PHOTOSELECTIVE CHEMISTRY

PART 2

ADVANCES IN CHEMICAL PHYSICS

VOLUME XLVII

PHOTOSELECTIVE CHEMISTRY

PART 2

Edited by

JOSHUA JORTNER
Tel-Aviv University

RAPHAEL D. LEVINE
Hebrew University of Jerusalem

STUART A. RICE
University of Chicago

ADVANCES IN CHEMICAL PHYSICS

VOLUME XLVII

Series editors

Ilya Prigogine
University of Brussels
Brussels, Belgium
and
University of Texas
Austin, Texas

Stuart A. Rice
Department of Chemistry
and
The James Franck Institute
University of Chicago
Chicago, Illinois

AN INTERSCIENCE® PUBLICATION

JOHN WILEY AND SONS
NEW YORK · CHICHESTER · BRISBANE · TORONTO

An Interscience® Publication

***Library of Congress Catalog Card Number:* 58–9935**

ISBN 0-471-06274-x

Printed in the United States of America

10 9 8 7 6 5 4 3 2 1

INTRODUCTION

Few of us can any longer keep up with the flood of scientific literature, even in specialized subfields. Any attempt to do more, and be broadly educated with respect to a large domain of science, has the appearance of tilting at windmills. Yet the synthesis of ideas drawn from different subjects into new, powerful, general concepts is as valuable as ever, and the desire to remain educated persists in all scientists. This series, *Advances in Chemical Physics*, is devoted to helping the reader obtain general information about a wide variety of topics in chemical physics, which field we interpret very broadly. Our intent is to have experts present comprehensive analyses of subjects of interest and to encourage the expression of individual points of view. We hope that this approach to the presentation of an overview of a subject will both stimulate new research and serve as a personalized learning text for beginners in a field.

ILYA PRIGOGINE
STUART A. RICE

PREFACE

Photoselective chemistry is concerned with the influence of selective optical excitation on the acquisition, storage and disposal of energy and on the reactivity of molecules, in both gaseous and condensed phases. The very considerable recent progress in this field is largely due to the introduction of lasers, both as pumping and as probing devices, and to the intense theoretical effort stimulated by the many new and intriguing experimental findings. Photoselective chemistry is an interdisciplinary research area, blending concepts and techniques from a wide variety of fields. The articles assembled in these volumes describe many of the theoretical and experimental results now in hand, with the goal of contributing to the synthesis of a conceptual framework with which one can understand a broad spectrum of photophysical and chemical processes. The field is too young, and too little is known, to permit compilation of a definitive treatise. Instead, the contributions in these volumes reflect our opinion of the kinds of information which must be accumulated before it is possible to develop an integrated approach to the interpretation of observations already made and to the development and exploitation of these for new approaches to photochemistry. The same (opinionated) point of view guided the Organizing Committee [J. Jortner (Chairman), S. Kimel, A. Levin, R. D. Levine] of the Laser Chemistry Conference, which took place on December 15–22, 1979 at Ein Bokek, Israel, under the auspices of the National Council for Research and Development, Jerusalem, Israel. Indeed the decision to compile these volumes was triggered by the intensive and exhaustive problem-oriented discussions which took place during that Conference. In addition to contributions by participants at the Ein Bokek meeting, we have included many others, the overall goal being to provide an up-to-date set of authoritative reviews spanning the various aspects of photoselective chemistry.

The first article is meant to serve as an introduction to the entire field; it provides an overview of the relevant concepts, problems, ideas and experiments described in the following papers. This introductory article was written after receipt of the other contributions so that cross references could be made. The following articles have been organized in topical groups. Where it was desirable and practical, an introductory article to the given topic is placed first. The general organization of the material is:

1. Aspects of Intramolecular Dynamics
2. Multiphoton Induced Chemistry
3. Studies of Collision Effects
4. Studies in Condensed Media
5. Other Aspects of Photoselective Chemistry

We have followed the general policy of the *Advances in Chemical Physics* in that the authors have been given complete freedom, our point of view being that the person who pioneered the topic is the best judge of the appropriate mode for its presentation. We believe that the results have more than vindicated our approach and hope that the reader concurs. These volumes offer the newcomer a review of the entire field, yet in each and every direction reach the forefront of the current research effort and even attempt to explore the perspectives and future of photoselective chemistry.

We are grateful to the participants of the Ein Bokek Conference and to numerous colleagues and friends, whose lively and probing discussions convinced us of the merits of this project, and to the staff of Wiley-Interscience for welcoming and supporting it. We thank the authors for their willingness to contribute to this endeavor and for their adherence to a timetable which enabled us to send the manuscripts to the publisher in the Fall of 1979. The wide range of subjects touched on in these volumes bears witness to the scope of photoselective chemistry and to the contagious enthusiasm of its practitioners.

JOSHUA JORTNER
R. D. LEVINE
STUART A. RICE

Tel-Aviv, Isreal
Jerusalem Isreal
Chicago, Illinois
January 1981

CONTRIBUTORS TO VOLUME XLVII, PART 2

URI AGAM, Department of Chemistry, Tel-Aviv University, Tel-Aviv, Israel

MICHA ASSCHER, Department of Physical Chemistry, The Hebrew University, Jerusalem, Israel

P. F. BARBARA, Bell Laboratories, Murray Hill, New Jersey

A. BARONAVSKI, Chemical Diagnostics Branch, Chemistry Division, Naval Research Laboratory, Washington, D.C.

A. BEN-SHAUL, Department of Physical Chemistry and Institute of Advanced Studies, The Hebrew University, Jerusalem, Israel

V. E. BONDYBEY, Bell Laboratories, Murray Hill, New Jersey

MARK EYAL, Department of Chemistry, Tel-Aviv University, Tel-Aviv, Israel

GEORGE FLYNN, Department of Chemistry and Columbia Radiation Laboratory, Columbia University, New York, New York

KARL F. FREED, The James Franck Institute, Department of Chemistry, University of Chicago, Chicago, Illinois

FREDERICK R. GRABINER, Department of Chemistry, Tel-Aviv University, Tel-Aviv, Israel

M. GUTMAN, Department of Biochemistry, Tel-Aviv University, Ramat Aviv, Israel

YEHUDA HAAS, Department of Physical Chemistry, The Hebrew University, Jerusalem, Israel

PAUL L. HOUSTON, Department of Chemistry, Cornell University, Ithaca, New York

D. HUPPERT, Department of Chemistry, Tel-Aviv University, Ramat Aviv, Israel

A. KALDOR, Exxon Research and Engineering Company, Corporate Research Laboratories, Linden, New Jersey

K. J. KAUFMANN, Department of Chemistry, University of Illinois, Urbana, Illinois

GERALDINE A. KENNEY-WALLACE, Department of Chemistry, University of Toronto, Toronto, Canada

M. C. Lin, Chemical Diagnostics Branch, Chemistry Division, Naval Research Laboratory, Washington, D.C.

A. Nitzan, Department of Chemistry, Tel-Aviv University, Tel-Aviv, Israel

David W. Oxtoby, Department of Chemistry, James Franck Institute, University of Chicago, Chicago, Illinois

P. M. Rentzepis, Bell Laboratories, Murray Hill, New Jersey

Stuart A. Rice, The Department of Chemistry and The James Franck Institute, The University of Chicago, Chicago, Illinois

R. Schmiedl, Fakultät für Physik, Universität Bielefeld, Bielefeld, Federal Republic of Germany

A. Tramer, Laboratoire de Photophysique Moleculaire CNRS, Université Paris-Sud, Paris, France

M. E. Umstead, Chemical Diagnostics Branch, Chemistry Division, Naval Research Laboratory, Washington, D.C.

Stephen C. Wallace, Department of Chemistry, University of Toronto, Toronto, Canada

M. R. Wasielewski, Argonne National Laboratory, Chemistry Division, Argonne, Illinois

Eric Weitz, Department of Chemistry, Northwestern University, Evanston, Illinois

K. H. Welge, Fakultät für Physik, Universität Bielefeld, Bielefeld, Federal Republic of Germany

Douwe A. Wiersma, Picosecond Laser and Spectroscopy Laboratory of the Department of Physical Chemistry, State University, Nijenborgh, Groningen, Netherlands

R. L. Woodin, Exxon Research and Engineering Company, Corporate Research Laboratories, Linden, New Jersey

CONTENTS

PHOTOSELECTIVE CHEMISTRY

PART 2

ADVANCES IN CHEMICAL PHYSICS

VOLUME XLVII

Section 3

ONE-PHOTON AND TWO-PHOTON PHOTOSELECTIVE CHEMISTRY

ENHANCEMENT OF CHEMICAL REACTIONS BY INFRARED LASERS

R. L. WOODIN AND A. KALDOR

Exxon Research and Engineering Company, Corporate Research Laboratories, Linden, New Jersey 07036

CONTENTS

I. INTRODUCTION

The detailed effect of reactant internal energy on chemical reactions has been a subject of considerable interest in chemical kinetics for many years. Classic work of Evans and Polanyi in the 1930s illustrated the importance of vibrational energy in reaction dynamics, and the intimate nature of the coupling of vibrational and translational energy with a reaction potential energy surface.[1,2] Although a topic of central importance in reaction dynamics, explicit effects of reactant vibrational energy on a chemical reaction have been difficult to assess because of problems associated with preparing molecules in well-defined initial states by traditional chemical or thermal activation techniques. Even so, molecular beam[3] and infrared chemiluminescence[4] experiments have provided a great deal of experimental data on the role of internal energy in chemical reactions.

A powerful new tool that has been added recently to the experimentalist's arsenal is the infrared laser. The advent of tunable infrared lasers now makes possible population of well-defined initial internal energy states, and the concomitant ability to examine more closely the role of vibrational energy in chemical reactions. Several recent review articles discuss the use of lasers in promoting chemical reactions and in probing reaction dynamics.[5–8] In this chapter we shall be concerned only with the effects of laser excitation on enhancing chemical reactivity.

The ability of both pulsed and cw infrared lasers to induce chemical reactions is well known. CO_2 lasers are now common equipment in many laboratories. The infrared laser-induced process studied most extensively is multiplephoton excitation of molecules (using megawatt CO_2 laser radiation) to high vibrational states from which reaction, usually dissociation, may occur. This field is the subject of intense effort by many research groups, and a number of excellent review articles have been written about multiplephoton excitation.[9] At lower laser intensities it is possible to prepare molecules in specific initial vibrational states below the dissociation threshold and to study their subsequent bimolecular and unimolecular (isomerization) reactions. In this chapter we shall restrict ourselves to considering only the results of low-level vibrational excitation on chemical reactions.

II. VIBRATIONAL ENHANCEMENT OF BIMOLECULAR REACTIONS

It was recognized early in the study of chemical kinetics (particularly with regard to unimolecular reactions) that increasing the energy of reactants increased reaction rates. This was usually accomplished by simply raising the reaction temperature. As experimental techniques were refined, it became possible to answer the question of whether vibrational or translational energy was more important in enhancing a bimolecular reaction. Evans and Polanyi presented the first qualitative description of how reactant initial energy is coupled to details of the potential energy surface.[1] Although strictly concerning three-atom reactions (1) the concepts are useful for more complex systems as well. For exothermic reactions, the

$$A + BC \rightarrow AB + C \tag{1}$$

ability of vibrational or translation energy to enhance a reaction depends on whether the potential energy surface is attractive (late barrier) or repulsive (early barrier). For an attractive surface trajectory studies show that reactant vibrational energy is most effective at promoting reaction. Conversely, for a repulsive surface reactant translational energy enhances reaction. Molecular beam studies (generally for reactions of hydrogen halides) have confirmed these predictions.

Endothermic reactions may be considered as the reverse of the exothermic reactions just discussed. Considering both attractive and repul-

TABLE I
Bimolecular Reactions of Infrared Laser Excited Molecules

Reaction	Pump laser[a]	$\Delta H^{\circ b}_{298}$	Comments[c]	Ref.
$HCl(v=1)+H \rightarrow H_2+Cl$	HCl	−1.2	No enhancement	34
$HCl(v=1)+K \rightarrow KCl+H$	HCl	1.5	$k_1/k_{th}=100$	13
$HCl(v=1)+O \rightarrow OH+Cl$	HCl	0.7	$k_1/k_{th}=300$	41
$HCl(v=2)+Br \rightarrow HBr+Cl$	HCl	15.5	$k_1/k_{th}=10^{11}$	34,42
$HF(v=1)+Ba \rightarrow BaF(v)+H$	HF	$\geqslant -2.6$	$k_1/k_{th}=0.1$–300^d	11
$HF(v=1)+Ca \rightarrow CaF+H$	HF	9.2	$k_1/k_{th}>10^4$	14
$HF(v=1)+Sr \rightarrow SrF+H$	HF	8.0	$k_1/k_{th}>10^4$	14
$DF(v=1)+Ca \rightarrow CaF+D$	DF	9.2	No reaction	14
$DF(v=1)+Sr \rightarrow SrF+D$	DF	8.0	No reaction	14
$HCl(v=6)+D_2 \rightarrow HD+DCl$	Visible dye	~0	No reaction	6
$O_3(v)+NO \rightarrow NO_2^*+O_2$	CO_2	−4.8	$k_1/k_{th}=4$	15,16,18
$\rightarrow NO_2^{\ddagger}+O_2$	CO_2	47.	$k_1/k_{th}=17$	15,16,18
$O_3(v)+O \rightarrow 2O_2$	CO_2	−94	No enhancement	33
$O_3(v)+O_2 \rightarrow 2O_2+O$	CO_2	25.4	$k_1/k_{th}=40$	15
$O_3(v)+SO \rightarrow SO_2+O_2$	CO_2	−107	$k_1/k_{th}=2$	32
$O_3(v)+Ba \rightarrow BaO+O_2$	CO_2	−107	$k_1/k_{th}=5$–10	43
$O_3(v)+Pb \rightarrow PbO+O_2$	CO_2	−59	$k_1/k_{th}=10$–20	44
$NO(v=1)+O_3 \rightarrow NO_2^*+O_2$	CO	−4.8	$k_1/k_{th}=6$	17
$\rightarrow NO_2^{\ddagger}+O_2$	CO	47.6	$k_1/k_{th}\leqslant 22$	17
$OCS(v)+O \rightarrow CO+SO$	CO_2	−51	No enhancement	45
$C_2H_4(v)+O \rightarrow CH_3+CHO$	CO_2	6.2	No enhancement	45
$CH_3Br(v)+Cl \rightarrow CH_2Br+HCl$	CO_2	NA	$k_1/k_{th}\sim 1.05$	19
$CH_3F(v)+D \rightarrow CH_3+DF$	CO_2	−26	No reaction	46
$CH_3F(v)+O \rightarrow CH_2F+OH$	CO_2	5	No enhancement	46
$BH_3PF_3(v)+BH_3 \rightarrow B_2H_6+PF_3$	CO_2	NA	Enhancement observed	20
$BCl_3(v)+3C_2Cl_4 \rightarrow C_6Cl_6 + 3Cl_2+BCl_3$	CO_2	NA	Laser promoted	21
$B(CH_3)_3(v)+HBr \rightarrow B(CH_3)_2Br + CH_4$	CO_2	NA	Laser promoted	22
$B(CH_3)_2Br(v)+HBr \rightarrow BCH_3Br_2 + CH_4$	CO_2	NA	Laser promoted	22
$BCH_3Br(v)+HBr \rightarrow BBr_3+CH_4$	CO_2	NA	Laser promoted	22
$SF_6(v)+e^- \rightarrow SF_5^-+F$	CO_2	~4	$k_1/k_{th}\sim 3$	28
$Fe(CO)_4(v)+CH_4 \rightarrow Fe(CO)_4CH_4$	CO	NA	Laser promoted, matrix isolation	25
$Fe(CO)_4(v)+CO \rightarrow Fe(CO)_5$	CO, SFRL[e]	NA	Laser promoted, matrix isolation	24
$UF_6(v)+SiH_4 \rightarrow UF_5, UF_4, SiH_3F$	16 μm diode[f] laser	NA	Laser promoted matrix isolation	26

[a]HCl laser: 3.8 μm; HF laser: 2.8 μm; DF laser: 3.8 μm; CO_2 laser: 10.6 μm; CO laser: 5.6 μm.

[b]ΔH°_{298} calculated from data in Ref. 47. Values are kcal/mole. NA indicates data not available.

[c]k_1/k_{th} = ratio of rate constant with laser excitation to thermal rate constant.

[d]Value of k_1/k_{th} depends on v' state of BaF.

[e]SFRL = spin flip raman laser.

[f]This reaction also promoted by broadband IR excitation.

sive potential energy surfaces leads to the conclusion that for endothermic reactions the crest of the activation barrier lies in the reaction exit channel (i.e., a late barrier).[10] The expectation, then, is that reactant vibrational energy will be most effective at enhancing an endothermic reaction. A second consideration for endothermic reactions is whether the vibrational excitation is sufficient to raise the system above the activation barrier. In those reactions where this is the case, substantial rate enhancement should be realized.

Tunable, narrow bandwidth lasers provide a very convenient method for obtaining selective vibrational excitation. A number of reactions of infrared laser-excited species have been studied to date, and are listed in Table I. Also included is the frequency of the pump laser, the ground-state heat of reaction, and the effect of vibrational excitation on reaction rate. Reactions in Table I are listed according to increasing molecular complexity.

A. Atom Plus Hydrogen Halide Reactions

The reactions of excited hydrogen halides are convenient to study because of the availability of hydrogen halide chemical lasers for excitation. The correlation between ground-state heats of reaction and vibrationally induced reaction rate enhancement is striking for these processes (Table I). As expected, reaction enhancement is observed for all reactions except for those that are exothermic. In the case of reaction (2) the reaction rate is

$$\mathrm{HF}(v=1)+\mathrm{Ba}\rightarrow\mathrm{BaF}(v')+\mathrm{H} \tag{2}$$

highly state specific.[11] In fact, as the reaction is made endothermic (higher v') substantial rate enhancements are observed.

In an elegant series of experiments, Brooks and co-workers explored reaction (3) as a function of both relative translational energy[12]

$$\mathrm{HCl}+\mathrm{K}\rightarrow\mathrm{KCl}+\mathrm{H} \tag{3}$$

and HCl vibrational energy.[13] While translational energy is capable of enhancing the reaction rate by up to one order of magnitude, HCl vibrational energy is found to enhance the reaction by two orders of magnitude. The preferential enhancement of reaction (3) by vibrational energy is evident.

Reactions (4) and (5) are both examples of endothermic

$$\mathrm{HF}+\mathrm{Ca}\rightarrow\mathrm{CaF}+\mathrm{H} \tag{4}$$

$$\mathrm{DF}+\mathrm{Ca}\rightarrow\mathrm{CaF}+\mathrm{D} \tag{5}$$

reactions, yet only the rate of reaction (4) is enhanced by reactant excitation.[14] This is because one quanta of HF vibration (3600 cm^{-1}) is enough to surmount the activation barrier, while one quanta of DF vibration (2700 cm^{-1}) leaves the reactants below the activation barrier. The same situation is true if calcium is replaced by strontium.[14]

B. Polyatomic Bimolecular Reactions

While it is tempting to apply the concepts developed for atom plus diatomic reactions to polyatomic reactions in general, it becomes rapidly apparent that the situation is much more complex. One of the first reactions studied was the reaction of O_3 with NO, reaction (6).[15] Two channels are available, one producing electronically excited NO_2 [exothermic, reaction

$$O_3(v) + NO \longrightarrow \begin{cases} NO_2^*(^2B_1) + O_2 & (6a) \\ NO_2^\dagger(^2A_1) + O_2 & (6b) \end{cases}$$

(6a)] and one producing vibrationally excited NO_2 [endothermic, reaction (6b)]. The rate for reaction (6a) is enhanced by a factor of 4, while the rate for reaction (6b) is enhanced by a factor of nearly 20. Because it is not known precisely what initial and final states are involved, and which O_3 vibrations are coupled to the reaction coordinate, it is not possible to correlate the observed rate enhancement with the reaction endo- or exothermicity. Initial studies indicated that reaction was dependent on intramolecular energy transfer from the O_3 ν_3 mode to the ν_2 mode. Subsequent studies show the O_3 ν_3 mode is involved in the reaction.[16] As a further probe of this reaction, a CO laser was used to excite NO, reaction (7).[17] Enhancement of reaction (7a) was comparable to enhancement in reaction (6a).

$$NO(v=1) + O_3 \longrightarrow \begin{cases} NO_2^*(^2B_1) + O_2 & (7a) \\ NO_2^\dagger(^2A_1) + O_2 & (7b) \end{cases}$$

Only an upper bound was put on the rate for reaction (7b), indicating it could be comparable to reaction (6b) or much slower.

In related experiments[18], SiF_4 was added to the O_3–NO system. Irradiation with a CO_2 laser at the 9.6μm P(30) transition excites O_3 while the 9.6μm P(32) transition excites SiF_4. Enhancement of reaction (6) is observed in both cases, either by direct O_3 excitation or by near resonant V–V transfer from SiF_4 to O_3. In the latter case the $O_3(v)$ concentration is much lower than when excited directly due to rapid redistribution of energy among SiF_4 vibrations not resonant with O_3. This technique allows the enhancement of reaction (6) to be studied both directly and via a V–V excitation step. While it is clear that vibrational energy enhances the reaction of O_3 with NO, the complexities of this reaction illustrate some of the difficulties in dealing with polyatomic systems.

Enhanced reaction rates are seen for several more complex reactions. In isotope separation experiments, CH_3Br reacting with Cl [reaction (8)] shows

$$CH_3Br(v)+Cl \rightarrow CH_2Br+HCl \tag{8}$$

an enhanced reaction rate when pumped with a CO_2 laser.[19] For process (9)

$$BH_3PF_3(v)+BH_3 \rightarrow B_2H_6+PF_3 \tag{9}$$

reaction of BH_3 with BH_3PF_3 has an activation energy of about 8 kcal/mole. Vibrational excitation of BH_3PF_3 is observed to enhance the reaction rate for reaction (9).[20] It is postulated that this reaction involves sequential absorption of two or three CO_2 laser photons for significant rate enhancement. In the reaction of BCl_3 and C_2Cl_4, process (10),

$$BCl_3(v)+3C_2Cl_4 \rightarrow C_6Cl_6+3Cl_2+BCl_3 \tag{10}$$

the products with lower activation energy are among those found for a thermal process.[21] However, reaction (10) is found to be specific for BCl_3; use of either SF_6 or BBr_3 as sensitizer leads to reaction products characteristic of higher temperature processes, with little formation of C_6Cl_6. Related experiments were carried out with brominated boron compounds, reactions (11)–(13).[22] These are all known thermal reactions, with

$$B(CH_3)_3+HBr \rightarrow B(CH_3)_2Br+CH_4 \tag{11}$$

$$B(CH_3)_2Br+HBr \rightarrow BCH_3Br_2+CH_4 \tag{12}$$

$$BCH_3Br_2+HBr \rightarrow BBr_3+CH_4 \tag{13}$$

reaction (11) characteristic in the 150–180°C range, reaction (12) at temperatures >250°C, and reaction (13) at temperatures >450°C. What is particularly interesting is that irradiating BCH_3Br_2 in a mixture of BCH_3Br_2 and $B(CH_3)_2Br$ produces only BBr_3, and no loss of $B(CH_3)_2Br$. Laser excitation selectively heats one component of the mixture, while the bulk sample remains at a low temperature.

Diborane is an interesting molecule for studying laser-enhanced reactions because a CO_2 laser may be used to excite three distinct vibrational modes of the molecule. Reactions (14)–(16) are those found to

$$B_2H_6(h\nu_1)+CO\rightarrow B_2O_3+H_2CO \tag{14}$$

$$B_2H_6(h\nu_2)+CO\rightarrow B_2O_3+H_2CO+CH_3OH \tag{15}$$

$$B_2H_6(h\nu_3)+CO\rightarrow BH_3CO \tag{16}$$

occur when a mixture of B_2H_6 and CO is irradiated with a CO_2 laser exciting the boron bridging mode, BH_2 scissors mode, and terminal BH stretching mode, respectively.[23] The only reaction resulting in products characteristic of the thermal process is reaction (16), where the terminal BH stretch is excited. Excitation at the other frequencies result in a proton stripping reaction. Thus the reaction is markedly influenced by which mode is excited.

Two experiments have dealt with the effect of higher levels of vibrational excitation on a bimolecular reaction. Overtone absorption from $v=0$ to $v=6$ can be induced in HCl using visible light from a dye laser. Due to the small cross-section associated with overtone transitions, the high laser intensities obtainable intracavity are needed. For reaction (17), the activation energy

$$HCl(v=6)+D_2\rightarrow HD+DCl \tag{17}$$

is estimated to be 34 kcal/mole. No reaction could be observed with irradiation at room temperature.[6] Since overtone excitation is expected to overcome the activation barrier, it is not clear why enhanced reaction is not observed in process (17). It may be that estimates of the activation energy are in error, and the four-center reaction requires more excitation than that available from $HCl(v=6)$.

The use of matrix isolation techniques has led to several interesting infrared laser promoted reactions. The reactions of $Fe(CO)_4$ with CH_4 and CO, processes (18) and (19), have been observed when the $Fe(CO)_4$ is excited

$$Fe(CO)_4+CH_4\rightarrow Fe(CO)_4CH_4 \tag{18}$$

$$Fe(CO_4)+CO\rightarrow Fe(CO)_5 \tag{19}$$

by a CO laser.[24–25] The mechanism for these reactions is unclear at this time. In similar experiments Catalano and Barletta have observed reactions of UF_6 and SiH_4 in matrices, to produce a variety of products, including UF_5, UF_4, and SiH_3F.[26] In the gas phase they estimate an activation energy of 30 kcal/mole yet in the matrix experiments, the reaction is apparently induced by single-photon absorption by UF_6 (3 kcal/mole). In fact, the reactions can be promoted by use of a broadband infrared source (Nernst glower). It has been concluded that the drastic change in activation barrier is due to the fixing of reactant configurations in the matrix. Recent attempts to duplicate the infrared photochemical reaction of UF_6 with SiH_4 in a matrix have failed to reproduce the original data. There is some speculation that the reaction is induced instead by small amounts of UV light from the broadband infrared source used for either excitation or product analysis.[27]

A somewhat different type of reaction for which vibrational enhancement of the reaction rate is observed is reaction (20).[28]

$$SF_6 + e^- \rightarrow SF_5^- + F \tag{20}$$

An estimate of the activation barrier for this reaction is 0.2 eV, comparable to the energy of a CO_2 laser photon. Excitation of the 10 μm ν_3 mode of SF_6 is seen to enhance the dissociative electron attachment process.

C. Laser-Enhanced Reactions at High Pressure

In order to maintain state-specific populations, low pressures are required. If pressures are raised to the point where collisions are frequent, then specific effects of laser excitation are lost and intermolecular energy transfer heats the entire sample. Reactions induced in this fashion then yield products characteristic of thermal processes. A particularly interesting aspect of this pyrolysis technique is the homogeneous nature of the reaction. Since heating is confined to the region near the laser beam, heterogeneous effects due to hot walls in a conventional pyrolysis reactor are avoided.

Table II lists a number of the high-pressure laser-induced reactions that have been studied. (References 29 to 30 contain more extensive tables of similar reactions not included here.) These reactions are carried out either directly or with a second component added as a sensitizer. Reaction (21)[31] proceeds only in the presence of a sensitizer such as SiF_4, since

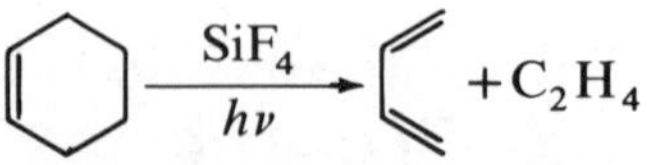

(21)

cyclohexene is transparent to CO_2 laser radiation. SiF_4 absorbs light, heating the cyclohexene via intermolecular energy transfer. The reaction is particularly clean from the standpoint of not involving heterogeneous wall effects. A sensitizer is not needed if the reacting species absorbs the laser

TABLE II
High-Pressure Laser-Enhanced Reactions

Reaction	Pump laser	Pressure[a]	Ref.
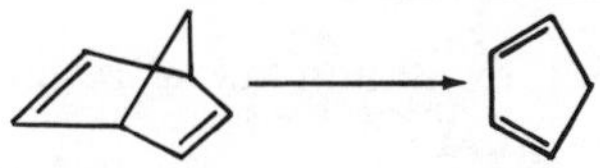 $+C_2H_2$	CO_2 [b]	10–50	31
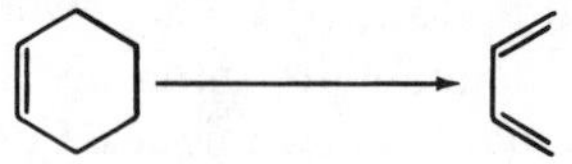 $+C_2H_4$	CO_2 [b]	10–50	31
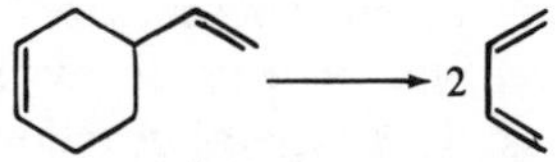	CO_2 [b]	10–50	31
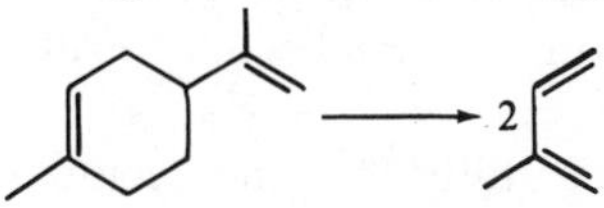	CO_2 [b]	10–50	31, 48
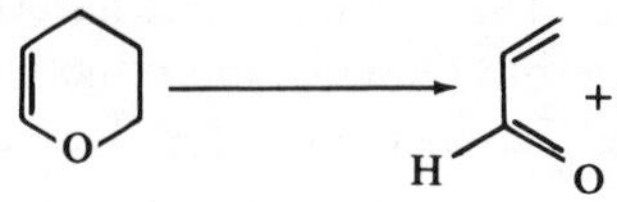 $+C_2H_4$	CO_2 [b]	10–50	31
$CF_2ClCF_2CL \rightarrow C_2F_4 + Cl_2$	CO_2	200	49
$CH_3CF_2Cl \rightarrow CH_2CF_2 + HCl$	CO_2	50–400	32
$B_2H_6 \rightarrow B_5H_9$, B_5H_{11}, $B_{10}H_{14}$,$(BH)_n$	CO_2 [c]	50–500	50
$CH_3F \rightarrow C_2F_4$, HF, C_3F_6, X[d]	CO_2 [c]	50–100	30
$CHClF_2 \rightarrow C_2F_4$, HCl, C_3F_6, X[d]	CO_2 [c]	50–100	30
$c-C_4F_8 \rightarrow C_2F_4$, C_3F_6, X[d]	CO_2 [c]	50–100	30
$CH_4 \rightarrow C_2H_4$, C_2H_6, C_2H_2	CO_2 [c]	50–100	30
$C_2H_4 \rightarrow CH_4$, C_2H_6, C_3H_6, C_3H_8, C_4, C_5 hydrocarbons	CO_2	100–500	30, 51, 52
$C—C_3H_6 \rightarrow C_3H_6$, CH_4, C_2H_4	CO_2 [c]	50–100	30
$C_2H_6 \rightarrow C_2H_4$, CH_4, C_2H_2	CO_2 [c]	50–100	30
trans-2-butene$\rightarrow CH_4$, C_2H_4, C_3H_6, *cis*-2-butene, C_4H_6	CO_2 [c]	50–100	30
cis-2-butene$\rightarrow CH_4$, C_2H_4, C_3H_6, *trans*-2-butene, C_4H_6	CO_2 [c]	50–100	30
$B_2H_6 + O_2 \rightarrow$boron oxides, X[d]	CO_2 [c]	50–100	30
$Fe(CO)_5 + O_2 \rightarrow$iron oxides, X[d]	CO_2 [c]	50–100	30
$Fe(CO)_5 \rightarrow$Fe, CO, X[d]	CO_2 [c]	50–100	30

[a] Units are torr.
[b] SiF_4 added as sensitizer.
[c] SF_6 added as sensitizer.
[d] X indicates unidentified products.

radiation. This is the case for reaction (22)[32]. The products are

$$CH_3CF_2Cl \rightarrow CH_2CF_2 + HCl \quad (22)$$

characteristic of the thermal pyrolysis, with excitation as a result of infrared absorption by CH_3CF_2Cl.

The reactions of various hydrocarbons listed in Table II are noteworthy in that they yield dehydrogenated products in amounts differing from those obtained by conventional heterogeneous pyrolysis techniques. For all hydrocarbon reactions listed in Table II, SF_6 was added as a sensitizer. An additional example, listed in Table II, of a novel type of reaction by laser pyrolysis is the production of metal particulates. This type of reaction, if selective for only one component of the starting material, could lead to production of very pure products.

The reactions listed in Table II illustrate the utility that this technique may have in conventional pyrolysis reactions. The homogeneous nature of the laser-induced reaction minimizes undesirable secondary products and results in cleaner processes with higher yields. Laser-induced pyrolysis may prove to be a useful synthetic technique.

D. Collisional Deactivation of Vibrationally Excited Species

It is appropriate here to point out that, in addition to the reactive collisions discussed so far, the possibility of inelastic nonreactive collisions is also present. In many instances the nonreactive channel may be dominant.

In the case of atom plus molecule reactions, the atom may very efficiently deactivate the excited molecule. Reactions (23)[33] and (24)[34]

$$O_3(v) + O(^3P) \rightarrow O_3 + O(^3P) \quad (23)$$

$$HCl(v=1) + H \rightarrow HCl(v=0) + H \quad (24)$$

are examples of rapid deactivation by reactive atomic species. The ability of a reactive atom to form a long-lived complex with the molecule allows the atom to remove energy efficiently when the complex decomposes. In looking at enhancement of reaction rates by laser excitation it is necessary to observe both decay of the excited species and also product formation to distinguish correctly between reactive and nonreactive pathways.

Vibrational deactivation by polyatomic species takes place predominantly by V–V relaxation rather than V–T relaxation.[35,36] The facility of one molecule for deactivating another molecule depends on many factors, for example, the energy mismatch between vibrational frequencies of the two species. When a near-resonance exists, the collisional deactivation process may be rapid, requiring 10–100 collisions.[36] For reactions of molecules with only one quantum of vibrational excitation, collisions that remove 2–5 kcal/mole are particularly effective at canceling any rate enhancement, since the amount of energy removed is roughly the same as the amount of energy absorbed by the molecule from the laser. These nonreactive collisions may be more efficient than reactive encounters and need to be considered if an experiment is done under collisional conditions.

III. VIBRATIONAL ENHANCEMENT OF ISOMERIZATION REACTIONS

Isomerization reactions induced by vibrational excitation are listed in Table III. Probably the first infrared photochemical reaction to be observed was the isomerization of nitrous acid in a nitrogen matrix, reaction (25).[37]

$$cis\text{-HONO} \rightarrow trans\text{-HONO} \tag{25}$$

The excitation source was a filtered Nernst glower. At the power levels

TABLE III
Isomerization Reactions of Infrared Laser-Excited Molecules

Reaction	Pump laser	Comments	Ref.
cis-HONO→*trans*-HONO	Broadband[a] 3000–4000 cm^{-1}	Infrared promoted, matrix isolation	37
asym-N_2O_3→*sym*-N_2O_3	Broadband near-IR	Infrared promoted, matrix isolation	38
$Fe(CO)_4 \xrightarrow[\text{exchange}]{\text{ligand}} Fe(CO)_4$	CO, SFRL[b]	Infrared induced, matrix isolation	24, 39
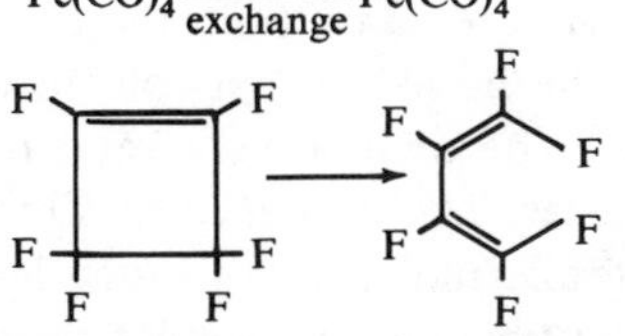	CO_2	Gas phase	53
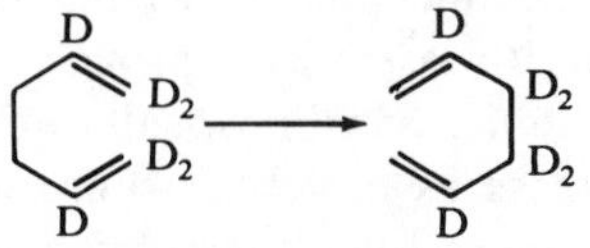	CO_2	Gas phase	54
trans-2-butene→*cis*-2-butene	CO_2	Gas phase	30, 55
trans-$C_2H_2Cl_2$→*cis*-$C_2H_2Cl_2$	CO_2	Gas phase	56

[a]Radiation from infrared spectrometer source.
[b]SFRL = spin flip Raman laser.

involved, only single-photon processes can occur. It was postulated that rapid intramolecular energy transfer into the torsional mode led to isomerization. In a similar reaction using near-infrared radiation from a broadband source, isomerization of N_2O_3 was observed in a matrix, reaction (26).[38]

$$\textit{asym}\text{-}N_2O_3 \rightarrow \textit{sym}\text{-}N_2O_3 \tag{26}$$

Recent experiments have been carried out on isomerization of $Fe(CO)_4$ in an Ar matrix. By using labeled CO groups, it is possible to distinguish between the various isomers.[24,39] It was found that, by tuning either a CO laser or spin flip Raman laser to a particular ligand absorption, isomerization could be induced. In addition, it was possible to distinguish between $Fe(CO)_4$ molecules in different orientations within the matrix, and to selectively excite those molecules in a particular orientation. Experiments with polarized light indicate that the observed isomerization is not due to simple heating of the matrix.[24]

Gas-phase isomerization reactions have also been induced with infrared lasers. Examples of these are included in Table III. In general, these reactions require more than one photon to overcome the isomerization barrier, but less photons than required for a dissociation process. The gas-phase isomerization reactions are unique in that the product species has enough internal energy to react back to starting material. The ratio of reactants to products will thus depend on the relative isomerization rates and the rates of deactivation of excited molecules.

IV. CONCLUSION

From Tables I–III it can be seen that there is a great deal of interest in a variety of infrared laser-enhanced reactions. At the same time there are many questions that remain to be answered. The basic concepts put forth in the 1930s by Evans and Polanyi are seen to be qualitatively valid for laser-excited hydrogen halide plus atom reactions. The influence of vibrational energy on reactions of complex polyatomic molecules is much less clear, especially in those cases where the initial and final molecular states are not well known. Tunable, narrow bandwidth lasers have proven to be valuable as tools in probing the reaction dynamics of vibrationally excited molecules. Their value will continue to increase as new laser sources are discovered that cover the infrared spectrum more fully. Combining these new laser sources with molecular beam techniques should allow the effects of vibrational vs translational energy to be sorted out for complex reactions, much as in the case of reaction (3). It will also be possible to explore the effects of exciting a molecule at various different frequencies. As intramolecular energy relaxation is thought to be slow at low levels of excitation, it may be that some of the current questions regarding energy localization in polyatomic molecules[40] will be answered by bimolecular reactions of vibrationally excited species. It is clear that much work remains to be done and will provide new insights into reaction dynamics in the years to come.

References

1. M. G. Evans and M. Polanyi, *Trans. Faraday Soc.*, **35**, 178 (1939).
2. J. C. Polanyi, *Acct. Chem. Res.*, **5**, 161 (1972).
3. R. D. Levine and R. B. Bernstein, *Molecular Reaction Dynamics*, Oxford University Press, New York, 1974.
4. K. G. Anlauf, D. M. Maylotte, J. C. Polanyi, and R. B. Bernstein, *J. Chem. Phys.*, **51**, 5716 (1969); J. C. Polanyi and D. C. Tardy, *J. Chem. Phys.*, **51**, 5717 (1969); K. G. Anlauf, P. J. Kuntz, D. H. Maylotte, P. D. Pacey, and J. C. Polanyi, *Discuss. Faraday Soc.*, **44**, 183 (1967), K. G. Anlauf, P. E. Charters, D. S. Horne, R. G. MacDonald, D. H. Maylotte, J. C. Polanyi, W. J. Skirlac, D. C. Tardy, and K. B. Woodall, *J. Chem. Phys.*, **53**, 4091 (1970).
5. M. J. Berry, *Ann. Rev. Phys. Chem.*, **26**, 259 (1975).
6. J. Wolfrum, *Ber. Bunsenges. Phys. Chem.*, **81**, 114 (1977).
7. S. Kimel and S. Speiser, *Chem. Rev.*, **77**, 437 (1977).
8. C. B. Moore, *Discuss. Faraday Soc.*, **67**, 146 (1979).
9. R. V. Ambartzumian and V. S. Letohkov, in *Chemical and Biological Applications of Lasers*, Vol. III, C. B. Moore, ed., Academic Press, New York, New York, 1977; C. D. Cantrell, S. M. Freund, and J. L. Lyman, in *Laser Handbook*, Vol. III, M. L. Stitch, ed., North-Holland Publishing, Amsterdam, 1979; J. G. Black, E. Yablonovitch, and N. Bloembergen, *Phys. Rev. Lett.*, **38**, 1131 (1977).
10. J. C. Polanyi, *J. Chem. Phys.*, **31**, 1338 (1959); M. H. Mok and J. C. Polanyi, *J. Chem. Phys.*, **51**, 1451 (1969).
11. J. G. Pruett and R. N. Zare, *J. Chem. Phys.*, **64**, 1774 (1976).
12. T. J. Odiorne, P. R. Brooks, and J. V. V. Kasper, *J. Chem. Phys.*, **55**, 1980 (1971).
13. J. G. Pruett, F. R. Grabine, and P. R. Brooks, *J. Chem. Phys.*, **60**, 3335 (1974).
14. Z. Karny and R. N. Zare, *J. Chem. Phys.*, **68**, 3360 (1978).
15. M. J. Kurylo, W. Braun, A. Kaldor, S. M. Freund, and R. P. Wayne, *J. Photochem.*, **3**, 71 (1974); R. J. Gordon and M. C. Lin, *Chem. Phys. Lett.*, **22**, 262 (1973); R. J. Gordon, and M. C. Lin, *J. Chem. Phys.*, **64**, 1058 (1976); W. Braun, M. J. Kurylo, C. N. Xuan, and A. Kaldor, *J. Chem. Phys.*, **62**, 2065 (1975); W. Braun, M. J. Kurylo, A. Kaldor, and R. P. Wayne, *J. Chem. Phys.*, **61**, 461 (1974).
16. J. Moy, E. Bar-Ziv, and R. J. Gordon, *J. Chem., Phys.*, **66**, 5439 (1977).
17. J. C. Stephenson and S. M. Freund, *J. Chem. Phys.*, **65**, 4303 (1976).
18. W. Braun, M. J. Kurylo, and A. Kaldor, *Chem. Phys. Lett.*, **28**, 440 (1974).
19. T. J. Manuccia, M. D. Clark, and E. R. Lory, *J. Chem. Phys.*, **68**, 2271 (1978).
20. E. R. Lory, S. H. Bauer, and T. Mannucia, *J. Phys. Chem.*, **79**, 545 (1975).
21. H. R. Bachmann, R. Rinck, H. Noth, and K. L. Kompa, *Chem. Phys. Lett.*, **45**, 169 (1977).
22. H. R. Bachmann, H. Noth, R. Rinck, and K. L. Kompa, *Chem. Phys. Lett.*, **33**, 261 (1975).
23. A. Kaldor, Electro-Optical Systems Design Conference, New York, N.Y., 1976.

24. B. Davies, A. McNeish, M. Poliakoff, M. Tranquille, and J. J. Turner, *Chem. Phys. Lett.*, **52**, 477 (1977).
25. A. McNeish, M. Poliakoff, K. P. Smith, and J. J. Turner, *J. Chem. Soc., Chem. Commun.*, **1976**, 859.
26. E. Catalano and R. E. Barletta, *J. Chem. Phys.*, **66**, 4706 (1977); E. Catalano, R. E. Barletta, and R. K. Pearson, *J. Chem. Phys.*, **70**, 3291 (1979).
27. L. H. Jones and S. A. Ekberg, paper presented at the National Meeting of the American Chemical Society, Sept. 9, 1979, Washington, D. C.
28. A. Kaldor U.S. Patent 4,000,051 (1976), P. J. Chantry, C. L. Chen, *J. Chem. Phys.*, **71**, 3897 (1979); R. L. Woodin, D. S. Bomse, and J. L. Beauchamp, unpublished results.
29. N. G. Basov, A. N. Oraevsky, and A. V. Pankratov, in *Chemical and Biochemical Applications of Lasers*, Vol. I, C. Bradley Moore, ed., Academic Press, New York, New York, 1974.
30. W. M. Shaub and S. H. Bauer, *Int. J. Chem. Kinetics*, **7**, 509 (1975).
31. D. Garcia and P. M. Keehn, *J. Am. Chem. Soc.*, **100**, 6111 (1978).
32. R. N. Zitter and D. F. Koster, *J. Am. Chem. Soc.*, **100**, 2265 (1978).
33. G. A. West, R. E. Weston, Jr., and G. W. Flynn, *Chem. Phys. Lett.*, **56**, 429 (1978).
34. D. Arnoldi and J. Wolfrum, *Ber. Bunsenges. Phys. Chem.*, **80**, 892 (1976).
35. G. W. Flynn, in *Chemical and Biological Applications of Lasers*, Vol. I, C. Bradley Moore, ed., Academic Press, New York, New York, 1974.
36. J. D. Lambert, *Vibrational and Rotational Relaxation in Gases*, Clarendon Press, Oxford, 1977.
37. R. T. Hall and G. C. Pimentel, *J. Chem. Phys.*, **38**, 1889 (1962).
38. L. Varetti and G. C. Pimentel, *J. Chem. Phys.*, **55**, 3813 (1972).
39. B. Davies, A. McNeish, M. Poliakoff, and J. J. Turner, *J. Chem. Soc.*, **99**, 7573 (1977).
40. I. Oref and B. S. Rabinovitch, *Acct. Chem. Res.*, **12**, 166 (1979).
41. D. Arnoldi and J. Wolfrum, *Chem. Phys. Lett.*, **24**, 234 (1974); Z. Karny, B. Katz, and A. Szoke, *Chem. Phys. Lett.*, **35**, 100 (1975); R. D. H. Brown, G. P. Glass, and I. W. M. Smith, *Chem. Phys. Lett.*, **32**, 517 (1975).
42. D. Arnoldi, K. Kaufmann, and J. Wolfrum, *Phys. Rev. Lett.*, **34**, 1597 (1975).
43. A. Kaldor, W. Braun, and M. J. Kurylo, *J. Chem. Phys.*, **61**, 2496 (1974).
44. M. J. Kurylo, W. Braun, S. Abramowitz, and M. Krauss, *J. Res. Nat. Bur. Std.*, **80A**, 167 (16).
45. R. G. Manning, W. Braun, and M. J. Kurylo, *J. Chem. Phys.*, **65**, 2609 (1976).
46. M. Kneba and J. Wolfrum, *Ber. Bunsenges. Phys. Chem.*, **81**, 1275 (1977).
47. D. R. Stull and H. Prophet, National Bureau of Standards, NSRDS-NBS 37, 1971; M. W. Chase, J. L. Curnutt, W. T. Hu, H. Prophet, A. N. Syvernd, and L. C. Walker, *J. Phys. Chem. Ref. Data*, **3**, 311 (1974); M. W. Chase, J. L. Curnutt, H. Prophet, R. A. MacDonald, and A. N. Syverud, *J. Phys. Chem. Ref. Data*, **4**, 1 (15).
48. A. Yogev., R. M. J. Loewenstein, and D. Amar, *J. Am. Chem. Soc.*, **94**, 1091 (1972).
49. R. N. Zitter and D. F. Koster, *J. Am. Chem. Soc.*, **100**, 2265 (1978).
50. C. Riley and R. Shatas, *J. Phys. Chem.*, **83**, 1679 (1979).
51. J. P. Bell, R. V. Edwards, B. R. Nott, and J. C. Angus, *Ind. Eng. Chem. Fundam.*, **13**, 89 (1974).
52. J. Tardieu de Maleissye, F. Lempereur, and C. Marsal, *Chem. Phys. Lett.*, **42**, 472 (1976).
53. A. Yogev and R. M. J. Benmair, *Chem. Phys. Lett.*, **46**, 290 (1977).
54. I. Glatt and A. Yogev, *J. Am. Chem. Soc.*, **98**, 7087 (1976).
55. A. Yogev and R. M. J. Loewenstein-Benmair, *J. Am. Chem. Soc.*, **95**, 8487 (1973).
56. R. V. Ambartzumian, N. V. Chekalin, V. S. Doljikov, V. S. Letokhov, V. N. Lokhman, *Opt. Commun.*, **18**, 220 (1976).

TWO-PHOTON EXCITATION AS A KINETIC TOOL: APPLICATION TO NITRIC OXIDE FLUORESCENCE QUENCHING

YEHUDA HAAS AND MICHA ASSCHER

Department of Physical Chemistry, The Hebrew University, Jerusalem, Israel

CONTENTS

I. MULTIPHOTON VERSUS ONE-PHOTON SPECTROSCOPY

The polarization **P** of an atom or a molecule can be expressed as a power series in the field intensity **E**

$$\mathbf{P} = \alpha\mathbf{E} + \alpha'\mathbf{E}^2 + \alpha''\mathbf{E}^3 + \cdots \tag{1}$$

One-photon spectroscopy is due to the linear term, whereas the nonlinear terms lead to the simultaneous absorption of two or more photons. Although the theory was worked out almost 50 years ago,[1] observation of multiphoton absorption was made feasible only after the development of lasers. This chapter deals with the application of two-photon excitation (TPE) to kinetic studies in low-pressure gas-phase samples. For a systematic, extensive discussion of spectroscopic applications, one of the excellent reviews available should be consulted.[2,3]

The two-photon transition rate, W_{gf}, defined as the probability for absorption of a photon per molecule may be written as[2,3]

$$W_{gf}=\delta_{gf} I(\omega_1) I(\omega_2) \tag{2}$$

Here $I(\omega_i)$ is a laser intensity (photons s^{-1} cm^{-2}) at frequency ω_i and δ_{gf} is the two-photon absorption cross-section (cm^4 s $photon^{-1}$ $molecule^{-1}$) for transition from state g (at energy E_g) to state f (at energy E_f). The resonance condition is

$$\hbar(\omega_1+\omega_2)=E_f-E_g \tag{3}$$

In the special case of two photons of the same frequency, ω, the final state is populated at a rate

$$\frac{dN_f}{dt}=(1/2)W_{gf}N_g=(1/2)N_g\delta_{gf}I(\omega)^2 \tag{4}$$

with N_g and N_f the number densities (molecules cm^{-3}) of the initial and final state, respectively. δ_{gf} is the analog of the one-photon absorption cross-section, and may be written as

$$\delta_{gf}=128\pi^3\cdot\alpha^2\cdot\omega^2\cdot S_{gf}\cdot g(2\omega) \tag{5}$$

Here α is the fine-structure constant ($\sim 1/137$), $g(2\omega)$ the line shape function, and S_{gf} the two-photon transition probability tensor, which is the counterpart of the transition vector in one-photon spectroscopy. For electric dipole transitions, by far the most important ones, S_{gf} may be written as

$$S_{gf}=\sum_k \frac{(\mathbf{e}\langle g|\mathbf{r}|k\rangle)(\langle k|\mathbf{r}|f\rangle\cdot\mathbf{e})}{\omega_{kg}-\omega} \tag{6}$$

where $\mathbf{r}$ is the displacement vector operator and $\mathbf{e}$ is a complex polarization vector of unit length ($\mathbf{e}\cdot\mathbf{e}^*=1$).[3] $|k\rangle$ are intermediate states such that

$\hbar\omega_{kg} = E_k - E_g$, summation must be extended over all k states. By (3) the transition is also subject to an overall resonance condition $E_f - E_g = 2\hbar\omega$.

The operator governing δ is thus not a transition vector, but rather a transition tensor, which may be represented by a 3×3 matrix. Matrix elements are of the form

$$\mathcal{S}_{gf}^{xy} = \sum_{K} \frac{\langle g|x|k\rangle\langle k|y|f\rangle}{\omega_{kg} - \omega} + \frac{\langle g|y|k\rangle\langle k|x|f\rangle}{\omega_{kg} - \omega} \tag{7}$$

From (2)–(7) one can deduce the properties of two-photon spectroscopy. First and foremost is the fact that selection rules are different from those pertaining to one-photon absorption. For instance, in centrosymmetric molecules the selection rules are $g\leftrightarrow g$, $u\leftrightarrow u$ and $g\not\leftrightarrow u$, that is, the opposite of the one-photon selection rules. Thus, two-photon spectroscopy in this case is complementary to one-photon spectroscopy much in the same way as infrared and Raman spectroscopy are in the domain of molecular vibrations. Another property is the fact that δ depends on the polarization of the radiation even in fluid samples, that is, for complete orientation randomization. The form of δ for all point groups at various polarization combinations of the two photons has been derived and tabulated.[3] Consequently, polarization studies may be used to obtain structural information. Another unique feature of two-photon absorption is the possibility to observe spectra free of Doppler broadening by using two counter propagating beams.[4, 5]

II. EXPERIMENTAL METHODS

A. The Need for High-Power Lasers

Power requirements for experimental observation of two-photon absorption may be estimated by rewriting (2) as (deleting subscripts from δ_{gf} for brevity):

$$\frac{\Delta I}{I} = \delta N_g I \tag{8}$$

ΔI is the intensity decrease on traversing unit distance in a sample with ground-state population density of N_g. A typical value of δ is 10^{-50} cm^4 s photon^{-1} molecule^{-1} and at 1 torr pressure N_g is about 10^{16} molecules cm^{-3}. In this case a relative attenuation $\Delta I/I$ of 10^{-6} is obtained at $I = 10^{28}$ photons cm^{-2} s^{-1}, i.e., about 10^9 watt cm^{-2} s^{-1} for visible radiation. Powers of these orders of magnitude, at narrow frequency bandwidths, are practical only with laser sources. For TPE applications, lasers are often (not always) operated in a pulsed mode, allowing easy attainment of high peak powers while avoiding optical damage to cell windows and associated optics by keeping average power as low as possible. In kinetic studies laser-pumped dye lasers offer both spectral tunability and resolution on the one hand, and good time resolution on the other hand. With typical available dye laser systems (pumped by a nitrogen or a Nd–YAG laser) pulse duration and energy are 10^{-8} s and 10^{-3} J, respectively. Using the data of the above example, about 10^{10} photons are absorbed per laser pulse, creating 5×10^{9} excited molecules. Detection of such small changes in either laser intensity or excited molecule population is a considerable experimental challenge. In the following section we briefly describe some of the methods currently used to monitor TPE.

B. Monitoring Two-Photon Absorption

1. *Direct Attenuation of the Laser Beam*

This is conceptually the most straightforward technique. In view of the small relative change in laser intensity, it is mainly used in cases where two different laser beams are crossed in the sample. Often one of the lasers is a high-power, fixed-frequency device; attenuation is measured on the second, weaker laser appropriately termed the probe, or monitor, laser.[6] The method was used in early days of two-photon absorption spectroscopy and applied mostly to high-density and liquid samples. Its obvious drawback is poor sensitivity, as the desired quantity is obtained as a small difference between two large numbers.

2. *Detection by Thermal Effects*

Rather than measuring the relative change in laser intensity, one can monitor the absolute of energy deposited in the sample. This energy may be reemitted as fluorescence (see below) or degraded into heat. This in turn can be detected by thermal lensing methods [7] or by optoacoustic techniques.[8] In the first method, a low-power laser collinear with the pump beam is used to monitor changes in the refraction index of the substrate (or the solvent) due to heating. In the second, a microphone picks up the acoustic wave created by the sudden heat change. We are not aware of a report using this technique for TPE applications, but it should be sensitive enough for many cases.

A major disadvantage of thermal methods for kinetic applications is their poor time resolution—determined by the rate of gas density changes. This usually limits the resolution to 10^{-3}–10^{-6} s, washing out many interesting temporal effects.

3. *Detection by Fluorescence Excitation*

As noted above, fluorescence emission often follows two-photon absorption. This one-photon process occurs at a higher frequency than that of the exciting laser. Usually the spectral separation between the frequencies is large, minimizing scattered light problems. The method is highly sensitive, as shown below, and has been used in many TPE applications.[9] Offering excellent time resolution (better than 10^{-9} s), it has so far been the only one used in kinetic studies.

The method's sensitivity can be estimated by following up the previous example. Assuming a fluorescence quantum yield of 0.1 and a spatial collection efficiency of 10^{-3}, a population of 5×10^9 excited molecules leads to 5×10^5 photons reaching the detector per pulse. Single-photon counting techniques easily allow the observation of 0.1–1 photon per pulse. Evidently fluorescence methods can easily be used to probe submillitorr samples.

4. *Detection by Ionization*

Being promoted to an electronically excited state, a molecule (atom) may be further energized by absorption of more photons. The process often leads eventually to ionization, making it easily observable.[10] If the rate-determining step of the overall process is the initial coherent two-photon absorption rate, multiphoton ionization spectra are dominated by two-photon resonances. The method is highly sensitive, as demonstrated by recent molecular beam studies.[11] It has obvious kinetic applications, mostly in determining branching ratios and appearance rates of ionic fragments.

5. *Other Nonlinear Techniques*

Four-wave mixing has been used to generate VUV laser radiation.[12] In some cases (such as NO) two-photon resonances are decisive factors in overall yield, and laser output may be related to the two-photon absorption cross-section. A similar technique involves mixing of two dye laser frequencies ω_1 and ω_2 to yield ω_3 by $\omega_3=2\omega_1-\omega_2$. This process is resonance enhanced when $2\omega_1$ coincides with a molecular transition.[13,14] Keeping ω_2 constant and scanning ω_1, one obtains the spectrum of ω_3. This spectrum is strongly enhanced in two-photon resonances and may be used to derive the two-photon cross-section of the medium. The technique does not depend on subsequent processes, such as fluorescence or ionization, for detection. Its sensitivity, however, is limited by the ever-present nonresonant third-order nonlinear susceptibility $\chi^{(3)}$,[15] limiting its use at present to condense phases and relatively high-pressure gases.

6. *Chemical Methods*

In some favorable cases TPE was efficient enough to form measurable amounts of photodecomposition products. Examples are the dissociation of iodoform by a ruby laser[16] and of water by a doubled dye laser.[17] In the first, the reaction was followed by titrating the liberated iodine, while in the second, OH radicals were monitored by laser-induced fluorescence, using the same laser frequency for both two-photon dissociation of H_2O and one-photon fluorescence excitation of OH.

7. *Null Methods*

Two-photon absorption can lead to changes in the polarization of the exciting beam. The changes can be described, in the case of elliptically polarized light, as rotation of the polarization ellipse. One can detect this rotation, by placing a two-photon absorption cell between two linear polarizer–quarterwave retarder combinations. In the absence of a sample, the combination is set so that the retardation caused by the first quarter-wave plate is exactly canceled by the second so that no light can pass through the crossed polarizers. With a sample in the cell, any slight change in the polarization ellipse destroys the exact match, and some light passes on through the second polarizer onto a detector. The method was termed ELLIPSA (elliptical polarization state alteration)[18] and should prove to be of general applicability, being potentially quite sensitive.

The fate of two-photon excited neutral molecules and atoms is most easily followed in real time by fluorescence monitoring. In the next sections we show that in spite of the small two-photon absorption cross-section, the method can be used for quantitative kinetic measurements. For some applications, it is more straightforward and leads to less ambiguous interpretation than one-photon excitation.

III. KINETIC STUDIES

A. Motivation

Thus far, the use of TPE for kinetic studies was limited to relatively few cases [see, e.g., (19)–(23)]. This is hardly surprising in view of the smallness of δ and the difficulties encountered in even detecting TPE. It turns out, however, that the small absorption cross-section can actually be of advantage in some instances, as discussed below. Furthermore, present dye laser technology (using doubling crystals) provides tunable, narrowband radiation in the range from about 200 nm to about 1000 nm. The time resolution is 10^{-8} s with most pump lasers currently used (Q-switched Nd–YAG, excimer, or nitrogen laser) but can be readily extended to 10^{-11}

s using mode-locked lasers.[24] TPE extends the wavelength range to 100 nm, without using expensive vacuum UV optics, being supplemental to frequency doubling and summing techniques that are not universally available. These properties, coupled with the different selection rules, make TPE a tool for populating a range of excited states not readily accessible by one-photon excitation.

In quantitative photokinetic studies one must control the spatial and temporal population of the initially excited species. The discussion in Section IV centers on the special case of fluorescence quenching. Energy transfer to an acceptor (quencher) results in changes in fluorescence lifetime and intensity. Typically these changes are monitored as a function of quencher pressure, keeping the donor pressure constant. It is attempted to keep the initial conditions (concentration and spatial distribution of excited donor molecules) constant as well. Difficulties arise when donor and acceptor absorption spectra overlap: inner filter effects decrease initial excited donor concentration, resulting in lower emission intensity. The same problem is encountered in the determination of self-quenching rate constants, which must be known for correct analysis. Taking NO as an example, it was found necessary to use ternary mixtures[25] to extract self-quenching rate constant. Cases of considerable photochemical interest were not studied, apparently because of overlapping donor and acceptor absorption. Examples pertinent to NO photochemistry are NO–SO_2, NO–CH_3NO_2, and NO–organic amines mixtures. Complications due to acceptor's absorption are avoided when optically thin conditions hold, namely, when the absorbed fraction ΔI_0 is much smaller than the incident light intensity I_0. In that case (optically thin conditions) Beer's law leads to the relation

$$\Delta I_0 = I_0 \alpha P l \tag{9}$$

with α the absorption coefficient and p the absorber pressure. For several absorbing species

$$\Delta I_0 = I_0 l \sum_i \alpha_i P_i = \sum \Delta I_i \tag{10}$$

i.e., light attenuation is additive. Excited state population of the donor per unit length, n_D^*, is given by

$$n_D^* = I_0 \alpha_D P_D \tag{11}$$

being proportional to the incident intensity and the donor pressure P_D, even in the presence of other absorbing species. In the case of two-photon absorption, we get

$$n_{\mathrm{D}}^{*}=(1/2)I_0^2\delta_{\mathrm{D}}P_{\mathrm{D}} \tag{12}$$

the small value of δ ensuring optically thin conditions at practically any pressure of donor and acceptor(s). Thus, quenching cross-sections can be extracted for donor–acceptor pairs whose one-photon absorption spectra strongly overlap. The method is of special interest in the energy range above 45,000 cm^{-1}, where many molecular absorption spectra are very intense, and where tunable, pulsed laser sources are not readily available.

B. Excited States of Nitric Oxide

A partial energy level diagram of NO is shown in Fig. 1. The ground state, $X^2\Pi$, has the configuration

$$KK(\sigma_g 2s)^2(\sigma_u^* 2s)^2(\sigma_g 2p)^2(\pi_u 2p)^4(\pi_g^* 2p) \tag{13}$$

One can thus roughly consider the molecule as consisting of a core, composed of 14 electrons in closed shells and a lone electron orbiting outside. Transitions of the lone electron to states with principal quantum numbers exceeding 2 lead to Rydberg series, converging to the ionization potential. Examples of Rydberg states are $A^2\Sigma^+, C^2\Pi, D^2\Sigma^+$.

Transitions of core electrons into the antibonding π^* orbital are called valence transitions. Examples of valence excited states are $a^4\Pi, b^4\Sigma^-, B^2\Pi$. It is noted that excitation into Rydberg states moves the antibonding electron away from the nuclei, while in valence state transition a bonding electron is moved to an antibonding orbital. Therefore, one expects the internuclear distance to decrease for Rydberg excitation and increase for valence state excitation. These expectations are borne out by experiment (cf. Fig. 1).

C. Quenching Kinetics

For states below the dissociation limit of the molecule, the decay rate constant in the presence of a quencher Q may be written as

$$k_{\mathrm{TOT}}=k_f+k_{\mathrm{SQ}}P+k_Q Q \tag{14}$$

with $k_f=\tau_f^{-1}$ the radiative rate constant, k_{Q} and k_{SQ} the quenching rate constants by Q and NO, respectively, and P and Q the pressures of NO and the quencher, respectively.

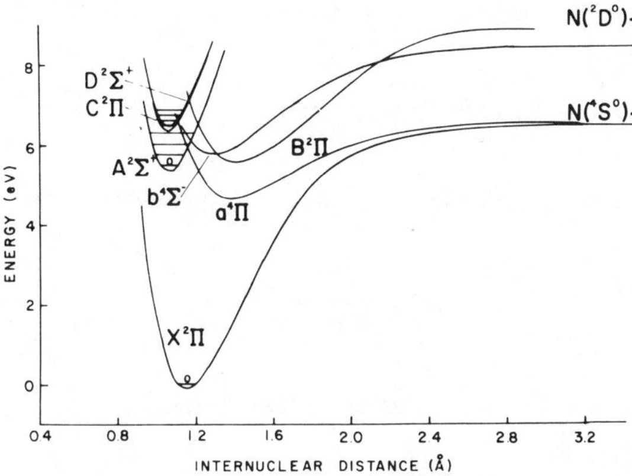

Fig. 1. Potential energy diagram of some low-lying states of NO (after Ref. 26). The energies of some vibrational states belonging to the *A*, *C*, and *D* electronic states are also shown.

The straightforward way to obtain k_{SQ} is to use the Stern–Volmer relation

$$\frac{\tau_f}{\tau(P)} = 1 + \frac{k_{SQ}}{k_f} P \tag{15}$$

where $\tau(P)$ is the measured lifetime at pressure P. Similarly

$$\frac{\tau(P)}{\tau(Q)} = 1 + k_Q \cdot \tau(P) \cdot Q \tag{16}$$

yields k_Q ($\tau(Q)$ is the measured lifetime in the presence of the quencher). In cases where direct lifetime measurements are not practical, one may use the integrated intensity as a measure. For exponential decay, $I(Q)$, the emission intensity at quencher pressure Q is proportional to $\tau(Q)$. One can thus write

$$\frac{I_0}{I(Q)} = 1 + k_Q \tau_0 Q \tag{17}$$

where I_0 and τ_0 are the intensity and lifetime, respectively, in the absence of Q (They are, of course, P-dependent.)

The case of self-quenching is slightly more complicated, as increasing the pressure causes an increase in the signal due to a larger excited-state population (12), even though the lifetime decreases. The quantum yield of fluorescence may be written as

$$\phi^f = \frac{k_f}{k_f + k_{SQ} P} \tag{18}$$

and from (12) fluorescence intensity is proportional to I_P^f:

$$I_P^f = \frac{k_f P I_0^2}{k_f + k_{SQ} P} \tag{19}$$

as $P \to \infty$, a constant value, I_∞^f, is approached, given by

$$I_\infty^f = \frac{k_f I_0^2}{k_{SQ}} \tag{20}$$

Experimental intensity ratios may be used to obtain k_{SQ} by plotting I_∞^f / I_P^f vs P^{-1} (21)

$$\frac{I_\infty^f}{I_P^f} = 1 + \frac{k_f}{k_{SQ} P} \tag{21}$$

For states beyond the dissociation limit, k_f is replaced by $k_f + k_{pr}$, where k_{pr} is the unimolecular predissociation rate constant.

D. Quenching Mechanisms

Very large quenching cross-sections were reported for Rydberg states of NO,[27] approaching and sometimes exceeding gas kinetic (GK) rate constants. Mechanisms leading to large cross-sections are

1) Resonant dipole–dipole energy transfer.[28]
2) Nonresonant dipole – dipole or dipole-induced dipole energy transfer.[29,30]
3) Charge transfer controlled energy transfer.[31]

For states beyond the dissociation limit, a collision-induced predissociation mechanism is also possible. In the case of NO, dissociation is energetically possible for the following states $A(v \geqslant 4)$, $C(v=0, J > 5\frac{1}{2}, v \geqslant 1)$ and $D(v \geqslant 0)$. Strong predissociation is known to take place in the $C^2\Pi$ state, possibly by strong coupling to the $a^4\Pi$ state.[32]

TPE is uniquely helpful in discerning between these mechanisms, as shown below. It turns out that the results appear to correlate well with mechanism 3), which was previously mostly discussed in relation to excited

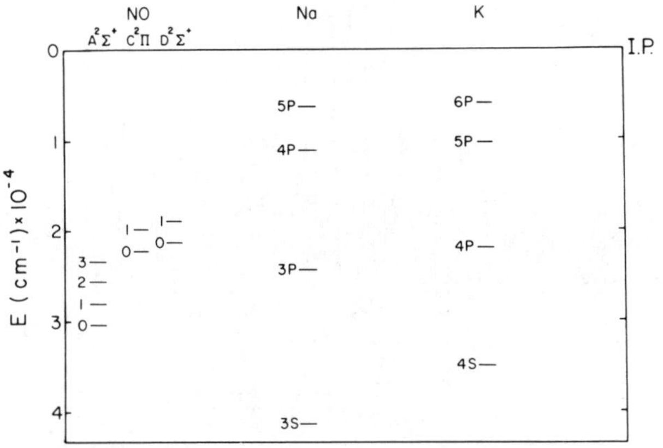

Fig. 2. Comparison of excited state energies of nitric oxide, sodium and potassium. Energies are measured with respect to the ionization potential (IP) of each species. It is seen that the ionization potentials of the *A*, *C*, and *D* states of NO are similar to those of the sodium 3*P* and the potassium 4*P* states.

atom quenching.[33,34] The process is envisioned as an electron transfer from the electronically excited species to an electron acceptor, *Q*, followed by charge transfer recombination leading to ground-state species (22)

$$NO^* + Q \rightarrow NO^+ + Q^- \qquad (22a)$$

$$NO^+ + Q^- \rightarrow NO + Q \qquad (22b)$$

Process (22a) requires a crossing between the ionic curve of the NO^+–Q^- pair and the covalent (largely repulsive) curve due to NO–Q interaction. This crossing is followed by rapid charge recombination. Quenching cross-sections may be determined from the distance *R* at which the crossing takes place. Similarity to alkaline atoms is suggested by Fig. 2, in which the ionization potentials of Na, K, and NO are taken as a common reference potential. Effective ionization potentials of the *A*, *C*, and *D* states of NO are seen to be close to those of the Na(3*P*) and K(4*P*) states.

IV. RESULTS

A. NO Self-Quenching

Self-quenching of NO is very efficient.[27] Rate constants for $A(v=0$, 1 and 2) were recently remeasured using two-photon excitation.[21] We have repeated those experiments, and extended them to the $A(v=3)$ $C(v=0,1)$ and $D(v=0)$ states.[22] Easy assignment of the initially excited state is

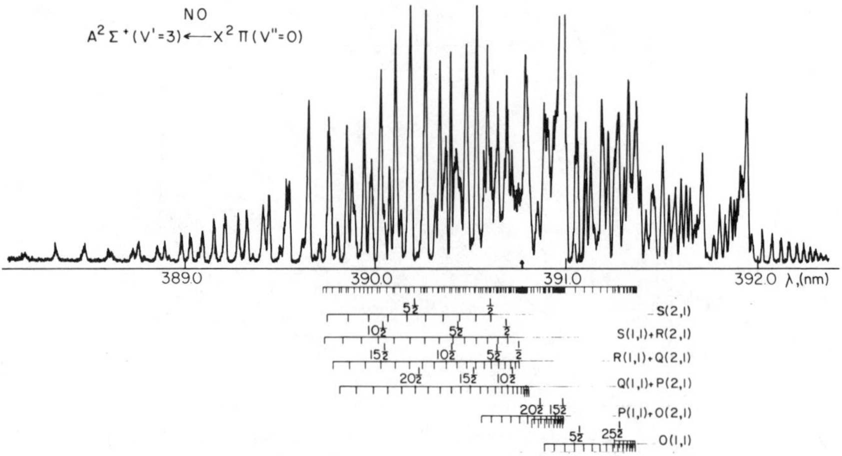

Fig. 3. A moderately high-resolution recording of the $A(v=3)$ two-photon excitation spectrum. Laser wavelength is given in the abscissa; resolution was about 0.15 Å. Partial rotational assignment is also shown.

ensured by the laser's narrow bandwidth. In fact, single rotational lines may be selectively excited, as shown in Fig. 3 for the $A(v=3)$ state. Self-quenching may result in population of emitting states other than the one directly reached by light absorption. The identity of the actual emitting state is established by dispersing the emission, using a low-resolution monochromator. Examples of the resulting spectra are shown in Fig. 4. Attention is drawn to the spectrum, assigned to $C(v=1)$. It has been stated[35] that this state is not observed in emission, due to fast predissociation. A recent synchrotron study reported emission on excitation to $C(v=1)$, but in the absence of spectral resolution of the emission, a definitive assignment could not be made.[36] Two-photon excitation leads to straightforward assignment: The structure shown in Fig. 4 agrees with calculated Franck–Condon factors and with band positions. These results and the fact that the spectrum is obtained only on excitation to the $C(v=1)$ state unambiguously determine the assignment. All the kinetic data reported below were obtained by monitoring the emission of a single vibronic transition that could be assigned to a well-defined upper state.

Data for the A state (with a zero-pressure lifetime of ~200 ns[32]) were obtained either from the decrease in fluorescence lifetime or from time integrated intensity attenuation. Agreement between the two methods was always better than 10%. The D state zero-pressure lifetime is about 25 ns[32] leading to a large experimental error in our data handling system. The problem is even more pronounced for the predissociative C state. Intensity

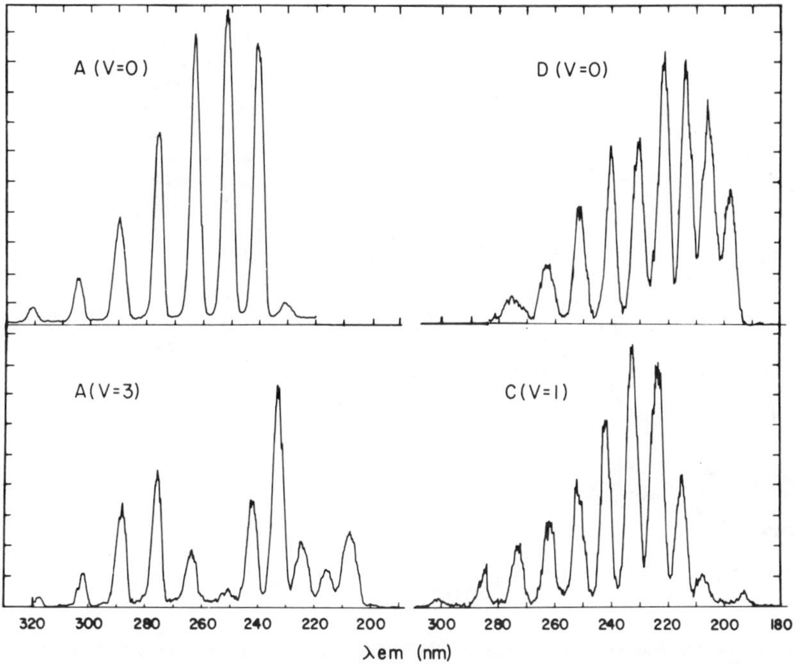

Fig. 4. Low resolution emission spectra of some vibronic states of NO, obtained by two-photon excitation. Relative intensities of the bands of a given progression agree with those calculated from known Frank–Condon factors,[44] except for the 0–0 bands which are attenuated by self-absorption. Intensities of bands due to different upper states were recorded at arbitrary sensitivity levels. The spectra shown are direct box car output recordings. Comparison with calculated spectra was made by taking into account the spectral response of the detection optics and the calibration curve of the monochromator.

measurements were thus employed, cf. (21), and they are summarized in Fig. 5 and in Table I.

Inspection of Table I shows that agreement between one- and two-photon excitation results is reasonable for the *A* state, and poor for the *C* and *D* states. As mentioned above, the *C* state predissociates on excitation beyond the $J=11/2$ level of $v=0$. In the absence of a reliable value for the zero-pressure lifetime for the $C(v=1)$ state, only a lower limit for the quenching rate constant can be obtained. It is seen to be much larger than for all other states, possibly indicating a collision-induced predissociation mechanism.

The $v=0$ level of the *D* state is not predissociative,[32] making the anomalously large self-quenching rate constant previously reported[37] rather

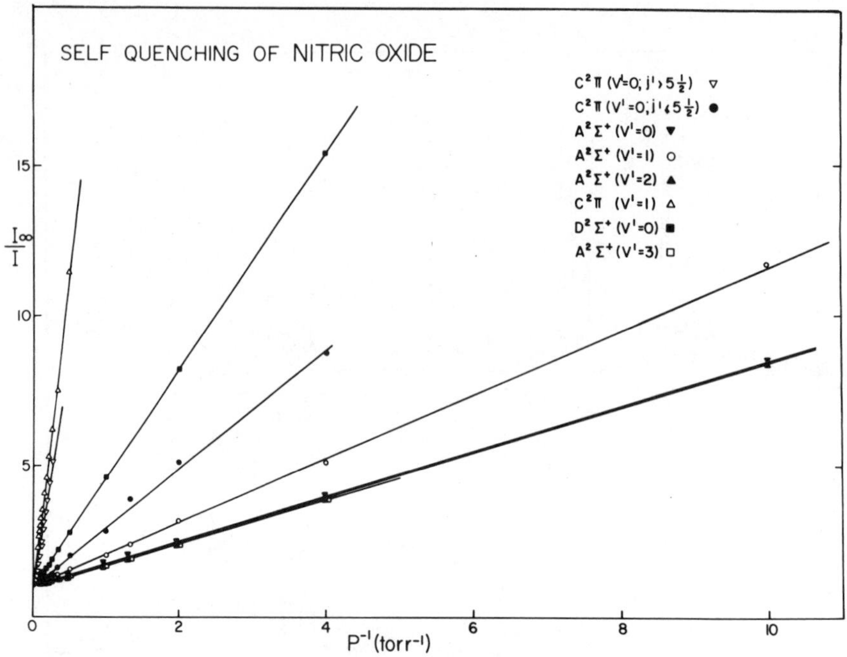

Fig. 5. Stern–Volmer plots of the self-quenching of several vibronic states of NO. For definition of I_∞/I see text.

TABLE I
Self-Quenching Rate Constants for Some Vibronic States of NO[a]

State	Zero-pressure lifetime[b] τ_f(ns)		Quenching rate constant, K_{SQ} (cm^3 s^{-1} molecule^{-1} × 10 — Two-photon excitation[b]		One-photon excitation[b]	
$A(v=0)$	205	[32]	1.9±0.1	[21, 22]	1.5±0.1	[25]
$A(v=1)$	200	[32]	1.4±0.1	[21, 22]	1.6±0.1	[25]
$A(v=2)$	195	[32]	2.0±0.2 1.5±0.15	[22] [21]	1.4±0.2	[25]
$A(v=3)$	213	[32]	1.9±0.1	[22]	1.7±0.1	[36]
$C(v=0)$, $J<11/2$	32	[32]	4.6±0.6	[22]	~4	[36]
$C(v=1)$	≤0.3	[36]	≥47	[22]	—	
$D(v=0)$	25.7	[32]	3.1±0.3	[22]	8±1 35	[36] [37]

[a] The gas kinetic constant is 2.5×10^{-10} cm^3 s^{-1} molecule^{-1}.[38]
[b] Reference numbers in brackets.

surprising. We have obtained a much smaller value, 3.1×10^{-10} cm^3 s^{-1} $molecule^{-1}$, being only slightly bigger than gas kinetic. It is found that collision-induced $D \rightarrow A$ transfer is very effective,[22] leading to strong A-state emission after D-state excitation. Since A-state lifetime is much longer than that of the D-state, steady-state intensity measurements may introduce an error in calculating k_{SQ}. In view of the rather indirect method used in Ref. 37, it is likely that the two-photon result is more dependable. As detailed below, comparison with other quencher molecules also lends support to the lower value reported herewith (see Tables II and III).

B. Quenching of NO Rydberg States by Other Molecules

One-photon absorption spectra of NO and CH_3NO_2 are shown in Fig. 6. It is evident that, in a mixture of the two, absorption of nitromethane dominates. The figure serves to illustrate the usefulness of two-photon excitation, as discussed in Section III. TPE allows selective excitation of NO in the presence of nitromethane without needing to correct for the latter's attenuation of the laser light. A remaining necessary precaution, as far as inner filter effects are concerned, is attenuation of NO fluorescence by the tail of nitromethane absorption. The problem can be minimized by using long wavelength vibronic emission bands (cf. Fig. 6). Rate constants were extracted from Stern–Volmer plots, such as shown in Fig. 7. They

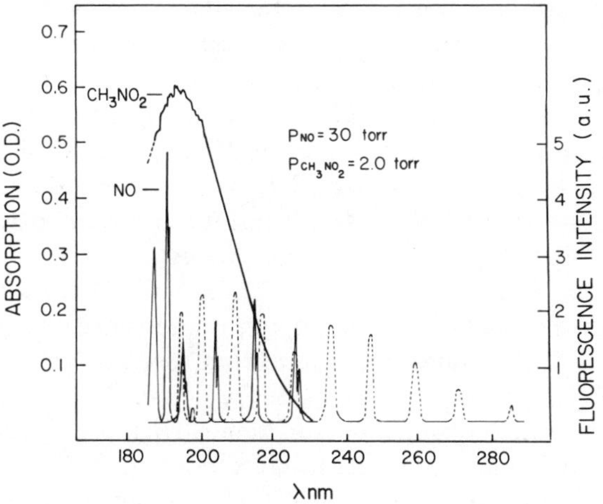

Fig. 6. Absorption spectra of nitric oxide and nitromethane between 180 and 240 nm. Note that nitromethane absorption would dominate the absorption of a mixture. The dashed curve (right-hand ordinate scale) shows the calculated emission spectrum after excitation to the $D(v=0)$ state of NO.

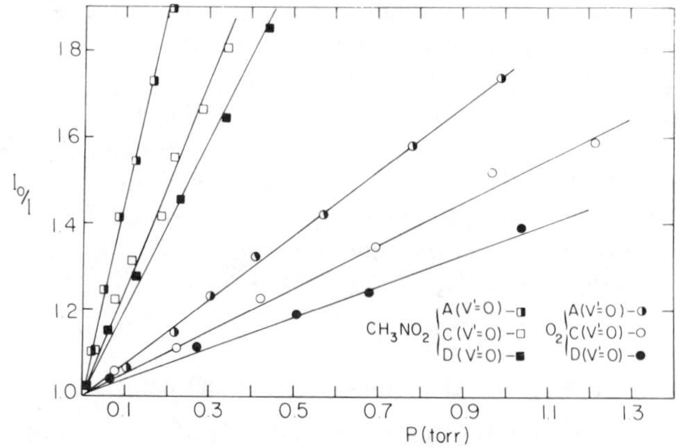

Fig. 7. Stern–Volmer plots of NO fluorescence quenching by oxygen and nitromethane. Data are shown for the ground vibrational states of the $A^2\Sigma^+$, $C^2\pi$ and $D^2\Sigma^+$ of NO, as indicated in the figure. Square symbols refer to CH_3NO_2, circles to 0_2.

were converted to energy transfer cross-sections, σ, using

$$k_Q = \sigma\langle V\rangle \tag{23}$$

where $\langle V\rangle$ is the average relative velocity given by $\langle V\rangle = (8kT/\pi\mu)^{1/2}$, μ being the reduced mass of the collision pair.

A case of special interest is that of SO_2, whose one-photon absorption is also known to overlap that of NO. In distinction with CH_3NO_2, the excited states are bound, with predissociation setting in at about 45,600 cm^{-1}.[39] At lower energies, excitation leads to fluorescence, which may introduce complexities and ambiguities to the kinetic analysis. It turns out that, while one-photon absorption of SO_2 is about one order of magnitude more intense than that of NO ($A \leftarrow X$ transition) the trend is reversed in the corresponding two-photon transitions. We estimate δ(NO) to be about 3 orders of magnitude larger than $\delta(SO_2)$ at around 4400 Å. This estimate is based on our observation of two-photon fluorescence excitation spectra of SO_2, taken with the same apparatus as that used for NO. Unambiguous assignment of the absorption was made by comparing the observed frequencies with known SO_2 single-photon bands. High pressures of SO_2 were required to obtain moderate signal-to-noise ratio, making it impossible to obtain an accurate comparison with NO. (The results were obtained with linearly polarized light.) This situation exemplifies the different selection rules controlling one- and two-photon transitions. Thus, even in the case of accidental energetic overlap, direct two-photon excitation of SO_2 in the presence of NO may be neglected.

TABLE II

Quenching Cross-Sections for Some NO Vibronic States (Å^2)

Quencher	$A(v=0)$	$A(v'=1)$	$A(v'=2)$	$A(v'=3)$	$D(v'=0)$	$C(v'=0)$	Gas kinetic
NO	29±3	20±2	31±3	29±3	48±5	71±7	38
O_2	22±2	22±2	—	23±2	67±6	78±8	37
N_2	—	<0.01	—	<0.01	6.8±0.7	—	40
CO_2	—	39±4	—	—	106±10	—	44
SO_2	140±14	125±12	150±15	120±12	228±23	217±21	49
CH_4	0.4±0.04	—	—	0.77±0.08	21±2	—	42
CH_3Cl	64±6	80±8	—	80±8	99±9	—	45
CH_2Cl_2	96±9	96±9	—	120±10	107±10	—	53
$CHCl_3$	110±11	130±13	—	90±9	175±17	—	58
CCl_4	102±10	142±14	—	80±8	188±18	—	64
CH_3NO_2	148±15	143±14	154±15	137±14	430±40	440±40	50

A second complication can arise from reemission of radiation by SO_2 after being populated by energy transfer. It turns out that such emission, if present, is quite weak: The emission spectrum of NO–SO_2 mixtures, for which extensive quenching took place, was observed to be essentially due to NO. As a further precaution, detection was limited, as always, to one of the well-known emission bands of NO.

The results for a number of quenchers are summarized in Table II. It is seen that in many instances the quenching cross-sections are considerably larger than gas-kinetic. In general, *C*- and *D*-state quenching probabilities are comparable for a given molecule, and are larger by a factor of 2–3 than the *A*-state quenching probabilities. A notable exception are N_2 and CH_4 for which a dramatic increase in the rate is observed on going from the *A* to the *D* state. Chlorination of methane with a single chlorine atom leads to a much larger rate constant for all states studied, while further chlorination has a relatively small effect. As shown below, this fact lends support to a nonresonant mechanism.

Vibrational excitation in the *A* state has a comparatively small effect on the overall cross-section. It should be recalled that the method used monitors total depletion of the emitting level, that is, the electronic and vibrational deactivation. Our results indicate that electronic energy transfer is faster than vibrational relaxation. Further support of this contention was provided by using the deuterated analogs of $CHCl_3$ and CH_2Cl_2: Within experimental error, the cross-sections obtained were the same as the protonated compounds, for all $A(v)$ states.

Finally, $C(v=1)$ quenching rate was measured for SO_2. As in the case of self-quenching, a huge value was obtained (over 30 times larger than gas kinetic), in line with the assumption of a different mechanism for this state.

V. QUENCHING MECHANISM OF NO RYDBERG STATES

In this section we attempt to provide a unified picture of the data presented in Table II and the discussion following it. While it is in principle possible that the quenching mechanism varies from one collision partner to the other, it will be shown that the data can reasonably well be accounted for by the charge transfer mechanism. Other mechanisms will be first briefly considered, in order to point out their inadequacies.

Resonant energy transfer[28] (Section III.D, mechanism 1), can be very efficient in favorable cases, leading to cross-sections of the order of 50–200 $Å^2$. The mechanism requires a small energy gap between donor and acceptor levels and favorable Franck–Condon factors. Figure 8 shows the approximate energies of known electronic states of the quenchers used in this study. Triplet states, which might play an important role in the case of substituted methanes, are not shown as their position is unknown. As Fig. 8 shows, a resonant mechanism can account for the large cross-sections in the case of O_2 and SO_2. However, it should lead to a sharp rise in quenching efficiency of N_2 between $A(v=2)$ and $A(v=3)$. In fact, above $A(v=2)$ O_2 and N_2 should exhibit similar efficiencies. Likewise, CO_2 should be a much poorer acceptor for $A(v=0)$ than $A(v \geqslant 1)$, a trend not substantiated by experiment.[27] Resonant transfer is expected to lead to observation of fluorescence from SO_2, in the case of $A(v=0)$, as it lies below the dissociation limit of SO_2. No such fluorescence was detected; while this negative result cannot be taken as a proof that electronically excited SO_2 was not produced, it is compatible with such a statement. A Förster mechanism requires strong overlap between donor emission and acceptor absorption spectra. This condition does not hold for methane and its chlorine-substituted derivatives.

It may be concluded that the overall picture strongly suggests that if a universal mechanism prevails in NO Rydberg state quenching, it is not the resonant energy transfer mechanism.

Mechanism 2 of Section III.D was often used to correlate quenching cross-sections, σ_Q, obtained for a common donor and a line of acceptors. An approximate result for σ_Q is[30]

$$\sigma_Q = A\mu^{\frac{1}{2}} I_P^2 I_Q^2 (I_P + I_Q)^{-2} \alpha_Q^2 R_C^{-9} + C \tag{24}$$

Here σ_Q is the quenching cross-section of molecules P by molecule Q, μ is the collision-reduced mass, α_Q the quencher polariabilizty, I_P and I_Q are the ionization potentials of P and Q, respectively, and R_C the Lennard–Jones collision diameter. C is a constant, and A is given by

$$A = 0.38\pi^{5/2} (2h^2 kT)^{-\frac{1}{2}} \rho F_{if} \alpha_P^2 \tag{24a}$$

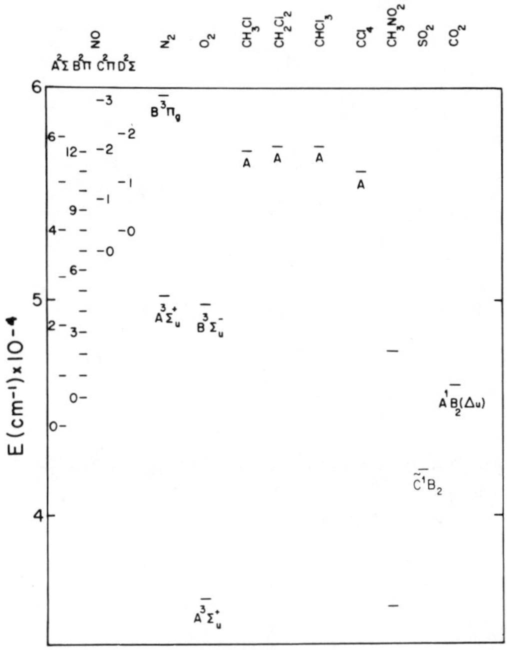

Fig. 8. Energy levels of NO and some acceptors. The vibrational states of NO are also shown. Data are from Ref. 40 except for CH_3Cl, CH_2Cl_2, $CHCl_3$, CCl_4 and CH_3NO_2, taken from Ref. 41. The lowest known excited state of methane is above 75,000 cm^{-1}.[41]

where ρ is the density of states of P at the excitation energy, F_{if} are Franck–Condon factors between initial and final states of P and α_P is its polarizability.

Equation (24) should hold for nonpolar quenchers. In the case of polar quenchers a term of the form $B\mu^{1/2} D_Q R_C^{-3}$ is added, D_Q being the quencher's dipole moment. These expressions proved to be useful in the analysis of energy transfer trends of, for example, propynal and glyoxal.[30,43]

A plot of σ_Q vs the Thayer–Yardley parameter is shown in Fig. 9 for the $A(v=1)$ state. A clear trend emerges from the figure—lending support to the validity of nonresonant energy transfer. A difficulty arises, however, on trying to estimate absolute values according to the model. Equation (24) may be used to calculate the quenching cross-sections for the $A(v=0)$ or $A(v=1)$ states. The process may be written as

$$NO(A, v') + Q \rightarrow NO(X, v'') + Q \tag{25}$$

that is, conversion of electronically excited NO molecule to a highly

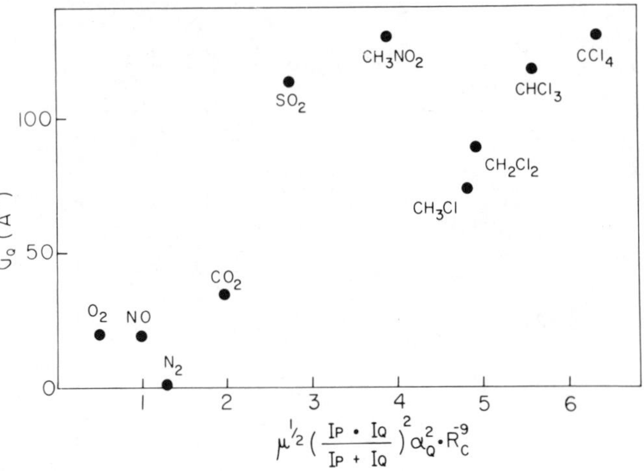

Fig. 9. A plot of observed rate constants vs the Thayer–Yardley parameter (Eq. 24), for quenching of the $A(v=1)$ state.

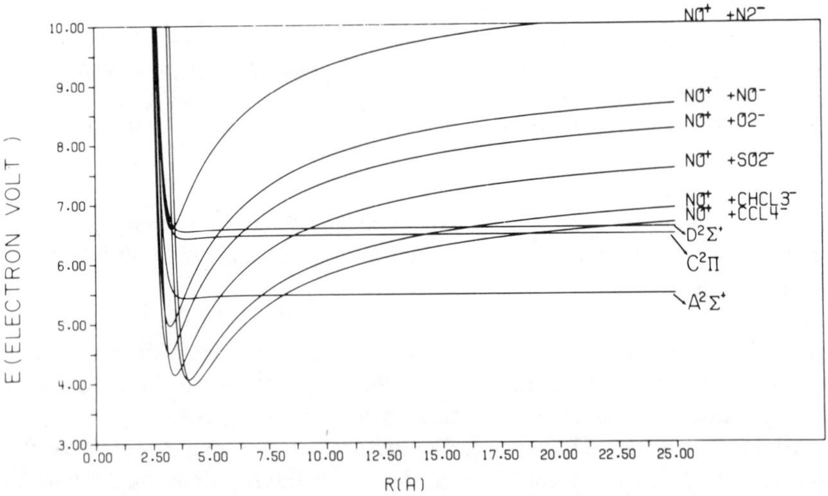

Fig. 10. Covalent and ionic potential curves for some NO-quencher pairs. Equation (26) was used to calculate the ionic curves, using the electron affinities of Table III. The covalent curves are Lennard–Jones potential of NO and the respective quencher, displaced by the electronic excitation energy of the appropriate Rydberg state. See text for further details.

vibrationally excited one. Dissociation is precluded for the $A(v=0)$ and $A(v=1)$ states. All the necessary data for obtaining the absolute value of σ_Q are available for the case in question. In particular the density of states and the Franck–Condon factor can be obtained.[40,44] It turns out that for molecules such as O_2 or CCl_4, the resulting cross-section is several orders of magnitude smaller than the experimental one. It thus seems that the correlation of Fig. 9 could be accidental.

The third mechanism proposed in Section III.D involved a charge transfer process. Figure 10 shows the potential curves calculated for the acceptors whose electron affinities are known. The ionic curve was calculated from the expression

$$V(r)=I(\mathrm{NO}^*)-EA(\mathrm{Q})-e^2/r+V_{\mathrm{LJ}} \tag{26}$$

where $I(\mathrm{NO}^*)$ is the effective ionization potential of an electronically excited NO molecule (cf. Fig. 2), $EA(\mathrm{Q})$ is the electron affinity of the quencher, r the intermolecular distance and V_{LJ} the Lennard–Jones potential $V_{\mathrm{LJ}}=4\varepsilon[(r_0/r)^{12}-(r_0/r)^6]$. ε was taken as 0.05 eV throughout and r_0 obtained from Ref. 38. The covalent curves in Fig. 10 are simply given by $V_{\mathrm{cov}}=E_{\mathrm{EL}}+V_{\mathrm{LJ}}$, with E_{EL} the electronic energy of the excited state of NO above ground-state energy. The quenching cross-section σ_Q can be obtained from the crossing distance R derived from Fig. 10, using the relation[42]:

$$\sigma_Q=\pi R^2 e^{-\alpha R} \tag{27}$$

α is a correction factor resulting from the radial dependence of the electronic wavefunction.

Table III summarizes the results. It is seen that, using a single α parameter, all calculated cross-sections agree with experimental values within a factor of 2. The credibility of the model is further enhanced by the following results.

The increased cross-section for a given quencher on going from the A to the C or D states is a natural consequence of the model. The special case of N_2 is interesting: For the A state, no crossing between the ionic and covalent curves is obtained, indicating that charge transfer mechanism is not applicable. In distinction, a finite crossing is obtained for the D state: Thus the large increase in cross-section observed for N_2 may be correlated with a change in the quenching mechanism. A similar situation possibly holds in the case of methane, which is expected to have a low, or negative electron affinity.

TABLE III

Calculated Quenching Cross-Sections

		$\sigma(\text{Å})^2$					
		$A(v'=0)$		$D(v'=0)$			
Molecule	Electron Affinity (eV)	Calc. (πR^2)	Calc. ($\pi R^2 e^{-\alpha R}$)[f]	Expr.	Calc. (πR^2)	Calc. ($\pi R^2 e^{-\alpha R}$)[f]	Expr.
NO	0.02[a]	45.3	33.4	29 ± 3	95.0	61.2	48 ± 5
O_2	0.44[b]	58.1	41.2	22 ± 2	132.7	78.9	67 ± 6
N_2	−1.5[c]	NO CROSSING			36.3	27.6	6.8 ± 0.7
SO_2	1.1[d]	91.6	59.5	140 ± 14	271.6	129.1	228 ± 23
$CHCl_3$	1.76[e]	162.8	91.5	110 ± 11	854.9	228.4	175 ± 17
CCl_4	2.0[e]	206.0	107.8	102 ± 10	1133.5	247.9	188 ± 18

[a] M. W. Siegel, R. J. Celotta, J. L. Hall, J. Levine, and R. A. Bennet, *Phys. Rev.*, **A6**, 607 (1972).
[b] R.J. Celotta, R. A. Bennett, J. L. Hall, M. W. Siegel, and J. Levine, *Phys. Rev.*, **A6**, 631 (1972).
[c] A. Lofthus and P. H. Krupenic, *J. Phys. Chem. Ref. Data*, **6**, 113 (1977).
[d] R. J. Celotta, R. A. Bennett, and J. L. Hall, *J. Chem. Phys.*, **60**, 1740 (1974).
[e] A. F. Gains, J. K. Kay, and F. M. Page, *Trans. Faraday Soc.*, **62**, 874 (1966).
[f] With $\alpha = 0.08$.

The model also accounts for the trend observed in the CH_xCl_{4-x}(x=0–4) series. Introduction of one chlorine atom is expected to lead to a large increase in *EA*. More chlorine atoms (or deuteration) are expected to have a much smaller effect. Indeed, comparison of $CHCl_3$ and CCl_4 is in line with this contention, as well as the negligible deuteration effect. A more detailed discussion of the model is given elsewhere.[45]

VI. SUMMARY

Two-photon excitation is complementary and occasionally superior to one-photon excitation for excited states kinetic study because of the following:

1. Selection rules are different from one-photon selection rules, leading to facile selective excitation of one component in a mixture of compounds with overlapping one-photon absorption spectra.
2. Small absorption cross-sections ensure optically thin conditions, and limit excitation to only a small volume (near the focal spot of the lens used to focus the beam).
3. It offers easy access to energy levels above 45,000 cm^{-1}, i.e., regions not readily accessible to one-photon tunable sources.

These properties were utilized in the study of energy transfer from NO to a variety of accepting molecules. Acceptors whose one-photon absorption spectrum strongly overlaps that of NO were investigated for the first time. Efficiency of resonant and nonresonant processes could thus be empirically compared. A unified mechanism, involving a charge transfer intermediate, was found to account reasonably well for all the observed rate constants.

A natural extension of the method consists of using it to monitor the nascent vibrational and rotational energy distribution in the products of a chemical reaction. The use of nanosecond laser sources and low enough substrate pressure can make the method most useful in the case of laser-induced dissociation (single or multiphoton). This application has recently been demonstrated for the photodissociation of CF_3NO.[46]

Acknowledgements

We thank the Israel Commission for Basic Research for financial support. We are grateful to Professor R. D. Levine for suggesting the curve–crossing model and for his enlightening discussions.

References

1. M. Gŏppert Mayer, *Ann. Phys. (Leipzig)*, **9**, 273 (1931).
2. H. Mahr, in *Quantum Electronics: A Treatise*, Vol. 1, part A, H. Rabin and C. L. Tang, eds., Academic Press, New York, 1975, pp. 285–361.

3. W. M. McClain and R. A. Harris, in *Excited States*, Vol. 3, E. C. Lim, ed., Academic Press, New York, 1978, pp. 1–56.
4. (a) F. Biraben, B. Gagnac, and G. Grynberg, *Phys. Rev. Lett.*, **32**, 643 (1974). (b) M. D. Levenson and N. Bloemberger, *Phys. Rev. Lett.*, **32**, 645 (1974). (c) T. W. Hänsch, K. C. Harvey, G. Meisel, and A. L. Schawlow, *Opt. Comm.*, **11**, 50 (1974).
5. R. Wallenstein and H. Zacharias, *Opt. Comm.*, **25**, 363 (1978).
6. (a) R. Swofford and W. M. McClain, *Rev. Sci. Inst.*, **46**, 246 (1975). (b) U. Fritzler, Ph. Kellerrand, and G. Schaak, *J. Phys. E.*, **8**, 530 (1975).
7. A. J. Twarowski and D. S. Kligler, *Chem. Phys.*, **20**, 253, 259 (1977).
8. W. R. Harshbarger and M. B. Robin, *Acct. Chem. Res.*, **6**, 329 (1973).
9. For an early application to nitric oxide, see, e.g., R. G. Bray, R. M. Hochstrasser, and J. E. Wessel, *Chem. Phys. Lett.*, **27**, 167 (1974).
10. (a) P. M. Johnson, *J. Chem. Phys.*, **64**, 4638 (1976). (b) D. Zakheim and P. M. Johnson, *J. Chem. Phys.*, **68**, 3644 (1978). (c) W. M. Jackson and C. S. Lin, Int. *J. Chem. Kinetics*, **10**, 945 (1978).
11. L. Zandee, R. B. Bernstein, and D. A. Lichtin, *J. Chem. Phys.*, **69**, 3427 (1978).
12. S. C. Wallace, in *Tunable Lasers and Applications*, A. Mooradian, T. Jaeger, and P. Stokseth, eds., Springer-Verlag, Berlin, 1976, p. 1.
13. R. T. Lynch and H. Lotem, *J. Chem. Phys.*, **66**, 1905 (1977).
14. R. B. Hochstrasser, G. R. Meredith, and H. P. Trommsdorf, *Chem. Phys. Lett.*, **53**, 423 (1978).
15. N. Bloembergen, *Nonlinear Optics*, Benjamin, New York, 1965.
16. S. Speiser and S. Kimel, *J. Chem. Phys.*, **51**, 5614, 1969; and **53**, 2392 (1970).
17. C. C. Wang and L. I. Davis, Jr., *J. Chem. Phys.*, **62**, 53 (1975).
18. R. J. M. Anderson, T. M. Stachelek, and W. M. McClain, *Chem. Phys. Lett.*, **59**, 100 (1978).
19. L. Wunsch, H. J. Neusser, and E. W. Schlag, *Chem. Phys. Lett.*, **31**, 1 (1975).
20. U. Boesl, H. J. Neusser, and E. W. Schlag, *Chem. Phys. Lett.*, **42**, 16 (1976).
21. H. Zacharias, J. B. Halpern, and K. H. Welge, *Chem. Phys. Lett.*, **43**, 41 (1976).
22. M. Asscher and Y. Haas, *Chem. Phys. Lett.*, **59**, 231 (1978).
23. D. S. Kligler and C. K. Rhodes, *Phys. Rev. Lett.*, **40**, 309 (1978).
24. J. N. Eckstein, A. I. Ferguson, and T. W. Hänsch, *Phys. Rev. Lett.*, **40**, 847 (1978).
25. A. B. Callear and I. W. M. Smith, *Trans. Faraday Soc.*, **59**, 1720 (1963).
26. R. F. Gilmore, *J. Quant. Spectr. Rad. Trans.*, **5**, 369 (1965).
27. See e.g., A. B. Callear, and M. J. Pilling, *Trans. Faraday Soc.*, **66**, 1618, 1886, (1970) and references therein.
28. Th. Forster, *Ann. Phys.*, **2**, 55 (1948).
29. J. I. Steinfeld, *Acct. Chem. Res.*, **3**, 313 (1970).
30. C. A. Thayer and J. T. Yardley, *J. Chem. Phys.*, **51**, 3992 (1972).
31. A. Bjerre and E. E. Nikitin, *Chem. Phys. Lett.*, **1**, 179 (1967).
32. (a) J. Brzozowski, N. Elander, and P. Erman, *Phys. Scripta*, **9**, 99 (1974). (b) J. Brzozowski, P. Erman, and M. Lyyra, *Phys. Scripta*, **14**, 290 (1976).
33. B. L. Earl and R. R. Herm, *J. Chem. Phys.*, **60**, 4568 (1974).
34. R. Bersohn and H. Horowitz, *J. Chem. Phys.*, **63**, 48 (1975).
35. F. Ackerman and E. Miescher, *J. Mol. Spectry*, **31**, 400 (1969).
36. O. Benoist d'Azy, R. Lopez Delgado, and A. Tramer, *Chem. Phys.*, **9**, 327 (1975).
37. A. B. Callear, M. J. Pilling, and I. W. M. Smith, *Trans. Faraday Soc.*, **64**, 2296 (1968).
38. J. O. Hirschfelder, C. E. Curtiss, and R. B. Bird, *Molecular Theory of Gases and Liquids*, Wiley, New York, 1964.
39. H. Okabe, *J. Am. Chem. Soc.*, **93**, 7095 (1971).

40. (a) G. Herzberg, *Spectra of Diatomic Molecules*, Van Nostrand, Princeton, N.J., 1950. (b) G. Herzberg, *Electronic Spectra of Polyatomic Molecules*, Van Nostrand, Princeton, N.J., 1966.
41. M. B. Robin, *Higher Excited States of Polyatomic Molecules*, Academic Press, New York, 1975.
42. R. E. Olson, F. T. Smith, and E. Bauer, *Appl. Opt.*, **10**, 1848 (1971).
43. R. A. Beyer and W. C. Lineberger, *J. Chem. Phys.*, **62**, 4024 (1975).
44. R. W. Nicholls, *J. Res. Natl. Bur. Std. U.S.*, **68A**, 535 (1964).
45. M. Asscher and Y. Haas, *J. Chem. Phys.*, **71**, 2724 (1979) and to be published.
46. M. Asscher, Y. Haas, M. P. Roellig and P. L. Houston, *J. Chem. Phys.*, **72**, 768 (1980).

INFRARED LASER-ENHANCED DIFFUSION CLOUD REACTIONS

MARK EYAL, URI AGAM, AND
FREDERICK R. GRABINER

Department of Chemistry, Tel-Aviv University, Tel-Aviv, Israel

CONTENTS

I. INTRODUCTION

The diffusion cloud (flame) technique developed by Hartel and Polanyi[1, 2] in the 1930s is one of the early methods of studying rapid bimolecular chemical reactions under pseudo-first-order, steady-state conditions. This method is the source of most measured rates for reactions of alkali metals with halogenated compounds[3] and still serves as a basis for experimental[4] and theoretical[5] studies. In most applications of the technique, sodium metal is heated in an oven, mixed with an inert carrier gas, and allowed to diffuse into a background of a reactant gas. In very exothermic reactions the sodium "flame" is chemiluminescent; otherwise the "cloud" is illuminated with a sodium resonance lamp. The reaction rate can be measured either by determining the distance the sodium diffuses until it all reacts or by spectroscopically measuring the total amount of sodium in the cloud.[3]

We have modified the diffusion cloud technique to study the vibrational energy dependence of reactions of alkali metals with polyatomic molecules. A steady-state reaction is prepared by the standard method, and the vibrational energy of the polyatomic reactant is then perturbed by absorption of energy from a *Q*-switched CO_2 laser. The effect of the added energy is determined by following the rate of departure from equilibrium sodium concentration after the laser pulse, on a time scale when the energy is still localized within the vibrational modes. The main reaction studied to date is that of sodium with sulfur hexafluoride:

$$Na + SF_6 \rightarrow NaF + SF_5$$

The relative importance of vibrational and translational energy in promoting chemical reactions is of both theoretical and practical interest. In reactions of diatomic molecules with atoms it has been substantiated both experimentally and theoretically that for endothermic reactions vibrational energy is more important, while for exothermic reactions the opposite is true.[6, 7] For polyatomic molecules, however, there is insufficient experimental and theoretical evidence to draw conclusions. The major work on laser-excited polyatomic reactions has involved the vibrational excitation of ozone in its exothermic reaction with nitric oxide.[8, 9] Although the vibrational energy increased the reaction rate, comparison with statistical models[10] and the temperature dependence of the thermal reaction[9] indicate about equal importance for vibrational and translational energy. On the other hand, a molecular beam study of the temperature dependence of the reaction of potassium with sulfur hexafluoride[11] has shown a definite preference for vibrational energy of the SF_6.

II. EXPERIMENTAL

The diffusion cloud apparatus is shown schematically in Fig. 1. A steady-state reaction is maintained in a flow cell containing a temperature-stabilized oven for heating sodium. Argon flowing through the oven carries sodium vapor into a background of the polyatomic reactant, and the mixture is pumped by a mechanical vacuum pump. Pressure and flow conditions are monitored with a capacitance manometer and rotameter flow meters. Since only the bottom part of the oven is heated, the temperature of the reaction zone is considerably lower than that of the oven. To measure the effect of additional vibrational energy the reaction zone is illuminated with the output from a *Q*-switched (pulsed) CO_2 laser. The laser can be tuned with a grating to coincide with the wavelength of maximum absorption of the gas. Since the laser is polarized, an infrared polarizer is used as a variable attenuator, and the laser output can be measured with a power meter either before or after the cell. The sodium concentration at a particular height above the nozzle is monitored by measuring the absorption from the collimated output of a hollow cathode sodium lamp. The change in sodium concentration following the laser pulse is measured with a transient recorder connected to a computer-based signal averager.

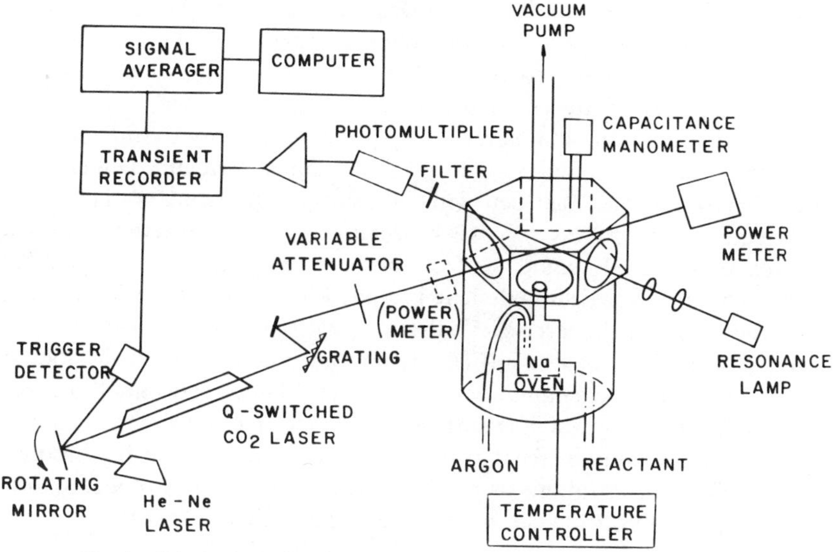

Fig. 1. Schematic diagram of the laser-excited diffusion cloud apparatus.

Steady-state reactions can be observed in this apparatus by causing the sodium cloud to fluoresce with light from a sodium resonance lamp. When the oven is heated, and the carrier gas is flowed through the oven, the sodium cloud is seen to fill the cell. The addition of a reacting gas contracts the cloud to a sphere or cone, depending on the flow rate and pumping speed. We have observed steady-state reactions with a number of gases which are known to also absorb radiation from a CO_2 laser: SF_6, CH_3F, CF_2Cl_2, CCl_4, CH_3Cl, CO_2, OCS, and SO_2. The laser-excited reaction has been observed with the first three of these molecules, and the reaction with SF_6 has been studied extensively.

Initial laser-excited experiments were performed at a relatively slow pumping rate achieved by pumping through a throttling valve, using CH_3F or SF_6 as reactants. In general, the sodium concentration decreased after the laser pulse, and returned to equilibrium at a slower rate. However, the signals also contained a large sinusoidal modulation, which at high pressures completely obliterated the desired signal. When helium was added to the gas mixture, the sinusoidal frequency increased in accordance with the difference in the velocity of sound. This implies that the modulation is due to changes in the sodium density due to acoustic waves. This is a phenomenon commonly observed in laser-induced fluorescence when the gas absorbs a significant amount of energy from the laser.[12] When ethylene was used as a "reactant" a similar signal was seen, which implies that even the initial signal can be due to a density change caused by laser heating of the gas. This is very similar to the thermal lensing effect,[13] which has been used to study laser-induced energy transfer.

When the flow rate was increased by pumping directly by the mechanical pump, the sinusoidal signals were no longer evident. Under these conditions no signal was seen with ethylene as the reactant. The signal with methyl fluoride was very weak because of its slow reaction rate. The fast reaction with SF_6 could easily be observed even at low pressures. The absence of any signal with ethylene is a good indication that what is being observed is in fact the laser-excited rate. Another interesting feature of the reaction with SF_6 is that at sufficiently high oven temperatures the sodium "cloud" becomes a chemiluminescent sodium "flame." This implies that a reaction is occurring which is sufficiently exothermic to excite the sodium *D* line. The chemiluminescent reaction, which occurs only at high sodium and low SF_6 concentrations, is probably due to the removal of a second or third fluorine atom from the sulfur by a sodium dimer. That is,

$$Na_2 + SF_5 \rightarrow SF_4 + NaF + Na^*$$

in addition to the normal reaction. The chemiluminescence intensity is also affected by the laser.

III. RESULTS

Typical rise and decay curves for the $Na + SF_6$ reaction are shown in Fig. 2, in which a positive signal implies a decrease in Na concentration. The time-dependent signal is a small fraction of the steady-state absorption and can be considered to be linear in concentration. The sodium concentration is seen to decrease after the laser pulse in a time determined by the reaction rate, and then to return to equilibrium on a longer time scale, as determined by the vibrational relaxation of SF_6, diffusion of Na from the oven, and flow of the reaction system past the region illuminated by the hollow cathode lamp. Since all the competing physical processes involved in the reaction, such as energy transfer and acoustic effects, have rates that depend on the SF_6 concentration, while the chemical effect of interest, the excited-state reaction, is the only effect that has a rate dependent on the level of SF_6 excitation, the reaction was measured as a function of laser power, with all other parameters fixed. The laser power was varied by passing the beam through a rotatable polarizer.

Experimental results for the reaction carried out under two different sets of conditions are shown in Fig. 3 and 4. Figure 3 represents an experiment performed at a temperature of 384 K with 87 mtorr of SF_6, ~10 μtorr of

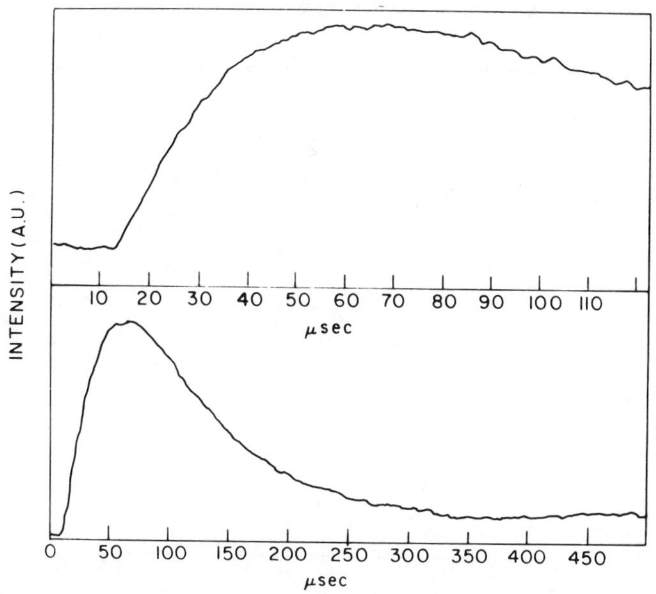

Fig. 2. Sample signal averager output for the reaction of sodium with sulfur hexafluoride, shown on two time scales. The transient recorder is triggered several microseconds before the laser pulse. A positive signal indicates a decrease in sodium concentration.

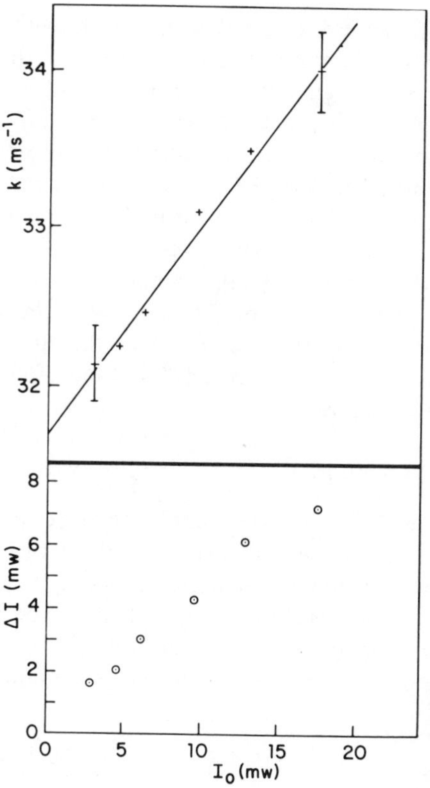

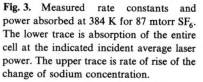

Fig. 3. Measured rate constants and power absorbed at 384 K for 87 mtorr SF_6. The lower trace is absorption of the entire cell at the indicated incident average laser power. The upper trace is rate of rise of the change of sodium concentration.

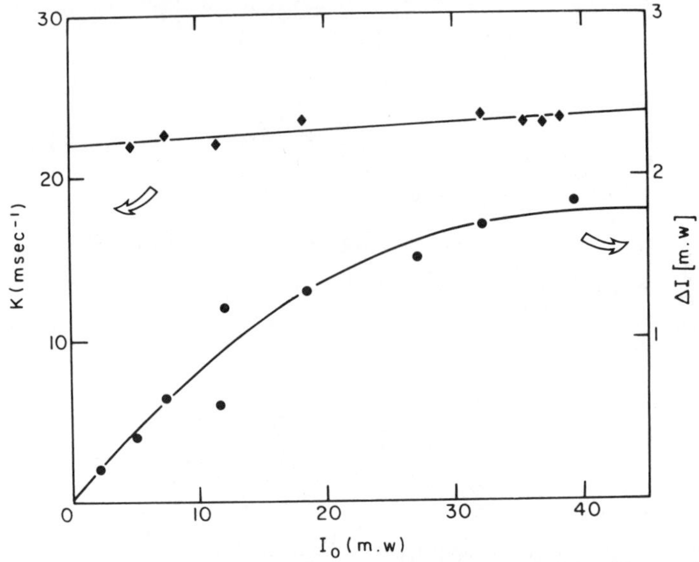

Fig. 4. Measured rate constants and power absorbed at 400 K for 40 mtorr SF_6.

Na, and about 1 torr of Ar. Figure 4 represents the data at 400 K with 40 mtorr SF_6, ~10 μtorr Na, and about 1 torr Ar. The top curve in both cases shows the measured rate of rise of the signal as a function of incident laser average power. The laser repetition rates are 122 Hz and 132 Hz, respectively. Since there are saturation effects in the absorption, the power absorbed is measured concurrently and plotted as the lower curve. In the second experiment the cell has been shortened and the pressure is lower, so the measured absorption is considerably lower. The slope of the rate constant graphs plotted against incident power is approximately linear. When plotted against the absorbed power the slopes increase exponentially. The measured rates are computed as the reciprocal of the time required for the signal to reach within $1/e$ of its maximum value after extrapolating the decay back to time equals zero.

IV. DISCUSSION

The first problem in analyzing the data is to determine the physical meaning of the measured rate and its intercept at zero added energy. Before the laser pulse, the sodium concentration at the point probed by the sodium lamp reaches a steady-state value determined by a balance between the reaction rate, diffusion of reactants, and linear flow of the reacting system. Under these conditions, the rate of diffusion of sodium into the probed region, R_D, is equal to the removal rate of sodium by the reaction:

$$\frac{d[\mathrm{Na}]_s}{dt} = -k_s[\mathrm{Na}]_s + R_D = 0 \tag{1}$$

or

$$R_D = k_s[\mathrm{Na}]_s \tag{2}$$

where k_s is the steady-state reaction rate constant and $[\mathrm{Na}]_s$ is the steady-state sodium concentration. Since $[SF_6]$ is considered to be constant and in excess, it is included in the rate constants.

At the temperature of the reaction (384 K), the vibrational partition function of SF_6, Q_v, is equal to 7.85, and the average energy per molecule, E_v, is 1094 cm^{-1}, computed using the accepted harmonic energy levels.[11] Under these conditions it is not realistic to think of the laser as exciting a particular fraction of ground-state molecules to a particular excited state. Since the time scale of the laser excited experiment is after vibrational equilibration (V–V) and before vibrational relaxation (V–T), we will assume that the function of the laser pulse is to increase the average energy of the molecules by an amount ΔE, which is equal to the average energy absorbed per molecule. This increases k_s by an amount k_E, which decays exponentially with the rate constant k_d, which is in turn determined by the V–T rate, diffusion of SF_6, and linear flow. That is;

$$\Delta k = k_E e^{-k_d t} \tag{3}$$

This approach defines the vibrational relaxation of a large polyatomic molecule as the rate of decay of the average molecular energy rather than the rate of decay of a particular vibrational state. For the purpose of describing the excited-state reaction it is assumed that the concentration of SF_6 remains constant but that the rate constant increases with the average energy.

The advantage of this approach is that it allows the determination of the total vibrational energy dependence of the reaction by varying the laser power. This is in contrast to the treatment of reactions of definite excited states of smaller molecules, in which it is considered that the laser excitation promotes a certain fraction of molecules from the ground state to an excited vibrational state.

Since under the experimental conditions the sodium concentration is changed by no more than 1%, we can make the approximation that the rate of supply of sodium to the probed region is constant. Thus,

$$\frac{d[\mathrm{Na}]}{dt} = -(k_s + k_E e^{-k_d t})[\mathrm{Na}] + k_s[\mathrm{Na}]_s \tag{4}$$

For the limiting case of a slow decay rate, it can be seen that if $e^{-k_d t} \approx 1$, this equation has a solution with rate constant $k_{\mathrm{exp}} = k_s + k_E$. When the full equation is solved numerically it reproduces the experimental curves fairly well. Analysis of the computed curves with rate constants comparable to the experimental values yields a rate of rise approximately equal to $k_s + k_E$. The change in measured rate as k_E is changed is proportional to k_E. Since the final data analysis is logarithmic this proportionality factor is unimportant. In the first experiment (384 K) the intercept is within 5% of the reported reaction rate[14] corrected for temperature. However, in the second experiment (400 K) the difference is 20%. Since there are other effects that can affect the intercept, mainly a slight disalignment of the laser and probe beams, the values for k_E are computed with respect to the experimental intercept, but the experimental values used in the data analysis are referenced to the literature value of k_s instead of the intercept. The reported value for the rate is $10^{14.7} \exp(-E_a/RT)$ cm^3 mole^{-1} s^{-1}, as measured at 520 K. In the units used here, $k_s/[\mathrm{SF_6}] = 350$ msec^{-1} torr^{-1} at 384 K and $E_a = 1155$ cm^{-1}.

The change in average energy, ΔE, is computed from the energy absorbed per laser pulse. A Beer–Lambert formula modified for saturation effects is used to calculate the absorption of SF_6 above the nozzle. The maximum absorption reached was on the order of 100 cm^{-1}, or 0.1 photon per molecule. The bottom panels in Fig. 5 and 6 show the logarithmic dependence of k_{exp} on additional vibrational energy.

The slope of these graphs gives the "conversion efficiency," α[15, 16], of the reaction, which can be considered to be a measure of the efficiency of vibrational energy utilization in the reaction. At 384 K,

$$\alpha \equiv \left[\frac{\partial \ln k_{\mathrm{exp}}}{\partial \Delta E} \right] \cdot \mathbf{k}T = 0.40$$

while at 400 K the value is 0.27. (**k** is Boltzmann's constant.)

A desired result in an experiment with laser-excited reactants is the determination of the relative efficiency of the vibrational excitation compared to translational excitation. If the translational excitation data are not

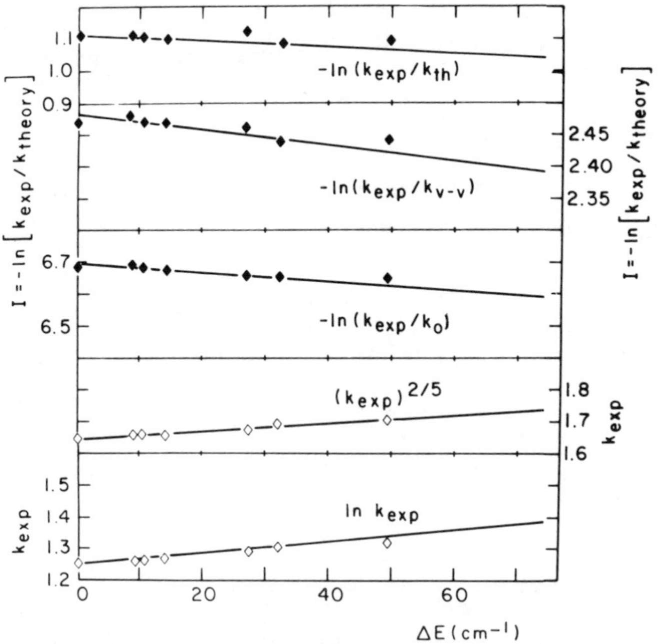

Fig. 5. Vibrational energy dependence of reaction of sodium with sulfur hexafluoride at 384 K. The lower two traces show the logarithmic and power law dependence of the experimental rate constant. The upper three traces are suprisal plots with reference to the different prior rates, as described in the text. The negative slope indicates a preference for vibrational over translational energy in promoting the reaction. All curves are plotted against the vibrational energy absorbed per molecule. 50 cm^{-1} is equivalent to 6.2 meV or 0.14 kcal/mole.

available, as in the present case, then the results must be compared with a statistical theory, which assumes equal effectiveness for vibrational and translational energy. Information theory can be applied to compare the statistical and experimental results.[17]

The reaction of sulfur hexafluoride with sodium is exothermic, but nevertheless has a reported activation energy. Either of these features can be used to define a priori reaction rate for comparison with the experimental results. Levine and Manz[17] have estimated that for a diatomic reaction with large exothermicity, $-\Delta E_0$, the reaction rate for a vibrational level with energy E_v should be proportional to $(E_v - \Delta E_0)^{5/2}$. This form can be considered for the present case if one assumes that when the S$-$F bond is broken, the SF_5 fragment does not carry off any excess internal energy. It is shown in Fig. 5 and 6 that a plot of $(k_{exp})^{2/5}$ vs. ΔE is linear.

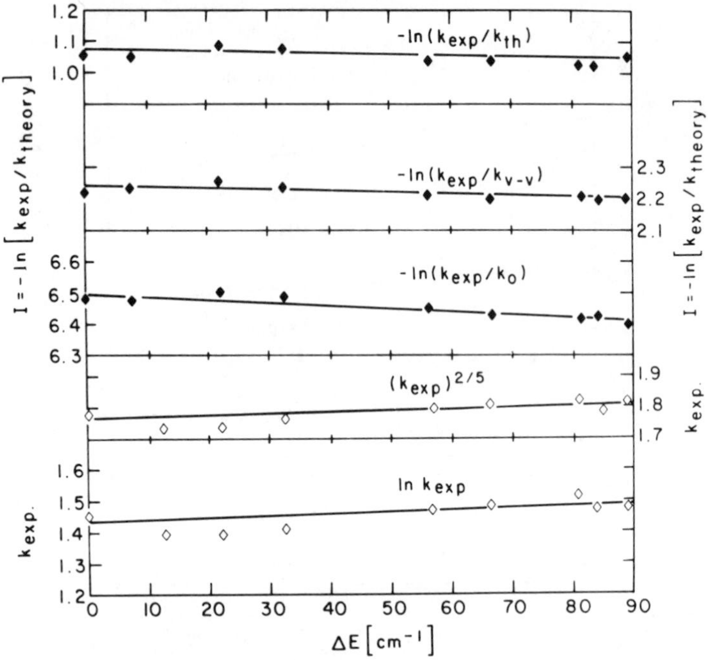

Fig. 6. Vibrational energy dependence of reaction of sodium with sulfur hexafluoride at 400 K.

In order to compute a theoretical reaction rate k_0 we first assume that the reaction rate, k_i^0, from vibrational level i with energy E_i is given by:

$$k_i^0 \propto \left(\frac{E_i - \Delta E_0}{\mathbf{k}T} \right)^{5/2} \tag{5}$$

If x_i is the fraction of molecules in level i, then

$$k_0 = \sum_{i=1}^{n} k_i x_i \tag{6}$$

or

$$k_0 = \frac{1}{Q_v(T_v)} \sum_{i=1}^{n} k_i^0 g_i \exp\left[\frac{-E_i}{\mathbf{k}T_v} \right] \tag{7}$$

where g_i is the degeneracy of level i, and T_v is the vibrational temperature after vibrational equilibration. k_0 was computed for each ΔE for which there was an experimental value. The vibrational temperature was determined by linear interpolation to calculate the temperature and partition function, which yield the given average energy per molecule. The exothermicity of the reaction was taken to be 5400 cm^{-1}, based on published values.[10, 18] The summation was carried out for all fundamental levels, their overtones up to about 2500 cm^{-1}, and all combinations thereof, using accepted values of SF_6 harmonic energy levels and degeneracies. This direct count yields a vibrational partition function within 1% of that computed by the method of infinite series of energy-level overtones.

Given this theoretical prior rate, k_0, the surprisal[17] of the experimental values is given by

$$I_v(E|T) = -\ln\left[\frac{k_{\text{exp}}(E|T)}{k_0(E|T)}\right] \tag{8}$$

and is plotted in Fig. 5 and 6. The slope of these graphs determines the "consumption potential",[17] $-\lambda_v$, of the reaction, given by

$$\lambda_v = \mathbf{k}T\left(\frac{\partial I_v}{\partial E}\right)_T \tag{9}$$

A negative value of λ_v implies a greater than statistical preference for vibrational energy. Our experimental values yield λ_v of -0.23 at 384 K and -0.17 at 400 K. These values are of the same order of magnitude as those for diatomic reactions, and show a preference for vibrational energy over translational energy.

Another possible prior rate for comparison is given by Cantrell et al.[10] and is based on the activation energy of the reaction. We assume the rate constant for each energy level to be given by

$$k_i \propto T^{1/2}\exp\left[\frac{-E_a+E_i}{\mathbf{k}T}\right] \qquad E_i \leqslant E_a \tag{10}$$

$$k_i \propto T^{1/2} \qquad E_i > E_a \tag{11}$$

If we assume vibrational equilibration as before, then this model used in (7) yields k_{v-v}. The surprisal compared to this model is shown in Figs. 5 and 6. In this case λ_v is equal to -0.19 at 384 K and -0.12 at 400 K. These results are virtually the same as for the previous prior rate.

A surprisal can also be computed using a thermal rate constant, k_{th}, for which it is assumed that the added energy is completely equilibrated. A uniform temperature is computed by dividing the absorbed energy by the heat capacity, and the rate is computed as before, with equal vibrational and translational temperatures. The surprisal for this model is also negative, and is equal to -0.14 and -0.17 for the respective temperatures.

The surprisals calculated from all three of these models thus indicate a preference for vibrational energy in promoting the reaction. Work is now progressing on a temperature analysis of both the steady-state and laser-excited reactions in order to provide further evidence. The results to date are in agreement with the molecular beam study of the equivalent reaction of SF_6 with potassium.[11]

The diffusion cloud method can thus be seen to be a potentially useful technique for studying the reactions of laser-excited polyatomic molecules. Since the reactant sodium is monitored, the same technique can be used for a large number of molecular reactants. By measuring the laser power dependence of the reaction rate information can be obtained on both the vibrational energy requirements and the steady-state value of the reaction rates.

Acknowledgment

This research was supported by a grant from the United States-Israel Binational Science Foundation (BSF), Jerusalem, Israel.

References

1. H. V. Hartel and M. Polanyi, *Z. Phys. Chem.*, **B11**, 97 (1930).
2. M. Polanyi, *Atomic Reactions*, Williams and Northgate, London, 1932.
3. E. Warhurst, *Q. Rev. Chem. Soc.*, **5**, 44 (1951).
4. U. C. Sridharan, T. G. Digiuseppe, D. L. McFadden, and P. Davidovits, *J. Chem. Phys.*, **70**, 5422 (1979).
5. D. G. Keil, J. F. Burkhalter, and B. S. Rabinovitch, *J. Phys. Chem.*, **82**, 355 (1978).
6. J. C. Polanyi, *Acct. Chem. Res.*, **5**, 161 (1972).
7. H. Kaplan, R. D. Levine, and J. Manz, *Chem. Phys.* **12**, 447 (1976).
8. M. J. Kurylo, W. Braun, C. N. Xuan, and A. Kaldor, *J. Chem. Phys.*, **62**, 2065 (1975).
9. R. J. Gordon and M. C. Lin, *J. Chem. Phys.*, **64**, 1058 (1976).
10. C. D. Cantrell, S. M. Freund, and J. L. Lyman, *Laser Handbook III(b)*, M. L. Stitch, ed., North-Holland Publishing, Amsterdam, 1979.
11. T. M. Sloane, S. Y. Tang, and J. Ross, *J. Chem. Phys.*, **57**, 2745 (1972).
12. R. D. Bates, Jr., G. W. Flynn, J. T. Knudtson, and A. M. Ronn, *J. Chem. Phys.*, **53**, 3621 (1970).
13. F. R. Grabiner, D. R. Siebert, and G. W. Flynn, *Chem. Phys. Lett.*, **17**, 189 (1972).
14. A. F. Trotman-Dickenson, *Gas Kinetics*, Butterworths, London, 1955, p. 219.
15. J. H. Birely and J. L. Lyman, *J. Photochem.*, **4**, 269 (1975).
16. R. B. Bernstein, State-to-state chemistry, *A.C.S. Symp. Ser.*, **56**, 3 (1977).
17. R. D. Levine and J. Manz, *J. Chem. Phys.*, **63**, 4280 (1975).
18. A. G. Gaydon, *Dissociation Energies and Spectra of Diatomic Molecules*, 3rd ed., Chapman and Hall, London, 1968, p. 277.

CHEMICAL LASER KINETICS

A. BEN-SHAUL

Department of Physical Chemistry and Institute of Advanced Studies, The Hebrew University, Jerusalem, Israel

CONTENTS

I. INTRODUCTION

Chemical lasers are complex nonequilibrium molecular systems governed by an intricate interplay between a variety of chemical, radiative, and collisional relaxation processes.[1, 2] Many of their kinetic properties are reflected by the temporal, spectral, and power characteristics of the out-coupled laser radiation. For example, threshold time measurements and other gain experiments[3, 4] have provided detailed information on vibrational distributions of nascent reaction products. Another, more qualitative, example: Single-line and simultaneous multiline operation indicate, respectively, whether the lasing molecules are rotationally equilibrated or not. Besides their practical applications, chemical lasers are widely used as means of selective excitation in state-to-state kinetic studies. On the other hand, many experimental and theoretical studies have been motivated by the wish to understand and improve the mechanism of chemical laser operation.

The preparation of nonequilibrium level or species populations is the first step in any kinetic experiment. The introduction of lasers to chemical research has opened up new possibilities for preparing, often state-selectively, the initial nonequilibrium states. However, the subsequent time evolution of the molecular populations occurs almost invariably along several relaxation pathways. Some of which, like intra- and intermolecular vibrational energy transfer in infrared multiphoton absorption experiments, may interfere with the exciting laser pulse and/or with the specific process investigated. In such cases, as in chemical laser research, one has to interpret the behavior of complex nonequilibrium molecular systems in which the laser radiation plays of course a major role. This establishes the link between the present article and the general subject of this volume.

The principal tool for analyzing and predicting the kinetic behavior of chemical lasers are the photon and level rate equations[5–7] (Section II). The complexity arising from the large numbers of significantly populated levels and active transitions and the multitude of rate processes sets two types of limitations on this approach, commonly known as kinetic modeling.[7] First, there are very few systems for which the list of relevant kinetic data is reasonably known. A more technical difficulty is associated with the numerical solution of large sets of nonlinear rate equations. Both types of problems are gradually diminished. The advent of fast computers, efficient integration algorithms, and approximation techniques help in overcoming the numerical difficulties. The gaps of missing kinetic information are constantly narrowed due to the rapid accumulation of reliable detailed kinetic information from various experimental and theoretical methods.[8] An important example demonstrating these developments is the question of rotational nonequilibrium in chemical lasers, which has several interesting theoretical and practical implications (Section III): for example, the laser efficiency increases upon adding to the lasing mixture a buffer gas enhancing rotational relaxation. Rotational nonequilibrium effects such as multiline operation, R-branch, and pure rotational transitions have already been observed in the early chemical laser experiments[9–12] and are revealed in more detail later on.[13–18] Nevertheless, these effects could not be properly treated in chemical laser analyses until the first quantitative information on rotational relaxation rates from infrared chemiluminescence data became available.[19] Currently such data are provided by various techniques[20–24] and models,[19–28] and their incorporation in kinetic modeling studies is almost routine.[29–43]

By a judicious choice of the external conditions the energy stored in nonequilibrium chemical systems can be partly converted into thermodynamic work. The extraction of electrical current from nonequilibrium ionic mixtures is a well-known manifestation of this principle. A less thoroughly analyzed example is the conversion of the internal energy stored in inverted laser populations into laser radiation.[44–48] This form of energy, being highly monochromatic, unidirectional, and phase coherent carries essentially no entropy and is therefore equivalent to thermodynamic work. Hence, chemical lasers are suitable for thermodynamic analyses and interpretations. We shall briefly outline the thermodynamic description of these systems in Section IV.

The oxygen–iodine chemical transfer laser,[49] $O_2(^1\Delta)+I(^2P_{3/2})\rightarrow O_2(^3\Sigma)+I(^2P_{1/2})$, based on the same electronic transition as the iodine photochemical laser[50], $I(^2P_{1/2})\rightarrow I(^2P_{3/2})$, and a few systems operating on pure rotational transitions[14,15,17,18] are among the recent developments in chemical laser research. Other electronic lasers such as the iodine photochemical laser and the large group of excimer lasers[51] are also classified sometimes as chemical lasers. Yet, most chemical laser systems utilize vibrotational transitions, almost exclusively of diatomic molecules. Our discussion will be confined to this type of chemical lasers. To emphasize the nonequilibrium characteristics and the time factor we shall consider only pulsed lasers. We shall not discuss important subjects such as optical properties,[52] gas dynamic factors,[53] and computational methods.[7] As specific guiding examples we shall refer to the well-studied $F+H_2\rightarrow HF+H$ laser[54] and the relatively simple (only one active vibrational band) $Cl+HBr\rightarrow HCl+Br$ system.[55]

II. KINETICS

A. Rate Equations

The rate processes governing the time dependence of the vibrotational populations, N_{vJ}, (molecules/cm^3), and photon densities, $\phi_{v,J}^{v',J'}$, (photons/cm^3), corresponding to the radiative transitions $v,J\rightarrow v',J'$ are illustrated schematically in Fig. 1. The symbols P_{vJ}, V_{vJ}, and $R_J(v)$ represent, respectively, the net contributions of the pumping, vibrational relaxation, and rotational relaxation to the rate of change of N_{vJ}. $S_{v,J}^{v',J'}$ and $\chi_{v,J}^{v',J'}$ denote the rates of spontaneous and stimulated radiation in $v,J\rightarrow v',J'$. To simplify the kinetic description we shall consider only P-branch transitions, $v,J\rightarrow v-1,J+1$ and disregard R-branch, $v,J\rightarrow v-1,J-1$, and pure rotational, $v,J\rightarrow v,J'$, transitions. This simplification is supported by the fact that the gains of P-branch lines are almost invariably higher than those of the R-branch and pure rotational lines originating at the same level.[3,4] Consequently, the intense laser radiation developing immediately after a P-line reaches threshold quenches the gain on the competing lines. R-branch and pure rotational transitions are commonly observed in grating experiments.[4,54] They may appear in free-running lasers, usually as

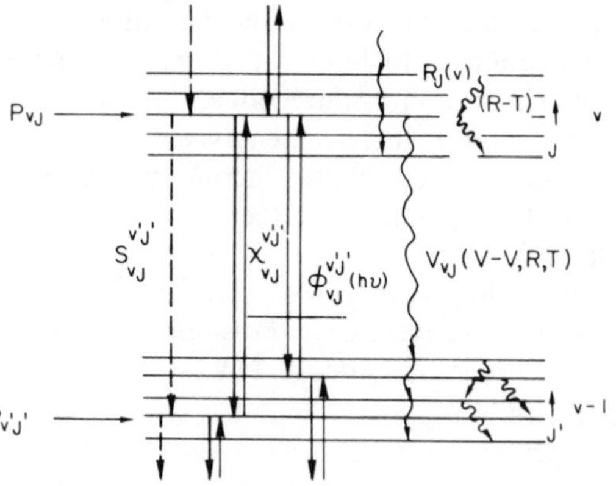

Fig. 1. Types of rate processes in vibrotational chemical lasers.

transients, only under extreme deviations from rotational equilibrium. It should be noted, however, that their inclusion in kinetic analyses is straightforward.

Assuming that the rate processes in the laser cavity are spatially uniform the rate equations are of the form

$$\dot{N}_{vJ} = P_{vJ} + V_{vJ} + R_J(v) + \chi_{v+1,J} - \chi_{v,J+1} + S_{v+1,J} - S_{v,J+1} \tag{1}$$

$$\dot{\phi}_{vJ} = \chi_{vJ} + \varepsilon S_{vJ} - \tau_c^{-1}\phi_{vJ} \tag{2}$$

In these equations and hereafter we use simplified symbols for quantities referring to the P-branch line $v, J-1 \rightarrow v-1, J$. Namely, χ_{v_J} instead of $\chi_{v,J-1}^{v-1,J}$ etc. The second term in (2) represents the rate of spontaneous emission into the oscillating cavity modes.[5, 6] $\varepsilon \ll 1$ is approximately the effective solid angle subtended by the mirrors after several reflections. (Alternatively, ε is the fraction of stable transverse modes.) After threshold εS_{vJ} is negligible, $\varepsilon S_{vJ} \ll \chi_{vJ}$. This term is important only before threshold —as a source of noise photons to trigger-on the lasing process. The spontaneous emission terms in (1) are given by $S_{v,J+1} = A_{v,J+1}N_{vJ}$; A_{vJ} is the Einstein coefficient. In infrared lasers where typically $A_{vJ}^{-1} \sim 10^{-2}$ s, S_{vJ} is generally very small compared to the other terms in (1). The last term in (2) is the total, useful, and useless rate of radiation losses from the cavity. The photon lifetime τ_c is given by[5,6] $\tau_c^{-1} = (c/2L)\ln(1/R_1R_2) + \alpha$ where the first term is the inverse lifetime for useful output coupling, while the second accounts for dissipative losses (absorption, scattering, diffraction); c is the speed of light, L is the length of the active medium, R_i is the reflectivity of mirror i, and α is the average dissipation rate. The total laser power (per unit active laser volume) is given by

$$\dot{W}=\sum_{v,J}\dot{W}_{vJ}=\tau_c^{-1}\sum_{v,J}h\nu_{vJ}\phi_{vJ} \tag{3}$$

where $\dot{W}_{vJ}$ and ν_{vJ} are the power and frequency of the $v, J-1\rightarrow v-1, J$ transition. $W=W(t)$ is the integrated laser energy in the time interval $0-t$.

The net amplification rate is

$$\chi_{vJ}=c\sigma_{vJ}\Delta N_{vJ}\phi_{vJ}=\gamma_{vJ}\phi_{vJ} \tag{4}$$

where γ_{vJ} is the gain coefficient (in s^{-1}) and

$$\sigma_{vJ}=c^{-1}g(\nu_{vJ},\nu_{vJ}^0)h\nu_{vJ}B_{vJ} \tag{5}$$

$$\Delta N_{vJ}=N_{v,J-1}-\left(\frac{g_J}{g_{J-1}}\right)N_{v-1,J} \tag{6}$$

are the cross-section for stimulated emission and the population inversion, respectively; $g_J=2J+1$. B_{vJ} is the Einstein coefficient for stimulated emission, $\nu_{vJ}^0=(E_{v,J-1}-E_{v-1,J})/h$ is the line-center frequency and $g(\nu_{vJ},\nu_{vJ}^0)$ is the normalized lineshape function. In kinetic modeling of chemical lasers it is common to assume that lasing occurs at line-center and to take the lineshape function as a convolution (Voigt profile) of the Gaussian and Lorentzian shapes describing Doppler and pressure broadening, respectively.[7] Based on the notion that the possibility of hole-burning in Doppler profiles is largely suppressed by fast translational (momentum transfer) relaxation it can be assumed that the gain curve saturates homogeneously even in the case of Doppler dominated broadening. Typical linewidths in HF or HCl lasers are $\Delta\nu\sim10^{-2}-10^{-3}$ cm^{-1}.

In writing rate equations only for N_{vJ} it is tacitly assumed that translational relaxation is instantaneous on the time scale of all the other rate processes. Hence, a well-defined temperature, T, characterizes the translational degrees of freedom of the lasing molecules and all degrees of freedom of the nonlasing species. This "heat bath" temperature appears as a parameter in the collisional rate constants. It also enters the gain coefficients via the linewidth and in the case of rotational equilibrium mainly via the population inversion. Thus (1) and (2) should be supplemented by a rate equation for T. Additional kinetic equations describe the time dependence of the nonlasing species concentrations.

B. Kinetic Scheme

To describe the pumping and relaxation terms in (1) we shall refer to a specific, but typical, example: the flash initiated CF_3I/H_2 system[54] based on the pumping reaction $F+H_2 \rightarrow HF(v,J)+H$. The kinetic scheme of this system is summarized in Table I.

The explicit form of the pumping term corresponding to Table I is

$$P_{vJ}=k(\rightarrow v, J; T)[F][H_2]-k(v, J\rightarrow; T)N_{vJ}[H] \tag{7}$$

where $N_{vJ}=[HF(v, J)]$ and $k(\rightarrow v, J; T)$ is the rate constant of reaction $(R-2)$; for thermal reactant distribution. The reversed rates are determined by detailed balancing: $k(v, J\rightarrow; T)=k(\rightarrow v, J; T)\exp[(\Delta E_0-kT+E_{vJ})/kT]$, where $\Delta E_0<0$ is the exoergicity of the reaction and $\sim kT$ is the average internal reactant energy, neglecting vibration. Typically $\Delta E_0 \gg kT$, e.g., for $(R-2)$ $\Delta E_0=32$ kcal/mole. Moreover, the highest accessible product states for which $E_{vJ}\sim\Delta E_0$ are barely populated. Hence the effect of the reverse reaction is negligible (except, of course, close to equilibrium where $P_{vJ}\rightarrow 0$).

The nascent product distribution $X_i(v, J)$ is defined by

$$X_i(v, J)=X_i(v)X_i(J|v)=\frac{k(\rightarrow v, J; T)}{k(\rightarrow; T)} \tag{8}$$

where $k(\rightarrow; T)$ is the total pumping rate constant, $X_i(v, J)$ is normalized by $\Sigma_{vJ}X_i(v, J)=1$. The vibrational distribution $X_i(v)=\Sigma_{J\in v}X_i(v, J)$ and the rotational distribution within v, $X_i(J|v)=X_i(v, J)/X_i(v)$ are normalized by $\Sigma_v X_i(v)=1$ and $\Sigma_{J\in v}X_i(J|v)=1$. The nascent distributions of $(R-2)$, as measured in ir chemiluminescence[56–58] and chemical laser experiments[4, 54] are shown in Fig. 2.

The rotational relaxation term $R_J(v)$ incorporates contributions from different reaction partners M_i, e.g.,

$$R^i_J(v)=-\sum_{J'}\left[k^i(J\rightarrow J'|v; T)X(J|v) - k^i(J'\rightarrow J|v; T)X(J'|v)\right]X(v)N[M_i] \tag{9}$$

where $X(v)$ and $X(J|v)$ are the instantaneous vibrational and rotational distributions, respectively;

$$X(v,J)=X(v)X(J|v)=\frac{N_v}{N}\frac{N_{vJ}}{N_v}=\frac{N_{vJ}}{N} \tag{10}$$

Experiments[19–24] and theoretical models[19–28] indicate that the $R-T$ rate constants can often be fitted to the exponential gap expression

$$k(J\rightarrow J'|v; T)=A\exp\{-C[E_J(v)-E_{J'}(v)]\} \quad J>J' \tag{11}$$

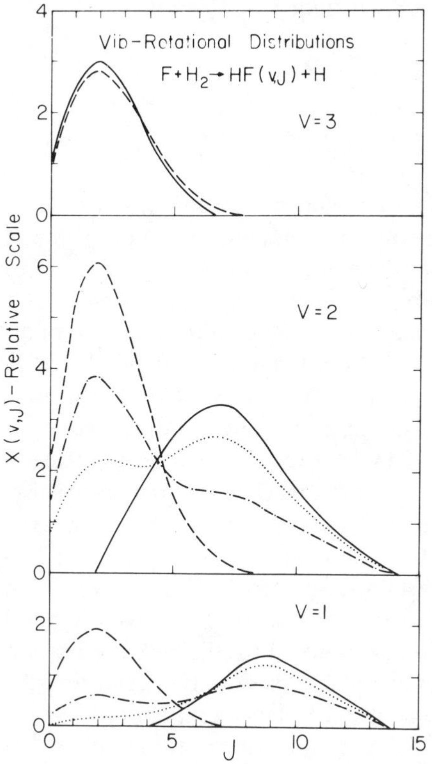

Fig. 2. The nascent vibrotational product distribution in the $F+H_2 \rightarrow HF(v, J)$ $+H$ reaction (full lines). The curves within each v represent $X_i(J|v)$. The areas under these curves are proportional to $X_i(v)$. The dashed lines describe thermal rotational populations $X_0(J|v)$. The doubly peaked (dotted and dash-dotted) curves correspond to partly relaxed distributions.

where A and C are empirical, temperature-dependent, parameters measuring, respectively, the absolute value of the rate constant and its J-dependency. The preexponential factor is sometimes[20,25] taken proportional to the final state degeneracy, i.e., $A=A'(2J'+1)$. $E_J(v)$ is the rotational energy of molecules in level v,J; $E_{vJ}=E_v+E_J(v)$; e.g., in the vibrating rotor model $E_J(v)=B_vJ(J+1)$. The excitation, $J'>J$, rate constants are obtained by detailed balance. Since rotational energy gaps increase with J, e.g., $E_J(v)-E_{J-1}(v)=2B_vJ$, (11) implies that the rate of rotational relaxation decreases exponentially with J. Hence, the rotational distribution in the high J regime is expected to relax relatively slowly and retain its original shape, though not its magnitude, long after the start of relaxation. This behavior is reflected in Fig. 2 by the doubly peaked intermediate distributions. The original peak is gradually decaying while the Boltzmann peak is building up, mainly due to population transfer from intermediate J levels. This relaxation pattern suggests that rotational nonequilibrium effects are more likely to be observed in the high J region.

A typical example to the V–R, T terms included in V_{vJ} is

$$V_{vJ}^{i} = -\sum_{v'J'} \left[k^{i}(v,J \rightarrow v',J';T) X(v,J) - k^{i}(v',J' \rightarrow v,J;T) X(v',J') \right] N[M_i] \tag{12}$$

Considerably more complex but just as straightforward to write down are the V–V and V–R, T terms due to diatom–diatom collisions. For example, HF–HF collisions are described by rate constants involving eight indices besides the translational temperature, i.e., $k(v,J,m,L \rightarrow v',J',m',L';T)$. Fortunately, under conditions ensuring rotational equilibrium—a condition often realized experimentally—the rotational state dependence is irrelevant and only averaged rate constants $k(v \rightarrow v';T)$ or $k(v,m \rightarrow v',m';T)$ are needed. (We shall discuss this in a more general context in the next section.) Vibrational rate constants are relatively easier to measure and to model and have been extensively tabulated as recommended lists for chemical laser modeling.[8] Yet many of them are associated with large uncertainties. Various methods have been suggested for testing the sensitivity of modeling results to these uncertainties but so far have rarely been applied.[59–62] The rotationally averaged rate constants still provide a good approximation when the deviations from rotational equilibrium are not too large. When the deviations are pronounced, or, for example, in testing the efficiency of V→R transfer as a significant secondary pumping mechanism of high J levels, consideration of the J dependency becomes crucial.[15, 17, 18] Currently, information about this dependency is only partly provided by classical trajectory studies, approximate dynamical models, and indirect, for example, temperature dependence, experimental observations (see, for example, Ref. 36 and references cited therein).

C. Model vs Experiment

A demonstration of the predictive potential (and the limitations) of kinetic simulations is provided in Fig. 3. The left column shows the experimental output pattern determined by M. J. Berry for the laser system[54] in Table I with initial gas mixture $CF_3I:H_2:Ar = 1:1:50$ torr. The high buffer gas pressure was used to enhance rotational relaxation. The central column in the figure shows the computed results obtained[42] by solving the rate equations (1) and (2) with R–T rate constants of the form (11), with recommended vibrational rate constants[8] and without attempting to optimize the agreement with experiment. In view of the complexity of the system the agreement reflected by Fig. 3 is good but clearly not perfect—a conclusion that should encourage additional kinetic analyses

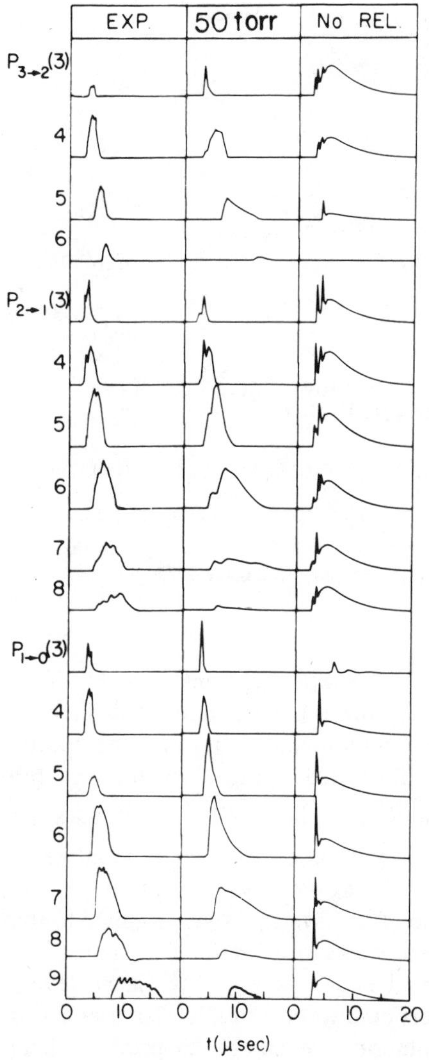

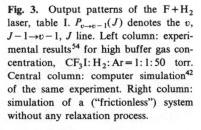

Fig. 3. Output patterns of the $F+H_2$ laser, table I. $P_{v\to v-1}(J)$ denotes the v, $J-1\to v-1$, J line. Left column: experimental results[54] for high buffer gas concentration, $CF_3I:H_2:Ar=1:1:50$ torr. Central column: computer simulation[42] of the same experiment. Right column: simulation of a ("frictionless") system without any relaxation process.

on the one hand and detailed chemical laser experiments on the other. It should be noted that at the high buffer gas pressure employed in the experiment the rotational equilibrium assumption[3,4,7] is almost fully justified. In this case the detailed, vibrotational equations (1) and (2) may be replaced by a few simple vibrational equations (next section). However the detailed solutions[42] shown in Fig. 3, yielded better agreement with experiment, especially (as could be expected) in the high J region.

TABLE I
Kinetic Scheme of the $CF_3I/H_2/Ar$ Laser System[a]

Initiation	
$CF_3I \underset{\text{flash}}{\overset{h\nu}{\rightarrow}} CF_2I + F$	(R-1)
Pumping, P_{vJ}	
$F + H_2 \rightleftarrows HF(v, J) + H$	(R-2)
Secondary reaction	
$F + CF_3I \rightarrow CF_3 + FI$	(R-3)
Stimulated and spontaneous radiation. (P-branch), χ_{vJ}	
$HF(v, J-1) \overset{h\nu_{vJ}}{\rightleftarrows} HF(v-1, J) + h\nu_{vJ}$	(R-4)
R–R, T transfer, $R_J(v)$	
$HF(v, J) + M_i \rightleftarrows HF(v, J') + M_i$	(R-5)
V–V transfer, V_{vJ}(V–V)	
$HF(v, J) + HF(v', J') \rightleftarrows HF(v \pm \Delta v, \bar{J}) + HF(v' \mp \Delta v, \bar{J}')$	(R-6)
$HF(v, J) + H_2(0) \rightleftarrows HF(v-1, J') + H_2(1)$	(R-7)
V–R, T transfer, V_{vJ}(V–R, T)	
$HF(v, J) + M_i \rightleftarrows HF(v', J') + M_i$	(R-8)
$M_i = F, H, HF, H_2, Ar, CF_3I, CF_3$	

[a] Based on the experiments in Ref. 54 with $CF_3I : H2 : Ar = 1 : 1 : 50$ torr. In the computations[42] reported in the text the Ar pressure which measures the R–T rate was taken as a variable.

In addition to radiation densities, ϕ, the rate equations provide kinetic information which is less easy to determine experimentally. This includes the time evolution of the level populations (Section III) and the specific effects of different rate processes, as illustrated in Fig. 4 for the time rate of change of [HCl ($v=1, J=9$)] in the flash-initiated ($Cl_2 \overset{h\nu}{\rightarrow} 2Cl$), $Cl + HBr \rightarrow HCl + Br$ laser.[39] The exoergicity of this reaction suffices to populate the $v=0,1,2$ levels of HCl, but effective lasing occurs only on the $P_{v=1\rightarrow 0}$ ($J=1-15$) lines.[55] The kinetic scheme of this system is very similar to that of Table I. The results in Fig. 4 correspond to a low-buffer gas-pressure condition; hence incomplete rotational equilibrium.[39] The figure reveals that the R–T rate is finite and proportional to that of stimulated emission —demonstrating the tendency of rotational relaxation to prevent "hole-burning" in the rotational profile. The trend displayed by the V–R,T processes is general, that is, to reduce the population of $v \geqslant 1$ levels, usually with greater efficiency as v increases. The increase in time of the V–R,T and V–V rates is due to the growth in the absolute value of N_{vJ}. These processes reduce the pumping efficiency and are the major cause for quenching the laser pulse. Under certain conditions the fast, near resonant, V–V transfer can be an efficient pumping rather than deactivating mechanism. This is the case in electrical,[63] and to a lesser extent chemical,[64] CO

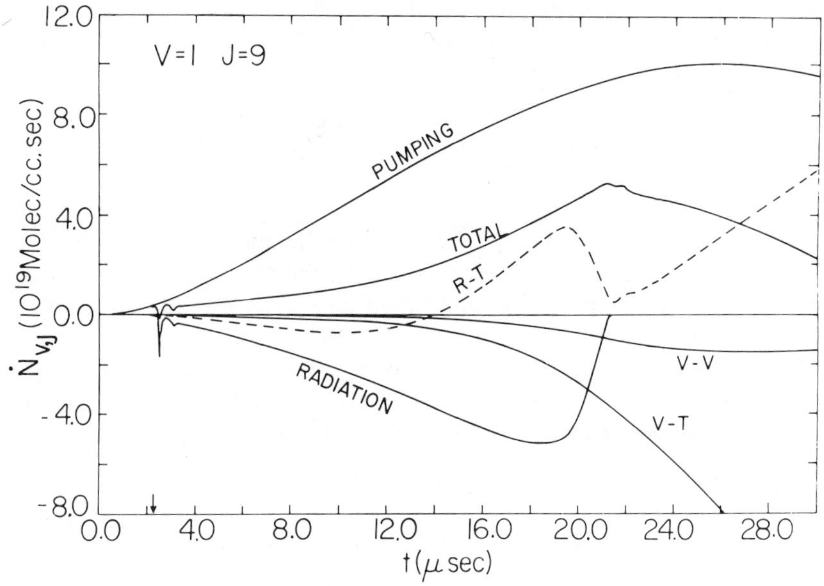

Fig. 4. The contributions of different rate processes to the temporal evolution of a given vibrotational level population; HCl($v=1, J=9$) molecules in the Cl + HBr laser[39].

lasers where the molar fractions of CO or CO and N_2 molecules are large. Here the anharmonic up-pumping (Treanor) mechanism[65] effectively promotes molecules to excited vibrational levels. The same mechanism is operative in HF lasers,[8,66,67] but its absolute contribution is usually minor[36,42] owing to the relative rarity of HF–HF collisions. In the Cl + HBr system of Fig. 4 [HCl]≪[HBr] and the major V–V route is the deactivation[39,55] process HBr ($v=0$) + HCl ($v=1$)→HBr ($v=1$) + HCl ($v=0$).

III. ROTATIONAL EFFECTS

The rotational equilibrium assumption implies substantial conceptual and technical simplifications of the kinetic description of chemical lasers. In this section we shall consider the validity and consequences of this assumption and compare the very different lasing mechanisms and efficiencies implied by fast and slow rotational relaxation.

A. Rotational Equilibrium

The physical content of the rotational equilibrium assumption is that rotational (like translational) relaxation is practically instantaneous on the time scale of the other rate processes in the laser cavity including, in particular, the stimulated emission. Its mathematical interpretation is that the $X(J|v)$'s are Boltzmann throughout the temporal evolution of the laser system (Fig. 2). Hence cf. (10):

$$\frac{N_{vJ}}{N_v} = X_0(J|v) = g_J \exp\left[\frac{-E_J(v)}{kT}\right] \Big/ Q_r \tag{13}$$

where Q_r is the rotational partition function, and T is the heat bath temperature. T is time-dependent. However, the excess buffer gas usually needed to ensure rotational equilibration also ensures moderate temperature variations.

Since N_{vJ} is uniquely determined by N_v and $X_0(J|v)$, the number of independent rate equations (1) is now reduced to the number of vibrational levels. The formal reduction is achieved by using (13) in (1) and summing over all J's in v. The result, is

$$\dot{N}_v = P_v + V_v + \sum_J (\chi_{v+1,J-1} - \chi_{v,J}) \tag{14}$$

where the spontaneous radiation terms have been neglected for simplicity. The rotational relaxation term does not appear in (14), since rotational equilibrium implies $R_J(v) \equiv 0$; as can be confirmed by setting (13) in (19) and using the detailed balance principle. The pumping and vibrational relaxation terms in (14), $P_v = \Sigma_J P_{vJ}$ and $V_v = \Sigma_J V_{vJ}$, involve now only the averaged, vibrational, rate constants; e.g., (7) is replaced by

$$P_v = k(\rightarrow v; T)[\mathrm{F}][\mathrm{H}_2] - k(v\rightarrow; T)N_v[\mathrm{H}] \tag{15}$$

where

$$k(\rightarrow v; T) = \sum_J k(\rightarrow v, J; T) = k(\rightarrow; T)X_i(v) \tag{16}$$

Similarly (12) reduces to

$$V_v^i = -\sum_{v'} \left[k^i(v\rightarrow v'; T)X(v) - k^i(v'\rightarrow v; T)X(v') \right] N[M_i] \tag{17}$$

where

$$k^i(v\rightarrow v'; T) = \sum_{J,J'} X_0(J|v)k(v, J\rightarrow v', J'; T) \tag{18}$$

In addition to simplifying the population rate equations the rotational equilibrium assumption also leads to fewer photon equations (2). Explicitly, instantaneous rotational relaxation implies that in each vibrational band, $v \to v-1$, there is only one active transition $v, J^*-1 \to v-1, J^*$ at any moment. $J^* = J^*(t)$, which specifies the highest gain transition in $v \to v-1$ at time t is an increasing function of time, i.e., the lasing is monotonically shifted toward higher J values. This is a well-known phenomenon, which will be described here again to prepare the basis for comparison with the lasing mechanism in the weak coupling limit.

The threshold for lasing oscillations in $v, J-1 \to v-1, J$ is reached when $\dot{\phi}_{vJ} \geqslant 0$ [cf. (2)] or equivalently (disregarding εS_{vJ}) when

$$\gamma_{vJ} = c\sigma_{vJ} \Delta N_{vJ} \geqslant \tau_c^{-1} \tag{19}$$

In the limit of rotational equilibrium we can use (5), (6), (10), and (13) to rewrite this condition in the form

$$\alpha_{vJ} \equiv c\sigma_{vJ} \left\{ X_0(J-1|v) - \frac{X(v)}{X(v-1)} \frac{g_{J-1}}{g_J} X_0(J|v-1) \right\} \geqslant (N_v \tau_c)^{-1} \equiv \alpha_v^0 \tag{20}$$

A graphical representation of the relative gain coefficient α_{vJ} corresponding to HF ($v=2$) at room temperature is given in Fig. 5 for different vibrational population ratios $X(v)/X(v-1)$. The strong J dependence of α_{vJ} is mainly due to the J dependence of the population difference term, $\{\ \}$, in (20) because of the sharp, exponential, decay of $X_0(J|v)$ with $E_J(v) = B_v J(J+1)$. Additional but much weaker J dependence results from[4,7,68] $\sigma_{vJ} \propto vJ/(2J-1)$. Note also that since B_v does not vary considerably from one v to another the behaviour displayed in Fig. 5 applies to all $v \to v-1$ bands.

At the early stages of the pumping process and before threshold is reached vibrational deactivation is negligible so that $X(v)/X(v-1) \simeq X_i(v)/X_i(v-1)$ and $N_v \simeq NX_i(v)$ increases like N (see Fig. 6). Hence α_{vJ} is constant while α_v^0 decreases. For $v=2 \to 1$ $X_i(v)/X_i(v-1) \simeq 1.6$ (Fig. 6). From Fig. 5 we find that the highest gain and therefore the first lasing transition in this band is $P_{2\to1}(3)$, ($J^*=3$), as confirmed by the experimental and the computed output patterns of Fig. 3. The intense laser radiation tends to reduce α_{vJ} to its threshold value α_v^0, thereby to "burn a hole" in $X_0(J|v)$ at v, J^*-1 (and simultaneously "build a hump" in $v-1, J^*$). However, the instantaneous population transfer by rotational relaxation prevents the hole-burning, and the Boltzmann profile in v is drained homogeneously. Moreover, since $\alpha_{vJ^*} = \alpha_v^0 > \alpha_{vJ \neq J^*}$ no lasing can take

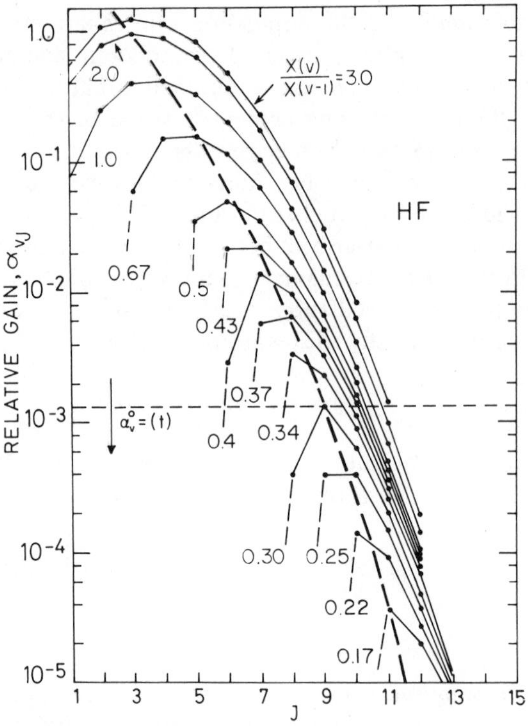

Fig. 5. Relative gain coefficients of HF for thermal rotational populations and different vibrational population ratios, $X(v)/X(v-1)$. The heavy line is the loci of highest gain transitions. At every vibrational population ratio lasing occurs only on the highest gain transition for which α_{vJ^*} is just above α_v^0. As $X(v)/X(v-1)$ decreases J^* increases provided that α_v^0 decreases too.[68]

place on $J \neq J^*$ and the upper vibrational population is "funneled" downward only through the highest gain transition. This picture is entirely equivalent to that of single-mode operation in homogeneously broadened lasing transitions.[5-7] The cavity modes are the analogs of the P-branch lines, the Lorentz lineshape corresponds to the Boltzmann profile and the dephasing mechanism (mainly elastic collisions), responsible for the homogeneous broadening is equivalent to the fast rotational coupling.

The lasing process reduces $X(v)/X(v-1)$, first sharply and then moderately, as shown in Fig. 6. According to Fig. 5 this implies a positive shift in J^*, as reflected in Fig. 3. Since α_{vJ^*} decreases as J^* increases and the lasing condition requires $\alpha_{vJ^*} \geqslant \alpha_v^0$ the J-shifting continues only as long as α_v^0 decreases or equivalently as long as $\dot{N}_v \geqslant 0$ [cf. (20)]. Now, $\dot{N}_v = \dot{X}(v)N + X(v)\dot{N}$ where $\dot{N} = \Sigma \dot{N}_v$ is the total pumping rate. At the beginning of the

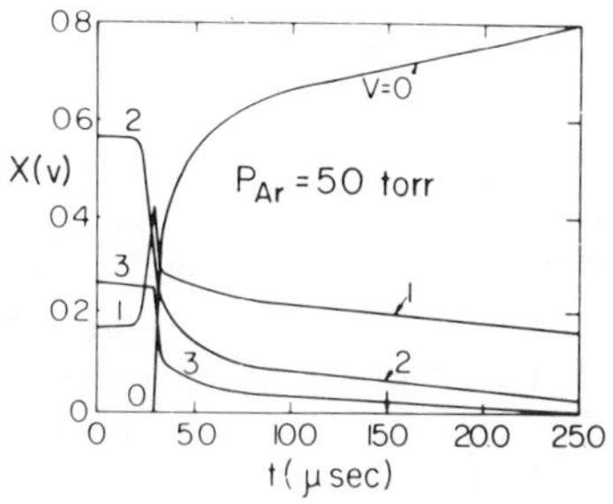

Fig. 6. Time evolution of the vibrational distribution, $X(v)$, in the $F+H_2$ laser at high buffer gas pressure, i.e., strong rotational coupling. The initial conditions are as for the central column in Fig. 3. Note the sharp changes in $X(v)$ just after threshold and that most of the laser radiation is extracted from partially inverted populations. The arrow indicates the end of the pulse.

lasing process $X(v)$, $(v \geqslant 1)$, falls steeply (Fig. 6), but N increases fast enough to ensure $\dot{N}_v > 0$. Later on $X(v)$ decreases more slowly but the pumping rate falls off rapidly (due to the consumption of reactants). As time evolves $\dot{N}$ becomes very small, while $X(v)$ continues to decrease because of lasing and vibrational relaxation. Eventually $\dot{N}_v < 0$, and lasing terminates. Using somewhat more quantitative arguments[68] it is not difficult to apply this criterion to determine $N_v^f = X_f(v)N$ and J_f^*, the respective values of the vibrational population and the last lasing line at the end of the laser pulse. Also, since J^* and $X(v)/X(v-1)$ are uniquely related by $\alpha_{vJ^*} = \alpha_v^0$ one can evaluate $X(v)$ from the measured output spectra. In particular this can be used to determine $X_f(v)$, which provides an upper bound to the photon yield ψ, (photons/molecule),

$$\psi = \sum_v v\left[X_i(v) - X_f(v)\right] \tag{21}$$

An estimate based on (21) yields[68] $\psi = 1.8$ photons/molecule for the $F+H_2$ system of Fig. 3 and Table I; in reasonable agreement with the value $\psi = 1.5$ obtained by solving the full set of rate equations.[42] Taking into account that the exoergicity of the $F+H_2$ reaction corresponds to $\sim$3.2 vibrational HF quanta, of which only two thirds ($\sim$2 quanta) appear as HF vibration the value $\psi = 1.5$ indicates very high chemical efficiency. This is a direct consequence of the fact that at rotational equilibrium most of the laser energy is extracted from partially inverted vibrational populations, $X(v)/X(v-1) < 1$. This conclusion is clear from Fig. 5 and can be confirmed by comparing Figs. 3 and 6. A simple qualitative explanation follows from the necessary (but not sufficient) lasing condition $\Delta N_{vJ} > 0$. Using (13) we find that $\Delta N_{vJ} = N_{v,J-1} - (g_{J-1}/g_J)N_{v-1,J} > 0$ yields

$$J > \frac{kT}{2B} \ln \frac{X(v-1)}{X(v)} \tag{22}$$

where we have used the rigid rotor approximation $E_J(v)=E_J=BJ(J+1)$. For HF $kT/2B \sim 5$ and, say, $X(v)/X(v-1)=0.3$ implies $J>7$.

B. Rotational Nonequilibrium

To assess the validity limits of (13) let us test its most crucial implication: The ability of rotational relaxation to prevent hole-burning by stimulated emission. To this end, suppose that $X(J|v)$ is either approximately or exactly given by (13) (otherwise we a priori assume rotational nonequilibrium). Let us further assume that stimulated emission in $v, J-1 \rightarrow v-1, J$ has already burnt a hole in $v, J-1$ (otherwise $R_{J-1}(v)\equiv 0$, cf. (9), and there is no net population transfer), so that $X(J-1|v)<X_0(J-1|v)$ but $X(J'\neq J-1|v)=X_0(J'|v)$. The hole is significant and will remain so if $X_0(J-1|v)-X(J-1|v)$ is appreciable and the net R–T rate into $v, J-1$ is slow compared to the rate of stimulated emission χ_{vJ}. To favor the R–T rate let us neglect the rate of population transfer out of $v, J-1$. That is, we decompose $R_J(v)$ as in (9), $R_J(v)=R_J^+(v)-R_J^-(v)$, and set

$$R_{J-1}(v)\simeq R_{J-1}^+(v)=\sum_{J'} k(J'\rightarrow J-1|v;T)X_0(J'|v)X(v)N[M] \quad (23)$$

The rotational equilibrium assumption will be valid when $\lambda \gg 1$, where

$$\lambda \equiv \frac{R_{J-1}^+(v)}{\chi_{vJ}} \quad (24)$$

The stimulated emission rate can be evaluated by replacing γ_{vJ} in $\chi_{vJ}=\gamma_{vJ}\phi_{vJ}$ by its threshold value $\gamma_{vJ}=\tau_c^{-1}$. An alternative estimate, based on the "gain–equal–loss" (steady-state) condition[5–7] and the neglect of vibrational relaxation is $\chi_{vJ}\simeq\Delta P_{vJ}=P_{v,J-1}-(g_J/g_{J-1})P_{v-1,J}$; the effective pumping rate of $v, J-1\rightarrow v-1, J$. Restricting the sum in (23) to $J'=(J-1)\pm 1$ and noting that for HF (and other molecules) the parameter C in (11) is $C \sim kT=5.10^{-3}$ cm^{-1}, $B\simeq 20$ cm^{-1} and $A\sim 5\cdot 10^{13}$ cm^3/mole·s we find $R_{J-1}^+(v)\simeq 10^{14}\exp[-0.4(J-1)]X(v)N[M]$. Typical values for the other parameters in the HF system[42, 54] are: $X(v)\sim 0.2$ (cf. Fig. 6) $N\sim 10^{-9}$ mole/cm^3 (0.2 torr), $\phi_{vJ}\sim 10^{-12}$ mole-photons/cm^3, $\tau_c=10^{-7}$ s (corresponding to $\Delta P_{vJ}\sim 10^{-5}$ mole/cm^3). Using these values and setting $[M]\simeq 5\cdot 10^{-8}P$, where P(torr) is the total gas pressure, we find

$$\lambda \simeq 10^2 P X_{J-1}^0(v)\exp[-0.4(J-1)] \quad (25)$$

At high buffer gas pressures, e.g., $P=50$ torr, like in the experiment[54] and calculation[42] corresponding to Fig. 3, we obtain $\lambda\sim 500$ for $J=3$ and $\lambda\simeq 2$

for $J=8$. This indicates, as mentioned earlier, that at the high J region rotational nonequilibrium effects may be observed even at high pressures. These effects are considerably more pronounced when P is lowered, e.g., for $P=2$ torr $\lambda<1$ already for $J\geqslant 4$. The limit of very weak rotational coupling, $\lambda \ll 1$, is favored by low pressures and large pumping rates but is practically impossible for experimental realization (over all J's).

The lasing characteristics in the (somewhat artificial but interesting) limit of weak rotational coupling are essentially opposite to those in the strong coupling limit. Here different transitions belonging to the same vibrational band reach threshold, lase, and decay independently from one another. The lasing mechanism may be appropriately classified as "inhomogeneous" or "individual" as opposed to the "homogeneous" or "cooperative" mechanism which dominates the strong coupling limit. The independence of the lasing transitions is clearly reflected by the output pattern shown in the right column of Fig. 3. This spectrum was derived by solving (1) and (2) with all the relaxation processes artificially switched off,[42] that is, with $R_J(v) = V_{vJ} \equiv 0$ and $P_{Ar}=0$. In this system, hereafter the "frictionless" laser, the efficiency is low, $\psi=1.0$ compared to $\psi=1.5$ in the rotationally equilibrated laser ($P_{Ar}=50$ torr), although in the latter all kinds of relaxation were included. This implies that the kinetic scheme and not the amount of heat losses by relaxation processes is the crucial factor in determining the laser efficiency. In the next section this phenomenon will be interpreted from a thermodynamic viewpoint. A simple kinetic explanation for the relatively low efficiency of the frictionless laser can be given as follows: Since radiative transitions are the only mechanism that modify $X(v,J)$, the entire manifold of v,J levels can be divided into disjointed groups. Each group contains the sequence of levels, $v,J-1; v-1,J; v-2,J+1;\ldots$. Population is exchanged within but not between groups. Moreover, the population flows only downward, i.e., $v,J-1 \rightarrow v-1,J$. The flow or, equivalently, the lasing terminates when $X(v,J-1) = (g_{J-1}/g_J)X(v-1,J)$. Since $g_{J-1}/g_J \lesssim 1$, the final (end of pulse) distribution is characterized by $X_f(v) \lesssim X_f(v-1)$, as illustrated in Fig. 7. This

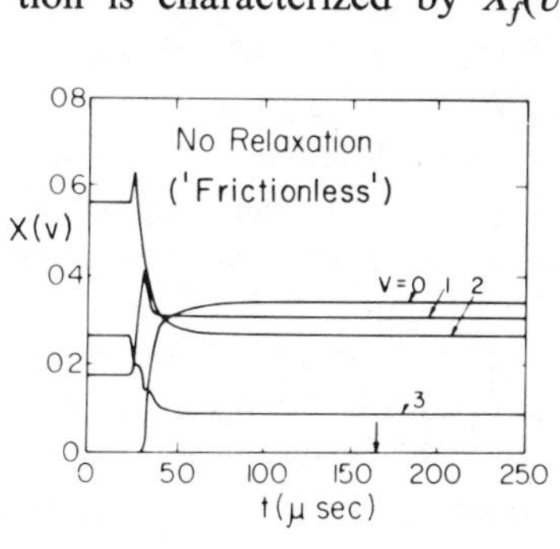

Fig. 7. Time evolution of the vibrational distribution of the "frictionless" (no relaxation) HF laser.[42] Lasing terminates (arrow) when the system is still energy-rich.

implies that lasing terminates when the energy-rich vibrational levels, $X_f(v \geqslant 1)$, are still heavily populated, i.e., low efficiency.

The buffer gas pressure can be regarded as the parameter measuring the gradual passage from the inhomogeneous-individual lasing mechanism predominating in the limit of week rotational coupling to the homogeneous –cooperative mechanism characterizing the lasing at rotational equilibrium. Model calculations show, as expected, that increasing the buffer gas pressure is accompanied by reduction in the extent of (time) overlap between lasing transitions in the same vibrational band and by an increase in the laser efficiency. For instance, the average photon yields for the $CF_3I:H_2:Ar$ system corresponding to constant initial reagent pressures $CF_3I:H_2=1$ torr and varying Ar pressure, $P_{Ar}=0$, 5, 25, and 50 torr are $\psi=0.97$, 1.28, 1.55, and 1.62, respectively.[42]

IV. THERMODYNAMICS

In this section we shall briefly review the thermodynamic description of chemical lasers. The analysis will be based on the fundamental laws of thermodynamics and the statistical-molecular definitions of entropy and energy. The approach outlined below does not intend to yield new detailed results, these have been supplied adequately in the kinetic analyses. On the contrary, it attempts to compact the detailed data by focusing attention on a few macroscopically significant observables, and by applying general thermodynamic relationships to shed a different light on the phenomena described in the previous sections.

A. Lasing Criteria

The key function in the thermodynamic analysis is the generalized entropy[69–72]

$$S=-k\sum_j P_j \ln P_j \tag{26}$$

where k is the Boltzmann constant and P_j is the probability of finding the system of interest in its quantum state j, $\Sigma P_j=1$. The definition (26) applies to any system whether in equilibrium or not, provided that P_j is the distribution that maximizes S subject to the appropriate thermodynamic constraints. For example, the equilibrium distribution of an isolated system is governed by no constraints except the normalization $\Sigma P_j=1$ and the maximal entropy distribution is the microcanonical one; $P_j^0=1/\Omega$, where Ω is the total number of accessible quantum states. Correspondingly, $S^0=k\ln\Omega$ is the familiar Boltzmann expression. If the system is coupled to a heat bath the normalization constraint is supplemented by the energy constraint

$$\langle E_j \rangle = \sum_j P_j E_j = \text{const} \tag{27}$$

and $S \rightarrow \max$ yields[69, 70] $P_j^0 = \exp(-\beta E_j)/Q(\beta)$, where $\beta = 1/kT$.

If the system is coupled to (heat, matter, mechanical) reservoirs the entropy of the composite system is $S_T = S + \Sigma_r S_r$ where S_r, the entropy of reservoir r, is given by (26) with j labeling the reservior states. The second law of thermodynamics states that in the composite system, which by definition is an isolated system, S_T cannot decrease, i.e.,

$$dS_T = dS + \sum_r dS_r \geqslant 0 \tag{28}$$

where the equality holds only at equilibrium (in accordance with the maximum entropy principle). Equation 28 is valid also for nonisolated systems provided the energy taken or given to the system is thermodynamic work, which by definition carries no entropy. In this case the equality in (28) holds for quasistatic (reversible) processes.

The systems of specific interest to us are gases of $N \cdot V$ weakly interacting molecules. Ignoring intermolecular interactions and disregarding quantum statistical effects which are irrelevant to our discussion the entropy of the gas (per unit volume) is[70–72]

$$S = -Nk \sum_n X(n) \ln\left[\frac{X(n)}{g_n}\right] = Ns \tag{29}$$

where $X(n) = N_n/N$ is the fraction of molecules in the g_n-fold degenerate energy level n. Similarly the energy of the gas (per unit volume) is

$$E = N \sum_n X(n) E_n = N\langle \varepsilon \rangle \tag{30}$$

The simplest application of thermodynamic arguments to laser systems is the derivation[44] of the necessary lasing condition $\Delta N > 0$. To this end consider the changes in E and S of a thermally isolated system when an infinitesimal fraction of molecules $dN_2 = -dN_1 = -N\,dX(2)$ drop from $n=2$ to $n=1$ by the radiative transition $2 \rightarrow 1$, $E_2 - E_1 = h\nu > 0$ (we assume $N = \text{constant}$). Setting $dX(2) = -dX(1) < 0$ and $dX(i \neq 1,2) = 0$ we find

$$dS = Nk\left\{ \ln\left[\frac{X(2)}{g_2}\right] - \ln\left[\frac{X(1)}{g_1}\right]\right\} dX(1) \geqslant 0 \tag{31}$$

where the inequality follows from the fact that the system is not coupled to any reservoir. Since $dX(1)>0$ we get the population inversion requirement

$$X(2)-\frac{g_2}{g_1}X(1)\geqslant 0 \tag{32}$$

The entropy associated with the internal degrees of freedom of the lasing molecules is given by (29) with $n=v, J$,

$$\begin{aligned} S=S[v,J]&=-Nk\sum_{v,J}X(v,J)\ln\left[\frac{X(v,J)}{g_J}\right]\\ &=S[v]+S[J|v]\\ &=-Nk\sum_{v}X(v)\ln X(v)-Nk\sum_{v}X(v)\sum_{J}X(J|v)\ln\left[\frac{X(J|v)}{g_J}\right] \end{aligned} \tag{33}$$

In the limit of weak rotational coupling the internal degrees of freedom form an isolated system and S increases during the lasing; more precisely each group $v, J-1$; $v-1, J$; $v-2, J+1\ldots$ is an isolated system whose entropy increases in the lasing process, as implied by (31) and (32). The situation is different when the lasing system is coupled to a heat bath. In this case the lasing process can go on even if $dS<0$ provided that $dS+dS_b\geqslant 0$; dS_b is the entropy change of the bath. This precisely is the reason for the high efficiency of the rotationally equilibrated lasers. The proof is simple. Consider the net change in the system entropy (33), and energy E,

$$\begin{aligned} E=N\sum_{v,J}X(v,J)E_{vJ}&=N\sum_{v}X(v)E_v+N\sum_{v}X(v)\sum_{J}X(J|v)E_J(v)\\ &=E_{\text{vib}}+E_{\text{rot}}, \end{aligned} \tag{34}$$

$[E_{vJ}=E_v+E_J(v)]$, when a small fraction of molecules are transferred from $v, J-1$ to $v-1, J$ by the lasing and simultaneously rotational relaxation ensures no hole-burning in $X(J|v)=X_0(J|v)$ and $X(J|v-1)=X_0(J|v-1)$. Clearly then, $dX(J|v)=dX_0(J|v)\equiv 0$, $dX(v)=-dX(v-1)<0$ and $dX(v'\neq v, v-1)=0$. Assuming for simplicity the rigid-rotor harmonic oscillator (RRHO) level scheme, $E_v=\hbar\omega v$ and $E_J(v)=E_J=BJ(J+1)$, we get $dS[J|v]\equiv 0$, $dE_{\text{rot}}\equiv 0$ and

$$dS=dS[v]=-Nk\ln\left[\frac{X(v)}{X(v-1)}\right]dX(v) \tag{35}$$

$$dE=dE_{\text{vib}}=N(E_v-E_{v-1})dX(v)=N\hbar\omega\, dX(v) \tag{36}$$

The frequency of a P-branch photon is $h\nu = \hbar\omega - 2BJ$. Thus only part of the vibrational quanta $\hbar\omega$ appears as thermodynamic work (laser radiation). The rest $\hbar\omega - h\nu = 2BJ$ is delivered to the heat bath so as to preserve the Boltzmann shapes of $X(J|v)$ and $X(J|v-1)$. In thermodynamic notation

$$-dE = dW + dQ = -Nh\nu\, dX(v) - 2NBJ\, dX(v) \tag{37}$$

The second term determines the entropy increase of the bath $dS_b = dQ/T$. Combining this result with the second law, $dS + dS_b \geqslant 0$, and using (35) and (36) with $dX(v) < 0$ we regain (22),

$$k \ln\left[\frac{X(v)}{X(v-1)}\right] + \frac{2BJ}{T} \geqslant 0 \tag{38}$$

The first term here is proportional to the entropy change of the system, which becomes negative when $X(v)/X(v-1) < 1$. This is allowed, however, provided the positive second term, which accounts for the entropy increase of the bath, has a larger absolute value. Similar arguments can be used to derive the lasing condition on R-branch lines, $v, J \to v-1, J-1$. Here, however, $h\nu = \hbar\omega + 2BJ$ and $dQ = T\,dS_b < 0$ and lasing requires $dS > 0$, which means [cf. (35)] that complete inversion $X(v)/X(v-1) > 1$ is a necessary condition; more precisely, instead of (38) we get[47] $k\ln[X(v)/X(v-1)] - 2BJ/T \geqslant 0$.

B. Maximal Work and Constraints

The different efficiencies of chemical lasers governed by different kinetic coupling schemes can be derived from a general statistical-thermodynamic approach to work processes in nonequilibrium molecular systems[72–74]. The two major components of this approach are the maximum entropy principle and the entropy deficiency function.[73] The entropy deficiency is a generalized thermodynamic potential (free energy). That is, it decreases monotonically in time in spontaneous relaxation processes and provides an upper bound to the thermodynamic work performed by the system in a controlled process.[75, 76] For systems of weakly interacting molecules the entropy deficiency $DS[X|X']$ is given by

$$DS[X|X'] = Nk \sum_n X(n) \ln\left[\frac{X(n)}{X'(n)}\right] \geqslant 0 \tag{39}$$

where $X(n)$ and $X'(n)$ are two molecular distribution functions. This function is always positive,[72–74] unless $X(n) \equiv X'(n)$, in which case it is identically zero. As a special but common example demonstrating the properties of DS consider a closed (N=const) system in contact with a heat bath T. Suppose that at time t the system is not at equilibrium and characterized by $X(n)$, while at $t=\infty$ it is equilibrated and specified by $X_0(n)=g_n\exp[-E_n/kT]/Q$. It can be shown[77] that if the passage from $X(n)$ to $X_0(n)$ is a spontaneous relaxation process described by a master equation, then $dDS[X|X_0]/dt \leqslant 0$ with equality only at $t=\infty$. Also, substituting the canonical expression for $X'(n)=X_0(n)$ into (39) we get after little algebra

$$DS[X|X_0]=(S_0-S)-T^{-1}(E_0-E)=-T^{-1}\Delta F=T^{-1}W_{\max} \quad (40)$$

where S and S_0 are the initial and final entropies [cf. (29)] and E and E_0 are the respective energies (30). ΔF is the change in Helmholtz free energy, which determines the maximal work available $W_{\max}$. It should be noted that (40) applies also for thermally isolated systems, if T is identified as the temperature specifying the single-molecule distribution in the final state. If the system is isolated (except mechanically) $S \geqslant S_0$ and $W_{\max}=-\Delta E=TDS$ is obtained in the isoentropic process $S=S_0$.

Any molecular distribution, equilibrium or not, can be represented in the generalized canonical form[78, 79] $X(n)=\exp(-\Sigma_r\lambda_r A_n^r)/Q(\{\lambda_r\})$, which obtains by maximizing (29) subject to the constraints $\Sigma_n A_n^r X(n)=\langle A^r\rangle$. The λ_r's are the appropriate Lagrange multipliers. (The canonical distribution is a special case corresponding to a single constraint $\Sigma_n E_n X(n)=\langle E\rangle$ with $\lambda_1=1/kT$). Based on this fact one can prove the following theorems[74]: (1) If $X_f(n)$ is obtained from $X_i(n)$ by removing one or more constraints then $W_{\max}=T_f DS[X_i|X_f]$ in the process $i\rightarrow f$; if the system is isolated then $W_{\max}$ obtains when $S_i=S_f$. (Mathematically a constraint r is removed when $\lambda_r=0$.) (2) $W_{\max}$ is a state function in the sense that $T_f DS[X_i|X_f]= T_j DS[X_i|X_j]+T_f DS[X_j|X_f]$, where X_j is any intermediate distribution governed by less (more) constraints than $X_i(X_f)$. Again, if the system is isolated $W_{\max}$ is obtained when $S_i=S_j=S_f$.

The different laser efficiencies in the limits of weak and strong rotational coupling can be naturally cast into the formal framework outlined above.[74] To this end the lasing molecules, together with the heat bath, should be regarded as one large thermally isolated system. In the absence of rotational relaxation no energy flows between neighboring transitions or between the lasing levels and the heat bath. This implies (physically "invisible" but realistic) constraints on $X(v, J)$, which restrict the efficiency of converting internal energy into laser radiation. These constraints are removed in the strong coupling limit giving rise to higher values of the available work.

C. Time Dependence

So far in the thermodynamic description we have not considered the explicit time dependence of the thermodynamic functions, we assumed that the number of lasing molecules is fixed and we disregarded dissipative losses by evaluating only upper bounds to the thermodynamic work. We shall conclude the discussion by showing briefly how all these factors can be properly incorporated into the thermodynamic analysis of chemical lasers.[80, 81] Let us visualize chemical lasers as heat engines in which the "working substance" are the internal degrees of freedom of the lasing molecules, henceforth the "system" with its entropy, S, and energy, E, given by (33) and (34), respectively. The time rates of change of E and S are[81]

$$\dot{E}=\sum_{v,J}\left[\dot{N}X(v,J)+N\dot{X}(v,J)\right]E_{vJ}=\sum_{v,J}\dot{N}_{vJ}E_{vJ} \tag{41}$$

$$\dot{S}=-Nk\sum_{v,J}\left[\dot{N}X(v,J)+N\dot{X}(v,J)\right]\ln\left[\frac{X(v,J)}{g_J}\right] \tag{42}$$

$$=-k\sum_{v,J}\dot{N}_{vJ}\ln\left[\frac{X(v,J)}{g_J}\right]$$

The reactants ("fuel") consumed by the pumping reaction are equivalent to a hot reservoir supplying energy E_p and entropy S_p to the system at rates $\dot{E}_p$ and $\dot{S}_p$,

$$\dot{E}_p=\dot{N}\sum_{v,J}X_i(v,J)E_{vJ} \tag{43}$$

$$\dot{S}_p=-\dot{N}k\sum_{v,J}X_i(v,J)\ln\left[\frac{X_i(v,J)}{g_J}\right] \tag{44}$$

where $\dot{N}=\Sigma\dot{N}_{vJ}$ is the total pumping rate and $X_i(v,J)$ is the nascent product distribution. E_p and S_p are given by (43) and (44) with N replacing $\dot{N}$.

Part of the system energy is converted into thermodynamic work W, coupled out as laser power $\dot{W}$ [cf. (3)]. Another part, Q, is transferred to the heat bath by the relaxation processes at a rate $\dot{Q}$. Correspondingly, the bath entropy increases at a rate $\dot{S}_b=\dot{Q}/T$ where $T=T(t)$ is the instantaneous bath temperature. The rest of the energy is stored in the system. The energy balance in the system is determined by the first law of thermodynamics

$$\dot{E}=\dot{E}_p-\dot{W}-\dot{Q} \tag{45}$$

From the second law, (28), we get

$$\dot{S}+\dot{S}_b-\dot{S}_p\equiv\dot{S}_{\rm ir}\geqslant 0 \tag{46}$$

with $-\dot{S}_p$ identified as the entropy change of the hot reservoir. The difference $\dot{S}_{\rm ir}$ accounts for the irreversible entropy production; it is identically zero in a reversible change. Analogous expressions to (45) and (46) relate the integrated quantities E, E_p, etc. Taking $t=0$ as the beginning of the pumping process and setting $X(t=0)=0$ for all the functions whose time derivatives appear in (45) and (46) imply that $E(t)$ and $S(t)$ are the instantaneous energy and entropy of the system, while all other quantities such as $E_p(t)$ or $S_b(t)$ represent accumulated changes.

All the thermodynamic functions considered above as well as their time rate of change can be computed using the solutions of the rate equations (1) and (2) as input data. This data compaction has been carried out[81] for the "frictionless" and the rotationally equilibrated HF laser systems and are shown in Figs. 8 and 9, respectively. The figures also show two types of efficiency. An integrated efficiency η defined as

$$\eta=\eta(t)=\frac{W(t)}{E_p(t)} \tag{47}$$

and an instantaneous efficiency[80]

$$\chi=\chi(t)=\frac{\dot{W}(t)}{\dot{E}_p(t)} \tag{48}$$

The ordinary efficiency is simply $\eta(t\to\infty)$. (Note that E_p is the energy pumped only into the internal degrees of freedom. Hence η is higher than the chemical efficiency $\eta_c=\langle f_{\rm int}\rangle\eta(\infty)$ where $\langle f_{\rm int}\rangle$ is the fraction of the reaction exoergicity released as products internal energy. $\langle f_{\rm int}\rangle\simeq 0.75$ for the $F+H_2$ reaction.) The instantaneous efficiency reflects "memory effects" of the laser system.[80] Since the lasing molecules can store energy and then suddenly release it χ may vary from low to high values or vice versa. In Q-switched lasers, for example, we would expect a very sharp peak in χ. The attainment of threshold also produces oscillations in χ (Figs. 8 and 9).

Figures 8 and 9 summarize and confirm quantitatively our qualitative inferences on the thermodynamic characteristics of chemical laser systems and thus do not deserve additional interpretation. Let us emphasize, however, one noteworthy feature: Despite the absence of relaxation

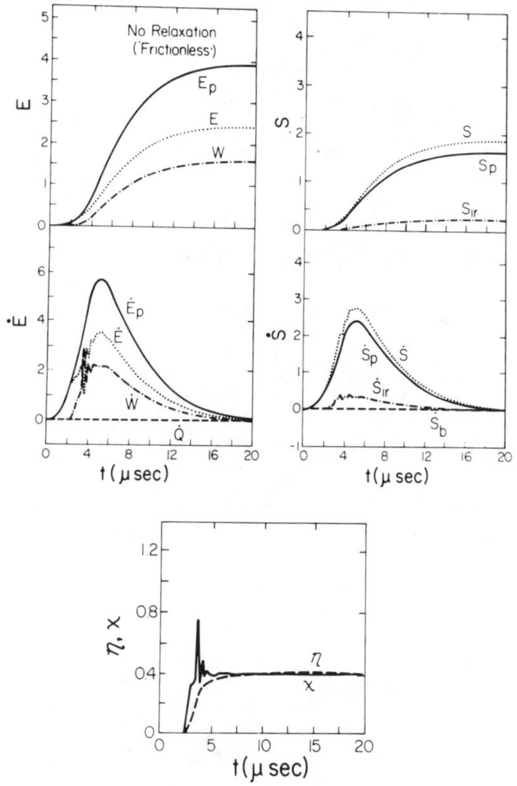

Fig. 8. Time dependence of the thermodynamic functions for the frictionless laser.[81] The entropies per unit volume S, are in units of 10^{-8} cal/cm^3 deg, $\dot{S}$ in 10^{-5} cal/cm^3 deg·s, E in 10^{-5} cal/cm^3, and $\dot{E}$ in cal/cm^3 s.

processes in the frictionless system as reflected by $\dot{Q}=\dot{S}_b=Q=S_b\equiv 0$, the irreversible entropy production S_{ir} is small but finite. A closer inspection reveals that $\dot{S}_{ir}\neq 0$ only when $\dot{W}\neq 0$, that is, only during the laser pulse; in accordance with the conclusion that (in the absence of coupling to a heat bath) the entropy of a lasing system increases due to the lasing tendency to equalize populations [cf. (31)]. Alternatively, note that in the frictionless laser the system entropy is always higher than the pumping entropy, $S>S_p$, reflecting the entropy production associated with the lasing process. On the other hand, in the rotationally coupled system $S<S_p$, corresponding to entropy decrease associated with the extraction of laser radiation from partially inverted population [cf. (35) and (38)]. Clearly this decrease is counterbalanced by the increase in the bath entropy, $S_b>S_p-S$.

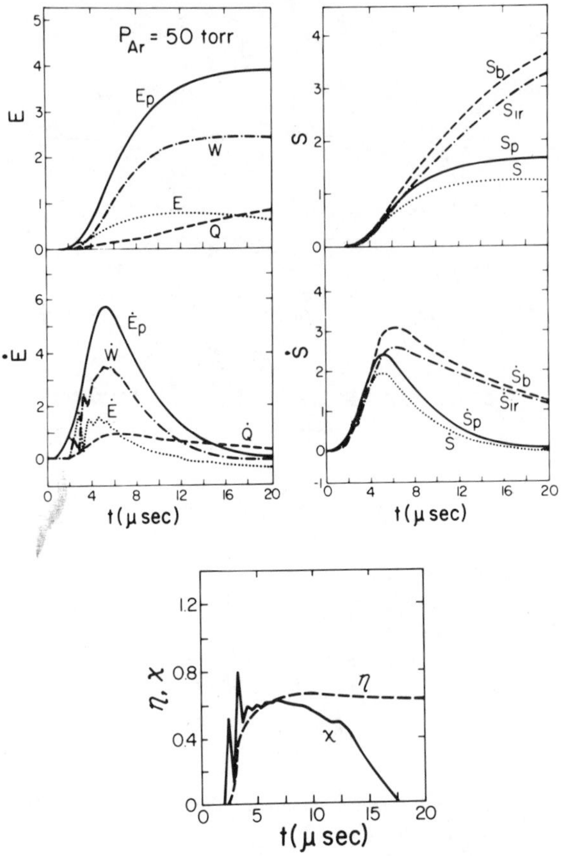

Fig. 9. Time dependence of the thermodynamic functions in the limit of strong rotational coupling,[81] $p_{Ar} = 50$ torr. Units as in Fig. 8.

V. CONCLUDING REMARKS

The aim of the descriptions and analyses presented in the foregoing sections was to illuminate some aspects of chemical lasers as molecular systems far from equilibrium. Particular emphasis was drawn on the limits of weak and strong rotational coupling since they represent extremely different kinetic schemes and consequently different kinetic behaviors. Thermodynamic considerations were employed to complement the detailed kinetic description. The thermodynamic approach can yield additional physical insights, but (at least so far) not new quantitative data. These are provided by the kinetic formalism. Nevertheless, further chemical laser experiments and state-to-state kinetic data are required in order to increase the reliability of kinetic analyses as useful predictive means.

References

1. R. W. F. Gross and J. F. Bott (eds.), *Handbook of Chemical Lasers*, Wiley, New York, 1976.
2. M. J. Berry, *Ann. Rev. Phys. Chem.* **26**, 259 (1975).
3. K. L. Kompa, *Chemical Lasers, Topics in Current Chemistry*, Vol. 37, Springer, Berlin, 1973.
4. M. J. Berry, in *Molecular Energy Transfer*, R. D. Levine and J. Jortner, eds., Wiley, New York, 1975, p. 114.
5. A. Yariv, *Quantum Electronics* 2nd ed., Wiley, New York, 1975.
6. A. E. Siegman, *An Introduction to Lasers and Masers*, McGraw-Hill, New York, 1971.
7. G. Emanuel, Chap. 8 in Ref. 1 and the references therein.
8. N. Cohen and J. F. Bott, Chap. 2 in Ref. 1 and the references therein.
9. T. F. Deutch, *Appl. Phys. Lett.*, **10**, 234 (1967); **11**, 18 (1967).
10. D. P. Akitt and J. T. Hardley, *IEEE J. Quant. Electron*, **QE-6**, 113 (1970).
11. H. L. Pummer and K. L. Kompa, *Appl. Phys. Lett.*, **19**, 356 (1972).
12. N. Skribanovitz, I. P. Hermann, R. M. Osgood Jr., M. S. Feld, and A. Javan, *Appl. Phys. Lett.*, **20**, 498 (1972).
13. T. D. Padrick and M. A. Gusinow, *Appl. Phys. Lett.*, **22**, 183 (1973).
14. O. D. Krogh and G. C. Pimentel, *J. Chem. Phys.*, **67**, 2993 (1977).
15. E. Cuellar, J. H. Parker, and G. C. Pimentel, *J. Chem. Phys.*, **61**, 422 (1974); E. Cuellar and G. C. Pimentel, *IEEE J. Quant. Electron.*, **QE-11**, 688 (1975).
16. H. L. Chen, R. L. Taylor, J. Wilson, P. Lewis, and W. Fyfe, *J. Chem. Phys.*, **61**, 306 (1974).
17. E. Sirkin and G. C. Pimentel, to be published.
18. J. H. Smith and D. W. Robinson, *J. Chem. Phys.*, **68**, 5474 (1978); G. D. Downey, D. W. Robinson, and J. H. Smith, *J. Chem. Phys.*, **60**, 1685 (1977).
19. J. C. Polanyi and K. B. Woodall, *J. Chem. Phys.*, **57**, 1547 (1972) and the references therein.
20. A. M. G. Ding and J. C. Polanyi, *Chem. Phys.*, **10**, 39 (1975).
21. J. J. Hinchen, *Appl. Phys. Lett.*, **27**, 672 (1975).
22. J. J. Hinchen and R. H. Hobbs, *J. Chem. Phys.*, **65**, 2732 (1976).
23. N. C. Lang, J. C. Polanyi, and J. Wanner, *Chem. Phys.*, **24**, 219 (1977).
24. P. Brechignac, *Opt. Comm.*, **25**, 53 (1978).
25. I. Procaccia, Y. Shimoni, and R. D. Levine, *J. Chem. Phys.*, **63**, 3181 (1975).
26. R. B. Bernstein, *J. Chem. Phys.*, **62**, 4570 (1975).
27. L. H. Sentman, *J. Chem. Phys.*, **67**, 966 (1977).
28. J. C. Polanyi, N. Sathyamurthy, and J. L. Schreiber, *Chem. Phys.*, **24**, 105 (1977).
29. A. Ben-Shaul, G. L. Hofacker, and K. L. Kompa, *J. Chem. Phys.*, **59**, 4664 (1973); U. Schmailzl, A. Ben-Shaul, and K. L. Kompa, *IEEE J. Quant. Electron.*, **QE-10**, 753 (1974); A. Ben-Shaul and G. L. Hofacker, Chap. 10 in Ref. 1.
30. T. D. Padrick and M. A. Gusinow, *Chem. Phys. Lett.*, **24**, 270 (1974).
31. L. H. Sentman, *J. Chem. Phys.*, **62**, 3523 (1975).
32. J. T. Hough and R. L. Kerber, *IEEE J. Quant. Electron.*, **QE-11**, 699 (1975).
33. J. R. Creighton, *IEEE J. Quant. Electron.*, **QE-11**, 699 (1975).

34. R. J. Hall, *IEEE J. Quant. Electron.*, **QE-12**, 453 (1976).
35. G. K. Vasil'ev, E. F. Makarov, A. G. Ryabenko, and V. L. Tal'roze, *Sov. Phys. JETP*, **44**, 690 (1976).
36. A. Ben-Shaul, U. Schmailzl, and K. L. Kompa, *J. Chem. Phys.*, **65**, 1711 (1976).
37. L. H. Sentman, *Appl. Opt.*, **15**, 744 (1976).
38. K. Daree and U. Hesse, *Opt. Comm.*, **19**, 269 (1976).
39. E. Keren, R. B. Gerber, and A. Ben-Shaul, *Chem. Phys.*, **21**, 1 (1977).
40. M. Baer, Z. H. Top, and Z. B. Alfassi, *Chem. Phys.*, **22**, 485 (1977).
41. R. L. Kerber and J. T. Hough, *Appl. Opt.*, **17**, 2369 (1978); **14**, 2960 (1975).
42. A. Ben-Shaul, S. Feliks, and O. Kafri, *Chem. Phys.*, **36**, 291 (1979).
43. Y. Reuven, A. Ben-Shaul, and M. Baer, *J. Appl. Phys.*, **51**, 130 (1980).
44. R. D. Levine and O. Kafri, *Chem. Phys. Lett.*, **27**, 175 (1974).
45. O. Kafri and R. D. Levine, *Opt. Comm.*, **12**, 118 (1974).
46. R. D. Levine and O. Kafri, *Chem. Phys.*, **8**, 426 (1975).
47. A. Ben-Shaul, O. Kafri, and R. D. Levine, *Chem. Phys.*, **10**, 367 (1975).
48. M. Garbuny, *J. Chem. Phys.*, **67**, 5676 (1977).
49. W. E. McDermot, N. R. Pchelkin, B. J. Bernar, and R. R. Bousek, *Appl. Phys. Lett.*, **32**, 469 (1978). See also *Electronic Transition Lasers*, L. E. Wilson, S. N. Suchard, and J. I. Steinfeld, eds., (MIT Press, Cambridge, Mass., 1976).
50. K. Hohla and K. L. Kompa, Chap. 12 in Ref. 1.
51. C. K. Rhodes ed., *Excimer Lasers, Topics in Applied Physics*, Vol. 30, Springer, Berlin, 1979.
52. A. N. Chester and R. A. Chodzko, Chap. 3 in Ref. 1.
53. G. Grohs and G. Emanuel, Chap. 5 in Ref. 1.
54. M. J. Berry, *J. Chem. Phys.*, **59**, 6229 (1973).
55. J. R. Airey, *J. Chem. Phys.*, **52**, 156 (1970).
56. J. C. Polanyi and K. B. Woodall, *J. Chem. Phys.*, **57**, 1574 (1972).
57. H. W. Chang and D. W. Setser, *J. Chem. Phys.*, **58**, 2298 (1973).
58. J. G. Moehlmann and J. D. McDonald, *J. Chem. Phys.*, **62**, 3061 (1975).
59. R. I. Cukier, C. M. Fortuin, K. E. Shuler, A. G. Petschek, and J. H. Schaibly, *J. Chem. Phys.*, **59**, 3873 (1973).
60. J. G. Schaibly and K. E. Shuler, *J. Chem. Phys.*, **59**, 3879 (1973).
61. H. B. Levine, System, Science and Software Report, SSS-R-75-2684 (1975).
62. J. T. Hwang, E. P. Dougherty, S. Rabitz, and H. Rabitz, *J. Chem. Phys.*, **69**, 5180 (1978).
63. S. D. Rockwood, J. E. Brau, W. A. Proctor, and G. H. Canavan, *IEEE J. Quant. Electron.*, **QE-9**, 94 (1973).
64. B. R. Bronfin and W. Q. Jeffers, Chap. 11 in Ref. 1 and references therein.
65. C. E. Treanor, J. W. Rich, and R. G. Rehm, *J. Chem. Phys.*, **48**, 1798 (1968).
66. S. N. Suchard and J. R. Airey, Chap. 6 in Ref. 1 and references therein.
67. R. W. F. Gross and D. J. Spencer, Chap. 4 in Ref. 1 and references therein.
68. A. Ben-Shaul, *Chem. Phys.* **18**, 13 (1976).
69. E. T. Jaynes, in *The Maximum Entropy Formalism*, R. D. Levine and M. Tribus, eds., MIT Press, Cambridge, Mass., 1978, p. 15.
70. J. C. Keck, in *The Maximum Entropy Formalism*, R. D. Levine and M. Tribus, eds., MIT Press, Cambridge, Mass., 1978, p. 219.
71. R. B. Bernstein and R. D. Levine, *Adv. Atom. Mol. Phys.*, **11**, 215 (1975).
72. R. D. Levine and A. Ben-Shaul in *Chemical and Biochemical Applications of Lasers*, C. B. Moore, ed., Academic Press, New York, 1977, Chap. 4.
73. I. Procaccia and R. D. Levine, *J. Chem. Phys.*, **65**, 3357 (1976).
74. A. Ben-Shaul and R. D. Levine, *J. Noneq. Therm.*, **4**, 363 (1979).

75. L. D. Landau and E. M. Lifshitz, *Statistical Physics*, Pergamon, Oxford, 1969.
76. H. B. Callen, *Thermodynamics* Wiley, New York, 1961.
77. M. Tabor, R. D. Levine, A. Ben-Shaul, and J. I. Steinfeld, *Mol. Phys.* **37**, 141 (1979).
78. I. Procaccia, Y. Shimoni, and R. D. Levine, *J. Chem. Phys.*, **65**, 3284 (1976).
79. R. D. Levine, *J. Chem. Phys.*, **65**, 3302 (1976).
80. O. Kafri and R. D. Levine, *Israel J. Chem.*, **16**, 342 (1978).
81. A. Ben-Shaul and O. Kafri, *Chem. Phys.*, **36**, 307 (1979).

LASER DIAGNOSTICS OF REACTION PRODUCT ENERGY DISTRIBUTIONS

A. BARONAVSKI, M. E. UMSTEAD, AND M. C. LIN

Chemical Diagnostics Branch, Chemistry Division, Naval Research Laboratory, Washington, D. C. 20375

CONTENTS

I. INTRODUCTION

One of the most successful applications of the laser to chemistry and physics lies in the area of diagnostics. Thanks to its intensity, monochromaticity and its continually expanding tunability, the internal states of chemical species can be probed readily to the concentration level of $\sim 10^8 - 10^9$ particles/cm^3. For several metal atoms which have large optical cross-sections, single-atom detection with the laser was achieved a few years ago.[1]

In this chapter we review in some detail the results of our reaction dynamics studies employing various types of lasers to probe the formation of reaction products in different internal quantum states—electronic, vibrational, and rotational, depending on the processes investigated and on the diagnostic techniques used. Two different laser probing methods are used in this work; they are resonance absorption and fluorescence methods using a cw CO laser and a tunable dye laser, respectively.

Three areas of our studies covering photodissociation (by the absorption of single and multiple photons), electronic-to-vibrational energy transfer and combustion related reactions are discussed. Because of space limitation, we present primarily results obtained in this Laboratory and related work.

II. EXPERIMENTAL DETAILS

A. The CO Laser-Probing Technique

A schematic diagram of the CO laser-probing apparatus is shown in Fig. 1. The cw-CO laser, similar to that designed by Djeu,[2] had an active length of about 1.7 m enclosed by a pair of BaF_2 windows and was cooled over its entire length by a flow of liquid N_2. The laser gases, He (5 torr), N_2 (5 torr), CO ($\sim$0.1 torr), and NO ($\sim$0.2 torr), were admitted through sidearms near electrodes at each end of the laser and were pumped out through a third sidearm near an electrode at the middle. By means of suitable current-limiting resistors in series with a Sorensen high-voltage power supply, an electric discharge was passed from each of the two end electrodes to the center one.

The laser cavity was formed by a 3-m output-coupling mirror (3% transmission between 4.7 and 6 μm) and a 300 line/mm grating blazed at 4 μm, both supported on a steel *I*-beam, which in turn was supported by thick rubber padding on a long table. The laser tube was separately mounted over the *I*-beam without touching it. Single transition operation was possible between the 1→0 and 26→25 bands, the upper limit imposed only by the spectral range of the grating used.

Single-mode operation was achieved by closing down the variable iris in front of the output coupling mirror. The mirror itself was mounted on a piezoelectric transducer, which is capable of converting an applied voltage into a proportional extension of the element (Lansing 80-214 Lock-In

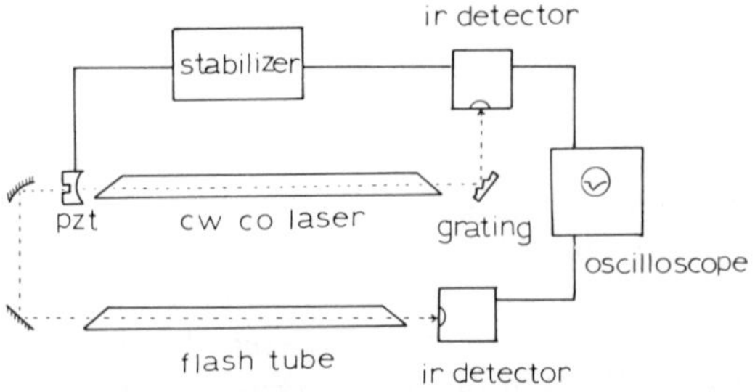

Fig. 1. Schematic diagram of the CO laser probing apparatus. PZT, piezoelectric transducer attached to a 3% transmitting, 3-m radius coupling mirror.

Stabilizer), thus continuously stabilizing the laser to line center. Under normal conditions the amplitude instability remained less than a few percent during the time required to make an absorption measurement. The output from the partially transmitting mirror was directed through the reaction tube, a focusing lens, and finally a suitable IR filter before impinging upon a Ge:Au detector maintained at 77 K. The detector signal was either displayed and photographed on a Tektronix 555 oscilloscope or fed into signal averaging equipment and plotted on an X–Y recorder.

For experiments involving flash photolysis, the reaction tube consisted of a coaxial flash tube, constructed similarly to that described previously.[3] The innermost tube, which contained the reaction mixture and which was surrounded by the flash gas (20 torr of 1% Xe in Ar), was readily replaceable allowing quartz, Pyrex, and Vycor tubes to be used interchangeably as filters. The temperature of the flash tube could be varied over a wide range and controlled to $\pm 1\,^{\circ}\mathrm{C}$ by circulating through the outer jacket of the tube, an ethylene glycol–water solution from a constant temperature bath, which could be either refrigerated or heated.

In a study of the relaxation of electronically excited species by CO, a Chromatix CMX-4 dye laser operating on 1.2×10^{-4} M rhodamine 590 (Exciton Chemical Co.) in a 1 : 1 solution of methanol and water was used for electronic excitation. The pulse width (fwhm) was approximately 1 μs. By means of a beam splitter, which transmits infrared and reflects visible radiation in the suitable spectral regions, the excitation radiation from the dye laser was brought to be colinear with the CO probing laser radiation. Both were directed through the center of a suitable reaction tube used in place of the flash tube. The signal from the Ge:Au detector was amplified by a Tektronix 1A7A differential amplifier, fed through a Biomation 610B transient recorder and signal averaged on a Nicolet 1072 signal averager. The final output from the signal averager was plotted on an X–Y recorder and displayed on an oscilloscope to be photographed.

Typically, reaction mixtures were either flash photolyzed or laser excited under conditions which led to the production of the species of interest in the presence of the CO laser beam, preset to the various vibrational–rotational lines accessible to the reaction. The initial vibrational population distribution of the CO formed in the reaction was determined from time-resolved absorption curves measured for all vibrational levels populated by the reaction under study. The $P(10)$ lines were used for most experiments because of their high output by the probing laser. However, no significantly different population distributions were obtained when other nearby lines were used randomly.

The absorption (or gain) coefficient, α, for a vibration–rotation transition was computed by the equation: $\alpha = \ln(I/I_0)/l$, where I_0 and I are the

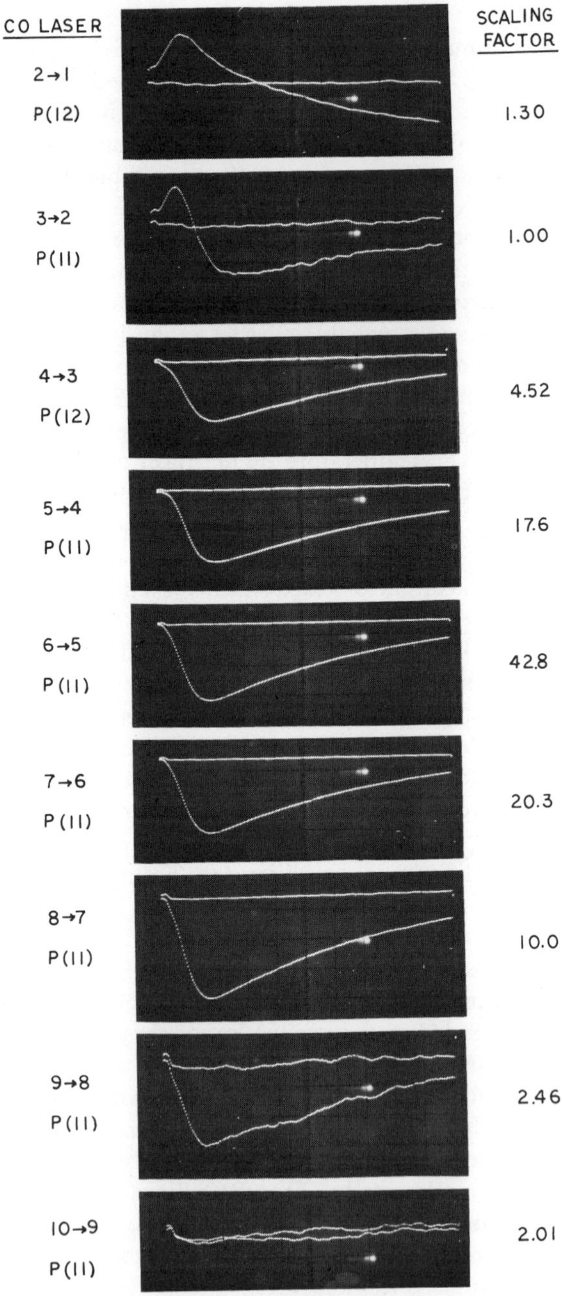

Fig. 2. Signal-averaged background and absorption traces for the Na (3^2P) + CO reaction at 528 K. Reprinted with permission from *Chem. Phys. Lett.*, **42**, 78 (1976). Copyright by North-Holland Publishing Company.

incident and the transmitted laser intensities, and l is the effective length of the absorbing medium. I_0 was measured by chopping the laser beam both before and after the flash. Any runs in which the two I_0 measurements differed by more than 10% were discarded. In the laser excitation experiments, the background CO laser intensity was first signal averaged over 1024 shots of the dye laser, which was tuned to one of the excited vibronic states of the species under study. The absorption measurement was then repeated in the presence of CO. The same procedure was employed for each of the CO laser transitions. As in the flash experiments, I_0 was measured by chopping the beam before and after absorption measurements for each transition. Typical signal averaged traces for the Na(3^2P)–CO system are shown in Fig. 2.

Since the rates of resonant V–V relaxation,

$$CO(v)+CO(0)\rightarrow CO(v-1)+CO(1)$$

are known to be very rapid,[4–10] only the early, rapidly rising portion of each absorption curve was evaluated. The time of the first appearance of absorption varied, depending on the rate of the reaction being studied and the reactant concentrations, but was on the order of a few microseconds. The CO vibrational population was determined by analyzing these initial portions of the time resolved absorption curves by means of a computer, according to the following equation:

$$\alpha_v(J)=\frac{8\pi^3}{3kT}\left(\frac{Mc^2}{2\pi kT}\right)^{1/2}|R_v|^2 JF_{vr}\left\{N_{v+1}B_{v+1}\exp\left[-B_{v+1}\frac{J(J-1)hc}{kT}\right]\right.$$
$$\left.-N_v B_v \exp\left[-B_v\frac{J(J+1)hc}{kT}\right]\right\}$$

where v and J represent the vibrational and rotational quantum numbers, respectively, of the lower level. B_v is the rotational constant for the vth vibrational level, $|R_v|^2$ is the rotationless matrix element for the $v\rightarrow v+1$ transition, and F_{vr} is the Herman–Wallis factor. All of these molecular constants are available in the literature and their sources can be found in Ref. 11.

The initial CO population distribution was evaluated by extrapolating the N_v/N_0 ratios to the appearance time of absorption (t_0). The absorption data obtained over the first several microseconds of the reaction were first fitted to a quadratic function of t by means of a least-squares method, and the N_v/N_0 ratios were then evaluated at t_0 for all v's from these best fitted equations. This extrapolation was carried out to eliminate the effect of both CO vibrational relaxation and any possible secondary reactions that might generate CO.

The amount of vibrational energy channeled into CO was calculated from the expression,

$$\langle E_v \rangle = \sum_{v>0} f_v E_v$$

where $f_v = N_v / \Sigma N_v$ is the normalized vibrational population distribution and E_v is the vibrational energy of CO at the vth level with the zero point excluded.

B. Laser-Induced Fluorescence Measurements

A detailed description of the theoretical aspects of LIF has recently been reviewed by Kinsey.[12] We will only briefly describe the experimental arrangement employed in this Laboratory.

Figure 3 shows a typical experimental arrangement for obtaining LIF spectra from photodissociation products. The first laser, which in our case is either a frequency quadrupled Nd^{3+}:YAG or an ArF excimer, is collimated and sent through a cell containing the parent molecule. Absorption of either a single photon or several photons causes this molecule to dissociate, forming radicals in specific electronic, vibrational, and rotational levels. The second laser, a tunable dye laser, is synchronized with the first and propagates collinearly with and in the opposite direction to the photolysis laser. As the wavelength of the second laser is scanned across resonances of the radical being probed, light at these specific wavelengths is absorbed and reemitted at longer wavelengths as fluorescence. For instance, if one excites the $v' = 4 \leftarrow v'' = 0$ level, fluorescence on the $v' = 4 \rightarrow v'' = 1, 2, 3$, etc. levels can be observed. In this way scattered light from the laser is minimized. This technique is much more sensitive than monitoring absorption. The fluorescence is observed at right angles to the propagating beams through either an appropriate set of glass and dielectric filters, or a monochromator set at a specific wavelength. Typically the photomultiplier used is an RCA 31034 due to its broad and relatively flat response, although other types have been used depending on specific experimental needs. The signal from the photomultiplier is smoothed by a boxcar integrator also synchronized with the lasers and is then available to be plotted on a strip chart recorder, or to be further processed in a signal averager or minicomputer. In all cases, in order to obtain correct population ratios, the power dependence of the probe laser as a function of wavelength, and the response curves of the detection system (filters or

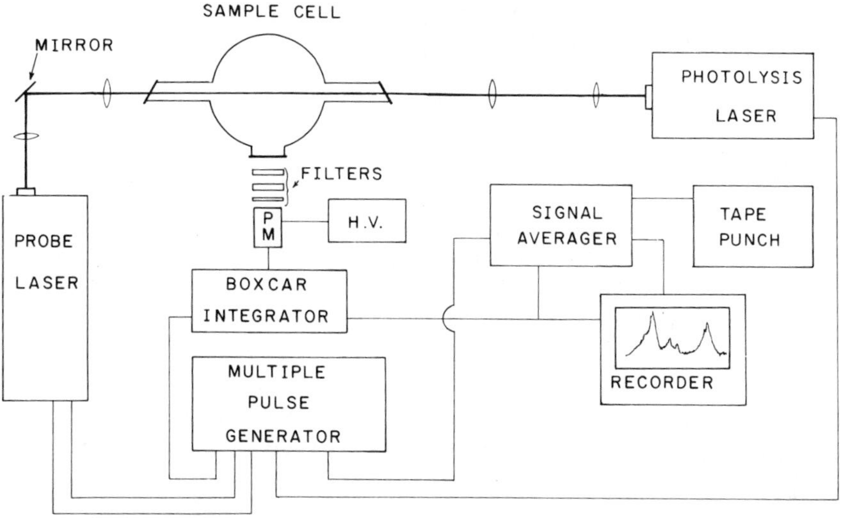

Fig. 3. Experimental apparatus for determining energy level populations following photodissociation. The photolysis laser can be either a Nd^{3+}:YAG laser quadrupled to 266 nm or an ArF excimer laser. The probe laser is a Chromatix CMX-4 tunable flashlamp pumped dye laser. The lasers propagate collinearly through the fluorescence cell and fluorescence is monitored by a photomultiplier tube through appropriate filters. In cases where direct chemiluminescence from fragments is observed following photolysis, the filters are replaced by a scanning 0.6 m Hilger–Engis monochromator. The multiple pulse generator allows the two lasers to be synchronized or, in some cases, allows the dye laser to be delayed in time.

monochromator and photomultiplier) must be known and the LIF spectrum corrected accordingly. In addition, the line strengths and Franck–Condon factors for the transitions being investigated must also be known. Once the spectrum is corrected for all these factors, the relative peak heights give the population ratios directly. In the case where fluorescence is directly observed from transients after photodissociation of the parent, no second laser is used; however, a monochromator is scanned over the region of interest and the signals are obtained and plotted in the same manner. It has been found that LIF can be used to detect transients in concentrations $< 10^9$ molecules/cm^3 and, if direct emission can be observed, concentrations $< 10^7$ molecules/cm^3 can be monitored.[13]

III. RESULTS AND DISCUSSION

A. Photodissociation Reactions

In the photodissociation of molecules one expects the initial energy deposited in the molecule to be partitioned into some combination of translational, rotational, vibrational, and electronic energies of the fragments. The conservation of energy requires

$$E_t + E_r + E_v + E_e = E_{h\nu} + E_{\text{int}} - D_0 \tag{1}$$

where

E_t = fragment translational energy
E_r = fragment rotational energy
E_v = fragment vibrational energy
E_e = fragment electronic energy

and

$E_{h\nu}$ = energy of absorbed photon
E_{int} = internal energy (distribution) of the parent molecule
D_0 = energy required to dissociate the parent molecule into the appropriate fragments

For the simplest molecules, diatomics, (1) can be written

$$E_t + E_e = E_{h\nu} + E_{\text{int}} - D_0$$

In a set of very elegant experiments, Wilson and co-workers have determined the translational energy distributions of atomic species following photodissociation of several diatomics including Cl_2,[14] Br_2,[15] IBr,[16] and I_2.[16, 17] In this simplest case the electronic energy distributions are easily found by measuring translational energies of the photofragments. In the case of diatomics such measurements provide information on the electronic states of the parent molecule as well as those of the atoms. An extension of this work to polyatomics has also been accomplished.[18–21] However, in this case, the internal energy distributions of the fragments must be inferred from their translational energy distributions. As a result, the internal energy content is much less definitive.

In the following section, two types of photodissociation will be discussed: single-photon dissociation and VUV multiphoton dissociation. The examples given are those with which we have worked and so is not a complete list of photodissociation reactions for which internal energies of photofragments have been determined. Several excellent reviews of a more general nature are available in the literature.[22–25]

1. Single-Photon Photodissociation

ICN. The photodissociation of halogen cyanides has been studied by several workers due to the ease with which the CN fragment may be probed.[18, 26–38] ICN, in particular, has been studied in several spectral regions, yet the longest wavelength absorption is still not well understood.[39–50] In order to probe the CN radical after dissociation, we have photolyzed ICN in the $\tilde{A}$ state using a frequency-quadrupled Nd:Yag laser.[13] Even though the output energy of the laser at 266 nm is typically only ~ 100 μJ/pulse, sufficient quantities ($\sim 10^9$ molecules/cm^3) of CN are produced to allow quantitative measurements by LIF. The YAG laser photolyzes ICN to form CN $X\,^2\Sigma^+$ $(v''=0,1,2\ldots)$ and those states are probed by LIF on the $A\,^2\pi_i \leftarrow X\,^2\Sigma^+$ (Red) system.

In this case the bandwidth of the probe laser is sufficiently broad to overlap several rotational lines. Still, we can estimate that $K_{max} \simeq 23-27$ corresponding to a Boltzmann temperature of >3000 K. In more recent experiments,[51] using a narrower bandwidth laser, we have been able to resolve individual rotational lines as shown in Fig. 4. By plotting $\ln(I_f/S_K \nu_K)$ against K, a bimodal distribution is found, which can adequately be fitted to two Boltzmann temperatures, 105 K and 3520 K, as shown in Fig. 5.

By adding 3 torr of He one can obtain the initial vibrational distribution since the He acts as an effective rotational quencher but is inefficient in relaxing vibrational levels.[13] Table I gives the vibrational distribution which has been observed. Direct emission from CN($A\,^2\pi_i$) was not observed under any conditions and must be <0.001 of the population of $X\,^2\Sigma^+$.

We have now totally characterized the internal energy of the CN fragment. However, Ling and Wilson have found a bimodal distribution in the time of flight spectrum for ICN photodissociation at 266 nm. Originally this was explained as electronically excited CN($A^2\pi_i$). However, it has been shown that this is not the case. Therefore, the electronically excited I atom, I($^2P_{1/2}$) must be formed. In a recent set of experiments[51] a Ge IR detector was used to observe I($^2P_{1/2}$) emission at 1.315 μm, using essentially the same apparatus as in Fig. 3 with the exception that no probe laser was needed and the cell used was a $1 \times 1 \times 5$ cm cuvette in order to collect emission efficiently. I($^2P_{1/2}$) emission was detected, and by using the photodissociation reaction $C_3F_7I + h\nu$ (266 nm)$\rightarrow C_3F_7 + I(^2P_{1/2})$(99%) as an actinometer we have found the following branching ratios:

$$\mathrm{ICN} + h\nu(266\ \mathrm{nm}) \overset{63\pm3\%}{\rightarrow} \mathrm{I}(^2P_{1/2}) + \mathrm{CN(X)}$$

$$\overset{37\pm3\%}{\rightarrow} \mathrm{I}(^2P_{3/2}) + \mathrm{CN(X)}$$

The quantum yields found in this case for the production of I($^2P_{1/2}$)(63%) are in agreement with the findings of Ling and Wilson[18] for the translational distribution peaks ($\sim$60% and 40%), which were originally ascribed

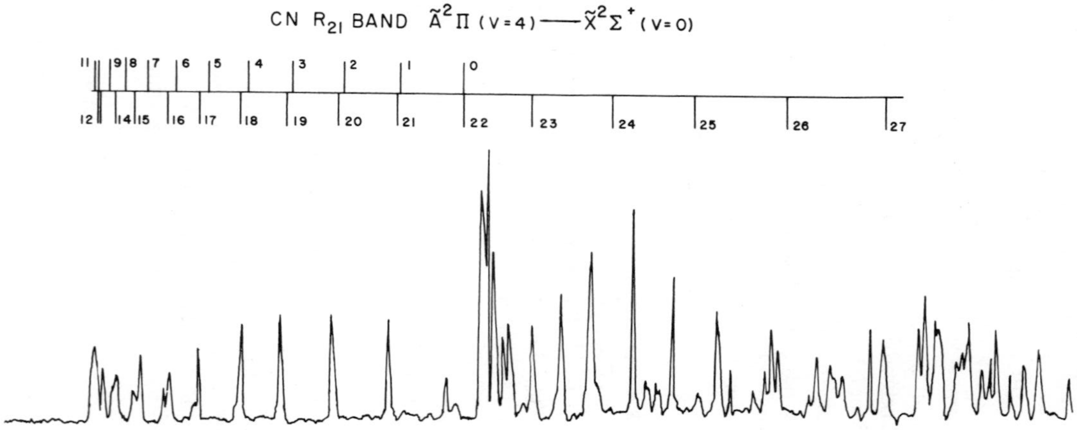

Fig. 4. A portion of the R_{21} branch of the CN($A^2\pi_i(v=4) \leftarrow X^2\Sigma^+(v=0)$) laser-induced fluorescence excitation spectrum. The rotational assignments are given above the spectrum.

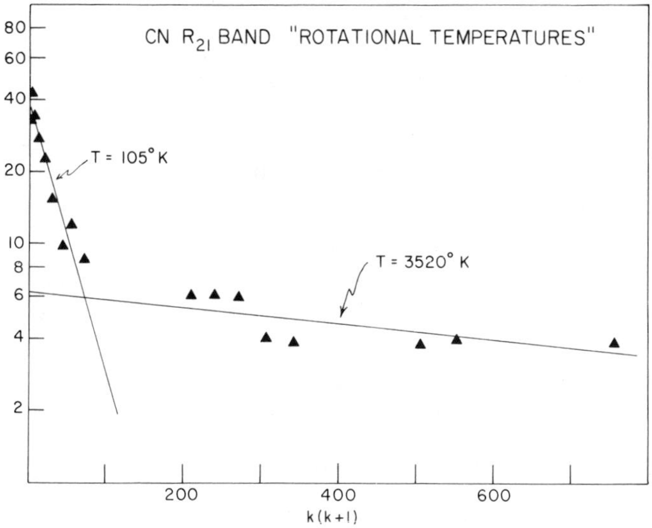

Fig. 5. Plot of ln $(I_f/S_K\nu_K)$ vs $K(K+1)$ where I_f is the corrected peak height of a given line, S_K is the rotational line strength and ν is the transition frequency. K is the rotational quantum number. The slope of each line is related to $1/T_R$ through the Boltzmann distribution equation.

to $CN(A^2\pi_i)$ and $CN(X^2\Sigma^+)$, respectively. We have, therefore, characterized the internal energy distribution for the photodissociation of ICN at 266 nm.

Several theories have been proposed to explain this photodissociation reaction. At present it is most reasonable to assume that at least two electronically excited states give rise to ICN absorption in the $\tilde{A}$ continuum,[42, 49, 50] one which correlates with $I(^2P_{1/2})+CN$ and one which correlates with $I(^2P_{3/2})+CN$. The very hot rotational temperature observed probably results from the steep repulsive curve of the potential

TABLE I

Vibrational Distribution of Primary CN Fragments[a]

v''	$N_{v''}$
0	$\equiv 1.00$
1	1.2×10^{-2}
3	$\ll 1\times10^{-4}$

[a]Reprinted with permission from *Chem. Phys. Lett.*, **45**, 172 (1977). Copyright by North-Holland Publishing Company.

surface, which also accounts for the relatively broad distribution of one of the translational energy peaks in Ling and Wilson's results. The colder vibrational temperature results from the peaked distribution in Frank–Condon overlaps between the excited ICN molecule and the I+CN fragments. Golden Rule calculations[52, 53] of vibrational distributions based on the ground-state CN stretch frequency and CN bond distance in ICN and those corresponding to the free CN radical are in excellent agreement with the experimentally observed distribution.[51]

HN_3. Using the apparatus of Fig. 3, the photodissociation of HN_3 has also been studied.[54–56] Previous work[57–61] has shown that the near UV excitation of HN_3 results in the formation of $NH(^1\Delta)+N_2(^1\Sigma_g^+)$ exclusively. However, the internal vibrational and rotational distributions had not been determined. The threshold for the reaction

$$HN_3+h\nu\rightarrow NH(^1\Delta)+N_2(^1\Sigma_g^+)$$

is 16,450 cm^{-1},[62, 63] and, for photolysis at 266 nm, E_{avl} is thus ~21,000 cm^{-1}. Reasonably, one could expect vibrational energy in $NH(^1\Delta)$ up to $v=6$ and ~2000 cm^{-1} of rotational energy. However, experimentally the $a(^1\Delta)$ state is vibrationally cold; >99.8% of the NH is formed in the $v=0$ level. No $X^3\Sigma^-$ or $b^1\Sigma^+$ was detected, in agreement with previous results for long wavelength photolysis. The rotational energy, on the other hand, is quite high, corresponding to a Boltzmann temperature of 1200 K as shown in Fig. 6. By measuring Doppler widths of individual rotational lines, a crude estimate of the translational energy can also be obtained. We have determined $E_T \leqslant 5200$ cm^{-1} for $NH(^1\Delta)$.[55] Therefore, ~15,000 cm^{-1} of energy must be partitioned into the N_2 fragment. From conservation of momentum we can estimate E_T for N_2; and we are left with ~11,000 cm^{-1} to be partitioned into rotational and vibrational energy of N_2.

The initially formed $NH(^1\Delta)$ undergoes a secondary reaction with HN_3 to form electronically excited $NH_2(^2A_1)$[54–56]: $NH(^1\Delta)+HN_3\rightarrow N_3+NH_2(^2A_1)$. The chemiluminescence can be monitored as a function of time and is shown in Fig. 7. The rise time corresponds to quenching of $NH_2(^2A_1)$ and the fall corresponds to the rate of reaction of

$$NH(^1\Delta)+HN_3\rightarrow NH_2+N_3 \qquad (2)$$

The kinetic mechanisms for the reaction and for the reaction of $NH(^1\Delta)$ with foreign gases are given elsewhere;[55] however, the Stern–Volmer plot for the reaction of $NH(^1\Delta)$ and HN_3 is given in Fig. 8 as an example. It has been determined that two types of reactions occur for $NH(^1\Delta)$, abstraction of a hydrogen atom and insertion into C–C bands.[56] Table II gives

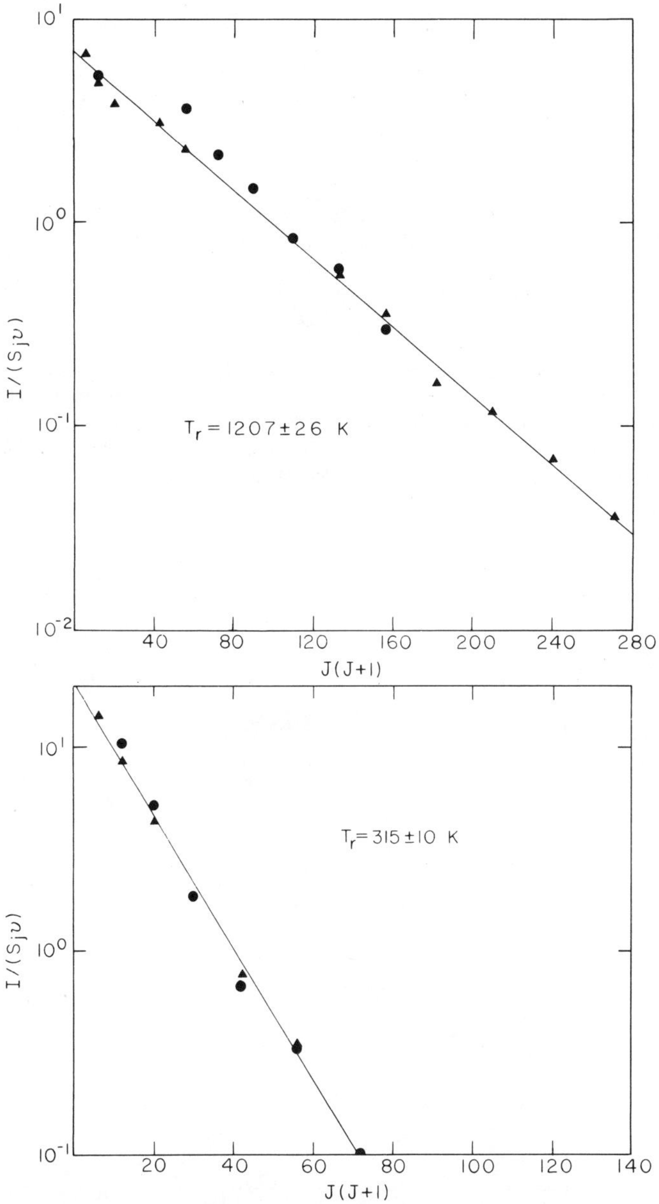

Fig. 6. Rotational distribution of NH($^1\Delta$, $v=0$) produced by photolysis of 30 mtorr of HN_3 at 266 nm, top, and in the presence of 95 torr of helium buffer base, bottom. *P*-branch line intensities are plotted as circles, while *Q*-branch lines are denoted by triangles. Reprinted with permission from *Chem. Phys.*, **30**, 119 (1978). Copyright by North-Holland Publishing Company.

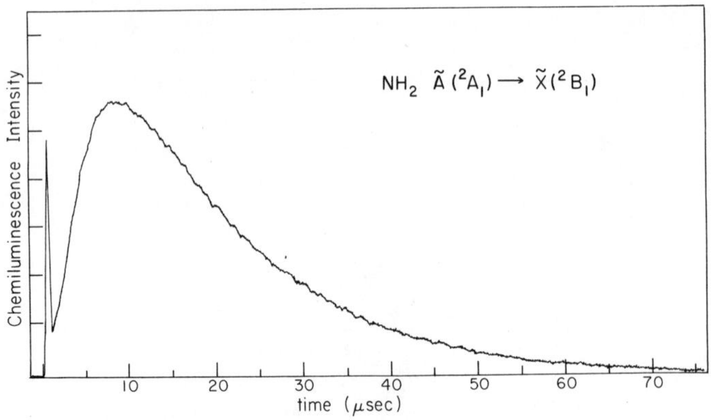

Fig. 7. Time resolved chemiluminescence of $NH_2(^2A_1) \rightarrow NH_2(^2B_1)$ resulting from the reaction of $NH(^1\Delta)$ with HN_3. The $NH(^1\Delta)$ is produced from the 266 nm photodissociation of 18.9 mtorr of HN_3. The sharp spike at the beginning of the trace results from scattered photolysis laser light and is not fluorescence. Reprinted with permission from *Chem. Phys.*, **30**, 119 (1978). Copyright by North-Holland Publishing Company.

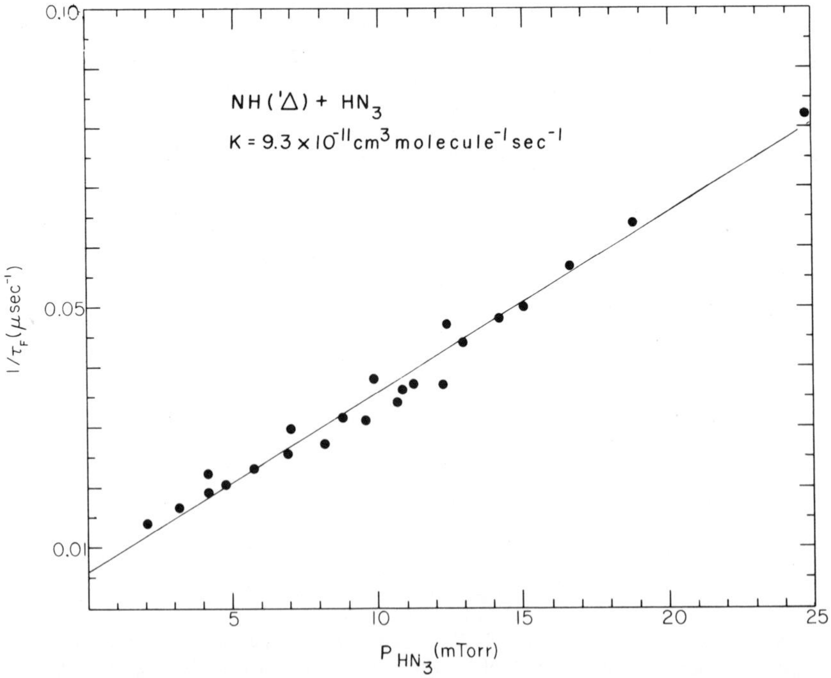

Fig. 8. A Stern–Volmer plot of $1/\tau$ vs HN_3 pressure. K is the rate constant for the reaction $NH(^1\Delta)+HN_3$. Reprinted with permission from *Chem Phys.*, **30**, 133 (1978). Copyright by North-Holland Publishing Company.

TABLE II
$NH(^1\Delta)$ Reaction Rates (in cm^3 $molecule^{-1}$ $s^{-1} \times 10^{11}$) and $NH_2(^2A_1)$ Quenching and Reaction Rates (in cm^3 $molecule^{-1}$ $s^{-1} \times 10^{10}$)[a]

Added gas	$NH(^1\Delta)$	$NH_2(^2A_1)$
HN_3	9.3 ± 0.9	3.2 ± 0.3
HCl	7.9 ± 0.8	4.0 ± 0.4
CH_4	1.2 ± 0.1	3.0 ± 0.3
C_2H_4	3.8 ± 0.4	3.1 ± 0.3
C_3H_6	3.6 ± 0.4	3.1 ± 0.3
C_6H_{12}	6.7 ± 0.7	

[a]Reprinted with permission from *Chem. Phys. Lett.*, **45**, 172 (1977). Copyright by North-Holland Publishing Company.

reaction rate constants for $NH(^1\Delta)$ and several gases and bimolecular quenching rates of $NH_2(^2A_1)$. It has been shown that the production of $NH_2(^2A_1)$ by reaction (2) is constant for photolysis wavelengths from 266 nm to 315.0 nm, indicating that:

1. HN_3 photolyzes at all wavelengths between 266 nm and 315 nm to produce $NH(^1\Delta)$ with approximately the same branching ratio.
2. $NH_2(^2A_1)$ is produced with nearly the same efficiency.
3. $NH_2(^2A_1)$ is produced with nearly the same internal energy.

These results show that the translational energy of $NH(^1\Delta)$ plays a small part in the formation of $NH_2(^2A_1)$.[55]

Based upon these observations and making reference to Fig. 9 we determine that previous assignments of the absorption spectrum of HN_3 in regions I and II (Fig. 9) must be reconsidered. It now appears that the continuous absorption at $\lambda > 280$ nm and the diffuse absorption for 236.5 nm$< \lambda <$280 nm both result in formation of $NH(^1\Delta)$ based upon the wavelength dependence of $NH_2(^2A_1)$ production.[55] Previous work has regarded these two regions as a single transition to a $^1A''$ state in HN_3 correlating with a $^1\Sigma^-$ linear molecule. However, this state would not correlate with $NH(^1\Delta)$. It now seems that there are, in fact, two transitions, one giving rise to a completely dissociative state that onsets at 330 nm and a second bound state that predissociates rapidly by coupling to the first dissociative state. Both states would then give $NH(^1\Delta)$ and $N_2(^1\Sigma_g^+)$ as products.

The most likely assignment for the totally dissociative state would be one of the components of a $^1\Delta$ derived HN_3 transition which is estimated to be a 1.5 eV higher in energy based on MO calculations.[64, 65] It is hoped

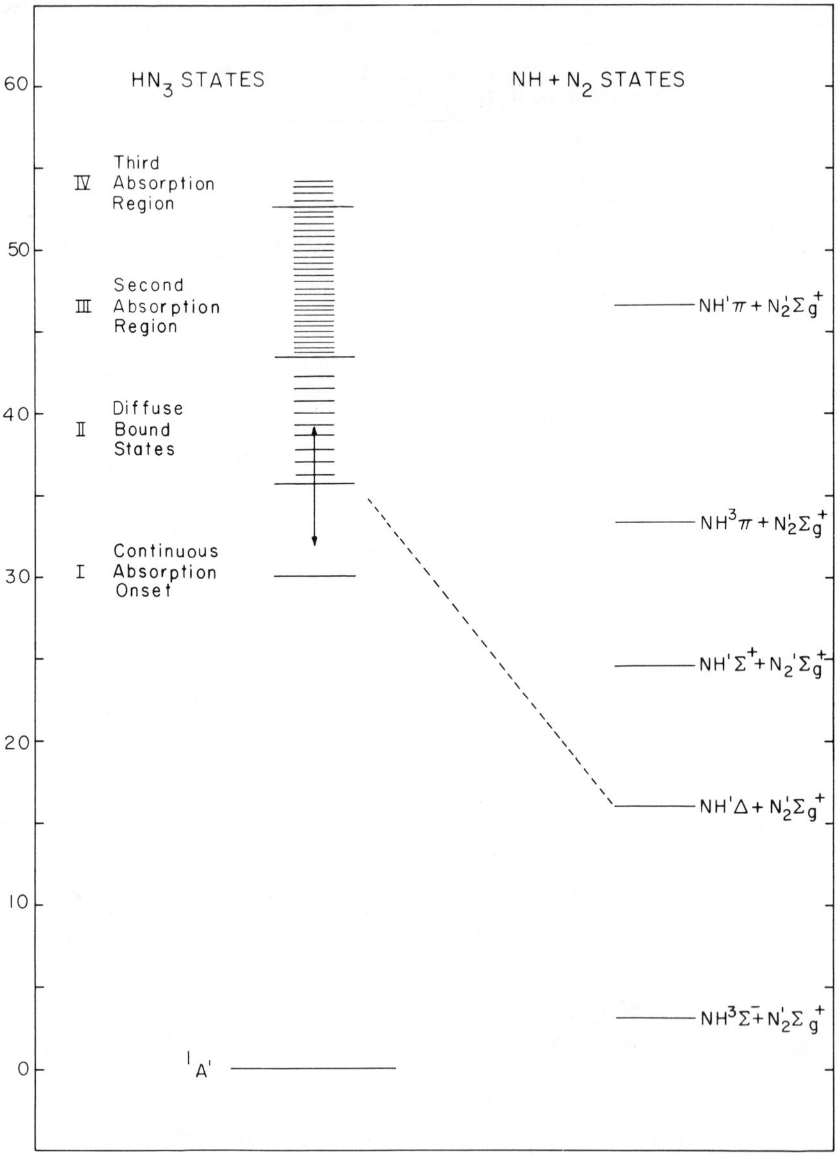

Fig. 9. Energy level diagram of HN_3 states and $NH + N_2$ states resulting from photodissociation. (From Ref. 55.) Reprinted with permission from *Chem. Phys.*, **30**, 119 (1978). Copyright by North-Holland Publishing Company.

that further work in the region from 194–265 nm can elucidate this assignment and also clarify the assignments of the shorter wavelength transitions.

NH_3. The single photon dissociation of NH_3 between 185 and 214 nm has been studied extensively[63, 66–70] and has been shown to result in predominantly $NH_2(\tilde{X}\,^2B_1)+H$. DiStefano et al.[71] have shown that photodissociation of NH_3 using a hydrogen discharge lamp results in a weak visible emission which they have assigned as $NH_2(\tilde{A}\,^2A_1 \rightarrow \tilde{X}\,^2B_1)$.

We have photolyzed NH_3 using an ArF excimer laser at 193 nm.[72] Both the $\tilde{X}\,^2B_1$ and $\tilde{A}\,^2A_1$ states of NH_2 are formed. Figure 10 gives a low-resolution scan of the emission observed between 620 and 900 nm. It should be noted that the emission extends almost to the thermochemical threshold at 619 nm. Assignments based on the work of Dressler and Ramsay[73] and Johns et al.[74] are also given. Since the NH_2 spectrum is complex due to extensive Renner–Teller coupling,[73] only approximate relative vibrational populations for the bending mode have been obtained for this emission, and this is shown in Fig. 11. Although there is a large

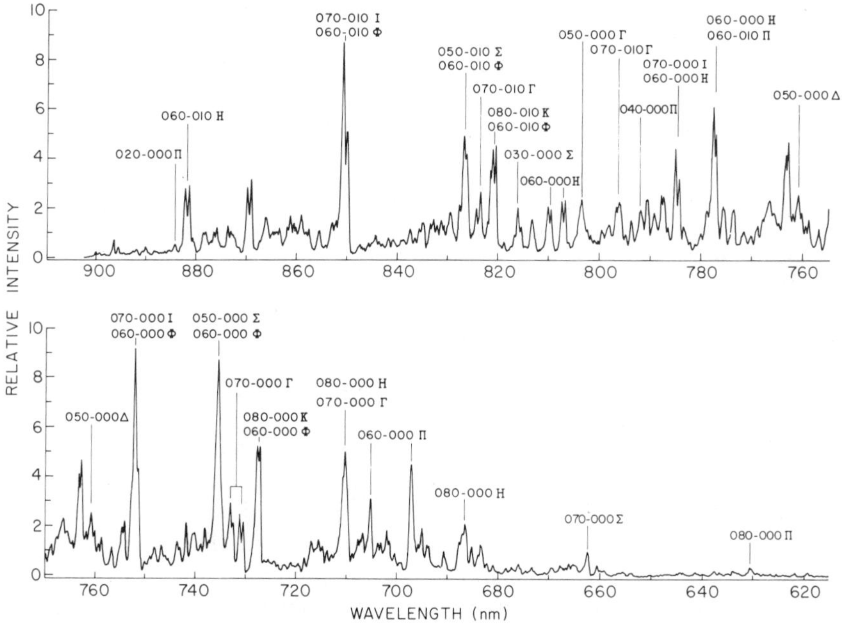

Fig. 10. Low resolution (0.7 nm) $NH_2(\tilde{A}^2A_1 \rightarrow \tilde{X}^2B_1)$ emission spectrum observed following NH_3 photodissociation at 193 nm $P_{NH_3}=30$ mtorr. Reprinted with permission from *Chem. Phys.*, **43**, 271 (1979). Copyright by North-Holland Publishing Company.

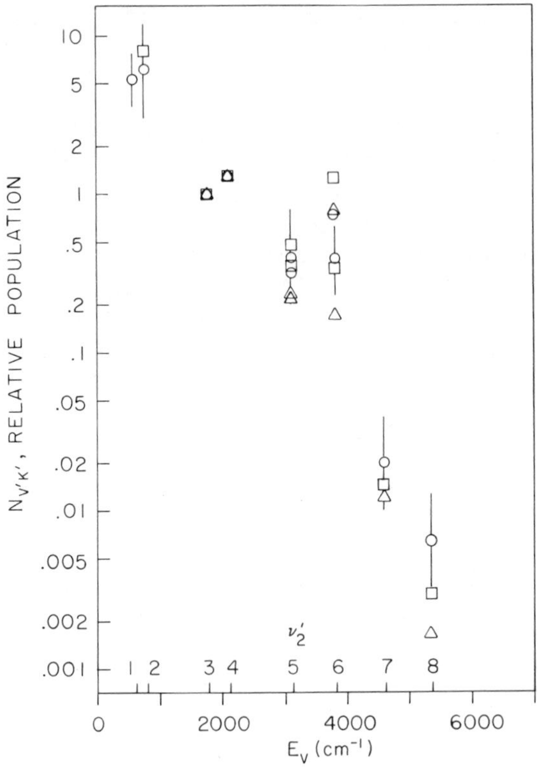

Fig. 11. Vibrational populations of $(Ov'O)\Sigma$ (odd v_2') and π (even v_2') levels of the $\tilde{A}^2A_1$ state of NH_2 vs vibrational energy for NH_3 pressures of 20 mtorr (0) and 1.2 mtorr (△ and □). Reprinted with permission from *Chem. Phys.*, **43**, 271 (1979). Copyright by North-Holland Publishing Company.

angle change in going from a planar $\tilde{A}\,^1A_2''$ state in NH_3 to the quasilinear $\tilde{A}\,^2A_1$ state in NH_2, there seems to be relatively little energy in the $^2A_1$ state of NH_2. This is indicative of a redistribution of energy in the dissociating complex.

The rotational "temperature" of the (050) Σ level has been determined. The rotational distribution is Boltzmann and corresponds to $T_r = 210 \pm 40$ K. This results from the fact that the projection of angular momentum about the *c*-axis (top axis) in NH_3 maximizes at $K=3$. This axis is parallel to the *c*-axis in NH_2. If C_{2v} symmetry is maintained in the NH_2 fragment during recoil, then rotation of NH_2 about its *c*-axis can arise only from rotation of NH_3 about its *c*-axis. Since NH_2 is quasilinear the *b* and *c* axes are indistinguishable[75] and one concludes that N' (in NH_2) $< K'$ (in NH_3) $=3$ and expects $T_r < 300$ K, which is borne out experimentally.

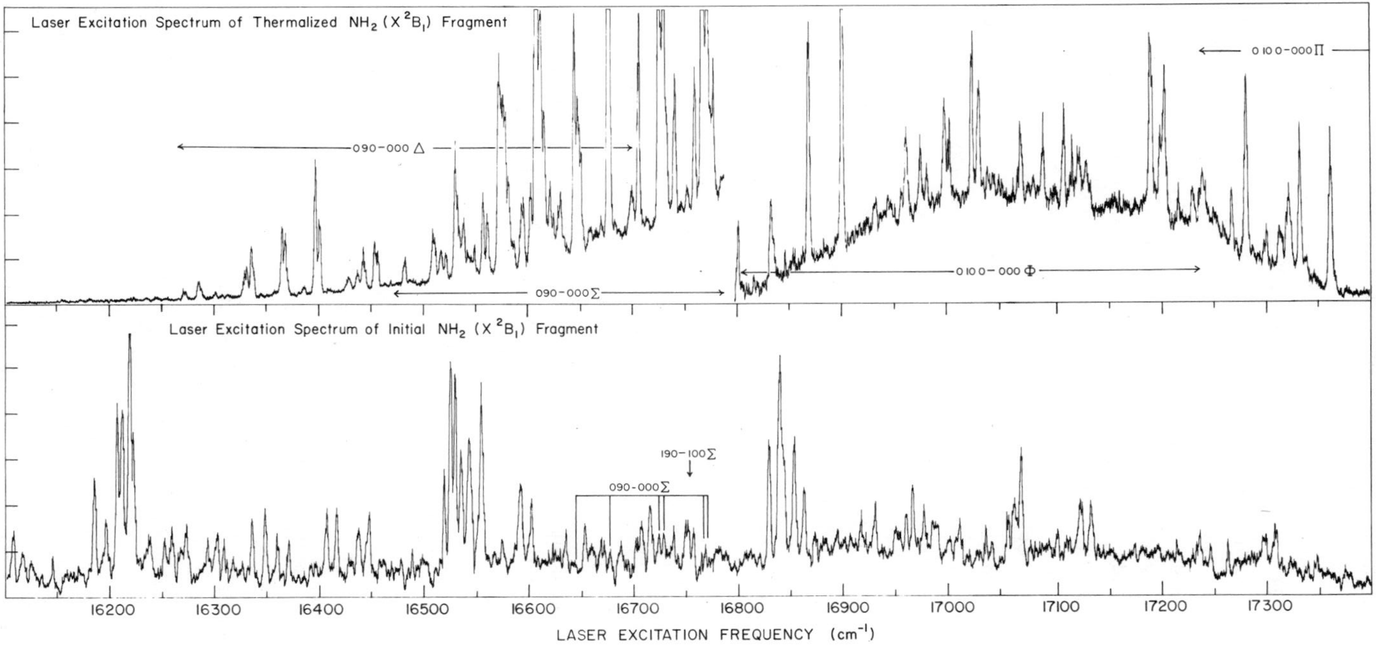

Fig. 12. Tunable dye laser-induced fluorescence excitation spectra of NH_2 ($\tilde{A}^2A_1 \rightarrow \tilde{X}^2B_1$). The lower spectrum is of initial NH_2 fragments formed < 1.5 μs after photolysis. The upper spectrum is of NH_2 thermalized with Ar. Reprinted with permission from *Chem. Phys.*, **43**, 271 (1979). Copyright by North-Holland Publishing Company.

In experiments using NO_2 fluorescence as an actinometer,[72] we have found that the $NH_2(\tilde{A}\,^2A_1 \rightarrow \tilde{X}\,^2B_1)$ emission accounts for ~2.5% of the total NH_3 dissociation. Therefore >97% of the NH_3 dissociates to give ground-state NH_2. The fluorescence excitation spectra for NH_2 $(\tilde{X}\,^2B_1)$ under two conditions are shown in Fig. 12. The upper panel shows the spectrum obtained when 2.2 torr of Ar is added to 130 mtorr of NH_3 and the probe laser is delayed by 400 μs. In this case, the ground-state NH_2 is totally thermalized and each line can be unambiguously assigned. Quite a different situation arises when the probe laser and the photolysis laser are coincident and the NH_3 pressure is 30 mtorr with no buffer gas. The resulting spectrum is shown in the lower panel. No assignments can be made in this case and we can eliminate all transitions arising from $v_2'' \leqslant 4$ and $K_a'' \leqslant 4$.[72] It is obvious that the initially formed $NH_2(\tilde{X}\,^2B_1)$ is rotationally and/or vibrationally very hot.

The production of $NH_2(\tilde{A}^2A_1)$ is surprising in that the excited planar state of NH_3 $(^1A_2'')$ correlates adiabatically with ground state NH_2. However, if NH_3 is nonplanar, this restriction is lifted and the $\tilde{A}$ and $\tilde{X}$ states of NH_3 belong to the same symmetry species. Therefore, an avoided curve crossing takes place and the excited NH_3 can dissociate to give $NH_2(\tilde{A}^2A_1)$. Evleth has treated this case for $H + NH_2$ recombination.[76]

It appears therefore that the dissociation of NH_3 occurs from both a planar and nonplanar geometry, with the planar geometry much more likely.

CH_3CHCO. The photodissociation of methylketene (CH_3CHCO) in the region of 240–370 nm has been reported to take place by the mechanism:[77, 78]

$$CH_3CHCO + h\nu \rightarrow CH_3CH^{\dagger} + CO^{\dagger}$$
$$CH_3CH^{\dagger} \rightarrow C_2H_4{}^{\dagger}$$
$$C_2H_4{}^{\dagger} \rightarrow C_2H_2 + H_2$$
$$C_2H_4{}^{\dagger} + M \rightarrow C_2H_4 + M$$

We have investigated this photodissociation by flash photolyzing CH_3CHCO in Vycor, ($\lambda \geqslant 220$ nm) and determining the CO vibrational population distribution by CO laser absorption measurements.[79] This reaction is very similar to the $O(^3P) + CH_3C{\equiv}CH$ reaction, which will be described in more detail later. Both involve the dissociation of an activated methylketene complex; in this case activation takes place photolytically, in the latter, chemically. The energetics of the two reactions are also very similar. In Vycor the CH_3CHCO absorption centers at about 226 nm,

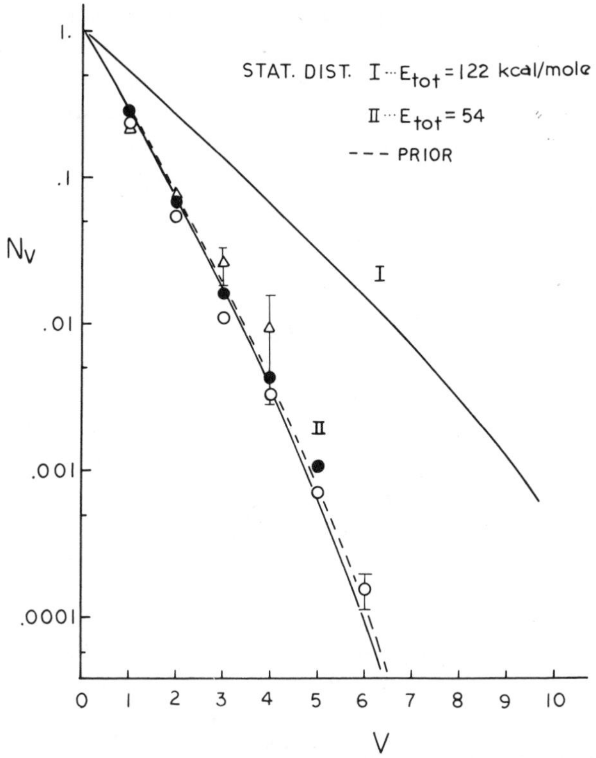

Fig. 13. Vibrational energy distribution of the CO formed in the methylketene photodissociation in Vycor: triangles, 2% CH_2CHCO in SF_6; open circles, 5% CH_3CHCO in SF_6; solid circles, $O+CH_3C_2H$ reaction. Reprinted with premission from *J. Phys. Chem.*, **82**, 1458 (1978). Copyright by the American Chemical Society.

which leads to the availability of 54 kcal/mole of reaction energy for CO excitation, essentially the same as in the $O+CH_3C{\equiv}CH$ reaction, 53.5 kcal/mole. As in the $O+CH_3C{\equiv}CH$ reaction, the isomerization energy $CH_3CH{\rightarrow}C_2H_4$ is not available for CO excitation, since it is not released until after the complex has already dissociated into CH_3CH and CO.

The CO vibrational population distribution resulting from the photolysis of CH_3CHCO is plotted in Fig. 13, along with that obtained for $O+$ $CH_3C{\equiv}CH$. The distributions are essentially the same. Also shown are the theoretical results obtained by randomizing in the CH_3CHCO complex, the total available reaction energy (upper solid curve), and the total available reaction energy less the CH_3CH isomerization energy (lower solid curve). The latter clearly fits the experimental results.

CH_3CHCO was also photolyzed in quartz ($\lambda \geqslant 200$ nm). The CO population distribution of the higher vibrational levels was distinctly hotter than was statistically expected. Photolysis of the isomeric acrolein (CH_2CHCHO) under similar conditions gave rise to a distribution very similar to that obtained for CH_3CHCO. (No dissocation of acrolein could be measured in Vycor.) These results have been tentatively explained by the occurrence of a new reaction channel at <200 nm, common to both isomers, which leads to the direct elimination of C_2H_4 from the complex.

2. *VUV Multiphoton Dissociation*

Multiphoton dissociation in the vacuum ultraviolet is a relatively new technique for the generation of high energy transients.[80–88] In this case, the first photon is used to excite a molecule to a state which may either be bound or dissociative, and one or more additional photons are absorbed by either the excited parent molecule or fragments from the parent. Such processes are possible due to the development of high-power excimer lasers, especially the ArF laser at 193 nm. Typically, $>10^{18}$ photons cm^{-2} $pulse^{-1}$ can be obtained in the focal region of such a laser.[81] The experimental apparatus is similar to Fig. 3 with the exception that the YAG laser is replaced by the ArF laser and the photomultiplier is attached to the exit slit of a Hilger–Engis 0.6-m monochromator. Fluorescence from excited fragments is focused onto the entrance slit of the monochromator.

CH_3Br. The absorption spectrum of CH_3Br is characterized by a continuum that extends from 285.0 nm to shorter wavelengths. The ArF laser was focused into a cell containing CH_3Br and the spectral region from 193.0 to 900 nm investigated.[81] Emissions from $CH(A^2\Delta \rightarrow X^2\pi)$ and $CH(^2\Sigma^- \rightarrow {}^2\pi)$ were observed. No attempt was made to observe LIF from $CH(X^2\pi)$ due to the strong emissions present and the fact that the $A^2\Delta \leftarrow X^2\pi$ transition is dominated by vertical (0–0, 1–1, etc.) promotions and results in a congested spectrum. From the observed intensities, the ratio $N(^2\Delta)/(^2\Sigma^-)$ is 19.6.

Thermodynamically, the threshold for $CH_3Br \rightarrow CH(A\,^2\Delta) + H_2 + Br$ is 10.57 eV and for $CH_3Br \rightarrow CH(B\,^2\Sigma^-) + H_2 + Br$ is 10.92 eV. The ArF photon is 6.4 eV, and the emissions are believed to be due to the following processes:

$$CH_3Br + h\nu(193) \rightarrow CH_3 + Br$$

$$CH_3 + h\nu(193) \rightarrow CH(^2\Delta, {}^2\Sigma^-) + H_2$$

Although it is energetically possible to form $CH(C\,^2\Sigma^+)$, we see no emission from this state.[81] Since the ionization potential of CH_3Br is 10.54

eV, it is likely that CH_3Br^+ is also formed. Recently, Rockwood et al. has reported the multiphoton ionization of C_6H_6 using a KrF laser laser at 249 nm.[88] At low-power levels, predominantly high mass ions are observed; however, as the laser intensity is increased to >50 MW/cm^2 only C^+ and H^+ ions are formed. It is possible that similar processes occur in CH_3Br.

C_2H_2. The multiphoton photolysis of acetylene at 193 nm results in emission from C_2 and CH fragments throughout most of the 190–1100 nm region.[82] The most intense emission is that from $C_2(A^1\pi_u \rightarrow X^1\Sigma_g^+)$ and accounts for $\sim$90% of the total emission observed. A portion of this spectrum is shown in Fig. 14. The relative vibrational populations for the $A^1\pi_u$ have been determined and are shown in Fig. 15. The population of $v'=0$ is unknown, since the emission falls outside the spectral region accessible to our photomultipliers. Since the reaction

$$C_2H_2 \rightarrow C_2(A^1\pi_u) + 2H$$

requires 11.75 eV, the process must involve at least two photons. The power dependence for this emission is shown in Fig. 16 and is clearly quadratic. However, there are several possible channels which are energetically allowed to give C_2 in the $A^1\pi_u$ state:

$$C_2H_2 + h\nu \rightarrow C_2H_2^* \tag{3}$$

$$C_2H_2^* + h\nu \rightarrow C_2(^1\pi_u) + H_2 \tag{4}$$

$$C_2H_2^* + h\nu \rightarrow C_2(^1\pi_u) + H + H \tag{5}$$

$$C_2H_2^* \rightarrow C_2H + H \tag{6}$$

$$C_2H(\tilde{X}\,^2\Sigma) + h\nu \rightarrow C_2(^1\pi_u) + H \tag{7}$$

For a two-photon process, the excess energy available in process (4) of 5.5 eV exceeds the $C_2(A\,^1\pi_u)$ dissociation energy. Since we observe vibrational levels in $^1\pi_u$ state up to only $v'=5$, we conclude that process 4 is unlikely.[82] The relative importance of (5) vs (6) followed by (7) depends on the lifetime of $C_2H_2^*$. Since this lifetime is estimated to be $\sim 10^{-11}-10^{-12}$ sec,[82] the dissociation of $C_2H_2^*$ to give C_2H+H followed by (7) is expected to compete favorably with process (5).

C_2 emission from $d^3\pi_g \rightarrow a\,^3\pi_u$ and $C\,^1\pi_g \rightarrow A\,^1\pi_u$ is also observed. The latter is very weak ($<1\%$ of the total emission). The $d\,^3\pi_g \rightarrow a\,^3\pi_u$ spectrum is shown in Fig. 17. This amounts to $\sim$1% of the total emission and from Fig. 16 the power dependence is found to be greater than quadratic. The production of $C_2\,d\,^3\pi_g + 2H$ by 3 photons is 49,000 cm^{-1} exothermic. Since vibrational levels up to at least $v'=6$ are observed and rotational energy

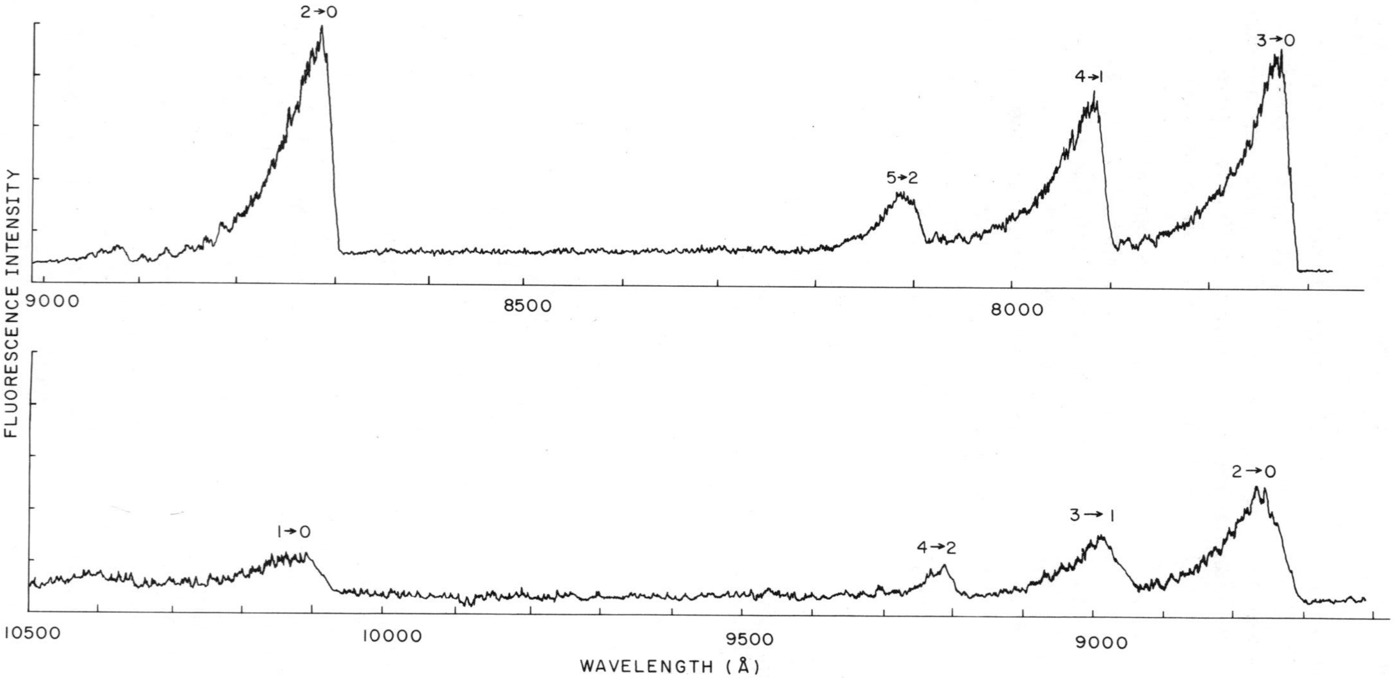

Fig. 14. A portion of the $C_2(A^1\pi_u \rightarrow X^1\Sigma_g^+)$ Phillips band emission. Upper trace: RCA 31034 photomultiplier tube. Lower trace: RCA 7102. Acetylene pressure = 30 mtorr. Reprinted with permission from *Chem. Phys.*, **33**, 161 (1978). Copyright by North-Holland Publishing Company.

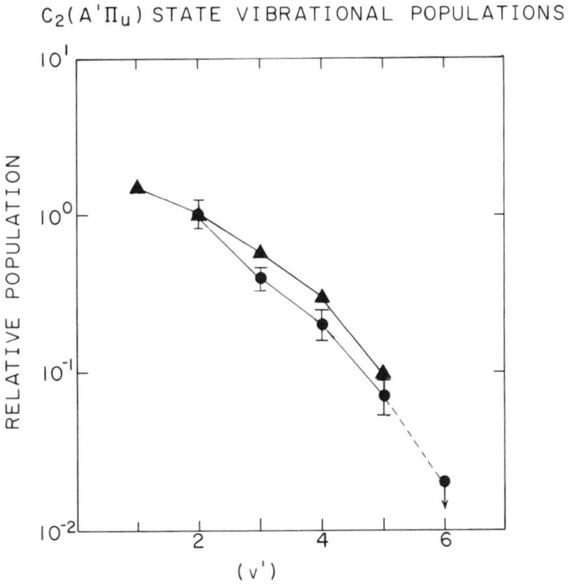

Fig. 15. Relative populations of $C_2(A^1\pi_u)$ vibrational levels for an acetylene pressure *f* 30 mtorr. ●: RCA 31034 photomultiplier tube, ▲: RCA 7102. Populations are normalized to $v'=2$ and $v'=6$ is an upper limit. Reprinted with permission from *Chem. Phys.*, **33**, 161 (1978). Copyright by North-Holland Publishing Company.

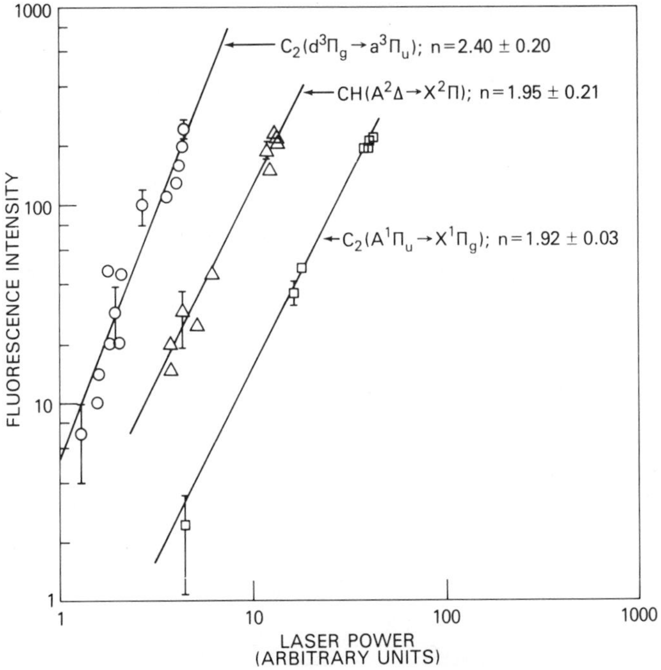

Fig. 16. Fluorescence intensity (F) vs laser power (I). Data are least squares fit to $F=I^n$. Acetylene pressure = 30 mtorr. Reprinted with permission from *Chem. Phys.*, **33**, 161 (1978). Copyright by North-Holland Publishing Company.

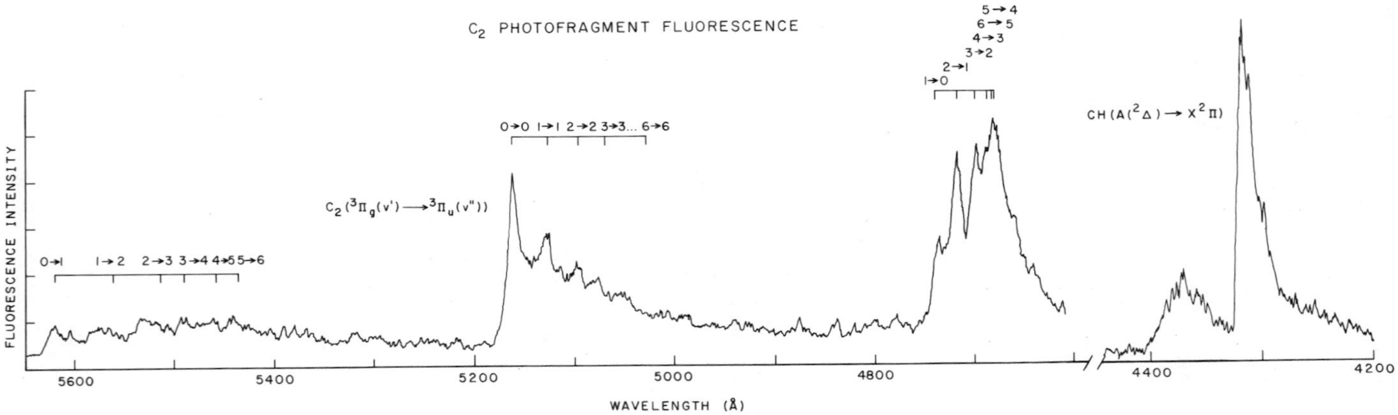

Fig. 17. $C_2(d^3\pi_g \rightarrow a^3\pi_u)$ emissions at an acetylene pressure of 30 mtorr. Reprinted with permission from *Chem. Phys.*, **33**, 161 (1978). Copyright by North-Holland Publishing Company.

can account for no more than ~2000–3000 cm^{-1}, more than 35,000 cm^{-1} must appear as translational energy.

Perhaps the most interesting observation is the $CH(^2\Delta \rightarrow {}^2\pi)$ emission shown in Fig. 17. The power dependence is shown to be quadratic (Fig. 16), and thus CH can be formed only by the two-photon absorption

$$C_2H_2 + h\nu \rightarrow C_2H_2^*$$

$$C_2H_2^* + h\nu \rightarrow CH(A^2\Delta, v=0) + CH(X^2\pi, v=0)$$

The CH emission accounts for ~1% of the total observed emission and requires 12.76 eV of energy which compares to the 12.84 eV provided by two ArF laser photons. It is interesting to note that Metzger and Cook observed visible emission in the 300–600 nm region as a function of acetylene photolysis wavelength between 103 and 58 nm.[89] Between 103 and 92 nm there is a sudden onset of visible fluorescence. This region corresponds very closely to the thermodynamic threshold for

$$C_2H_2(\tilde{X}^1\Sigma_g^+) \rightarrow CH(A\,^2\Delta) + CH(X\,^2\pi)$$

at 97.2 nm. It is likely that this emission is due to the $CH(^2\Delta \rightarrow {}^2\pi)$ transition at 431.5 nm, which infers that the 193 nm two-photon excitation is equivalent to absorption of a single 96.5 nm photon.

It is possible, therefore, that photolysis in the hard VUV at 96.5 nm and 64.3 nm can be done by using multiple photon photolysis with ArF laser photons at 193 nm. The production of high-energy fragments for use in kinetics measurements can easily be achieved and this technique has been used by other workers in the field.[90, 91]

B. E→V Energy Transfer Reactions Involving CO

Recently there has been considerable interest in both experimental and theoretical aspects of the quenching of electronically excited atoms.[92] One of the most important facets of these studies lies in the understanding of the mechanisms by which a relatively large amount of electronic energy carried by the excited atom is released to the quencher in the course of collision. Since a majority of these reactions involve changes in spin and orbital quantum numbers of excited atoms, reliable kinetic data, particularly information on detailed product energy distribution, are essential to the understanding of the dynamics of curve-crossing processes occurring in these quenching reactions. The CO laser absorption technique has proved to be a valuable tool in obtaining these data.

$O(^1D_2)+CO$. The quenching of $O(^1D_2)$ by CO was the first reaction investigated by the CO laser-probing technique.[11, 92] Earlier, it was attempted to use this reaction as a CO laser pumping source, but no stimulated emission could be detected.[93] It was concluded that either the CO was not vibrationally excited as a result of the reaction, or that it was produced with an uninverted energy distribution. The laser probing experiments were undertaken to distinguish between these two possibilities.

$O(^1D_2)$ atoms were produced by the flash photolysis of O_3 in quartz ($\lambda \geqslant 200$ nm) in the presence of the diluent SF_6, which has a negligibly small cross-section for $O(^1D)$ quenching and a high heat capacity for CO rotational thermalization.[94] The CO was found to be vibrationally excited up to $v=7$, the limit of the available electronic energy. A Boltzmann CO vibrational population distribution was obtained, which corresponded to a vibrational temperature of about 8000 K. At 298 K the average vibrational energy $\langle E_v \rangle$ of the CO was found to be 9.9 ± 0.5 kcal/mole, or 22% of the available electronic transfer energy, 45.4 kcal/mole. In order to calculate $\langle E_v \rangle$, it was necessary to know the vibrational population at $v=0$ (N_0). N_0, which represents the probability of quenching that leads to only translational and rotational excitation of CO, could not be determined experimentally because of the presence of a large number of unexcited reactant molecules. It was evaluated by extrapolation of the linear portion of the Boltzmann plot.

The results of simple statistical calculations based on the RRK and RRKM theories of unimolecular reactions[95] agree well with the observed vibrational population distribution. At 293 K, the computed average CO vibrational energy is 9.8 kcal/mole according to RRK and 10.2 according to RRKM calculations. Both agree quantitatively with the experimental value, 9.9 ± 0.5.

These results strongly support the complex formation mechanism: $O(^1D) + CO(X^1\Sigma^+) \rightarrow {}^1CO_2* \rightarrow {}^3CO_2* \rightarrow O(^3P) + CO(X^1\Sigma^+, v<7)$, where ${}^1CO_2*$ and ${}^3CO_2*$ represent electronically excited singlet and triplet CO_2 molecules, respectively. One of the singlet states may be the ground electronic state. There are several low-lying singlet and triplet states known to be present in the vicinity of the $O(^3P)+CO(X^1\Sigma^+)$ limit.[96] It is quite possible that all these low-lying states may be initially involved in the reaction prior to a successful crossing-over to the repulsive triplet states that correlate with $O(^3P_J)+CO(X^1\Sigma^+)$. However, if the lifetime of the ${}^1CO_2^*$ is as long as a period of several vibrations, then it is probably immaterial which initial states are involved.

The complex-formation mechanism was demonstrated quite conclusively by some experiments with isotopically labeled molecules in which $^{16}O(^1D)$ was quenched by $C^{18}O$.[92] The absolute rate of formation of $C^{16}O$ from

$^{16}O(^1D)+C^{18}O$ was found to be 50% of that measured for the $^{16}O(^1D)+C^{16}O$ reaction. Clearly the $^{16}OC^{18}O$ complex must be involved in the reaction which produces $C^{16}O$ and $C^{18}O$ with equal probabilities.

The observed E→V energy transfer efficiency, which was found to range from 17±1% at 246 K to 22.5±0.8% at 323 K, is in good agreement with the theoretically predicted values of 20% by Zahr et al.[97] and 30±10% by Tully[98] for the isoelectronic $O(^1D)+N_2$ reaction.

$Na(3^2P_J)+CO$. Extensive experimental data have been reported for the quenching of excited Na atoms by various gases, most of it dealing with the overall quenching cross-sections.[99] In order to understand the mechanism by which the electronic energy is released in the course of a collision, it would be extremely helpful to have detailed microscopic data. Theoretical developments have reached the stage where detailed dynamical properties can now be predicted or correlated. By means of CO laser absorption, the vibrational population distribution of the excited CO produced in the reaction $Na(3\,^3P_{1/2,3/2})+CO(v=0)\rightarrow Na(3\,^2S)+CO(v)$ has been measured and compared with some theoretical models.[99]

The reaction tube for the Na atom experiments consisted of a low-carbon stainless steel tube, enclosed by a pair of sapphire windows, with a side-arm reservoir for Na. It was in the center of an aluminum-block oven enclosed in high-porosity firebricks, and was maintained at 528±0.2 K. The reservoir temperature was controlled separately and was maintained 37°C cooler than the reaction tube during experiments. The Na atoms were excited by the dye laser tuned to one of the Na doublet lines at 589.1 or 589.7 nm. Typical signal averaged absorption curves obtained are shown in Fig. 2.

Absorption was observed from CO vibrational bands 1→2 through 8→9, indicating that the CO was vibrationally excited up to the limit of the available electronic energy, 48.5 kcal/mole. Figure 2 reveals that optical gain occurred in the 1→2 and 2→3 bands, suggesting a high degree of vibrational excitation in the $v=2$ and $v=3$ levels. Figure 18 shows the vibrational population distribution obtained. The E→V transfer is clearly nonresonant. The same distribution resulted from pumping either the $^2P_{1/2}$ (589.7 nm) or the $^2P_{3/2}$ (589.1 nm) state, which is consistent with the large cross-section for the doublet mixing through collisions with either Na or Ar.[100, 101] Both states are therefore effectively in equilibrium before appreciable energy transfer to CO has taken place, regardless of the state initially pumped. The average CO vibrational energy was computed to be 18.0 kcal/mole, which corresponds to an electronic to vibrational energy transfer efficiency of 35%.

The quenching of electronically excited alkali atoms in the ^2P state by diatomic molecules is believed to take place via successive crossings involving a covalent $M(^2P)+XY(^2\Sigma^+)$, an ionic $M^+(XY)^-$ and a covalent $M(^2S)+XY(^2\Sigma^+)$ surface.[102–110] Schematically, the quenching mechanism can be represented by $M^*(^2P)+XY(^1\Sigma^+, v=0) \rightarrow M^+(XY)^- \rightarrow M(^2S)+XY(^1\Sigma^+, v \geqslant 0)$, where M is the alkali atom and XY, the diatomic quencher. Bauer et al.[103, 104] applied this model to account for the relaxation of $Na(^2P)$ atoms by N_2, CO, and O_2. For the $Na(^2P)+CO$ reaction, they obtained a CO product vibrational distribution with a maximum at $v=5$, assuming an initial kinetic energy of 0.2 eV (~1500 K) and a value of 10 Å^3 for the sum of the polarizability of Na^+ and CO^-. More recently, Fisher and Smith[106] adopted a larger value of 40 Å^3 and obtained a

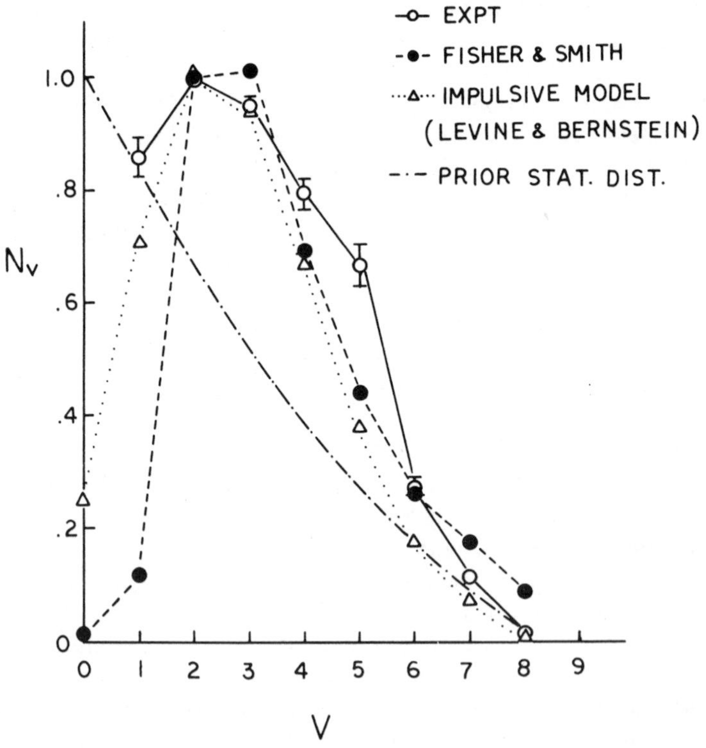

Fig. 18. CO vibrational population distributions obtained from the experiment (open circles and solid curve), the curve-crossing model by Fisher and Smith (solid circles and dashed curve), and the impulsive model by Levine and Bernstein (triangles and dotted curve). Reprinted with permission from *Chem. Phys. Lett.*, **42**, 78 (1976). Copyright by North-Holland Publishing Company.

distribution peaking at $v=3$. Their results for an initial relative kinetic energy of 0.2 eV are presented in Fig. 18 by the dashed curve. The theoretical distribution agrees quite nicely with our experimental results except the latter peaks at $v=2$ and has a slightly broader width.

The impulsive nature of the E→V transfer is also supported by the fact that our experimental results can be well correlated by a Poisson distribution based on the near-collinear impulsive model of Levine and Bernstein,[111] shown by the dotted curve in Fig. 18.

In conclusion, our results show unequivocably that the Na* + CO reaction is not a resonant E→V transfer process and also that the reaction, presumably taking place via an ionic $Na^+(CO)^-$ intermediate, produces CO with a nonstatistical distribution. This is in sharp contrast with the statistical distribution found in the $O(^1D)$ + CO reaction, in which nearly the same amount of electronic energy is involved. This implies that the transition of the assumed $Na^+(CO)^-$ intermediate to $Na(^2S)+CO(v)$ occurs approximately in the period of a collision. More detailed discussion of the relaxation of electronically excited Na atoms has recently been made by Hertel.[112]

I_2^*, ICl* and NO_2^* + CO. E→V energy transfer from electronically excited I_2, ICl, and NO_2 to CO has been investigated recently by a technique similar to that used for the Na* + CO reaction.[113] By means of CO laser absorption measurements, the relative populations at $v=1$ and $v=2$ (N_2/N_1) were determined to be 0.0072 ± 0.0018, 0.035, and 0.0058 ± 0.0043, respectively. Absorption due to transitions above 2→3 was too weak to be measured. The extent of CO vibrational excitation resulting from these reactions was considerably less than that observed for the $O(^1D)$ and $Na(^2P)$ + CO reactions even though the electronic excitation energies are comparable.

At first glance, the small extent of CO vibrational excitation appeared to be in line with the presence of additional degrees of freedom in the quenched molecules, I_2, ICl, and NO_2. However, a (prior) statistical calculation made for the I_2^* + CO reaction, for example, indicated that the N_2/N_1 ratio should be statistically as high as 0.60–0.79, depending on the quenching mechanism; viz., $I_2^* + CO(v=0) \rightarrow CO(v) + I_2^\dagger$ (or 2I). The collision-induced dissociation producing I atoms is quite likely because the state ($^3\pi_{O^+u}$) of the I_2 excited in the present study is considerably higher than the dissociation limit. A similar calculation for the relaxation of NO_2^*, $NO_2^* + CO(v=0) \rightarrow NO_2^\dagger(1, m, n,) + CO(v)$ gave rise to $N_2/N_1 = 0.45$, which is about two orders of magnitude greater than the observed value, 0.0058. Accordingly, the complexity of the quenched molecules alone is not sufficient to account for the observed E→V energy transfer efficiencies. It is worth noting that the efficiency of E→V transfer in the $ICl^* + CO$ reaction is five times higher than that observed in the I_2^* reaction. This is probably related to the fact that ICl is more polar and $ICl^*(A^3\pi)$ is more stable.

C. Combustion-Related Reactions

CO laser probing of the vibrationally excited CO formed in combustion-related reactions has proved valuable in gaining an understanding of the course of these reactions. Much useful data have been obtained pertinent to the determination of reaction mechanisms and the identity of intermediates taking part in the reactions. A number of reactions of $O(^3P)$ with allene, alkynes, carbon suboxide, and some free radicals are described below.

$O(^3P)$ + Allene and Alkynes. The reaction of $O(^3P)$ and allene leads primarily to CO and C_2H_4 as final products.[114, 115] This reaction is believed to proceed through an excited cyclopropanone complex which decomposes to CO and C_2H_4 directly:

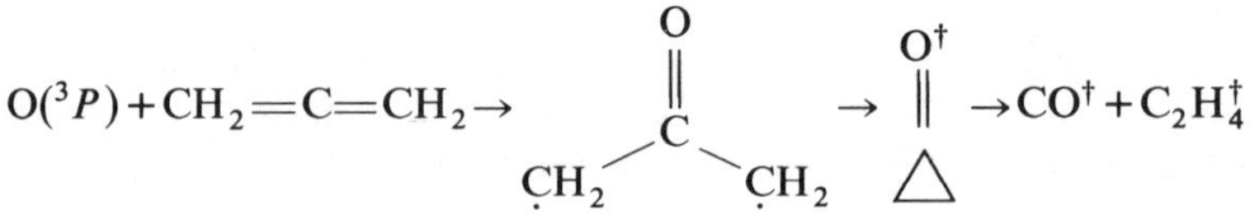

$$\Delta H° = -119 \text{ kcal/mole}$$

with CO being formed predominantly in the initial step of the reaction.[114]

By means of CO laser absorption measurements, the vibrational population distribution of the CO initially formed in the reaction has been determined.[116] The $O(^3P)$ atoms were produced by the flash photolysis of NO_2 in Pyrex ($\lambda \geqslant 300$ nm). Figure 19 shows the experimental results, along with a theoretically calculated distribution (upper solid curve) based on randomization of the total reaction energy, 119 kcal/mole, among the degrees of freedom of a loose cyclopropanone complex prior to its dissociation into CO and C_2H_4. The experimental average vibrational energy $\langle E_v \rangle$ carried by CO was calculated to be 6.8 ± 0.6 kcal/mole and is in excellent agreement with the value predicted by the model, 6.79 kcal/mole. These results strongly support the previously proposed reaction mechanism.

The reactions of alkynes with $O(^3P)$ are believed to proceed via excited ketene intermediates. Indeed, in the $O + C_2H_2$ reaction, $CH_2{=}C{=}O$ has been successfully isolated and identified in a low-temperature Ar matrix.[117]

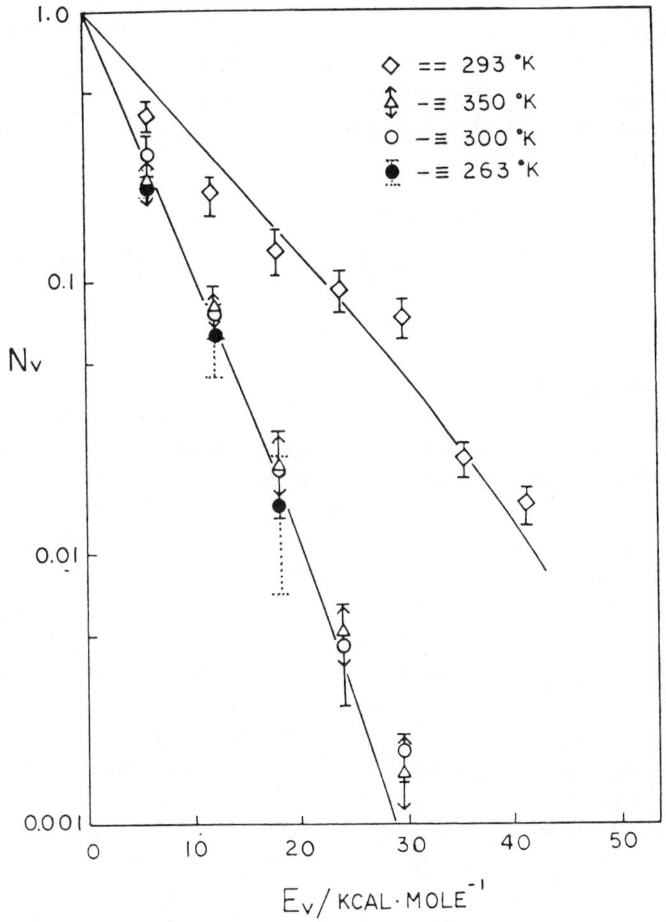

Fig. 19. Boltzmann plots for the CO formed in the $O(^3P)+CH_3C_2H$ and $CH_2=C=CH_2$ reactions. Solid lines: theoretically predicted population distributions. Reprinted with permission from *Chem. Phys.*, **20**, 271 (1977). Copyright by North-Holland Publishing Company.

The commonly accepted mechanism for the C_2H_2 reaction is:[118] $O(^3P)+HC{\equiv}CH \rightarrow [H\dot{C}{=}\overset{\overset{\displaystyle O\cdot}{|}}{C}H] \rightarrow CH_2{=}C{=}O^{\dagger} \rightarrow CH_2{}^{\dagger}+CO^{\dagger}$. In the case of higher alkynes, the migration of hydrogen atoms or alkyl groups leads to excited substituted ketene complexes, which subsequently decompose to form CO and olefins.[119–125]

We have investigated the reactions of $O(^3P)$ with a number of alkynes by measuring the vibrational populations of the nascent CO formed in the reactions.[116, 126, 127] The $O(^3P)$ atoms were again produced by the flash photolysis of NO_2 in Pyrex ($\lambda \geqslant 300$ nm). As an example, the O+ methylacetylene ($CH_3C{\equiv}CH$) reaction leads to CO and C_2H_4 as the major stable products and is believed to take place by the following mechanism:[119–122]

$$O(^3P) + CH_3C{\equiv}CH \rightarrow CH_3\dot{C}{=}\overset{\cdot O}{\overset{|}{C}}H \rightarrow CH_3CH{=}C{=}O^\dagger \rightarrow CH_3CH^\dagger + CO^\dagger$$
$$CH_3CH^\dagger \rightarrow C_2H_4{}^\dagger$$

The overall exoergicity of the reaction is 118 kcal/mole.

No direct evidence exists for the formation of the CH_3CH radical, but its presence has been presumed by analogy with the C_2H_2 reaction and by the presence of 2-butene and C_2H_2 as reaction products. The formation of 2-butene has been attributed to the recombination of CH_3CH radicals,[121] and the high yields of C_2H_2 to the decomposition via H_2 elimination of excited C_2H_4 produced by the isomerization of excited CH_3CH.[119–122] Recently, Blumenberg, Hoyermann, and Sievert reported that the C_2H_4 formed in the reaction had a different mass-spectral cracking pattern than admixed thermal C_2H_4, and attributed this to the initial formation of the CH_3CH radical.[125]

It is interesting that $CH_3C{\equiv}CH$ is isomeric with allene, and that the exoergicities of their reactions with $O(^3P)$ are very similar, 118 and 119 kcal/mole, respectively. If the $CH_3C{\equiv}CH$ reaction does indeed proceed via the CH_3CH radical, the isomerization energy for $CH_3CH \rightarrow C_2H_4$, 68 kcal/mole, will not be available for CO excitation, since this energy is not released until after the CO has dissociated from the CH_3CHCO complex. This should result in a considerably lesser extent of CO vibrational excitation than was found in the case of allene.

The vibrational population distribution measured for $CH_3C{\equiv}CH$ is also shown in Fig. 19.[116, 126] The lower solid curve represents the theoretical distribution based on a statistical model in which the total available energy less the CH_3CH isomerization energy, or a net of 53.5 kcal/mole, is randomized in a CH_3CHCO complex prior to its dissociation into CO and CH_3CH. At 300 K, the experimental value for $\langle E_v \rangle$ was found to be 2.3 ± 0.3 kcal/mole, in good agreement with the theoretical value based on the model, 2.20 kcal/mole. These results strongly support the formation of CO and CH_3CH as the major initial products of the reaction.

The reactions of $O(^3P)$ with 1- and 2-butyne[127] and trifluoromethylacetylene[128] have also been investigated. As in the case of methylacetylene, the CO vibrational population distributions obtained from these reactions

TABLE III
Average Vibrational Energy of the CO Formed in Some $O(^3P)$ + Alkyne/Allene and Some Photodissociation Reactions

		$\langle E_v \rangle^a$	
Reaction	Intermediate	Expt	Calc[b]
C_2H_2O	$CH_2CO^\dagger$	3.23 ± 0.69	4.51
$CH_3C{\equiv}CH + O$	$CH_3CHCO^\dagger$	2.3 ± 0.3	2.20
$CF_3C{\equiv}CH + O$	$CF_3CHCO^\dagger$	2.0 ± 0.3	~2.2
$C_2H_5C{\equiv}CH + O$	$C_2H_5CHCO^\dagger$	0.97 ± 0.04	1.18
$CH_3C{\equiv}CCH_3$	$(CH_3)_2CCO^\dagger$	1.14 ± 0.07	1.18
$n{-}C_3H_7C{\equiv}CH + O$	$C_3H_7CHCO^\dagger$	0.92 ± 0.04	1.04
$n{-}C_4H_9C{\equiv}CH + O$	$C_4H_9CHCO^\dagger$	0.73 ± 0.03	0.80
$CH_2{=}C{=}CH_2 + O$	▷$=O^\dagger$	6.8 ± 0.6	6.79
$CH_3CHCO + h\nu$	$CH_3CHCO^\dagger$	2.0 ± 0.2	2.20

[a]Average CO vibrational energy in kcal/mole.
[b]Statistically expected values.

were completely consistent with the formation of CO and a biradical as the initial reaction products. The results obtained are summarized in Table III, along with some preliminary results for C_2H_2 and some higher alkynes.[129]

$O(^3P)$ + Carbon Suboxide. The reaction of $O(^3P)$ with carbon suboxide is rather unique in that three identical molecular species are formed in the same reaction, $O(^3P) + C_3O_2(X^1\Sigma_g^+) \rightarrow 3CO(X^1\Sigma^+)$. Although apparently spin forbidden, the reaction takes place very rapidly. The spin-conserved process yielding $CO_2 + C_2O$ has been shown to be negligible ($\leqslant 0.4\%$).[130] The mechanism of the reaction has been proposed to be:

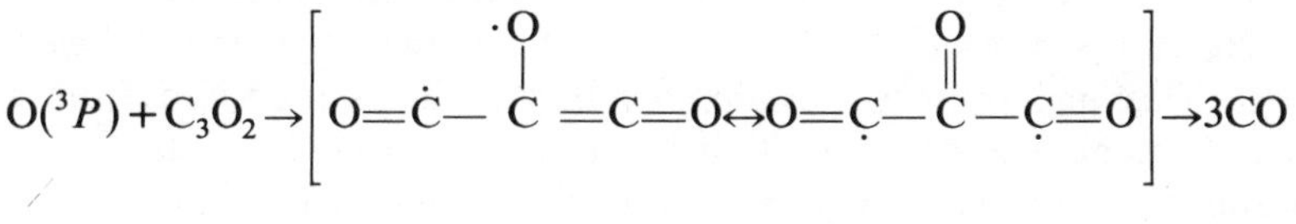

$$\Delta H° = -115 \text{ kcal/mole}$$

in which the O atom attacks predominantly the center C atom.[130–133] It thus bears similarities to the O + allene reaction (O=C=C=C=O cf. $H_2{=}C{=}C{=}C{=}H_2$).

Figure 20 shows the vibrational population distribution of the CO produced in this reaction.[133] Also shown (solid curve) are the computed results predicted by a simple statistical model based upon a C_3O_3 complex consisting of three indistinguishable CO stretching modes. The experimental points lie close to the theoretical curve.

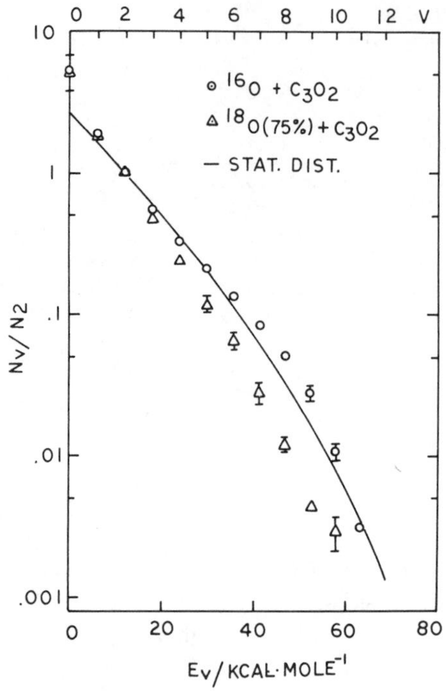

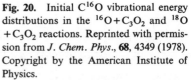

Fig. 20. Initial $C^{16}O$ vibrational energy distributions in the $^{16}O + C_3O_2$ and $^{18}O + C_3O_2$ reactions. Reprinted with permission from *J. Chem. Phys.*, **68**, 4349 (1978). Copyright by the American Institute of Physics.

Experiments were also carried out using isotopically labeled ^{18}O atoms.[133] Under these conditions, the $C^{16}O$ laser probe cannot detect the $C^{18}O$ produced by the attack on the center C atom, the entire CO absorption signal is due to the $C^{16}O$ formed from the ^{16}O already present in the C_3O_2 molecule. The $C^{16}O$ vibrational distribution thus obtained is also shown in Fig. 20, and is noticeably colder than that produced by the unlabeled O atoms. This implies that vibrationally hotter CO is produced from the center C atom. The deconvolution of these data showed that the average vibrational energy carried by CO from the center C atom amounted to about 15 kcal/mole, compared with only about 4 kcal/mole for the two CO's derived from the C=O bonds initially present in the C_3O_2.

The most likely mechanism by which the C_3O_2 complex may decompose involves a concerted decomposition process:

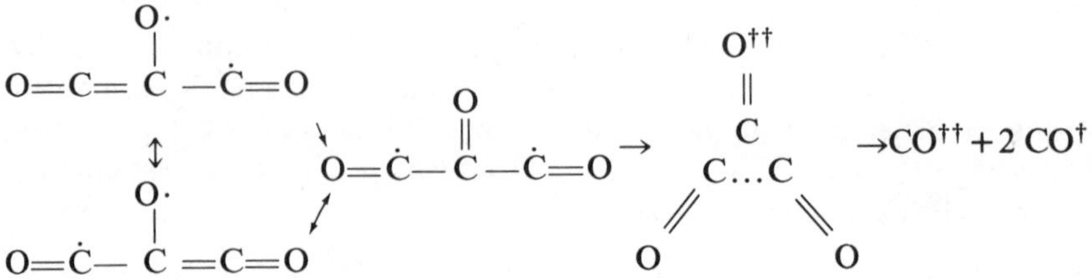

where the initial adduct may be long-lived because of resonance stabilization. Since the O-atom attack occurs from the side, the most likely reaction coordinate is the skeletal bending motion. This bending vibration, aided by the overlapping of the unpaired electrons after the formation of the new CO double bond, may lead to the formation of a highly energetic, *short-lived* cyclotrione, which rapidly decomposes concertedly into a vibrationally hotter CO and transient $(CO)_2$ that immediately dissociates into two CO's each carrying an equal amount of energy. Because of the short lifetime of the cyclotrione complex, the total available energy has not been completely randomized.

$O(^3P)+CS_2$. Mass-spectrometric studies by Slagle et al.[134] and by Graham and Gutman[135] have shown that the reaction of $O(^3P)$ with CS_2 takes place by three reaction channels:

$$O(^3P)+CS_2(X\,^1\Sigma_g^+) \begin{cases} \xrightarrow{a} CS+SO+21 \text{ kcal/mole} \\ \xrightarrow{b} CO+S_2+83 \text{ kcal/mole} \\ \xrightarrow{c} OCS+S+54 \text{ kcal/mole} \end{cases}$$

The exoergicities listed are those for the formation of ground electronic state products. The branching ratio for channel (c) was found to vary from 0.098 ± 0.004 at 249 K to 0.081 ± 0.007 at 500 K, and that for channel (a) was crudely estimated to be within the range, 0.7 ± 0.85. These values allow channel (b) to be roughly estimated as in the range of 0.05 ± 0.2.

Channel (a) has been studied quite extensively,[136–139] and the results have shown that SO production occurs by a direct stripping process probably involving the bent transition state O–S–C–S.

In addition to the evidence of Slagle et al.[134] for the existence of channel (b), some indirect but strong indications of its presence have been provided by the observed CO vibrational energy distributions obtained from some pulsed electrical discharge experiments.[140–143] The dynamics of CO formation from this channel was first investigated by Hudgens et al. by means of the chemiluminescence method.[144] They obtained a CO vibrational population distribution showing a second hump peaking at $v=7$, and attributed this dual maxima to the presence of two reaction paths, one occurring by the direct attack of $O(^3P)$ on the center C atom forming the

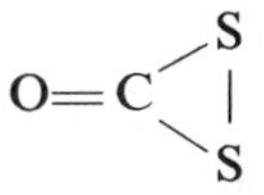

complex, and the other via an indirect attack on one of the S atoms with its subsequent migration onto the C atom.

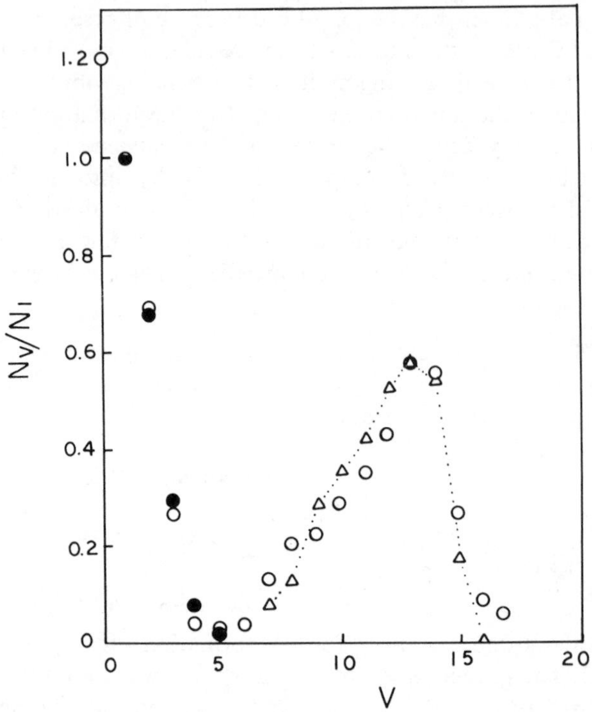

Fig. 21. Relative CO vibrational energy distributions from the $O(^3P)+CS$ and CS_2 reactions. Open circles: distribution obtained from the flash photolysis of SO_2, CS_2, N_2O, and Ar mixtures above 200 nm, filled circles: distribution obtained from the flash photolysis of NO_2, CS_2, and SF_6 mixtures above 300 nm. Both are normalized to $v=1$. (Ref. 145.) Triangles: Data of Hancock et al. obtained from the O+CS reaction, (Ref. 147) normalized to $v=13$ of the open-circle distribution. The distributions below $v=5$ are attributable to the $O+CS_2 \rightarrow CO+S_2$ reaction.

By CO laser absorption measurements, we have also determined the CO vibrational population produced by this reaction,[145] as well as the branching ratio for channel (*b*), 0.016 ± 0.002. $O(^3P)$ atoms were produced from NO_2 by flash photolyzing mixtures of NO_2 and CS_2 in Pyrex. The population distribution obtained (Fig. 21) was considerably colder than that reported by Hudgens et al., and was free from the second maximum at $v=7$. In another experiment performed at a higher flash energy, we detected the presence of a small residual tail, which peaked at $v=13$ and extended up to $v=16$, resembling that of the O+CS reaction, discussed below. At this higher flash energy, a very small fraction of CS_2 might be dissociated near 300 nm. Since the O+CS reaction is probably at least two orders of magnitude faster than the $O+CS_2$ reaction, this secondary

(O+CS) reaction thus could become prominent in this high-energy flash. This secondary reaction probably influenced the results of Hudgens et al. The C_{2v} OCS_2 complex may well be the intermediate involved in both channels (*b*) and (*c*), producing $CO+S_2$ and OCS+S, respectively. Our observed CO distribution from channel (*b*) is close to the prior statistical distribution.[142]

O(^3P)+CS. The reaction of O(^3P) with the CS radical,

$$O(^3P)+CS(X\,^1\Sigma^+)\rightarrow CO(X\,^1\Sigma^+)+S(^3P)$$
$$\Delta H^\circ=-85\text{ kcal/mole}$$

has been the subject of numerous studies[140, 141, 144–148] and was proposed to be responsible for the laser emission obtained from the flash photolysis of mixtures of CS_2 and O_2.[149–150] Figure 21 shows the initial vibrational energy distribution reported by Hancock et al. for the CO produced by the O+CS reaction.[147] It was obtained from chemiluminescence measurements from the O+CS_2 reaction carried out in a slow flow-discharge system. This distribution has been corroborated by laser absorption measurements,[146, 148] including our results,[145] also shown in Fig. 21.

Our experiments were carried out in a quartz flash system ($\lambda \geqslant 200$ nm) using Ar-diluted mixtures of SO_2, CS_2, and N_2O. Although N_2O was added to relax the $CS^\dagger$ initially formed in the photodissociation reaction,[146] the fact that $CO^\dagger$ ($v>15$) was still present (Fig. 21) indicates that the relaxation was not complete. The results obtained agree closely with those of Hancock et al.[147] above $v=7$. The $CO^\dagger$ formed in these higher vibrational levels results mainly from the O+CS reaction. However, in our experiment we observed a significant amount of $CO^\dagger$ present below $v=5$. Since the concentration of the CS radical formed in our low-energy flash (0.3 kJ) was quite small, these vibrationally colder CO molecules were most likely produced by the reaction of O atoms with undissociated CS_2 molecules. The population distribution from $v=5$ and below compares well with that obtained from the O(^3P)+CS_2 reaction.

O(^3P)+OCS. The reaction of O(^3P) with OCS,

$$O(^3P)+OCS(X\,^1\Sigma^+)\rightarrow CO(X\,^1\Sigma^+)+SO(X\,^3\Sigma^-)$$
$$\Delta H^\circ=-52\text{ kcal/mole}$$

has also been investigated by CO laser spectroscopy in the manner described above.[145, 151] $CO^\dagger$ was detected up to $v=5$, but because of the slow reaction rate at room temperature,[152] absorption for all levels was rather weak. The vibrationally excited CO was estimated to carry about 9%

of the total available energy, which is similar to that carried by the $CS^{\dagger}$ in the $O+CS_2$ reaction. Accordingly, the mechanism of $O(^3P)+OCS$ is probably similar to channel (*a*) in the $O+CS_2$ reaction which produces $CS+SO$ by a direct stripping process. Experiments using ^{18}O atoms showed that the CO was formed exclusively from the parent OCS molecule.[145]

$O(^3P)+CF$, CHF, CF_2 and CH_2. The reaction of $O(^3P)$ with the CF radical

$$O(^3P)+CF(X\,^2\pi)\rightarrow CO(X\,^1\Sigma^+)+F(^2P)$$

has been shown to generate stimulated CO emission.[153, 154] Although this reaction is considerably more exoergic ($\Delta H° = -126\pm2$ kcal/mole) than the $O+CS$ reaction, the intensity of the emission produced by CF was much weaker than that obtained from CS under similar experimental conditions.

From CO laser absorption measurements, the CO formed in the reaction of $O+CF$ was found to be vibrationally excited up to $v=24$ with a Boltzmann temperature of 1.4×10^4 K.[153, 154] The CF was produced by the flash photolysis of $CFBr_3$ in quartz ($\lambda \geqslant 200$ nm) or Suprasil ($\lambda \geqslant 165$ nm). A near-statistical vibrational population distribution was obtained, with partial inversion between all levels. The absence of a complete population inversion accounts for the weaker laser emission compared with the $O+CS$ system. A surprisal analysis of the observed CO distribution yielded a near straight line with a zero slope (i.e., $\lambda_v \simeq O$), which is markedly different from the value, $\lambda_v \simeq -9$, for the CO formed from the $O+CS$ reaction, based on the data of Hancock et al.[146]

The gross difference in the dynamics of these two reactions is attributable to the involvement of a relatively long-lived CFO intermediate in the CF reaction. The RRK lifetime of the CFO complex was estimated to be 10^{-12} to 10^{-13} s and corresponds to a period of several vibrations. This is quite sufficient for the available reaction energy to be completely randomized among all internal degrees of freedom of this three-atom intermediate.

The reactions of CHF, CF_2, and CH_2 with $O(^3P)$ all appear to take place via long-lived complexes also. CO laser absorption measurements indicate that the CHF reaction proceeds by two major pathways:[154, 155]

$$O(^3P)+CHF(\tilde{X}\,^1A')\rightarrow[^3HFCO^{\dagger *}\rightarrow {}^1HFCO^{\dagger}]$$

$$\xrightarrow{a} HF^{\dagger}+CO^{\dagger}+\sim190 \text{ kcal/mole}$$

$$\xrightarrow{b} H+F+CO^{\dagger}+\sim54 \text{ kcal/mole}$$

where $^3HFCO^{\dagger *}$ and $^1HFCO^{\dagger}$ represent the vibronically excited triplet and singlet HFCO molecules, respectively. The CHF radicals were produced by the flash photolysis of $CHFCl_2$ in Suprasil or $CHFBr_2$ in quartz. The experimental CO population distribution was found to lie between two statistical distributions based on paths (*a*) and (*b*), indicating that both paths are taking place simultaneously. A near-statistical vibrational energy distribution was observed for the HF produced by path (*a*).[155] It was determined by measuring the appearance times of each vibrational–rotational HF laser line selectively oscillating within a cavity formed by a grating and a mirror. By analysis of the computed statistical curves and the observed CO distribution, it was estimated that about 40% of the reaction occurs via the molecular-elimination channel (*a*) and 60% via the atom-formation channel (*b*). A study by Klimek and Berry of the photolytic decomposition of HFCO also concluded that the dissociation takes place by both molecular and atomic elimination channels, but no branching ratios could be determined.[156]

The $O(^3P)+CF_2$ reaction was investigated in order to determine the identity of the CF_2 radical produced in the $O(^3P)+C_2F_4$ reaction:

$$O(^3P)+C_2F_4 \rightarrow CF_2O+CF_2$$

$$CF_2+O(^3P) \rightarrow CO+F_2(\text{or } 2F)$$

Kinetic absorption spectroscopy measurements on the reaction carried out by flash photolysis detected only ground state $CF_2(^1A_1)$.[157, 158] Experiments based on product analysis suggested that the excited triplet $CF_2(^3B_1)$ radical was present in the reaction,[159, 160] as might be expected according to the spin conservation rule.

In an experiment attempting to resolve the identity of the CF_2, mixtures of C_2F_4 and NO_2 were flash photolyzed in Pyrex ($\lambda \geqslant 300$ nm) to avoid dissociating the C_2F_4, and the CO product was monitored by the CO laser.[154, 161] The CO was found to be vibrationally excited up to $v=11$, corresponding to as much as 63 kcal/mole of energy. By analysis of the observed population distribution and consideration of all possible paths and energetics associated with this reaction, it was concluded that $CO^{\dagger}$ excited to this extent could come only from the process:

$$O(^3P)+CF_2(^3B_1) \rightarrow CO^{\dagger}+F_2 \qquad \Delta H° = -89 \text{ kcal/mole}$$

The experimental distribution agreed closely with the statistically predicted one based upon the reaction proceeding via a CF_2O complex. The conclusion that CF_2 ($^3B_1$) is indeed produced has been confirmed recently by Koda,[162] who detected an emission from this reaction in the 470–720 nm region attributable to the CF_2 ($^3B_1 \rightarrow {}^1A_1$) phosphorescence.

The $O(^3P) + CH_2$ reaction, an important process in combustion chemistry, has been proposed to proceed by two major channels:[118, 119]

$$O(^3P) + CH_2 \rightarrow [^3CH_2O^{\dagger *} \rightarrow {}^1CH_2O^{\dagger}]$$

$$\xrightarrow{a} CO + H_2 + 179\ \text{kcal/mole}$$

$$\xrightarrow{b} CO + 2H + 75\ \text{kcal/mole}$$

Hydrogen is a known major product, but experimentally it has not been determined whether H_2 and H atoms are formed simultaneously, or if the H_2 results from H atom recombination. If H_2 is produced by molecular elimination, one would expect different reactions in combustion chemistry than if only H atoms were present initially.

This reaction was investigated by carrying it out in a quartz flash photolysis tube and measuring the absorption of the CO produced. $O(^3P)$ and CH_2 were produced by the photodissociation of SO_2 and either CH_2Br_2 or CH_2I_2. The CO was found to be vibrationally excited up to $v = 18$ with a vibrational temperature of $\sim 10^4$ K. Since channel (b) could only excite the CO up to $v = 13$, it was concluded that the molecular-elimination channel (a) was occurring simultaneously. Preliminary analysis of the data indicates that both channels are of approximately equal importance. Significant CO vibrational excitation has also been observed by chemiluminescence measurements of the $O(^3P) + C_2H_2$ reaction, in which $O + CH_2$ is a major secondary process.[163, 164]

IV. CONCLUDING REMARKS

We have reviewed in some detail the dynamics of photodissociation, E→V energy transfer and some combustion related reactions, studied either by laser induced fluorescence or by laser resonance absorption. We have shown that many photodissociation reactions can be studied in detail, using two lasers—one for parent molecular excitation and the other for probing the internal states of photofragments. Several examples have been presented to illustrate the techniques of single and multiple UV-photon dissociations. The latter technique, although developed rather recently, has already been used to generate highly reactive transients (such as CH, C_2,

C_2H, etc.) and to study their reaction kinetics.[91, 165] In the case of UV-multiphoton dissociation, the information concerning exact mechanisms of fragmentation is more sketchy and, at times, totally unknown. However, its usefulness as a means of producing important free radicals and ions warrants further thorough investigation.

With regard to E→V energy transfer reactions, we report primarily the vibrational (microscopic) energy content of the quencher molecules, and no effort was made to measure total (macroscopic) cross-sections. The data obtained for the relaxation of $O(2^1D_2)$ and $Na(3^2P_J)$ by CO have demonstrated unambiguously the mechanisms of these two important reactions. The former occurs by a complex-forming mechanism and the latter is impulsive in nature.

Those combustion reactions described here are best studied by the flash photolysis-laser resonance absorption technique. Other well-established techniques (such as chemiluminescence) are less useful for these cases because free-radical species cannot be cleanly produced by microwave and electrical discharge methods.

References

1. W. M. Fairbank, Jr., T. W. Hänsch, and A. L. Schawlow, *J. Opt. Soc. Am.*, **65**, 199 (1975). For review see, for example, W. M. Fairbank, Jr. and C. Y. She, *Opt. News*, **5**, 4 (1979).
2. N. Djeu, *Appl. Phys. Lett.*, **23**, 309 (1973).
3. M. C. Lin, *Int. J. Chem. Kinetics*, **6**, 173 (1973).
4. Y. Fushiki and S. Tsuchiya, *Chem. Phys. Lett.*, **22**, 47 (1973).
5. G. Hancock and I. W. M. Smith, *Appl. Opt.*, **10**, 1827 (1971).
6. H. T. Powell, *J. Chem. Phys.*, **59**, 4937 (1973).
7. P. B. Sackett, A. Horduih, and H. Schlossberg, *Appl. Phys. Lett.*, **22**, 367 (1973).
8. J. C. Stephenson, *Appl. Phys. Lett.*, **22**, 576 (1973).
9. J. K. Hancock, D. F. Starr, and W. H. Green, *J. Chem. Phys.*, **61**, 5421 (1974).
10. Y. S. Liu, R. A. McFarlane, and G. J. Wolga, *J. Chem. Phys.*, **63**, 228 (1975).
11. M. C. Lin and R. G. Shortridge, *Chem. Phys. Lett.*, **29**, 42 (1974).
12. A detailed presentation of the theory of laser-induced fluorescence (LIF) has been given by J. L. Kinsey, *Ann. Rev. Phys. Chem.*, **28**, 349 (1977).
13. A. P. Baronavski and J. R. McDonald, *Chem. Phys. Lett.*, **45**, 172 (1977).
14. G. E. Busch, R. T. Mahoney, R. I. Morse, and K. R. Wilson, *J. Chem. Phys.*, **51**, 449 (1969).
15. R. J. Oldman, R. K. Sander, and K. R. Wilson, *J. Chem. Phys.*, **63**, 4252 (1975).
16. G. E. Busch, R. T. Mahoney, R. I. Morse, and K. R. Wilson, *J. Chem. Phys.*, **51**, 837 (1969).
17. R. K. Sander and K. R. Wilson, *J. Chem. Phys.*, **63**, 4242 (1975).
18. J. H. Ling and K. R. Wilson, *J. Chem. Phys.*, **63**, 101 (1975).
19. G. E. Busch and K. R. Wilson, *J. Chem. Phys.*, **56**, 3626 (1972).
20. G. E. Busch and K. R. Wilson, *J. Chem. Phys.*, **56**, 3638 (1972).
21. G. E. Busch and K. R. Wilson, *J. Chem. Phys.*, **56**, 3655 (1972).
22. K. F. Freed and Y. B. Band, in *Excited States*, Vol. 3, E. C. Lim, ed, Academic Press, New York, 1977.

23. W. M. Gelbart, *Ann. Rev. Phys. Chem.*, **28**, 325 (1977).
24. J. P. Simons, *Chem. Soc. Specialist Period. Rep.*, **2**, 58 (1977).
25. H. Okabe, *Photochemistry of Small Molecules*, Wiley, New York, 1978.
26. M. N. R. Ashfold and J. P. Simons, *Chem. Phys. Lett.*, **47**, 65 (1977).
27. M. N. R. Ashfold and J. P. Simons, *J. C. S., Faraday Trans. II.*, **73**, 858 (1977).
28. D. D. Davis and H. Okabe, *J. Chem. Phys.*, **49**, 5526 (1968).
29. N. Basco, J. E. Nicholas, R. G. W. Norrish, and W. H. Vickers, *Proc. Roy. Soc.* **A272**, 147 (1963).
30. J. P. Simons and P. W. Tasker, *Mol. Phys.*, **27**, 1691 (1974).
31. M. J. Sabety-Dzvonik and R. J. Cody, *J. Chem. Phys.*, **66**, 125 (1977).
32. A. Jakovleva, *Acta Physicochim.*, **9**, 665 (1938).
33. A. Terenin and H. Neujmin, *Nature*, **134**, 255 (1934).
34. R. J. Donovan and J. Konstantatos, *J. Photochem.*, **1**, 75 (1972/73).
35. R. Engleman, Jr., *J. Photochem.*, **1**, 317 (1972/73).
36. A. Mele and H. Okabe, *J. Chem. Phys.*, **51**, 4798 (1969).
37. G. A. West and M. J. Berry, *J. Chem. Phys.*, **61**, 4700 (1974).
38. M. Sabety-Dzvonik and R. Cody, *J. Chem. Phys.*, **64**, 4794 (1976).
39. S. Mukamel and J. Jortner, *J. Chem. Phys.*, **60**, 4760 (1974).
40. S. Mukamel and J. Jortner, *J. Chem. Phys.*, **65**, 3735 (1976).
41. O. Atabek, J. A. Beswick, R. Leferbvre, S. Mukamel, and J. Jortner, *J. Chem. Phys.*, **65**, 4035 (1976).
42. J. A. Beswick and J. Jortner, *Chem. Phys.*, **24**, 1 (1977).
43. M. Shapiro and R. D. Levine, *Chem. Phys. Lett.*, **5**, 499 (1970).
44. U. Halavee and M. Shapiro, *Chem. Phys.*, **21**, 105 (1977).
45. O. Atabek and R. Lefebvre, *Chem. Phys. Lett.*, **52**, 29 (1977).
46. K. E. Holdy, L. C. Klotz, and K. R. Wilson, *J. Chem. Phys.*, **52**, 4588 (1970).
47. M. D. Morse, K. F. Freed, and Y. B. Band, *Chem. Phys. Lett.*, **44**, 125 (1976).
48. Y. B. Band, M. D. Morse, and K. F. Freed, *J. Chem. Phys.*, **68**, 2702 (1978).
49. M. D. Morse, K. F. Freed, and Y. B. Band, *J. Chem. Phys.*, **70**, 3604 (1979).
50. M. D. Morse, K. F. Freed, and Y. B. Band, *J. Chem. Phys.*, **70**, 3620 (1979).
51. A. P. Baronavski and J. R. McDonald, to be published.
52. M. J. Berry, *Chem. Phys. Lett*, **27** 73 (1974).
53. M. J. Berry, *Chem. Phys. Lett.*, **29**, 329 (1974).
54. J. R. McDonald, R. G. Miller, and A. P. Baronavski, *Chem. Phys. Lett.*, **51**, 57 (1977).
55. A. P. Baronavski, R. G. Miller, and J. R. McDonald, *Chem. Phys.*, **30**, 119 (1978).
56. J. R. McDonald, R. G. Miller, and A. P. Baronavski, *Chem. Phys.*, **30**, 133 (1978).
57. A. O. Beckman and R. G. Dickinson, *J. Am. Chem. Soc.*, **52**, 124 (1930).
58. R. S. Konar, S. Matsumoto, and B. DeB. Darwent, *Trans. Faraday Soc.*, **67**, 1698 (1971).
59. B. A. Thrush, *Proc. Roy. Soc.*, **235A**, 143 (1956).
60. R. J. Paur and E. J. Bair, *J. Photochem.*, **1**, 255 (1973).
61. R. J. Paur and E. J. Bair, *Int. J. Chem. Kinetics*, **8**, 139 (1976).
62. H. Okabe, *J. Chem. Phys.*, **49**, 2726 (1968).
63. H. Okabe and M. Lenzi, *J. Chem. Phys.*, **47**, 5241 (1967).
64. J. R. McDonald, J. W. Rabalais, and S. P. McGlynn, *J. Chem. Phys.*, **52**, 1332 (1970).
65. J. W. Rabalais, J. R. McDonald, V. Scherr, and S. P. McGlynn, *Chem. Rev.*, **71**, 73 (1971).
66. J. R. McNesby, I. Tanaka, and H. Okabe, *J. Chem. Phys.*, **36** 605 (1962).
67. J. R. McNesby and H. Okabe, *Adv. Photochem.*, **3**, 157 (1964).
68. W. E. Groth, U. Schurath, and R. M. Schindler, *J. Phys. Chem.*, **72**, 3914 (1968).
69. U. Schurath, P. Tiedman, and R. M. Schindler, *J. Phys. Chem.*, **73**, 456 (1969).

70. R. A. Back and S. Koda, *Can. J. Chem.*, **55**, 1387 (1977).
71. G. DiStefano, M. Lenz, A. Margani, and C. W. Xuan, *J. Chem. Phys.*, **67**, 3832 (1977).
72. V. M. Donnelly, A. P. Baronavski and J. R. McDonald, *Chem. Phys.*, **43**, 271 (1979).
73. K. Dressler and D. A. Ramsay, *Phil. Trans. Royal Soc., London*, **251**, 69 (1959).
74. J. W. C. Johns, D. A. Ramsay, and S. C. Ross, *Can. J. Phys.*, **54**, 1804 (1976).
75. R. N. Dixon, *Mol. Phys.*, **9**, 357 (1965).
76. E. M. Evleth, *Chem. Phys. Lett.*, **38**, 516 (1976).
77. G. B. Kistiakowsky and B. H. Mahan, *J. Am. Chem. Soc.*, **79**, 2412 (1957).
78. D. P. Chong and G. B. Kistiakowsky, *J. Phys. Chem.*, **68**, 1793 (1964).
79. M. E. Umstead, R. G. Shortridge and M. C. Lin, *J. Phys. Chem.*, **82**, 1455 (1978).
80. W. M. Jackson, J. B. Halpern, and C. S. Ling, *Chem. Phys. Lett.*, **55**, 254 (1978).
81. A. P. Baronavski and J. R. McDonald, *Chem. Phys. Lett.*, **56**, 369 (1978).
82. J. R. McDonald, A. P. Baronavski, and V. M. Donnelly, *Chem. Phys.*, **33**, 161 (1978).
83. Z. Karny, R. Nagman, and R. N. Zare, *Chem. Phys. Lett.*, **59**, 33 (1978).
84. C. L. Sam and J. T. Yardley, *J. Chem. Phys.*, **69**, 4621 (1978).
85. W. M. Jackson and C. S. Lin, *Int. J. Chem. Kinetics*, **10**, 745 (1978).
86. J. B. Halpern, W. M. Jackson, and V. McCrary, *Appl. Opt.*, **18**, 590 (1979).
87. W. M. Jackson and J. B. Halpern, *J. Chem. Phys.*, **70**, 2373 (1979).
88. S. Rockwood, J. P. Reilly, K. Hohla, and K. L. Kompa, *Opt. Comm.*, **28**, 175 (1979).
89. P. H. Metzger and G. R. Cook, *J. Chem. Phys.*, **41**, 642 (1964).
90. V. M. Donnelly and L. Pasternack, *Chem. Phys.*, **39**, 427 (1979).
91. L. Pasternack and J. R. McDonald, *Chem. Phys.*, **43**, 173 (1979).
92. R. G. Shortridge and M. C. Lin, *J. Chem. Phys.*, **64**, 4076 (1976) and references cited therein.
93. L. E. Brus and M. C. Lin (unpublished work).
94. H. F. Yamazaki and R. J. Cventanovic, *J. Chem. Phys.*, **41**, 3703 (1964).
95. W. Forst, *Theory of Unimolecular Reactions*, Academic Press, New York, 1973.
96. N. W. Winter, C. F. Bender, and W. A. Goddard, *Chem. Phys. Lett.*, **20**, 489 (1973).
97. G. E. Zahr, R. K. Preston, and W. H. Miller, *J. Chem. Phys.*, **62**, 1127 (1975).
98. J. C. Tully, *J. Chem. Phys.*, **61**, 61 (1974).
99. D. S. Y. Hsu and M. C. Lin, *Chem. Phys. Lett.*, **42**, 78 (1976), and references cited therein.
100. J. Pitre and L. Krause, *Can. J. Chem.*, **45**, 2671 (1967); **46**, 125 (1968).
101. H. L. Chen and S. Fried, *IEEE J. Quantum Electron.*, **QE-11**, 669 (1975).
102. A. Bjerre and E. E. Nikitin, *Chem. Phys. Lett.*, **1**, 179 (1967).
103. E. Bauer, E. R. Fisher, and F. R. Gilmore, *De-excitation of Electronically Excited Sodium by Nitrogen and Some Other Diatomic Molecules*, IDA p-471, March 1969.
104. E. Bauer, E. R. Fisher, and F. R. Gilmore, *J. Chem. Phys.*, **51**, 4173 (1969).
105. E. R. Fisher and G. K. Smith, *Chem. Phys. Lett.*, **6**, 438 (1970).
106. E. R. Fisher and G. K. Smith, *Appl. Opt.*, **10**, 1803 (1971).
107. E. R. Fisher and G. K. Smith, *Chem. Phys. Lett.*, **13**, 448 (1972).
108. E. A. Andreev, *High Temp. Phys.*, **10**, 637 (1972).
109. J. L. Magee and T. Ri, *J. Chem. Phys.*, **9**, 638 (1941).
110. K. J. Laidler, *J. Chem. Phys.*, **10**, 34 (1942).
111. R. D. Levine and R. B. Bernstein, *Chem. Phys. Lett.*, **15**, 1 (1972).
112. I. V. Hertel, to be published in *Adv. Chem. Phys.* on "The Excited States in Chemical Physics II," J. Wm. McGowan, ed.
113. D. S. Y. Hsu and M. C. Lin, *Chem. Phys. Lett.*, **56**, 79 (1978).
114. P. Herbrechtsmeier and H. G. Wagner, *Ber. Bungenges. Physik. Chem.*, **76**, 517 (1972).
115. J. J. Havel, *J. Am. Chem. Soc.*, **96**, 530 (1974).
116. M. C. Lin, R. G. Shortridge, and M. E. Umstead, *Chem. Phys. Lett.*, **37**, 279 (1976).

117. I. Haller and G. C. Pimentel, *J. Am. Chem. Soc.*, **84**, 2855 (1962).
118. R. E. Huie and J. T. Herron, *Prog. React. Kinetics*, **8**, 1 (1975), and references cited therein.
119. C. A. Arrington, W. Brennen, G. P. Glass, J. V. Michael, and H. Niki, *J. Chem. Phys.*, **43**, 525 (1965).
120. J. M. Brown and B. A. Thrush, *Trans. Faraday Soc.*, **63**, 630 (1967).
121. P. Herbrechtsmeier and H. G. Wagner, *Z. Physik. Chem. NF*, **93**, 143 (1974).
122. J. R. Kanofsky, D. Lucas, F. Pruss, and D. Gutman, *J. Phys. Chem.*, **78**, 311 (1974).
123. P. Herbrechtsmeier and H. G. Wagner, *Ber. Bunsenges. Physik. Chem.*, **79**, 461, 673 (1975).
124. H. E. Avery and S. J. Heath, *Faraday Trans. I*, **3**, 512 (1972).
125. B. Blumenberg, J. Hoyermann, and R. Sievert, *Sixteenth Symposium (Int.) on Combustion*, The Combustion Institute, Pittsburg, Pa., 1977, p. 841.
126. M. E. Umstead, R. G. Shortridge, and M. C. Lin, *Chem. Phys.*, **20**, 271 (1977).
127. M. E. Umstead and M. C. Lin, *Chem. Phys.*, **25**, 353 (1977).
128. D. S. Y. Hsu, L. J. Colcord, and M. C. Lin, *J. Phys. Chem.*, **82**, 121 (1978).
129. W. M. Shaub, T. L. Burks, and M. C. Lin, *Chem. Phys.*, **45**, 455 (1980).
130. F. Pilz and H. G. Wagner, *Z. Phys. Chem.*, **92**, 323 (1974).
131. G. Liuti, C. Kunz, and S. Dondes, *J. Am. Chem. Soc.*, **89**, 5542 (1967).
132. D. G. Williamson and K. D. Bayes, *J. Am. Chem. Soc.*, **89**, 3390 (1967).
133. D. S. Y. Hsu and M. C. Lin, *J. Phys. Chem.*, **68**, 4347 (1978).
134. I. R. Slagle, J. R. Gilbert, and D. Gutman, *J. Chem. Phys.*, **61**, 704 (1974).
135. R. E. Graham and D. Gutman, *J. Phys. Chem.*, **81**, 207 (1977).
136. I. W. M. Smith, *Disc. Faraday Soc.*, **44**, 194 (1967).
137. J. Geddes, P. N. Clough, and P. L. Moore, *J. Chem. Phys.*, **61**, 2145 (1974).
138. P. A. Gorry, C. V. Nowikow, and R. Grice, *Chem. Phys. Lett.*, **49**, 116 (1977).
139. P. N. Clough, G. M. O'Neil, and J. Geddes, *J. Chem. Phys.*, **69**, 3128 (1978).
140. S. Tsuchuya, N. Nielsen and S. H. Bauer, *J. Phys. Chem.*, **77**, 2455 (1973).
141. H. T. Powell and J. D. Kelley, *J. Chem. Phys.*, **60**, 2191 (1974).
142. J. D. Kelley, *Chem. Phys. Lett.*, **41**, 7 (1976).
143. R. D. Levine and Ben-Shaul, *Chemical and Biological Applications of Lasers*, Academic Press, New York, 1977, p. 145.
144. J. W. Hudgens, J. T. Gleaves, and J. D. McDonald, *J. Chem. Phys.*, **64**, 2528 (1976).
145. D. S. Y. Hsu, W. M. Shaub, T. L. Burks, and M. C. Lin, *Chem. Phys.*, **44**, 143 (1979).
146. G. Hancock, C. Morley, and I. W. M. Smith, *Chem. Phys. Lett.*, **12**, 193 (1971).
147. G. Hancock, B. A. Ridley, and I. W. M. Smith, *JCS Faraday Trans. II*, **68**, 2117 (1972).
148. N. Djeu, *J. Chem. Phys.*, **60**, 4109 (1974).
149. M. A. Pollock, *Appl. Phys. Lett.*, **8**, 237 (1966).
150. G. Hancock and I. W. M. Smith, *Chem. Phys. Lett.*, **3**, 573 (1969).
151. R. G. Shortridge and M. C. Lin, *Chem. Phys. Lett.*, **35**, 146 (1975).
152. K. Schofield, *J. Phys. Chem. Ref. Data*, **2**, 25 (1973).
153. D. S. Y. Hsu and M. C. Lin, *Int. J. Chem. Kinetics.*, **10**, 839 (1978).
154. D. S. Y. Hsu, M. E. Umstead, and M. C. Lin, in *Flourine-Containing Free Radicals, Kinetics and Dynamics of Reactions*, J. W. Root, ed., ACS Symposium Series, No. 66, 1978, Chap. 5.
155. D. S. Y. Hsu, R. G. Shortridge, and M. C. Lin, *Chem. Phys.* **38**, 285 (1979).
156. D. E. Klimek and M. J. Berry, *Chem. Phys. Lett.*, **20**, 141 (1973).
157. R. C. Michell and J. P. Simons, *J. Chem. Soc.*, **13**, 1005 (1968).
158. W. J. R. Tyerman, *Trans. Faraday Soc.*, **65**, 163 (1969).
159. J. Heicklen, N. Cohen, and D. Saunders, *J. Phys. Chem.*, **69**, 1774 (1965).

160. D. Saunders and J. Heicklen, *J. Phys. Chem.*, **70**, 1950 (1966).
161. D. S. Y. Hsu and M. C. Lin, *Chem. Phys.*, **21**, 235 (1977).
162. S. Koda, *Chem. Phys. Lett.*, **55**, 353 (1978).
163. P. N. Clough, S. E. Schwartz, and B. A. Thrush, *Proc. Roy. Soc.*, **A317**, 575 (1970).
164. D. M. Creek, C. M. Melliar-Smith and N. Jonathan, *J. Chem. Soc. (A)*, **1970**, 646.
165. J. E. Butler, L. P. Goss, M. C. Lin and J. W. Hudgens, *Chem. Phys. Lett.*, **63**, 104 (1979).

DOPPLER SPECTROSCOPY OF PHOTOFRAGMENTS*

K. H. WELGE AND R. SCHMIEDL

Fakultät für Physik, Universität Bielefeld, Bielefeld, FRG

CONTENTS

I. INTRODUCTION

This chapter is concerned with experimental investigations of the dynamics of the dissociation of polyatomic neutral molecules carried out by the technique of laser Doppler spectroscopy, in bulk and under crossed-beam condition. Photodissociation is a basic process in the interaction of light with molecules, of interest in itself as an elementary molecular process and also with respect to a variety of applications in different fields. The interest has increased considerably in recent years, first, because the experimental investigation of photodissociation is rapidly advancing by the use of the laser, and second, because the laser makes possible to achieve photodissociation, state, and isotope selectively, by new excitation mechanisms. These are, aside from the common one-photon absorption, stepwise

*Work supported by Deutsche Forschungsgemeinschaft. Lecture delivered by K. H. Welge at the Laser Chemistry Conference, Ein Bokek, Israel, December 1978.

excitation through vibrational plus electronic states or solely electronic states, multiphoton excitation through virtual states, and finally the multiple-photon excitation in the infrared, either by infrared light alone or also in combination with electronic state absorption.

Most detailed experiments on dissociation dynamics of a molecule would include preparation (or knowledge) of the initial state of the molecule prior to absorption and complete analysis of the energy and momentum distribution of the fragments as they evolve into the excited channel states. For polyatomic molecules this includes, particularly, the rotational and vibrational fragment states, aside from the translational degree of freedom. Experiments to such an extent and in such detail have practically not yet been performed, even with triatomics.

The experimental investigation of polyatomic neutral molecule dissociation has been considerably advanced by the technique of photofragment spectroscopy with mass spectroscopic fragment detection combined with the time-of-flight measurement applied to one-photon dissociation first by Diesen et al.[1] and Busch et al.[2] The potential of this method has recently been exploited, for instance, in studies of the infrared multiple-photon dissociation (MPD) process, particularly by Lee and co-workers, examples being given in Refs. 3–6. A recent example for application to one-photon dissociation, is a work with SO_2.[7] The mass spectroscopic time-of-flight photofragment technique, in principle universally applicable, can deliver interfragment recoil energy and momentum distribution. It demands, however, a very high degree of experimental complexity and sophistication and, furthermore, because of practical limitation of velocity resolution, detailed investigations on the quantum state-specific basis are normally not feasible for small molecules.

This is different with optical spectroscopy, as it is in principle quantum-state selective. Its application depends of course on the condition that spectroscopically suitable fragments are produced. Using laser-induced fluorescence (LIF), state-selective detection has been increasingly employed in photodissociation studies. However, practically all previous applications of LIF have been concerned with *intrafragment energy distribution* measurement. First examples have been one-photon dissociation experiments.[8,9] LIF investigations have been carried out recently also on the infrared MPD process; see, for instance, Refs. 10–16. Laser excitation spectroscopy also offers the possibility to investigate the *interfragment energy and momentum distribution*, with the special and attractive property that this can be done state selectively. State-selective optical photofragment recoil experiments can be performed in two ways: (1) by fragment *time-of-flight spectroscopy*, and (2) by *Doppler spectroscopy*. To exploit the full potential of these techniques crossed-beam configurations are necessary, particularly where slow fragments are involved, as will generally be the case, for example, at excitation energies close to dissociation limits and in the IR multiple-photon dissociation process.

Laser-induced fluorescence photofragment spectroscopy studies with state-selective interfragment recoil measurement have not been performed to any appreciable extent. A preliminary kind of experiment has recently been carried out in an MPD case by using the time-of-flight technique under bulk condition and with rather limited spectral resolution.[17] A Doppler spectroscopy study has been carried out in a one-photon dissociation case where electronically excited fragments were produced under bulk condition, and linewidth measurement was made in emission interferometrically.[18] A first experiment with tunable LIF detection has been made recently by McDonald et al.,[19] who dissociated HN_3 at 266 nm and observed the recoil of the NH fragment to yield a linewidth of $\sim$0.5 cm^{-1}, compared to $\sim$0.2 cm^{-1} bandwidth of the probing laser.

II. SUBJECT OF REPORT

Here we report Doppler experiments carried out recently in our laboratory. Two kinds of studies have been made, both with the main objective to establish and investigate the potential of the method: (1) first multiple-photon dissociation Doppler experiments under crossed-beam condition with C_2H_3CN[20] and CH_3NH_2,[21] and (2) first linewidth measurement in single vibrational–rotational levels of NO produced by one-photon dissociation of NO_2 in bulk with unpolarized and polarized light,[24] in connection with complete internal energy distribution measurements of the NO fragment.[22,23]

In previous MPD experiments with C_2H_3CN[17,25] $C_2(a^3\Pi_u)$ has been observed by LIF as product, resulting with all likelihood from a sequential dissociation mechanism. Time-of-flight photofragment experiments in bulk have indicated that $C_2(a^3\Pi_u)$ is produced with little recoil energy, of the order of 0.05 eV.[17] Because of this previous investigation, C_2H_3CN has been chosen as a first case for the present Doppler experiments. The MPD of CH_3NH_2 produces NH_2, as previously observed by LIF.[26-29] A reason for choosing this molecule was that NH_2 can be rather easily monitored by laser-induced fluorescence,[30-32,26] since it has a relatively open and well-studied spectrum in the visible.[33] Furthermore, the fragment is likely to be produced by direct decay into $CH_3 + NH_2$, interestingly to an appreciable extent in excited vibrational levels.[26] Whereas the experiments with C_2H_3CN were of relatively crude nature, the CH_3NH_2 experiments have been considerably improved.

The NO_2 spectroscopy and photochemistry has attracted great attention for a long time. The dynamics of the one-photon dissociation of NO_2 into $NO + O$ has previously been investigated at the doubled ruby-laser frequency (347.15 nm) by Busch and Wilson,[34,35] using the mass spectroscopic photofragment technique. The results have been explained theoretically,[36–38] indicating a decay mechanism with statistical energy distribution. In the work reported here, the dissociation has been investigated at the N_2-laser wavelength (337 nm), and the intrafragment energy distribution of NO has been fully analyzed and Doppler measurements have been performed for the first time in selected states of the NO fragment.

III. METHOD

The employment of Doppler spectroscopy with laser-excited fluorescence for the investigation of the dynamics of molecular scattering processes has previously been discussed by Kinsey.[39] Kinsey and co-workers have recently also reported a first application to a reaction, $H + NO_2 \rightarrow NO + OH$, with observation of OH in a selected state.[40] Another recent application to an inelastic scattering process [41] has been reported by Phillips et al. The theoretical analysis [39] shows that the three-dimensional velocity distribution, $F(\mathbf{v}) = f(v;\theta,\phi)$, of a scattered particle can be obtained from the measurement of Doppler line profiles $D(v,\theta,\phi)$ as a function of the spatial angle (θ,ϕ) of the probing light beam with respect to a given laboratory frame of reference. $F(v;\theta,\phi)$ can be deduced unambiguously from the set of Doppler profiles $D(v;\theta,\phi)$ by a Fourier transformation procedure. If the angular distribution contains some form of symmetry, and if the symmetry is known, the theoretical derivation of F from D is correspondingly simplified, and of course also the experimental procedure. A most simple situation is given for instance when the observed particles are scattered with spherical symmetry, that is F does not depend on the velocity direction, but on the speed only. Then it is sufficient to make the Doppler measurement in one direction only, and $D(v)$ and $F(v)$ are related by

$$\frac{dD(v)}{dv} = -2\pi v \cdot F(v) \tag{1}$$

with the Doppler relation $\nu = \nu_0[1 + v/c]$.

This case is of interest for the MPD process, since a previous investigation[41] has indicated that this dissociation type will normally lead to spatially randomized fragments. Therefore, spherical symmetry has been assumed in the MPD experiments reported here, though the limited amount of work should be kept in mind for the angular distribution in MPD. In general photodissociation will produce more or less spatially nonisotropic fragment distributions, particularly where the dissociation occurs from electronically excited states, as in one- and multiphoton absorption. Also it should be noted that in experiments with molecular beam configuration nonspherical symmetry always exists with respect to the laboratory frame of reference. Anisotropic fragmentation is observed in the case of NO_2 dissociation.

Figure 1 illustrates the situation as applicable in the case of spherically symmetric fragment distribution and dissociation in a molecular beam. In this arrangement, used in the Doppler MPD experiments, the molecular beam is crossed at right angles by the probe laser beam. The direction of the dissociating light beam is not specified, since it can be freely chosen, assuming spherical symmetry for the dissociation process. Because of the perpendicular directions of the probe laser and molecular beams, the velocity of the parent molecules, $\mathbf{v}_P$, does not contribute to the Doppler component, $\mathbf{v}_D$, in the direction of the probe laser beam. Hence, $\mathbf{v}_D$ is determined, within experimental limits, solely by the fragment recoil velocity. This means the measurement is done in the center-of-mass system of the parent molecules, an essential and practically important difference to the time-of-flight technique which always fully incorporates $\mathbf{v}_p$ and accordingly probes in the laboratory system.

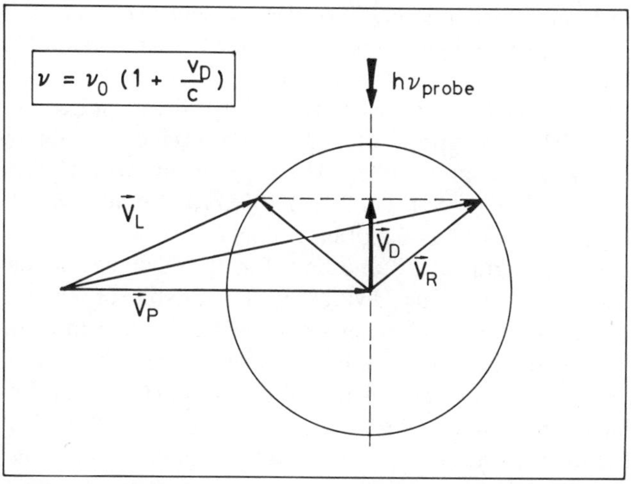

Fig. 1. Doppler spectroscopy; velocity vector diagram: $\mathbf{v}_p$ parent molecule velocity, $\mathbf{v}_R$ fragment recoil velocity in CM-system, $\mathbf{v}_L = \mathbf{v}_p + \mathbf{v}_R$ fragment velocity in lab-system, $\mathbf{v}_D$ Doppler component in probe laser direction.

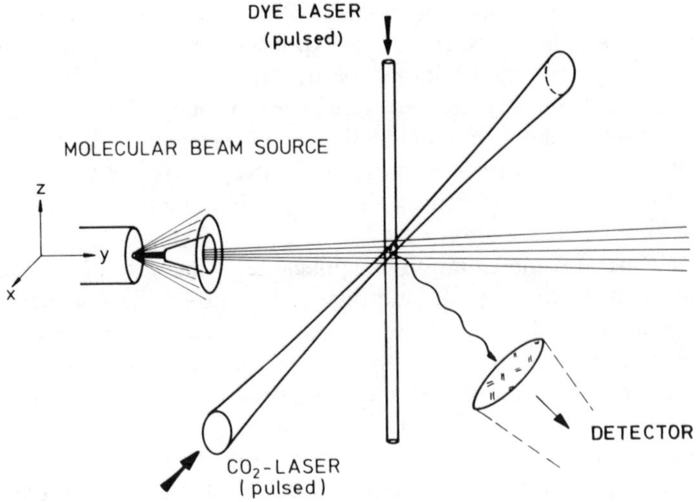

Fig. 2. Experimental arrangement for Doppler photofragment spectroscopy.

IV. MULTIPLE-PHOTON DISSOCIATION IN BEAM

A. Experimental

The crossed-beam arrangement (Fig. 2) used in the MPD experiments[20, 21] consisted of nozzle beam system, multimode pulsed TEA-CO_2 laser, pulsed tunable probe laser, fluorescence detection system. The CO_2 laser beam of about rectangular cross-section, 2×3 cm^2, was focused onto the molecular beam with a lens of 25 cm focal length. Most of the laser output ($\sim$70%) of $\sim$1 J/pulse was contained in a 0.2-μs pulse, the rest in a 2-μs long tail. The repetition rate of 10 Hz was compatible with the repetition of the pulsed probe laser. The experiments with C_2H_3CN have been carried out at the 10.591 μm $P(20)$, $(00^\circ1-10^\circ0)$ and with CH_3NH_2 at the 9.586 μm P(24), $(00^\circ1-02^\circ0)$ laser line.

The essential experimental condition for the Doppler method is of course a sufficiently small bandwidth of the tunable probe laser. Since dissociation was done in a pulsed mode, probing also had to be done with a pulsed laser. Aside from this condition pulsed probing has the advantage that time-resolved studies are possible and, particularly, that the spectral region of detection is much wider than with cw laser. On the other hand, it is more difficult to obtain sub-Doppler bandwidth with pulsed lasers. The tunable dye laser system used in the present experiments has been described in detail.[42] It had 10 Hz repetition frequency and 7 ns pulse duration and could be operated in high- and low-resolution modes. At low resolution the bandwidth was ~1.5 GHz and at high resolution less than 0.2 GHz. The output was, for instance at 493 nm, ~3 mJ/pulse and ~1 mJ/pulse in the two modes, respectively. The probe beam diameter was kept at about 3 mm at the intersection with the molecular and CO_2-laser beams.

The molecular beam, produced with a nozzle and skimmer arrangement, had a divergence of ~5° at FWHM, which was sufficient to fully neglect the parent molecule velocity component in the probe laser beam direction. According to the vapor pressure of the gases used the nozzle was heated to allow stagnation pressures of up to several hundred torr. At the crossing of the beams (Fig. 2) the overlapping, of the CO_2- and probe-laser beams resulted in an effective fluorescence excitation volume of 3 mm^3 on the molecular beam axis. Because of the small extension of this volume it was possible to work without the skimmer, nevertheless retaining practically the crossed-beam condition at the intersection, as with the skimmer. This "optimal skimming" simplified naturally the experimental procedure very much. The probe-laser pulse was delayed with respect to the CO_2-laser pulse peak by 1 μs, and the fluorescence was received subsequently through a time gate of 10 μs.

B. Results

1. *Doppler Measurements of $C_2(a^3\Pi_u)$ from C_2H_3CN*

As in previous LIF studies with $C_2(a^3\Pi_u)$ the fragment was monitored by excitation in the Swan band system, $d^3\Pi_g$–$a^3\Pi_u$, in the visible [see, for example, Refs. 43, 17, 44, 11]. Figure 3a shows a section of the excitation spectrum from about 516.3 to 516.5 nm obtained at the dissociating CO_2-laser line at 10.591 μm. The beam of C_2H_3CN was produced without carrier gas. The probe laser bandwidth was $\Delta\nu_L = 2$ GHz, compared to a FWHM linewidth of 1.3 GHz of thermal C_2 at room temperature. Another, narrower part of the spectrum around 516.2 nm, taken also with 2 GHz bandwidth, is shown in Fig. 3b.

No isolated lines are present in this spectrum, preventing a precise analysis of the lineshape. The linewidth of 5 GHz FWHM exceeds, however, substantially the spectral resolution and clearly indicates the fragment recoil. As shown in Fig. 3b, the lines may be approximated by a Gaussian profile so that the width of 5 GHz can be related to a most probable recoil speed $\bar{v} = 1430$ m/s of $C_2(a^3\Pi_u)$ in a respective vibrational–rotational level. $\bar{v}$ corresponds to an average kinetic recoil energy of ~2 CO_2 photons of 10.591 μm, compared to 0.4 photons obtained in the previous experiment[17] with the time-of-flight technique.

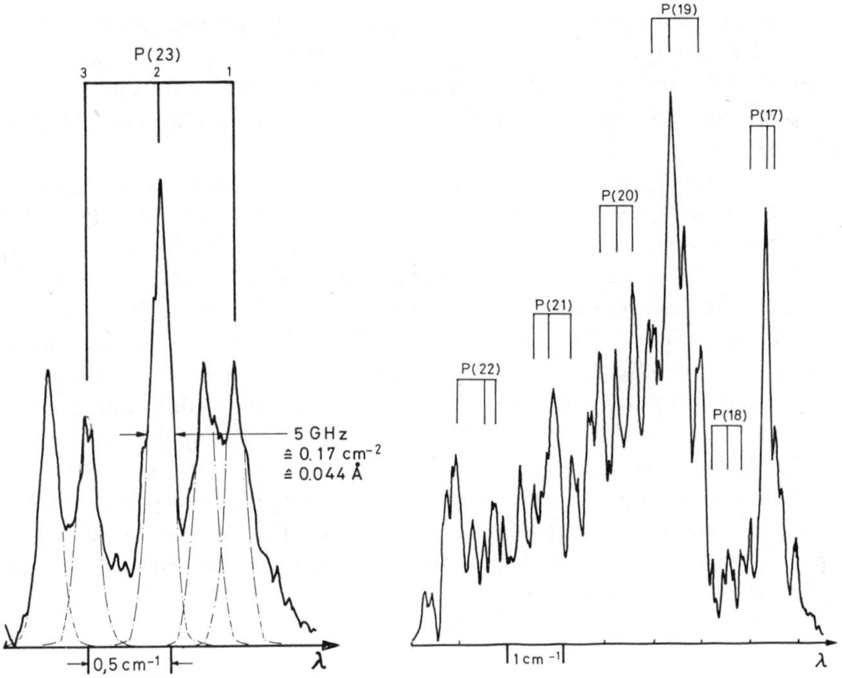

Fig. 3. Excitation spectra of $C_2(a^3\Pi_u)$ produced by MPD of C_2H_3CN in a beam. Probe laser bandwidth ~2 GHz.

2. *Doppler Measurements of NH_2 from CH_3NH_2*

More detailed and improved experiments have been made with NH_2 from CH_3NH_2. As in previous MPD studies [for instance, Refs. 10, 26, 29] NH_2 has been observed by laser excitation in the $NH_2(\tilde{A}^2A_1\Pi_u–\tilde{X}^2B_1)$ transition system, which exhibits a many-line spectrum reaching from the near-UV into the near-IR.[33] Figure 4 shows in the lower part a small section around 493 nm of the whole NH_2 excitation spectrum taken with a probe laser bandwidth of ~1.5 GHz and a scanning speed of 4 GHz/min. The CH_3NH_2 beam was formed without carrier gas at a stagnation pressure of 500 torr. The section covers transitions from rotational levels $N''=1$ to 6 in the $v''=(0,0,0)$ state of NH_2. The MPD in beams with laser-induced fluorescence is obviously a powerful technique to produce and observe radicals with sufficient amounts for state-selective experiments of various kinds under rather refined conditions.

Linewidth measurements at high resolution, i.e., 0.2 GHz, have been made so far in the two rotational levels 6_{16} of the $v''=(0,0,0)$ and $(0,1,0)$ vibrational states (see Fig. 6). The energy of both levels is different by 0.18

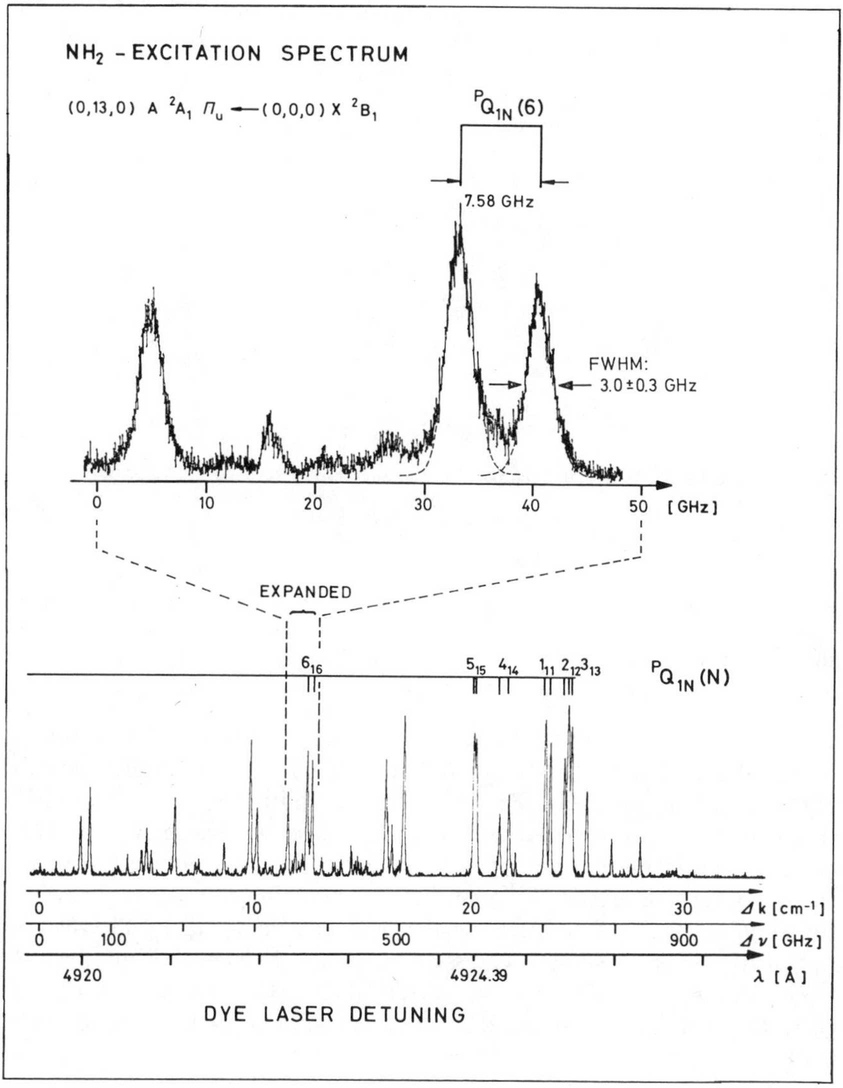

Fig. 4. Excitation spectra of $NH_2(\tilde{X}^2B_1)$ produced by MPD of CH_3NH_2 in a beam. Probe laser bandwidth: lower spectrum ~1.5 GHz, upper spectrum ~0.2 GHz. NH_2 rotational level 6_{16}, $v''=(0,0,0)$. Broken line shows fitted Gaussian profile.

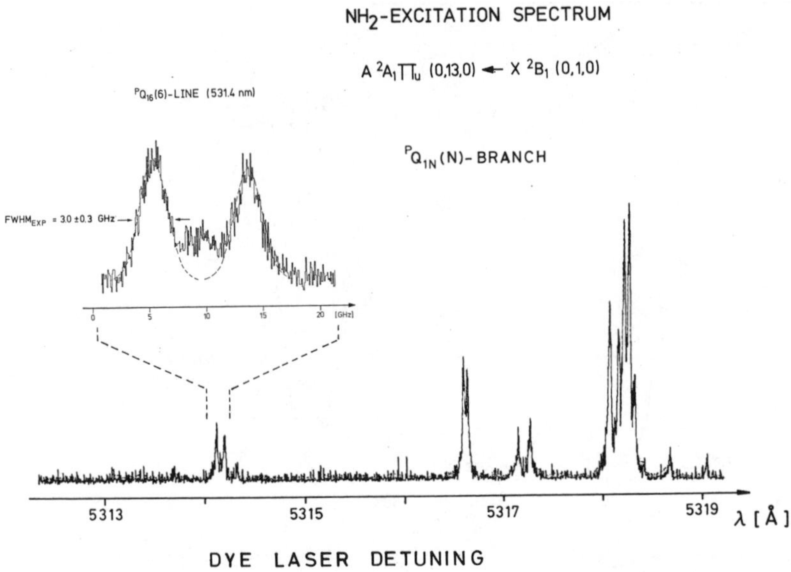

Fig. 5. Excitation spectrum of $NH_2(\tilde{X}^2B_1)$ produced by MPD of CH_3NH_2 in a beam. Probe laser bandwidth ~0.2 GHz. NH_2 rotational level 6_{16}, $v''=(0,1,0)$. Broken line shows fitted Gaussian profile.

eV corresponding to about 1.5 CO_2 photons. The results for $v''=(0,0,0)$, $N''=6$ and that for $v''=(0,1,0)$, $N''=6$ are shown in the upper parts of Figs. 4 and 5, respectively. The measured linewidth $\Delta\nu_{exp}$ of about 3 GHz is obviously large compared with the probe laser linewidth $\Delta\nu_L$, and the observed lineshape can therefore be taken to be practically equal to the true Doppler profile $\Delta\nu_0$. For quantitative estimation of the laser linewidth contribution one can deconvolute the observed linewidth by the relation $\Delta\nu_{exp}^2=\Delta\nu_0^2+\Delta\nu_L^2$, assuming Gaussian contours. Even with a probe laser bandwidth of 1.5 GHz and an observed FWHM line of 3 GHz the correction amounts to about 25% only and with $\Delta\nu_L=0.2$ GHz to about 0.4%.

3. *Speed Distribution of NH_2*

Assuming spherically symmetric fragment distribution, the speed distribution can be obtained from the lineshape by (1). This procedure poses a problem as the relation becomes formally undetermined for the line center and the error increases as the line center is approached, the degree of error depending of course on the precision of the measurement. Instead of the formal analyses a more direct procedure can be applied in the present case, namely, the lines can be fitted quite well by Gaussian profiles for, as can

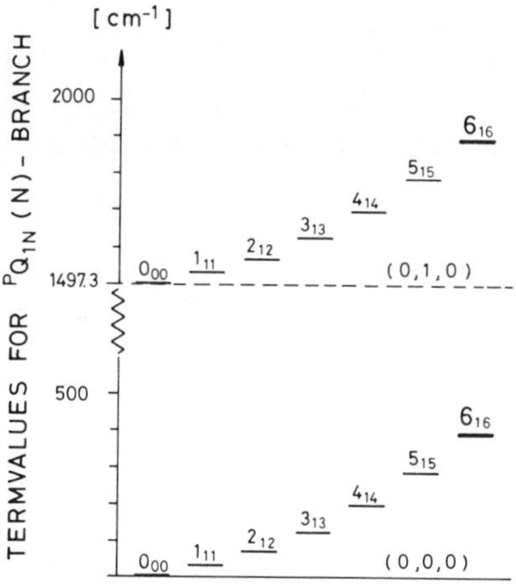

Fig. 6. Rotational levels of $NH_2(X^2B_1)$ in $v''=(0,0,0)$ and $(0,1,0)$. Doppler measurements made in levels indicated by strong lines.

be seen in Figs. 4 and 5, application of the formal procedure of (1) yields the same result,[21] the deviation is well within the experimental error limits. Corresponding to the Gaussian line profile the velocity distribution of the NH_2 radicals in the observed levels can be represented by a Maxwell–Boltzmann distribution. Hence, the FWHM linewidth can be converted to most probable speed v. For the two rotational levels in $v''=(0,0,0)$ and $(0,1,0)$ the values are $v=890$ m/s and 960 m/s, respectively. The average translational energy is correspondingly 0.1 eV and 0.11 eV, that is the recoil energy distribution and the average value of the recoil energy is thus practically the same in the two levels compared to the relatively much larger energy difference of the two levels (Fig. 6). This finding, the meaning of which is not yet clear, has to be substantiated by further systematic investigations.

V. ONE-PHOTON DISSOCIATION OF NO_2

A. Experimental

The NO_2 dissociation experiments[23] have been carried out under bulk condition at 337 nm with N_2-laser light of 8 ns pulse duration. Part of the N_2-laser light was used to pump a tunable dye laser for LIF detection of NO, as has been previously carried out.[45,46] The probe and dissociation beams crossed each other at right angle in the center of a fluorescence cell, filled with NO_2 at typically 0.01 to 0.1 torr. Fluorescence was observed at right angle to both laser beams. The probe laser pulse of ~5 ns duration was optically delayed to the dissociation pulse by 10 ns. In the intrafragment energy experiments the dissociation laser radiation was unpolarized and the probe laser bandwidth was 0.015 nm. The Doppler experiments have been made with unpolarized and also polarized N_2-laser light and high-resolution probe laser bandwidth of ~0.2 GHz, which is negligibly small in comparison with the linewidth of NO in the (X–A) transition at room temperature.

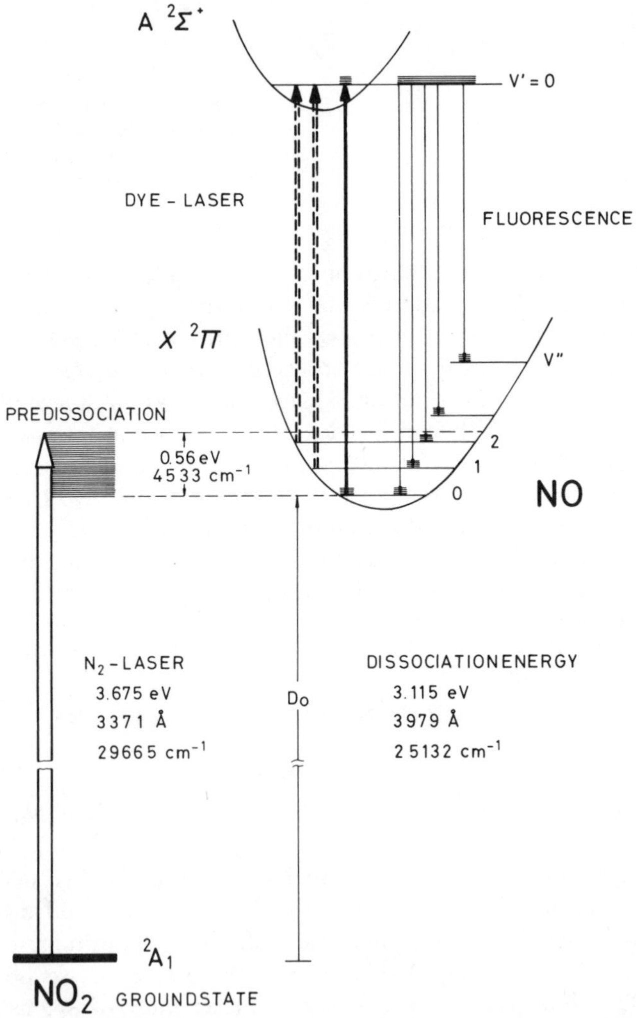

Fig. 7. Energetics of NO_2 dissociation at 337 nm and procedure of laser excitation spectroscopy of NO($X^2\Pi$, v, J).

B. Results

1. Intrafragment Energy Distribution

At 337 nm the dissociation can yield NO in vibrational levels $v''=0, 1, 2$ (Fig. 7). Taking excitation spectra of the $(v'', v')=(0,0), (1,0), (2,0)$ bands in the $NO(X^2\Pi - A^2\Sigma^+)$ transition, rotational level populations have been obtained in each vibrational level. Figure 8 represents, as an example, excitation spectra of the (1,0) band, once for thermal NO at room temperature (lower spectrum) and once for NO from the dissociation (upper spectrum). The nonthermal character of the spectrum from the dissociation is clearly observable. Figure 9 shows the rotational energy distribution of NO in $v''=0,1,2$ plotted versus the total internal NO energy of vibration and rotation, however, for each level separately. The dotted line indicates the maximum possible internal NO energy, that is, the excess energy at 337 nm. The population beyond this limit is attributed to internal energy of the NO_2 molecules. The energetics of the NO_2 dissociation is given by $E(h\nu)+E_i(NO_2)-D_0=E_i(NO)+E_i(O)+E_{ki}$, where $E(h\nu)$ is the photon energy at 337 nm, $E_i(NO_2)$ is the internal parent molecule energy, D_0 is the dissociation energy with NO and O in their ground states, $E_i(NO)$ is the internal NO and $E_i(O)$ that of the O atom. E_{ki} is the recoil energy of NO and O in the respective internal states. The energy of the $O(^3P_{2,1,0})$ sublevels (0; 157; 227 cm^{-1}) is small compared to the excess energy, $E(h\nu)+E_i(NO_2)-D_0$, so that $E_i(O)$ may be neglected to a first approximation. $E_i(NO_2)$ is of the order of 400 cm^{-1} and may also be neglected for the same reason.

Most of the excess energy obviously appears as internal energy of NO, with strongly nonthermal and inverted rotational distribution. As to the vibrational distribution, summation over the rotational population shows that it also is strongly nonthermal and inverted (Fig. 10). These results are essentially different from those previously obtained by Busch and Wilson[34,35] at 347.15 nm, who obtained a much less pronounced nonthermal distribution, with most of the excess energy in translational motion of the fragments. The decay at 337 nm obviously occurs via a nonstatistical mechanism, which is in contrast to the theoretical analysis of the previous experiments at 347 nm by Quack and Troe.[36–38] A theoretical treatment of the nonstatistical behavior at 337 nm has not yet been made. The rotational and vibrational distributions can, however, be qualitatively understood by a model where dissociating NO_2 states are involved with equilibrium bond angles and bond lengths substantially different from those of the NO_2 ground state. This appears to be the case for the $^2B_2$ state,[34] which may be excited at 337 nm.

It is interesting to observe the effect of resolution on the energy distribution measurement. This is illustrated in Fig. 11, which shows in

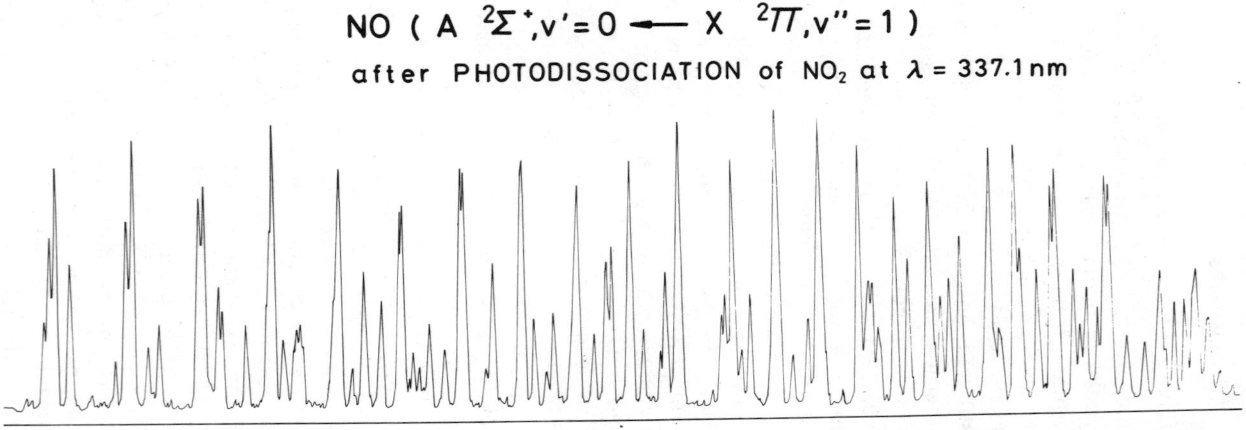

Fig. 8. NO excitation spectra of the $v''=1 \rightarrow v'=0$ band in the $(X^2\Pi - A^2\Sigma^+)$ transition. Spectrum (a): Thermal NO at room temperature. Spectrum (b): NO from NO_2 photodissociation at 337 nm.

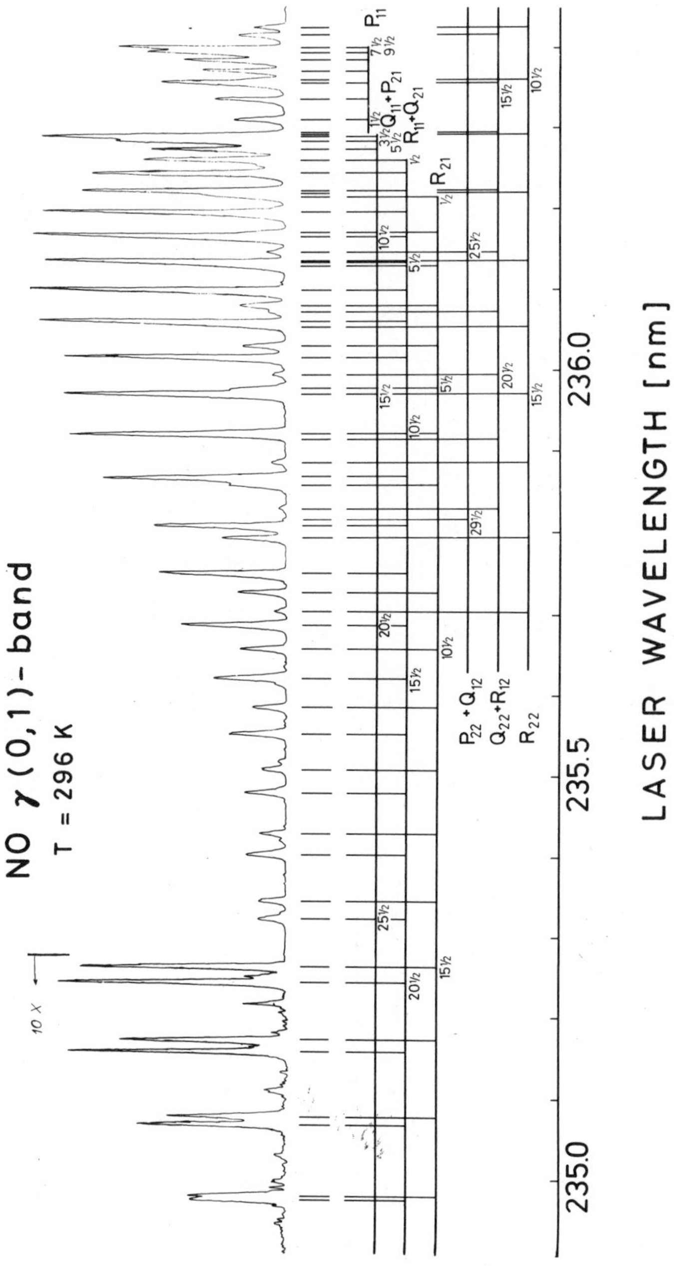

Fig. 8. *(Continued)*

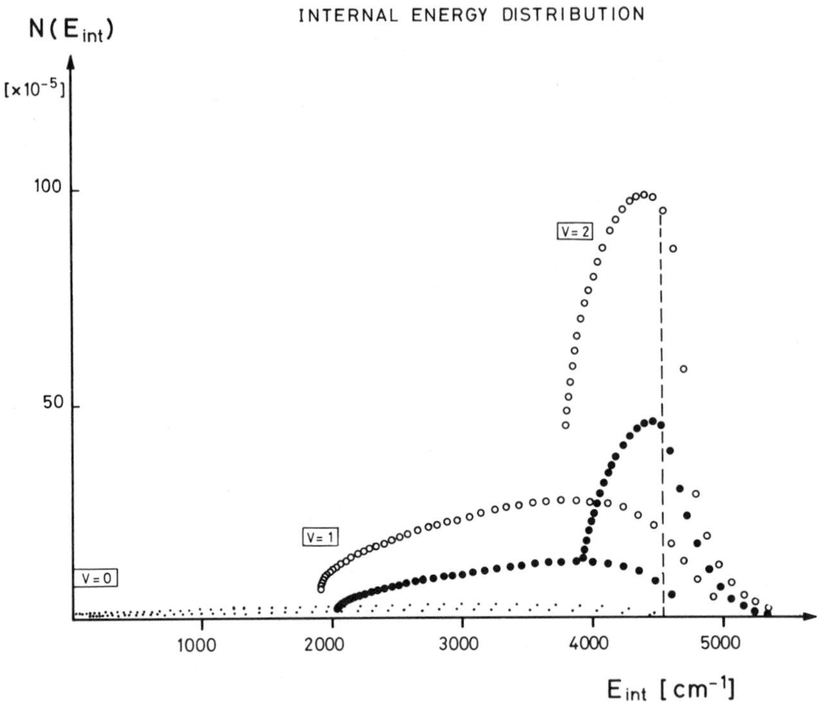

Fig. 9. Rotational energy population of NO($X^2\Pi$) in $v=0,1,2$ plotted vs internal energy for each rotational level individually. Open circles and dots refer to the two spinstates, $\Pi_{1/2}$ and $\Pi_{3/2}$, respectively.

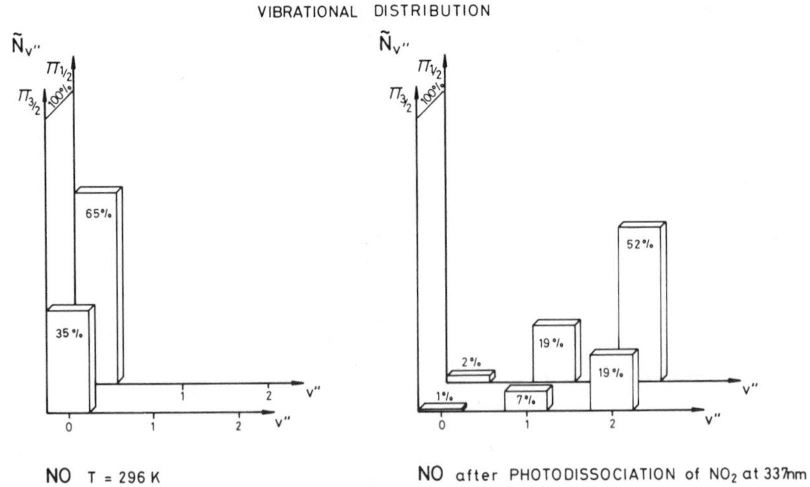

Fig. 10. Vibrational energy distribution in $v=0,1,2$ separately for $^2\Pi_{1/2}$ and $^2\Pi_{3/2}$.

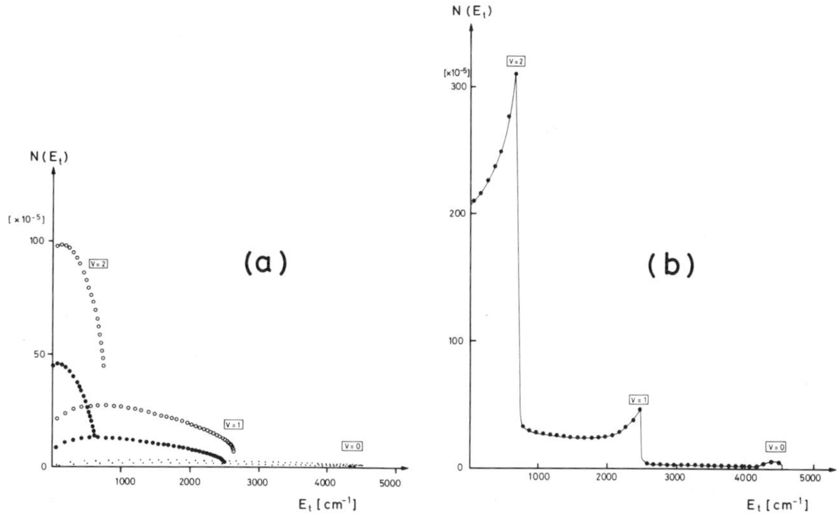

Fig. 11. (a) Translational energy distribution of NO in $v=0,1,2$ derived from the internal energy distribution (Fig. 9) and plotted for each level separately. (b) The same distribution for the case of an assumed energy resolution of 100 cm^{-1}.

part (a) the kinetic energy distribution of NO, straightforwardly derived from the internal energy distribution (Fig. 9), and plotted again for each rotational state separately. In a direct translational energy measurement, as by the mass spectroscopic time-of-flight technique, the limited resolution normally covers a range of internal quantum states. Taking the true distribution in Fig. 11a, the result one would obtain from a translational energy measurement with a resolution of, for example, 100 cm^{-1} is shown in Fig. 11b.

2. *Interfragment Doppler Recoil Experiments*

Results of the Doppler experiments [24] are shown in Fig. 12, representing shapes of the $R_{21}(9\frac{1}{2})$ line from the ($v''=1$, $J''=9\frac{1}{2}$) rotational level of the $NO(X^2\Pi_{\frac{1}{2}})$ ground state, measured under collisionless condition. The line shown in Fig. 12b was measured with unpolarized dissociating N_2-laser light, whereas the lines in Fig. 12c and 12d have been obtained with linearly polarized N_2-laser light. In one case (c) the electric field vector, $\boldsymbol{E}$, was perpendicular and in the other (d) parallel to the probe laser beam direction. Figure 13 represents another NO line, R_{21} $(25\frac{1}{2})$, again from the same vibrational level, $v''=1$, however with a substantially higher rotational level, $J''=25\frac{1}{2}$. In this case the N_2-laser light was again unpolarized and the line can thus be compared with the R_{21} $(9\frac{1}{2})$ line (Fig. 12b) taken

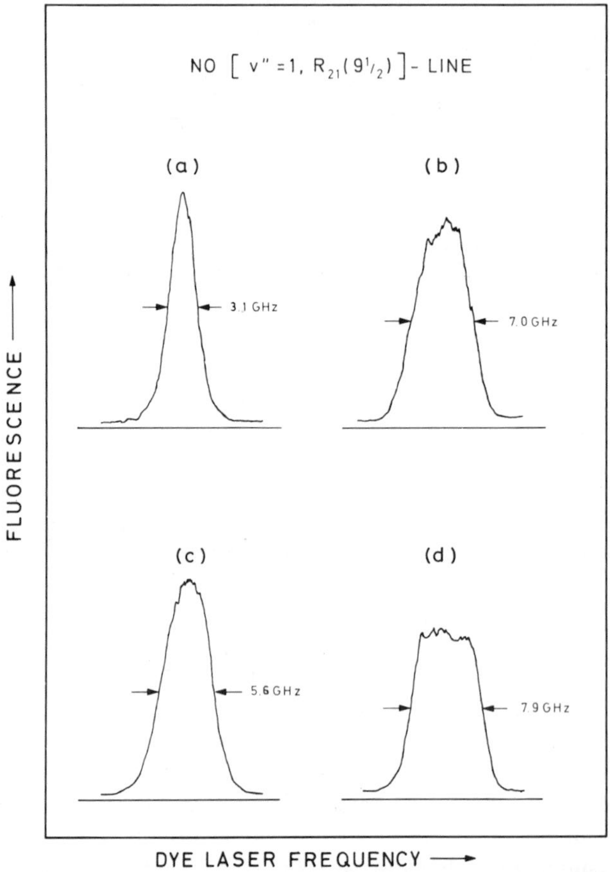

Fig. 12. Doppler measurement of the $R_{21}(9\frac{1}{2})$ NO line in the $(X^2\Pi - A^2\Sigma^+)$, (1,0) band from ($v''=1, j''=9\frac{1}{2}$) level (Ref. 24). (a) Thermal NO line at room temperature. (b) NO line from NO_2 photodissociation at 337 nm, unpolarized. (c) NO from NO_2 photodissociation of 337 nm, polarized with electric field vector perpendicular to probe laser beam. (d) NO from NO_2 photodissociation at 337 nm, polarized with electric field vector parallel to probe laser beam.

under identical condition. In order to estimate the contribution of the thermal motion in bulk to the linewidth, the R_{21} $(9\frac{1}{2})$ line has been measured with NO in bulk under thermal condition at room temperature (Fig. 12a). Taking the mass ratio of NO and NO_2 into account the motion of NO_2 will contribute to the linewidth of NO from the dissociation roughly 2.6 GHz FWHM. Despite this considerable amount the effects of the recoil energy and nonisotropic spatial distribution of the NO fragment in the selected levels can clearly be seen, demonstrating the power of the technique.

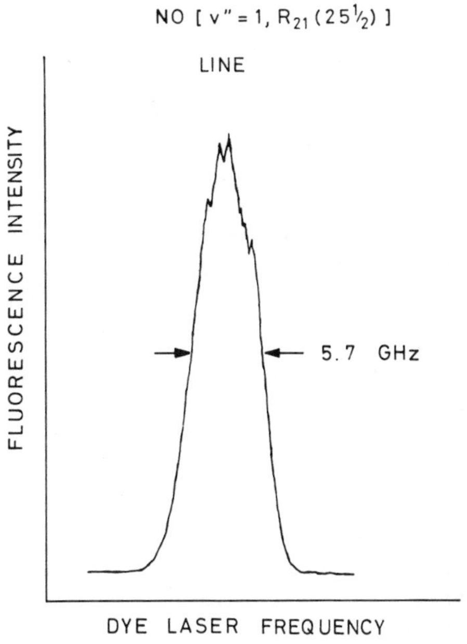

Fig. 13. Doppler measurement of the R_{21} $(25\frac{1}{2})$ NO line in the $(X^2\Pi - A^2\Sigma^+)$ (1,0) band transition from the $v'' - 1$, $J'' = 25\ 1/2$ level. NO_2 dissociation at 337 nm with unpolarized light.

The present observations can be qualitatively readily understood: Comparison of the unpolarized R_{21} $(9\frac{1}{2})$ and R_{21} $(25\frac{1}{2})$ lines (Figs. 12b and 13) shows that, as to be expected, the linewidth decreases as the internal fragment energy increases. The difference in linewidth, and the equivalent kinetic energy, corresponds also quantitatively with the energy difference of the levels. The two polarized lines, (c) and (d) in Fig. 12, nicely exhibit the spatial anisotropy of the departing fragments. The dissociation obviously proceeds predominantly parallel to the electric field vector, a finding that is qualitatively consistent with the previous work by Busch and Wilson,[34,35] who also found an angular fragment distribution peaking in the "parallel" direction.

References

1. R. W. Diesen, J. C. Wahr, and S. E. Adler, *J. Chem. Phys.*, **50**, 3635 (1969).
2. G. E. Busch, R. T. Mahoney, R. I. Morse, and K. R. Wilson, *J. Chem. Phys.*, **51**, 449 (1969).
3. M. J. Coggiola, P. A. Schulz, Y. T. Lee, and Y. R. Shen, *Phys. Rev. Lett.*, **38**, 17 (1971).
4. E. R. Grant, M. J. Coggiola, Y. T. Lee, P. A. Schulz, Aa. S. Sudbo, and Y. R. Shen, *Chem. Phys. Lett.*, **52**, 595 (1977).
5. Aa. Sudbo, P. Schulz, D. Krajnovich, Y. R. Shen, and Y. T. Lee, *Advances in Laser Chemistry*, *Springer Series in Chemical Physics*, Vol. 3, A. H. Zewail, ed., Springer-Verlag, Berlin-Heidelberg-New York, 1978, p. 308.
6. F. Brunner, T. P. Cotter, K. L. Kompa, and D. Proch, *J. Chem. Phys.*, **67**, 1547 (1977).
7. A. Freedman, S. Ch. Yang, and R. Behrson, *J. Chem. Phys.*, **70**, 5313 (1979).

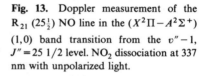

8. R. J. Cody, M. J. Sabety-Dzvonik, and W. M. Jackson, *J. Chem. Phys.*, **66**, 2145 (1977).
9. M. J. Sabety-Dzvonik and R. J. Cody, *J. Chem. Phys.*, **66**, 125 (1977).
10. J. D. Campbell, G. Hancock, J. B. Halpern, and K. H. Welge, *Opt. Comm.*, **17**, 38 (1976).
11. N. V. Chekalin, V. S. Dolzihikov, V. S. Letokhov, V. N. Lokham, and A. N. Shibanor, *Appl. Phys.*, **12**, 191 (1977).
12. M. L. Lesiecki and W. A. Guillory, *J. Chem. Phys.*, **66**, 4239 (1977).
13. M. L. Lesiecki and W. A. Guillory, *J. Chem. Phys.*, **69**, 4572 (1978).
14. S. E. Bialkowski and W. A. Guillory, *J. Chem. Phys.*, **68**, 3339 (1978).
15. D. S. King and J. C. Stephenson, *Chem. Phys. Lett.*, **51**, 48 (1977).
16. J. C. Stephenson, and D. S. King, *J. Chem. Phys.*, **69**, 1485 (1978).
17. J. D. Campbell, M. H. Yu, M. Mangir, and C. Wittig, *J. Chem. Phys.*, **69**, 3854 (1978).
18. H. G. Hansen, *J. Chem. Phys.*, **47**, 4773 (1967).
19. J. R. McDonald, R. G. Miller, and A. P. Baronavski, *Chem. Phys. Lett.*, **51**, 57 (1977).
20. R. Schmiedl, R. Boettner, H. Zacharias, U. Meier, D. Feldmann, and K. H. Welge, *Laser Induced Processes in Molecules, Springer Series in Chemical Physics*, Vol. 6, K. L. Kompa and S. D. Smith, eds., Springer-Verlag, Berlin-Heidelberg-New York, 1979, p. 186.
21. R. Schmiedl, R. Boettner, H. Zacharias, U. Meier, and K. H. Welge, *Opt. Comm.*, **31**, 329 (1979).
22. M. Geilhaupt, Diplomarbeit, Universität Bielefeld, 1979.
23. H. Zacharias, M. Geilhaupt, K. H. Welge, to be published.
24. R. Wallenstein (private communication).
25. M. H. Yu, H. Reisler, M. Mangir, and C. Wittig, *Chem. Phys. Lett.*, **62**, 439 (1979).
26. S. V. Filseth, J. Danon, D. Feldmann, J. D. Campbell, and K. H. Welge, *Chem. Phys. Lett.*, **63**, 615 (1979).
27. G. Hancock, R. J. Hennessy, and T. Villis, *J. Photochem.*, **9**, 197 (1978).
28. G. Hancock, R. J. Hennessy, and T. Villis, *J. Photochem.*, **10**, 305 (1979).
29. M. N. R. Ashfold, G. Hancock, and G. Ketley, *Faraday Disc.* **67**, 204 (1979).
30. M. Kroll, *J. Chem. Phys.*, **63** 319 (1975).
31. J. B. Halpern, G. Hancock, M. Lenzi, and K. H. Welge, *J. Chem. Phys.*, **63**, 4808 (1975).
32. G. Hancock, W. Lange, M. Lenzi, and K. H. Welge, *Chem. Phys. Lett.*, **33**, 168 (1975).
33. K. Dressler, and D. A. Ramsay, *Phil. Trans. Roy. Soc. (London)* **A 251**, 553 (1959).
34. G. E. Busch, and K. R. Wilson, *J. Chem. Phys.*, **56**, 3626 (1972).
35. G. E. Busch, and K. R. Wilson, *J. Chem. Phys.*, **56**, 3638 (1972).
36. M. Quack, and J. Troe, *Ber. Bunsenges. Physik. Chem. Bd.*, **79**, 170 (1975).
37. M. Quack, and J. Troe, *Ber. Bunsenges. Physik. Chem. Bd.*, **79**, 469 (1975).
38. M. Quack, and J. Troe, *Ber. Bunsenges. Physik. Chem. Bd.*, **80**, 1140 (1976).
39. J. L. Kinsey, *J. Chem. Phys.*, **66**, 2560 (1977).
40. E. J. Murphy, J. H. Brophy, G. S. Arnold, W. L. Dimpfl, and J. L. Kinsey, *J. Chem. Phys.*, **70**, 5910 (1979).
41. W. D. Phillips, J. A. Serri, D. J. Ely, D. E. Pritchard, K. R. Way, and J. L. Kinsey, *Phys. Rev. Lett.*, **41**, 937 (1978).
42. H. Zacharias, R. Schmiedl, R. Wallenstein and K. H. Welge, *Laser 79–Opto Electronik*, IPC Science and Technology Press, Guiltfort, 1979, pp. 74–80.
43. J. D. Campbell, M. H. Yu, and C. Wittig, *Appl. Phys. Lett.*, **32**, 413 (1978).
44. J. H. Hall, M. L. Lesiecki, and W. A. Guillory, *J. Chem. Phys.*, **68**, 2247 (1978).
45. H. Zacharias, A. Anders, J. B. Halpern, and K. H. Welge, *Opt. Comm.*, **19**, 116 (1976).
46. R. Wallenstein, and H. Zacharias, *Opt. Comm.*, **25**, 363 (1978).

NONLINEAR OPTICS AND LASER SPECTROSCOPY IN THE VACUUM ULTRAVIOLET

STEPHEN C. WALLACE*

Department of Chemistry, University of Toronto, Toronto, Canada M5S 1A1

CONTENTS

Over the past five years we have witnessed dramatic strides in the theoretical framework and experimental development of lasers and quantum electronics, such that sources of coherent radiation now span an enormous spectral region, beginning in the far-infrared and extending down through the vacuum ultraviolet to approach the soft X-ray region. This chapter will focus on the recent developments in vacuum ultraviolet (VUV) laser sources and nonlinear optics which have led to novel studies in laser spectroscopy in the VUV. It is this short wavelength spectral region that has been the last frontier in the laser revolution in chemistry and physics and that still presents the final challenge, that is, the achievement of a laser source in the X-ray region. Moreover, even in the spectral range between 1000 and 2000 Å, in which the most recent progress has

*Alfred P. Sloan Foundation Fellow.

been made, there is still a great deal of scope for further research. Nevertheless the feasibility and significance of high-resolution laser spectroscopy in the VUV have been clearly demonstrated using a readily accessible level of experimental sophistication, thus finally opening the VUV spectral region to the wide range of studies intrinsic to laser spectoscopy and photoselective chemistry.

The development of laser sources in the vacuum ultraviolet has followed two distinct but quite complementary paths. On the one hand, we have the family of high-power gas discharge[1] and electron-beam pumped lasers,[2–4] while on the other there is nonlinear optics or harmonic generation from visible and ultraviolet dye lasers. As we will discuss later, only through the latter has high-resolution laser spectroscopy and state-selective studies been possible. However, both approaches will play an important role in laser-induced chemistry, as the VUV is integrated into the spectral domain normally accessible for study with laboratory-scale laser systems. A further objective of this article is to balance both the exciting prospects, which are at hand for laser studies in the VUV, and the realities of venturing into such a challenging area of research endeavor. In treating the subject, we will follow a historical perspective: first, briefly outlining the evolution of VUV lasers, then tracing the more recent application of nonlinear optics to produce widely tunable coherent radiation in the VUV, and finally describing the application of VUV radiation to contemporary problems in chemical physics. These two examples, chosen from photophysics and reaction dynamics, illustrate the ultimate implementation of harmonic generation techniques to photoselective studies in VUV laser spectroscopy, and therefore establish the role of quantum optics in revealing novel vistas in molecular dynamics.

I. VACUUM ULTRAVIOLET LASERS—A RETROSPECTIVE

In 1970 the first report of the molecular hydrogen laser[1] opened up a decade of activity in VUV laser development, which included the appearance of rare gas excimer and exciplex lasers and the achievement of tunable coherent radiation in the Lyman-α region[5] via harmonic generation. The surge of activity in the development of VUV lasers arose in part from the uniqueness of the VUV region, in part from the ultimate interest in X-ray lasers and, from our perspective, from the exciting prospects in spectroscopy and molecular dynamics promised by narrow linewidth, tunable, high-power VUV laser pulses for state-selective studies. Here we review the principles on which VUV lasers are based.

It was always recognized that the major obstacles to producing a short wavelength laser would be classified as operational rather than fundamental. The principle difficulty lies in the unfavorable dependence of the laser gain resulting from the frequency scaling of the Einstein A factor as ν^3. Thus for a given oscillator strength the rate of depopulation of the upper laser level by spontaneous emission becomes a serious problem at short wavelengths, and hence the rate of excitation necessary to obtain a net gain in each round trip of the optical cavity increases rapidly with decreasing wavelength. Simple considerations based on the Schalow–Townes equations show that the pumping power necessary to achieve some arbitrary gain scales as ν^4 for a Doppler-broadened transition and ν^6 for a radiatively broadened line! Thus it is not surprising that the only practical means of pumping VUV lasers has been either with high-power gas discharge circuits (Blumlein) or high-intensity relativistic electron beams. The second, nontrivial problem has been in constructing the appropriate optical cavity. Important improvements in producing high reflectivity optical coatings[6] for the VUV have helped eliminate many of the problems of catastrophic laser ablation of the mirrors in optical resonators and thus permitted peak output powers in excess of 10^9 W to be observed. Despite this, continuously tunable, narrow linewidth lasers, analogous to visible wavelength dye lasers, are far from being a reality for the reasons discussed below. In contrast, many lasers in the VUV (notably H_2 and CO) are excited in a traveling-wave configuration, where the actual optical emission closely follows the excitation pulse as it propagates through the laser medium, so that in fact an optical cavity would be superfluous. In such designs, therefore there is no phase coherence in the "laser" output, which should be correctly classified as amplified spontaneous emission.

The rare gas excimer lasers are based on bound–continuum transitions from an excited diatomic species to its dissociative ground state. The observed continuum emission is a superposition of the Franck–Condon factors from the vibrational levels of the upper state. Thus these molecular dissociation lasers display relatively broad fluorescence as a consequence of the steeply repulsive ground-state potential, and there is always a population inversion on such transitions. However, the net gain is significantly lower than that for a bound–bound transition because of the distribution of oscillator strength over the broad fluorescence band. Figure 1 illustrates schematic potential energy curves for such transitions in the excimer and exciplex lasers.

The identity of the lasing states and the kinetics of these laser systems have been the subject of interesting and lively discussion over the past few years, which is not appropriate to reproduce in detail here. The upper laser level is generally accepted to be populated in three body collisions following excitation of atomic states in the discharge, but the complex kinetics involved in completely modeling the pressure dependence of the laser

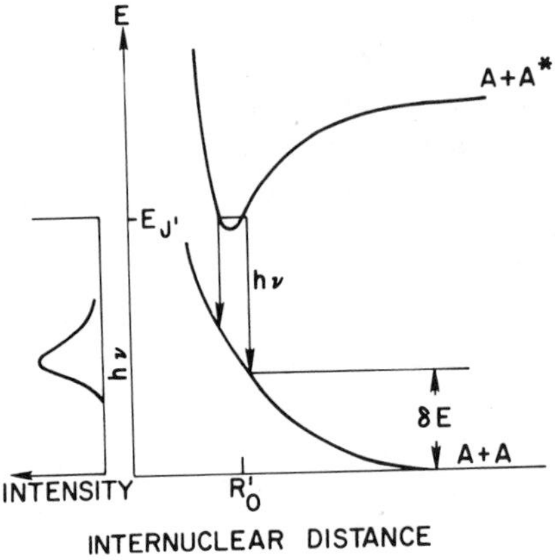

Fig. 1. Schematic potential energy curves for the bound-continuum emission characterizing the laser transition in excimer lasers.

intensity are far from being solved. Many more detailed studies of the individual kinetic processes and measurements of absorption and gain cross-sections will be necessary to fully model the behavior of excimer and exciplex lasers.

Salient parameters for all types of VUV lasers reported to date are summarized in Table I. As comprehensive and recent reviews[7, 8] have discussed in detail both the traveling wave systems, such as H_2, and the electron-beam pumped excimers, such as Xe_2, Kr_2, and Ar_2, we will forego further commentary. More recent developments, namely the ArF exciplex system[11–13] and the molecular fluorine laser, hold much promise for the future. This is primarily due to the ease by which these molecules may be excited in a simple, volume-preionized discharge scheme, which was previously developed for CO_2 TEA lasers. Hence these VUV lasers exhibit high output powers with relatively low pumping energies, although they do not achieve the same efficiency as their forebearers. Already we have seen an impressive range of photofragmentation studies[28, 50] reported for such molecules as CH_3OH, C_2N_2, and C_2H_2 using the ArF laser at 1932 Å. The F_2 laser will permit parallel studies at 1550 Å. It is not expected, however, that any of these exciplex (ArF) or eximer (Xe_2) lasers, although proven tunable[4, 16] will ever yield a sufficiently narrow linewidth $\leqslant 0.1\ cm^{-1}$ for state-selective studies in the VUV, as indeed dye lasers have already permitted in the visible region of the spectrum. Thus it is a general feature of these gas-phase systems, where the optical gain is low, that the constraints of geometrical optics make it very difficult to achieve linewidths $\leqslant 5\ cm^{-1}$. Perhaps a hybrid scheme, in which one amplifies the narrow

TABLE I
Vacuum Ultraviolet Lasers

System	Wavelength	Excitation source	Peak power (MW)	Pulse duration (ns)	Linewidth (cm^{-1})	Ref.	Remarks
CO	1811–1970 Å	Gas discharge	10^{-5}	1.5	$\leqslant 0.1$	9	
		Electron beam	1.5×10^{-2}	3	$\leqslant 0.1$	10	Discrete spectrum
		Gas discharge	10	10–20	1000 cm^{-1}	11,12	
ArF	1932 Å	Gas discharge	~2	20	150 cm^{-1}	13	600 cm^{-1} tuning range
		Electron beam	1.6×10^3	55	1000 cm^{-1}	14	
Xe_2	1730 Å	Electron beam	2	5	600 cm^{-1}	15	
		Electron beam	2×10^{-2}	5	150 cm^{-1}	4	1000 cm^{-V} tuning range
		Electron beam	1	50	10 cm^{-1}	16	
Kr_2	1457 Å	Electron beam	$\leqslant 10^{-2}$	50	600	3	
Ar_2	1261 Å	Electron beam	$\leqslant 10^{-2}$	50	600	17	
H_2	1250–1613 Å	Gas discharge	$\leqslant 10^{-1}$	1–2	$\leqslant 0.1$	1,18–22	Discrete spectrum
H_2, HD, D_2, $P-H_2$	1098–1613	Electron beam	$\leqslant 10^{-1}$	1–2	$\leqslant 0.1$	2,23	
H_2	1613 Å	Electron beam	0.3	1	$\leqslant 0.1$	24	
F_2	1580 Å	Gas discharge	~1	20		25	Discrete spectrum
C IV	1548.2, 1550.8	Gas discharge	10^{-3}	$\leqslant 1$	$\leqslant 50$	26	
Kr IV	1950.3, 1756.1	Gas discharge	10^{-3}	100	$\leqslant 0.1$	27	

linewidth coherent radiation produced by harmonic generation (see below) in an exciplex amplifying medium, will ultimately provide a high-brightness, megawatt intensity, and tunable laser source in the VUV. Of course, in this latter case, the finite bandwidth over which the exciplex systems displays gain will still limit the tuning range to typically less than 500 cm^{-1}.

II. GENERATION OF HARMONIC RADIATION IN THE VACUUM ULTRAVIOLET

Nonlinear optical phenomena provided the next and in a sense more flexible approach to producing coherent radiation in the vacuum ultraviolet. Advances in producing VUV radiation via harmonic generation and frequency mixing from visible and near ultraviolet lasers followed closely behind the development of VUV lasers. The impetus for this work has been largely provided by the interest in obtaining sources of coherent VUV radiation with all of the excellent spectral and spatial properties that can be acheived with visible wavelength lasers (i.e., single longitudinal and transverse modes). As discussed in the previous section, such performance is yet to be realised for any laser operating in the VUV, and it is now questionable whether it will ever be worth the effort to accomplish such a feat. Indeed, harmonic generation schemes involving higher order processes such as the 5th or 7th harmonic have now even surpassed VUV lasers in the quest for coherent radiation in the X-ray region, and the lowest wavelength achieved to date is 380 Å.[30]

From the perspective of high-resolution spectroscopy in the VUV, frequency-tripling from a visible laser provides a very high-brightness source of radiation, which may be readily tuned over a wide spectral region by tuning the dye laser, which provides the fundamental radiation. Extremely narrow linewidths ($\Delta\nu \leqslant 0.1$ cm^{-1}) suitable for state-selective studies[31–33] are presently realized, and following the demonstration of cw harmonic generation[34] we can anticipate several orders of magnitude further improvement in spectral resolution. A complete summary of the various wavelengths and spectral regions which have been covered up to the present time is given in Tables II and III.

The results to-date have really only sampled a small fraction of the possible atomic and molecular systems that could be used for nonlinear materials in VUV harmonic generation. Thus, there is no longer any doubt that all of the spectral region between 1000 Å and 2000 Å can be spanned by means of sum-frequency mixing from the visible and near-ultraviolet. We will now outline the theory on which these processes are based.

TABLE II
VUV Harmonic Generation—Fixed Frequency

Nonlinear material	Wavelength (Å)	Peak power	Pulse duration (ps)	Linewidth (cm^{-1})	Ref.	Remarks
Cd	1773	7 kW	50	≤ 5	35	
	1520	70 W	50	≤ 5	35	
Xe	1182	300 kW	25	≤ 5	36	
Hg	896	—	50	≤ 10	37	
Ar	887	—	25	≤ 5	38	
He	532	300 kW	30	≤ 5	30, 39	
	380	3 W	30	≤ 5	30, 39	28th harmonic of Nd:Yag

TABLE III
VUV Harmonic Generation—Continuously Tunable Sources

Nonlinear material	Wavelength range (Å)	Photons/pulse F (normalized)[a]	Linewidth (cm^{-1})	Ref.	Remarks
Sr	1778–1817 1883–1957	$10^5 \leq F \leq 10^7$	≤ 0.3	40,41,42	Tuning range determined by autoionizing spectrum
	1700	$10^8 \leq F \leq 10^{10}$	≤ 0.3	41	Autoionizing states
	1920	3×10^{11}	≤ 0.3	43	Autoionizing states, 500 ns pulse
	1700	10^7 photons/s	≤ 0.2	34	Continuous wave
	1712	10^8 photons/s	≤ 50	44	Continuous wave-mode-locked (10 ps pulses)
Ca	1760–1770	(unnormalized) 4×10^8	≤ 0.1	42, 46	Autoionizing states
	2000	3×10^{11} at 2.5×10^7 W (unnormalized)	40	45	4 ps pulses
Eu	1855		0.3	46	
Mg	1390–1600	$10^9 \leq F \leq 10^{11}$	0.1	31, 47, 5	Tunable over 10,000 cm^{-1} includes Lyman$-\alpha$
	1200–1300	$10^7 \leq F \leq 10^8$	0.1		
Kr	1215.6	3×10^{11} at 10^6 W	12	48	Requires high power laser
NO	1520, 1430 1360, 1300	$10^6 \leq F \leq 10^7$	0.1	49	Continuous tunability (600 cm^{-1}) over each band, using *one* dye laser
Ar	570		10	50	

[a] For purposes of comparison reported powers have been converted to photons/pulse at fundamental dye laser intensities of 10 kW.

A. THEORY OF THIRD-ORDER NONLINEAR OPTICAL PHENOMENA

In nonlinear optics the theoretical framework falls naturally into two parts. First, one must calculate a nonlinear susceptibility (χ) which describes the microscopic (i.e., atomic contribution to the polarization induced by the propagation of the laser beam at frequency ω_i). For the general case, this polarization is written as a power series in the electric field $E(\omega_i)$ so that

$$P=\chi^{(1)}_{\omega_i}E(\omega_i)+\chi^{(2)}_{2\omega_i}E(\omega_i)E(\omega_i)+\chi^{(3)}_{3\omega_i}E(\omega_i)E(\omega_i)E^*(\omega_i) \tag{1}$$

In isotropic media, only the odd-order susceptibilities are nonzero because of inversion symmetry, hence the lowest order, nonlinear term is the third-order susceptibility. If (as in the gas phase) the number density N is low enough so that local field effects are small, then the macroscopic nonlinear polarization induced by the laser field is simply $N\chi^{(3)}$ and this polarization is now introduced in the wave equation as a nonlinear source term. Thus for the second part of the calculation, one may revert to classical electrodynamics to describe the evolution of the new third harmonic wave $\omega=3\omega_i$. The important physical feature of this latter stage is the phase matching between the fundamental ($\omega=\omega_i$) and harmonic wave ($\omega=3\omega_i$). The relevant parameters in expression for the intensity of the harmonic radiation are as follows:

$$I_{3\omega_i}\propto N^2|\chi^3|^2 I^3_{\omega_i} f(\Delta k) \tag{2}$$

where I_{ω_i} is the laser intensity at ω_i, Δk is the wavevector mismatch ($k=2\pi n/\lambda$) and f is a function determined by the actual geometrical conditions of the experiment (a comprehensive analysis of phasematching for various third order effects may be found in Ref. 51). For plane waves $f(\Delta k)$ is maximized at $\Delta k=0$ or $n_{\omega_i}=n_{3\omega_i}$; but for a focussed beam, where there is a phase slip between the fundamental and harmonic waves through the focus, f is maximized for $n_1<n_3$ and $f=0$ for $n_1=n_3$! Since all materials show significant dispersion in their transparent regions, it might appear to be difficult to achieve perfect phasematching. However, Harris[35] recognized that, in atomic systems, by choosing the harmonic radiation to lie in the region of anomalous dispersion on the high-energy side of a resonance line then $n_{3\omega_1}<n_{\omega_1}$, and hence by simply adding a second normally dispersive gas, one could compensate for the phase mismatch. This is schematically illustrated in Fig. 2. Thus even though the nonresonant susceptibility per unit volume is low in gases, because the nonlinear material may be made of arbitrary dimension and has perfect optical

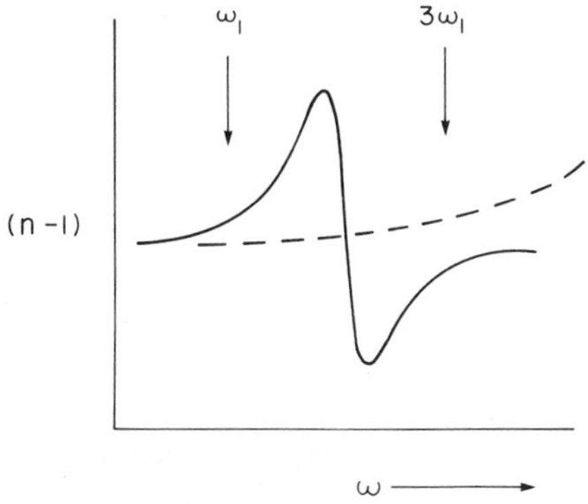

Fig. 2. Schematic representation of the compensation of refractive index for phase-matching in atomic vapors.

homogeneity, as well as a high threshold for laser-induced breakdown, significant fractions of the incident fundamental radiation can be converted to the third harmonic. For example, in rubidium vapor 3% conversion efficiencies have been reported.[52] Moreover, up to seventh-order effects have been observed in helium,[30] yielding radiation at 380 Å and approaching the soft X-ray region.

So far we have not considered the importance of the magnitude or the frequency dependence of the susceptibility $\chi^{(3)}_{3\omega}$. Although it was recognized early in the development of nonlinear optics[53, 54] that the third-order nonlinear susceptibilities were frequency dependent (in an analogous manner to the resonant Raman effect), the potential for resonant enhancement in harmonic generation lay untapped until third-order effects were explored in gases. The important feature, which may be extracted from the perturbation expansion[53, 55] for the susceptibility $\chi^{(3)}_{3\omega_i}$, is the existence of terms corresponding to a two-photon resonant enhancement process, which permits a large increase in $\chi^{(3)}_{3\omega i}$ without an associated linear absorption. Such effects were analyzed theoretically by Harris and Bloom[56] and first observed experimentally in strontium vapor by Hodgson, Sorokin, and Wynne.[40] Resonant enhancement effects can increase the susceptibility by as much as 10^3-10^4 in favorable cases, so that it is feasible to produce VUV radiation, starting with tunable dye lasers of moderate power, ~10 kW.

A detailed analysis of the atomic and molecular physics, which is contained in third-order susceptibilities for atoms and molecules, is beyond the scope of this article. However, it is possible to illustrate most of the important factors governing harmonic generation by selecting examples from several different experimental studies in atomic systems. We first approximate the susceptibility by a single term (there are a total of 24), which is selected because it alone contains all three resonant enhancement terms in the denominator,

$$\chi^{(3)}_{3\omega_i} \sim \sum_{a,b,c} \frac{\langle g|u|c\rangle\langle c|u|b\rangle\langle b|u|a\rangle\langle a|u|g\rangle}{(\Omega_{cg}-3\omega_i)(\Omega_{bg}-2\omega_i)(\Omega_{ag}-\omega_i)} \tag{3}$$

where Ω_{cg} is the energy of state c, ω_1 is the laser frequency, and the $\langle i|u|j\rangle$ are the dipole matrix elements between state i and j. Clearly, in any real atomic system only one term in the denominator can be made to approach zero (at which stage a linewidth is introduced phenomenologically by adding a term $-i\Gamma$). But, having chosen this term to be due to $(\Omega_{bg}-2\omega_i)$, then the summation may be contracted to that over a single index:

$$\chi^{(3)}_{3\omega_i} \sim \sum_{c} \frac{\langle g|u|c\rangle\langle c|u|b\rangle\langle b|u|a\rangle\langle a|u|g\rangle}{(\Omega_{cg}-3\omega_i)(\Omega_{bg}-2\omega_i-i\Gamma)(\Omega_{ag}-\omega_i)} \tag{4}$$

The two-photon transition strength to state b is approximated by a single sequence of matrix elements $\langle b|u|a\rangle\langle a|u|g\rangle$. It is this term, which is chosen for the principal resonant enhancement of $\chi^{(3)}$ in all four-wave sum-mixing schemes because, in contrast to its single-photon analog, the resonant Raman effect, there is no linear absorption associated with it. By itself, the two-photon resonant enhancement term provides only a fixed frequency source (corresponding to the Doppler width of the transition of state b) of harmonic radiation. However, by using two laser frequencies, ω_1 fixed so that $2\omega_1$ gives rise to two-photon resonant enhancement and ω_2 variable, one may produce tunable harmonic radiation at the sum-frequency $2\omega_1+\omega_2$.

This is illustrated for the case of Mg in Fig. 3, where $2\omega_1$ is set to be equal to the $3s4s$ state and ω_2 is chosen to span the np Rydberg series. In this case the susceptibility reduces to a single term.

$$\chi^{(3)}2\omega_1+\omega_2 \sim \frac{\langle 3s^2|u|np\rangle\langle np|u|3s4s\rangle\langle 3s4s|u|3s3p\rangle\langle 3s3p|u|3s^2\rangle}{(\Omega_{np}-(2\omega_1+\omega_2)-i\Gamma_{np})(\Omega_{3s4s}-2\omega_1-i\Gamma_{3s4s})(\Omega_{3s3p}-\omega_i)} \tag{5}$$

and we have both two- and three-photon resonant enhancement terms in $\chi^{(3)}_{2\omega_1+\omega_2}$. This is the situation that is normally encountered. The experimental results from Mg vapor for this particular choice of energy levels is shown in Fig. 4. The experimental variable is the intensity of the harmonic

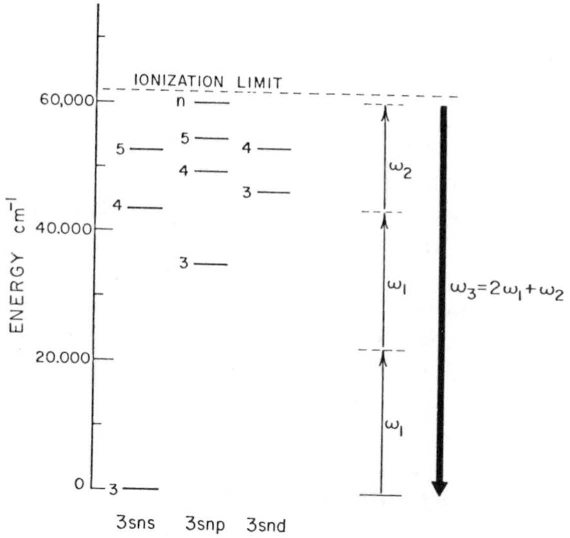

Fig. 3. Energy level diagram for magnesium showing resonant enhancement by frequencies $2\omega_1$ and ω_2.

radiation at the sum-frequency $\omega = 2\omega_1 + \omega_2$, which, as noted above, is proportional to $|\chi^{(3)}|^2$. This figure clearly illustrates the resonant enhancement, which is due to the *np* Rydberg series, but the actual lineshapes also show the importance of phasematching. These linewidths are not determined by the bandwidth of the dye lasers at ω_1 and ω_2, but by the dispersion around each *p* level, which changes the phasematching as one

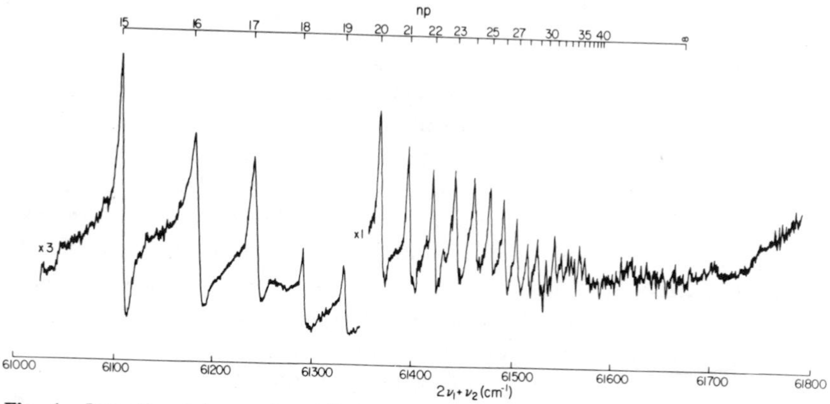

Fig. 4. Intensity of harmonics radiation in magnesium vapor, produced under double resonance conditions shown in Fig. 3.

tunes through each line. Moreover, as the principle quantum number increases, the oscillator strength decreases, and so the variations in the refractive index become smaller as is indeed observed.

B. Experimental Considerations

Before proceeding further in our discussion of different harmonic generation schemes, a brief diversion into some experimental details for producing such tunable VUV radiation is in order. The basic components required are (1) tunable dye lasers for frequencies ω_1 and ω_2, and (2) a metal vapor heat-pipe[57] oven to contain the nonlinear material. These are shown in Fig. 5, within the larger context of the block diagram for our laser-induced fluorescence studies[32] in carbon monoxide. The beams from the two dye lasers are spatially and temporally overlapped using a Glan–Thompson-type polarizer (the escape window is cut so that it is perpendicular to the direction of propagation of the orthogonal polarization) and then focused to a confocal beam parameter $b=1$ cm in the center of the Mg vapor zone (length $\leqslant 5$ cm) in the heat pipe. Both the quality of the optics and the spatial properties of the dye laser beams are important at this stage in determining the ultimate intensity in the VUV harmonic radiation. Generally, one can use a good quality archromatic lens for focusing, but a system of *two* spherical mirrors as shown in Fig. 5 provides the most flexible arrangement for shorter wavelengths. Some care is also necessary in setting up the two dye lasers so that they have the same divergence and hence focus to approximately the same spot size at the same displacement from the lens. The harmonic radiation emerging from the heat pipe is then simply recollimated in vacuum by a LiF lens (note that the solid angle subtended by this VUV radiation is less than that of the incident fundamental[51]) before entering the sample cell or molecular beam apparatus. This VUV radiation will, of course, consist of a mixture of the third harmonic of ω_1, that is, $\omega=3\omega_1$, and the sum frequency $2\omega_1+\omega_2$. Generally speaking, the presence of the fixed frequency harmonic is not a problem, since the probability of a chance overlap with other electronic transitions in most atoms or molecules is small. However, it also may be completely eliminated by using circularly polarized beams of opposite senses of polarization for ω_1 and ω_2.[40] The constraints of angular momentum conservation rigorously disallow the third harmonic of circularly polarized light in isotropic media, but, since ω_1 and ω_2 have opposite senses of polarization, the sum-frequency $2\omega_1+\omega_2$ is permitted, thus yielding a strictly monochromatic source of tunable VUV radiation. Experimentally this latter step is elegantly accomplished by introducing a quarterwave retarder after the Glan–Thompson prism, as shown in Fig. 5. Thus one can control the sense of circular polarization in the VUV by

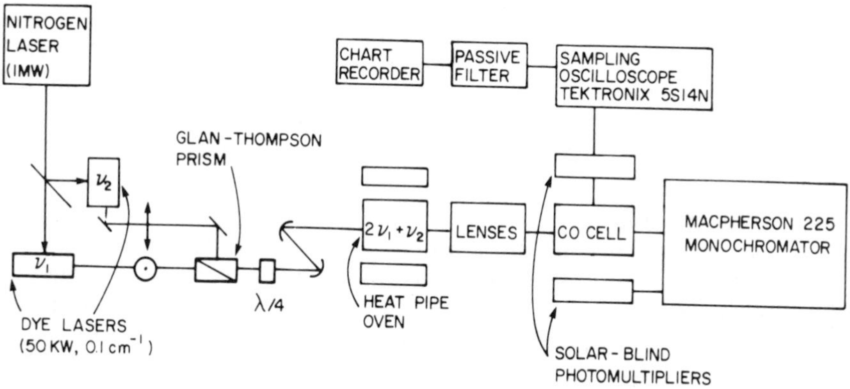

Fig. 5. Block diagram of apparatus for VUV laser spectroscopy. See text for discussion.

changing the polarization of the fundamental in the visible wavelength dye laser beams. Furthermore, ultrahigh-resolution circular dichroism studies in the VUV also would become possible by means of this technique, which opens up a new area for the study of molecular state symmetries.

Some brief comments on the practical details of heat-pipe operation are important. Detailed articles describing the theory of heat-pipe operation may be found in Ref. 57. First, one must combat the metallurgy of what turn out to be highly reactive materials, that is, molten metals at up to 1200 °K, both in finding inert materials to contain them and in simultaneously choosing a material for the wick, which the metal will wet. Second, because it is frequently necessary to add sufficient quantities of buffer gas to achieve good phasematching, the partial pressure of this buffer gas often exceeds that of the metal vapor and so the system no longer strictly functions as a heat pipe. In other words, the metal vapor boundaries are no longer sharply defined. Finally, realizing that our ultimate goal is to produce VUV photons and not heat pipes, some experimental compromise must be reached between a buffered gas cell and a heat pipe. Thus, even for an element with a very high vapor pressure, such as Mg, stable operation of the "heat-pipe containment system" can be maintained successfully for periods of several hundred hours.

C. Strontium and Autoionizing States

The distinguishing feature between various atomic systems, which are used as nonlinear media for harmonic generation, is whether or not there are discrete energy levels at the wavelength of the harmonic radiation. Generally speaking, the states in question are autoionizing levels converging on the higher ionization potentials. Thus we may compare strontium, in which several Rydberg series overlap the ionization continuum, and magnesium where the autoionizing states lie at much higher energy and do not apparently interact. In strontium the effect of the configuration interaction, between these bound states and the continuum in which they are embedded, is to produce a dramatic three-photon resonant enhancement in the susceptibility, so that the harmonic radiation mimics the linear absorption spectrum as shown in Fig. 6. A detailed analysis[58] of the frequency dependence of the third-order nonlinear susceptibility shows how the Fano q-parameters for the bound–continuum transition from the intermediate state responsible for the two-photon resonant enhancement may be obtained and hence illustrates the power of such studies for nonlinear spectroscopy. However, at the same time the consequence of such strong resonant enhancement is to significantly restrict the range of

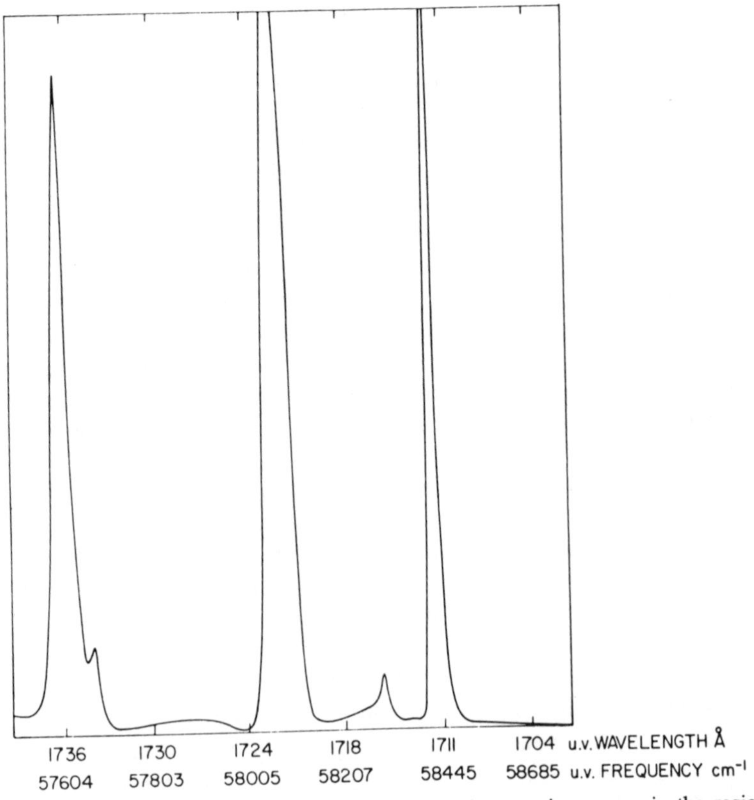

Fig. 6. Frequency dependence of harmonic radiation in strontium vapor in the region of autoionizing states.

tunability as also illustrated in Fig. 6. Moreover, it has recently been shown[43] that only under the conditions where the atomic metal vapor is optically thin, will the three-photon resonant enhancement be particularly beneficial, when one is attempting to scale the total harmonic intensity by increasing the pathlength of the nonlinear medium.

It turns out that the much more important role of configuration interaction is to increase the transition probabilities for selected two-photon transitions and so significantly improve the two-photon resonant enhancement term.[41] This effect was recently capitalized on by Freeman et al.,[34] who, by using the $5s^2(^1S_0)$–$5s7s(^1S_0)$ transition in strontium, produced continuous wave (cw) harmonic generation in the VUV for the first time. Although the actual intensity of VUV radiation was modest (10^6 photon/s) this achievement marks the beginning of ultrahigh-resolution ($\sim$1 MHz) laser spectroscopy in the VUV. Other atomic media of similar atomic structure to strontium, such as calcium[42, 46] and europium,[46] have also been explored and some initial results are listed in Table II. Again, the autoionizing states are observed to dominate the frequency dependence of the susceptibility. Comprehensive experimental studies have so far been restricted only to strontium. However, with the recent and extremely detailed studies on configuration interaction in calcium and barium,[59] it would be of considerable interest to attempt to locate the optimum two-photon resonant enhancement effects as in strontium.

D. Studies in Magnesium Vapor

In contrast to the heavier alkaline earths, magnesium shows a strong, unperturbed ionization continuum, with the odd parity autoionizing states shifted far enough away in energy that there is little interaction. In this case, the formal expression for the susceptibility must include an integral over the entire continuum as follows [taking the state responsible for two-photon resonant enhancement as the $3s3d(^1D_2)$];

$$\chi^3 \sim \frac{\langle 3s3d|u|3s3p\rangle\langle 3s3p|u|3s^2\rangle}{(\Omega_{3s3d}-2\omega_i-i\Gamma_{3s3d})(\Omega_{3s3p}-\omega_i)}\int_\epsilon \frac{\langle 3s^2|u|\acute{\epsilon}p\rangle\langle\acute{\epsilon}p|u|3s3d\rangle d\acute{\epsilon}}{(\acute{\epsilon}-3\omega_i-i\Gamma)} \tag{6}$$

Thus, interference effects between bound–continuum matrix elements will play an important role in determining the magnitude of the susceptibility. However, since these terms vary slowly with energy, the tuning range for harmonic generation will still be very large.

For magnesium, both the nonlinear susceptibility and tuning range for harmonic generation are significantly greater than reported for any other

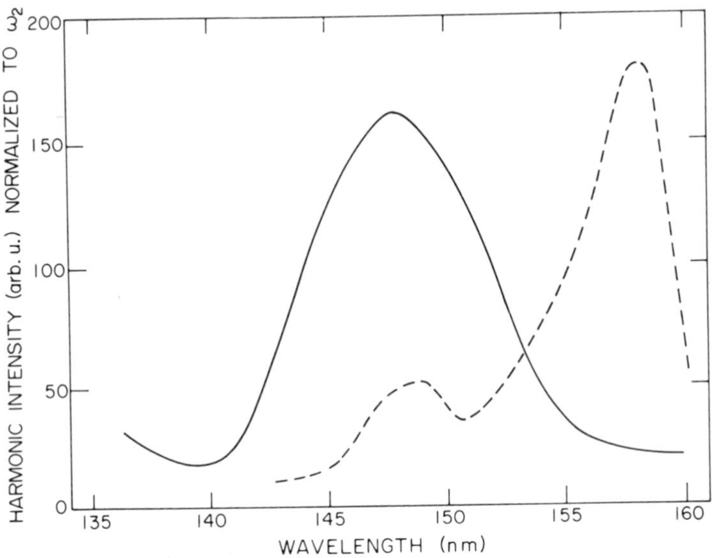

Fig. 7. Frequency dependence of harmonic radiation in magnesium vapor in the region of the ionization continuum.

equivalent set of experimental conditions. Thus, we are able to obtain peak power of up to 100 W/ pulse at 1460 Å (1.5×10^{12} photons per pulse at 30 kW dye laser incident powers) and continuous tunability, which spans the region from 1360 to 1600 Å. Figure 7 shows the wavelength regions that are covered by choosing the two lowest even parity states for two-photon resonant enhancement [$3s3d(^1D_2)$ and $3s4s(^1S_0)$]. It should be noted that these curves are obtained by the sequential superposition of the data from the tuning ranges for some 10 different dyes, so that the actual frequency dependence of the susceptibility may deviate by up to a factor of 2 at either extreme of the wavelengh region shown. More recently, we have been able to achieve[5] the landmark wavelength of Lyman-α radiation (1215.61 Å), by choosing higher energy, even parity states for two-photon resonant enhancement and tuning over the broad profile of the $3p4s(^1S_0)$ autoionizing level.

So far, the combination of a large susceptibility with a wide region of wavelength tunability over the ionization continuum seems to be unique to magnesium vapor. Two factors are important. First, there is constructive interference in the bound–continuum matrix elements, which reflect the continuum electronic structure. Second, configuration interaction with the even parity doubly excited np^2 series brings extra two-photon transition probability to the $3s3d(^1D_2)$ state and hence an extra degree of resonant enhancement analogous to the case of the $5s7s$ level in strontium.

E. Molecules vs Atoms

Although molecular systems clearly offer a rich field of study for resonantly enhanced (electronic) four-wave sum-frequency mixing, so far there has been only one example[49, 60] that has been reported in any detail. This is in nitric oxide, where we have used the resonant enhancement due to rovibronic energy levels of the lowest electronically excited state $A^2\Sigma^+$ to produce tunable VUV radiation at 1520, 1430, 1360, and 1300 Å. Of particular interest is the fact that it is possible to surpress the rotational structure of each vibronic band by pressure broadening and hence obtain continuously tunable VUV radiation by simple third harmonic generation. Typical data for the (0,0) band at both 50 torr and 10 atm pressure is given in Fig. 8. Hence with nitric oxide (NO) we have in the VUV the analog to the doubling crystals used in the ultraviolet and, moreover, the eventual possibility that NO could be used for cw third harmonic generation into the VUV. Perhaps the only drawback of such molecular nonlinear materials is that the magnitude of the nonlinear susceptibility is still considerably less than for the *best* atomic metal vapors. The principle features which account for the lower susceptibilities in molecules are the distribution of a given electronic oscillator strength over many vibrational and rotational energy levels (hence reducing the resonant enhancement) and the random orientation of the molecular axis with respect to the linearly polarized radiation at ω_1. In nitric oxide, the estimated (using known oscillator strengths and Franck–Condon factors) nonlinear susceptibility for the (0,0) band of $A^2\Sigma$ is $\leqslant 10^{-4}$ of that obtained in magnesium vapor in the 1400 Å region. The measured intensity of the harmonic signal is in agreement with this order of magnitude calculation. Nevertheless, molecules offer some interesting possibilities, first, because of the obvious advantage of avoiding heat-pipe ovens and having a fixed gas for the nonlinear material, and second, because with the higher powers from the latest generation of dye lasers we can compensate for the lower intrinsic susceptibility. Moreover, by using both two- and three-photon resonant enhancement effects in NO, we have observed a factors of up to 100 times increased susceptibility. We have also now observed[47] harmonic generation in benzene, IBr, and N_2O, and clearly more experimental work will continue to reveal a rich nonlinear spectroscopy in molecules as well as useful molecular systems for VUV harmonic generation. Perhaps molecules will eventually supplant atomic vapors as the desirable nonlinear material.

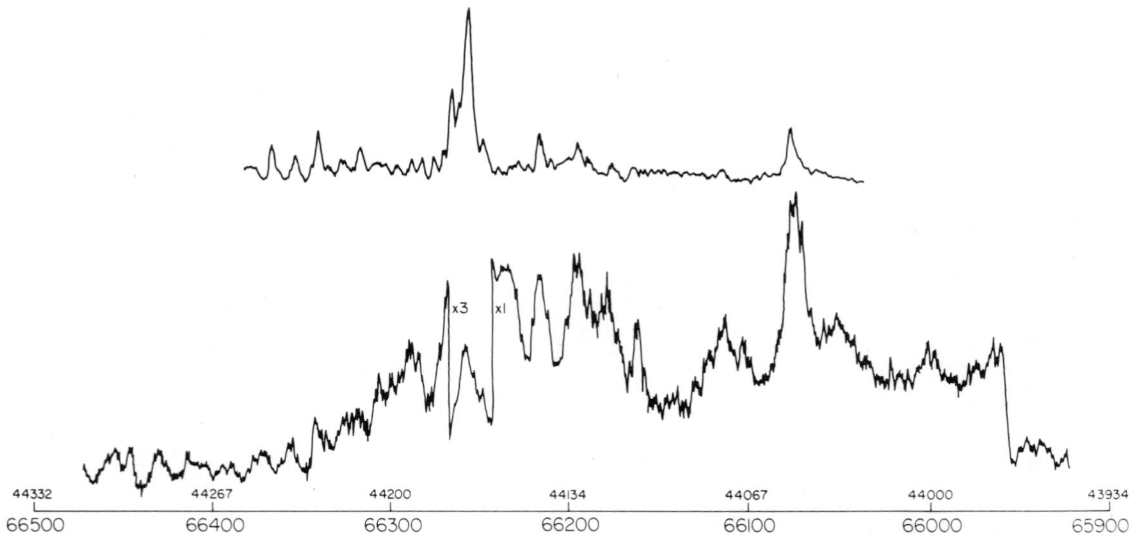

Fig. 8. Pressure dependence of harmonic radiation in NO; upper trace 50 torr pressure, lower trace 10 atm.

F. Higher Order Nonlinear Effects

The present record for producing the shortest wavelength laser radiation was achieved by generating[30] the seventh harmonic of 2660 Å (this is the fourth harmonic of Nd:Yag) in helium. The relevant energy level diagram is shown in Fig. 9, and there is no appreciable resonant enhancement of the susceptibility. Such higher order effects are only observed with extremely intense laser radiation at the fundamental wavelength, in this case 10^{15} W·cm^{-2}! Since these high-power pulses can only be produced from fixed-frequency lasers, the prospects for a laboratory scale, tunable source naturally fall far in the future. In order to avoid dielectric breakdown, such pulses are always of picosecond duration, which imposes a limit on the ultimate spectral resolution as well as providing a rather miniscule number of photons per pulse. Although, in general, the intensity of the various harmonics would be expected to fall off rapidly with increasing order, this

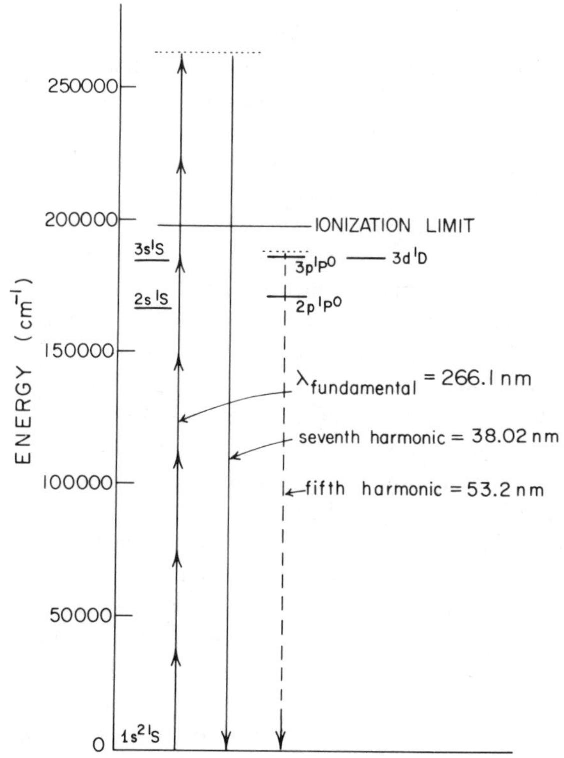

Fig. 9. Energy level diagram for He, showing fifth- and seventh-order harmonic generation processes.

may not always be the case. This is illustrated by comparing the conversion efficiency to the third with that to the fifth harmonics in helium.[39] With sufficient pump intensity at 2660 Å, the power of the fifth harmonic significantly exceeds that at the third. Even though $\chi^{(5)}$ is much less than $\chi^{(3)}$, at the extremely high laser intensities used in this work the fifth order term becomes dominant, and this interesting feature is in fact borne out by theory.[39]

G. Saturation Phenomena

There are fundamental limitations in the scaling of both resonantly enhanced and nonresonant harmonic generation to high powers, and such effects have generally been observed in all attempts to obtain high harmonic conversion efficiencies in gases. These saturation phenomena arise through a variety of different nonlinear effects, including multiphoton ionization, field-dependent susceptibilities resulting from dynamical Stark shifts and depletion of ground-state species, and field-induced changes in phase matching arising from the nonlinear refractive index. For resonantly enhanced harmonic generation, multiphoton ionization seems to be the principal limiting factor. That is, along with a resonantly enhanced third-order nonlinear susceptibility comes resonantly enhanced three-photon ionization. These multiphoton ionization rates can become large enough that the subsequent number density of free electrons gives rise to significant Stark broadening and so reduces the resonant enhancement in $\chi^{(3)}$. Therefore three-photon resonant enhancement terms resulting from strong autoionizing states present an additional disadvantage to power scaling. In contrast, for the case of magnesium, where there is no odd parity autoionizing state overlapping the continuum, the form of the multiphoton ionization cross-section differs from $\chi^{(3)}$ by the cross-term in the integral over bound–continuum matrix elements. Hence the susceptibility can be large, while the multiphoton ionization effects are small. In magnesium vapor the onset of saturation effects is observed only for harmonic intensities $\geqslant 100$ W (corresponding to dye laser incident powers $\sim$40 kW) so that even if the power density is kept constant, with current megawatt dye lasers this would scale the VUV to $\geqslant 10^4$ W. Note that this is the same power level as would be expected of visible dye lasers less than seven years ago! Nevertheless, it should be emphasized that scaling to high powers in resonantly enhanced harmonic generation is not the general case, and the three-photon ionization (and perhaps other related effects) may be expected to limit output VUV powers to the 1–10 W range.

Even though saturation effects in nonresonant harmonic generation begin to set in at high powers and power densities, it is interesting to observe that the harmonic intensity at the onset of these limiting conditions is approximately the same as for the best results obtained using resonant enhancement. Thus in studies on nonresonant frequency trebling in krypton gas to produce radiation around Lyman-α, the intensity of the harmonic radiation saturates at a VUV power of <100 W, which is quite comparable to the resonantly enhanced case discussed above. The one important difference is that for the nonresonant case, the dye laser powers required are greater by two or three orders of magnitude.

It is quite difficult to identify any one specific effect as the principal source of saturation phenomena for nonresonant harmonic generation. More likely it is a combination of a number of the competing nonlinear processes described above that determine the limiting intensities for these harmonics. So far, the best result for nonresonant tripling into the VUV has been obtained in Xe where with high-power picosecond pulses the conversion efficiency is 0.3%.[36]

III. APPLICATION TO CHEMICAL PHYSICS

Perhaps the ultimate test of the success of any harmonic generation scheme is in its final application to problems in chemistry and physics. So far, in Toronto, we have been fortunate to have been in the forefront of exploring and exploiting the unique spectral properties of tunable laser sources in the VUV, particularly to problems in chemical physics. Two recent examples serve to illustrate the potential of these techniques. The first is the direct measurement of the radiative lifetimes of individual rotational energy levels in a given vibrational level of the first excited state of CO ($A^1\pi\ v=0$), in the region of strong interstate perturbation by neighboring triplet states ($d^3\Delta$ and $e^3\Sigma^-$). The second is the use of VUV laser-induced fluorescence as a probe of molecular beam reactive scattering experiments, in particular the reaction of $H+Br_2 \rightarrow HBr+Br$.

The carbon monoxide experiment is shown schematically in the block diagram given in Fig. 5. The intensity (and the net absorption) of the exciting light was monitored through the VUV monochromator by one photomultiplier while the fluorescence was viewed perpendicular to the exciting radiation, through a quartz filter. The measured fluorescence intensity was consistent with that estimated from the incident VUV flux, and even though the collection optics and spectral response of the particular VUV photomultiplier available were far from optimum the signal-to-noise ratio was good enough that fluorescence decays were recorded in a single (internal) scan of the sampling oscilloscope time-window. Excitation spectra were obtained by simple charge integration using an electrometer, and a typical example is shown in Fig. 10 for pressures of 0.4 and 4 torr. As the carbon monoxide begins to become optically thick at the higher

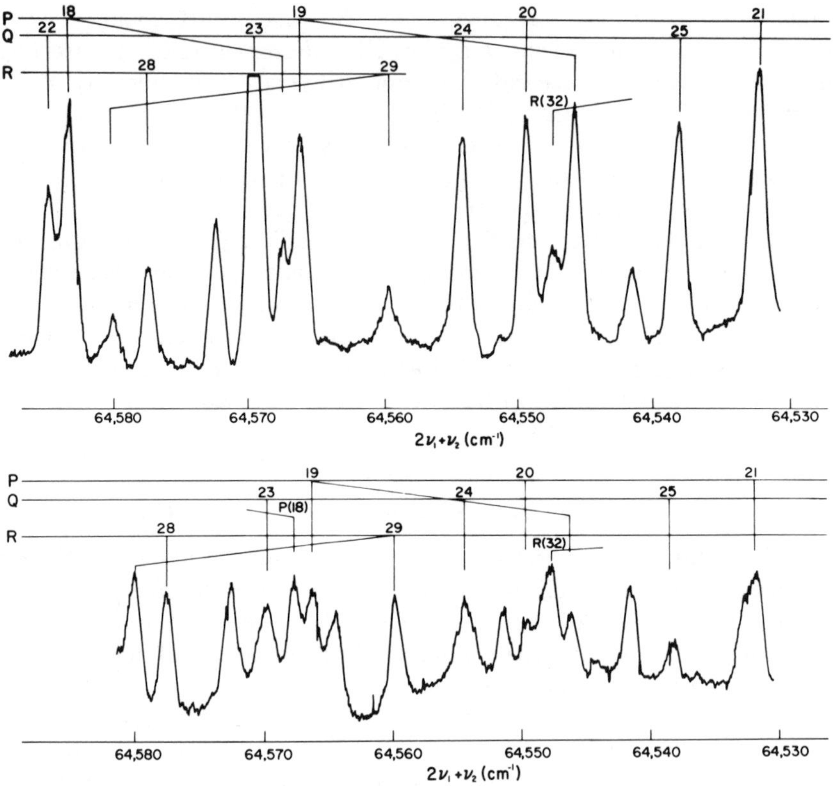

Fig. 10. Excitation spectra in CO at 0.4 torr (upper trace) and 4 torr (lower trace).

pressures ($\geqslant 4$ torr) very weakly allowed singlet–triplet transitions begin to be observed as noted in Fig. 10. Two contrasting examples of fluorescence decays are given in Fig. 11. The first trace is for an unperturbed level of the $A^1\Pi$ state $Q(24)$, while the second is for a weak, triplet-like level (interacting with the $d^3\Delta$ state) $R(29)$. By measuring the radiative lifetimes of the "singlet-like" levels of the $A^1\Pi$ state as a function of rotational quantum number we are able to quantitatively monitor the strength of the interaction between singlet and triplet states in isolated CO molecules

The unusual time dependence shown in the second trace of Fig. 11 is due to collision-induced triplet–singlet energy transfer from the triplet-like levels of the $d^3\Delta$ to the $A^1\Pi$ state. The sequence of events is indicated in Fig. 12. Although the initial absorption is very weak because of the triplet-like character ($\geqslant 95\%$) of the upper state, these levels can be probed with a tunable laser source. Following excitation, collisions with other CO

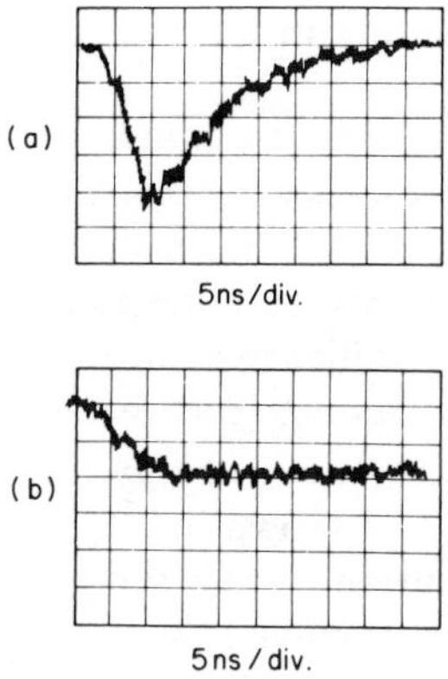

Fig. 11. Fluorescence decay curves in CO; see text for discussion.

molecules result in population transfer to nearby levels in $A^1\Pi$. Since these states have close to unperturbed lifetimes similar to trace (a) in Fig. 11, the actual time evolution of the fluorescence in trace (b) is a direct measure of the triplet–singlet energy transfer

These studies are the first and unique example of state-selective photophysical studies with tunable VUV lasers and will make an important

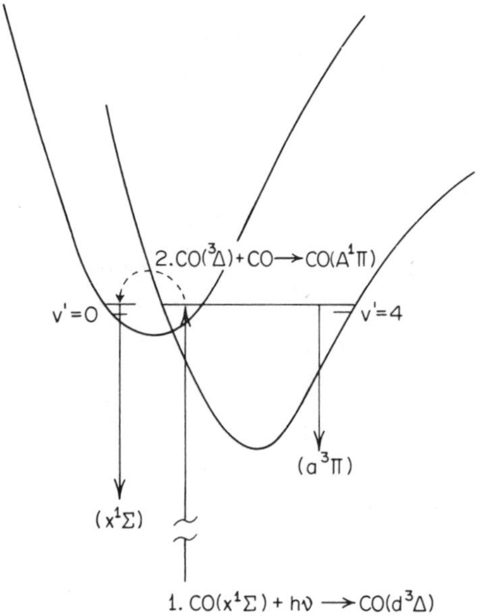

Fig. 12. Potential energy curves for CO showing the energetics of collision-induced, singlet–triplet energy transfer.

contribution to our understanding of molecular dynamics in the limit of a sparse density of states.

In chemical dynamics we would like to be able to detect a number of simple atoms and molecules by laser-induced fluorescence in the VUV under molecular beam conditions. With these tunable VUV laser sources we are able for the first time to make such measurements for a large class of light atoms ranging from H to the halogens. Our initial experiments[29] have been directed to measuring the total cross-sections for reactive scattering of $H + Br_2$ as a function of hydrogen atom kinetic energy. The crossed molecular beam apparatus used for this work is shown in Fig. 13. The source of atomic hydrogen is a supersonic expansion from a tungsten oven, heated to 2800°C, with variable kinetic energy achieved by inverse seeding techniques. The other reactant is introduced as an effusive beam from a collimated hole structure. The crossing region is illuminated by the VUV laser beam and fluorescence is detected perpendicular to the three intersecting beams. At present, the estimated lower limit of detection for ground-state Br atoms is $\leqslant 10^6\ cm^{-3}$, but with improved dye laser intensities this figure will likely be reduced by several orders of magnitude. Results for the reaction of D, $H + Br_2$ are given in Fig. 14. These data are

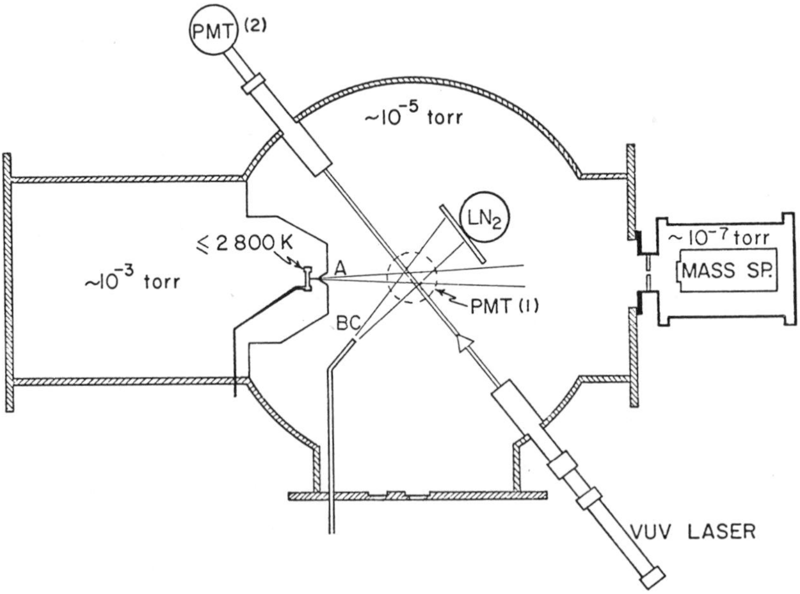

Fig. 13. Schematic diagram of molecular beam apparatus used to study the total cross-section for the process $H + Br_2 \rightarrow HB + H$.

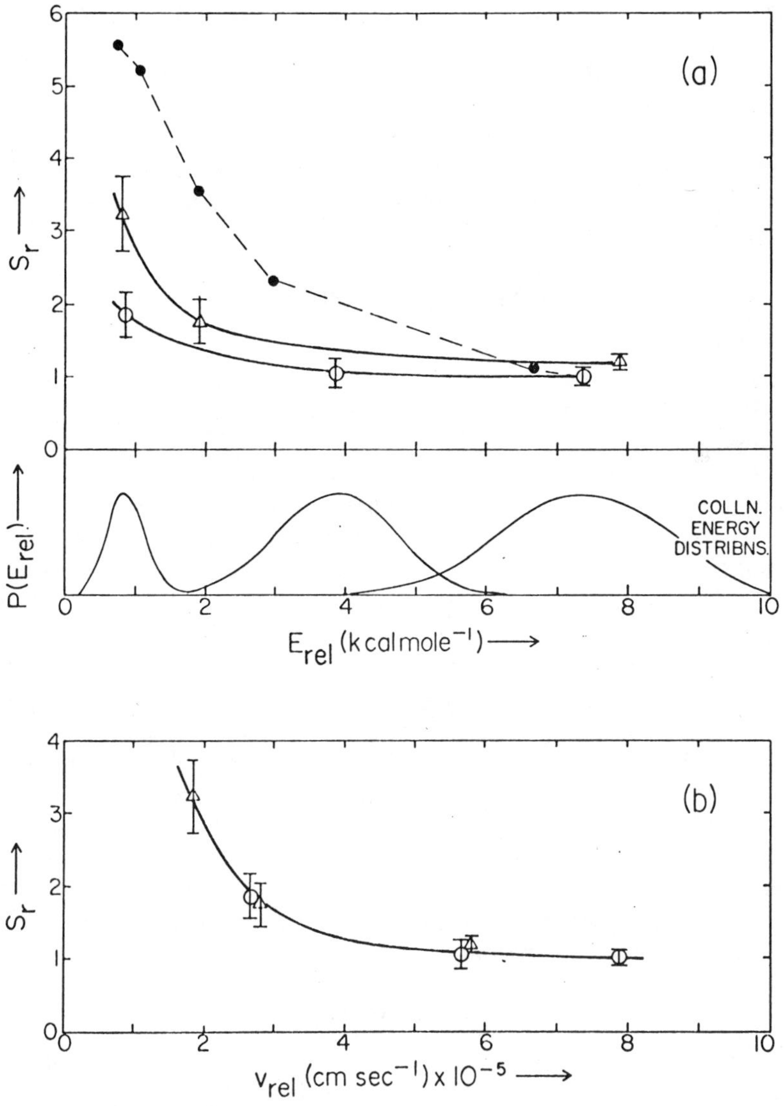

Fig. 14. Total cross-sections for the reaction of $H, D + Br_2 \rightarrow HBr(DBr) + Br$ plotted as a function relative kinetic energy and relative velocity.

consistent with a very small barrier to reaction, namely, $\leqslant 1$ kcal/mole. Moreover, cross-sections for D and H become superimposable when plotted against relative velocity, as would be expected in a simple classical picture[61] for the reaction of light and heavy heavy in the absence of any barrier to reaction. Finally, it should be noted that we are not restricted to detecting only ground-state species, for by simply changing the dye laser wavelength it has also been possible to detect excited state bromine atoms produced by photodissociation.

IV. CONCLUSION

In this chapter we have attempted to describe the enormous strides, which have taken place over the last ten years, in the development of lasers and laser spectroscopy in the VUV. Although a great deal has been accomplished in extending the wavelength span of sources of coherent radiation to the VUV and even to the soft X-ray region, this area of research still has great scope for further studies. We anticipate that significant progress will be made in extending the range of wavelength tunability, in scaling nonlinear optical effects to higher intensities, and in using optical coherent transient phenomena to minimize saturation effects. There is even greater potential for laser spectroscopy in the VUV. The two examples presented here represent the only studies to date and hence sample a tiny fraction of the range of effects that could be investigated with tunable laser sources produced by harmonic generation into the VUV. With current advances in dye laser technology we can expect a further increase in intensities so that even nonlinear optical phenomena will be accessible to study. Furthermore, recent work in synchronous pumping of picosecond dye lasers now permits tunable picosecond VUV pulses to be produced as well. The future for laser spectroscopy in the VUV looks good from all perspectives, both in the time and frequency domains, as we continue to exploit the rich field of nonlinear optical effects.

V. APPENDIX

Considerable improvements in both the intensity and tunability of VUV laser sources have been recently possible as the new generation of tunable dye laser systems and excimer lasers, which provide significantly higher intensity input radiation, are used for nonlinear mixing. However, it is important to note at the outset that only a selected group of nonlinear media simultaneously satisfy the requirements in electronic properties necessary for scaling the VUV output ot larger values already obtained. In fact, recent work[62] on nonresonant third harmonic generation in krypton clearly shows how limiting effects such as Kerr–induced index mismatch and dielectric breakdown reduce harmonic conversion efficiences at high powers. Thus, in this case the intensity of Lyman–α radiation which can be produced by third harmonic generation is less than 50 W for any laboratory scale laser system (in a slightly different experimental arrangement intensities of ~5 W for the same process have also been reported[63]). Finally, Lyman–β radiation has also been produced[64] (10^{-3} W, 10^{-8} efficiency), again by a non-resonant process in Argon, but in this latter case using tunable radiation from a XeCl laser around 3080 Å.

The highest photon flux of tunable coherent VUV radiation is currently avaliable from resonantly enhanced four-wave sum-mixing in Mg vapor. As noted earlier (see article), simply increasing the dye laser power and the spot size would permit at least linear scaling. This has been verified[65] by using high–power Nd:Yag pumped dye lasers[66] (~3 MW), where the resulting VUV intensity is increased by a factor of $\geqslant 100$ times over that reported for nitrogen pumped dye lasers (~30 kW). This result implies peak intensities on the order of the dye laser intensities used in the initial studies!

Resonantly enhanced third harmonic generation at 830 Å has been observed in atomic Xenon, using a unique, very high–brightness ultraviolet laser source (injection locking of a KrF laser) at 2490 Å for the fundamental radiation. The efficiencies reported $\leqslant 10^{-7}$ are suprisingly small, but can be expected[68] to be significantly improved at longer wavelengths below the ionization limit of Xenon.

A new source of tunable radiation in the far-ultraviolet and near–VUV, which has recently been reported is sequential antistokes raman generation in molecular hydrogen.[69] This is also a four–wave mixing process acting through $\chi^{(3)}$, the third–order nonlinear susceptibility, but of course consists of a cascade of CARS type processes. By using the second harmonic of a tunable dye laser as the input radiation, wavelengths as short as 1800 Å are obtained so that this source covers an important new spectral region. Since it now possible[70] to servo–lock second harmonic crystals so that the phase–matching angle follows the tuning of the dye laser, this becomes a useful, continuously tunable source.

Applications of these tunable VUV sources continue to be mostly in the detection of atoms and small molecules by laser–induced fluorescence in molecular beam scattering studies. Of particular importance has been the improved intensities available from Mg vapor, so that it will be possible, for example, to study the internal energy distributions in CO molecules following scattering from surfaces. This capability for both very sensitive and state–selective detection of small molecules will lead to important advances in our understanding of molecular interactions at surfaces.

References

1. R. T. Hodgson, *Phys. Rev. Lett.*, **25**, 494 (1970).
2. R. T. Hodgson and R. W. Dreyfus, *Phys. Rev. A*, **9**, 2635 (1974).
3. P. W. Hoff, J. C. Swingle, and C. K. Rhodes, *Appl. Phys. Lett.*, **23**, 245 (1973).
4. S. C. Wallace and R. W. Dreyfus, *Appl. Phys. Lett.*, **25**, 498 (1974).
5. T. J. McKee, B. P. Stoicheff, and S. C. Wallace, *Opt. Lett.*, **3**, 207 (1979).
6. Acton Research Incorporated.
7. S. C. Wallace and G. A. Kenney-Wallace, *Int. J. Radiation Phys. Chem.*, **7**, 345 (1975).
8. C. K. Rhodes, *IEEE JQE*, **10**, 153 (1974).
9. R. T. Hodgson, *J. Chem. Phys.*, **55**, 5378 (1971).
10. G. A. Kenney-Wallace, unpublished data.
11. R. Barnham and N. Djeu, *Appl. Phys. Lett.*, **29**, 707 (1976).
12. V. N. Ischenko, V. N. Lisitsyn, and A. M. Razhev, *Appl. Phys.*, **12**, 55 (1977).
13. T. R. Loree, K. B. Butterfield, and D. L. Barker, *Appl. Phys. Lett.*, **32**, 171 (1978).
14. J. M. Hoffman, A. K. Hays, and G. C. Tisone, *Appl. Phys. Lett.*, **28**, 538 (1976).
15. S. C. Wallace, unpublished data.
16. D. J. Bradley, M. H. R. Hutchinson, and C. C. Ling, "Tunable Lasers and Applications," Proceedings of the Loen Conference, Norway 1976, A Mooradian, T. Jaeger, and P. Stokseth, eds., Springer-Verlag, 1976, p. 40.
17. W. M. Hughes, J. Shannon, and R. Hunter, *Appl. Phys. Lett.*, **24**, 488 (1974).
18. R. W. Waynant, J. D. Shipman, R. C. Elton, and A. W. Ali, *Appl. Phys. Lett.*, **17**, 383 (1970).
19. R. W. Waynant, *Phys. Rev. Lett.*, **28**, 533 (1972).
20. V. S. Antonov, I. N. Knyazev, V. S. Letokhov, and V. G. Movshev, *Zh. ETF Pis. Red.*, **17**, 545 (1973).
21. R. W. Waynant, J. Shipman, R. C. Elton, and A. W. Ali, *Proc. IEEE*, **59**, 679 (1971).
22. S. A. Borgstrom, *Opt. Comm.*, **11**, 105 (1974).
23. R. T. Hodgson and R. W. Dreyfus, *Phys. Lett.*, **38A**, 213 (1972).
24. R. W. Dreyfus, *Appl. Phys. Lett.*, **29**, 348 (1976).
25. H. Pummer, K. Hohla, M. Diegelmann, and J. P. Reilly, *Opt. Comm.*, **28**, 104 (1979).
26. R. W. Waynant, *Appl. Phys. Lett.*, **22**, 419 (1973).
27. J. B. Marling and D. B. Lang, *Appl. Phys. Lett.*, **31**, 181 (1977).
28. W. J. Jackson, J. B. Halpern, and C-S. Lin, *Chem. Phys. Lett.*, **55**, 254 (1978).
29. J. W. Hepburn, K. Klimek, K. Liu, J. C. Polanyi, and S. C. Wallace, *J. Chem. Phys.*, **69**, 4311 (1978).
30. J. Reintjes, C. Y. She, R. C. Eckardt, N. W. Karangelen, R. A. Andrews, and R. C. Elton, *Appl. Phys. Lett.*, **30**, 480 (1977).
31. S. C. Wallace and G. Zdasiuk, *Appl. Phys. Lett.*, **28**, 449 (1976).
32. A. C. Provorov, B. P. Stoicheff, and S. C. Wallace, *J. Chem. Phys.*, **67**, 5393 (1977).
33. R. R. Freeman, G. C. Bjorkland, N. P. Economou, P. F. Liao, and J. E. Bjorkholm, *Appl. Phys. Lett.*, **33**, 739 (1978).
35. A. H. Kung, J. F. Young, G. C. Bjorklund, and S. E. Harris, *Phys. Rev. Lett.*, **29**, 985 (1972).
36. A. H. Kung, J. F. Young, and S. E. Harris, *Appl. Phys. Lett.*, **22**, 301 (1973).
37. V. L. Slabka, A. K. Popov, and V. F. Lukinyh, *Appl. Phys.*, **15**, 239 (1977).
38. S. E. Harris, J. F. Young, A. H. Kung, D. M. Bloom, and G. C. Bjorklund, *Laser Spectroscopy*, Plenum, New York, 1973, p. 59.
39. J. Reintjes and C. Y. She, *Opt. Comm.*, **27**, 469 (1978).
40. R. T. Hodgson, P. P. Sorokin, and J. J. Wynnes, *Phys. Rev. Lett.*, **32**, 343 (1974).

41. G. C. Bjorklund, J. E. Bjorkholm, R. R. Freeman, and P. F. Liao, *Appl. Phys. Lett.*, **31**, 330 (1977).
42. S. C. Wallace and G. Zdasiuk, unpublished data.
43. H. Scheingraber, H. Puell, and C. R. Vidal, *Phys. Rev. A*, **18**, 2585 (1978).b
44. R. R. Freeman, G. C. Bjorklund, N. P. Economu, P. F. Liao, and J. E. Bjorkholm, *Appl. Phys. Lett.*, **33**, 739 (1978).
44. N. P. Economou, R. R. Freeman, J. P. Heritage, and P. F. Liao, to be published.
45. A. I. Ferguson and E. G. Arthurs, *Phys. Lett.*, **58A**, 298 (1976).
46. P. P. Sorokin, J. J. Wynne, J. A. Armstrong, and R. T. Hodgson, *N. Y. Acad. Sci.*, **267**, 30 (1976).
47. S. C. Wallace, unpublished data.
48. R. Mahon, T. J. McIlrath, and D. W. Koopman, *Appl. Phys. Lett.*, **33**, 305 (1978).
49. K. K. Innes, B. P. Stoicheff, and S. C. Wallace, *Appl. Phys. Lett.*, **29**, 715 (1976).
50. M. H. R. Hutchinson, C. C. Ling, and B. J. Bradley, *Opt. Comm.*, **18**, 203 (1976).
51. G. J. Bjorkland, *IEEE JEQ*, **11**, 287 (1975).
52. H. Puell, K. Spanner, W. Falkenstein, W. Kaiser, and C. R. Vidal, *Phys. Rev. A*, **14**, 2240 (1976).
53. J. A. Armstrong, N. Bloembergen, J. Ducing, and P. S. Pershan, *Phys. Rev.*, **127**, 1918 (1962).
54. P. H. Maker and R. W. Terhune, *Phys. Rev. A*, **137**, 801 (1965).
55. B. J. Orr and J. Ward, *Mol. Phys.*, **20**, 513 (1971).
56. S. E. Harris and D. M. Bloom, *Appl. Phys. Lett.*, **27**, 229 (1974).
57. C. R. Vidal and F. B. Haller, *Rev. Sci. Instrum.*, **42**, 1779 (1971).
58. L. Armstrong and B. L. Bears, *Phys. Rev. Lett.*, **34**, 1290 (1975).
59. J. A. Armstrong, J. J. Wynne, and P. Esherick, *J. Opt. Soc. Am.*, **69**, 211 (1979).
60. S. C. Wallace and K. K. Innes, *J. Chem. Phys.*, **72**, 4805 (1980).
61. C. A. Parr, J. C. Polanyi, and W. H. Wong, *J. Chem. Phys.*, **58**, 5 (1973).
62. R. Mahan and Y. M. Liu, *Opt. Lett.*, **5**, 279 (1980).
63. D. Cotter, *Opt. Comm.*, **31**, 397 (1979).
64. J. Reintjes, *Opt. Lett.*, **5**, 342 (1980).
65. J. C. Polanyi, communicated.
66. Quanta-Ray Inc.
67. S. C. Wallace and G. Zdasiuk, *Appl. Phys. Lett.*, **28**, 449 (1976).
68. S. C. Wallace, unpublished.
69. V. Wilke and W. Schmidt, *Appl. Phys.*, 177 (1979).
70. G. J. Bjorklund and R. A. Storz, *IEEE JQE-15*, 228 (1979).

Section 4

STUDIES OF COLLISION EFFECTS

VIBRATIONAL ENERGY FLOW IN THE GROUND ELECTRONIC STATES OF POLYATOMIC MOLECULES

ERIC WEITZ*

Department of Chemistry, Northwestern University, Evanston, Illinois 60201

GEORGE FLYNN

Department of Chemistry and Columbia Radiation Laboratory, Columbia University, New York, New York 10027

CONTENTS

*Alfred P. Sloan Fellow

I. INTRODUCTION AND SCOPE

The transfer of energy among vibrational, rotational, and translational degrees of freedom has been a topic of both fundamental and practical importance for many years. An understanding of the processes that control the rates, pathways, and cross-sections for energy exchange between different degrees of freedom and the flow of energy between various modes of polyatomic molecules is crucial to the development of improved theories of chemical reactivity. In addition experimental energy-transfer data can be used to test the quality of intermolecular potentials as well as the accuracy of approximate scattering theories. A knowledge of the efficiency of energy-transfer processes is also of fundamental importance in developing new, more versatile, and more powerful laser systems. In fact, it is the detailed changes in the populations of various quantum states, controlled largely by relaxation processes, that determine the overall population inversion and efficiency of most laser systems.

Studies of vibrational relaxation using lasers have expanded dramatically in the past fourteen years due largely to the increased availability of versatile, high-power, pulsed-laser devices. Increased demand for energy-transfer parameters has resulted from attempts to develop new lasers and to push existing systems to their practical limits. Thus the development of lasers and the study of energy-transfer phenomena have been mutually synergistic processes.

Several excellent reviews in related and parallel areas have appeared recently.[1–14] Though much early work in the field concentrated on vibration–translation/rotation relaxation phenomena using ultrasonic methods, a remarkable resurgence of interest in the area occurred with the advent of infrared lasers and their application to laser-induced infrared fluorescence.[15,16] Laser methods have been particularly successful in providing data concerning intermode vibration–vibration energy-transfer processes, which have themselves taken on increasing importance as interest in infrared laser-driven chemical reactions has grown. In the last few years a greater depth of understanding of the factors that govern vibrational energy-transfer rates and pathways in polyatomic molecules has resulted both from improved experimental methods and equipment as well as from a more detailed look at the microscopic processes that govern overall vibrational energy-transfer pathways and how information about these processes might be obtained.

The present review has been primarily limited to an in-depth survey of energy-transfer processes in five selected molecules in an attempt to illustrate a number of important unifying features regarding vibrational energy-transfer phenomena in polyatomics. In addition we have reviewed progress in the development of diagnostic techniques for unraveling energy-transfer mechanisms, an area of crucial importance as experimental studies of more and more complex molecular systems become possible. Finally, a set of qualitative "collisional propensity rules" are presented, which appear to summarize in a general way the features that control the efficiency of intermode energy transfer in the ground electronic state of most of the small molecules studied so far. Many, many excellent paper have appeared in the field, which are not covered in this present focused review, but which can be found in a number of the most recent summaries of the subject.[7,10–14]

II. ENERGY TRANSFER IN METHANE, DEUTEROMETHANES, AND METHYL FLUORIDE

A. CH_4

Methane was the first polyatomic molecule other than CO_2 for which vibrational energy-transfer pathways and rates were investigated.[16] This study of methane involved the utilization of the phase-shift method for determining the lifetimes of vibrationally excited states. The asymmetric stretching vibration, ν_3, of methane at 3010 cm^{-1} (see Fig. 1) was excited by a chopped He–Ne laser operating on the 2947.9 cm^{-1} Ne transition. Fluorescence was detected from both the ν_3 mode and the ν_4 bending mode of methane at 1306 cm^{-1}. Rates were extracted from phase-shift measurements, and Ref. 16 provides an excellent discussion of the background for, and use of the phase-shift method.

As a result of these measurements the authors conclude that the vibration–translation/rotation (V–T/R) relaxation time in CH_4 is 0.69 ms^{-1}

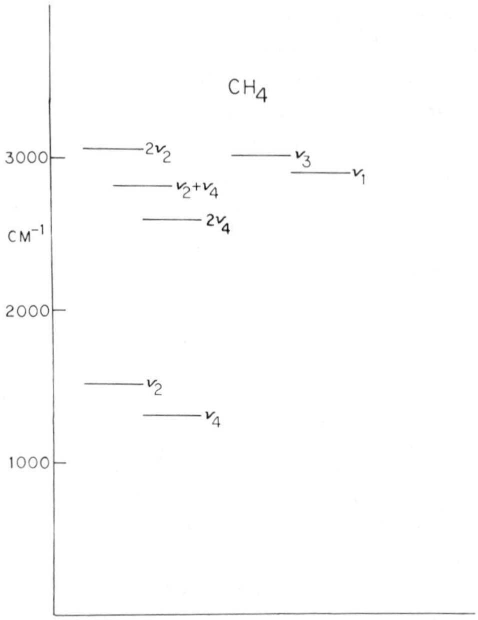

Fig. 1. Partial vibrational energy level diagram for CH_4.

$torr^{-1}$, which is in good agreement with the results of ultrasonic studies of CH_4.[17] The deactivation rate due to vibration–vibration (V–V) energy-transfer processes involving the initially excited ν_3 mode was measures as 188 ± 27 ms^{-1} $torr^{-1}$, while the rise rate of the ν_4 bending mode was measured to be between 440 and 165 ms^{-1} $torr^{-1}$. (For a typical small molecule, a rate of 10^4 ms^{-1} $torr^{-1}$ corresponds to a gas kinetic process.) While this measurement had large error bars due to a number of experimental difficulties, the authors indicated that they believed the actual value for the above process was most likely between 330 ms^{-1} $torr^{-1}$ and 220 ms^{-1} $torr^{-1}$, which would, surprisingly, make the rise of this state faster than the fall of ν_3.

An equilibration pathway that involved the sequential flow of energy from

$$CH_4(\nu_3) \rightarrow CH_4(\nu_4) \rightarrow CH_4(\nu_2) \tag{1}$$

where the transfer between ν_3 and ν_4 may actually proceed through an overtone or combination band of ν_4, could provide an explanation for the fact that ν_4 rises faster than ν_3 falls. Since in the above picture of the system ν_4 would be an intermediate state, it would in general possess a fluorescence signal that would be multiexponential in character and could have significant contributions from both the $\nu_3 \rightarrow \nu_4$ and the $\nu_4 \rightarrow \nu_2$ equilibrations, provided that the above two processes occurred on a comparable time scale.[18] While the ν_3 state will also be multiexponential in character, in many cases an intermediate state such as ν_4 can exhibit a faster rise, due to further equilibration with ν_2, than the fall of the initially pumped state, ν_3.[19] A result that bears on the equilibration of ν_2 and ν_4 is reported in Ref. 20 where, as a result of ultrasonic studies, ν_2 and ν_4 were found to be in equilibrium on the V–T/R time scale.

Since the original study of CH_4 a number of other studies have been performed by the phase-shift technique using a chopped He-Ne laser, and by pumping of the ν_3 mode of CH_4 using a tunable infrared pulsed laser.[21–23] While these additional studies have resulted in the measurement of a number of other rates in the methane system, including the rare-gas dependence of the V–T/R energy-transfer process, the rare-gas dependence of the fall of ν_3 and the rise of ν_4, and the rate of vibrational energy transfer to oxygen from methane, they have basically confirmed the overall results of Ref. 16 with regard to V–V, V–T/R rates and vibrational energy-transfer pathways. The fall rate of ν_3 and rise of ν_4 have been refined and are now reported as 260 ± 40 ms^{-1} torr^{-1} and 370 ± 80 ms^{-1} torr^{-1}, respectively. While these rates are within the reported error bars of each other, the possibility that the ν_4 rise is faster than the ν_3 fall has again been discussed.[21] There is also evidence to indicate that the ν_4 rise signal is not a single exponential. The above results, coupled with the direct observation via cold gas filter studies of the rise and fall of $2\nu_4$ and $\nu_2+\nu_4$, suggest a pathway for vibrational energy equilibration involving

$$CH_4(\nu_3)+CH_4 \rightleftarrows CH_4(2\nu_4)+CH_4 \tag{2}$$

followed by

$$CH_4(2\nu_4)+CH_4(0) \rightleftarrows 2CH_4(\nu_4) \tag{3}$$

and with

$$CH_4(\nu_4)+CH_4 \rightleftarrows CH_4(\nu_2)+CH_4-227\text{cm}^{-1} \tag{4}$$

occurring on roughly the same time scale as that for the combined processes of (2) and (3). The role of IR inactive ν_1 in the overall equilibration of the vibrational manifold is still not well understood; however, either the $\nu_1 \leftrightarrow \nu_3$ equilibration process is very rapid ($\sim$3 collisions), or else it is slow on the time scale of energy transfer from ν_3 to ν_4.[21] Another interesting feature in the CH_4 system is that the effect of rare gases on the rate of rise of ν_4 and the fall of ν_3 decreases as lighter rare gases are used. The rise of ν_4 again appears to be faster than the fall of ν_3 for all rare gases over the range of rare-gas pressures studied.[21]

An interesting experiment to test the hypothesis that the $\nu_4 \leftrightarrow \nu_2$ equilibration leads to the apparent faster rise of ν_4 than fall of ν_3 would involve a study of the fall of ν_3 and rise of ν_4 at high rare-gas pressure. In the region of high rare-gas pressure, ν_4 and ν_2 should be rapidly equilibrated, while the rate of resonant steps such as that in (3) should not be affected. Thus in the high rare-gas pressure region, the fall of ν_3 and rise of ν_4 should be identical if the existing apparent discrepancy in the rise of ν_4 and fall of ν_3 is due to the $\nu_3 \leftrightarrow \nu_2$ equilibration step of (4).

B. *CD_3H*

Vibrational energy transfer in CD_3H has been dealt with in two studies.[24,25] The rare-gas dependence of the V–T/R deactivation rate of CD_3H has been reported. Perhaps surprisingly, at least in comparison with the methyl halides, the dependence of the deactivation rate of CD_3H on the square root of the reduced mass of the collision pair yields a straight line. This is quite similar to what has been observed for CH_4 and CH_2D_2;[24,26] however, whether the slope for this line is compatible with SSH predictions has not been dealt with in detail. An energy level diagram for CD_3H is shown in Fig. 2. The rates of activation of the various vibrational modes in CD_3H have been measured and pathways postulated for equilibration steps. Excluding the ν_5 mode at 1291 cm^{-1}, the six vibrational modes of the molecule group into three regions, and this grouping of modes, as will be discussed below, bears on the possible energy-transfer pathways in the system.

Subsequent to pumping the ν_6 mode of CD_3H at 1037 cm^{-1} with the *P*(30) line of the 9.6 μ CO_2 laser transition, fluorescence was observed in the 3000 cm^{-1} region which was ascribed to the ν_1 mode, in 2000 cm^{-1} region ascribed to the tightly coupled and coriolis mixed (ν_2, ν_4) C—D stretching modes and in the 1291 cm^{-1} region, which is attributed to the ν_5, CD_3 rocking mode. The pumped state is mixed with the ν_3 mode at 1003 cm^{-1} and is assumed to be tightly coupled to it.

Equilibration of the ν_5 state with the tightly coupled (ν_3, ν_6) state is assumed to occur via a direct endothermic step of the type

$$CD_3H(\nu_3, \nu_6) + CD_3H \rightleftharpoons CD_3H(\nu_5) + CD_3H \; - \sim 250 \; cm^{-1} \qquad (5)$$

with a rise rate for ν_5 fluorescence of 180 ± 30 ms^{-1} $torr^{-1}$. Assignment of

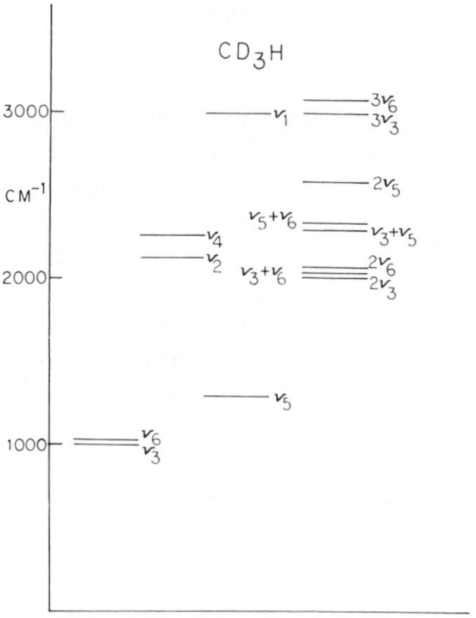

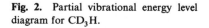

Fig. 2. Partial vibrational energy level diagram for CD_3H.

the rate of this process appears to be on firm ground, since a thermal lensing study indicates that translational cooling of CD_3H occurs immediately after absorption of the laser pulse. This cooling should be indicative of an endothermic step involving equilibration of a significant fraction of the vibrational population initially pumped into ν_6.[25]

Equilibration of states in the 2000 cm^{-1} region at a rate of 300 ± 50 ms^{-1} $torr^{-1}$ is postulated to occur via an "up-the-ladder" step involving overtones or combination bands of the initially pumped (ν_3, ν_6) state and subsequent crossover to the (ν_2, ν_4) state. A process requiring a combination band involving the initially pumped state and ν_5 followed by subsequent transfer to (ν_2, ν_4) appears unlikely, since this could not produce a single exponential rise of (ν_2, ν_4) at a rate of 300 ms^{-1} $torr^{-1}$. Only a lower limit of 250 ms^{-1} $torr^{-1}$ for the rise of the ν_1 state could be measured due to signal quality constraints. Energy transfer from CD_3H to SF_6, SiF_4 and CF_4 were also measured and reported.[25]

Energy transfer from CO_2 and N_2O (001) to the molecules CH_nD_{4-n} where $n = 0$–4 have been measured and discussed.[27] The deactivation of CO_2 (001) or N_2O (001) appears to correlate well with the number of stretching modes in the 2200 cm^{-1} region, which in turn correlates with the number of deuterium atoms in the collision partner. This correlation is discussed with regard to near-resonant long-range energy-transfer theory as is the role of the large rotational constant of the collision partners.[28]

C. CD_4

CD_4 whose energy level diagram is shown in Fig. 3 has been the subject of a number of vibrational energy-transfer studies.[29-31] In the most extensive study the ν_4 mode at 966 cm^{-1} was pumped with a *Q*-switched CO_2 laser and fluorescence was observed from the only other infrared active mode, ν_3 at 2259 cm^{-1}.[30] The rise rate of fluorescence was reported as $133 \pm 30\ ms^{-1}\ torr^{-1}$. The fall rate of ν_3 corresponding to V–T/R processes was measured in added rare gas to eliminate heating effects and is reported as $0.313 \pm 0.05\ ms^{-1}\ torr^{-1}$, which is in good agreement with both ultrasonic studies[31] and a study using vibrationally excited HCl to collisionally excite CD_4.[32]

A study[29] involving the use of thermal lensing to monitor vibrational energy flow in CD_4 revealed some interesting variations of the magnitude of the thermal lensing signal as a function of the laser line that was used to excite CD_4; however, the overall conclusion regarding V–V processes was that a fast endothermic V–V event, assigned to the pathway

$$CD_4(\nu_4) + CD_4 \rightleftharpoons CD_4 + CD_4(\nu_2) - 96\ cm^{-1} \tag{6}$$

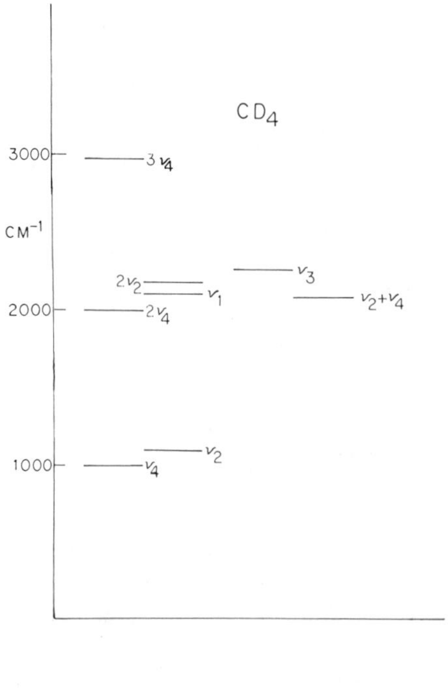

Fig. 3. Partial vibrational energy level diagram for CD_4.

rapidly coupled the ν_4 and ν_2 modes. The rapidity of this process as well as the rare gas dependence of the rise of the ν_3 state has been used to argue[31] for a ladder-climbing mechanism as a possible path leading to the equilibration of ν_3 via an overtone or combination band.

D. CH_2D_2

CH_2D_2 has been investigated in a number of studies involving pulsed CO_2 laser excitation.[19,26,33] Most of these investigations have been performed via excitation of the ν_7 mode with the $R(22)$ 9.6 μm CO_2 laser line at 1080 cm^{-1}. Since CH_2D_2 is an asymmetric top, it possess nine nondegenerate normal modes, which leads to a more complex energy level diagram (Fig. 4) than that exhibited by C_{3v} or higher symmetry molecules with 5 atoms. Fluorescence has been observed from all infrared active fundamentals in the molecule. The only noninfrared active fundamental is ν_5, since $A_2 \leftrightarrow A_1$ transitions are not dipole allowed under C_{2v} symmetry.[19] The observed V–T/R rate for the parent molecule and all rare gases is the same independent of which fluorescing state is observed. The deactivation rate for pure CH_2D_2 is 1.0 ms^{-1} $torr^{-1}$. The rare gas deactivation rates vary by a little over an order of magnitude in going from He^4 to Xe, and, while the rates fall on a straight line when a plot is made of the log of the deactivation probability vs the square root of the reduced mass of the collision partners, the line predicted by simple SSH theory has a much steeper slope than observed experimentally. The SSH calculated deactivation probabilities vary over almost four orders of magnitude in going from He^4 to Xe.

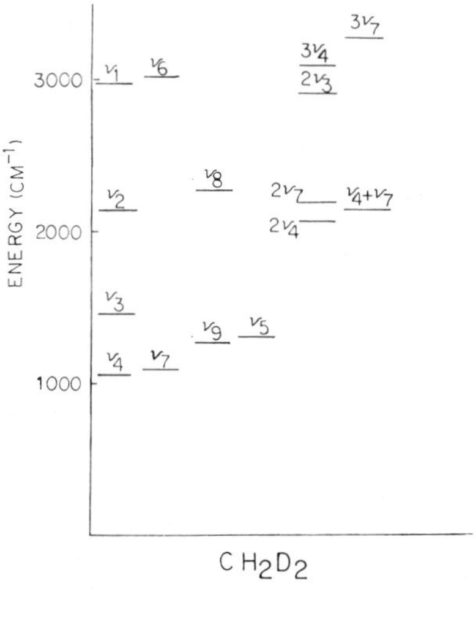

Fig. 4. Partial vibrational energy level diagram for CH_2D_2.

The fluorescent rise times were also measured for a variety of states in the system. As in previous studies of other molecules, a number of states that are close in energy to each other and for which fluorescence from an individual state could not be resolved were considered to be a single tightly coupled state. The rate of rise of the ν_9, ν_3, (ν_2, ν_8), and (ν_1, ν_6) states have been measured. With the rise signals analyzed as single exponentials, the reported rates are 263 ± 12, 212 ± 21, 291 ± 29, and 130 ± 13 ms^{-1} $torr^{-1}$, respectively. In the original study there was some speculation as to the likely paths for vibrational equilibration of these states;[19] however, much more specific and detailed information is now available with regard to vibrational equilibration pathways.[26,33]

Extensive use was made of the rare-gas dependence of the rise rate for various states in an attempt to determine the vibrational energy transfer pathways. These studies were extended over a much larger range of rare-gas pressure than normally used in the study of the rare-gas dependence of intermode V–V rates. As previously mentioned in regard to methane, if a state is filled via a consecutive mechanism such as

$$CH_2D_2(\nu_x)+CH_2D_2(\nu_x)\rightleftharpoons CH_2D_2(2\nu_x)+CH_2D_2(0) \tag{7}$$

followed by

$$CH_2D_2(2\nu_x)+CH_2D_2\rightleftharpoons CH_2D_2(\nu_y)+CH_2D_2 \tag{8}$$

the observed filling rate of y will normally be rate limited by the second step in this mechanism, since resonant processes, such as illustrated in (7), are normally quite fast relative to steps such as (8) if (8) is accompanied by a significant energy gap. On addition of rare gas, the rate of the second step will speed up, while the rate of the first step will be largely unaffected by the addition of rare gas. At low parent-gas pressure and upon addition of sufficient rare gas to make (8) faster than (7), a plot of the rise rate of state y vs rare-gas pressure will start to level off. This behavior is indicative of a resonant process being important in the filling of the observed state.[19] Furthermore, a knowledge of the energy and nature of states x, $2x$, and y provide sufficient information to determine the magnitude of the rate constants in (7) and (8). Extensions of this procedure can be used to treat systems in which there are one or more states intermediate between $2\nu_x$ and ν_y. In addition, the procedure can also be used to treat the reverse case, where ν_y is the initially excited state.

The aforementioned technique was applied to CH_2D_2 with excellent success. The rare-gas dependence of the rise rate of the ν_9 state 144 cm^{-1} above the initially pumped level, ν_7, was linear vs added rare gas. This implies that the ν_9 state is filled by a direct endothermic process of the type

$$CH_2D_2(\nu_7)+CH_2D_2 \rightleftharpoons CH_2D_2(\nu_9)+CH_2D_2-144\ \mathrm{cm}^{-1} \qquad (9)$$

The rare gas dependence of the rise rate of ν_3 is also linear upon the addition of significant quantities of rare gas, which implies that this state is filled by a nonresonant endothermic process that could be either a direct step

$$CH_2D_2(\nu_7)+CH_2D_2 \rightleftharpoons CH_2D_2(\nu_3)+CH_2D_2-360\ \mathrm{cm}^{-1} \qquad (10)$$

or more likely a path involving an intermediate state such as ν_9 and/or ν_5. In this case the equilibration process would be

$$CH_2D_2(\nu_7)+CH_2D_2 \rightleftharpoons CH_2D_2(\nu_i)+CH_2D_2+(\nu_7-\nu_i) \qquad (11)$$

where $i=9,5$, followed by

$$CH_2D_2(\nu_i)+CH_2D_2 \rightleftharpoons CH_2D_2(\nu_3)+CH_2D_2+(\nu_i-\nu_3) \qquad (12)$$

The fact that ν_3 is filled by an endothermic process was confirmed by another set of experiments.[33] Oxygen was added to CH_2D_2 in sufficient quantities to guarantee that the vibrational heat capacity of the oxygen becomes significant compared to the vibrational heat capacity of the excited states of CH_2D_2. The fractional fall in the amplitude of the fluorescence signal from ν_3 due to equilibration of CH_2D_2 with the oxygen vibrational manifold can then be measured. Assuming that the large majority of vibrational energy flows through ν_3 to the nearly resonant oxygen $v=1$ level at 1556 cm^{-1}, the fractional fall in amplitude can be used to fingerprint a given energy-transfer pathway for filling ν_3. This fingerprinting procedure is discussed in Section III.C. The fingerprinting procedure is consistent with a direct or consecutive nonresonant endothermic pathway for filling ν_3.

The rates of filling of both (ν_2, ν_8) and (ν_1, ν_6) exhibit very different behavior on addition of rare gas (neon) than that observed for the ν_9 and ν_3 states. Upon addition of rare gas, the rise rate of the (ν_2, ν_8) states remained constant over the range from 0→30 torr of rare gas for an initial CH_2D_2 pressure of 1 torr. When the (ν_1, ν_6) states were monitored, the rise rate at first increased linearly with rare gas and then increased more and more slowly until at a rare-gas pressure ~30 times the CH_2D_2 pressure the rise rate reached a plateau. The behavior of both these sets of states, (ν_2, ν_8) and (ν_1, ν_6), is characteristic of a resonant process involving the collision of two excited molecules. On the addition of sufficient rare gas to make this resonant process rate limiting, further rare gas will not speed up the rise rate of the state being monitored.

For the (ν_2, ν_8) state, the implication is that the resonant process is already rate limiting, slower than the crossover rate from the overtones or combination bands of the initially pumped state, ν_7, and those states that may be tightly coupled with ν_4 and ν_7. This rapid coupling of states in the 2000 cm^{-1} region may be a consequence of both the number of states in that spectral region and the extensive mixing that has been observed in spectroscopic investigation of this region.[34]

The behavior of the (ν_1, ν_6) state on addition of rare gas is indicative of the situation that is expected to be generally observed when a state is filled by consecutive steps involving a resonant process requiring two excited state molecules and a nonresonant process. An example of such a consecutive mechanism is depicted in (7) and (8). The nonresonant step will normally be rate limiting at low rare-gas pressure, while at high rare-gas pressure, the resonant process would be expected to become rate limiting.

Two pathways have been postulated as likely possibilities for the filling of (ν_1, ν_6). One involves the use of the initially pumped state and states that are assumed tighly coupled to it as a ladder for the filling of (ν_1, ν_6). For example

$$CH_2D_2(\nu_7)+CH_2D_2(\nu_7)\rightleftharpoons CH_2D_2(2\nu_7)+CH_2D_2(0) \qquad (13)$$

and

$$CH_2D_2(2\nu_7)+CH_2D_2(\nu_7)\rightleftharpoons CH_2D_2(3\nu_7)+CH_2D_2(0) \qquad (14)$$

followed by

$$CH_2D_2(3\nu_7)+CH_2D_2\rightleftharpoons CH_2D_2(\nu_1, \nu_6)+CH_2D_2 \qquad (15)$$

or similar processes involving overtones of ν_4 or combination bands of ν_4 and ν_7. An alternative pathway involves the ν_3 manifold where (ν_1, ν_6) would be filled from $2\nu_3$ via

$$CH_2D_2(2\nu_3)+CH_2D_2\rightleftharpoons CH_2D_2(\nu_1, \nu_6)+CH_2D_2 \qquad (16)$$

preceded by the step

$$CH_2D_2(\nu_3)+CH_2D_2(\nu_3)\rightleftharpoons CH_2D_2(2\nu_3)+CH_2D_2(0) \qquad (17)$$

with the filling of ν_3 occurring via transfer from the lower states in the vibrational manifold as discussed previously. Consideration of the effect of the degree of exo- or endothermicity of the nonresonant step in the consecutive processes of (13) to (15) shows that for a given resonant rate, the rate for filling of a level decreases when a large amount of population must be funneled through a state with a relatively small population.[19] Thus small energy gap endothermic nonresonant steps may be favored over exothermic steps. This consideration along with the rate of filling of the (ν_2, ν_8) state leads to the conclusion that, while the (ν_1, ν_6) state could be filled by an "up-the-ladder process" involving the ν_4 or ν_7 manifolds in the absence of rare gas, on addition of rare gas, up-the-ladder processes involving the ν_3 manifold must also become important in filling (ν_1, ν_6). Within this model, actual rate constants have been reported for a number of the steps involved in the equilibration of the (ν_1, ν_6) and (ν_2, ν_8) states.[19,33]

Since the majority of the vibrational population in the CH_2D_2 system is contained within the five lowest states, effects of equilibration processes involving the higher states on the lowest five states can be neglected.[35] Additionally, if the ν_4 and ν_7 states and the ν_9 and ν_5 states are considered to be rapidly coupled due to their proximity to each other, the lower part of the vibrational manifold in CH_2D_2 can be modeled as a three-level system composed of (ν_4, ν_7), (ν_9, ν_5), and ν_3. The mathematical description of the kinetic behavior of a three-level system has been well worked out and has been presented in detail.[18] Signals governing population equilibration in a three-level system will in general be double exponentials with each eigenvalue being given by a combination of kinetic rate constants. Thus to determine actual kinetic rate constants, both eigenvalues must be measured. These measurements have been performed in CH_2D_2 and the results of this analysis presented in a recent publication,[33] which also addresses the validity of the approximation of rapid coupling within the (ν_4, ν_7) and (ν_9, ν_5) composite states. Input regarding the validity of these approximations is also gained via experiments involving pumping of the ν_4 state as well as pumping of a hot-band transition.

Another interesting aspect of CH_2D_2 energy-transfer dynamics involves the part of the intermolecular potential that is important for near-resonant V–V energy transfer of the type

$$CH_2D_2(\nu_i) + CH_2D_2(\nu_i) \rightleftharpoons CH_2D_2(2\nu_i) + CH_2D_2(0) \tag{18}$$

CH_2D_2 is unlike CH_3F and a number of other systems that possess modes that have large dipole-moment derivatives for fundamental transitions. In CH_2D_2 the dipole-moment derivative of the ν_7 mode is $\sim 1/4$ of the dipole-moment derivative of the ν_3 mode of CH_3F.[36,37] In addition, virtually all the fundamentals in CH_2D_2 have dipole-moment derivatives that are of the same order.[37] Therefore, CH_2D_2 ladder-climbing processes of the type given in (18) would not be expected to be dominated by long-range energy-transfer processes, which have a fourth-power dependence on the dipole-moment derivative.[28] This, of course, assumes that higher order multipole moments do not dominate the long-range energy-transfer mechanism. Unfortunately, virtually no reliable information exists on the magnitude of quadrupole-moment derivatives and higher order moments in CH_2D_2 that could be used to check this assumption. In the absence of significant contributions to (18) by long-range energy transfer, the process should be dominated by hard collisions that sample the repulsive part of the intermolecular potential. The theory that describes such processes has been formulated by Schwartz, Slawsky, and Herzfeld.[38] Use of this theory with modifications introduced by Tanczos and Stretton allows for the calculation of energy transfer probabilities for V–V processes in this system.[39] The agreement between calculation and experiment is quite good, though this may be fortuitous.

An interesting experiment to check the above predictions as to the part of the intermolecular potential that is dominant in controlling process (18) would be a measurement of the temperature dependence of V–V rates in the system. Near-resonant energy-transfer theory predicts a T^{-1} behavior, while SSH theory predicts a $T^{1/3}$ behavior.

E. CH_3F

Methyl fluoride has been the subject of a large number of investigations over the last decade, which have led to a complete eluciation of the vibrational energy-transfer processes that result after low-level excitation of this molecule.[53, 40–53] It is one of a very few molecular systems where measured rates can be deconvoluted into actual kinetic rate constants with a high degree of certainty. Actual kinetic rate constants are essential for the testing of theoretical predictions with regard to vibrational energy-transfer probabilities. Additionally, CH_3F is probably the only molecular system larger than 3 atoms for which a full energy-transfer map can now be drawn.

The initial investigation of energy transfer in methyl fluoride dates back to 1972.[40] In the beginning, rapid energy transfer from the initially pumped ν_3,C—F stretch at 1050 cm^{-1} to the C—H stretching mode at 3000 cm^{-1} was reported. An actual rate for the population of the 3000 cm^{-1} states could not be measured, but the process was seen to be very rapid on the time scale of the observed V–T/R deactivation rate which was measured as 0.59 ms^{-1} $torr^{-1}$ (15,000 gas kinetic collisions). Due to the rapidity with which states 2000 cm^{-1} removed from the pumped state are populated, very efficient energy-transfer processes clearly must dominate the dynamics of filling of the (ν_1, ν_4) states at 3000 cm^{-1}. Resonant up-the-ladder energy-transfer processes were postulated for the filling of these states. At the time of this publication,[40] the most likely process leading to filling of the (ν_1, ν_4) states seemed to be rapid energy transfer up the ν_3 manifold with crossover from $3\nu_3 \rightarrow (\nu_1, \nu_4)$. The possibility of direct pumping of (ν_1, ν_4) by hot-band transitions was considered and discarded. The results of all subsequent investigations are consistent with collisional pumping and equilibration of all excited states in CH_3F.[35,40–52] Hot-band transitions, if they exist, do not seem to play a significant role in the equilibration behavior of CH_3F when it is pumped via the $P(20)$ 9.6 μm unfocused Q-switch CO_2 laser transition.

The next study of the vibrational equilibration of the CH_3F system reported observation of all the remaining fundamentals in the CH_3F molecule except the initially pumped ν_3 fundamental.[42] Though the reported observation of ν_6 level fluorescence proved to be in error due to leakage of (ν_2, ν_5) emission through the interference filter used to observe ν_6 (see Fig. 5 for energy level diagram)[35,48] the general conclusion of this study proved valid; all the vibrational fundamentals and the $2\nu_3$ overtone

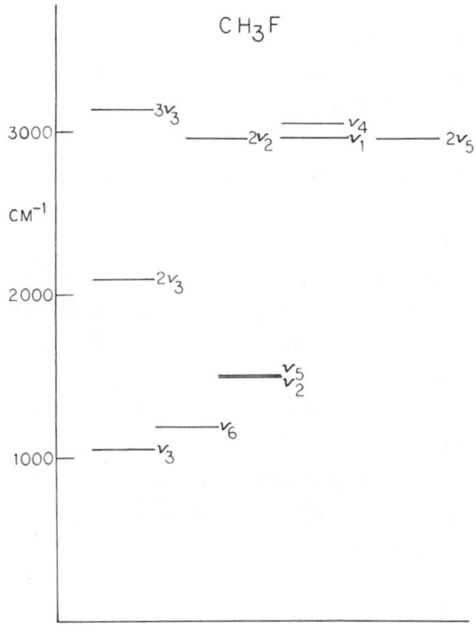

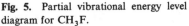

Fig. 5. Partial vibrational energy level diagram for CH_3F.

in CH_3F equilibrate on a time scale corresponding to ~100 gas kinetic collisions. Additionally, as might be expected from the previous results, the V–T/R time of the system was found to be the same for all states observed, which included $2\nu_3$ and all the fundamentals except ν_3. Subsequently the V–T/R time of the ν_3 fundamental was also measured to be the same as that for all other states in the system.[45]

At this time the hypothesized pathways for filling of various CH_3F states were the following:[27]

(ν_1, ν_4)

$$CH_3F(\nu_3)+CH_3F(\nu_3) \rightleftharpoons CH_3F(2\nu_3)+CH_3F \quad (19a)$$

$$CH_3F(2\nu_3)+CH_3F(\nu_3) \rightleftharpoons CH_3F(3\nu_3)+CH_3F \quad (19b)$$

$$CH_3F(3\nu_3)+CH_3F \rightleftharpoons CH_3F(\nu_1,\nu_4)+CH_3F+155\ cm^{-1} \quad (19c)$$

ν_6

$$CH_3F(\nu_3)+CH_3F \rightleftharpoons CH_3F(\nu_6)+CH_3F-133\ cm^{-1} \quad (20)$$

(ν_2, ν_5)

$$CH_3F(\nu_6)+CH_3F \rightleftharpoons CH_3F(\nu_2,\nu_5)+CH_3F-284\ cm^{-1} \quad (21)$$

$2\nu_3$

$$CH_3F(\nu_3)+CH_3F(\nu_3) \rightleftharpoons CH_3F(2\nu_3)+CH_3F+17\ cm^{-1} \quad (22)$$

The three lowest states in the CH_3F system, ν_3, ν_6, (ν_2, ν_5) were recognized as forming a subsystem that could be isolated from the remaining vibrational manifold and treated independently.[42] A solution of the kinetic rate equations for this three-level system will yield expressions for the population evolution that are double exponentials; however, experimentally signal quality and apparatus constraints precluded a full double exponential analysis of fluorescence signals.

While the pathways filling ν_6, (ν_2, ν_5), and $2\nu_3$ [(20)–(22)] have stood up to a whole host of experimental investigations, which will be described later, (ν_1, ν_4), rather than filling through $3\nu_3$, is now believed to fill from (ν_2, ν_5) via $(2\nu_2, 2\nu_5)$. The coupling between $(2\nu_2, 2\nu_5)$ and (ν_1, ν_4) is aided by Fermi resonance between $2\nu_5$ and ν_1.[36] The $(\nu_2, \nu_5) \rightarrow (\nu_1, \nu_4)$ pathway is supported by the fact that the signals from these states are virtually identical in either pure parent gas or at low pressures of rare gas, which implies that they are filled via the same rate-limiting step. This would be the case provided that the rate for the process

$$CH_3F(2\nu_2,2\nu_5)+CH_3F \rightleftharpoons CH_3F(\nu_1,\nu_4)+CH_3F \quad (23)$$

and the up-the-ladder process involving the (ν_2, ν_5) manifold are fast compared to the equilibration of ν_3 with (ν_2, ν_5). Additional support for this pathway is provided by a study of the rare-gas dependence of (ν_1, ν_4) and (ν_2, ν_5) fluorescence at high pressures of rare gas. By variation of the rare-gas pressure over the region where at low rare-gas pressure the filling of (ν_2, ν_5) is rate limiting, to high rare-gas pressure where the resonant "(ν_2, ν_5) ladder process" is rate limiting, estimates of the rate constants for a number of the processes involved in filling (ν_1, ν_4) can be obtained, all of which are consistent with the major filling pathway involving the (ν_2, ν_5) manifold.[19,33] Additionally, since the filling of (ν_1, ν_4) from $3\nu_3$ is an exothermic process which requires funneling many times the ambient population of $3\nu_3$ through $3\nu_3$ to (ν_1, ν_4) such a process would be expected to be overall less efficient than the filling of (ν_1, ν_4) through an endothermic process requiring much less population transfer.[19] Finally, in the first measurement of its kind, the vibrational temperature T_{14} of the (ν_1, ν_4) modes of strongly excited CH_3F was found to be the same as the vibrational temperature T_{25} for the (ν_2, ν_5) modes, consistent with filling via the $(2\nu_2, 2\nu_5)$ overtones.[47] As discussed in detail in a later section of this chapter, energy-transfer steps (19–21) would require $T_{14} > T_3 \gg T_6 \gg T_{25}$. Thus while filling of (ν_1, ν_4) may have some contribution from the $3\nu_3$ pathway, it is clearly dominated by the (ν_2, ν_5) channel.[19,35]

The fact that the lower vibrational manifold $\nu_3, \nu_6, (\nu_2, \nu_5)$ fills in a sequential fashion has now been established by a variety of studies. Thermal lensing experiments in CH_3F indicated that a rapid translational cooling accompanied vibrational equilibration in the CH_3F system.[43] The obvious process to which this translational cooling could be attributed was the direct endothermic process

$$CH_3F(\nu_3) + CH_3F \rightleftharpoons CH_3F(\nu_6) + CH_3F - 133\ \text{cm}^{-1} \tag{24}$$

Due to the double degeneracy of ν_6, the populations of ν_3 and ν_6 are expected to be almost equal at room temperature. The relative amplitude of the cooling signal when compared to the heating signal, which appears on a longer time scale and is due to V–T/R relaxation of the equilibrated vibrational manifold, is consistent with an endothermic step sufficient to populate ν_6.

Further information regarding this pathway and the subsequent filling of (ν_2, ν_5) from ν_6 came from studies of energy transfer in $CH_3F{:}O_2$ mixtures.[49,50] In one study oxygen was added to CH_3F until the heat capacity of the oxygen excited-state manifold was essentially equal to the vibrational heat capacity of the CH_3F excited vibrational manifold. Rare gas was added to this mixture to insure that CH_3F vibrational states (particularly ν_3, ν_6 and (ν_2, ν_5) which possessed the majority of the heat capacity) equilibrated more rapidly than any vibrational state of CH_3F equilibrated with O_2. Equilibration with O_2 is assumed to take place via the (ν_2, ν_5) states so that the process can be written

$$CH_3F(\nu_2, \nu_5) + O_2(v=0) \rightleftharpoons CH_3F(v=0) + O_2(v=1) - 90.4 \text{ cm}^{-1} \quad (25)$$

The fractional fall in the fluorescence amplitude of the equilibrated CH_3F manifold can be monitored as energy is transferred to O_2. The magnitude of the fall can then be correlated with possible energy-transfer pathways that lead to the filling of (ν_2, ν_5). This concept, called "vibrational quanta conservation," is developed in detail in Section III.C, but, briefly, if (ν_2, ν_5) were filled by a step that involved the conversion of one quantum initially pumped into ν_3 and $\sim$450 cm^{-1} of translational energy, then the magnitude of the decrease in the fluorescence signal from $CH_3F(\nu_2, \nu_5)$ as the CH_3F manifold equilibrates with O_2 would be less than if (ν_2, ν_5) were filled via a process, which involved two quanta initially pumped into ν_3 and the conversion of $\sim$550 cm^{-1} of vibrational energy into translational energy.[49]

Similar conclusions were reached from a study involving thermal lensing in CH_3F/O_2 mixtures, which indicated that (ν_2, ν_5) was filled from ν_3 by an endothermic process or processes.[50] The thermal lensing signal was monitored in the presence of added O_2. As the O_2 concentration and thus the oxygen heat capacity increased, the thermal lens cooling signal also increased in amplitude. Provided the assumption of coupling of CH_3F to O_2 via process (25) is valid, this study provides strong evidence that (ν_2, ν_5) is filled via an endothermic process or processes.

Though the above experiments clearly indicate that (ν_2, ν_5) is filled from ν_3 by an endothermic pathway, the experiments are not sensitive to whether (ν_2, ν_5) is filled in a direct step of the type

$$CH_3F(\nu_3) + CH_3F \xrightarrow{k_{32}} CH_3F(\nu_2, \nu_5) + CH_3F - 417 \text{ cm}^{-1} \quad (26)$$

or via a sequential process of the type

$$CH_3F(\nu_3) + CH_3F \xrightarrow{k_{36}} CH_3F(\nu_6) + CH_3F - 133 \text{ cm}^{-1} \quad (27a)$$

$$CH_3F(\nu_6) + CH_3F \xrightarrow{k_{62}} CH_3F(\nu_2, \nu_5) + CH_3F - 284 \text{ cm}^{-1} \quad (27b)$$

An experiment that can distinguish between these pathways has recently been performed.[35] The rise of fluorescence from (ν_2, ν_5) was carefully monitored and analyzed as a double exponential. The signal was found to have both an induction period and a point of inflection. Both of these features are characteristic of a state being filled by consecutive processes rather than a direct filling. By considering the maximum magnitude that a parallel direct process of the type illustrated in (26) could have if it acted in conjunction with the process of (27), the upper limit for k_{32} was found to be 24 ms^{-1} $torr^{-1}$. Additionally k_{36} and k_{62} were determined as 292 ± 27 ms^{-1} $torr^{-1}$ and 70 ± 17 ms^{-1} $torr^{-1}$, respectively, if $k_{32} = 0$.

The equilibration of ν_3 with ν_6 and (ν_2, ν_5) was also monitored via another experimental procedure.[48] Though ν_3 and $2\nu_3$ were known to equilibrate rapidly, a reliable rate seems only recently to have been established (2580 ± 107 ms^{-1} $torr^{-1}$) for the forward rate constant in (19a). This rate constant was determined by the direct observation of the $2\nu_3$ fluorescence rate of rise.[35] Upon careful observation of the $2\nu_3$ fluorescence signals two rapidly falling exponentials were found to be superimposed on the slower V–T/R fall.[48] Though the rates of these falling exponentials are rapid compared to V–T/R processes, they are slow compared to the rise of $2\nu_3$ fluorescence. Due to this time-scale separation, the rates of the two rapidly falling exponentials can be measured accurately. These rapidly falling exponentials are interpreted as being due to the equilibration of ν_3 with ν_6 and subsequent equilibration with (ν_2, ν_5). $2\nu_3$ acts as an accurate monitor of these processes, since it is in dynamic equilibrium with ν_3 on a time scale faster than any other equilibration process in the system involving substantial population movement. The rates for the processes illustrated in (27) obtained by monitoring $2\nu_3$ fluorescence agree within experimental error with those determined by direct observation of (ν_2, ν_5).[35]

Recently the rise rate of (ν_1, ν_4) has also been analyzed as a double exponential.[35] The rates obtained from this analysis yield an identical rate for the slow eigenvalue to that measured when observing (ν_2, ν_5); however, the fast eigenvalue for (ν_1, ν_4) is slower than the fast eigenvalue for (ν_2, ν_5). A small rate constant for the process

$$CH_3F(\nu_2, \nu_5) + CH_3F(\nu_2, \nu_5) \rightarrow CH_3F(2\nu_2, 2\nu_5) + CH_3F(0) \qquad (28)$$

could also lead to a slowing of the rise of (ν_2, ν_5), but this appears unlikely as a result of the studies in Ref. 19. Additionally, the rise rate of (ν_1, ν_4) is expected to contain at least 4 exponentials which may appear with different preexponential factors for the (ν_2, ν_5) states possibly leading to a different effective fast eigenvalue.[19]

When a significant number of rates are known for a system, numerical modeling of the system can be quite valuable as a check on the validity of assumed pathways. Some numerical modeling of the CH_3F system has been performed in Ref. 35 and the results agree quite well with expectation. More extensive numerical modeling of the CH_3F system has been performed in Ref. 19. The role of the ν_3 manifold in filling (ν_1, ν_4) and the effect of that filling on the shape of the (ν_1, ν_4) signals were considered. By looking at the effect of energy transfer from $3\nu_3$ on the (ν_1, ν_4) signal shape and comparing that to the experimentally observed signal shape an upper limit has been set on the contribution of $3\nu_3$ to the filling of (ν_1, ν_4).[19] A number of other interesting features that have been observed experimentally were duplicated, such as the almost identical rise behavior of (ν_2, ν_5) and (ν_1, ν_4) as well as the fall behavior of $2\nu_3$ and the information contained therein regarding the ν_6 and (ν_2, ν_5) rise times. A plot of the time evolution of the 8 states that go into the model of Ref. 19 is presented in Fig. 6.

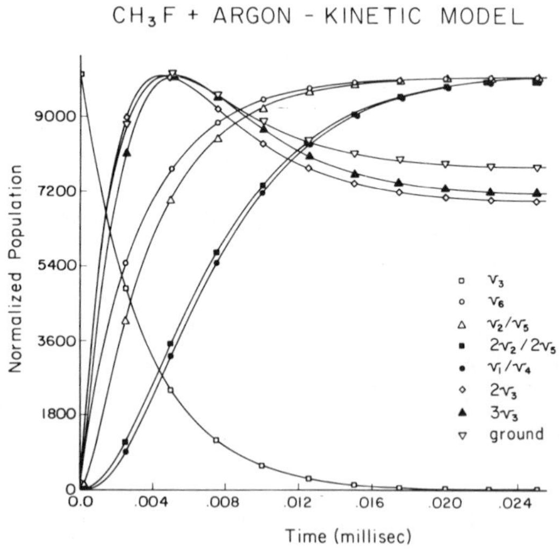

Fig. 6. Population evolution of eight coupled states in a CH_3F (0.17 torr)–argon (10 torr) mixture. The time dependence of the population is calculated from a numerical integration of the differential equations governing the rate processes of equations 19a, 19b, 20, 21, 28. The magnitudes of the rate constants are taken from Refs. 19, 35, and 55. All populations are shown normalized to 10,000 for the peak of the population of a given state. The laser pulse occurs at $t = 0$. Only V–V processes are included in the model.

1. *Rare Gas Behavior of CH_3F Rates*

In the study of the variation of the V–T/R deactivation rate of CH_3F vs the square root of the reduced mass of the CH_3F–rare gas collision pair, simple SSH theory was not adequate for predicting the observed variation of rates.[41] By inclusion of V–R processes good agreement with qualitative trends in the CH_3F–rare gas system could be obtained though quantitative predictions regarding the variation of the V–T/R rates were still substantially at variance with experiments. Recently an integrated theory including both V–T and V–R effects has had good success in matching the observed trends for deactivation rates in the CH_3F–rare gas system.[54]

Now that the vibrational energy-transfer pathways in CH_3F have been deconvoluted and the measurement of individual rate constants is a possibility, the time appears ripe for the measurement of the rare-gas dependence of the the rate constants for V–V processes in the CH_3F system. Some progress has already been made in this area. The rare-gas dependence of the eigenvalues for the rate processes governing the $\nu_3 \rightarrow \nu_6 \rightarrow (\nu_2, \nu_5)$ pathways (27) have been measured,[55] but as discussed in Ref. 19 the exact deconvolution of the rare-gas dependence of V–V rise rates that govern the process of (27) may prove difficult, since the observed rare-gas variation of the V–V rates may not be very sensitive to the magnitude of some of the rate constants in the system. Reference 19 also discusses the dependence of a number of CH_3F rate constants on added argon.

Though deconvolution of the rare gas-dependent rate constants may prove difficult in some cases, the effort appears to be worthwhile, since these data should prove to be very valuable as checks of theoretical calculations.

2. *Behavior of CH_3F in the High-Excitation Region*

The high-excitation region is characterized by the situation where population deviations in a state are no longer small compared to the ambient population of that state. Energy transfer behavior in CH_3F has been investigated in this regime and is reported on in Refs. 35 and 53. Under high-excitation conditions the risetime of 3 μm fluorescence becomes more rapid than the rise time of 3 μ fluorescence under low-excitation conditions. This rapid rise is followed by a partial decay of the fluorescence signal and then a second slower rise.[35] An explanation for this behavior could result from the process

$$CH_3F(\nu_3) + CH_3F(\nu_1) \rightleftharpoons CH_3F(\nu_1 + \nu_3) + CH_3F(0) \tag{29}$$

followed by emission

$$CH_3F(\nu_1+\nu_3) \rightarrow h\nu + CH_3F(\nu_3) \tag{30}$$

where the emission wavelength in (30) could not be spectrally resolved in a typical fluorescence experiment from (ν_1, ν_4) emission. $(\nu_1+\nu_3)$ emission would be expected to rise rapidly as ambient ν_1 molecules collide with laser-pumped ν_3 molecules.

This hypothesis was at least partially confirmed by direct observation of $(\nu_1+\nu_3)$ and $(\nu_3+\nu_4)$ fluorescence, which was not observable in the low-excitation regime. The $(\nu_1+\nu_3)$ emission signal is seen to rise rapidly and decay more rapidly than the normal V–T/R rate, presumably due to nonlinear V–T/R effects, which can occur in the high excitation regime, and/or V–V processes that will affect this signal, since not all of the vibrational manifold is equilibrated on the time scale of the $(\nu_1+\nu_3)$ rise time. The second rise rate that is observed may be due to normal filling of (ν_1, ν_4) via processes (20), (21), (23) and (28).

These conclusions, however, must be considered highly tentative until further experiments can be performed to elucidate the behavior of CH_3F under high excitation conditions. The possibility remains that the fast rise of 3 μm fluorescence, which accompanies strong laser excitation, may signal the presence of a new dominant energy-transfer mechanism valid under high excitation.[53] Furthermore, signals observed with focused *Q*-switch CO_2 laser excitation[35] and TEA CO_2 laser excitation[53] may be probing very different processes, even though both represent strong excitation conditions.

F. Conclusions

A number of factors are apparent when the dynamics of energy transfer processes in CH_4, CD_3H, CH_2D_2, and CH_3F are considered. One of these is the complexity of even these "simple" systems. Fluorescence signals from individual or tightly coupled levels in these molecules will in general exhibit multiple exponential behavior, which reflects the fact that the kinetic processes involved in the filling of a given state will be represented by a coupled set of differential equations. However, under certain circumstances individual eigenvalues of the kinetic rate matrix can be measured. This can occur when all exponentials of significant amplitude can be separated and their rates measured, or when one exponential dominates the behavior of a given state. Under these circumstances with the aid of a kinetic model, deconvolution of actual kinetic rate constants from the observed rise and fall rates for the system may be possible. These points are discussed in detail in the next section. Methyl fluoride represents an impressive illustration of this procedure. An energy-transfer map for the system is well established and a variety of kinetic rate constants have been determined.

A feature that stands out in all of the studies discussed in this chapter is that vibrational energy transfer within a given normal mode is rapid. This is the origin of the up-the-ladder energy-transfer mechanism, though as manifested in methane it can also lead to rapid energy transfer to states below the initially pumped state. The up-the-ladder mechanism is a fast, efficient process for transferring energy to states thousands of wavenumbers from an initially pumped level. However, the role of up-the-ladder processes in populating other vibrational modes depends on a number of factors such as the mixing between the overtones of the "ladder mode" and nearby fundamentals, the amount of population that must be funneled through the overtone to fill other states, and the number of quanta that must change in the process leading to filling of other modes.

Both small energy-gap endothermic and exothermic processes can be important in filling modes. In particular, states that lie between levels that correspond to "rungs of the ladder" will in general be filled by nonresonant processes. A powerful method has been developed for the determination of whether a state fills by resonant or nonresonant processes or a combination of both.[19] Endothermic events may be relatively more important in filling states than might have been intuitively thought because of the less severe requirement on the magnitude of population transfer compared to that necessary for exothermic processes.

The value of numerical modeling of a system has been discussed. Such an approach is particularly useful for examining the validity of assumptions and simplifications that have been made for a given system, for relating observed rates to eigenvalues, and for deconvoluting these eigenvalues into actual kinetic rate constants. This approach can also be used as a powerful predictive tool for the behavior of as yet unobserved states or for systems in the high excitation region where the rate constants obtained from studies in the low excitation region can be used as input data.

A variety of other molecules similar to those discussed in this chapter have been recently reviewed in Ref. 52. The general pathways for filling of vibrational states that are discussed in that article conform well to those discussed here. V–T and V–R effects in CH_3F and the other methyl halides are discussed in detail with the same overall conclusion: in general V–T and V–R effects must both be taken into account, preferably in a generalized theory that has the dynamics of both effects included at a fundamental level.[52] The results of the theoretical treatments based on the aforementioned principle are encouraging.[54]

III. TECHNIQUES FOR DETERMINING DOMINANT ENERGY-TRANSFER MECHANISMS

A. General Comments

When vibrational excitation is supplied to a polyatomic molecule by a technique such as pulsed-laser pumping, the return of the system to equilibrium via collisions is an enormously complex process for a molecule with more than just a few energy levels as should be evident from the data presented for CH_4, CD_4, CD_3H, CH_2D_2, and CH_3F in the preceding section. For example, let $N_i(t) = N_i^0 + n_i(t)$ be the population density of level i at time t after a laser pulse where N_i^0 is the equilibrium (preexcitation) density and $n_i(t)$ the deviation from equilibrium. Provided $n_i \ll N_i^0$ for all levels, the rates of change, $\dot{n}_i$, for the deviations obey a set of coupled, linear differential equations, which can be expressed by

$$\dot{\mathbf{n}} = A\mathbf{n} \tag{31}$$

where $\dot{\mathbf{n}}$ and $\mathbf{n}$ are vectors consisting of the $\dot{n}_i$ population deviation derivatives and n_i population deviations, respectively. A is a matrix whose entries are kinetic rate constants which are the quantities of interest. For a molecule such as CH_3F, considering only the 17 lowest vibrational states (energy $\lesssim 3200\ cm^{-1}$) and ignoring all rotational structure, the matrix A contains 136 independent rate constants corresponding to the physical collision processes (scattering events), which connect these 17 levels! Because of the constraint $\Sigma_i n_i = 0$, the solution to the matrix equation (31) for each level i takes the form

$$n_i(t) = \sum_{k=1}^{16} C_{ik} \exp[-\lambda_k t] \tag{32}$$

where the C_{ik} are constant amplitudes. The λ_k are the eigenvalues of the matrix A and the sum is limited to 16 terms only because we have chosen to restrict ourselves to the 17 lowest vibrational levels of CH_3F.

The complexity, even apparent hoplessness, of the situation begins to become clear from this example. To recover the full matrix A would require extracting all 16 eigenvalues λ_k and the many amplitudes C_{ik} for all n_i from experimental data! Furthermore, CH_3F is a relatively ideal case with a very small number of vibrational levels below 3200 cm^{-1}. Finally, the restriction to levels below 3200 cm^{-1} is arbitrary, though this limitation is usually not a serious one as long as the linear coupling approximation ($n_i \ll N_i^0$) is valid. *Experimentally, the extraction of more than 4 kinetic eigenvalues from a time-dependent infrared fluorescence curve is very difficult, and accurate measurements of eigenvalue amplitudes* (C_{ik}) *are rare.*

Fortunately, the practical situation is not as bleak as the general picture above would suggest. For example, usually only one collision process is responsible for overall vibration–translation/rotation (V–T/R) relaxation on a time scale $\tau_{V-T/R}$. Kinetic rate or collision events that are considerably slower than $1/\tau_{V-T/R}$ will *usually* appear in the fluorescence eigenvalue spectrum with negligible amplitude. Though there are exceptions to this rule (SO_2 is a clear case where the rule fails) [57,58] the fact remains that many of the possible kinetic rate processes contribute minimally to the eigenvalue amplitudes. Thus only some limited subset of kinetic collision events, the *dominant energy-transfer path or mechanism*, is required to describe the population deviations n_i and to invert the eigenvalue/amplitude spectrum in order to obtain the kinetic rate constants in A. Despite this fortunate circumstance, which leads to enormous simplifications, a knowledge of the subset of collision events that make up the dominant path is required to take advantage of this phenomenon and to properly interpret experimental data. Fortunately, several very powerful experimental techniques, which show great promise as diagnostic tools for determining energy-transfer mechanisms, have appeared in the past few years. These techniques take advantage of the concepts of vibrational temperature,[59–62] vibrational quanta conservation,[49] "kinetic sensitivity analysis,"[35,63] and rare-gas engineering[19] in the interpretation of kinetic data. In addition to their usefulness in the analysis of kinetic mechanisms, such concepts pay added dividends in that they provide exceptionally simple pictures of metastable population and energy distributions in laser-pumped polyatomic molecules.[62,64–66] Such descriptions should be very helpful in predicting chemical reactivity and population inversions[67] in polyatomic molecules. These general concepts are considered separately below.

B. Vibrational Temperatures

1. *Definitions*

For a molecular system completely at equilibrium, Boltzmann statistics immediately predict

$$\frac{N_i}{N_j} = \frac{g_i}{g_j} \exp\left[-\frac{(E_i - E_j)}{kT} \right] \tag{33}$$

where g_k, E_k, N_k are, respectively, the degeneracy, energy, and population density of level k, and T is the uniform single temperature for the molecule. In general a laser-pumped molecule is not at equilibrium for some time after pulsed excitation; nevertheless, we may still write

$$\frac{N_i}{N_j} = \frac{g_i}{g_j} \exp\left[-\frac{(E_i - E_j)}{kT_{ij}} \right] \tag{34}$$

where T_{ij} is the "temperature" specific to only levels i and j. Ordinarily $T_{ij} \neq T_{kl}$ and all temperatures are changing with time, though often very slowly. Unfortunately, (34) provides a description which is nearly useless, since the number of T_{ij}'s required is essentially equal to the number of state pairs. An enormous simplification arises if we take as the definition of vibrational temperature[59–62]

$$\frac{N_i}{N_0} = g_i \exp\left[-\frac{E_i}{kT_i} \right] \tag{35}$$

where T_i gives the ratio of the population N_i of state i to that of the *ground-state* N_0. Thus all T_i's use the ground state as a reference. In a metastable laser-pumped polyatomic molecule there exist relationships between different T_i's, and these relationships are controlled by the dominant energy-transfer path or mechanism for the molecule.[59] In the harmonic oscillator limit the number of vibrational temperatures is equal to the number of modes, and all of these temperatures are interrelated through the instantaneous translational temperature T' and the dominant energy-transfer path.[62]

2. *Intermode Collisional Coupling*

To illustrate some of these features consider just two energy-transfer processes for CH_3F that occur following laser excitation[35,40,41,48–50]

$$\text{Excitation: } CH_3F(0) + h\nu \rightarrow CH_3F(\nu_3) \tag{36}$$

$$\text{V–V Transfer } CH_3F(\nu_3) + M \underset{k_{63}}{\overset{k_{36}}{\rightleftharpoons}} CH_3F(\nu_6) + M - 133\ \text{cm}^{-1} \tag{37}$$

$$\text{V–T/R Relaxation } CH_3F(\nu_3) + M \underset{k_{03}}{\overset{k_{30}}{\rightleftharpoons}} CH_3F(0) + M + 1049\ \text{cm}^{-1} \tag{38}$$

Equation (37) is a fast intermode V–V energy-transfer event, which reaches steady state in ~30 gas kinetic collisions, while (38) is a very slow V–T/R decay, which requires ~15,000 gas kinetic collisions in pure CH_3F.

Thus shortly after pulsed laser excitation of the CH_3F ν_3 state, the ν_3 and ν_6 levels may be considered as a very slowly decaying metastable pair with

$$\frac{k_{36}}{k_{63}} = \frac{N_6}{N_3} \tag{39}$$

However, the kinetic rate constants are functions only of the instantaneous translational temperature T'. Using this fact and the definition (35), yields the parametric relationship[59–62]

$$\exp\left[-\frac{(E_6 - E_3)}{kT'}\right] = \frac{\exp[-E_6/kT_6]}{\exp[-E_3/kT_3]}$$

or

$$\frac{E_3 - E_6}{T'} = \frac{E_3}{T_3} - \frac{E_6}{T_6} \tag{40}$$

Equation. (40) is a most fundamental, crucial result, which provides the following insights:

1. In the case of energy transfer between degenerate levels (e.g., $E_3 = E_6$) equal vibrational temperatures must result ($T_3 = T_6$).
2. If $E_3 < E_6$, $T_3 > T_6$. (Translationally endothermic coupling leads to a lower temperature for the collisionally excited mode.)
3. If $E_3 < E_6$, $T_6 > T_3$. (Translationally exothermic coupling leads to a higher temperature for the collisionally excited mode.)

These results, which apply as long as $T' < T_3, T_6$ (the case for laser-pumped metastable polyatomics) are independent of whether the laser excites ν_3 or ν_6, or in fact any other mode.

3. Ladder-Climbing Effects

These arguments can be applied to ladder-climbing collisions, which are responsible for carrying vibrational energy to higher levels in both polyatomics and diatomics. For example in CH_3F the event:

$$2CH_3F(\nu_3) \underset{k_d}{\overset{k_f}{\rightleftharpoons}} CH_3F(2\nu_3) + CH_3F(0) + 17\ \text{cm}^{-1} \tag{41}$$

reaches steady state in ~3.3 gas kinetic collisions.[35] Application of the same reasoning used for the $\nu_3 - \nu_6$ intermode energy transfer [(37) and

(38)], immediately yields the parametric equation[60,61]

$$\frac{E_{2\nu_3}-2E_{\nu_3}}{T'}=\frac{E_{2\nu_3}}{T_{2\nu_3}}-\frac{2E_{\nu_3}}{T_{\nu_3}} \tag{42}$$

analogous to (40). This equation provides the following insights:

1. For harmonic oscillators, where, e.g., $E_{2\nu_3}=2E_{\nu_3}$, $T_{2\nu_3}=T_{\nu_3}$.
2. For anharmonic oscillators in which $E_{2\nu_3}<2E_{\nu_3}$, $T_{2\nu_3}>T_{\nu_3}$.
3. For anharmonic oscillators in which $E_{2\nu_3}>2E_{\nu_3}$, $T_{2\nu_3}<T_{\nu_3}$.

These arguments can be extended to higher levels such as $E_{3\nu_3}$, $E_{4\nu_3}$ by considering collisional coupling of the form

$$CH_3F(n\nu_3)+CH_3F(0)\rightleftharpoons CH_3F[(n-1)\nu_3]+CH_3F(\nu_3)$$

As long as the energy level spacing for successive pairs of levels within a mode is *decreasing*, the temperature of the levels will be *increasing*.

The origin of these multiple temperature effects lies in the dependence of the kinetic rate constants on only the translational (or more generally the "bath") temperature. For vibrational states, these results were first derived for diatomic anharmonic oscillators[60, 61] and used to explain laser action in CO via "anharmonic pumping."[60, 61, 68] Multiple vibrational temperature concepts were later applied to the polyatomic molecule SF_6 to explain discrepancies between relaxation rates measured by ultrasonic and laser techniques.[59] The close correspondence between the temperature picture for vibrational levels weakly coupled to a thermal bath and that for nuclear or electron spin levels also weakly coupled to a thermal bath will be immediately obvious to those familiar with spin temperature concepts.[69,]

4. *The Harmonic Oscillator Limit for Polyatomics*

The vibrational state temperature differences predicted by (40) and (42) are maximized when the magnitude of the terms $[(E_3-E_6)/T']$ or $[(E_{2\nu_3}-2E_{\nu_3})/T']$ are made as large as possible. Figure 7 shows how dramatic these temperature differences can become when the bath temperature is shrunk to very low values. This plot is a hypothetical one for OCS, which assumes that the ν_2 mode and the ν_3 mode (2062 cm^{-1}) reach steady state through the kinetic process

$$OCS(4\nu_2)+OCS\rightleftharpoons OCS(\nu_3)+OCS+43\ cm^{-1} \tag{43}$$

which can proceed via long-range dipole–dipole coupling.[71] The temperatures for the two modes are related by $43/T'=4\nu_2/T_2-\nu_3/T_3$ with

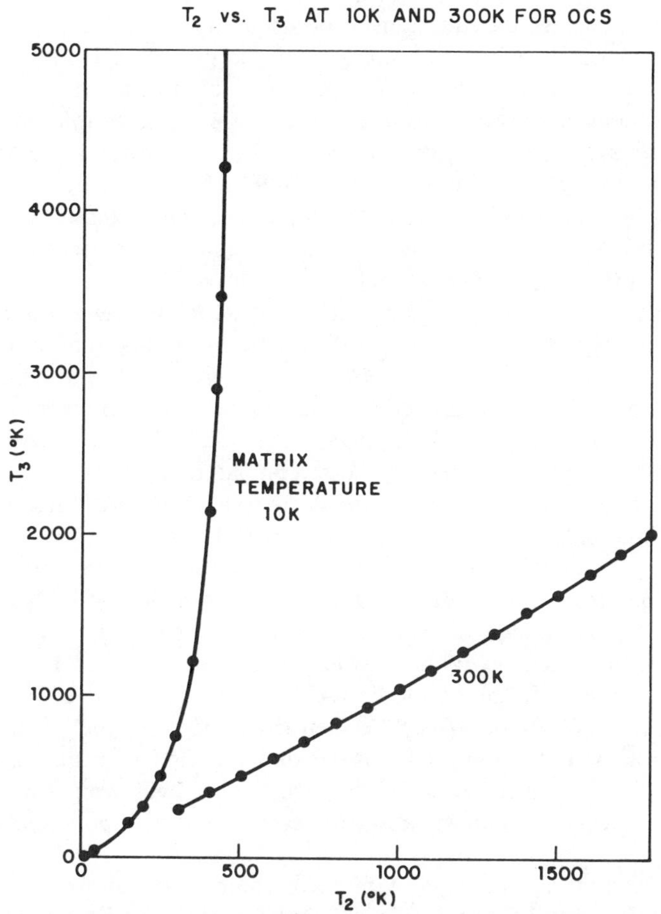

Fig. 7. Hypothetical temperature variations for the $\nu_3(T_3)$ and $\nu_2(T_2)$ modes of OCS. A metastable steady state has been assumed to occur via energy transfer between $4\nu_2$ and ν_3 [see Eq. (43)]. Calculations are included for both gaseous (300 K) and matrix-isolated species (10 K). The temperatures obey the equation $(2105/T_2)-(2062/T_3)=(43/T')$ with $T'=300$ K or 10 K.

$4\nu_2=2105$ cm^{-1}. As can be seen, when $T'=10$ °K (matrix isolated case), the temperature T_3 of mode 3 shoots up rapidly, while T_2 reaches a limiting value $T_2=(4\nu_2/43)10=489.5$ °K. On the other hand, when the bath temperature is 300 °K, T_2 and T_3 are very similar for large ranges of T_2 values. This illustrates an important feature of the multiple vibrational temperature description of molecules. For bath temperatures near 300 °K, very small temperature differences occur in modes coupled by kinetic processes in which only small amounts of energy are exchanged with the thermal bath. On the other hand, very low bath temperatures lead to dramatic mode temperature differences even for levels coupled by kinetic events with small energy discrepancies. Of course even for $T' = 300$ °K, T_2 reaches a limiting value, but this value (14,686 °K!) is 30 times higher than in the 10 °K matrix isolated case. Many real features of molecular systems cause a breakdown in the simple model calculations presented here for OCS; nevertheless, the general features remain valid.

Another way to obtain large vibrational state temperature differences is to couple the levels of interest via kinetic processes that require the exchange of relatively large amounts of energy with the bath. Such coupling processes typically occur with rather high efficiency between modes in many relatively small polyatomics [e.g., (37)]. In such cases, the intermode energy gaps are very large compared to the anharmonicities within a mode. Hence, the vibrational mode temperature differences can be large, and only small temperature differences occur for successive levels within a given mode. For this reason, many of the most interesting effects of multiple temperature distributions in polyatomic molecules can be obtained by assuming that each mode is harmonic, thus neglecting the smaller temperature variations within a mode arising from the small anharmonicities.[62] In the harmonic oscillator limit only one temperature is required for each mode as can be seen from (42) with the left-hand side equal to zero. Extension to levels beyond the first overtone is straight forward. For the remainder of this section we will consider only the harmonic oscillator approximation for each mode of a polyatomic molecule.

Since each vibrational state within a harmonic oscillator is at the same temperature, only $3N-6$ vibrational temperatures and one bath temperature T' are required to describe an N atom metastable polyatomic. For CH_3F this corresponds to 9 vibrational temperatures and T'; however, degenerate or nearly degenerate modes are generally tightly coupled by kinetic collision process such as (e.g.)

$$CH_3F(\nu_2) + M \rightleftharpoons CH_3F(\nu_5) + M + 4\ \text{cm}^{-1} \tag{44}$$

which guarantees $T_2 \simeq T_5$. Thus for CH_3F, which has E symmetry, doubly degenerate modes ν_4, ν_5, ν_6 and in addition $\nu_2 \simeq \nu_5$, $\nu_1 \simeq \nu_4$, only $\underline{4}$ vibrational temperatures are required to describe the molecule. These are T_3, T_6, T_{25}, and T_{14}. Furthermore, for laser-pumped CH_3F the relationships between these temperatures are defined via equations such as (40) once the dominant energy-transfer path can be identified. This reduces the problem of describing metastable CH_3F to one of determining only T' and one vibrational temperature, a remarkably simple task compared to what might have been expected!

Fortunately, the energy and population distributions of a metastable polyatomic molecule can be described by equations familiar from ordinary one temperature statistical thermodynamics. A multiple temperature vibrational partition function Q_{vib} can be derived, which has the form[62]

$$Q_{vib} = q_1(T_1)q_2(T_2)\ldots q_{3N-6}(T_{3N-6}) \tag{45}$$

with

$$q_i = \left[1 - \exp\left(\frac{-h\nu_i}{kT_i}\right)\right]^{-1} \tag{46}$$

Q_{vib} is thus identical to the ordinary one temperature result, except that q_i for each particular mode is evaluated at its own specific vibrational temperature T_i. Energy, enthalpy, entropy, and population distributions can be obtained from Q_{vib} for multiple temperature metastable states using standard techniques, but ambiguities are left for the free energy, work function, and heat capacity functions, which depend for their definition explicitly on the temperature.[62]

5. *Calculations of Energy and Population Distributions for Metastable Laser Pumped Polyatomics in the Harmonic Oscillator Limit*

In the harmonic oscillator approximation a knowledge of the dominant intermode V–V energy-transfer pathway provides a relationship between the vibrational temperatures T_i and the instantaneous bath temperature via equations of the type.(40) This leaves only two parameters one T_i and T', to be determined in order to fully describe a polyatomic at V–V steady state. These two parameters can be related to measureable initial conditions (e.g., the initial uniform preexcitation temperature T) by employing additional constraint equations.[62,67,72] For a gas these are conservation of excited molecules (vibrational quanta) and translational/rotational energy changes. The first constraint arises because a true steady state must conserve the number of excited molecules, while the second constraint arises because nonresonant energy transfers between modes [see, e.g., (37)] must be compensated by changes in translational/rotational energy. Combining a knowledge of the energy-transfer path [to define a set of relations such as (40)] with the above two conservation relations allows the steady-state energy, temperature, and population distributions to be computed using an iterative, inexpensive computer program.[62,67]

6. *Summary of Results for Polyatomic Molecules in the Harmonic Oscillator Limit*

Using the above procedure, the steady-state parameters for any polyatomic *which actually achieves a steady state* (fast V–V, slow V–T/R relaxation) after laser pumping can be computed for any assumed dominant V–V energy transfer pathway in any molecule.[62] This approach has yielded the following results:

1. The vibrational state population or temperature distribution is a sensitive function of the dominant intermode vibrational energy-transfer mechanism or pathway. Thus *measurements of steady-state populations can conversely be used to determine energy-transfer mechanisms*. This is a remarkably useful result. Consider CH_3F where the dominant mechanism is believed to be

$$\langle \text{Pathway A} \rangle$$

$$CH_3F(\nu_3) + M \rightleftharpoons CH_3F(\nu_6) + M - 133\ \text{cm}^{-1}$$

$$CH_3F(\nu_6) + M \rightleftharpoons CH_3F(\nu_2, \nu_5) + M - 284\ \text{cm}^{-1}$$

$$2CH_3F(\nu_2, \nu_5) \rightleftharpoons CH_3F(2\nu_2, 2\nu_5) + CH_3F(0)$$

$$CH_3F(2\nu_2, 2\nu_5) + M \rightleftharpoons CH_3F(\nu_1, \nu_4) + M - 60\ \text{cm}^{-1}$$

This scheme predicts $T_3 > T_6 > T_{25} \sim T_{14}$. The alternate, very physically reasonable mechanism (first suggested to explain the rapid upconversion of vibrational energy in CH_3F), which couples (ν_1, ν_4) to $3\nu_3$ but not to $(2\nu_2, 2\nu_5)$

$$\langle \text{Pathway B} \rangle$$

$$2CH_3F(\nu_3) \rightleftharpoons CH_3F(2\nu_3) + CH_3F(0)$$

$$CH_3F(2\nu_3) + CH_3F(\nu_3) \rightleftharpoons CH_3F(3\nu_3) + CH_3F(0)$$

$$CH_3F(3\nu_3) + CH_3F \rightleftharpoons CH_3F(\nu_1, \nu_4) + CH_3F + 155\ \text{cm}^{-1}$$

$$CH_3F(\nu_3) + M \rightleftharpoons CH_3F(\nu_6) + M - 133\ \text{cm}^{-1}$$

$$CH_3F(\nu_6) + M \rightleftharpoons CH_3F(\nu_2, \nu_5) + M - 284\ \text{cm}^{-1}$$

predicts $T_{14} > T_3 > T_6 > T_{25}$. Many other paths can be suggested and have been investigated, but only pathway A agrees with the experimentally measured vibrational temperature distribution.[47] There are, of course, "degenerate" mechanisms that predict the same temperature distribution.[50] These are considered in Section III. B. 8.

2. The steady state differs *significantly* from that predicted for a Boltzmann distribution with equivalent total energy.[62,64,65] Depending on the energy-transfer path, localization of energy in one mode can occur. A corollary to this statement is that energy saturation of some modes does occur. This is a condition in which the energy of a mode does not increase with increasing pump power. An illustration of these effects is given in Fig. 8, which compares the CH_3F laser-pumped steady-state

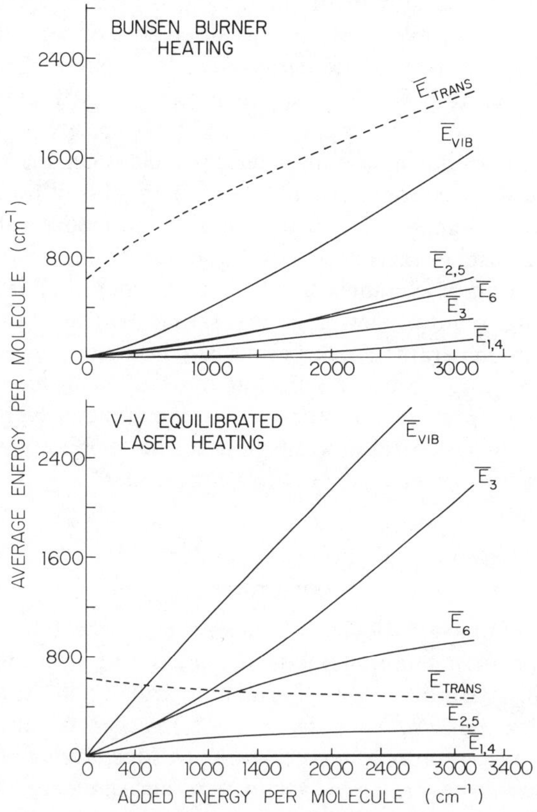

Fig. 8. (Upper) Plot of the average energy per molecule in the translational, rotational, and vibrational degrees of freedom of CH_3F vs added energy. A single temperature, Boltzmann distribution with harmonic oscillator states was assumed. $\bar{E}_{TRANS}$ is the sum of the translational and rotational energies. $\bar{E}_{VIB}$ is the total average vibrational energy and $\bar{E}_i$ is the mean energy of mode i. (Lower) Plot of the average energy distribution in CH_3F vs laser input energy for a system at vibration–vibration steady state achieved through Pathway A of the text. The calculated curves for the lower plot neglect vibration–translation/rotation relaxation.

energy distribution with that for a Boltzmann distribution at the same energy.

3. The steady-state distribution is dependent on the identity of the vibrational level initially pumped.
4. Large steady-state population inversions can occur between vibrational modes with minimum laser excitation energy.[67,73,74]
5. A stable, well-defined energy and population distribution can only be achieved through a restricted energy-transfer mechanism[62,72] (see Section III. B. 8).
6. Because of the sensitivity of population distributions to the energy-transfer pathway, experimental measurements of vibrational populations during the steady-state period after laser pumping can serve as a diagnostic tool to probe for changes in the energy-transfer mechanism as a function of laser pump power.[62] Ordinarily, for weak laser pumping, intermode energy-transfer channels occurring between levels high up in the electronic ground-state potential well do not affect the population or temperature distribution in the modes, since very little population reaches these levels. However, as the laser excitation power is increased, these channels should become more important. From an experimental point of view emphasis should be placed on the fact that vibrational temperatures can be measured on slow time scales (of the order of $\tau_{V-T/R}$, which can be hundreds of microseconds for some molecules at pressures in the 1–100 torr range) even through the processes which control or contribute to the temperature distributions may be occurring on very, very fast time scales.[47]

7. *Experimental Techniques for Measuring Vibrational Temperatures*

Given the diagnostic power of steady-state vibrational temperature distributions, experimental methods for measuring vibrational temperatures are highly desirable. For a harmonic oscillator the ratio of any two population densities $N_i/N_j = \exp[-(E_i - E_j)/kT_{\text{vib}}]$ is sufficient to determine the mode vibrational temperature provided levels i and j are within the same mode. In principle absorption measurements in either the infrared, visible, or ultraviolet range are capable of providing relative populations of successive levels in a harmonic oscillator. Visible/UV laser-induced fluorescence probing of vibration–rotation population distributions has been used for some time in molecular beam and IR photofragmentation studies,[75–77] while IR absorption probing requires a laser source to be reasonably useful. An extremely simple IR fluorescence technique, however, has been devised which is applicable to vibrational

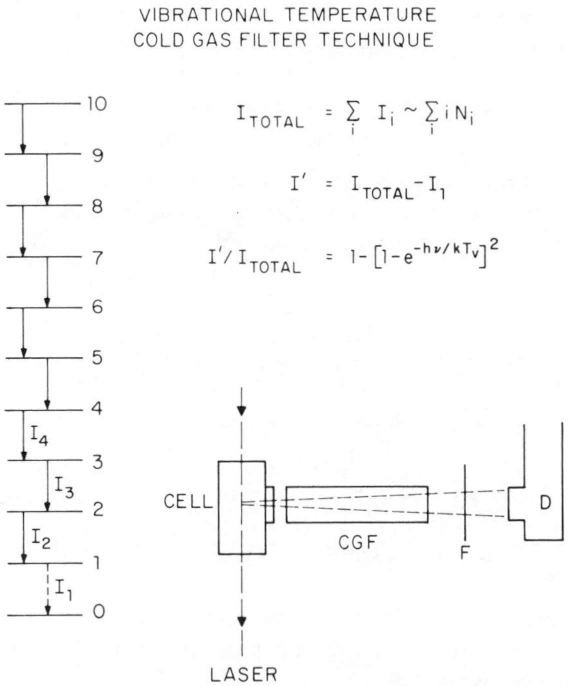

Fig. 9. Schematic diagram of cold gas filter technique for determining mode vibrational temperatures. The cell contains a sample gas of interest which is excited by a pulsed laser. Fluorescence from the sample is viewed through CGF, the cold gas filter cell. Initial intensity measurements are performed with the CGF empty to give the total infrared fluorescence I_{Total}. $I_{\text{Total}} - I_1 \equiv I'$ is then measured by filling the CGS with the same sample gas as contained in the cell. This unexcited (cold) gas absorbs radiation being emitted from level 1 to 0. In the harmonic oscillator limit I'/I_{Total} depends only on the vibrational temperature T_V. F is a filter used to select a single emitting mode of the sample and D is an infrared detector.

temperature measurements in bulk gases.[47] This method employs a cold gas filter to remove emission from the $v = 1 \rightarrow 0$ transition of a given mode (see Fig. 9). Measurements of fluorescence emission are made on a steady-state time scale after laser excitation with the cold gas filter full and empty. The ratio of the IR fluorescence emission for the two cases is a function of only the vibrational temperature for that mode in the harmonic oscillator limit. Though some care must be exercised in the use of cold gas filters for polyatomics, this technique is extremely simple and can provide very useful data for many molecular systems. Application of this method to CH_3F has provided an important corroborative piece of evidence to support pathway A above as the dominant intermode energy-transfer mechanism for this molecule.[47] Measurements of vibrational temperatures in COF_2 have also yielded promising initial results.[78]

8. *Degenerate Mechanisms, "Catastrophic" Cyclic Paths, and the Unique Path Assumption*

Degenerate energy-transfer mechanisms may be thought of as collections of physically distinct kinetic processes that lead to the same vibrational temperature distribution. An example of such a degenerate set for CH_3F is

$$\langle \text{Set 1} \rangle$$

$$CH_3F(\nu_3)+M \rightleftharpoons CH_3F(\nu_6)+M-133\ \text{cm}^{-1}$$

$$CH_3F(\nu_6)+M \rightleftharpoons CH_3F(\nu_2,\nu_5)+M-284\ \text{cm}^{-1}$$

$$\langle \text{Set 2} \rangle$$

$$CH_3F(\nu_3)+M \rightleftharpoons CH_3F(\nu_6)+M-133\ \text{cm}^{-1}$$

$$CH_3F(\nu_3)+M \rightleftharpoons CH_3F(\nu_2,\nu_5)-417\ \text{cm}^{-1}$$

Both these cases lead to identical temperature distributions. Fortunately, kinetic techniques can frequently be used to distinguish between such degenerate mechanisms.[35]

If sets 1 and 2 are simultaneously important, they constitute a *stable* cyclic path in which population can move from ν_3 to (ν_2, ν_5) along one path but return to ν_3 along a distinct second path *without loss of total vibrational population*.[62,72] A single temperature distribution results.

Catastrophic cyclic paths are those in which a loss of population can occur in a single cycle.[62,67,72] A possible case where this might happen is the molecule OCS

$$OCS(\nu_2)+M \rightleftharpoons OCS(\nu_1)+M-339\ \text{cm}^{-1}$$

$$OCS(\nu_1)+M \rightleftharpoons OCS(2\nu_2)+M-179\ \text{cm}^{-1}$$

$$OCS(2\nu_2)+OCS(0) \rightleftharpoons OCS(\nu_2)+OCS(\nu_2)$$

A careful consideration of the stoichiometry for this cyclic path reveals that a ν_2 molecule can travel from ν_2 to ν_1 to $2\nu_2$ without loss of vibrational population, but that return from $2\nu_2$ to ν_2 requires that a molecule be removed from the ground state thus increasing the total number of vibrationally excited molecules. The reverse cyclic path obviously leads to a loss of vibrationally excited molecules. Two vibrational temperature relationships of the type in (40) can be written for this cyclic system, but the only simultaneous solution occurs when $T_1 = T_2 = T'$ (complete one-temperature equilibrium).[62,67,72] In such cases the molecular system cannot reach a steady state. "V–T/R" relaxation is actually taking place on each cycle by exchanging 339 cm^{-1} + 179 cm^{-1} = 518 cm^{-1} between the vibrational and translational/rotational degrees of freedom and by returning one molecule to the ground state. Fundamentally, the difference between a stable and a catastrophic cyclic path is the presence or absence of a constraint *in addition to* the usual $\Sigma_i n_i = 0$ which guarantees that all of the population deviations must sum to zero.[62] For the CH_3F case above the additional constraint is $n_3 + n_6 + n_{2,5} =$ constant, or the sum of the vibrationally excited molecules is a constant. This is of course, the essence of the assumption that CH_3F reaches a V–V steady state in which the total vibrationally excited population decays away very slowly. In OCS, a constraint on $n_{\nu_2} + n_{\nu_1} + n_{2\nu_2}$ is not justified by the kinetic path. The presence of an additional constraint in CH_3F forces one of the eigenvalues of the rate matrix to zero leading to a metastable V–V equilibrium.[62] In a certain sense the experimental test for the presence of "catastrophic" cyclic paths in a polyatomic is the presence or absence of a metastable vibrationally excited steady state. This is a somewhat misleading statement, however, since a "catastrophic" cyclic path can certainly lead to a long-lived vibrationally excited population distribution if one or more of the kinetic rate processes in the cycle is slow. Of particular importance in the overall decay of a catastrophic cyclic path is the rate of exchange between the two highest states in the system (ν_1 and $2\nu_2$ for OCS). These states will contain on the average the smallest population deviations and relaxation will be rate limited or bottlenecked by the flow of population through these states. Such bottlenecking will occur even if the kinetic rate constants connecting states of small population are large.[19, 79] The situation is somewhat analogous to water flow in a system of pipes where a single capillary restriction can limit the total fluid flow.

In the most general sense a "catastrophic" cyclic path is one where the two possible pathways actually exchange different numbers of vibrational quanta as has been discussed in some detail in several places.[62,67,72] The probability for this happening should increase with the level of vibrational excitation where increasing anharmonic coupling makes the collisional or noncollisional exchange of multiple quanta between modes more probable. Nevertheless, as noted above for OCS, if the cyclic path is completed by states of very high energy, the population of these states will almost always be small enough (except at very, very high vibrational temperatures) to limit the rate of overall relaxation thus producing in effect a metastable state.[79]

The definition of a "unique" path for the vibrational temperatures purposes can now be developed. Such paths are simply those that produce a stable "long-lived" vibrational temperature distribution. These include all "degenerate" mechanisms leading to the same temperature distribution. "Catastrophic" cyclic paths are in principle excluded from such a unique path; nevertheless, if there is a slow kinetic step in such a "catastrophic" path, this step can be identified and removed thus producing a stable "unique" temperature distribution. While the steady-state vibrational temperature picture can correctly identify the presence of a "catastrophic" cycle, such a description provides no insight into the time scale required for catastrophe (or decay) to occur.

9. Chemical Reactivity of Vibrationally Metastable Distributions, Continuous Wave, and Discharge Pumping

The vibrational temperature picture provides an added bonus for experiments involving chemical reactivity of vibrationally excited molecules.[56,62,64–66,72] A casual glance at Fig. 8 shows the remarkable differences that can occur between the laser-pumped, V–V equilibrated, and Boltzmann vibrational energy distributions. The multiple temperature picture may be useful for describing the bimolecular reactivity of laser-pumped molecules excited to energies where the harmonic oscillator picture is still a reasonably good approximation. Since unimolecular decompositions a priori require pumping to very anharmonic molecular levels, extension of the multiple temperature picture to such reactions is less clear.

There is no reason in principle why the multiple temperature picture should be restricted to pulsed laser-pumped molecules. Dissociation of CO by a combination of V–V energy transfer and optical pumping with a continuous wave CO laser has been reported.[80] A modified version of the anharmonic pumping equations, appropriate for systems with a long-lived or continuous source of excitation,[81] has been used to at least partially explain these observations.[80] Even nonlaser-pumped systems can be described using the multiple temperature approach. For example, supersonic nozzles[82] and electric discharges[83] can produce metastable molecules with long-lived vibrationally excited states. The first recognition of these effects in the vibrational states of molecules, of course, was for systems that did not involve laser excitation at all.[60,61,84]

Some efforts have also been made to extend the strictly steady-state picture described here and to take into account the dynamic effects of excitation pulse width, rotational relaxation rates, and V–V pumping rates on the vibrational energy distribution.[47,85,86] Given the interest and practical importance of such descriptions, efforts in this direction seem sure to increase.

C. Vibrational Quanta Conservation

1. General Comments

Though the vibrational temperature concept is extremely powerful and useful, it cannot be easily employed with weakly excited gases studied near room temperature unless techniques for measuring vibrational temperatures accurately can be dramatically improved. For example, in the case of CH_3F $\nu_3-\nu_6$ equilibration (40) gives $-133/T'=1049/T_3-1182/T_6$. For $T'=300$ K, $T_3=350$ K (weak excitation) we find $T_6=343$ K. The differences between T_3 and T_6 are too small to measure, at least by the cold gas filter method. Thus the gas must either be strongly excited (provide a large T_3) or translationally cooled.

A technique that can be used for gases at room temperature under weak excitation where population deviations can be kept small ($n_i \ll N_i^\circ$) is "vibrational quanta conservation."[33,49] This method provides pathway information mainly by concentrating on relatively large changes in the eigenvalue amplitude coefficients C_{ik} [see (32)], which depend intimately on the particular intermode energy-transfer path followed to establish the V–V steady state. In a certain sense "vibrational quanta conservation" supplies information redundant to that available from vibrational temperature measurements; nevertheless, quanta conservation can be applied in the region where linear response assumptions are valid and all population deviations are coupled by linear differential equations. Experimentally, such measurements can be done with high repetition rate relatively low-power lasers where signal averaging provides excellent signal quality.

2. Energy Conservation Conditions

To illustrate the vibrational quanta conservation concept, consider the equilibration of CH_3F/O_2 mixtures after pulsed laser pumping of the $CH_3F(\nu_3)$ mode.[49,50] Energy transfer between CH_3F and O_2; which is rapid, occurs via

$$CH_3F(\nu_2,\nu_5)+O_2(0) \rightleftarrows CH_3F(0)+O_2(1)-90.4\ \text{cm}^{-1}$$

For simplicity, suppose that transfer from the laser-pumped ν_3 state to (ν_2,ν_5) can occur via the two pathways

$$\langle\text{Pathway C}\rangle$$

$$CH_3F(\nu_3)+M \rightleftarrows CH_3F(\nu_2,\nu_5)+M-417\ \text{cm}^{-1}$$

$$\langle\text{Pathway D}\rangle$$

$$2CH_3F(\nu_3) \rightleftarrows CH_3F(0)+CH_3F(2\nu_3)+17\ \text{cm}^{-1}$$

$$CH_3F(2\nu_3)+M \rightleftarrows CH_3F(\nu_2,\nu_5)+M+615\ \text{cm}^{-1}$$

Assume that these V–V steps for CH_3F are fast compared to CH_3F/O_2 equilibration and to overall V–T/R relaxation (this is an excellent assumption for CH_3F/O_2). Conservation of energy allows us to immediately calculate the change in CH_3F vibrational energy (ΔE_{vib}) and in translational/rotational energy (ΔE_{trans}) required to "convert" a $CH_3F(\nu_3)$ molecule into an $O_2(v=1)$ molecule by the mechanisms C and D. For path C, $\Delta E_{vib} = -1049$ cm^{-1} and $\Delta E_{trans} = -507.4$ cm^{-1} while for path D, $\Delta E_{vib} = -2098$ cm^{-1} and $\Delta E_{trans} = +541.6$ cm^{-1}. For both paths, of course, $\Delta E_{vib}(O_2) = 1556.4$ cm^{-1}. Note that path C leaves CH_3F vibrational energy *rich* at the expense of translational energy, while path D leaves CH_3F vibrational energy *poor* and significantly increases the translational energy. In path C, *one* ν_3 vibrational quantum is required to excite an $O_2(v=1)$ while in path D, *two* ν_3 vibrational quanta are needed. Thus if CH_3F equilibrates with O_2 by path D, the decrease in fluorescence intensity during the CH_3F–O_2 equilibration process will be significantly greater than the corresponding decrease for path C. Note also that the translational energy changes, which have been investigated in a separate study, are quite distinct for the two paths.[50]

3. *Quantitative Results*

The above qualitative features can be put on a quantitative basis to predict the ratio of the CH_3F fluorescence intensity before and after CH_3F–O_2 equilibration for each fluorescing state. Let N_i^0 be the ambient population of level i of CH_3F *before* laser excitation. The corresponding ambient populations for O_2 are $N_{v=1}^0$ and $N_{v=0}^0$. The deviations of the populations from their equilibrium values due to laser pumping and/or relaxation are taken to be n_i for CH_3F and $n_{v=1}, n_{v=0}$ for O_2. Let n_i^0 be the deviation of the population of a given CH_3F level *after* laser pumping of CH_3F *and* internal (V–V) equilibration of the CH_3F modes but *before* CH_3F–O_2 (V–V) equilibration. Finally, let n_i^f be the deviation of a given CH_3F level from ambient after complete V–V equilibration of all CH_3F modes with each other *and* with O_2. We assume that n_i^f describes the deviation of any level before V–T/R relaxation sets in. The fractions $f_i = n_i^f/n_i^0$ give the ratio of population deviations in CH_3F before and after equilibration with O_2. As shown below, *these fractions depend intimately on the path or mechanism which the* CH_3F *molecule takes to equilibrate its own modes*. The reason for this is that *some* V–V *equilibration schemes require more vibrational energy than others to reach the steady-state condition*.

It is a simple matter to show that for a path of type C above the population deviations are related as follows *at steady-state*

$$\frac{n_i}{n_j} = \frac{g_i}{g_j}\exp\left[\frac{E_j - E_i}{kT}\right] \tag{47}$$

The temperature T appearing in (47) is quite rigorously the translational temperature. In a similar manner the population deviations for O_2 are related to those for CH_3F by

$$\frac{n_i}{n_{v=1}}=\left(\frac{N_0^0}{N_{v=0}^0}\right)g_i\exp\left[\frac{E_{v=1}-E_i}{kT}\right] \tag{48}$$

which yields to an excellent approximation

$$\frac{n_i}{n_{v=1}}=\left(\frac{1-X_{O_2}}{X_{O_2}}\right)g_i\exp\left[\frac{E_{v=1}-E_i}{kT}\right] \tag{49}$$

The fractions f_i can now be easily determined. If n is the number of CH_3F molecules initially excited by the laser, simple stoichiometric considerations for path C yield immediately

$$n=n_3^0+n_{2,5}^0=n_3^f+n_{2,5}^f+n_{v=1}^f \tag{50}$$

(note that $n_{v=1}^0=0$). Using (47)–(49) in (50) gives

$$n_3^0\left\{1+3\exp\left[\frac{E_3-E_{2,5}}{kT}\right]\right\}=n_3^f\left\{1+3\exp\left[\frac{E_3-E_{2,5}}{kT}\right]+\left[\frac{X_{O_2}}{1-X_{O_2}}\right]\exp\left[\frac{E_3-E_{v=1}}{kT}\right]\right\}$$

For $kT=208\ \text{cm}^{-1}$ and $X_{O_2}=0.82$. $f_3=n_3^f/n_3^0=0.78$. Thus path C predicts a 22% drop in the $CH_3F(\nu_3)$ population when CH_3F comes into equilibrium with O_2.

Similar arguments can be used to construct the equations which give f_3 for path D. All steps in D which are linear (no excited state–excited state collisions) obey relationships identical to (47)–(49). The crucial difference between paths C and D is contained in the steps that bring ν_3 and $2\nu_3$ in CH_3F into equilibrium through excited state–excited state collisions. Because of this, in the low-excitation limit where the vibrational temperature of the ν_3 mode is essentially ambient,

$$n_{2\nu_3}=2n_3\exp\left[\frac{E_3-E_{2\nu_3}}{kT}\right] \tag{51}$$

The factor of 2 that appears in (51) is quite important and reflects the stoichiometry of path D. If n is again the number of CH_3F molecules initially pumped by the laser, the population balance condition for path D becomes

$$n = n_3^0 + 2\left(n_{2\nu_3}^0 + n_{2,5}^0\right) = n_3^f + 2\left(n_{2\nu_3}^f + n_{2,5}^f + n_{v=1}^f\right) \tag{52}$$

Using (47)–(49) and (51) transforms (52) to

$$n_3^0\left\{1 + (4)(3)\exp\left[\frac{E_3 - E_{2,5}}{kT}\right] + 4\exp\left[\frac{E_3 - E_{2\nu_3}}{kT}\right]\right\}$$

$$= n_3^f\left\{1 + (4)(3)\exp\left[\frac{E_3 - E_{2,5}}{kT}\right] + 4\exp\left[\left(E_3 - E_{2\nu_3}\right)kT\right]\right.$$

$$\left. + 4\left[\frac{X_{O_2}}{1 - X_{O_2}}\right]\exp\left[\frac{E_3 - E_{v=1}}{kT}\right]\right\} \tag{52a}$$

Inserting the constants and $X_{O_2} = 0.82$ gives $n_3^f / n_3^0 = f_3 = 0.625$. Thus path D predicts a 37.5% drop in the $CH_3F(\nu_3)$ population when the initially excited and equilibrated CH_3F comes into equilibrium with O_2. Under low-level laser excitation, where all population deviations are linearly related, the f_i for all states are the same.[49]

In a similar way each energy-transfer path is fingerprinted by the amount of CH_3F vibrational energy required to bring O_2 into equilibrium with CH_3F. The vibrational energy requirements in turn define the value for f_i.

4. *Results, Extension to Pure Systems, and Limitations*

Careful measurements of fluorescence intensities emanating from laser pumped CH_3F before and after equilibration with O_2 have been used to distinguish various paths in CH_3F.[49] The experimental results differ from those illustrated here because for simplicity the above treatment ignores the existence of the $CH_3F(\nu_6)$ level, which stores considerable vibrational population. The mole fraction of oxygen (or any other added "impurity" molecule) can obviously be adjusted for experimental convenience. Using impurity molecules to probe intermode energy-transfer paths in a "parent" polyatomic works best if (1) the intermode collisional coupling of the parent is fast compared to energy crossover to the impurity (this can usually be guaranteed by addition of a heavy rare gas to the mixture); (2) energy crossover to the impurity is fast compared to overall V–T/R relaxation of the combined parent/impurity system; and (3) the impurity is a diatomic.

This technique also does not strictly require an impurity molecule, though variations in the mole fraction of one species provide tremendous experimental convenience. Consider, for example, COF_2 which can be laser pumped to excite the ν_2 C–F symmetric stretch mode.[87,88] Two possible paths exist for energy transfer to lower levels:

Path 1

$$COF_2(\nu_2)+M \rightleftarrows COF_2(\nu_4)+M-280\ cm^{-1}$$

$$COF_2(\nu_4)+M \rightleftarrows COF_2(2\nu_5)+M+\sim 15\ cm^{-1}$$

$$COF_2(2\nu_5)+COF_2(0) \rightleftarrows 2COF_2(\nu_5)-9\ cm^{-1}$$

$$COF_2(\nu_5)+M \rightleftarrows COF_2(\nu_6)+M-147\ cm^{-1}$$

$$COF_2(\nu_5)+M \rightleftarrows COF_2(\nu_3)+M+37\ cm^{-1}$$

Path 2

$$COF_2(\nu_2)+M \rightleftarrows COF_2(\nu_6)+M+199\ cm^{-1}$$

$$COF_2(\nu_6)+M \rightleftarrows COF_2(\nu_5)+M+147\ cm^{-1}$$

$$COF_2(\nu_5)+M \rightleftarrows COF_2(\nu_3)+M+37\ cm^{-1}$$

Using the same procedure as employed for CH_3F/O_2 above, and assuming $n_2^0=n$ (number of molecules pumped by the laser to ν_2) with all other $n_i^0=0$, the values of f_2 are $\sim$0.20 for path 1 and $\sim$0.06 for path 2. Experimental measurements are presently underway to determine which of these paths is the correct one. In this case the experimental studies can be performed on a single species because population loss from ν_2 requires $\sim$500 collisions, while overall V–T/R relaxation of ν_3 via

$$COF_2(\nu_3)+M \rightarrow COF_2(0)+M+583\ cm^{-1}$$

requires $\sim$2400 collisions. Thus experimental criteria for applying the vibrational quanta conservation technique actually exist in the pure COF_2 species. In general errors will occur in the vibrational quanta conservation technique if the rates are not separated, since the simple "equilibrium" arguments used here to determine amplitudes will fail. The method depends for its success on being able to identify various segments of a fluorescence curve as belonging to a specific intermode V–V process or to a specific overall relaxation event. Whenever a system becomes tightly coupled and the rate constants for some of the processes are similar, the kinetic rate constants begin to have a serious effect on the fluorescence amplitudes for different parts of the curve (different eigenvalues).

D. KINETIC SENSITIVITY ANALYSIS AND RARE GAS ENGINEERING

1. *Sensitivity Analysis*

Very recently a simple concept which can be appropriately called "kinetic sensitivity analysis" has been employed to simplify the interpretation of vibrational energy-transfer data.[35,48] Again, using CH_3F as an example, this phenomenon can be understood by considering just the two energy-transfer processes

$$CH_3F(\nu_3)+M \rightleftharpoons CH_3F(\nu_6)+M-133\ \text{cm}^{-1} \tag{53}$$

$$2CH_3F(\nu_3) \rightleftharpoons CH_3F(2\nu_3)+CH_3F(0) \tag{54}$$

Under weak pumping conditions where $n_i \ll N_i^{\circ}$ for any level i, the population deviations of levels ν_3 and ν_6 together (n_3+n_6) are nearly 1000 times larger than that of $2\nu_3$ $(n_{2\nu_3})$ because $E_{2\nu_3}-E_{\nu_3} \simeq 1000\ \text{cm}^{-1}$ and $kT_{\text{vib}} \simeq 200$ cm^{-1}. Thus level $2\nu_3$ drains away a negligible fraction of total vibrationally excited population via (54). Since fluorescence amplitudes in laser excitation experiments are roughly controlled by population densities, the equilibration $2\nu_3$ with ν_3 is essentially impossible to detect *if fluorescence from only levels* ν_3 *and* ν_6 *is observed*. The amplitude of the kinetic eigenvalue which most closely corresponds to process (54) simply has negligible amplitude in the expressions for n_3 and n_6.[48] Conversely, however, this same eigenvalue amplitude is large for $n_{2\nu_3}$. Furthermore, the exchange of population between ν_3 and ν_6 will drastically affect the fluorescence from $2\nu_3$.[48] The overall principle is that high-energy states with very small populations do not affect low-energy states with large population while variations in low-energy state populations generally have significant affects on high-energy states. As a result, a reduced kinetic matrix can often be used to accurately describe energy-transfer processes among a small number of low-lying vibrational levels. This approach has been used successfully to describe intermode energy transfer in CH_3F[35] and CH_2D_2.[19]

There are, of course, exceptions to this general rule. For example, if levels ν_3 and ν_6 were not *directly* coupled via (53) but rather equilibrated through (54) and

$$CH_3F(2\nu_3)+CH_3F(0) \rightleftharpoons CH_3F(\nu_3)+CH_3F(\nu_6)-150\ \text{cm}^{-1} \tag{55}$$

The variations in $n_{2\nu_3}$ could not be neglected when observing ν_3 and ν_6. In such a case the $2\nu_3$ level is a bottleneck through which population must be funneled from ν_3 to ν_6. Recently, quantitative efforts have been made to develop a rigorous sensitivity test for the analysis of chemical kinetic data.[63] This approach should be extendable to energy transfer kinetics in a straightforward manner.

2. *Rare Gas Engineering of Kinetic Rates*

The idea of using rare gases to control various types of kinetic processes in polyatomic molecules has been described in detail in the review of CH_2D_2 data[19] [see (7)–(17) and associated discussion]. The basic idea is to distinguish events which *require* excited state–excited state collisions of parent species [e.g., ladder-climbing events such as (54)] from those that can be induced by a heavy rare gas. Generally, excited state-excited state collisions quickly lead to very highly excited species through resonant V–V energy exchange, while heavy rare gases only exhibit significant cross-sections for V–V energy transfer processes having energy gaps of the order of kT. Thus rare-gas engineering of kinetic rates is an extremely powerful technique for unraveling energy transfer mechanisms in molecules with mode vibration frequencies that are at least several times kT.

IV. COLLISIONAL PROPENSITY RULES FOR VIBRATIONAL ENERGY TRANSFER

Based on the data reviewed here plus a large quantity of experimental evidence that now exists in the literature, a set of *qualitative* propensity rules for intermode V–V collisional energy transfer can be formulated. These rules appear to be useful for making rough guesses about the relative efficiency of some energy-transfer processes. Molecules which undergo "sticky" collisions (hydrogen bonding, pseudomolecule complex formation), molecules with low-frequency torsional modes, or "large" molecules may not be well represented by these rules. Nevertheless, for molecules with relatively dilute vibrational levels (e.g., less than ~50 vibrational states below 3000 cm^{-1}) under conditions of low to moderate vibrational excitation, propensity rules do appear to be operative in vibrational energy-transfer processes.

Rule 1. Within a mode, "up-the-ladder" energy transfer is "fast." This simply means that nearly resonant V–V energy transfer events of the type

$$X(v)+X(1) \rightleftharpoons X(v+1)+X(0)+\Delta E \tag{56}$$

are very efficient. For high v states where anharmonicity becomes sizable (ΔE of the order of kT) the efficiency is expected to drop off. A compensating factor which slows this drop off is the increase in the square of the matrix element for $v \rightarrow v+1$ processes, which scales like v for a harmonic oscillator.

Rule 2. Energy transfer between 2 modes is "fast" if the $v=1$ levels of the different modes are close to each other in energy (within $\sim kT$). This rule can be illustrated by the equation

$$X(\nu_i)+M \rightleftharpoons X(\nu_j)+M+(\nu_i-\nu_j) \tag{57}$$

If $h(\nu_i-\nu_j) \lesssim kT$, such V–V energy transfer steps generally occur rapidly ($\lesssim 100$ gas kinetic collisions where M is the parent gas). There is some slight evidence that indicates such intermode exchanges are somewhat slower (100–200 gas kinetic collisions) for molecules which contain all heavy atoms (no hydrogen or deuterium).

Rule 3. If the $v=1$ levels of two modes are *not* close in energy, V–V exchange between the modes is enhanced by mechanical Fermi mixing of the modes, or by a large mechanical or electrical anharmonicity for the mode of lowest frequency. To illustrate this rule consider the equation

$$X(n\nu_i)+M \rightleftharpoons X(\nu_j)+M+(n\nu_i-\nu_j) \tag{58}$$

where $n\nu_i$ is the nth level of mode i (the level of mode i which is closest to the $v=1$ level of mode j) and $h(n\nu_i-\nu_j)$ is $\lesssim kT$. Rule 3 simply indicates that a process of the type (58) will be enhanced if there is some way of coupling $n\nu_i$ to the ground state by a process that is *first order* in some normal coordinate. Alternatively, such energy-transfer events will be enhanced if the potential field of the molecule is *not* a rapidly converging function of the normal coordinates for mode i (large mechanical or electrical anharmonicity). Energy exchange involving the $SO_2(2\nu_2)$ state appears to "anti"-demonstrate this rule quite well. For example, transfer of the type[57]

$$SO_2(\nu_1)+SO_2 \rightleftharpoons SO_2(2\nu_2)+SO_2+120\ \text{cm}^{-1}$$

and[89]

$$CH_3F(\nu_3)+SO_2(0) \rightleftharpoons CH_3F(0)+SO_2(2\nu_2)+18\ \text{cm}^{-1}$$

are both very slow ($\gtrsim 2500$ collisions). The $SO_2(2\nu_2)$ level, which has never been observed spectroscopically via a direct $0 \rightarrow 2\nu_2$ transition,[90] appears to be largely unaffected by any mechanical or electrical anharmonicities. On the other hand, efficient energy exchange between the bending and stretching modes of OCS has been observed and attributed to the large electrical anharmonicity of the bending mode. For example, the process

$$OCS(4\nu_2)+M \rightleftharpoons OCS(\nu_3)+M+43\ cm^{-1}$$

is rather efficient[71] (~120 collisions when M=OCS) and the OCS(ν_2) mode has a well-documented, large electrical anharmonicity.[91]

These rules are quite obviously what one would expect on the basis of a first-order time-dependent perturbation theory approach to collisional energy transfer in which the interaction potential is expanded only to first order in the molecular normal coordinates. There will certainly be many situations where such a simple picture is expected to fail. One may ask why these simple rules are even qualitatively successful in many cases. The answer may be that V–V energy-transfer processes which have been observed to date using laser techniques, have almost always been faster than V–T/R relaxation or they could not have been detected. (Among the obvious exceptions to this statement are CO_2[15] and SO_2[57] where laser excitation produces vibrationally excited population in states that only slowly relax to levels where V–T/R energy transfer can occur.) Since V–T/R relaxation probabilities per collision are typically 10^{-2}–10^{-4}, only rather efficient V–V processes can be observed. Rules 1–3, based as they are on a first-order perturbation picture, clearly emphasize collision events with large cross-sections. A more insidious possibility exists, however. At this early state in the study of intermode V–V energy transfer, the data base may be so restricted that these rules are valid for the cases studied but have no general long-range validity. The answers to this question will become clear as more data are collected.

V. CONCLUDING REMARKS

While much progress has been made in the area of vibrational energy transfer over the past decade, a detailed knowledge of energy-transfer rates and mechanisms is still only available for a handful of small, rigid molecules. Data for large molecules or for molecules with internal rotors is sparse. Studies of systems with very high levels of vibrational excitation are just now beginning to appear in the literature. So far rates for energy transfer from a specific vibration–rotation state $|vJ\rangle$ to another state $|v'J'\rangle$ are not available for polyatomic molecules, since almost all studies to date have measured only rotationally averaged cross-sections. Experiments which can probe detailed, rotationally specific cross-sections seem to be within sight and will probably be performed in the near future. Much of the present effort in the field is being directed toward the development of a general picture of polyatomic molecule energy transfer, which can be used to describe metastable states, very highly excited systems, and the interaction between intense laser fields and complex polyatomics. These efforts seem certain to have a strong effect on the study of chemically reacting molecular species, particularly those in vibrationally excited states.

Acknowledgments

We wish to thank our many students, collaborators, and scientific colleagues who have contributed to the work described here, which comes from our own laboratories. Support for energy-transfer research at Columbia has been provided by the National Science Foundation under grants MPS-75-04118 and CHE 77-24343, and by the Joint Services Electronics Program (U.S. Army, U.S. Navy, and U.S. Air Force) under contract DAAG29-77-C-0019. At Northwestern support has been provided by the National Science Foundation under grant CHE 76-10333 and the Donors of the Petroleum Research Fund administered by the American Chemical Society.

References

1. (a) K. T. Herzfeld and T. A. Litovitz, *Absorption and Dispersion of Ultrasonic Waves*, Academic Press, New York, 1959. (b) T. L. Cottrell and J. C. McCoubrey, *Molecular Energy Transfer in Gases*, Butterworths, London, 1961. (c) J. L. Stretton, *Transfer and Storage of Energy by Molecules*, Vol. 2, *Vibrational Energy*, G. M. Burnett, A. M. North, Wiley-Interscience, New York, 1969. (d) J. D. Lambert *Vibrational and Rotational Relaxation in Gases*, Clarendon Press, Oxford, 1977.
2. R. G. Gordon, W. Klemperer, and J. I. Steinfeld, *Ann. Rev. Phys. Chem.*, **19**, 215 (1968).
3. (a) C. B. Moore, *Ann. Rev. Phys. Chem.* **22**, 387 (1971). (b) C. B. Moore, *Ann. Rev. Chem. Phys.*, **23**, 41 (1973).
4. A. B. Callear, J. D. Lambert, in *Comprehensive Chemical Kinetics*, Vol. 4, C. H. Bamford and C. F. H. Tipper, eds., Elsevier, Amsterdam, 1969, Chap. 3, pp. 182–273.
5. B. Stevens, *International Encyclopedia of Physical Chemistry and Chemical Physics*, Vol. 3, Pergamon, Oxford, 1967.
6. P. Borrell, *Adv. Mol. Relaxation Proc.*, **1**, 69 (1967).
7. R. L. Taylor, *Can J. Chem.*, **52**, 1436 (1974).
8. D. Secrest, *Ann. Rev. Phys. Chem.*, **24**, 379 (1973).
9. T. Oka, *Adv. At. Mol. Phys.*, **9**, 127 (1973).
10. E. Weitz and G. W. Flynn, *Ann. Rev. Phys. Chem.*, **25**, 275 (1974).
11. M. J. Berry, *Ann. Rev. Phys. Chem.*, **26**, 259 (1975).
12. V. S. Letokhov, *Ann. Rev. Phys. Chem.*, **28**, 133 (1977).
13. S. Lemont and G. W. Flynn, *Ann Rev. Phys. Chem.* **28**, 261 (1977).
14. (a) S. Kimel and S. Speiser, *Chem. Rev.*, **77**, 437 (1977). (b) S. H. Bauer, *Chem. Rev.*, **78**, 147 (1978). (c) B. L. Earl, L. A. Gamss, and A. M. Ronn, *Acct. of Chem. Res.*, **11**, 183 (1978).
15. L. O. Hocker, M. A. Kovacs, C. K. Rhodes, G. W. Flynn, and A. Javan, *Phys. Rev. Lett.*, **17**, 233 (1966).

16. (a) J. T. Yardley and C. B. Moore, *J. Chem. Phys.*, **45**, 1066 (1966). (b) J. T. Yardley and C. B. Moore, *J. Chem. Phys.*, **49**, 1111 (1968).
17. (a) T. L. Cottrell and A. J. Matheson, *Trans. Faraday Soc.*, **58**, 2336 (1962). (b) P. D. Edmonds and J. Lamb, *Proc. Roy. Soc.*, **72**, 940 (1958).
18. T. M. Lowry and W. T. John, *Chem. Soc.* 97, 2634 (1910).
19. V. A. Apkarian and E. Weitz, *J. Chem. Phys.*, **71**, 4349 (1979).
20. W. M. Madigosky, *J. Chem. Phys.*, **39**, 2704 (1963).
21. P. Hess and C. B. Moore, *J. Chem. Phys.*, **65**, 2339 (1976).
22. J. T. Yardley and C. B. Moore, *J. Chem. Phys.*, **48**, 14 (1968).
23. J. T. Yardley and M. N. Fertig, and C. B. Moore, *J. Chem. Phys.*, **52**, 1450 (1970).
24. W. S. Drozdoski, A. Fakhr, and R. D. Bates, Jr., *Chem. Phys. Lett.*, **47**, 309 (1977).
25. W. S. Drozdoski, R. D. Bates, Jr., and D. R. Siebert, *J. Chem. Phys.*, **69**, 863 (1978).
26. V. A. Apkarian and E. Weitz, *Chem. Phys. Lett.*, **59**, 414 (1978).
27. R. Mehl, S. A. McNeil, L. Napolitano, L. M. Portal, W. S. Drozdoski, and R. D. Bates, Jr., *J. Chem. Phys.*, **69**, 5349 (1969).
28. (a) R. D. Sharma and C. A. Brau, *J. Chem. Phys.*, **50**, 924 (1969). (b) R. D. Sharma, *Phys. Rev.*, **177**, 102 (1969).
29. D. R. Siebert, F. R. Grabiner, and G. W. Flynn, *J. Chem. Phys.*, **60**, 1564 (1974).
30. F. R. Grabiner, D. R. Siebert, and G. W. Flynn, *Chem. Phys. Lett.*, **17**, 189 (1972).
31. D. R. Siebert, Ph.D. Thesis, Columbia University, 1963.
32. P. Zittel and C. B. Moore, *J. Chem. Phys.*, **58**, 2004 (1973).
33. M. Moser, A. Apkarian, and E. Weitz, "A Complete Determination of Vibrational Energy Transfer Pathways in CH_2D_2 for states below 3000 cm^{-1}; A Laser Induced Fluorescence Study" *J. Chem. Phys.*, (1981).
34. L. M. Sverdlov, *Opt Spektr.*, **10**, 33 (1961).
35. R. S. Sheorey and G. W. Flynn, "Collision Dynamics of Intermode Energy Flow in Laser Pumped Polyatomic Molecules: CH_3F," to be published.
36. W. L. Smith and I. M. Mills *J. Mol. Spectn.*, **11**, 11 (1963).
37. J. C. Deroche, *J. Phys.*, **34**, 559 (1973).
38. R. N. Schwartz, Z. I. Slawsky, and K. F. Herzfeld, *J. Chem. Phys.*, **20**, 1591 (1952).
39. F. R. Tanczos, *J. Chem. Phys.*, **25**, 439 (1956).
40. E. Weitz, G. W. Flynn, and A. M. Ronn, *J. Chem. Phys.*, **56**, 6060 (1972).
41. E. Weitz and G. W. Flynn, *J. Chem. Phys.*, **58**, 2679 (1973).
42. E. Weitz and G. W. Flynn, *J. Chem. Phys.*, **58**, 2781 (1973).
43. F. R. Grabiner, D. R. Siebert, and G. W. Flynn, *Chem. Phys. Lett.*, **17**, 189 (1972).
44. F. R. Grabiner, G. W. Flynn, and A. M. Ronn, *J. Chem. Phys.*, **59**, 2330 (1973).
45. G. T. Fujimoto and E. Weitz, *J. Chem. Phys.*, **65**, 3795 (1976).
46. J. M. Preses and G. W. Flynn, *J. Chem. Phys.*, **66**, 3112 (1977).
47. R. E. McNair, S. F. Fulghum, G. W. Flynn, M. S. Feld, and B. J. Feldman, *Chem. Phys. Lett.*, **48**, 241 (1977).
48. R. S. Sheorey, R. C. Slater, and G. W. Flynn, *J. Chem. Phys.*, **68**, 1058 (1978).
49. J. M. Preses, G. W. Flynn, and E. Weitz, *J. Chem. Phys.*, **69**, 2782 (1978).
50. I. Shamah and G. W. Flynn, *J. Chem. Phys.*, **70**, 4928 (1979).
51. B. L. Earl, P. C. Ysolani, and A. M. Ronn, *Chem. Phys. Lett.*, **39**, 95 (1976).
52. B. L. Earl, L. A. Gamss, and A. M. Ronn, *Acct. Chem. Res.*, **11**, 183 (1978).
53. M. Kneba and J. Wolfrum, *Ber Bunsenges Phys. Chem.*, **81**, 1275 (1977).
54. A. Miklavc and S. Fischer, *J. Chem. Phys.*, **69**, 281 (1978).
55. R. S. Sheorey and G. W. Flynn, unpublished results.
56. G. W. Flynn, ACS Symposium Series #56, "State to State Chemistry", Chapter 18, p. 145, P. R. Brooks and E. F. Hayes, Ed.

57. D. R. Siebert and G. W. Flynn, *J. Chem. Phys.*, **62**, 1212 (1975).
58. B. L. Earl, A. M. Ronn, and G. W. Flynn, *J. Chem. Phys.*, **9**, 307 (1975).
59. W. D. Breshears and L. S. Blair, *J. Chem. Phys.*, **59**, 5824 (1973); W. D. Breshears, *Chem. Phys. Lett.*, **20**, 429 (1973).
60. J. D. Teare, Seventh AGARD Colloquium, Oslo, 1966; J. D. Teare, Semi-Annual Program Progress Report Re-entry Physics (REP) Program, Avco Everett Research Laboratory, 1966; J. D. Teare, R. L. Taylor, and R. L. Von Rosenberg, Jr., *Nature*, **255**, 240 (1970).
61. C. E. Treanor, J. W. Rich, and R. G. Rehm, *J. Chem. Phys.*, **48**, 1798 (1968).
62. I. Shamah and G. W. Flynn, *J. Chem. Phys.*, **69**, 2474 (1978).
63. J. -T. Hwang, E. P. Dougherty, S. Rabitz, and H. Rabitz, *J. Chem. Phys.*, **69**, 5180 (1978).
64. I. Shamah and G. W. Flynn, *J. Am. Chem. Soc.*, **99**, 3191 (1977).
65. S. Mukamel and J. Ross, *J. Chem. Phys.*, **66**, 5235 (1977).
66. I. Procaccia and R. D. Levine, *J. Chem. Phys.*, **63**, 4261 (1975); R. D. Levine and A. Ben-Shaul, in *Chemical and Biochemical Applications of Lasers*, C. B. Moore, ed., Academic, New York; 1977.
67. I. Shamah, Ph.D. Thesis, "Vibrational Steady States Produced by the Vibrational Relaxation of Laser Pumped Polyatomic Molecules", Department of Chemistry, Columbia University, New York, September 1978.
68. R. E. Center, R. L. Taylor, C. A. Brau, and G. E. Caledonia, 6th International Conference on Quantum Electronics: Digest of Technical Papers, Kyoto, 1970, pp. 356–367.
69. A. W. Overhauser, *Phys. Rev.*, **92**, 411 (1953).
70. A. Javan, in *Quantum Optics and Electronics*, Dewitt, Blandin, and Cohen-Tannoudi, eds., Gordon and Breach, London, 1965.
71. D. R. Siebert and G. W. Flynn, *J. Chem. Phys.*, **64**, 4973 (1976).
72. A. Ben-Shaul and K. L. Kompa, *Chem. Phys. Lett.*, **55**, 560 (1978).
73. Columbia Radiation Laboratory, Progress Report #28, pp. 20–23, March 31, 1978.
74. Irwin Shamah and George Flynn, "Vibrational Relaxation Induced Population Inversions and Randomization Effects in Laser Pumped Polyatomic Molecules," *Chem. Phys.*, in press (1981).
75. A. Schultz, H. W. Cruse, and R. N. Zare, *J. Chem. Phys.*, **57**, 1354 (1972).
76. M. L. Lesiecki and W. A. Guillory, *Chem. Phys. Lett.*, **49**, 92 (1977).
77. D. S. King and J. C. Stephenson, *Chem. Phys. Lett.*, **51**, 48 (1977).
78. M. I. Lester and G. W. Flynn, work in progress.
79. M. L. Mandich and G. W. Flynn, work in progress.
80. J. W. Rich and R. C. Bergman, Dissociation of Carbon Monoxide by Optically Initiated Vibration-Vibration Pumping, Calspan Corporation Report #WG-6005-A-1, Buffalo New York, 14221.
81. V. F. Gavrikov, A. P. Dronov, V. K. Orlov, and A. K. Piskunov, *Sov. J. Quantum Electron.*, **6**, 938 (1976).
82. N. G. Basov, V. G. Mikhailov, A. N. Oraevskii, and V. A. Shcheglov, *Sov. Phys.-Tech. Phys.*, **13**, 1630 (1969).
83. A. N. Oraevskii, A. F. Suchkov, and Yu. N. Shebeko, *High Energy Chem.*, **12**, (2) 135 (1978).
84. C. T. Hsu and F. H. Maillie, *J. Chem. Phys.*, **52**, 1767 (1970).
85. L. P. Kudrin and Yu. V. Mikhailova, *Sov. Phys. JETP*, **41** (6), 1049 (1975).
86. R. A. Forber, R. E. McNair, S. F. Fulghum, G. W. Flynn, M. S. Feld, and B. J. Feldman, *J. Chem. Phys.*, **72**, 4463 (1980).

87. K. R. Casleton and G. W. Flynn, *J. Chem. Phys.*, **67**, 3133 (1977).
88. Y. V. Chalapati Rao, K. R. Casleton, and G. W. Flynn, "Intermode Energy Transfer in COF_2," manuscript in preparation.
89. R. C. Slater and G. W. Flynn, *J. Chem. Phys.*, **65**, 425 (1976).
90. K. Fox, G. D. T. Tejwani, and R. J. Corice, Jr., *Chem. Phys. Lett.*, **18**, 365 (1973).
91. A. G. Maki, E. K. Plyler, and E. D. Tidwell, *J. Res. Natl. Bur. St., Ser. A*, **66**, 163 (1962).

COLLISION-INDUCED INTRAMOLECULAR ENERGY TRANSFER IN ELECTRONICALLY EXCITED POLYATOMIC MOLECULES

STUART A. RICE*

The Department of Chemistry and The James Franck Institute, The University of Chicago, Chicago, Illinois 60637

CONTENTS

I. INTRODUCTION

The title of this chapter seems to promise a general discussion of the nature of collision-induced intramolecular energy transfer in electronically excited polyatomic molecules. If interpreted as just stated, the title promises more than can be delivered at this time. It is only recently that advances in experimental technique have permitted the study of the pathways of intramolecular energy redistribution following collision, and the few results now available were neither anticipated nor can they yet be fully accounted for by the available theories of collision-induced energy transfer. This chapter describes a preliminary synthesis of the limited experimental and theoretical information in hand and discusses some of its implications. It will be seen that more questions are raised than are answered.

*During 1979: Sherman Fairchild Distinguished Scholar, California Institute of Technology, Pasadena, California.

Studies of collision-induced mode-to-mode energy transfer in the ground electronic states of several small polyatomics reveal two qualitative features of the process.[1,2] First, the energy-transfer cross-sections are, with only a few exceptions, small compared with the cross-section for collision. Second, there are strong propensity rules governing the pathways of mode-to-mode transfer so that not all levels within the range defined by the distribution of collision energies are equally accessible from the initial level.

Much less is known about collision-induced energy transfer in the excited states of molecules. For diatomic molecules it is generally the case that vibrational relaxation in an excited state is very rapid compared to vibrational relaxation in the ground state.[3] A similar pattern has emerged from the very few studies of polyatomic molecules that have been made. In the earliest of these studies only a vibrational relaxation rate averaged over many levels was determined. For example, Stockburger measured the steady-state fluorescence spectra of naphthalene and benzene after excitation at 2537 Å and in the presence of varying amounts of added gas.[4] Changes in the spectra as a function of pressure suggested that vibrational relaxation requires about 1 collision. In other work, vibrational relaxation rates have been deduced from the observation that the lifetime of total emission depends on the directly excited vibronic level at low pressure, but is independent of the initial level at high pressure. A simple model is then used to infer an average relaxation rate from measurements of the total emission lifetime as a function of pressure. As one example of the use of this method, S_1 β-naphthylamine was studied by exciting vibronic levels in groups widely separated in energy. It was found that relaxation requires about 6 collisions with argon.[5] Similarly, when single vibronic levels in S_1 methyltriazoline dione are excited vibrational equilibration requires 12 collisions with argon or 4 collisions with CCl_4.[6] More detailed results have been obtained from studies in which the initial state is a single vibronic level and the fluorescence spectrum is dispersed. Lineberger and co-workers have used time- and wavelength-resolved emission to make a careful study of the total depletion of many individual vibronic levels in S_1 glyoxal, glyoxal-*hd*, and glyoxal-d_2 due to collisions with the S_0 molecule.[7–9] They concluded that intersystem crossing was the major process quenching high vibrational levels, having a cross-section six times the collision cross-section at 2000 cm^{-1} of vibrational energy. For the zero-point level they found that vibrational relaxation within the S_1 manifold required about 2 collisions, and for the torsional mode ($\tilde{\nu}_7 = 233$ cm^{-1}), they found that the vibrational relaxation cross-section was about 1.7 times the collision cross-section. The latter observation is consistent with a direct measurement showing that population of the zero point level from the 7^1 level requires slightly more than one collision with argon.[10]

Much more detailed information concerning vibrational relaxation in an excited electronic state of a polyatomic molecule comes from studies of $^1B_{2u}$ benzene[11] and $^1B_2$ aniline.[12] In both studies wavelength resolved fluorescence spectroscopy was used to monitor the steady-state populations of different vibronic levels populated from a given initial level as a function of added gas pressure. In the benzene experiments the initial state prepared was, in all cases, the 6^1 level of the $^1B_{2u}$ state, lying 522 cm^{-1} above the origin. Ground-state benzene (S_0) and 13 other gases were used as collision partners. In contrast, in the aniline experiments vibrational energy transfer was traced from 8 vibronic levels in the $^1B_2$ electronic state, but only with argon as the collision partner. It is the data obtained from these experimental studies that we discuss in the following sections. Briefly put, vibrational relaxation in $^1B_{2u}$ benzene and $^1B_2$ aniline is very efficient, the cross-section being an appreciable fraction of the hard-sphere collision cross-section. The vibrational relaxation process is also very selective in the new levels which are populated. In the case of $^1B_{2u}$ benzene, depletion of the level 6^1 is accounted for by growth of population in only 6 levels, which in fact represent only four channels of decay. Yet there are 20 levels which lie within 500 cm^{-1} of 6^1. The branching behavior depends on the vibrational structure of the collision partner. For example, only two decay channels are important when He or N_2 is the collision partner, while all four are important when *n*-heptane is the collision partner. Relaxation in $^1B_2$ aniline has similar features, namely, a very large cross-section and strong propensity rules govern the pathways of decay. But more interesting and more important insofar as collision-induced energy transfer is concerned, the vibrational modes of $^1B_2$ aniline can be separated into at least two nearly noncommunicating groups of levels. One group contains the inversion mode and the ring skeleton mode 6a, the other contains the origin, the out-of-plane bending modes 10b and 16a, and the in plane bending mode 15 (see Fig. 1). A mapping of the energy flow in $^1B_2$ aniline shows that it will take a surprisingly large number of collisions to achieve full randomization of the energy given an initial distribution in which only one vibrational mode is excited.

Another common feature of relaxation in $^1B_{2u}$ benzene and $^1B_2$ aniline is that a very frequent first step in the collision-induced vibrational energy exchange is endoergic "up-pumping" of the excited molecule, even when exoergic channels are available. The ubiquity and overall importance of this first endoergic step must be explained by any plausible mechanism for the processes observed.

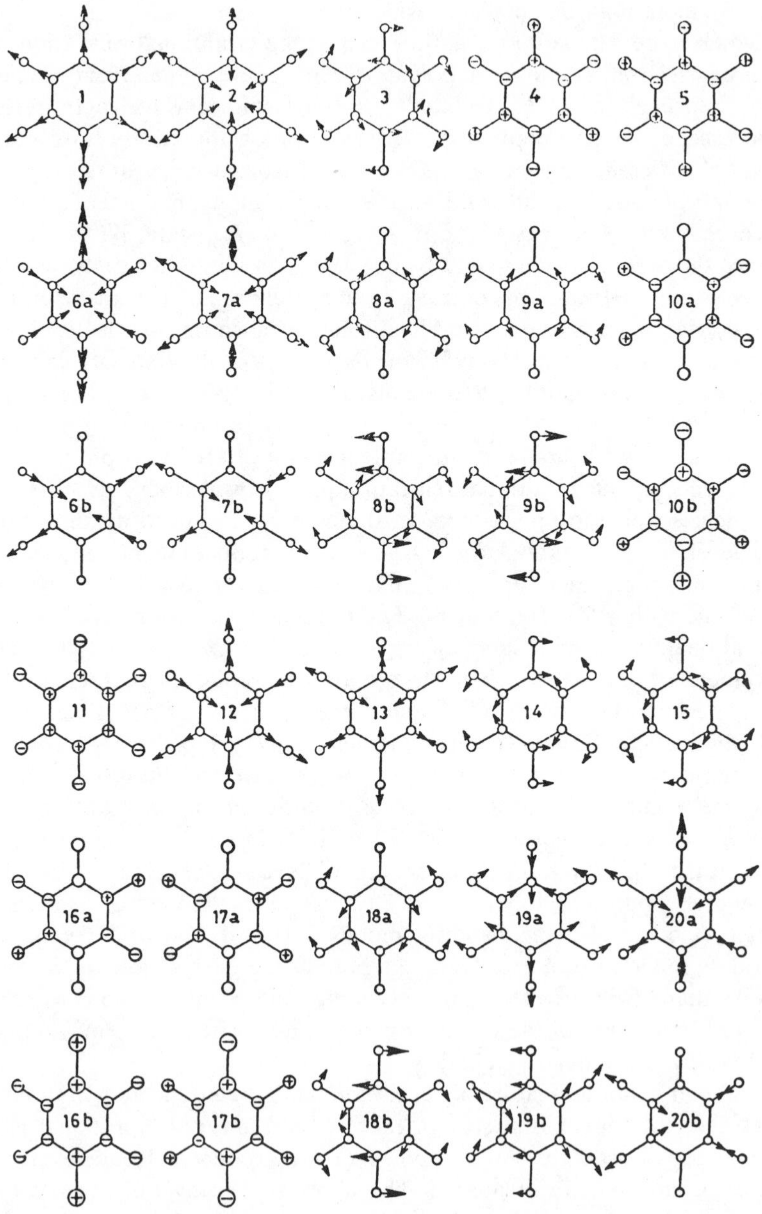

Fig. 1. The normal modes of vibration of benzene.

II. OVERVIEW OF THEORETICAL DEVELOPMENT

There exists an elegant formal theory of molecule–molecule scattering.[13] Unfortunately, current knowledge of the relevant potential energy surfaces is meager, and under many conditions of interest there are a vast number of open channels, representing different distributions of translational, rotational, and vibrational energy in the final state of the system. Given this complication, it is not surprising that the reduction of the scattering theory formalism to a useful algorithm has not yet been accomplished.

A few computer simulations of classical trajectories in atom–polyatomic molecule systems have been reported.[14] Leaving aside any questions related to the adequacy of the potential energy surfaces used, these studies have been directed toward understanding the overall vibrational relaxation of the polyatomic partner in the collision, with little attention focused on the nature of the accompanying intramolecular energy exchange.

To date, the theoretical results most useful for the discussion of collision induced intramolecular energy exchange have come from grossly oversimplified models of the collision. The most widely used of these models was proposed by Schwartz, Slawsky, and Herzfeld (SSH)[15,16] and later modified by Tanczos[17] and Stretton.[18]

The SSH model, although nominally three-dimensional, reduces the collision dynamics to a collinear interaction of two particles in an effective potential, which includes the centrifugal potential. It is argued that the centrifugal potential plays no essential role, and that translation-to-vibration energy exchange and its reverse are determined by the short-range repulsive interaction between the collision partners. The overall transition probability for any energy transfer process is represented as the product of independent transition probabilities for the several possible excitation and deexcitation steps. The only ways in which the analyses of atom–diatomic molecule and atom–polyatomic molecule collisions differ are as follows:

1. Inclusion of a steric factor to account for the possible orientations of the polyatomic molecule in a collision.
2. Inclusion of the possibility of interaction of the atom with different normal modes of the molecule in proportion to the squares of the projections of the normal modes along the collision trajectory.
3. Allowing for all multiple transitions in the set of vibrational modes that satisfy the constraint of conservation of energy.

The SSH model completely suppresses the dynamical role of molecular rotation, neglects energy transfer to or from rotation, and ignores the possibility of internal coupling of vibrations, for example, by Fermi resonance. It is to be expected that some or all of these neglected aspects of the dynamics play an important role in various collision processes. Nevertheless, in a few favorable cases the SSH model provides a qualitative guide to the relative importance of transitions involving changes in the status of different numbers of vibrational modes, the dependence of the transition probability on the energy mismatch, and the effect of changing the mass of the colliding partner.[1]

Fischer and co-workers[19,20] have extended the SSH model. In their formulation the translation and rigid-body rotation of the molecule are taken to be classical, and only the internal vibrational motion is quantized. The atom–polyatomic molecule collision trajectory is approximated by a straight line; any such trajectory not forbidden by steric hindrance and which terminates at the equilibrium position of a peripheral atom of the polyatomic is given equal weight in the overall collision process. In addition, Fischer and co-workers explicitly consider the role of Fermi resonance in the collision-induced energy exchange process. The distribution of energy released into rotational and translational degrees of freedom when a vibrational transition occurs is determined from the consistency conditions imposed by the conservation of energy and angular momentum.

The Fischer modification of the SSH model leads to some interesting predictions. When the energy released by a vibrational transition is larger than $k_B T$ the dominant contribution to the transition probability comes from collisions with small angle of approach averaged effective masses, and for these collisions a large part of the released vibrational energy goes into the rotational energy of the molecule. In the case that the vibrational energy released is small relative to $k_B T$, the Fischer model gives smaller transition probabilities than the SSH model. The change arises from the different ways the two models treat the collision between the incoming atom and a peripheral atom on the molecule, and is not a consequence of the inclusion of rotational motion in one but not the other model. Finally, as is to be expected, Fischer finds that Fermi resonance can play a very important role in several kinds of collision-induced vibrational transitions.

The SSH and Fischer models are based on simple approximations to the collision dynamics, but neither one assumes the existence of a collision complex, postulates rapid randomization of vibrational energy in the polyatomic collision partner, or separates the vibrational degrees of freedom into groups categorized by their rates of relaxation, or ease of exchanging a quantum, etc. Other theories of atom–molecule collision dynamics make use of some or all of these ideas. Schatz,[21] Rabitz,[22] and Eu,[23] and their co-workers, have proposed different variations of a multiple-time-scale stochastic description of collision dynamics, and Forst[24] has described a model in which the collision complex lifetime is due to orbiting collisions and the internal degrees of freedom rapidly exchange energy.

In Schatz's formulation the classical mechanical equations of motion for the "slow" degrees of freedom, say translation and possibly also rotation, are rewritten, in the form of a generalized Langevin equation. The coupling of the degrees of freedom that are responsible for energy exchange is represented by "friction" and "random force" terms in the generalized Langevin equation. In principle a formulation of this type is exact if none of the correlations and so-called memory effects are neglected. In practice, the generalized Langevin equation cannot be solved unless approximations are made in the treatment of correlations and memory effects in the dynamics. Schatz assumes that the dynamical correlations between the fast and slow variables are weak, and uses this approximation to eliminate the memory terms from the generalized Langevin equation, so that it takes on a Markovian form. Some of the dynamical effects neglected in the Markovian limit are reintroduced by approximating the calculation of the friction term in such a way as to include the correlations between fast and slow degrees of freedom along one selected trajectory. For details of this and other approximations in the theory the reader is referred to the original paper.[21] It suffices for us to note that this form of analysis does not provide any simple physical description of the relative importance of different pathways of intramolecular energy exchange. In the one model calculation reported, the collinear collision of an atom and a linear triatomic molecule, the predicted total translation-to-vibration energy exchange as a function of translational energy is very close to that obtained from exact semiclassical calculations.[21]

Rabitz's use of a multiple-time-scale representation of the collision dynamics is somewhat different. The separation of timescales in this theory is based on the rate of phase accumulation, since in the semiclassical limit this is related to the time needed for transfer of a quantum of energy. When the interactions are such as to generate rapid-phase accumulation, as in the description of deflected translational motion, classical mechanics is appropriate, and when the interactions generate slow-phase accumulation, as in vibrational depopulation, quantum mechanics must be used. The effect of interactions on rotational motion spans the range of behavior between these two limits. The stochastic assumption introduced by Rabitz asserts that large and rapidly varying phases permit use of a random phase approximation. Reduction of the equations of motion to a useful form requires further approximations; the reader is referred to the original paper for full discussion of these.[22] The theory described has some very interesting features, and may give useful insights into the dynamics of atom–molecule collisions, but it has not yet been applied to cases in which polyatomic molecules are involved in the collision.

As already mentioned, Forst has described a model of energy exchange in atom–molecule collisions that is based on the assumption that a collision complex is formed and that energy exchange among the vibrations of the molecule is rapid. Forst's model differs from the simplest version of a collision-complex model by imposing three contraints. First, it is assumed that the collision complex arises from orbiting collisions, so that its lifetime must be just the collision duration. Second, it is necessary that the internal degrees of freedom of the collision complex that correlate with relative translation and intrinsic rotation of the separated atom and molecule can only have an energy less than the maximum energy for which orbiting can occur. Third, those internal degrees of freedom of the collision complex that correlate with the internal degrees of freedom of the isolated molecule must have a density of states such that, subject to the constraint on the maximum amount of relative translational and intrinsic rotational energy, the complex lifetime is equal to the collision duration. Forst finds that the appropriate density of states is equivalent to the participation in energy exchange by only one oscillator.[24] The nature of the model precludes the strong conclusion that this is one of the true modes of the molecule; it can be any combination of modes with a density of states close to that expected for only one oscillator at the same energy. Nevertheless, the model shows that the dynamics of the collision process greatly constrains the randomization of energy during a collision, even in the favorable case of orbiting collisions. Clearly, this model gives a first hint of the possibility of a propensity for collision-induced intramolecular energy transfer to involve only a subset of vibrational modes of the molecule.

At the other extreme from the statistical model of atom–molecule collision dynamics is the simplification of the multichannel quantum theory by the use of classical mechanics to describe some of the degrees of freedom of the atom–molecule system, followed by essentially exact solution of the resulting equations of motion. The "classical part" of the system generates a time-dependent perturbation on the "quantum-mechanical part." By using the time-dependent Schrödinger equation it is possible to formulate a set of coupled equations for the transition amplitudes in the quantized degrees of freedom. This method, originally developed for atom–diatom collisions, has been extended by Billing[25] to the description of molecule–molecule collisions. Billing's analysis is more complete than earlier, similarly motivated, calculations by Shin.[26] Unfortunately, the prediction of the behavior of particular systems remains numerically tedious; judging from the few approximate numerical calculations available, the method is capable of accurate prediction of energy transfer cross-sections.[26] However, the analysis does not lend itself to the deduction of simple propensity rules or generalization about the overall characteristics of different types of collisions.

Finally, we mention that a theory of collision-induced intramolecular vibrational energy exchange can be constructed using ideas like those employed in the theory of collision-induced electronic relaxation.[27] That approach is now being developed by Freed.[28] It is expected to be of greatest value in the formulation of general propensity rules descriptive of the atom–molecule energy transfer process, and not as an algorithm for the calculation of cross sections.

III. SELECTED EXPERIMENTAL DATA

A. Vibrational Energy Transfer

The data that we now discuss, from the experiments of Parmenter and Tang[11] and of Chernoff and Rice,[12] were obtained by the same method. Briefly, a dictionary of single vibronic level (SVL) fluorescence spectra was recorded under collision-free conditions. These reference spectra were compared with SVL fluorescence spectra recorded under conditions such that the excited molecule experienced no more than one collision during the lifetime of the prepared level. Differences between the two spectra were then used to determine the pathways of energy transfer and the magnitudes of the energy transfer cross-sections. Detailed descriptions of the apparatus used, the method of reducing the observed spectral intensities to cross-sections, the precautions needed in the experiments, and the errors involved in the interpretation of the observations can be found in the original papers.[11,12]

A quick overview of the nature of collision-induced intramolecular energy transfer in $^1B_{2u}$ benzene and $^1B_2$ aniline can be obtained from a diagram that maps the flow from the initially prepared state. Figures 2–19 display the results of the studies of Parmenter and Tang and of Chernoff and Rice. In each case the rate constants written next to the arrows are expressed as a fraction of the corresponding hard-sphere collision rate. It is immediately apparent that

1. The efficiency of collision-induced intramolecular energy transfer on an excited electronic state energy surface is very much higher than on the ground-state energy surface.
2. Collision-induced intramolecular vibrational energy transfer is restricted to a small subset of the levels in the energetically accessible range.

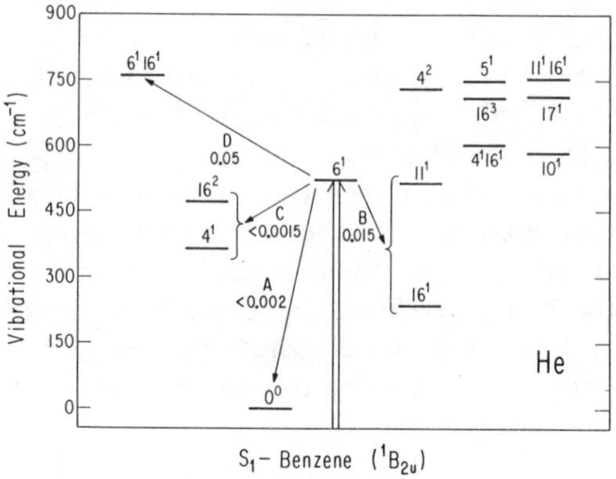

Fig. 2. The pathways for collision-induced intramolecular vibrational energy transfer from the 6^1 level of ${}^1B_{2u}$ benzene. The figures entered are the cross-sections relative to the hard sphere collision cross-section. Collision partner: He.

3. If the data for the Ar: ${}^1B_2$ aniline system are representative of the general case, the small set of collision-coupled modes is itself subdivided into several weakly interacting groups of modes. In particular, Chernoff and Rice find that collision of an Ar atom with ${}^1B_2$ aniline induces facile exchange of energy among the low-frequency modes 10b, 15, and 16a, but that modes 6a and I are only weakly coupled to the others (see Fig. 1 for the mode numbering used).

Figure 20 provides a further illustration of the pattern of collision-induced energy flow in aniline by diagramming the vibrational relaxation, which

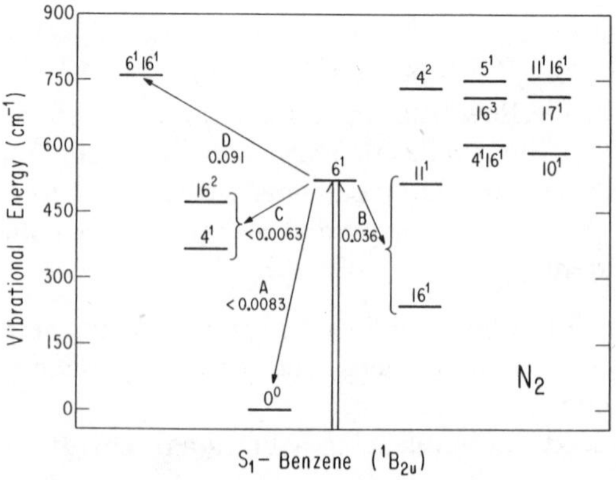

Fig. 3. Same as Fig. 2. Collision partner: N_2.

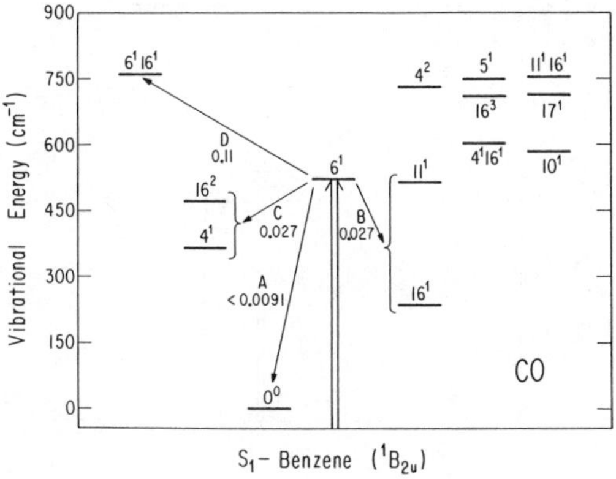

Fig. 4. Same as Fig. 2. Collision partner: CO.

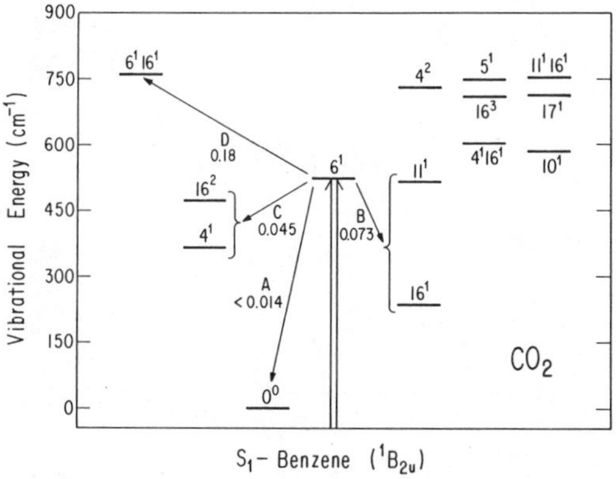

Fig. 5. Same as Fig. 2. Collision partner: CO_2.

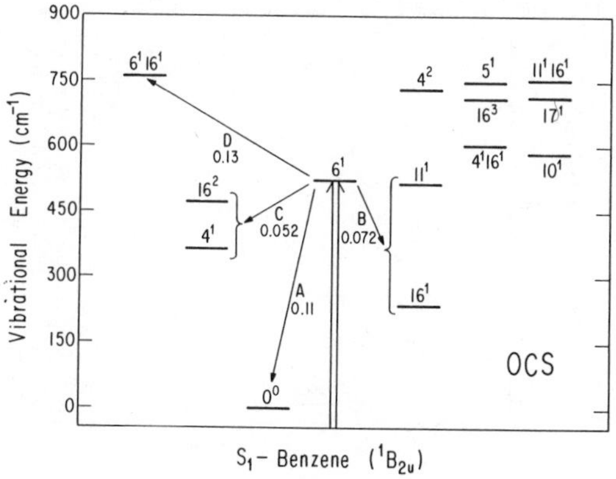

Fig. 6. Same as Fig. 2. Collision partner: OCS.

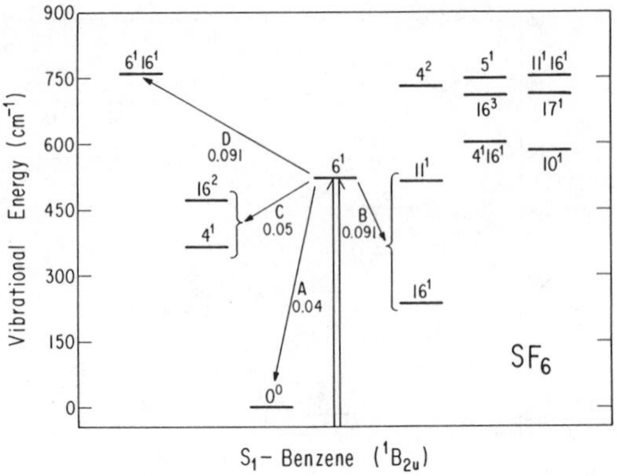

Fig. 7. Same as Fig. 2. Collision partner: SF_6.

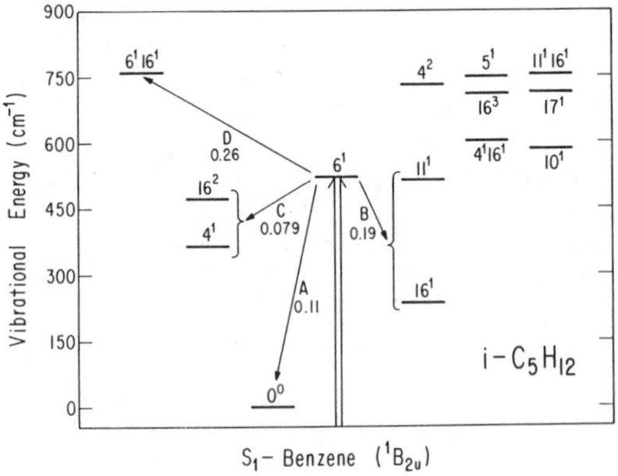

Fig. 8. Same as Fig. 2. Collision partner: i-C_5H_{12}.

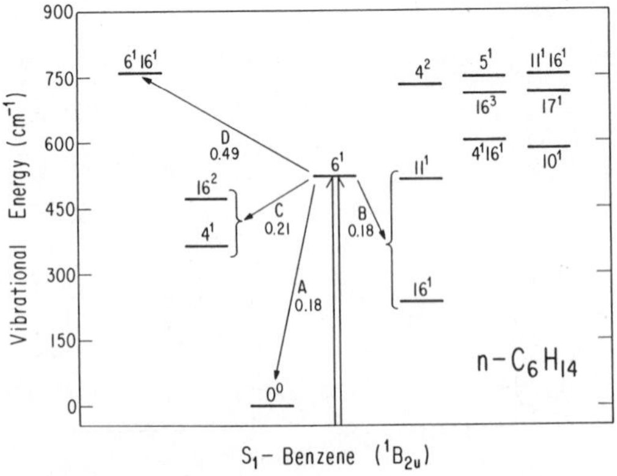

Fig. 9. Same as Fig. 2. Collision partner: n-C_6H_{14}.

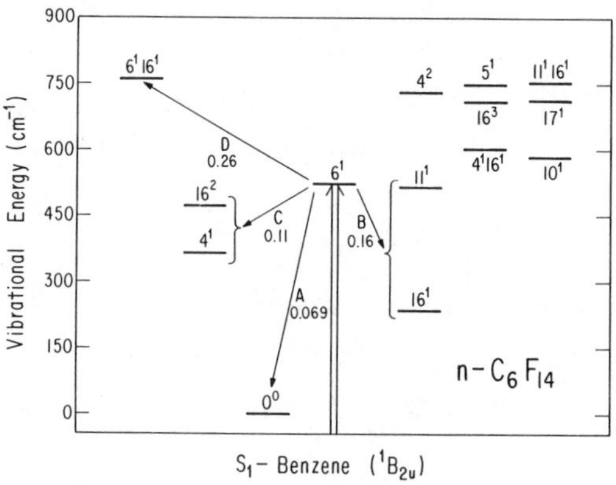

Fig. 10. Same as Fig. 2. Collision partner: n-C_6F_{14}.

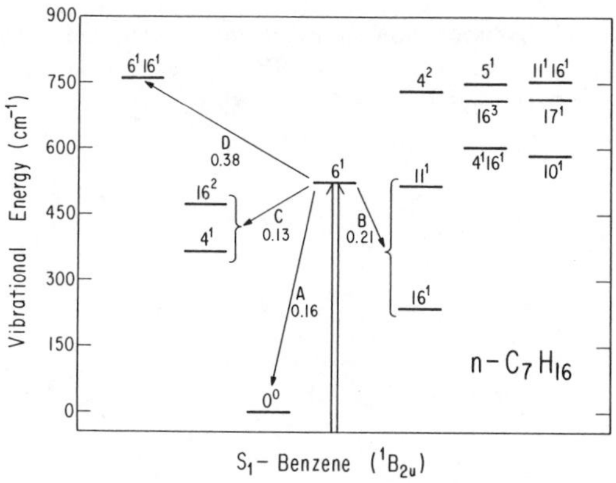

Fig. 11. Same as Fig. 2. Collision partner: n-C_7H_{16}.

occurs in two collisional steps after initial preparation of level $16a^1$. It is clear that the low-frequency modes are tightly coupled and that equilibration of all the vibrations must be a slow process.

There is some evidence that this is also true at high vibrational energies in $^1B_2$ aniline. As part of a study of electronic energy transfer in benzene–aniline gas mixtures, Lardeux and Tramer[29] measured the fluorescence spectra of aniline excited to produce $^1B_2$ species having vibrational energy at several nominal values in the range 2000–6000 cm^{-1}. The spectra are continuous and the maximum is shifted further to the red with increasing energy. In the presence of helium the spectra are essentially unchanged,

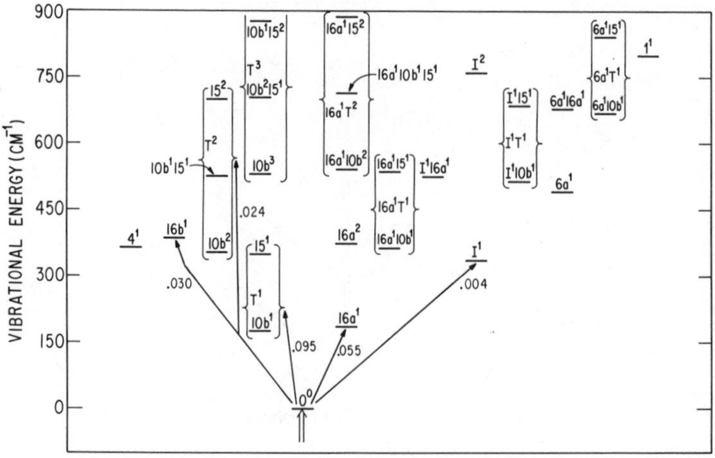

Fig. 12. The pathways for collision-induced intramolecular vibrational energy transfer from ${}^1B_2$ aniline levels. The collision partner is Ar. The figures entered are the cross-sections relative to the hard-sphere collision cross-section. Level: 0^0.

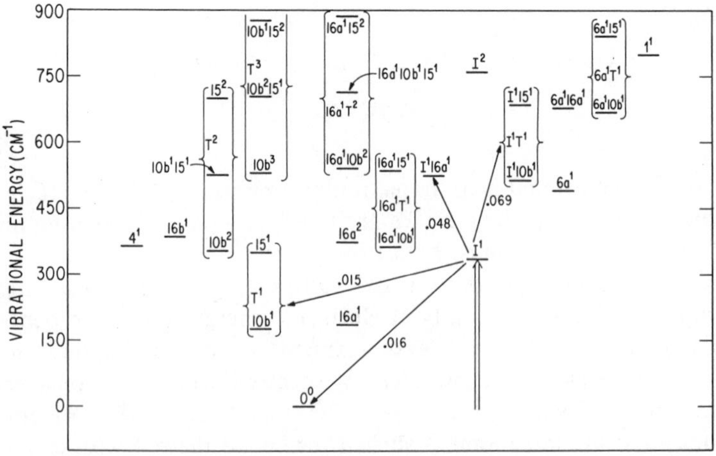

Fig. 13. Same as Fig. 12. Level: I^1.

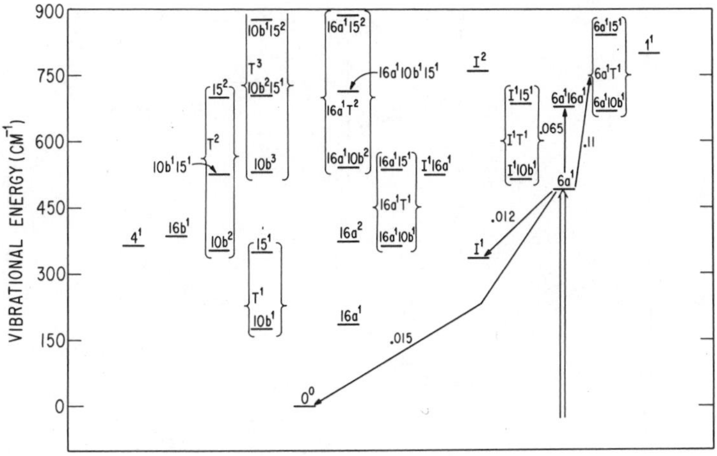

Fig. 14. Same as Fig. 12. Level: $6a^1$.

even with pressures high enough to result in 5 collisions during the lifetime of the excited electronic state, which suggests that these collisions leave the average vibrational energy essentially unchanged. Put another way, vibrational energy transfer down the ladder of the high-frequency modes which are excited is rather inefficient with helium as a collision partner. The work of Logan, Buduls, and Ross is also relevant to this point.[30] They studied the vibrational relaxation of $^1B_{2u}$ benzene excited at 2537 Å to levels with

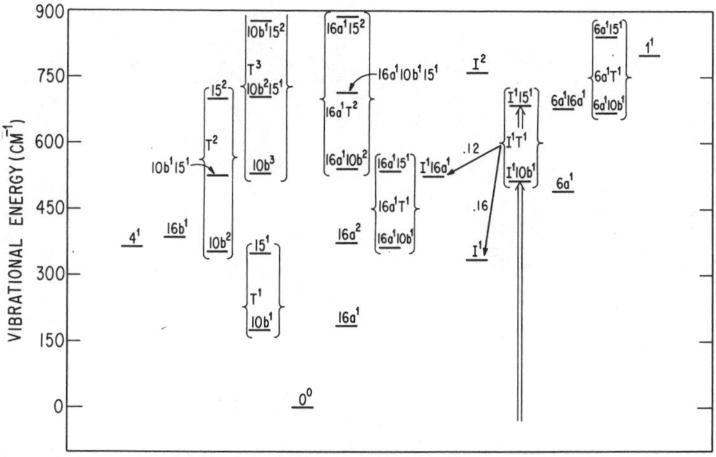

Fig. 15. Same as Fig. 12. Level: I^1T^1.

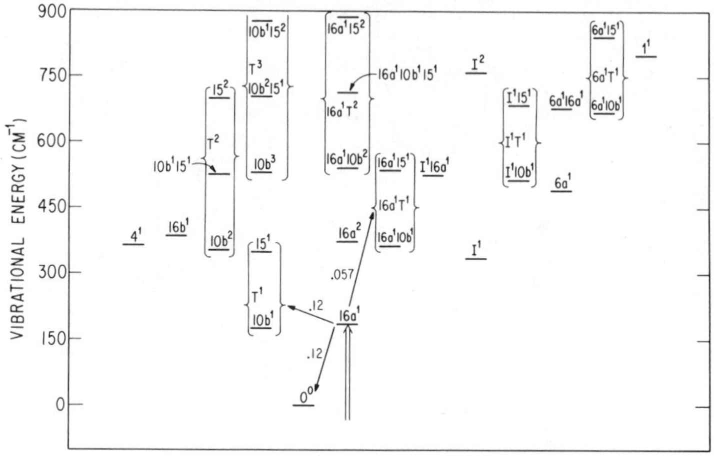

Fig. 16. Same as Fig. 12. Level: $16a^1$.

about 1700 cm^{-1} of vibrational energy. Polyatomic collision partners like ether and isopentane were one order of magnitude more effective than helium in depleting the initial levels and two orders of magnitude more effective than helium in achieving vibrational equilibrium, which reinforces the notion that V→T/R· energy transfer is ineffective for large energy gaps.

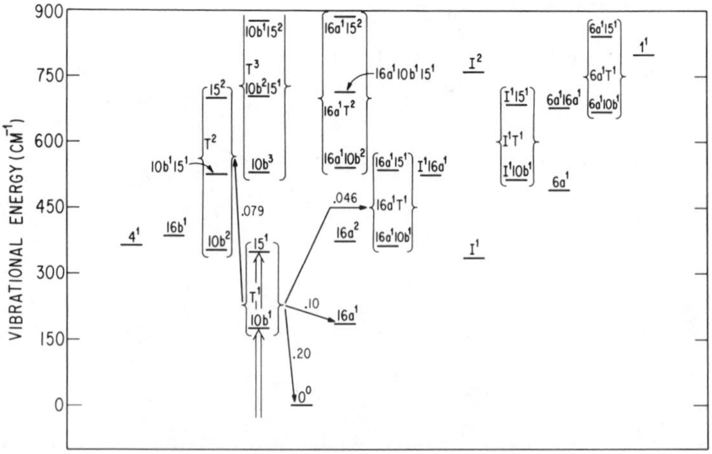

Fig. 17. Same as Fig. 12. Level: T^1.

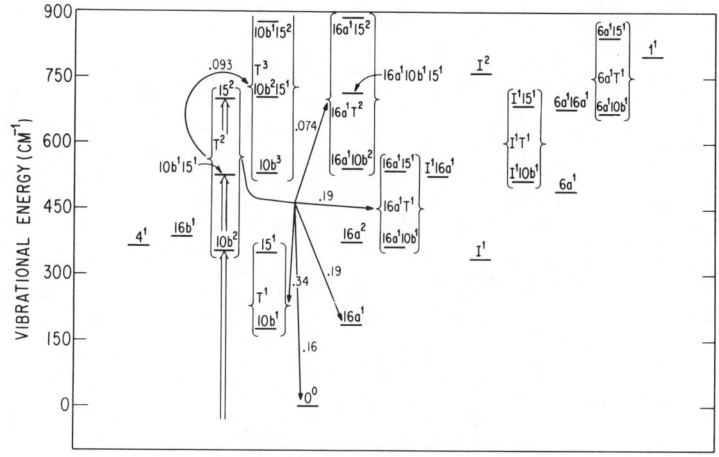

Fig. 18. Same as Fig. 12. Level: T^2.

An important clue concerning the nature of the constraints on collision induced intramolecular vibrational relaxation can be obtained from the data of Chernoff and Rice. Table I displays the rate constants for processes involving different changes in vibrational quantum number in the Ar:$^1B_2$ aniline system. Despite the uncertainties in the values of the entries it is clear that the rate of transfer of a vibrational quantum is independent of

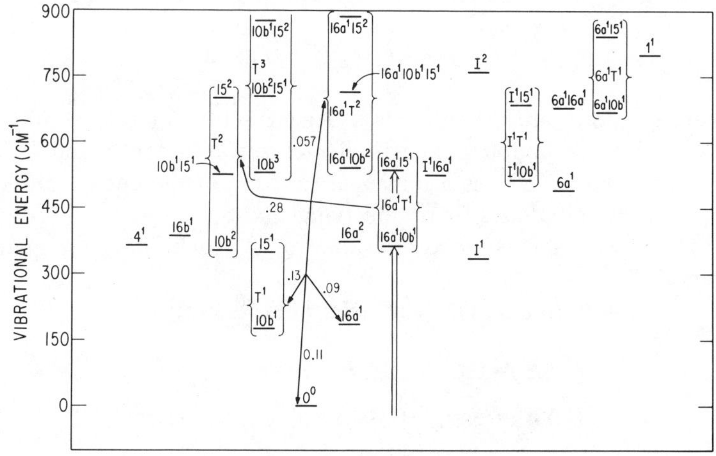

Fig. 19. Same as Fig. 12. Level: $16a^1T^1$.

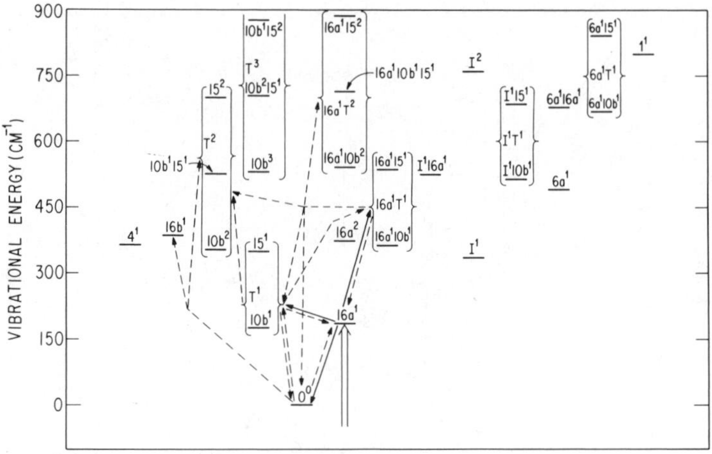

Fig. 20. Successive stages of Ar collision-induced relaxation of $^1B_2 16a^1$ aniline. Pathways for two successive collisional steps are indicated. Solid arrows show the first step, dashed arrows the second.

the state of excitation of the other vibrations provided the latter remains unchanged.

Parmenter and Tang proposed the following semiempirical propensity rules for collision-induced vibrational energy transfer:

1. The relative probability for a mode-to-mode transition, in which mode a undergoes the change $i \to j$ and mode b the change $k \to l$, is

$$P(i \to j | k \to l) \propto g_j g_l F_a^2 F_b^2 I(\Delta E)$$

   where g_j and g_l are the final state degeneracies of modes a and b, F_a and F_b are the coupling matrix elements within the a and b mode manifolds, and $I(\Delta E)$ is a universal function of the energy difference between the initial and final vibrational states.
2. F_a^2 and F_b^2 are each assigned the value $10^{-|\Delta v|}$ for a change in quantum number Δv.
3. When the vibrational energy transfer is exothermic

$$I(\Delta E) = 0.6 \qquad 0 \leqslant E \leqslant 50 \text{ cm}^{-1}$$
$$I(\Delta E) = \exp[-0.01|\Delta E|] \qquad E > 50 \text{ cm}^{-1}$$

   When the vibrational energy transfer is endothermic, $P(i \to j | k \to l)$

TABLE I
Classification of Vibrational Energy Transfer Processes in $^1B_2$ Aniline

Class	Process	$k_4(i)(10^6 \text{Torr}^{-1}\text{s}^{-1})$
$\Delta v(T) = +1$	$0^0 \to T^1$	0.95 ± 0.11
	$16a^1 \to 16a^1 T^1$	0.57 ± 0.13
	$I^1 \to I^1 T^1$	0.69 ± 0.17
	$6a^1 \to 6a^1 T^1$	1.07 ± 0.49
	$T^1 \to T^2$	0.79 ± 0.17
	$16a^1 T^1 \to 16a^1 T^2$	0.57 ± 0.26
	$T^2 \to T^3$	0.93 ± 0.15
	Mean	0.80 ± 0.20
$\Delta v(T) = -1$	$T^1 \to 0^0$	2.01 ± 0.22
	$I^1 T^1 \to I^1$	1.57 ± 0.27
	$16a^1 T^1 \to 16a^1$	0.9 ± 0.2
	$T^2 \to T^1$	3.38 ± 0.41
	Mean	1.97 ± 1.05
$\Delta v(T) = +2$	$0^0 \to T^2$	0.24 ± 0.04
$\Delta v(T) = -2$	$T^2 \to 0^0$	1.61 ± 0.14
$\Delta v(16a) = +1$	$0^0 \to 16a^1$	0.55 ± 0.05
	$I^1 \to I^1 16a^1$	0.48 ± 0.06
	$6a^1 \to 6a^1 16a^1$	0.65 ± 0.13
	$T^1 \to T^1 16a^1$	0.46 ± 0.06
	$T^2 \to T^2 16a^1$	0.74 ± 0.16
	Mean	0.58 ± 0.12
$\Delta v(16a) = -1$	$16a^1 \to 0^0$	1.18 ± 0.21
	$16a^1 T^1 \to T^1$	1.29 ± 0.34
	Mean	1.24 ± 0.08
$\Delta v(16a) = +1$ with	$T^1 \to 16a^1$	1.00 ± 0.15
$\Delta v(T) = -1$	$T^2 \to 16a^1 T^1$	1.93 ± 0.25
	$I^1 T^1 \to I^1 16a^1$	1.21 ± 0.22
	Mean	1.38 ± 0.49
$\Delta v(16a) = -1$ with	$16a^1 \to T^1$	1.24 ± 0.14
$\Delta v(T) = +1$	$T^1 16a^1 \to T^2$	2.77 ± 0.69
	Mean	2.0 ± 1.1

must be multiplied by the Boltzmann factor for that value of ΔE; the evaluation of $P(i \to j | k \to l)$ is otherwise the same as for an exothermic energy transfer.

The Parmenter–Tang rules combine in a very sensible manner the common analytic and numerical features of the SSH-type theories currently used to describe vibrational relaxation. They give a very good description of the pattern of energy flow as a result of collisions of $^1B_{2u}6^1$ benzene and a variety of partners, as shown in Fig. 21. However, the same rules give a much less accurate description of the patterns of energy flow for the

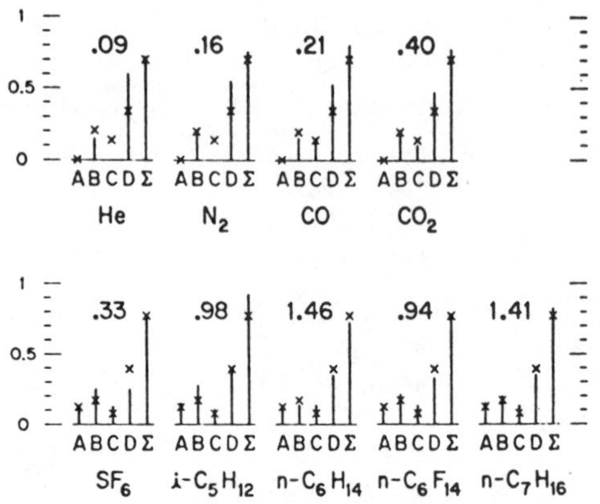

Fig. 21. A comparison of the observed and predicted collision efficiencies of various partners in the relaxation of $^1B_{2u}6^1$ benzene. The vertical bars indicate the observed fraction of vibrational energy transfer into specific channels, the crosses the values calculated from the Parmenter–Tang rules. The column labeled Σ is sum of the fractions in channels A, B, C, and D.

collisions of $^1B_2x^v$ aniline with Ar. For example, although the rate of the two-quantum change $0^0 \rightarrow T^2$ is less than of $0^0 \rightarrow T^1$, the values are not nearly as disparate as is suggested by application of the propensity rules of Parmenter and Tang. The observed ratio of the two-quantum and one-quantum exchange rates is 0.25, whereas that predicted is 0.073. Similarly, the nearly isoenergetic processes involving $\Delta v(16a) = +1$ with $\Delta v(T) = -1$ and $\Delta v(16a) = -1$ with $\Delta v(T) = +1$ are an order of magnitude faster than the estimates obtained from the propensity rules. The data of Chernoff and Rice also show that there is a general tendency for the rate of vibrational energy transfer to decrease as the energy gap increases, but not smoothly or as rapidly as implied by the propensity rules inferred from the study of benzene.

There are, as is to be expected, some striking similarities between the results of Chernoff and Rice and those of Parmenter and Tang. Consider the relative efficiencies of N_2 collision-induced transfer from the level 6^1 of $^1B_{2u}$ benzene to other levels, and of Ar collision-induced transfer from the level $6a^1$ of $^1B_2$ aniline to other levels. Parmenter and Tang find that the most efficient relaxation process in benzene is $6^1 \rightarrow 6^1 16^1$, which is endoergic by 237 cm^{-1}. The next most efficient process (down a factor of 2.5) is $6^1 \rightarrow 11^1$, 16^1; 11^1 is nearly isoergic to 6^1; 16^1 is 285 cm^{-1} below 6^1). Collision-induced transitions from 6^1 to 0^0, exoergic by 552 cm^{-1}, and from 6^1 to 4^1, 16^2, which are exoergic by 157 and 50 cm^{-1}, respectively, are very much slower than the other two mentioned. Chernoff and Rice find that relaxation from the $6a^1$ level of $^1B_2$ aniline is fastest to $6a^1T^1$ which is endoergic (T consists of 10b and 15, which lie at 177 cm^{-1} and 350 cm^{-1}, respectively). It is only a little slower, about a factor of 0.7, to $6a^1 16a^1$, which is endoergic by 187 cm^{-1}, and very slow to 0^0 and I^1, which are exoergic by 492 cm^{-1} and 155 cm^{-1}, respectively. The similarity in the relative importance of transfer from level 6^1 to level $6^1 16^1$ is immediately apparent. Moreover, in both molecules levels which lie below the initially populated level are only poorly coupled to it by collision despite the favorable energy balance. It seems fair to infer that in benzene, as in aniline, the vibrational levels are grouped into tightly coupled subsets, and that collisions redistribute energy between the several subsets much slower than between the levels within one subset. This inference, if generally valid, suggests that details of intramolecular coupling of vibrational modes have as much, or more, influence on the rate and pattern of collision-induced energy transfer as do the dependences of the collisional matrix element on the energy gap, potential, and related parameters. It is clear that at the grossest level of analysis the available data support the existence of propensity rules which define the dependence of the cross-section for intramolecular energy exchange on the vibrational energy gap and on the change in vibrational quantum number, as predicted by SSH and related theories and codified by Parmenter and Tang. However, it is also clear from the data that the available propensity rules do not cover all cases, and that the detailed dynamics of collision-induced intramolecular energy exchange is poorly represented by the extant theories.

Finally, a few words must be said about the magnitude of the cross section for collision-induced vibrational relaxation in $^1B_{2u}$ benzene and $^1B_2$ aniline. No fundamental explanation has been found for the greatly increased efficiency of vibrational relaxation in excited electronic states. It is tempting to argue that, since the excited molecule has an open-shell electron configuration and a smaller ionization potential than the ground-state molecule, there will be stronger long-range attractive forces between the excited polyatomic molecule and other molecules. Parmenter and Tang support this notion by presenting a correlation between the relaxation efficiency and the contribution of the collision partner to the intermolecular well depth; they also point out that no such correlation is found in ground-state vibrational relaxation data. On the other hand, spectroscopic studies of several van der Waals complexes indicate little change in binding energy between the ground and excited electronic states.[31] The largest observed increase in the binding energy is 22 cm^{-1}, for Ar-*s*-tetrazine, to be compared with a lower limit of 165 cm^{-1} for the binding energy of this electronically excited complex.[32]

It is possible that the electronic excitation of one of a pair of molecules increases both the attractive and repulsive forces between them so that the well depth is little changed. A correlation would then be found between the intermolecular attraction and the energy-transfer efficiency even if the repulsive part of the potential were responsible for the induced transition. Specific cases must be examined to test the applicability of this idea, but it seems more likely that our understanding of the long-range attractive forces between two complex molecules is inadequate for the prediction of differences ascribable to electronic excitation of one of them. Indeed, within the framework of the density functional theory of long-range forces,[33] any electronic excitation that does not change the radial distribution of electronic charge leaves the long-range force sensibly unchanged. Note that the π-electron excitations that lead to the $^1B_{2u}$ and $^1B_2$ states of benzene and aniline, respectively, are of this type. Unfortunately, if this argument is used to account for the very small difference in the binding energies of van der Waals molecules containing ground-state and excited molecules we are left with no explanation for the enhanced rate of collision-induced intramolecular energy transfer on the excited electronic surface.

B. Rotational Energy Transfer

Very little is known about the nature of rotational energy transfer in a collision between an electronically excited molecule and a ground-state atom or molecule. In the few reported studies[12, 34–39] the experimental method is fundamentally the same as that described at the beginning of Section III.A. An initial rotational distribution is established by narrow-band excitation. The fluorescence emission contour is recorded twice, under collision-free and thermal equilibrium conditions, and then again under conditions such that there is one collision during the lifetime of the excited state. The differences in the rotational contours of the three emission spectra are then used to infer the pathway of rotational energy transfer, and the rate of that transfer. Some examples of the emission spectra recorded under these conditions are shown in Fig. 22. Because of the small spacings between the rotational levels of polyatomic molecules most excitation sources prepare nonthermal superpositions of rotational states rather than pure rotational states, and this complicates interpretation of the observations.

The following features are found to be common to the rotational relaxation processes in ^1A_u glyoxal,[35–39] $^1B_{2u}$ benzene[34] and $^1B_2$ aniline:[12]

1. The cross-section for collision-induced rotational energy transfer is very large, of the order of magnitude of the hard sphere cross-section or larger.
2. Unlike the case of collision-induced vibrational energy transfer, collision induced rotational energy transfer seems to be free of strong restrictions on the changes in the rotational quantum numbers. When account is taken of the spectral widths of the excitation sources used, the nature of the rotation–vibration structure in the fluorescence and absorption spectra, and the possibility of resonant energy transfer in the collision, it is concluded that the studies of $^1B_2$ aniline are the weakest, those of $^1B_{2u}$ benzene better, and those of ^1A_u glyoxal the best available. With this hierarchy of quality of information kept in mind, the following weaker conclusions can also be obtained from the studies cited.
3. Collision-induced rotational relaxation is very much more efficient when resonant transfer of electronic energy is possible than when it is not possible. The cross-sections reported for rotational energy transfer in collisions of $^1B_{2u}$ and $^1A_{1g}$ benzene molecules,[34] and in collisions of ^1A_u and ^1A_g glyoxal molecules,[35–39] are of the order of magnitude of 10 times the hard-sphere cross-section. It is likely that resonance transfer of energy between electronically excited molecules with a nonthermal distribution of rotational states and ground-state molecules with a thermal distribution of rotational states is responsible for the large size of the rotational energy transfer cross-section.
4. Collision-induced intramolecular vibration-to-rotation energy transfer appears to be inefficient. The evidence for this inference comes from the study of rotational contours in the one collision-induced transition $7^1 \rightarrow 0^0$ in ^1A_u glyoxal.[39] It is found that the emission from 0^0 has a distribution over rotational transitions that is close to the thermal distribution. But the vibration ν_7 in glyoxal is a torsional motion, and the axis of torsion very nearly coincides with the smallest axis of inertia of the molecule, so if collision-induced intramolecular vibration-to-rotation transfer were efficient the emission from 0^0 should have a nonthermal distribution in the quantum number K (which describes quantization of the motion about the smallest axis of inertia). Note, however, that the collision partner used in this experiment was ^1A_g glyoxal. Since the ground-state glyoxal molecules have an equilibrium distribution of rotational states, resonance transfer of energy could obscure an intramolecular vibration-to-rotation energy conversion. The investigators report that "identical results" were obtained

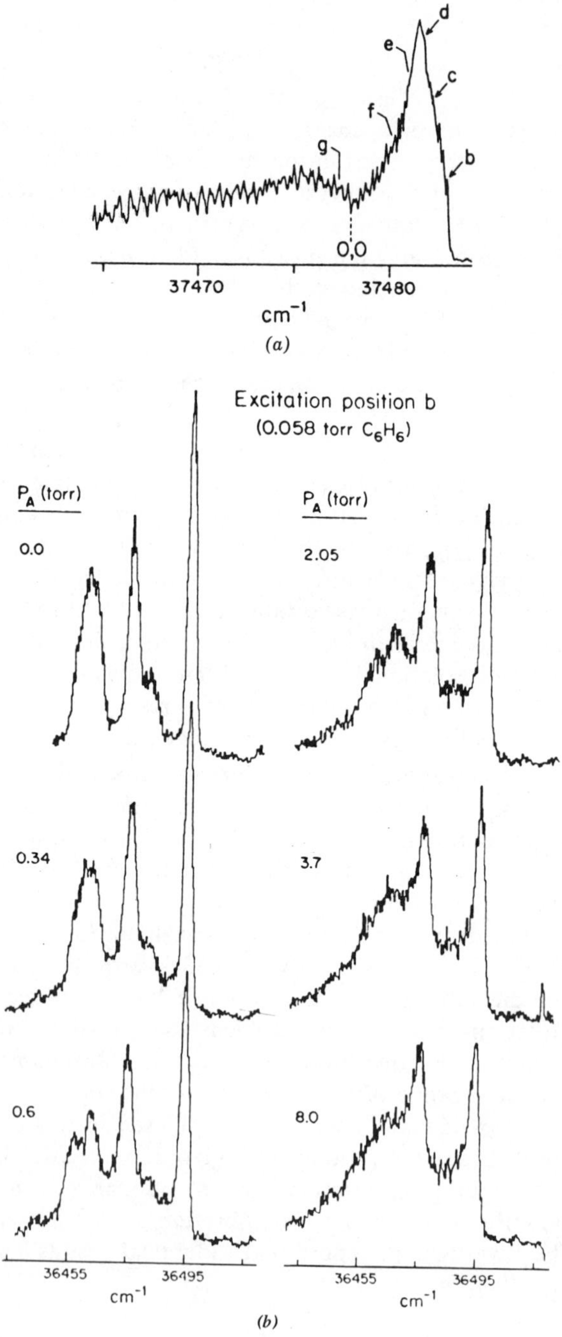
d
e
c
f
b
g
0,0
37470
37480
cm⁻¹
(a)
Excitation position b
(0.058 torr C_6H_6)
P_A (torr)
0.0
0.34
0.6
P_A (torr)
2.05
3.7
8.0
36455
36495
cm⁻¹
36455
36495
cm⁻¹
(b)

Fig. 22

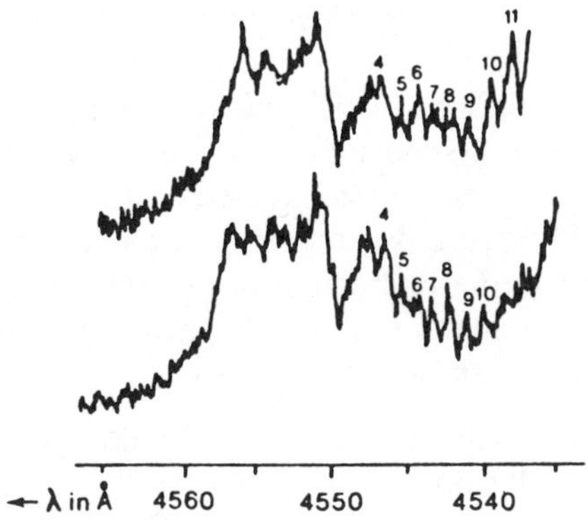

(c)

Fig. 22. (a) The rotational contour of the 6^0_1 absorption band of the ${}^1B_{2u} \leftarrow {}^1A_{1g}$ transition in benzene [from J. H. Callomon, T. M. Dunn, and I. M. Mills, *Phil. Trans. Roy. Soc.*, **A259**, 499 (1966)]. The absorption positions $b, c, \ldots$ are those used to excite groups of rovibronic states. (b) Benzene ${}^1B_{2u}1^0_16_1$ emission contours generated at position b with 0.10 torr benzene in the presence of various argon pressures. (From Ref. 34.) (c) Collision-induced emission of glyoxal from the 0^0 level after excitation of the 7^1 level. The recorded spectrum is the 0^0_0 emission band at a pressure of 0.012 torr. The K' values of the characteristic ^{r}R band heads are indicated on each band. Upper spectrum: excitation into $K'=9$. Lower spectrum: excitation into $K'=0$. (From Ref. 38.)

when the collision partner is CO_2, N_2, CO, or Ar,[39] but it is not clear whether this statement refers only to the overall property that one collision generates a broad distribution of final rotational states, or to that and the above-described behavior of the collisional relaxation of the torsional mode ν_7.

5. Although rotational relaxation accompanying the $7^1 \rightarrow 0^0$ and $8^1 \rightarrow 0^0$ vibrational transitions in ${}^1A_u$ glyoxal behave in the same way, for the $8^1 \rightarrow 6^1$ vibrational transition there appears to be some memory of the initial distribution of rotational levels.

It is relevant to note that studies of rotational relaxation in the collisions of electronically excited diatomic molecules and other atoms or molecules also indicate the absence of any strong constraints to the change in rotational quantum number. The most extensive data are for collisions of $B\,{}^3\Pi_{0_u^+}$ I_2 and other atoms and molecules, from the work of Steinfeld and collaborators.[40–44] For a given collision partner, say He, the rotational energy distributions generated by vibrationally inelastic collisions are found to be much broader than that generated by a vibrationally elastic encounter.

Classical trajectory calculations of vibrationally and rotationally inelastic collisions of $B\,^3\Pi_{O_u^+}$ I_2 and simple atoms successfully account for all of the observed phenomena.[45] The potential used in these calculations was the sum of three two-body terms, one referring to the intramolecular I—I bond, and the other two to the interaction of an I atom and the collision partner. A self-consistent scheme was used to generate realistic potential parameters. The calculations show that:

1. As the depth of the collider–I atom interaction increases the net energy transfer in a collision increases.
2. As the position of the minimum in the collider–I atom potential decreases the net energy transfer in a collision increases. The changes in collider–atom interaction generated by the changes in well depth and position described in (1) and (2) each lead to an increase in force at the classical turning point of a trajectory. Put this way, we have the simple result that the "harder" the collision, the more inelastic it will be. And, of course, as the magnitude of the well depth increases, so does the energy available to be transferred.
3. The cross-section for rotational energy transfer increases as the position of the minimum in the collider–I atom potential increases. The orbital angular momentum of the collision partner is, of course, $L = \mu v b$, where μ is the system reduced mass, v the relative velocity, and b the impact parameter. If we replace v with the thermal velocity $\langle v \rangle$, and note that $\langle v \rangle \propto \mu^{-1/2}$, we find that $L \propto \mu^{1/2}\, b$. Since size and mass are correlated, we expect to find that as the position of the minimum in the collider–I atom interaction increases so does μ, hence also the angular momentum. The energy changes accompanying even large changes in ΔJ are small for I_2, so there is no difficulty in transferring orbital to intrinsic angular momentum. Simply put, the more orbital angular momentum available, the larger the range of ΔJ generated in the collision.

The trajectory studies also give a useful picture of vibrationally inelastic encounters between the I_2 molecule and an atom. Large changes in vibrational level were generated by collisions in which the maximum force was exerted along the diatomic axis. When collisions occurred with the I_2 bond fully compressed energy transfer was from translation to vibration, and when with the I_2 bond fully extended energy transfer was from vibration to translation. The greater the initial relative velocity, the larger the amount of energy transferred in a collision. Collisions resulting in large losses of vibrational energy, say corresponding to four or more quanta, and collisions resulting in gains in vibrational energy, say larger than one quantum, depended on the initial relative velocity in the system being greater than the thermal velocity. On the other hand, collisions resulting in losses of vibrational energy corresponding to 1–3 quanta, and collisions leading to a gain of one quantum of vibrational energy, occurred when the relative velocity in the system was less than the thermal velocity. In many of these latter cases the interaction between the atom and I_2 was not impulsive in character (lingering collisions).

As for rotational energy transfer, this was favored in collisions that had large impact parameters. In general, impact parameters less than $0.7b_{max}$ were required before vibrational energy transfer could be achieved; the smaller the impact parameter the greater the possible vibrational energy transfer. Pure rotational energy transfer occurred with probability at least 0.25 up to impact parameters of $0.95b_{max}$, and reached a maximum probability of 0.5 at $0.5b_{max}$. Vibrational energy transfer occurred with probability 0.6 or greater for all impact parameters less than $0.45b_{max}$, with probability 0.2 at $0.7b_{max}$, and with only a few percent probability for impact parameters near b_{max}.

We shall see in Section IV how some of the results of the classical trajectory studies can be incorporated into a set of symmetry considerations that give propensity rules for atom-excited polyatomic molecule collisions.

C. Enhancement of Vibrational Relaxation by Orbiting Resonances

Thus far we have not discussed any experimental manifestation of the dependence of the rate of collision-induced energy transfer on the relative kinetic energy of the collision partners, although some comments based on trajectory calculations were made at the end of Section III.B. The canonical method for obtaining this information is to study the temperature dependence of the energy transfer process. Unfortunately, the averaging over the Maxwell–Boltzmann distribution inherent to this method makes it impossible to determine the detailed behavior of the energy-transfer cross-section as a function of relative kinetic energy of the colliding pair. And, at least at high temperatures, the fraction of molecular pairs with very low relative kinetic energy is too small to influence much the thermal average. Yet such low-energy collisions are of particular interest, since orbiting resonances can contribute to the rate of vibrational relaxation, and such resonances occur only for low relative kinetic energy.

At present there are no data available for the temperature dependence of the collision-induced intramolecular vibrational relaxation of electronically excited polyatomic molecules. As part of a program of obtaining information about the relative kinetic energy dependence of energy-transfer cross-section Tusa, Sulkes, and Rice[46, 47] have studied low energy collisional relaxation of $^3\Pi_{O_u^+}$ I_2 by He, Ne, and Ar. The experimental method takes advantage of the characteristics of supersonic free jets. In a supersonic jet, generated by adiabatic expansion through a nozzle, the translational temperature and density of the gas varies with distance from the nozzle. The residual collisions characteristic of the temperature at any given distance from the nozzle can be used to induce vibrational relaxation in an excited seeded molecular species. Thus, by changing the downstream location of the excitation region and dispersing the fluorescence it is possible to probe the effect of collision energy on vibration relaxation. Calculations based on the kinetic theory of the jet show that (Fig. 23) the collision energy range that can be studied is of the order 100–1 cm^{-1}, or less, for a gas originally at 300 K, and it can be extended upward by heating the nozzle. Furthermore, the range of collision energies selected is very narrow when the local temperature is low. Finally, by adjusting the carrier gas pressure it can be arranged that there are one or less collisions per lifetime, or if so desired that there are several collisions per lifetime, of the excited molecule.

In the experiments by Tusa, Sulkes, and Rice I_2 was seeded in He, Ne, or Ar, a particular vibrational level of $^3\Pi_{O_u^+}$ I_2 was excited, and the vibrational relaxation to other levels followed as a function of relative kinetic energy of the collision partners. Typical results for I_2 in He are shown in Figs. 24, 25, and 26. The conditions employed in obtaining the data of Fig. 24 correspond to there being one He:I_2^* collision per excited

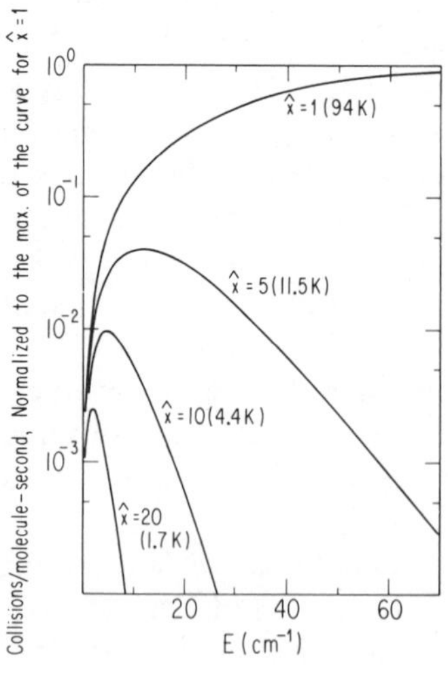

Fig. 23. Number of collisions per molecule per second in a free jet as a function of distance from the nozzle, measured in nozzle diameters.

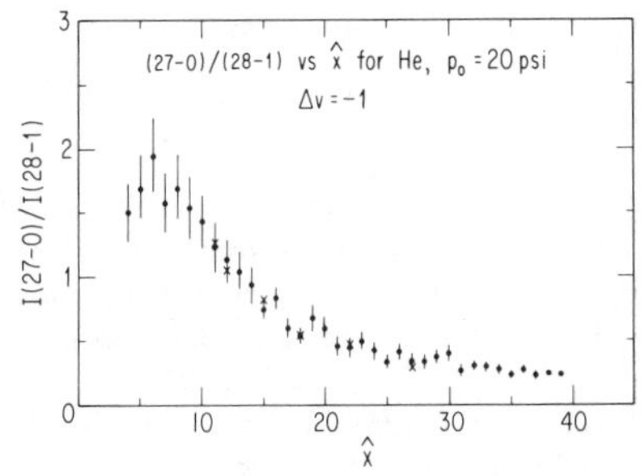

Fig. 24. Relative intensity of levels 27 and 28 of ${}^3\Pi_{0_u^+}I_2$ as a function of distance from the nozzle. The system was initially excited to level 28. The crosses represent the calculated ratio of intensities under the assumption that all collisions with $0 \leqslant E \leqslant 3$ cm^{-1} are equally effective. Collision partner: He.

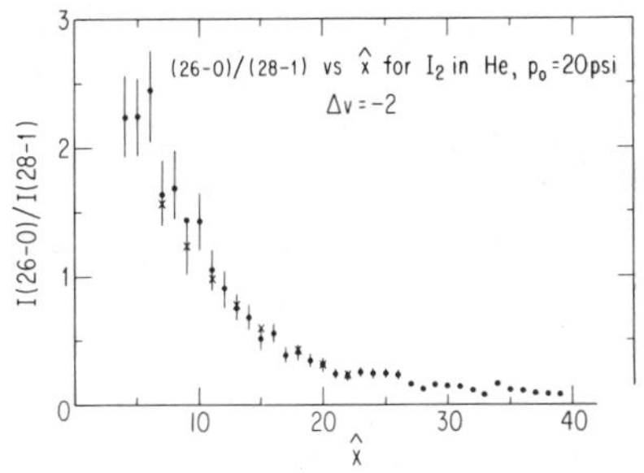

Fig. 25. Same as Fig. 24 except for monitoring of level 26.

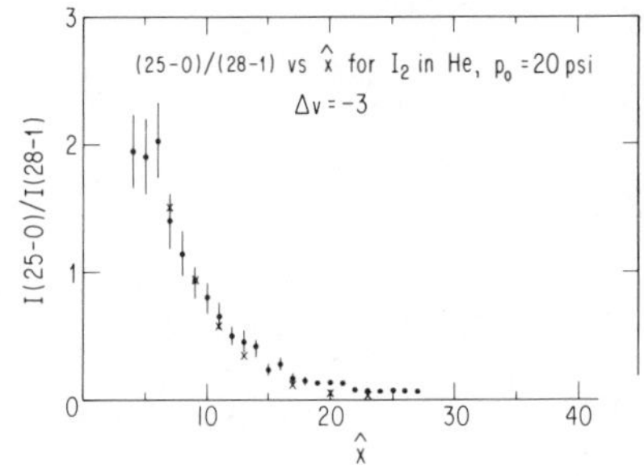

Fig. 26. Same as Fig. 24 except for monitoring of level 25.

state lifetime whereas there were, respectively, two and three collisions per lifetime for the data of Figs. 25 and 26. Results for I_2 in Ne, shown in Figs. 27–29, are similar in the dependence of depopulation of the initial level on relative kinetic energy, but differ in one detail, as will be made clear below.

There are two features of these observations that are striking. First, collisional depopulation persists even when the relative kinetic energy of the collision partners is sensibly zero, as shown by the existence of

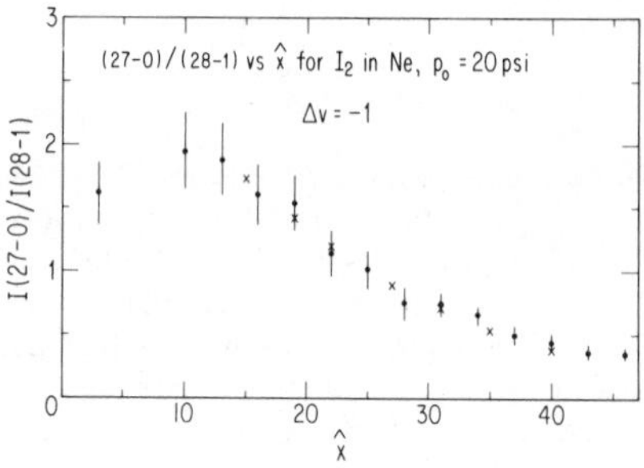

Fig. 27. Relative intensity of levels 27 and 28 of $^3\Pi_{O_u^+}I_2$ as a function of distance from the nozzle. The system was initially excited to level 28. The crosses represent the calculated ratio of intensities under the assumption that all collisions with $0 \leqslant E \leqslant 1.9\ \text{cm}^{-1}$ are equally effective. Collision partner: Ne.

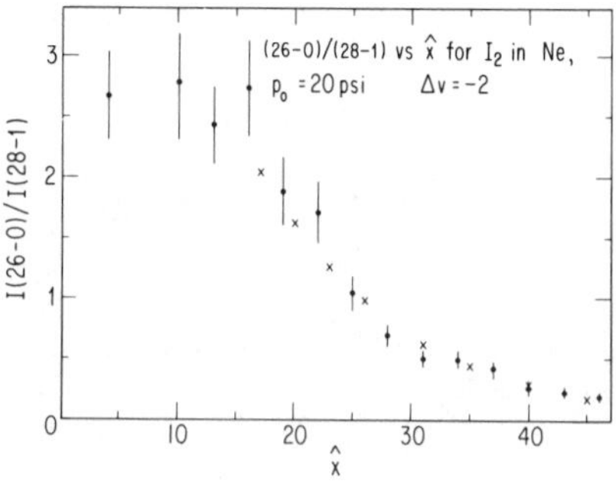

Fig. 28. Same as Fig. 27 except for monitoring of level 26.

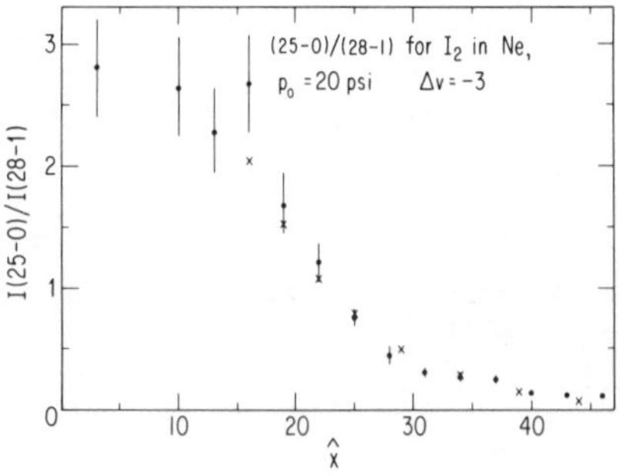

Fig. 29. Same as Fig. 27 except for monitoring of level 25.

emission from new levels even under conditions where $T \sim 1$ K. Second, the cross-section for this process is very large. The validity of this point will be demonstrated below.

A consistent interpretation of the observations can be constructed as follows. Suppose collisions with $E_{\min} \leqslant E \leqslant E_{\max}$ are equally effective in producing depopulation of the initial vibrational level, and consider the case that there is only one collision per excited state lifetime. The collision energy distribution per molecule-second obtained from the kinetic theory of the jet can be integrated over this range of E for each position along the jet axis. For the correct range (E_{min}, E_{max}) the computed ratio of populations as a function of distance from the nozzle should reproduce the experimental curve. If there is more than one collision per excited state lifetime similar calculations can be performed. Tusa, Sulkes, and Rice have examined the case of successive collisions, each inducing a one-quantum transition, and the case of a single collision inducing two, three, ...quantum transitions.

The data displayed in Fig. 24, which refer to one-quantum transitions, are quantitatively fit by the model calculation described if the range of energy for effective collisions is 0–3 cm^{-1}. For the cases that there are two and three collisions per excited state lifetime, the ranges of energies required to fit the observations are also 0–3 cm^{-1} provided the mechanism of depopulation is successive single-quantum transitions. A fit based on the dominance of collisions inducing two- or three-quantum changes requires a larger energy range, 0–7 cm^{-1}. Finally, for the case of I_2 in Ne, consideration of collision-induced one-, two-, and three-quantum transitions leads to the conclusion that the range of energy of the effective collisions is 0–1.9 cm^{-1}, and that the successive collision model is a better fit to the several-quantum transition data than is the single-collision model. Note that the successive collision model contains the assumption that the cross-section for energy transfer is essentially the same as the hard-sphere cross-section. Although this implication is also contained in the analysis of single-quantum transitions, the use of ratios in the fitting process makes the inference weak in that case. In contrast, in the multiple-collision model the quality of the fit strongly supports the model assumption that every collision in the energy range $E_{min} \leqslant E \leqslant E_{max}$ leads to a single-quantum transition in the diatomic molecule. Moreover, the intensities of transitions corresponding to $\Delta v = -1$, $\Delta v = -2, \cdots$ support this conclusion.

The observed "zero-energy" collision-induced vibrational depopulation of I_2^* indicates the existence of a relaxation mechanism that is qualitatively different from that dominant at higher relative kinetic energy. To begin, we note that at the midpoints of the energy ranges in which the collisions are effective, the de Broglie wavelengths of He, Ne, and Ar are 11, 7, and 4 Å, respectively. Accordingly, we must go beyond the semiclassical description of atom-diatomic molecule collisions to pick out the important physical processes involved in the vibrational depopulation mechanism. However, it is worthwhile to consider some of the aspects of the semiclassical description of an I_2^*:M collision since, despite its shortcomings, this description provides useful insights.

Suppose, for simplicity, that the collision dynamics is dominated by the central component of the interaction; we will later remove this restriction. In this case the effective potential is

$$V_{\text{eff}}(R)=V_0(R)+\frac{\hbar^2 l(l+1)}{2\mu R^2}=V_0(R)+\left(\frac{b^2}{R^2}\right)E_{\text{c.m.}}$$

where l is the orbital angular momentum quantum number and $E_{\text{c.m.}}$ is the collision energy in the center of mass coordinate system. The effective potential has a maximum at some intermediate value of R if $l\neq 0$ and $E_{\text{c.m.}} \leqslant c\varepsilon$, where c is a constant and ε is the depth of the interaction potential $V_0(R)$; for $V_0(R)$ a Lennard-Jones potential, $c=0.8$. The relative motion behind the centrifugal barrier will, in general, support quasivibrational resonances; the widths of these resonances depend on the strength of the coupling, which for states with $E_{\text{c.m.}}$ less than the maximum of $V_{\text{eff}}(R)$ determines the rate of tunneling through the centrifugal barrier. There are also broad scattering resonances for collision energies greater than the maximum of $V_{\text{eff}}(R)$. Examples of the influence of orbiting resonances in atom–diatomic molecule scattering have been reported in a recent paper by Toennies, Wetz, and Wolf.[48] Earlier work by Levine[49] has established the importance of subexcitation resonances.

The appearance of the centrifugal potential in $V_{\text{eff}}(R)$ is a consequence of the conservation of angular momentum, which requires that the rotational energy depend on the separation R. Consider now the consequences of the deviation from spherical symmetry in the interaction between an atom and a diatomic molecule; for simplicity we assume that molecular angular momentum and orbital angular momentum can be exchanged, but that the central component of the force field still dominates. In this case conservation of angular momentum can be achieved by several different partitionings between molecular and orbital angular momentum, and it is the simultaneous conservation of energy that restricts accessible partitionings.

The smallest possible value for the maximum impact parameter clearly is the sum of the van der Waals radii of the collision partners, which we take to be 5.6 Å for I_2^*:He, 5.8 Å for I_2^*:Ne, and 6.4 Å for I_2^*:Ar. The corresponding maximum angular momenta, obtained from $b_{\max}$ and the velocities at the upper ends of the effective collision energy ranges found experimentally, are $5\hbar$, $6\hbar$, and $17\hbar$, respectively. It seems likely that impact parameters somewhat larger than these are still important for the collision process we describe. Using the known rotational constant for I_2^*, we conclude that a transfer from orbital to molecular angular momentum of $10\hbar$ is sufficient to convert all of the relative translational energy of I_2^*: He to rotational energy of the I_2^*; for I_2^*: Ne and I_2^*: Ar similar but smaller transfers do the same. Given the uncertainty in the estimate of b_{max}, we conclude it is reasonable to expect, in the low-energy collisions of concern to us, the existence of orbiting resonances in which there is internal rotational excitation of the I_2^* collision partner.

Both classes of scattering resonances mentioned above could be contributors to the vibrational relaxation process since both provide a means of prolonging the duration of a collision. Similarly, both types of resonances provide grounds for expecting a cutoff in the energy range of the effective collisions. When molecular rotation is not excited, the cutoff measures the energy above which resonances no longer exist or are so broad that the time delay in the collision is too small to promote vibrational relaxation. When molecular rotation is excited, it is the increase in the difficulty of transferring very large amounts of angular momentum as the energy increases that generates a cutoff in the effectiveness of collisions. Put another way, if the collision energy is too high, the accessible rotational states of I_2^* cannot soak up the translational energy.

In general, a vibrational relaxation process is efficient if there are Fourier components of the driving force that are close to the frequency of the driven oscillator. The preceding suggestion of the importance of orbiting resonances in vibrational relaxation does not at first sight seem to provide a means for driving I_2^* at an appropriate frequency. In fact, however, because of the delocalized nature of the atoms in the resonances, it is easy to see how the quantum dynamics leads to the existence of just such a Fourier component of the driving force. In the orbiting resonance the I_2^* is effectively embedded in the atom, which spreads over the entire molecule. Then, because of the strong repulsion between the molecule and atom at small separation, vibration of the I_2^* generates amplitude oscillations in the delocalized atomic spatial distribution, and these in turn create a reaction force at the driving frequency. This reaction force is clearly appropriate for promoting vibrational transitions.

If the important resonances involve excitation of I_2^* rotation, it is easy to see that when the vibrational transition $v' \to v' - 1$ occurs the product I_2^* will be in a broad distribution of rotational states. If the importance resonance does not involve excitation of I_2^* rotation, we still expect the product $I_2^*(v' - 1)$ to have a broad distribution of rotational states for the following reason. Although the atomic wavelengths, at the energies of interest, exceed the I_2 bond length, they are not so large that the anisotropy of the interaction can be neglected. Under the conditions used by Sulkes, Tusa, and Rice, the only available initial rotational states of I_2^* correspond to very low J. The spatial distribution of, say, the delocalized

He in the orbiting resonance is not spherical because of interference in the scattered waves from the I atom centers. And since I_2 rotates very slowly, no individual resonance is subject to rotational averaging. Of course, all angles of approach of the atom to the I_2^*, and all impact parameters, are represented in the ensemble of resonances. Therefore, in addition to generating a force along the I_2 bond, in any particular resonance the atom exerts different forces on the two I atoms; when the vibrational transition $v' \to v' - 1$ is generated, excited rotational states of the relaxed I_2^* are also generated.

Given the picture proposed, we must expect the product relaxed I_2^* to have a distribution of rotational states which include all angular momenta up to that corresponding to the maximum impact parameter and the velocity of the departing atom. This statement is not a tautology. It is intended to suggest that there is no selection rule restriction on the angular momentum of the product $I_2^*(v'-1)$; the moment of inertia of I_2^* is so large that the amount of energy transferred to rotation for even the highest value of J possible (maximum orbital angular momentum) is small relative to the vibrational spacing, hence also small compared to the relative kinetic energy of the He and $I_2^*(v'-1)$ fragments. It is, therefore, easy to satisfy the conservation of energy and angular momentum in the product states for a wide range of fragment intrinsic angular momentum.

The model described is in agreement with the major features of the observations and is consistent with the data available for the predissociation of HeI_2^* van der Waals complexes.[50] It remains to be seen if detailed close coupling calculations will provide quantitative verification of all of its features. Although there are not now data to support a generalization, it seems plausible that the "zero energy" orbiting resonance mechanism for efficient collision-induced vibrational relaxation will occur in all systems. It will be particularly interesting to see what selectivity of vibrational pathway exists in the case of relaxation of a polyatomic molecule.

We note in passing that the mechanism described is probably not confined to the case of collisions with an electronically excited molecule. Herschbach and co-workers[51] have reached a similar conclusion for ground-state vibrational relaxation of I_2 in He and other gases. Vibrational relaxation by low-energy resonances could be quite common.

IV. DISCUSSION

The experimental data surveyed in Section III have several striking features. Although the data base is too small to permit definitive conclusions to be drawn, it seems likely that the extremely large cross-sections and the existence of propensity rules will be characteristic of all collision-induced intramolecular energy transfer in electronically excited molecules.

Calculation of the cross section for collision-induced energy transfer requires knowledge of the potential energy surface on which the collision pair moves. At present such information is not generally available, and it is not known if simple models of the potential energy surface, such as that generated by the superposition of atomic interactions, are accurate enough to be useful. Given the success of the computer simulations of collision-induced energy transfer in systems with $^3\Pi_{O_u^+}I_2$,[45] there is reason to be hopeful that such a simple model will be adequate, at least for qualitative understanding. At a somewhat lower level of description, it is useful to divide the contributions to the cross-section into those from "direct collisions" and those from "lingering collisions." The simulations reported by Rubinson, Garetz, and Steinfeld[45] show that an important contribution to the energy transfer cross-section comes from lingering collisions. It seems reasonable to expect that in an atom–polyatomic molecule collision, for which each adiabatic potential energy surface is multidimensional and for which there are likely to be many potential energy surface crossings, such lingering collisions will be common. What we have called lingering collisions are nothing more than complicated orbiting resonances. Although the detailed structure of these resonances will depend on the energy of the collision, a combination of the results of the computer simulations of Rubinson, Garetz, and Steinfeld and the experimental observations of Tusa, Sulkes, and Rice[46, 47] leads us to conjecture that in an atom–polyatomic molecule collision they will dominate the energy dependence of the energy-transfer cross-section over a wide energy range.

Interpretation of the propensity rules in the collision-induced energy transfer is as yet incomplete. The Parmenter–Tang[11] rules are fundamentally sound as a qualitative guide, but cannot be trusted to give quantitative predictions outside the domain for which they were parameterized. A simpler, completely qualitative, approach to understanding the observed propensity rules can be based on the idea that the perturbation of a particular mode of a polyatomic molecule in a binary collision is proportional to the projection of the amplitude of that mode along the collision trajectory. This commonsense notion has been used before. It is inherent in the SSH[16–18] model, hence also in the Parmenter–Tang rules, and has been employed by Shobotake, Rice, and Lee[52] in a modified ITFITS[53] theory of energy transfer in collinear atom–polyatomic molecule collisions. Here we use it somewhat differently, and only in a qualitative sense, to guide the construction of a classification of the possible effects of a collision.

Simple considerations, as well as computer simulations, suggest that energy transfer in a collision is dominated by those trajectories that generate the largest force on the molecule, and that the number of types of important collisions is crudely proportional to geometric factors such as available solid angle of approach, etc. In turn, these observations suggest that a suitable sampling of collision trajectories can adequately map the most important features of all collisions vis-à-vis energy transfer. Mr. D. McDonald has formulated a realization of this approach.[54] We assume that, in the lowest order of approximation, we can suppress the coupling of orbital and intrinsic angular momenta for the purpose of examining how a collision generates atomic displacements in the polyatomic molecule. Then for any selected collision we construct a set of vector displacements for the atoms of the polyatomic molecule, representing the effect of the atomic impact. The vector displacements are undefined as to absolute magnitude, but are scaled relative to one another such that the center of mass position and the angular momentum of the molecule are conserved. To determine the symmetry classes to which these vector displacements belong we transform the displacement pattern under each symmetry operation belonging to the parent molecule, take the dot products of the vector displacements of the transformed and original patterns, and sum the products so generated by symmetry classes. A set of irreducible representations is then obtained by projection onto the irreducible representations of the parent molecule. Finally, some estimate of the relative phase space available for different classes of collisions is made, and the results used to prepare a *qualitative* prediction of the vibrational modes involved in an "average collision."

Details of this scheme and its applications are discussed elsewhere.[54] For the present it is sufficient to note that the predictions appear to be consistent with the observed characteristics of collision-induced intramolecular energy transfer in $^1B_2$ aniline and $^1B_{2u}$ benzene. For example, the major fraction of the collisions between an Ar atom and a benzene ring will have an appreciable component of force perpendicular to the plane of the ring. It is reasonable to expect, then, that out-of-plane modes of benzene will play an important role in the energy transfer process. Consider, as one example of such a collision, the one which generates the "boatlike" motion shown in Fig. 30a. This kind of motion can be generated by impact on a C atom or on the center of a CC bond. The suppression of the role of angular momentum permits us to consider atomic motions in a nonrotating benzene ring. Then in the case under consideration atoms 1 and 4, say, move up twice as far as atoms 2, 3, 5, and 6 move down. Application of the procedure described leads to the following for the dot products of the original and transformed sets of displacement vectors:

E	$2C_6$	$2C_3$	C_2	$3C_2'$	$3C_2''$	i	$2S_3$	$2S_6$	σ_h	σ_d	σ_v
12	−6	−6	12	0	0	−12	6	6	6	−12	0

This is the character set of an E_{2u} irreducible representation, and it corresponds to vibration 16b of Fig. 1. It takes only a small change in the nature of the collision considered to generate vibrations 16a and 10b on impact. The data of Chernoff and Rice[12] show that collisions between an Ar atom and the vibrationless level of $^1B_2$ aniline lead to population of only out-of-plane modes, the important three being 16a, 16b, and 10b (see Figs. 1 and 12; arguments based on the principle of detailed balance strongly suggest that the unresolved sequence T is, actually, 10b).

The example given is particularly simple because the postulated set of collision-induced displacements coincides with that of mode 16b. That this is so could have been seen by comparison of Figs. 30a and 1 and needs no analysis. In other cases even simple atomic displacement patterns generate superpositions of normal mode displacements, and these must be decomposed systematically. A collision that leads to the displacement pattern shown in Fig. 30b generates excitation of B_{2g}, E_{2u}, E_{1g} and A_{2u} modes, which are modes 5, 17, 10, and 11 of Fig. 1. The analysis suggests that the amplitudes of 5 and 17 are much greater than those of 10 and 11, and that

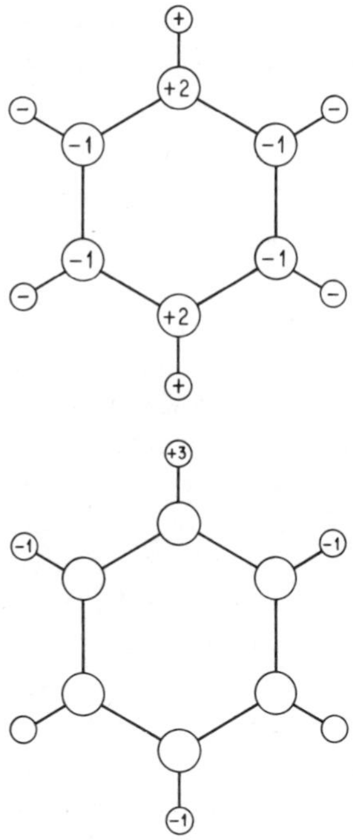

Fig. 30. (a) (top) Schematic representation of a possible collision-induced displacement pattern. (b) (bottom) Schematic representation of a different possible collision-induced displacement pattern.

if coupling of orbital and intrinsic angular momenta were not suppressed, and recoil were not suppressed, then 10 and rotation will be coupled and 11 and translation will be coupled, the detailed tradeoffs depending on properties of the collision not considered in this scheme.

The scheme described is a kind of correlation diagram approach to the classification of the effects of collisions on a polyatomic molecule. As such, it has attractive features of directness and simplicity. There are, however, inadequacies which remain to be worked out, assuming that the method is capable of improvement. Among these are the problem of verifying that the sample set of collisions is representative of all collisions, finding a way of properly incorporating the coupling of orbital and intrinsic angular momenta and of recoil motion, and fitting into the scheme the possible generation of overtone excitations.

The survey of theory and data presented in this chapter shows that we are barely on the threshold of an understanding of collision-induced intramolecular energy transfer in electronically excited molecules. So few experimental data are available that there are likely many qualitatively new phenomena as yet undiscovered. And, at least at present, the theory of collisions has not provided predictions or qualitative concepts which can be used to guide experimental studies, nor has it been developed to the point that it can fully account for the observations in hand. Clearly, much remains to be done in this subject area.

Note Added in Proof

Recent studies pertinent to the topics discussed in this paper include the following.

McDonald and Rice[55] have investigated the intramolecular vibrational redistribution induced by collisions between ${}^1B_{3u}$ (n, π^*) pyrazine and Ar atoms. Their data, in summary form, are displayed in Figs. 31–38; in these figures the transfer rates are specified per 1000 hard sphere collisions. As can be seen by comparison with the collision-induced processes in ${}^1B_{2u}$ benzene and ${}^1B_2$ aniline, energy transfer rates in ${}^1B_{3u}$ pyrazine are generally smaller than those previously observed for electronically excited polyatomic molecules. Among the single quantum changes possible, those involving mode 16b (see Fig. 31) predominate. Note that the vibronic coupling to the upper ${}^1B_{3u}$ (π, π^*) state, induced by mode 10a, increases the rate of energy transfer dramatically, especially when both initial and final states contain at least one quantum of 10a. Note also that the patterns of energy transfer are selective, so that collisions couple only subsets of levels, as observed for benzene, aniline, and glyoxal.

McDonald and Rice have completed an analysis of collision-induced intramolecular energy transfer based on the correlation diagram model

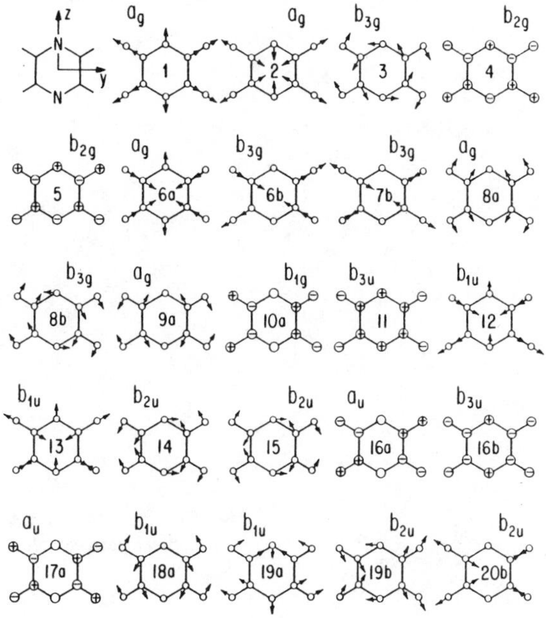

Fig. 31. The normal modes of pyrazine.

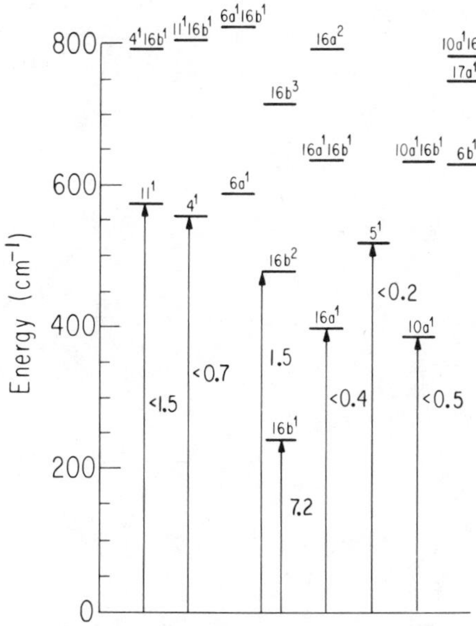

Fig. 32. Pyrazine: collision-induced energy transfer pathways from 0^0.

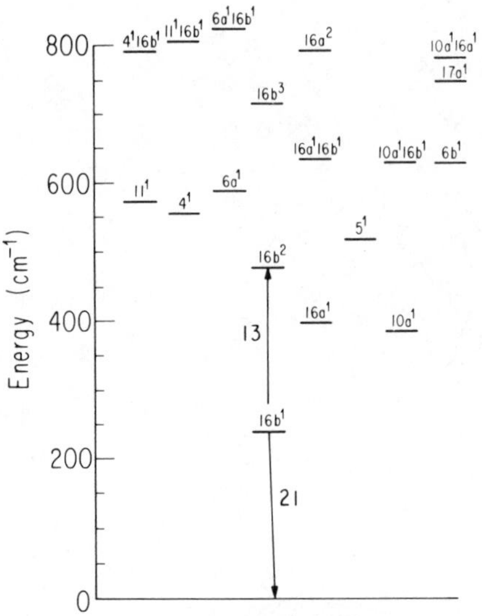

Fig. 33. Pyrazine: collision-induced energy transfer pathways from $16b^1$.

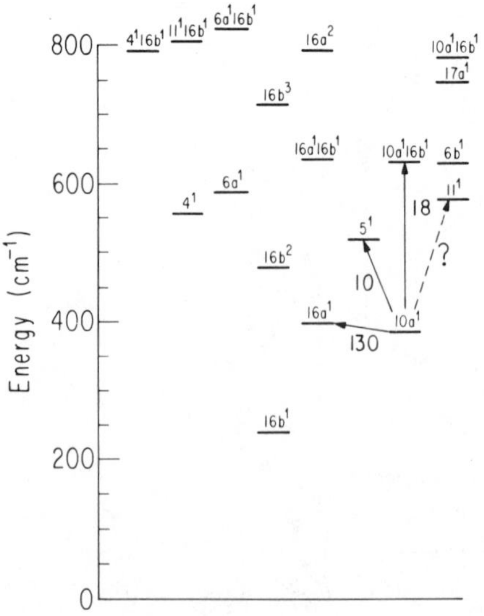

Fig. 34. Pyrazine: collision-induced energy transfer pathways from $10a^1$.

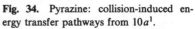

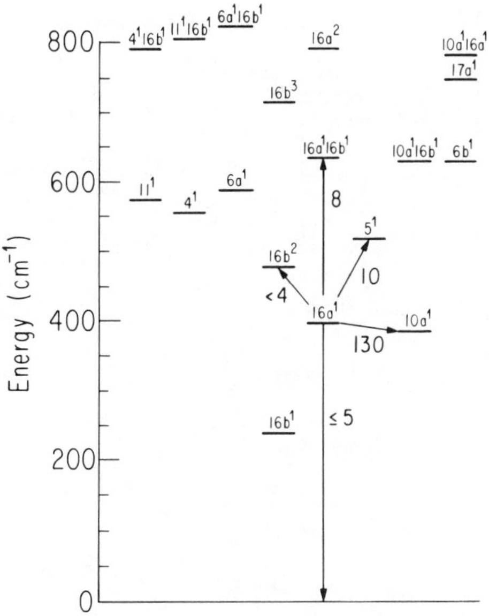

Fig. 35. Pyrazine: collision-induced energy transfer pathways from $16a^1$.

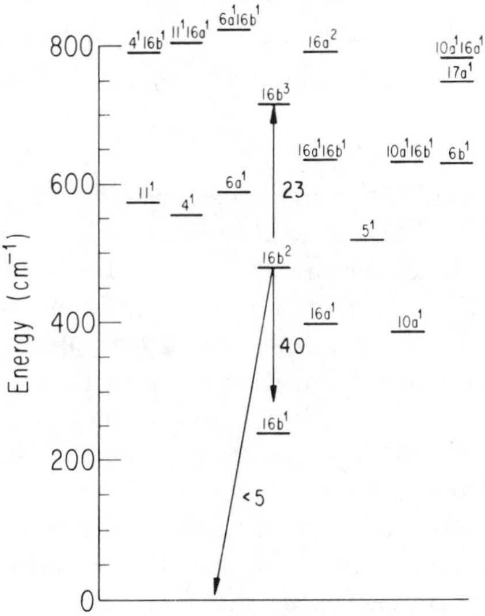

Fig. 36. Pyrazine: collision-induced energy transfer pathways from $16b^2$.

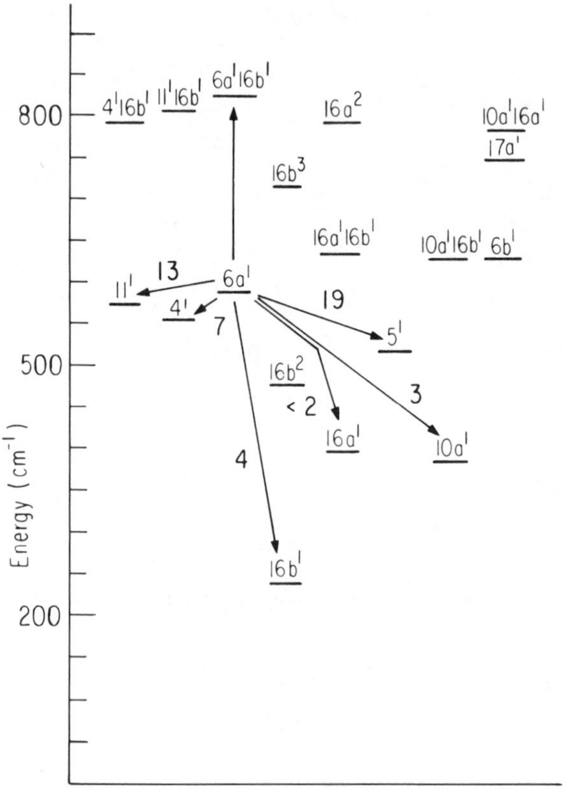

Fig. 37. Pyrazine: collision-induced energy transfer pathways from $6a^1$.

described in the text.[54] A rough quantification of the predictions was accomplished by an ad hoc generalization of the modified SSH breathing sphere model. As is shown by the entries in Tables II, III, and IV, predicted relative rates agree qualitatively, and, in some cases, quantitatively with those observed for collisions of benzene, aniline, and pyrazine with inert atoms. McDonald and Rice also provide a sketch of the possible relationships between their correlation diagram model and the symmetry rules applicable to atomic force correlation functions which determine the rate of vibrational relaxation of a molecule.[56] They regard the qualitative aspects of the correlation diagram model to be most valuable, and the particular set of crude approximations that were added for the purpose of quantifying the predictions as only a first step and merely a guide to generating a suitable algorithm for detailed calculations.

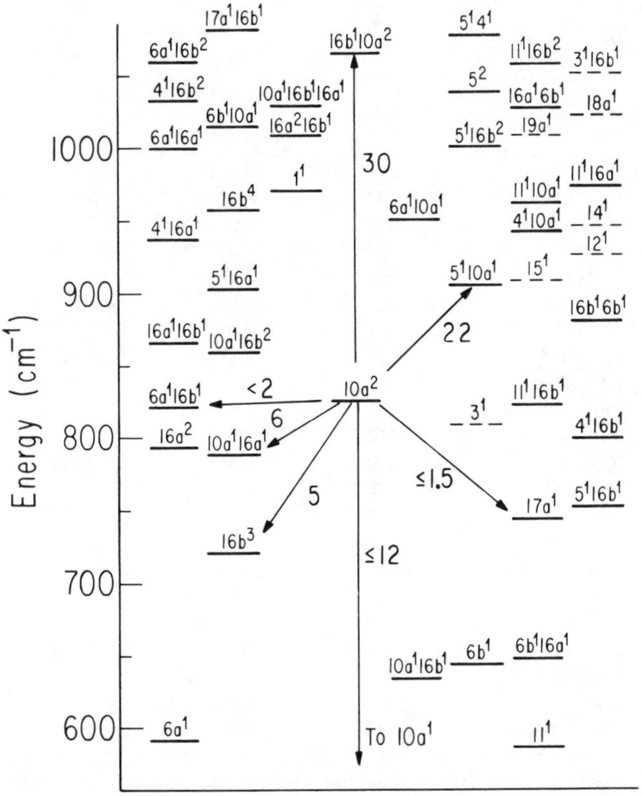

Fig. 38. Pyrazine: collision-induced energy transfer pathways from $10a^2$.

TABLE II
Observed and Predicted Transfer Rates for Benzene Collisions with Helium

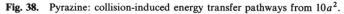

Decay channel	Observed rate (relative to maximum)	Predicted rate: Relative to maximum: McDR[54]	PT[11]	Relative to Σ: McDR	PT
$6^1 \rightarrow 6^1 16^1$	1.0	1.0	1.0	1.13	0.59
$6^1 \rightarrow 16^1$ or 11^1	0.3	0.23	0.6	0.27	0.35
$6^1 \rightarrow 16^2$ or 4^1	<0.03	0.04	0.4	0.05	0.23
$6^1 \rightarrow O^0$	<0.04	0.11	0.04	0.12	0.02
All others	0.4	0.28	1.31	0.32	0.77
Σ	1.7	1.5	2.9	1.7	1.7

TABLE III
Calculated Transfer Rates for Aniline Collisions with Argon[54]

Process	ΔE (cm^{-1})	Rate (10^6 $Torr^{-1}$ s^{-1}) Calculated	Observed[a]
$0^0 \to 10b^1$	177	0.84	0.84
$0^0 \to 16a^1$	187	0.48	0.55
$0^0 \to 15^1$	350	<0.01	—
$0^0 \to I^1$	337	0.01	<0.04
$0^0 \to 4^1$	365	<0.01	—
$0^0 \to 16b^1$	387	<0.01	(≤0.3)[b]
$0^0 \to 6a^1$	492	<0.01	(≤0.001)[b]
$0^0 \to 10b^2$	354	<0.01	(≤0.37)
$10b^1 \to 16a^1$	10	0	1.49
$10b^1 \to 10b^2$	177	1.7	1.46
$10b^1 \to 16b^1$	210	0.01	—
$10b^1 \to I^1$	160	0.35	(≤0.07)[b]
$10b^1 \to 4^1$	188	<0.01	—
$16a^1 \to I^1$	150	0.0	—
$16a^1 \to 16a^2$	187	1.0	Obscured
$I^1 \to 15^1$	13	<0.01	—
$I^1 \to 4^1$	28	<0.01	—
$I^1 \to 16b^1$	50	<0.01	—
$I^1 \to 6a^1$	155	<0.01	(≤0.06)[b]
$10b^1 \to 6a^1$	315	0.0006	—
$16a^1 \to 6a^1$	305	0.0	—
$10b^2 \to 10b^1 16a^1$	10	0.0	3.64

[a] The values in parentheses are those required to reproduce experimental results.
[b] These observed rates result from data highly subject to noise.

Vandersall, Chernoff, and Rice[57] have examined vibrational redistribution in $^1B_2$ aniline induced by collisions with H_2O and CH_3F. The observed rates and patterns of energy transfer from the levels 0^0, $10b^1$ and I^1 show both a high collision efficiency and mode-to-mode specificity; they also indicate that hydrogen bonding interactions are unimportant. The greater efficiency of energy transfer induced by collisions of $^1B_2$ aniline with H_2O and CH_3F as compared to collisions with Ar is evident from the data in Table V. In general, both H_2O and CH_3F depopulate vibronic levels of $^1B_2$ aniline at rates that are in excess of the hard sphere collision rate, whereas for Ar the decay rate is, on average, only a third of the hard-sphere collision rate. The patterns of relaxation induced by collisions with H_2O and CH_3F are quite similar (see Table VI and Figs. 39–41); in many cases the probabilities of particular mode-to-mode transfers are the same within the experimental error. However, the inversion mode is

TABLE IV

Calculated Transfer Rates for Pyrazine Collisions with Argon[54]

Process	ΔE (cm^{-1})	Rate (per 1000 collisions) Calculated	Observed
$0^0 \rightarrow 16b^1$	237	2.5	2.5(6.9)[a]
$0^0 \rightarrow 10a^1$	383	0.5	<0.15
$0^0 \rightarrow 16a^1$	397	0.06	<0.2
$0^0 \rightarrow 16b^2$	474	<0.01	<0.3
$0^0 \rightarrow 5^1$	517	<0.01	<0.1
$0 \rightarrow 4^1$	557	<0.01	<0.3
$0 \rightarrow 11^1$	571	<0.01	<0.5
$0 \rightarrow 6a^1$	585	<0.01	Obscured
$16b^1 \rightarrow 10a^1$	146	2.3	<1.0
$16b^1 \rightarrow 16a^1$	160	0	<0.2
$16b^1 \rightarrow 16b^2$	237	5.0	4.3
$16b^1 \rightarrow 5^1$	280	0.08	1.0
$10a^1 \rightarrow 16a^1$	14	18.0	45.0
$10a^1 \rightarrow 16b^2$	91	1.1	<0.4
$10a^1 \rightarrow 5^1$	134	8.0	3.0
$10a^1 \rightarrow 4^1$	174	0.6	<0.2
$10a^1 \rightarrow 11^1$	188	2.0	Obscured
$10a^1 \rightarrow 6a^1$	202	0.1	<0.9
$16a^1 \rightarrow 16b^2$	77	0	<1.0
$16a^1 \rightarrow 5^1$	120	4.0	3.2
$16a^1 \rightarrow 4^1$	160	0.4	<0.6
$16a^1 \rightarrow 11$	174	0.3	1.5
$16a^1 \rightarrow 6a$	188	0	
$16b^2 \rightarrow 5$	43	1.1	1.1
$5^1 \rightarrow 4^1$	40	8.0	
$5^1 \rightarrow 11^1$	54	27.0	
$5^1 \rightarrow 6a^1$	68	3.0	4.5(32)[a]
$4^1 \rightarrow 11^1$	14	2.0	
$4^1 \rightarrow 6a^1$	28	0.3	2.4(42)[a]
$11^1 \rightarrow 6a^1$	14	5.0	4.1(<8)[a]
$16b^2 \rightarrow 16b^3$	237	7.5	7.5

[a] Results in parentheses are in cases where quanta of mode 10a are present in both initial and final levels.

TABLE V
Overall Collision-Induced Decay Rate, k_4, Expressed in Units of $10^6\ s^{-1}\ torr^{-1}$ and as Probability per Hard Sphere Collision (parentheses)

Emitting level	Ar[a]	CH_3F	H_2O
0^0	1.1±0.1 (0.11)	6.1±0.2 (0.58)	13.8±1.0 (1.19)
$10b^1$	4.1±0.5 (0.41)	16.2±1.1 (1.53)	
I^1	2.6±0.1 (0.26)	19.0±0.7 (1.79)	23.4±0.5 (2.02)

[a] D. A. Chernoff and S. A. Rice, *J. Chem. Phys.* **70**, 2521 (1979).

coupled to other low-energy ring modes more effectively by H_2O and CH_3F than by Ar, and it is noteworthy that exoergic processes dominate in the case of relaxation from I^1 induced by H_2O and CH_3F, whereas the predominant relaxation pathways induced by collision with Ar involve addition of a quantum of mode 16a or 10b. Table VII compares the probabilities for collisional transitions in particular modes of $^1B_2$ aniline induced by the three collision partners mentioned.

It is not surprising that the overall rate of depopulation of an SVL of $^1B_2$ aniline is larger for collisions with H_2O and CH_3F than for collisions with Ar, since the attractive component of the intermolecular force is almost certainly larger for H_2O and CH_3F than for Ar, irrespective of the electronic state of the aniline. On the other hand, it is very surprising that hydrogen bonding between water and aniline plays no discernible role in

TABLE VI
Mode-to-Mode Energy Transfer Rates, Expressed as Probability per Hard Sphere Collision[a]

		Final level								
Initial level	Collider	0^0	$10b^1$	$16a^1$	I^1	$10b^2$	$16a^1 10b^1$	$16b^1$	$I^1 10b^1$	$I^1 16a^1$
0^0	Ar[b]		0.095	0.055	0.004	0.024		0.030		
	CH_3F		0.42	0.38	0.087	0.19		0.061	0.068	0.051
	H_2O		0.72	0.42	0.10	0.22		0.095		0.076
$10b^1$	Ar[b]	0.20		0.10		0.079	0.046			
	CH_3F	0.71		0.27	0.11		0.24		0.12	0.061
	H_2O	1.10		0.65	0.17		0.41		0.15	0.12
I^1	Ar[b]	0.016	0.015						0.069	0.048
	CH_3F	0.35	0.26	0.27		0.20		0.14	0.41	0.18
	H_2O	0.53	0.51	0.30					0.42	0.086

[a] Average uncertainty in these numbers is ±25%.
[b] D. A. Chernoff and S. A. Rice, *J. Chem. Phys.* **70**, 2521 (1979).

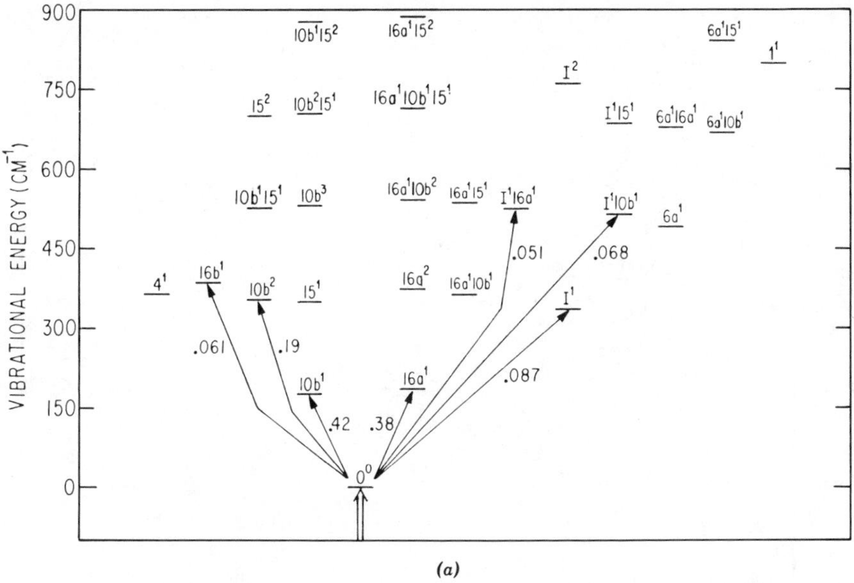

(a)

Fig. 39a. Vibrational energy transfer from 0^0 with CH_3F as the collision partner. The double arrow marks the level excited by the laser. Single arrows connect this level with levels populated by collisions. The rate constants written next to the arrows are in units of probability per hard-sphere collision.

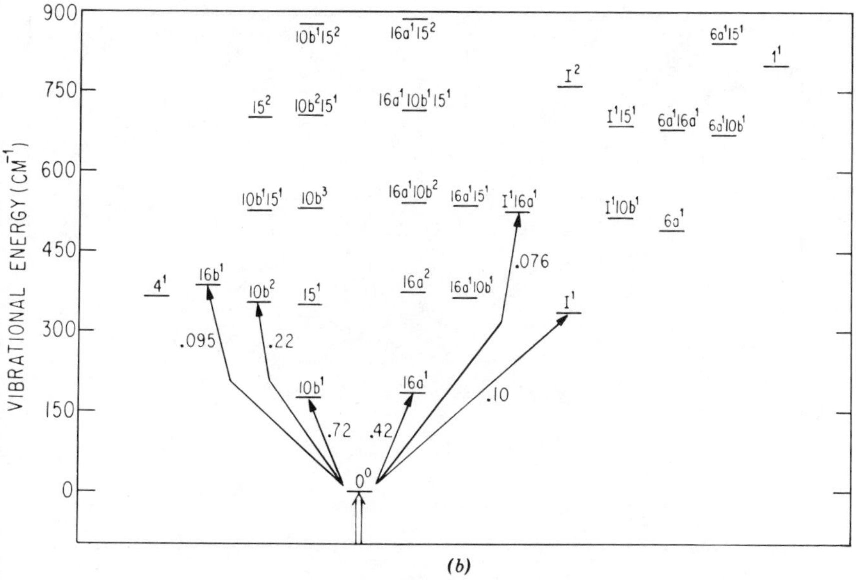

(b)

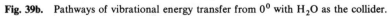

Fig. 39b. Pathways of vibrational energy transfer from 0^0 with H_2O as the collider.

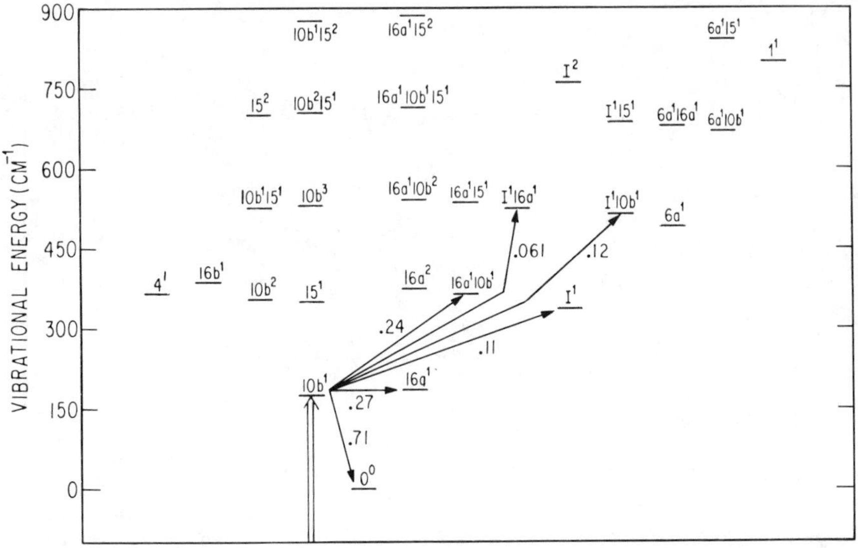

(a)

Fig. 40a. Pathways of vibrational energy transfer from $10b^1$ with CH_3F as the collider.

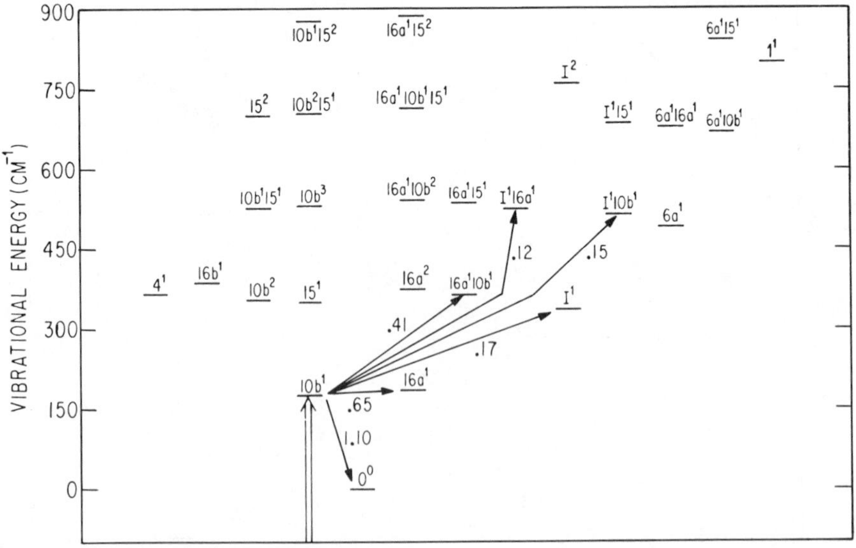

(b)

Fig. 40b. Pathways of vibrational energy transfer from $10b^1$ with H_2O as the collider.

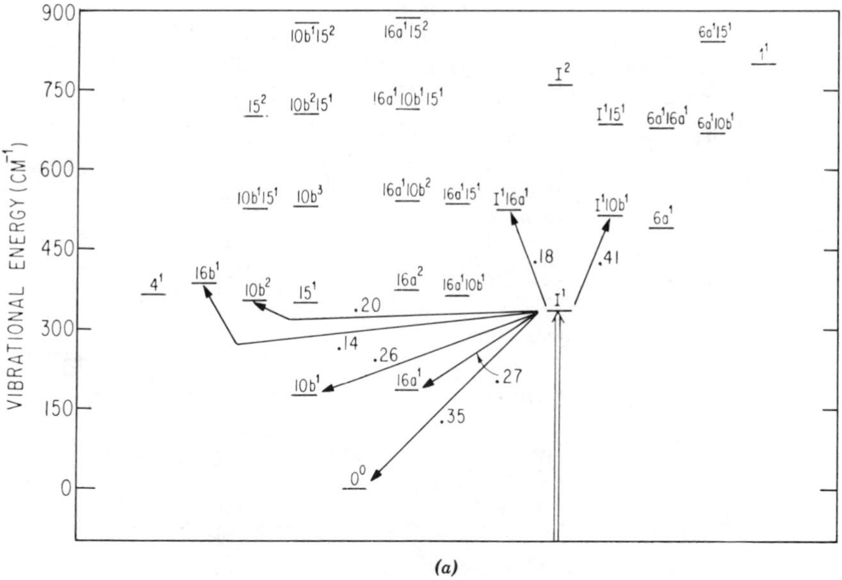

Fig. 41a. Pathways of vibrational energy transfer from I^1 with CH_3F as the collider.

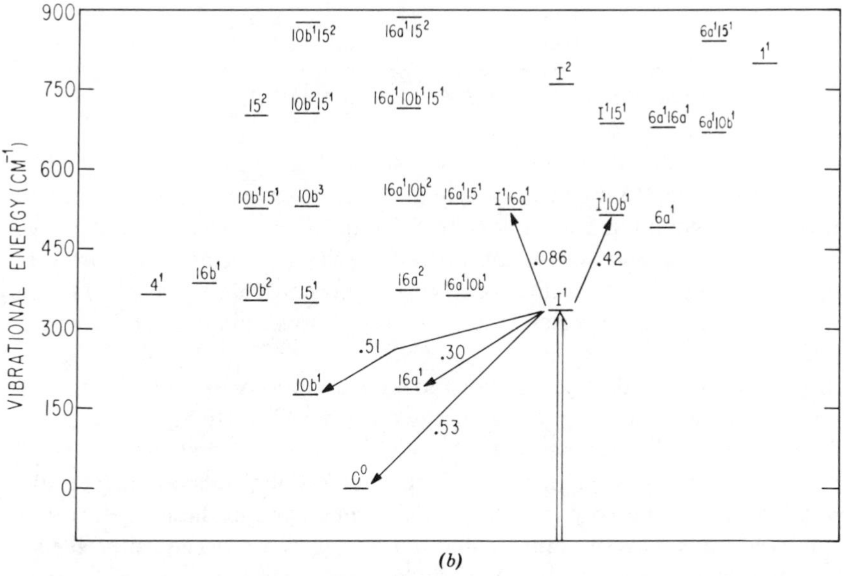

Fig. 41b. Pathways of vibrational energy transfer from I^1 with H_20 as the collider.

TABLE VII
Classification of Vibrational Energy Transfer Processes, and $k_4(i)$, as Probability per Hard Sphere Collision

Class	Process	Ar[a]	CH_3F	H_2O
$\Delta v(10b) = +1$	$0^0 \rightarrow 10b^1$		0.12	0.72
	$I^1 \rightarrow I^1 10b^1$		0.41	0.42
	Mean	0.08	0.42	0.57
	Relative probability	1	5.3	7.1
$\Delta v(10b) = -1$	$10b^1 \rightarrow 0^0$	0.20	0.71	1.10
	Relative probability	1	3.6	5.5
$\Delta v(16a) = +1$	$0^0 \rightarrow 16a^1$		0.38	0.42
	$I^1 \rightarrow I^1 16a^1$		0.18	0.086
	$10b^1 \rightarrow 10b^1 16a^1$		0.24	0.41
	Mean	0.058	0.27	0.31
	Relative probability	1	4.7	5.3
$\Delta v(I) = +1$	$0^0 \rightarrow I^1$	0.004	0.087	0.10
	$10b^1 \rightarrow I^1 10b^1$		0.12	0.15
	mean	0.004	0.10	0.13
	Relative probability	1	25	33
$\Delta v(I) = -1$	$I^1 \rightarrow 0^0$	0.016	0.35	0.53
	Relative probability	1	22	33

[a] D. A. Chernoff and S. A. Rice, *J. Chem Phys*. **70**, 2521 (1979).

the collision-induced energy-transfer process. Surely any incipient hydrogen bond formation ought to grossly alter the inversion mode motion, and thereby the coupling and energy transfer to other modes of aniline. An after-the-fact accounting for the observed behavior seems to require that hydrogen bounding does not occur, possibly because only a small fraction of the collision trajectories result in orientations suitable for hydrogen bonding.

Tusa, Sulkes, and Rice[58] have studied the vibrational redistribution in $^1B_2$ aniline induced by very-low-energy collisions with He atoms. As in the case of such collisions between I_2^* and He, they find that "zero-energy" encounters are very efficient in effecting vibrational relaxation (see Fig. 42). Because of the very low temperatures employed in these experiments, hot band transitions of aniline could not be pumped. Consequently, the set of initial states prepared by Tusa, Sulkes, and Rice has only one overlap with the set of initial states studied by Chernoff and Rice: $^1B_2 6a^1$. For this

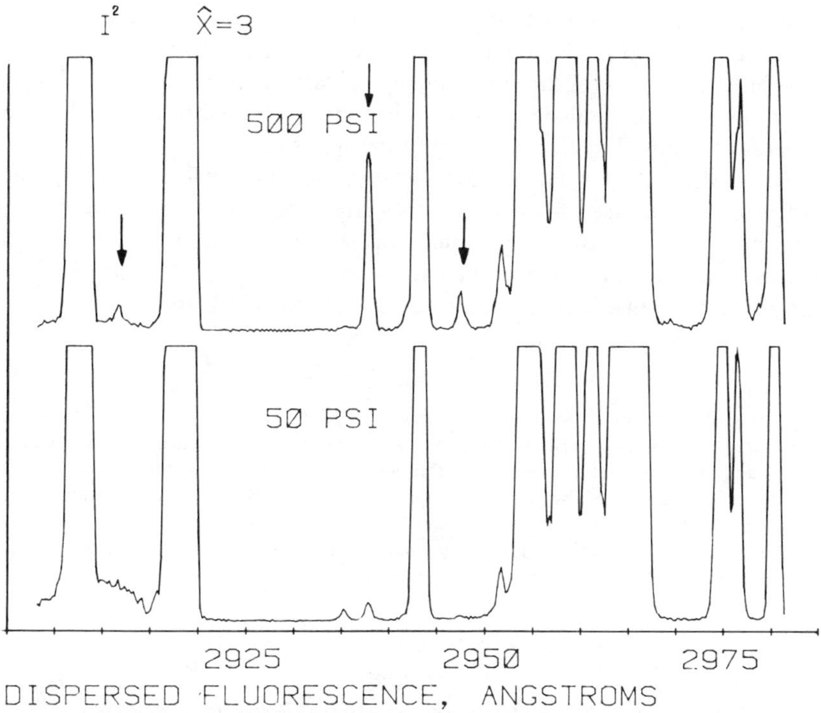

Fig. 42. Dispersed fluorescence spectra from the level I^2 of ${}^1B_2$ aniline, taken in a pulsed supersonic jet at $\hat{x}=3$, where the temperature is 25°K. Note the bands that grow in when the stagnation pressure is increased, indicating that very low energy collisions with the He atoms of the carrier gas are efficient at redistributing the vibrational energy of the aniline molecule.

one level, collisions of ${}^1B_2$ aniline with Ar atoms at 300 K lead to the transitions $6a^1 \rightarrow I^1$, $6a^1 \rightarrow 0^0$, $6a^1 \rightarrow 6a^1 16a^1$ and $6a^1 \rightarrow 6a^1 10b^1$, whereas collisions of ${}^1B_2$ aniline with He atoms at 1–5 K lead only to $6a^1 \rightarrow I^1$ and $6a^1 \rightarrow 0^0$. As expected, redistribution pathways involving endoergic steps are blocked when the relative kinetic energy of the collision partners is "zero."

Tusa, Sulkes, and Rice find that even when the distribution of relative collisions energies is peaked at 75 cm^{-1} (corresponding to one nozzle diameter downstream in the supersonic jet) transitions involving up-pumping of the aniline molecule cannot be observed. Given the population distribution of collisions with energy in excess of 75 cm^{-1}, the clear implication is that up-pumping is generated only in collisions in which the relative kinetic energy exceeds the energy gap by severalfold.

Tusa, Sulkes, and Rice have also obtained data for the zero-energy collisional relaxation of the levels I^1, I^2, $6a^2$, and 12^1 of $^1B_2$ aniline. In each case only a very few of the energetically accessible levels are coupled to the initial level by the collision, so the selectivity of the vibrational relaxation process is retained in the zero-energy mechanism.

There should be a correlation between the nature of the vibrational redistribution induced by a zero-energy collision and the vibrational distribution in the polyatomic product partner of a predissociated van der Waals molecule, since the latter can be viewed as a half collision. There are, at present, insufficient data to elucidate that correlation.

Acknowledgments

I am grateful to my co-workers D. Chernoff, D. McDonald, M. Sulkes, and J. Tusa for their many contributions to the subject discussed in this paper. I have had fruitful discussions on a number of the points covered with Professor R. D. Levine and Professor K. F. Freed, and have been influenced by as yet incomplete theoretical and experimental studies by my co-workers V. Sethuraman and M. Vandersall.

This research has been supported by grants from the Air Force Office of Scientific Research and the National Science Foundation.

References

1. B. L. Earl, L. A. Gamss, and A. M. Ronn, *Acct. Chem. Res.* **11**, 183 (1978).
2. E. Weitz and G. W. Flynn, *Ann. Rev. Phys. Chem.* **25**, 275 (1974).
3. G. Ennen and C. O. Hirger, *Chem. Phys.* **3**, 404 (1974).
4. M. Stockburger, *Z. Phys. Chem. (Leipzig)* **35**, 179 (1962).
5. H. von Weyssengoff and E. W. Schlag, *J. Chem. Phys.* **59**, 729 (1973).
6. A. V. Pocius and J. T. Yardley, *J. Chem. Phys.* **61**, 3587 (1974).
7. R. A. Beyer, P. F. Zittel, and W. C. Lineberger, *J. Chem. Phys.* **62**, 4016 (1975).
8. R. A. Beyer and W. C. Lineberger, *J. Chem. Phys.* **62**, 4024 (1975).
9. P. F. Zittel and W. C. Lineberger, *J. Chem. Phys.* **66**, 2972 (1977).
10. C. S. Parmenter, private communication.
11. C. S. Parmenter and K. Y. Tang, *Chem. Phys.* **27**, 127 (1978).
12. D. A. Chernoff and S. A. Rice, *J. Chem. Phys.* **70**, 2521 (1979).
13. W. A. Lester, The *N*-coupled channel problem, in *Dynamics of Molecular Collisions*, Part A, W. H. Miller, ed., Plenum, New York, 1976, p. 1, and references cited therein.
14. See, for example, R. N. Porter and L. M. Raff, Classical trajectory methods in molecular collisions, in *Dynamics of Molecular Collisions*, Part B, p. 1, and P. J. Kuntz, Features of potential energy surfaces and their effect on collisions, in *Dynamics of Molecular Collisions*, Part B, p. 53, W. H. Miller, ed., Plenum, New York 1976; S. A. Jayich, A. Study of the Collisional Energy Transfer of Methyl Isocyanide with Monatomic and Diatomic Collision Partners, Ph.D. Thesis, University of California, Irving, Calif., 1975.
15. R. N. Schwartz, Z. I. Slawsky, and K. F. Herzfeld, *J. Chem. Phys.* **20**, 1591 (1952).
16. R. N. Schwartz and K. F. Herzfeld, *J. Chem. Phys.* **22**, 767 (1954).
17. F. I. Tanczos, *J. Chem. Phys.* **25**, 439 (1956).
18. J. L. Stretton, *Trans. Faraday Soc.*, **61**, 1053 (1965).
19. A. Miklavc and S. Fischer, *J. Chem. Phys.* **69**, 281 (1978).

20. R. Zygan-Maus and S. Fischer, *Chem. Phys.*, **41**, 319 (1979).
21. G. C. Schatz, *Chem. Phys.* **31**, 295 (1978).
22. S. D. Augustin and H. Rabitz, *J. Chem. Phys.* **70**, 1286 (1979).
23. B. C. Eu, *Chem. Phys. Let.*, **47**, 555 (1977); *Chem. Phys.*, **27**, 301 (1978).
24. R. C. Bhattacharjee and W. Forst, *Chem. Phys.*, **30**, 217 (1978).
25. G. D. Billing, *Chem. Phys.* **33**, 227 (1978).
26. H. K. Shin, Vibrational energy transfer, in *Dynamics of Molecular Collisions*, Part A, p. 131, W. H. Miller, ed., Plenum, New York, 1976, and references cited therein.
27. K. F. Freed and C. Tric, *Chem. Phys.* **33**, 249 (1978).
28. K. F. Freed, private communication.
29. C. Lardeux and A. Tramer, *Chem. Phys.* **18**, 363 (1976).
30. L. M. Logan, I. Buduls, and I. G. Ross, Pressure dependence of the fluorescence spectrum of benzene, in *Molecular Luminescence*, E. C. Lim, ed., Benjamin, New York, 1969.
31. D. H. Levy, *Adv. Chem. Phys.* **47**, 323 (1981).
32. R. E. Smalley, L. Wharton, D. H. Levy, and D. W. Chandler, *J. Chem. Phys.* **68**, 2487 (1978).
33. R. A. Harris, *Chem. Phys. Lett.*, **33**, 495 (1975).
34. R. A. Covaleskie and C. S. Parmenter, *J. Chem. Phys.* **69**, 1044 (1978).
35. B. F. Rordorf, A. E. W. Knight, and C. S. Parmenter, *Chem. Phys.* **27**, 11 (1978).
36. A. Frad and A. Tramer, *Chem. Phys. Lett.* **23**, 297 (1973).
37. R. A. Beijer and C. W. Lineberger, *J. Chem. Phys.* **62**, 4024 (1975).
38. R. P. H. Rettschnick, H. M. ten Brink, and J. Langelaar, *J. Mol. Struct.*, **47**, 261 (1978).
39. H. M. ten Brink, J. Langelaar, and R. P. H. Rettschnich, *Chem. Phys. Lett.*, **62**, 263 (1979).
40. J. I. Steinfeld and W. Klemperer, *J. Chem. Phys.* **42**, 3475 (1965).
41. R. B. Kurzel and J. I. Steinfeld, *J. Chem. Phys.* **53**, 3293 (1970).
42. R. B. Kurzel, J. I. Steinfeld, D. A. Hatzenbuhler, and G. E. Leroi, *J. Chem. Phys.* **55**, 4822 (1971).
43. J. I. Steinfeld and A. N. Schweid, *J. Chem Phys.* **53**, 3304 (1970).
44. J. I. Steinfeld, in *Molecular Spectroscopy: Modern Research*, K. N. Ras and C. W. Mathews, eds., Academic, New York, 1972, p. 223.
45. M. Rubinson, B. Garetz, and J. I. Steinfeld, *J. Chem. Phys.* **60**, 3082 (1974).
46. J. Tusa, M. Sulkes, and S. A. Rice, *J. Chem. Phys.*, **70**, 3136 (1979).
47. M. Sulkes, J. Tusa, and S. A. Rice, *J. Chem. Phys.*, **72**, 5733 (1980).
48. J. P. Toennies, W. Wetz and G. Wolf, *J. Chem. Phys.*, **71**, 614 (1979).
49. R. D. Levine, *J. Chem. Phys.* **46**, 331 (1967). R. D. Levine, B. R. Johnson, J. T. Muckerman, and R. B. Bernstein; *J. Chem. Phys.* **49**, 56 (1969). R. D. Levine, *Acct. Chem. Res.* **3**, 273 (1970).
50. M. S. Kim, R. E. Smalley, and D. H. Levy, *J. Chem. Phys.*, **65**, 1216 (1976). R. E. Smalley, L. Wharton, and D. H. Levy, *J. Chem. Phys.* **66**, 2750 (1976).
51. G. M. McClelland, L. L. Saenger, J. J. Valentine, and D. R. Herschbach, *J. Phys. Chem.* **83**, 947 (1979).
52. K. Shobatake, S. A. Rice and Y. T. Lee, *J. Chem. Phys.* **59**, 2483 (1973).
53. F. E. Heidrich, K. R. Wilson, and D. Rapp, *J. Chem. Phys.*, **54**, 3885 (1971). B. H. Mahan, *J. Chem. Phys.* **52**, 5221 (1970).
54. D. McDonald and S. A. Rice, *J. Chem. Phys.*, in press.
55. D. McDonald and S. A. Rice, *J. Chem. Phys.*, in press.
56. S. Velsko and d. Oxtoby, *J. Chem. Phys.*, **72**, 4583 (1980).
57. M. Vandersall, D. Chernoff and S. A. Rice, *J. Chem. Phys.*, in press.
58. J. Tusa, M. Sulkes and S. A. Rice, *J. Chem. Phys.*, in press.

COLLISION-INDUCED INTERSYSTEM CROSSING

KARL F. FREED

The James Franck Institute and The Department of Chemistry, The University of Chicago, Chicago, Illinois 60637

CONTENTS

I. INTRODUCTION

Considerable experimental effort has been aimed at elucidating the collision-free unimolecular dynamics of excited molecules. Processes of interest include the dynamics of highly excited vibrational states, which have been reached by multiphoton absorption, and the various electronic relaxation processes that can occur in electronically excited states of moderate to large molecules,[1–4] etc. The idealized collision-free limit is approached either by extrapolating data to the limit of zero pressure or by performing experiments in molecular beams. Alternatively, estimates of expected collisional effects are made by using collision cross-sections that are computed from hard-sphere collision rates. These estimates are then utilized to determine whether the experiments are performed in the "collision-free" domain.

Often it is impossible to carry out the experiments at sufficiently low pressures to be in the predicted collision-free regime. On the other hand, collisional events provide the possibility of producing states that cannot be populated directly by photon absorption and/or intramolecular decay mechanisms. Thus, the collisional processes can qualitatively alter the dynamics of the system, and it is this aspect of the collisional processes that is of central interest to some researchers.

This review is concerned with the process of collision-induced intersystem-crossing cases in which collisions alter the spin multiplicity of the molecule. This phenomenon is closely related to collision-induced internal conversion, where the spin multiplicity is unaltered by the collision, and to collision-induced vibrational relaxation,[5] where it is the nature of the excited vibrations that is changed by the collision process. Our discussion is centered about the spin-changing collision-induced intersystem-crossing case, but some passing references are made to the related internal conversion and vibrational relaxation.

In small polyatomic molecules and intermediate case molecules (large molecules with small singlet–triplet spacing), there are many instances in which irreversible singlet–triplet intersystem-crossing processes are found to be absent in the isolated molecules. This situation arises because the small and intermediate case molecules have too low a triplet density of states to drive the irreversible decay from the singlet into the triplet. It is then found that collisions can, in fact, induce the intersystem crossing.[6] Experimentally this is verified by virtue of the fact that the collisions quench the fluorescence and lead to an enhancement of the phosphorescence with longer wavelength emission and considerably longer emission lifetimes. Some of the molecules in which collision-induced intersystem crossing has been found to date are methylene,[7] glyoxal,[6, 8–10] propynal,[11] primidine,[12] carbon monoxide,[13] sulfur dioxide,[14] pyrazine,[15] benzophenone,[16] and biacetyl.[17] It is quite clear that during a collision the translational continuum provides the continuous density of states necessary to drive the irreversible electronic relaxation. Thus, there is no surprise that collisions can produce intersystem crossing when the intersystem crossing cannot proceed in the isolated molecule.

Collision-induced intersystem crossing sounds like an ideal example for the application of standard quenching theory. For instance, if a singlet excited state (or group thereof) is initially excited and if a collision causes a transition to triplet levels, then it is expected that the perturbers' spin–orbit or exchange interaction induces the singlet–triplet transition in the initially excited molecule. This quenching mechanism would be expected to produce a collision induced intersystem crossing rate of the form,

$$k_{S\to T} \propto Z^2_{\text{perturber}} \tag{1.1}$$

where $Z_{\text{perturber}}$ is the atomic number of the perturbing system in the simple case that the collision partner is just an atom. By analogy with intramolecular intersystem-crossing processes in large molecules,[1-4] it is also to be expected that the singlet–triplet collision–induced crossing rates should somehow depend on the Franck–Condon factors for the singlet–triplet transition,

$$k_{S\to T} \propto |\langle \phi^S_{\text{vib}} | \phi^T_{\text{vib}} \rangle|^2 \tag{1.2}$$

where ϕ^S_{vib} and ϕ^T_{vib} are respective vibrational wavefunctions for the initially prepared singlet (S) level and the triplet level (T), which is produced by the collisional process. The expected Franck–Condon variation in (1.2) would then lead to the observation of large hydrogen–deuterium isotope effects on the collision-induced intersystem-crossing rates.

Since quenching occurs in a single collision, the kinetics of the collision-induced intersystem-crossing process are expected to follow simple Stern–Volmer quenching kinetics so long as the density of triplet states is considerably larger than that for the singlet, so that we can safely ignore the triplet to singlet reverse collision processes. (In small, e.g., diatomic, molecules the densities of states in the singlets and triplets are comparable, and this assumption must be modified to incorporate any of the effects of back transition.[71])

Our first interest[18] in collision-induced intersystem crossing had been stimulated by the important case of methylene.[7] Methylene is usually produced by the photolysis of diazomethane or ketene, which produces the molecules primarily in the lowest excited singlet state. The singlet methylene reacts via an insertion mechanism, whereas the ground-state triplet methylene reacts through an addition mechanism. The effects of collision-induced intersystem crossing from the lowest singlet to the ground triplet states have to be incorporated in order to properly describe the kinetics of the singlet and triplet reaction mechanisms. Earlier calculations, based on the above simple quenching mechanism, had led to predictions of the dependence of the collision-induced intersystem-crossing rates on the masses of rare-gas atom collision partners that are in disagreement with experimental values. We introduced an alternative mechanism for the collision-induced intersystem crossing to provide a different dependence on the mass of the perturbing atoms in an attempt to qualitatively explain the experimental collision-induced intersystem-crossing rates.[18] Subsequently, collision-induced intersystem crossing or internal conversion has been observed in a number of molecules.

A. Review of Experimental Data

We may investigate the accuracy of a simple quenching mechanism for collision-induced intersystem crossing by examining some of the experimental data. Thayer and Yardley[11] introduced a perturber-dependent expression for quenching rates and applied this correlation to collision-induced intersystem-crossing rates in propynal and glyoxal. This theory implies that the cross-section for collision-induced intersystem crossing varies as[11]

$$\sigma_{S\to T}=A\mu^{1/2}\left(\frac{I_{\mathrm{M}}I_{\mathrm{Q}}}{I_{\mathrm{M}}+I_{\mathrm{Q}}}\right)^2\alpha_{\mathrm{Q}}/R_c^q+\frac{B\mu^{1/2}D_Q^2}{R_c^3} \tag{1.3}$$

where for nonpolar molecules like glyoxal I_{M} is the ionization potential of the molecule; I_{Q} is the ionization potential of the quencher, α_{Q} is the dipole polarizability of the quencher, D_{Q} is the dipole moment of the quencher, R_c is a critical impact parameter, μ is the reduced mass for the collision pair, and A and B are dependent on the molecule of interest. Equation (1.3) is derived by Thayer and Yardley under the assumption of straight-line, constant velocity trajectories because of the assumed dominance of long-range dipole–dipole interactions. Figure 1 presents a Thayer–Yardley plot of the data of Beyer and Lineberger[8] on the cross-sections for collision-induced intersystem crossing from the lowest vibrational level in the first excited singlet state of glyoxal with a wide variety of collision partners. The Thayer–Yardley correlation is seen to provide a good representation of the experimental data. Oxygen as perturber fits the

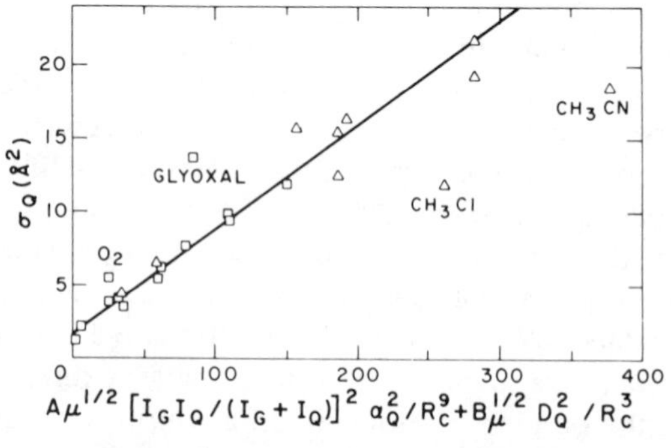

Fig. 1. Cross-section for collision-induced intersystem crossing in glyoxal presented in a Thayer–Yardley plot.[11] Figure taken from Beyer and Lineberger.[8] Symbols defined following Eq. (1.3). Note that CH_3Cl and CH_3CN have large quadrupole moments and 0_2 has a magnetic moment which are not included in the Thayer–Yardley correlation.

correlation fairly well despite the fact that its permanent spin would be expected to greatly enhance the glyoxal spin-flipping necessary to cause collision-induced intersystem crossing. The case of methyl chloride and methyl cyanide deviate markedly from the Thayer–Yardley correlation. However, these molecules have very nearly compensating bond–dipole moments and a very large quadrupole moment. Thus, estimates of the collision rates would have to incorporate the quadrupole moment of the perturber in order to adequately represent these two molecules in a fashion consistent with the treatment of the other perturbers.

The striking aspect of the Thayer–Yardley correlation in (1.3) and Fig. 1 is that it works at all. Equation (1.3) refers only to quantities associated with *long-range attractive interactions between molecules*. There is no reference to atomic number, spin–orbit, or exchange interactions! Furthermore, the cross sections for collision-induced intersystem crossing are considerably different from those anticipated on the basis of the quenching mechanism. For instance, helium provides a cross-section of 1.3 Å^2, while H_2 provides a cross-section of 2 Å^2, on the order of 5% of gas kinetic values. Helium and hydrogen have the smallest possible spin–orbit and exchange interactions, and if these interactions were required to flip the spin of the glyoxal molecule, the cross-sections for collision-induced intersystem crossing would be many orders of magnitude smaller than the observed values. Thus, our initial assumption that the collision-induced intersystem-crossing process is dominated by perturber–induced spin–orbit and exchange interactions is totally incorrect; the experimentally determined important forces emanate from long-range attractive interactions.

The above feature of the experimental data is also displayed by a correlation due to Parmenter and Seaver,[19] where the cross-section is taken to depend in an exponential form upon the well depth in a Lennard–Jones interaction between the molecule and the perturber,

$$\sigma_{S\to T} \propto \exp\left(-\frac{\epsilon_{MQ}}{k_B T}\right) \tag{1.4}$$

where $k_B T$ is the temperature in energy units and ϵ_{MQ} is the Lennard–Jones well depth that is estimated from the geometric combining rule

$$\epsilon_{MQ} \approx (\epsilon_M{}^* \epsilon_Q)^{1/2} \tag{1.5}$$

with $\epsilon_M{}^*$ the well depth for the electronically excited state of the molecule and ϵ_Q that for the quencher–quencher interaction. The Parmenter–Seaver plot for glyoxal is reproduced in Fig. 2, and the statistical correlation is

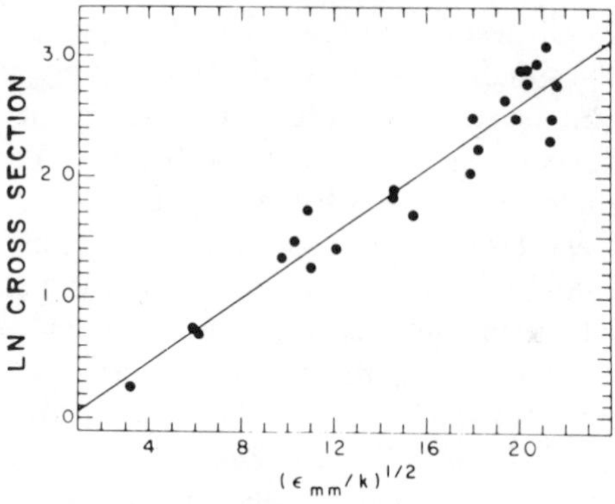

Fig. 2. Data of Fig. 1 replotted according to the Seaver–Parmenter plot where ϵ_{mm} is the perturber–perturber Lennard–Jones well depth. (From Ref. 19.)

found to be a slight improvement over the Thayer–Yardley plot. Again, (1.4) does not involve spin–orbit or exchange interactions; it relates the cross-section for collision-induced intersystem crossing to the long-range attractive nature of the molecule–quencher interaction.

Because of the singlet–triplet nature of the transition, it is to be expected that the coefficients A and B in (1.3) or the proportionality factor in (1.4) depend on the Franck–Condon factors as in (1.2). This expectation has been tested by Zittel and Lineberger[9] for the glyoxal molecule where they studied the isotope effect on the collision-induced intersystem crossing process. The vibrationless level of the first excited singlet state of glyoxal displays a negligible isotope effect in going from the dihydride to the monohydride to the dideutero glyoxal in marked constrast to the large isotope effects obtained in intramolecular intersystem-crossing process.[1–4]

Zittel and Lineberger further investigated the dependence of the collision-induced intersystem-crossing rates on the particular vibration, which is excited in the first excited singlet state. Theoretical analyses of the dependence of intersystem-crossing rates on initial vibrational states[1, 2] indicate that this dependence is a strong function of the Franck–Condon factors as in (1.2). However, the experimental results,[9] displayed in Fig. 3, demonstrate a negligible isotope effect for all the vibrational states studied as well. The data in Fig. 3 exhibit a slight increase in quenching rate with excess vibrational energy, E_{vib}, which correlates with a collision-induced intersystem-crossing rate that increases as the density of triplet states at the energy of the initially decaying singlet state,

$$k_{S\to T}^{\text{glyoxal}} \propto \rho_T(E_{vib}) \tag{1.6}$$

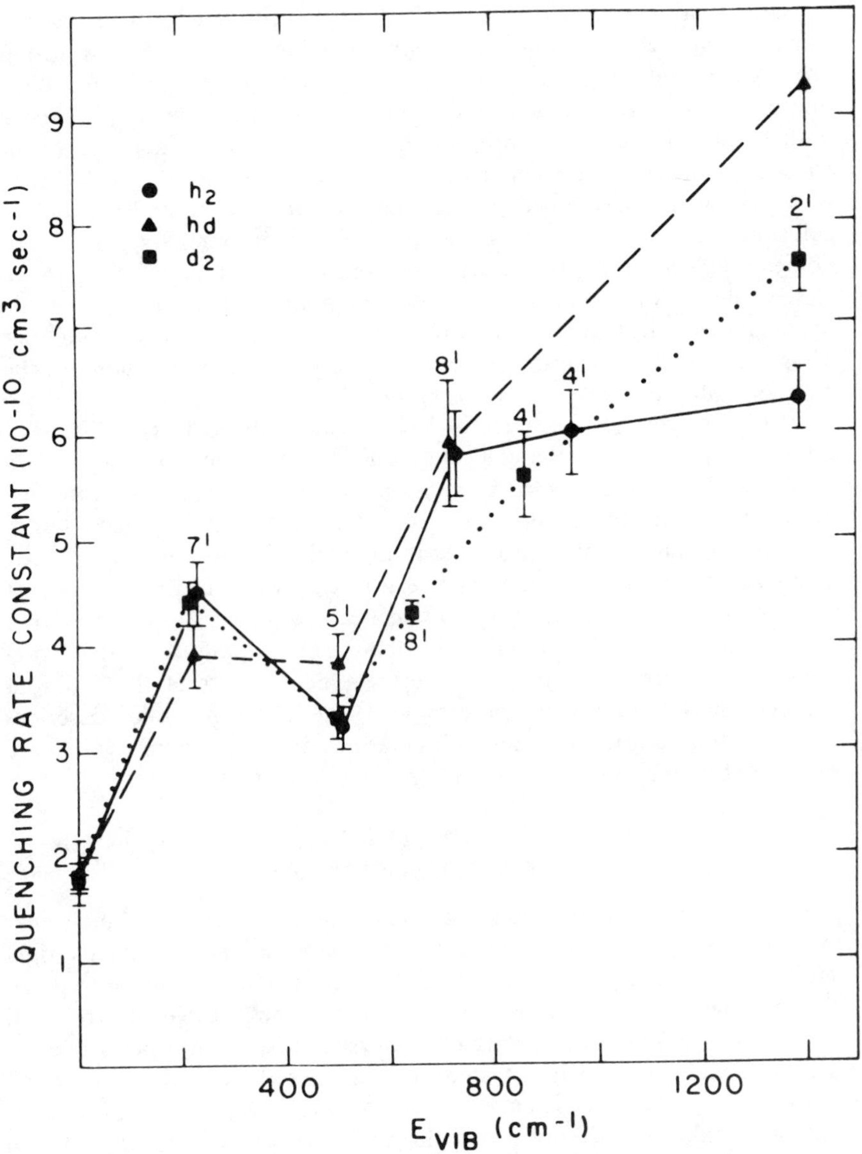

Fig. 3. Deuterium and vibrational state dependence of glyoxal collisional quenching rates as taken from Zittle and Lineberger.[9]

The glyoxal data presented above are sufficient to demonstrate that the standard quenching model of collision-induced intersystem-crossing processes is entirely wrong. The puzzling data on glyoxal have been interpreted through a generalization and extension of the simple theory initially introduced to explain the rare-gas atom mass dependence of collision-induced intersystem-crossing rates in methylene. The new theory was then further developed to describe the pressure dependence of the collision-induced intersystem-crossing process,[20–22] predicting new phenomena, which were subsequently observed experimentally.[14, 23] Here we review the major facets of the theory in a qualitative fashion, as the detailed theoretical analysis has recently been provided elsewhere.[22, 24] Thus, emphasis is placed on providing a physical description of the collision-induced intersystem-crossing process.

Despite the conflicts between the lack of isotope dependence of Franck–Condon dependence of the collision-induced intersystem-crossing rates in glyoxal and those anticipated from radiationless transition theory, the key to understanding the collision-induced intersystem-crossing process lies in the theory of radiationless processes in isolated collision-free molecules.[1, 2] Consequently, in Section II, a brief review is provided of those aspects of radiationless transition theory that are conceptually necessary to introduce the description of collision-induced intersystem crossing. Section III presents a qualitative description of the theoretical results of the collision-induced intersystem-crossing process, while the following section describes the predictions and experimental observations on the pressure dependence of the collision-induced intersystem-crossing process.

II. BASIC FACETS OF RADIATIONLESS TRANSITIONS IN ISOLATED MOLECULES

Early studies of the photophysical radiationless processes[25] of molecular systems were carried out on molecules in condensed media, liquids, rigid matrices, and high-pressure gases. This experimental situation introduces the complication associated with the presence of the possible occurrence of a number of different competing photophysical relaxation processes in the same molecular system in a fashion that mimics the complexity of a full photochemical reaction scheme. In order to study the primary photophysical radiationless transitions, it is optimal to consider experiments in which only the elementary individual processes of interest appear. Such investigations often involve the experimental determination of radiationless transition rates in isolated collision-free molecules.[26–29] For instance, collision-free experiments enable the consideration of the important phenomena of electronic relaxation and intramolecular vibrational redistribution.[30] Studies on isolated molecules have greatly contributed to our understanding of these processes, and continued progress in these areas is to be anticipated.

A previous review provides a description of the theory of electronic relaxation in polyatomic molecules with particular emphasis on the vibronic state dependence of radiationless transition rates.[1, 2] A sequal review[24] considers the general question of collisional effects on electronic relaxation, while the present one covers only the special phenomenon of collision-induced intersystem crossing. It departs from the other collisional effects review in presenting only a qualitative description of the theory; the full theoretical details can be obtained from the previous review[24] and the original papers.[20–23] As a review of the basic concepts of radiationless transitions theory is necessary as a prelude to a discussion of collision-induced intersystem crossing, considerable overlap exists between this section and Section II of the previous collision effects review.[24] However, since many concepts from radiationless transition theory, such as the nature and criteria for irreversible decay, the role of the preparation of the initial state, the occurrence of intramolecular vibrational relaxation, etc. pervade the other papers on laser chemistry in these volumes, it is useful to recall the primary results of the theory of electronic relaxation in isolated molecules and its relevance to the material in the present volume as well as to this review.

If the adiabatic Born–Oppenheimer approximation were exact, photochemistry and photophysical processes would be rather straightforward to describe. Molecules would be excited by the incident radiation to some upper electronic state. Once in this electronic state, the molecules could radiate to a lower electronic state, or they could decompose or isomerize on the upper electronic potential energy surface. No transitions to other electronic states would be possible. The spectroscopy of the systems would also be greatly simplified, as there would no longer be any phenomena such as lambda doubling,[31] etc., which lifts degeneracy of some energy levels of the clamped-nucleus electronic Hamiltonian, H_{el}.

There are numerous interactions which are ignored by invoking the Born–Oppenheimer approximation, and these interactions can lead to terms that couple different adiabatic electronic states. The full Hamiltonian, H, for the molecule is the sum of the electronic Hamiltonian, H_{el}, the nuclear kinetic energy operator, T_N, the spin–orbit interaction, H_{SO}, and all the remaining relativistic and hyperfine correction terms. The adiabatic Born–Oppenheimer approximation assumes that the wavefunctions of the system can be written in terms of a product of an electronic wavefunction, $\phi_n(r, R)$, a vibrational wavefunction, $\chi_{ni}(R)$, a rotational wavefunction, $\Theta_{JM}\cdots$, and a spin wavefunction, χ_{spin}. However, such a product wavefunction is not an exact eigenfunction of the full Hamiltonian for the system. One consequence, noted above, is the lifting of degeneracies[31] of energy levels of these simple product wavefunctions due to couplings ignored by the Born–Oppenheimer approximation. A second consequence[25] involves the possibility for the occurrence of radiationless transition processes, in particular, electronic relaxation processes.[1-4]

A. Description of the Molecular Model

The adiabatic Born–Oppenheimer approximation provides a general picture of the energy levels in a polyatomic molecule, which is presented schematically in Fig. 4. A molecule has a ground electronic state, ϕ_0, with a collection of vibrational sublevels. The photophysical experiment begins with the molecule in some thermally accessible vibrational level of the ground state (representation of the rotational and spin sublevels is omitted at this juncture.) There are also a host of electronically excited states of the system. Figure 4 depicts one excited state of the system, ϕ_s. Isoenergetic with the low-lying levels of ϕ_s is a dense manifold of vibrational levels arising from two possible sources. The first comes from high-lying vibrational levels of the ground electronic state. There may also be vibrational states belonging to other excited electronic states with origins lower than ϕ_s. It is assumed that electric dipole-allowed (or vibronically induced) transitions between the thermally accessible ground-state vibronic levels, $\phi_{0\lambda}$, and these high-lying states $\{\phi_l\}$, are not possible because of either

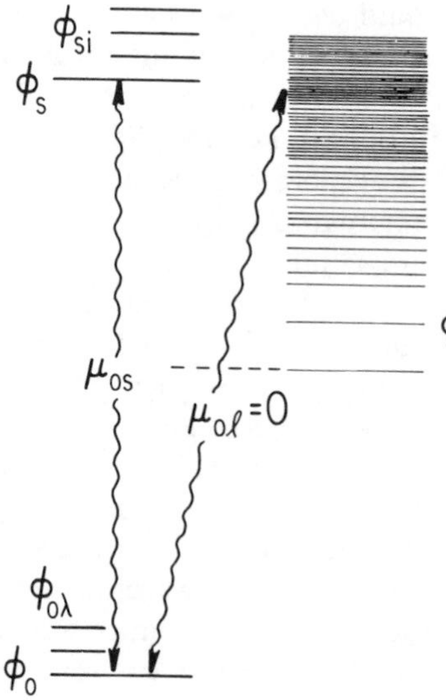

Fig. 4. Molecular energy level diagram used to discuss radiationless processes in polyatomic molecules. ϕ_0 is the ground electronic state, and $\phi_{0\lambda}$ denotes a thermally accessible vibronic component of this state. Electric dipole transitions from $\phi_{0\lambda}$ to the electronic state ϕ_s are allowed (or vibronically induced), while those to $\{\phi_l\}$ are forbidden. ϕ_{si} designates a vibronic component of ϕ_s, and ϕ_{lj} is a component of ϕ_l. The electronic states $\{\phi_0, \phi_s, \phi_l\}$ are obtained from the adiabatic Born–Oppenheimer approximation.

highly unfavorable Franck–Condon factors or because of spin selection or other symmetry considerations. Thus the primitive adiabatic Born–Oppenheimer representation of the energy levels of a polyatomic molecule in Fig. 4 would lead to the expectations of the observation of an absorption spectrum characterized by the energy levels of ϕ_s.

The presence of the vibronic states, $\{\phi_l\}$, can and generally does modify these simple expectations. This "dense" manifold of vibronic levels leads to a series of phenomena, which depend, in part, on the densities of levels, ρ_l, within this set. For instance, the small-molecule limit is characterized by a very low density of levels in this set $\{\phi_l\}$. In this case the non-Born–Oppenheimer couplings can lead to the observation of perturbations in the spectrum of ϕ_s arising from the vibronic state $\{\phi_l\}$. Such perturbations are associated with the displacement of levels of ϕ_s from their anticipated positions and possibly in the emergence of additional lines in the absorption spectra. This small molecule limit of low-level densities is an ideal case for the assignment and analysis of all spectroscopic lines.

The other extreme involves the large-molecule statistical limit wherein the density of $\{\phi_l\}$ levels is extremely high. Then the set of levels, $\{\phi_l\}$, can behave *as if* these levels were a continuum of levels on the time scales relevant to the photophysical processes in the excited electronic state. Thus these $\{\phi_l\}$ levels can act as a dissipative quasicontinuum and produce *irreversible* electronic relaxation, from ϕ_s to $\{\phi_l\}$.

Between these two extremes there resides the very interesting intermediate case that covers all other possibilities. First, the intermediate case may display some characteristics of both the small- and large-molecule limits because of strong variations in the ϕ_s–ϕ_l couplings arising from symmetry considerations, etc. A second intermediate case situation occurs where the density of $\{\phi_l\}$ levels becomes too high to hope to ever resolve and assign each of these levels individually, but where the level density is still not sufficiently high to lead to electronic relaxation in isolated molecules. This intermediate case situation is termed the "too-many level" small molecule case and is one of the interesting situations for further consideration of collisional effects on electronic relaxation.

One of the most striking facets of radiationless transition processes is the fact that they are *all* governed by the same general molecular energy level scheme as is given in Fig. 4.[1, 2] Thus, this figure encompasses the molecular model for the description of the small-molecule limit of perturbations in molecular spectroscopy, the large-molecule limit of irreversible electronic relaxation, as well as everything intermediate between these two limits. The phenomena are governed primarily by the details of the energy level density, ρ_l, as noted above, by the decay rates, $\{\Gamma_l\}$, of the zeroth order levels $\{\phi_l\}$ and by the $\phi_s - \{\phi_l\}$ couplings. Important other considerations center about the nature of the initially prepared electronic state. The theoretical explanation of all of these phenomena involve a description of the interaction of radiation with a coupled many-level molecular system and of the subsequent time evolution of the prepared state of the system. A treatment of many of these facets of the theory along with discussions of much of experimental results can be found in a variety of review articles.[1-4, 25] Here we merely state some of the conditions that delineate and explain the nature of the different small-, intermediate-, and large-molecule limits.

B. The Small, Intermediate, and Statistical Limits

The important parameter characterizing the molecular limit and the nature of the observed "radiationless processes," is given by

$$x_l = \frac{\hbar \Gamma_l}{\epsilon_l} \tag{2.1}$$

Here $\hbar\Gamma_l$ represents the energy uncertainty of a level with a decay rate of Γ_l, and ϵ_l is the average spacing between the $\{\phi_l\}$ levels. The condition for the small-molecule limit is

$$x_l \ll 1, \text{ small-molecule limit} \tag{2.2}$$

whereas for the large molecule case it is

$$x_l \gg 1, \text{ large-molecule, statistical limit.} \tag{2.3}$$

These limits are rather simply understood by the consideration of even a simple diatomic molecule example. Suppose first that ϕ_s and ϕ_l represent two bound electronic states of the diatomic molecule. Then under isolated molecule conditions Γ_s corresponds to the radiative decay rate of the individual ϕ_s level. Γ_s is typically on the order of $10^{-6}-10^{-9}$ sec^{-1}. However, in the diatomic molecule the average spacing between vibronic levels, ϵ_l, is on the order of hundreds or thousands of cm^{-1}. Thus the simple diatomic molecule example is definitely within the small molecule limit of (2.2).

On the other hand, we may consider a model in which ϕ_s represents a bound electronic state of the diatomic molecule, whereas $\{\phi_l\}$ represents a dissociative electronic state of the diatomic molecule. The dissociative state then has an energy level spacing that is zero, $\epsilon_l \rightarrow 0$, while the radiative lifetime is nonzero (albeit possibly rather small in many cases). Thus the parameter x_l of (2.1) is infinite for this example, and the conditions for the "large-molecule" limit, (2.3), are, perforce, automatically satisfied. In this situation we can observe a radiationless transition from ϕ_s to ϕ_l, which is an *irreversible nonradiative process*. Here the process of photon absorption and subsequent decomposition is simply represented (in simplifying limits) as follows: First, the molecule absorbs the incident radiation and undergoes a transition from the ground electronic state, ϕ_0, to the excited electronic state, ϕ_s, which carries oscillator strength in the appropriate spectral region. The quasibound state then undergoes a nonradiative transition to the dissociative $\{\phi_l\}$ levels.

By contrast, in the small molecule limit the situation corresponds to the spectroscopist's description of perturbations in the spectra of small molecules. The molecular Hamiltonian is represented in the basis set of adiabatic Born–Oppenheimer functions, $\{\phi_s, \phi_l\}$. Coupling among only a few levels need be considered, namely, those nearby coupled levels with couplings that satisfy the relationship

$$\frac{|v_{sl}|}{|E_s - E_l|} \gtrsim 1 \tag{2.4}$$

where E_s and E_l are the Born–Oppenheimer energies of ϕ_s and ϕ_l, respectively, and v_{sl} are the nonadiabatic matrix elements,

$$v_{sl} = \langle \phi_s | H | \phi_l \rangle$$

Those "nonresonant" levels $\{\phi_l\}$, grossly violating (2.4), provide a small contribution and can be treated separately by perturbation theory. The diagonalization of the molecular Hamiltonian within the resonantly coupled states, satisfying (2.4), then leads to the molecular eigenstates, ψ_n, with energies ϵ_n, which are a linear superposition of the adiabatic Born–Oppenheimer states. In our two bound states simple diatomic molecule example, ignoring the minor modifications necessary to include nuclear hyperfine structure, in the usual weak coupling limit the only resonant mixing can involve a pair of levels at a time for the case that ϕ_s and ϕ_l are both singlet states. In this simple case the molecular eigenstates are represented as

$$\psi_n = a_{sn}\phi_s + a_{ln}\phi_l \tag{2.5}$$

whereas when more levels of ϕ_l may be coupled, (2.5) involves the linear superposition of these few levels,

$$\psi_n = a_{sn}\phi_s + \sum_{j=1}^{N} a_{ljn}\phi_{lj} \tag{2.6}$$

where N is typically a small number.

In the absence of accidental degeneracies among the molecular eigenstates, $\{\psi_n\}$, (2.5) or (2.6) are the appropriate solutions of the Schrödinger equation for the isolated molecular system in a real world where spontaneous emission processes are admissible. Assuming that the ϕ_s and $\{\phi_l\}$ do not radiate to any set of common levels, it is possible to evaluate the radiative decay rates of the molecular eigenstates, $\{\psi_n\}$, as

$$\Gamma_n = |a_{sn}|^2\Gamma_s + \sum_{j=1}^{N} |a_{ljn}|^2\Gamma_l \tag{2.7}$$

For many cases of interest the zeroth order radiative decay rate of ϕ_s is much greater than that for the $\{\phi_l\}$. So under the condition that

$$\Gamma_s \gg \Gamma_l \tag{2.8}$$

the radiative decay rates of the molecular eigenstates are equal to

$$\Gamma_n \simeq |a_{sn}|^2\Gamma_s \tag{2.9}$$

The condition that the original zeroth order states, ϕ_s, be distributed amongst the molecular eigenstate implies the normalization condition,

$$\sum_{n=0}^{N} |a_{sn}|^2 = 1 \tag{2.10}$$

It follows from (2.10) that if more than one of the a_{sn} is nonzero, we must have

$$|a_{sn}|^2 < 1 \tag{2.11}$$

Equations (2.9) and (2.11) yield the result

$$\Gamma_n < \Gamma_s \qquad (\text{for } \Gamma_s \gg \Gamma_l) \tag{2.12}$$

which implies that the radiative decay rates of the mixed, molecular eigenstates, $\{\psi_n\}$, are less than the radiative decay rate of the parent Born–Oppenheimer zeroth order state, ϕ_s, containing all of the original oscillator strength. In fact, the radiative decay rates of more than one of the molecular eigenstates are often appreciable enough to enable observation of absorption or emission to more than one of the mixed molecular eigenstates. In practice this results in the observation of perturbations in the absorbtion spectrum where additional spectral lines are present.[32]

When the spacing between molecular eigenstates is large compared to the uncertainty widths of these levels,

$$\min|E_n - E_{n'}| \gg \tfrac{1}{2}\hbar(\Gamma_n + \Gamma_{n'}) \tag{2.13}$$

monochromatic excitation can only lead to the preparation of individual molecular eigenstates. Pulsed excitation, on the other hand, can lead to the excitation of a coherent superposition of a number of nearly molecular eigenstates. However, when the pulse duration, τ, satisfies in addition to (2.13), the inequality

$$\min|E_n - E_{n'}| > \frac{\hbar}{\tau} \tag{2.14}$$

this pulsed excitation cannot excite more than one of the individual molecular eigenstates. If the pulse is sufficiently short or the spacing between molecular eigenstates is sufficiently small that (2.14) and/or (2.13) is violated, then a coherent superposition of these nearby molecular eigenstates is possible. Apart from isolated cases of accidental degeneracies, the general situation (ignoring hyperfine structure) in diatomic molecules is found to satisfy (2.14) for nanosecond pulse durations, so even pulsed excitation is taken to prepare the molecule in single excited molecular eigenstate.

The above examples of electronic couplings in diatomic molecules are quite elementary and obvious. The basic principles, illustrated by these examples, however, apply to larger molecules where considerable confusion has arisen in the literature as to the applicability of either the small, large, or intermediate case limits in specific experimental situations.

As we pass to triatomic molecules, the level densities in bound electronic states increase. Thus there may be a large number of levels $\{\phi_l\}$ that can be resonantly [cf. (2.4)] coupled to a single level, ϕ_s. Nevertheless, when all electronic states are bound, it is quite often true that the level density is still very low, so that (2.2) is still safely satisfied. Then the small-molecule limit ensues, and the molecular eigenstates, (2.6), may contain a fairly large number of terms, N, on the order of 10–100 for large couplings, v_{sl}. In this case there are many additional spectral lines that may emerge in the absorption spectrum. One prime example of this complicating feature is the NO_2 molecule where the large couplings $v_{sl} \sim 150$–$200\ \text{cm}^{-1}$ and the moderate density of states leads to the appearance of a considerable number of additional levels.[32] Again, under the assumption of (2.8), all of the results (2.9)–(2.14) are equally valid for these small-molecule limit triatomic molecules.

There are then no possibilities for the occurrence of irreversible radiationless decays in such small-molecule limit triatomics. However, interesting effects, arising from the coherent superposition of many levels, may still appear when (2.14) is violated. The presence of hyperfine structure makes this possibility very likely. For instance, Demtroder[33] has observed nonexponential decays of excited states of NO_2 in a molecular beam where the spacing between hyperfine levels is claimed to be sufficient to excite a single hyperfine component with his MHz bandwidth laser. Demtroder then has no recourse but to explain the nonexponential decays in terms of some elusive "radiationless decay" despite the fact that the conditions (2.2) for the small-molecule limit are obeyed and prohibit *irreversible* decays. It should, however, be recalled that when traveling along with the molecule in the molecular beam, the molecule encounters a pulse of radiation whose duration is given by the laser spatial extent divided by the molecular velocity. For a laser spot size of 10^{-2} cm and a molecular velocity of 10^5 cms^{-1}, the pulse duration is 10^{-7} s. This yields an effective pulse frequency width of 10 MHz which could yield a coherent superposition of a number of hyperfine levels. The nonexponential decay of such a superposition is discussed in Section II. C.

The above general considerations may apply in an extremely large molecule. For instance, if the origins of the electronic states ϕ_s and ϕ_l are very close so that the density of states of $\{\phi_l\}$ is small at the energies E_s, the "small-molecule" limit conditions (2.2) are still obeyed. This situation applies to naphthalene, where the S_1–S_2 splitting is ca. 3500 cm^{-1}, and to 3, 4-benzepyrene, where it is ca. 3800 cm^{-1}. In these cases the respective observed lifetimes of S_2, $\tau_{S_2} \approx 4\times10^{-8}$ s and 7×10^{-8}s, are longer than those deduced from the absorption oscillator strengths, $S_2{}^{\mathrm{rad}}$, for both molecules. The intermediate case can also arise for small S_1–T_1 splittings[17, 34], e.g., benzophenone where the splitting is ca. 3000 cm^{-1} and $\tau_{S_1} \approx 10^{-5}$ s $> \tau_{S_1}{}^{\mathrm{rad}} \approx 10^{-6}$ s. Thus the behavior of the molecule is still governed by the small-molecule case, (2.9) through (2.14), when the condition (2.8) is obeyed. [If (2.8) is not met, all of the above arguments are readily modified to incorporate radiative decay processes from the $\{\phi_l\}$ levels.[1]]

When, on the other hand, the spacing between the origins of ϕ_s and ϕ_l become large enough and/or the size of the molecule is sufficiently large so there are a large number of vibrational modes and therefore a large enough density of states, the conditions for the emergence of the large molecule statistical limit, (2.3), are obtained. In this situation the molecular eigenstates could, in principle, be evaluated, but they are not the most appropriate eigenfunctions for the description of the radiative decay processes of the isolated molecule. This is because the condition,

(2.13), is violated for large sets of levels $\{\psi_n\}$, so these groups of levels no longer decay radiatively in an independent fashion.[1, 35–39] A theoretical analysis of the radiative and nonradiative decay of a group of closely coupled levels shows that the relevant quantity to be diagonalized is the effective molecular Hamiltonian,[1, 35–39]

$$H_{\text{eff}} = \frac{H - i\hbar\Gamma}{2} \tag{2.15}$$

where H is the full molecular Hamiltonian and $\mathbf{\Gamma}$ is the damping matrix. In the basis set of Born–Oppenheimer functions the latter is written as

$$\Gamma_{ij} = \Gamma_s \delta_{si} \delta_{sj} + \Gamma_l \delta_{lj} \delta_{li} \tag{2.16}$$

where it is assumed that the damping matrix is diagonal among the $\{\phi_l\}$ levels, for simplicity. Note that (2.15) is a nonhermitian Hamiltonian, so its eigenvalues, E_r, are complex numbers

$$E_r = \frac{\epsilon_r - i\hbar\Gamma_r}{2} \tag{2.15a}$$

The real parts of the eigenvalues, ϵ_r, give the energies of the quasistationary states, while the negative of the imaginary parts of the eigenvalues give one half the uncertainty widths, $\hbar\Gamma_r$, of these quasistationary states or the lifetimes from $\tau_r = \Gamma_r^{-1}$.

In the small-molecule limit the effective Hamiltonian, (2.15), is diagonalized to a good approximation by the molecular eigenstates, $\{\psi_n\}$, of (2.6) with the decay rates given by (2.7) (in the absence of accidental degeneracies). Thus equation (2.15) reduces to the small-molecule limit when condition (2.2) is obeyed. In the large-molecule limit, (2.3), in contrast, the nature of the eigenfunctions of the effective Hamiltonian, (2.15), is entirely different. The solutions correspond very much to the emergence of an eigenfunction which is very similar to ϕ_s and which is now a *resonant state* having a radiative decay rate equal to Γ_s and a nonradiative decay rate Δ_s, given in the weak coupling limit by the familiar golden rule expression

$$\Delta_s = \frac{2\pi}{\hbar} \sum_j |v_{slj}|^2 \rho_{lj} \tag{2.17}$$

The condition (2.3) implies that *if* a molecule begins in state ϕ_s and then crosses over to the $\{\phi_l\}$, the final states $\{\phi_l\}$ decays with rates Γ_l before the molecule ever has a chance to cross back to the original state ϕ_s. Thus it is the decay of the final manifold of levels $\{\phi_l\}$ that *drives* the irreversible electronic relaxation from ϕ_s to $\{\phi_l\}$.

There are, however, many situations in which the inequality (2.8) is obeyed for these large molecules, and the condition (2.3) might therefore appear to be inapplicable for these systems. However, in these cases we are primarily interested in whether or not the radiationless transition from ϕ_s to $\{\phi_l\}$ *appears* to be *irreversible* on the time scales of the experiment.[1, 36, 37] The time scales for the experiment may be limited by the radiative lifetime, Γ_s, of the state ϕ_s. It might be limited by the detection apparatus or by the presence of collisional or other competing relaxation processes. Thus it is often of interest to consider the nonradiative decay properties of a system on a particular time scale, τ_e, for the experiment. The parameter

$$x_l' = \hbar \frac{[\Gamma_l + (\tau_e)^{-1}]}{\epsilon_l} \tag{2.18}$$

then enables us to characterize the decay characteristics of a molecule on this experimental time scale τ_e. When x_l' obeys the condition

$$x_l' \ll 1 \tag{2.2'}$$

the small-molecule behavior is still manifest on the time scale τ_e. However, if

$$x_l' \gg 1 \tag{2.3'}$$

is obeyed on the time scale τ_e, the decay characteristics of the molecule correspond to the statistical limit of states, ϕ_s, with both radiative and nonradiative decay *on the time scales*, τ_e, of the experiment. On these time scales for which (2.3′) is obeyed, the nonradiative decay rate of ϕ_s is given by (2.17) in the weak coupling limit.

C. Dephasing in an Isolated Molecule

One interesting manifestation of the intermediate case occurs when the density of $\{\phi_l\}$ states is quite large but still insufficient for the statistical limit (2.3) to be obeyed. Then the molecular eigenstates in (2.6) have a large number of contributing terms, and it would be extremely difficult to unravel all of the individual molecular eigenstates to determine the zeroth order Born–Oppenheimer energies and their couplings v_{sl}. This is the "too-many level" small-molecule situation. It is possible to encounter cases in which the inequalities,

$$\Delta_s \gtrsim \Gamma_s \gtrsim \Gamma_l \tag{2.19}$$

are found. The quantity Δ_s of (2.17) is not the nonradiative decay rate of the state ϕ_s because the molecule conforms to the small-molecule limit. Broadband excitation can, in principle, produce an initially prepared nonstationary state of the molecule, which closely approximates the Born–Oppenheimer state ϕ_s. The subsequent time evolution of the system is governed by the time evolution of the molecular eigenstates. For the illustrative example where the initial state of the system is ϕ_s, the state of the system at time t can be shown in this small-molecule limit to be

$$\Psi(t)=\sum_n a_{ns}\exp\left[\frac{-i\epsilon_n t}{\hbar}-\frac{\Gamma_n t}{2}\right]\psi_n \tag{2.20}$$

The probability of observing radiative emission characteristic of the state ϕ_s at time t is proportional to the probability of finding the molecule in the zeroth order state ϕ_s at t, which is given by

$$\begin{aligned} P_s(t) &\equiv |\langle\phi_s|\Psi(t)\rangle|^2 \\ &= \left|\sum_n |a_{ns}|^2\exp\left[\frac{-i\epsilon_n t}{\hbar}-\frac{\Gamma_n t}{2}\right]\right|^2 \end{aligned} \tag{2.21}$$

Note that the probability of being in ϕ_s involves the absolute value squared of a complicated sum over molecular eigenstates of oscillatory damped factors.

The general behavior of (2.21) could, in principle, be extremely complicated. However, in the too-many level, small-molecule case the large number of contributing terms in the summation does provide a simplification. Again for illustrative purposes assume that the condition (2.8) is obeyed for this molecule and that the reason for its classification in the small-molecule limit is the small value of Γ_l. However, on a short enough time scale the condition (2.3′) may be satisfied for time scales on the order of or less than the radiative decay rates, Γ_n, of the individual molecular eigenstates. Then, on the time scales, τ_e, the molecule *appears as if it conforms to the large-molecule statistical limit*. Consequently, on these time scales the molecule appears to undergo a nonradiative decay from ϕ_s to $\{\phi_l\}$ with a nonradiative decay rate Δ_s of (2.17). Thus, as shown by Lahmani et al.[40] $P_s(t)$ of (2.21) for short enough times, such that (2.3′) is satisfied, displays an exponential decay with a decay rate given by

$$\Gamma_s^{\text{tot}}(\tau_e)=\Gamma_s+\Delta_s \tag{2.22}$$

On longer time scales, where (2.3′) is violated, the behavior of the system is then again given by the small-molecule limit and the complicated expression (2.21). However, in almost all experiments to date the long time behavior of these molecules, that is, the long time behavior of (2.21), is found to be adequately represented by a simple exponential decay with some average molecular eigenstate decay rate, $\bar{\Gamma}_n$. Between the short time exponential decay of (2.22) and the long time average molecular eigenstates exponential decay of $\bar{\Gamma}_n$, in principle, the decay given by (2.21) could be very complicated.[41] It could exhibit oscillations that result from quantum-mechanical interference effects (quantum beats). Recent molecular beam experiments by Chaiken et al.[42] have observed quantum beats in biacetyl. Previous bulb experiments could not unambiguously observe these oscillatory effects, and this is presumably due to the fact that large numbers of rotational levels are involved, so that the expression (2.21) appropriate to these experiments presumably involves a further weighted summation of this expression over all the individual rotational levels that are excited by the pulse of radiation, thereby washing out any beat pattern.[43] Thus, the net result is that the decay properties of these intermediate case molecules, the too-many level, small-molecule limit, is well represented by a biexponential decay involving the short time apparent radiationless transition and the long time decay of the individual molecular eigenstates.

This short time apparent decay is a particularly interesting intramolecular dephasing process, which merely accounts for the fact that the individual molecular eigenstates in the summation in (2.21) all have slightly different energies, $\{\epsilon_n\}$. When the molecule is initially prepared in the nonstationary state ϕ_s or with some other set of $\{a_{ns}\}$, there is a coherent superposition of all of these molecular eigenstates with relative phases which are fixed. However, because of the differences between the energies, $\{\epsilon_n\}$, these phases become different at a subsequent time, the $\exp[-i\epsilon_n t/\hbar - \Gamma_n t/2]$ factors are different for each of the ψ_n. The intramolecular dephasing of the different molecular eigenstates is what leads to the apparent exponential decay for short times governed by (2.3′), *making it look as if the initial state* ϕ_s is decaying into the $\{\phi_l\}$. The subsequent time evolution then governed by that of the individual molecular eigenstates. This process is not a truly irreversible electronic relaxation since, in principle, a multiple pulse sequence could be applied to the system to produce the state ϕ_s (with diminished overall amplitude) at a subsequent time $t>0$. The presence of a finite number of levels is what makes the dephasing (in principle) reversible. In the statistical limit the condition (2.3) implies the participation effectively of an infinite number of levels (i.e., continuous density of states) that can produce irreversible electronic relaxation.

When an intermediate case molecule, satisfying (2.2) conforms to (2.3′) on some time scales τ_e, this molecule has not had enough time to realize that there are only a finite number of levels $\{\phi_l\}$; it cannot resolve energy differences less than τ_e^{-1}. Because of time-energy uncertainty it "thinks" there is a continuous density of $\{\phi_l\}$ levels on time scales τ_e, so it undergoes its exponential decay. At longer times for which (2.2′) becomes obeyed, the molecule "finds out" the $\{\phi_l\}$ states are discrete, and the molecule behaves as a small molecule. Hence, the initial exponential decay of $P_s(t)$ is not an irreversible one. The term "intramolecular dephasing" is useful to distinguish this process from irreversible electronic relaxation.

D. Intramolecular Vibrational Relaxation

As mentioned in the introduction, the above discussion of the small-, large-, and intermediate-molecule limits of electronic relaxation processes can also be utilized with very minor modifications to discuss the phenomena of intramolecular vibrational relaxation in isolated polyatomic molecules.[30] Figure 4 is still applicable to this situation. The basis functions are now taken to be either pure harmonic vibrational states, some local-mode vibrational eigenfunctions, or some alternative nonlinear mode-type wavefunctions. In the following the nomenclature of vibrational modes is utilized, but its interpretation as normal or local can be chosen to suit the circumstances at hand.

ϕ_0 is now the vibrationless level that carries electric dipole oscillator strength to some excited zero-order vibrational level ϕ_s. Again isoenergetic with ϕ_s is a dense manifold of other zeroth order vibrational states, $\{\phi_l\}$, which do not carry oscillator strength from the ground vibrational level ϕ_0 (or any of the other low-lying thermally accessible vibrational levels). The zeroth order functions $\{\phi_s, \phi_l\}$ are not exact eigenfunctions of the molecular vibrational Hamiltonian because of the presence of anharmonicities and Coriolis couplings in the normal-mode description, and because of interlocal-mode couplings and anharmonicities in the local-mode description, etc. Thus there are again couplings between ϕ_s and the dense isoenergetic manifold $\{\phi_l\}$. These couplings can lead to a wide variety of phenomena depending on (1) the nature of initially prepared states, (2) the value of x_l of (2.1), and (3) the number of levels satisfying the near-resonance criterion (2.4).

In (2.1), ϵ_l is again taken to be the average spacing between neighboring $\{\phi_l\}$ levels at the energy of ϕ_s. Γ_s is again the decay rate of the zeroth order level, ϕ_s, which in isolated molecules is due to infrared emission and has typical values on the order of $10^{-3}-10^{-5}\mathrm{s}^{-1}$. In the small molecule limit the condition (2.2) is obeyed, and the states of the system are characterized by the vibrational eigenstates of (2.6). In this limit the mixing among a few zeroth order vibrational levels produces the familiar Fermi resonance. Because typical values of the couplings, v_{sl}, are on the order of tens of cm^{-1}, the energy separation, observed in many instances of Fermi resonance between lower lying levels in polyatomic molecules, safely satisfies the inequality (1.3) and (2.14). Hence, nanosecond-type pulsed excitation can only excite single vibrational eigenstates, ψ_n, of mixed parentage as in (2.5) or (2.6). With weak Fermi resonance and picosecond pulses, perhaps there are cases in which the inequality (2.14) is violated, so, in principle, it may be possible to coherently excite a number of vibrational eigenstates and to produce an initial nonstationary vibrational state that closely resembles the zeroth order vibrational state, ϕ_s, which carries all of the oscillator strength from ϕ_0.

On the other hand, for large enough molecules and/or high enough total vibrational energy content, the density of vibrational levels can become rather high. First, consider the case where $\rho_l = \epsilon_l^{-1}$ is of the order of 1–100 cm. Because of the small values of Γ_l, the small-molecule limit, (2.2), is *still* obeyed for this situation. There are unfortunately far too many vibrational levels to ever permit a complete spectroscopic analysis and sorting of each of the individual vibrational levels, so we again may term this situation the "too-many level, small-molecule limit" of vibrational states (an intermediate case). At a vibrational level density of 10^3 cm^{-1} and for τ_e on the order of nanoseconds, we find that x_l' of (2.18) satisfies the large-molecule limit of (2.3′) on these time scales. Hence we again would expect to observe an exponential decay of the probability $P_s(t)$ of (2.21) for finding the system in the initial nonstationary state ϕ_s *if this initial nonstationary state were one that could be prepared experimentally*. Perhaps one means for producing such an initial nonstationary state would involve optical excitation from a lower electronic state to high-lying vibrational levels in the upper electronic state when the optical absorption spectra involves readily assignable peaks corresponding to the zeroth order mode description that would be utilized in the vibrational representation of Fig. 4. Another possibility involves high overtone absorption within the ground electronic manifold.[44] On longer time scales, the criterion (2.2′) becomes satisfied, so the system is again in the small-molecule limit with longer decay lifetimes characterized by the decay rates $\{\Gamma_n\}$. Thus, again, the biexponential decay is expected to characterize the intermediate case vibrational situations *when* it is possible to initially excite a zeroth order vibrational level, ϕ_s, of some particular mode parentage.

At much higher vibrational level densities the criterion (2.3) can become satisfied, whereupon the vibrational relaxation of an initially prepared ϕ_s

state now involves a dissipative irreversible decay into the effective continuum of levels $\{\phi_l\}$. Questions concerning the precise nature of the prepared state, the contribution from the reversible dephasing processes at intermediate energies, and the role of intramolecular vibrational relaxation at high enough energies are central to the understanding of infrared multiphoton dissociation processes in isolated molecules.[45]

The RRKM theory of unimolecular decomposition is generally expressed in terms of the assumption of very rapid intramolecular vibrational relaxation.[46] However, despite the apparent successes of this theory, a kinetically equivalent formulation[47] invokes the opposite assumption of the *absence* of intramolecular vibrational relaxation along with the alternative assumption that it is the collisional excitation which produces its statistical distribution amongst the relevant energy levels of the system. Collisional processes are discussed in Section III, but we can anticipate results by noting that at very high vibrational energy contents, corresponding to the amount of energy required to break chemical bonds, a statistical description of the initial and final states of collisions of highly vibrationally excited molecules may be quite a reasonable one. Thus, a study of intramolecular vibrational processes is of considerable interest in also furthering our understanding of unimolecular decomposition processes.

Smalley and co-workers[72] have probed intramolecular vibrational relaxation by viewing the yields and the time-dependence of the fluorescence from S_1 in alkylated benzenes. They focus attention on those ring modes whose vibrational frequencies are unshifted by alkylation; these are vibrations with nodes at the alkylated ring carbon atom. The absorption lines are sharp, but as the alkyl chain is lengthened, the emission spectrum develops a broad "relaxed" component, while the intensity of the sharp "unrelaxed" resonance fluorescence diminishes in intensity as the intensity of the "relaxed" spectrum increases. The time-dependence of the relaxed and unrelaxed emission is found to be a single exponential decay, so unfortunately, the rapid intramolecular dephasing decay has not yet been followed.

As the majority of cases studied by Smalley and co-workers fall into the too many level small molecule limit, the intermediate case, we [73] have generalized the intermediate case relaxation theory to describe the Smalley type experiments. This generalization involves the description of the radiative decay characteristics of the $\{\phi_{lj}\}$ states to evaluate the time dependence of the relaxed emission as well as the relaxed and unrelaxed quantum yields. The lack of observation of the rapid intramolecular dephasing in the alkylated benzenes may be due to an excitation pulse too long to coherently excite the zeroth order ring vibration. An alternative possibility arises from the fact that the intramolecular vibrational coupling is channeled through the motion of the substituted ring carbon and its adjacent alkyl carbon atom. The theory predicts [73] that this situation would produce an intramolecular dephasing rate which is roughly independent of alkyl chain length, but with relative relaxed and unrelaxed yields which vary greatly. A test of this assumption might be provided by considering di-alkyl substituted benzenes.

III. COLLISIONALLY INDUCED INTERSYSTEM CROSSING AT LOW PRESSURES

By definition irreversible electronic relaxation processes cannot occur in isolated small and too-many level small (intermediate) case molecules because of the insufficient density of final levels. For long times the molecule senses the presence of a finite number of possible final levels instead of the effective continuum that is required to drive irreversible electron relaxation. When collisional processes are appended, it is clear that the continuous density of states of the colliding pair can provide the necessary driving force for irreversible relaxation. The observed magnitudes of electronic relaxation rates as well as dependencies on the initial state, perturbing molecules, temperature, and so on, are the aspects of the processes that are of central interest.

The description of collision-induced electronic relaxation simplifies in the low- and high-pressure limits. For extremely low pressures the collision probability and, therefore, the probability of collision induced electronic relaxation is extremely small, so the general theoretical results of Section II provide guides to the mechanism for collision-induced electronic relaxation processes.

When condition (2.2) is satisfied, the $\phi_s-\{\phi_l\}$ interelectronic state coupling, v_{sl}, leads to molecular eigenstates that are of mixed electronic parentage, (2.5)–(2.6) for the isolated molecule. If, for instance, ϕ_s is a zeroth order singlet level, S, and $\{\phi_l\}$ are zeroth order triplet levels, $\{T_l\}$, the zero-pressure molecular eigenstates (2.6) for the small-molecule limit are of three general varieties as depicted in Fig. 5. The mixed singlet states have wavefunctions

$$\hat{S}=\left(1-\sum_l y_l^2\right)^{1/2} S+\sum_l y_l T_l \tag{3.1}$$

while the mixed triplets yield

$$\hat{T}_l=z_l S+\left(1-z_l^2-\sum_{l'\neq l} z_{l'}^2\right)^{1/2} T_l+\sum_{l'\neq l} z_{l'} T_{l'} \tag{3.2}$$

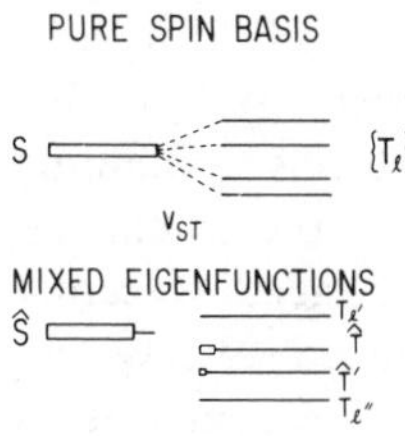

Fig. 5. In the pure-spin basis one singlet level is coupled by v_{ST} to a number of triplet levels $\{T_l\}$. When the molecular Hamiltonian is diagonalized, there results mixed eigenstates. $\hat{S}$ has primarily singlet character; $\hat{T}$ and $\hat{T}'$ are mostly triplets, while T_l', and T_l'' remain pure triplets. The observed low-pressure collisional processes are governed by the mixed states.

There may be some pure triplets, $T_{l''}$, which escape coupling to singlets because of symmetry selection rules.

Collisions between states of mixed electronic parentage are of a *wholly different character*[18, 20–22, 24] than those involving molecules in pure electronic states. This is a collisional generalization of familiar facets of isolated molecule spectroscopy where small mixings between different electronic states can lead to very sizable consequences. For instance, small nonadiabatic electronic mixings can lead to the vibronic inducement of otherwise electronically forbidden radiative transitions. Likewise, the presence of these zero-pressure interelectronic state couplings leads to the occurrence of highly efficient collision-induced electron relaxation processes. In particular, the theory discussed below shows how the mixed states S "borrow" some pure triplet state collision-induced vibrational and rotational relaxation processes to utilize for the collision-induced intersystem crossing in a manner that loosely parallels intensity borrowing in isolated molecules. Radiative intensity is generally borrowed from electronic states which are distant in energy. The collisional "borrowing," on the other hand, emerges from electronic states that have nearly degenerate levels.

Given that the small-molecule limits (2.2) and (2.2′) for relevant experimental time scales are both satisfied, the real difference between a very small diatomic molecule and an intermediate-sized one like glyoxal resides only in the number of coupled levels in (2.5) and (2.6), respectively. Hence it is convenient to pursue the general analysis for the case of a pair of coupled levels, S and T. Then the more general case of many levels (2.6) is readily generated.

Diatomic molecules can also exhibit strong collision-induced electronic relaxation. An old example is the CN molecule[48] that undergoes collision-induced internal conversion, $A^2\Pi \rightarrow B^2\Sigma^+$ due to the presence of perturbations involving $v=10$ of $A^2\Pi$ and $v=0$ of $B^2\Sigma^+$ for K values of 4, 7, 8, 11, and 15. At sufficiently low pressures in active nitrogen–organic vapor flames the $A^2\Pi$ levels have very large steady-state populations relative to the isoenergetic $B^2\Sigma^+$ levels. Most of the emission is observed in the

$A^2\Pi \rightarrow X^2\Sigma^+$ red-band system. The anomalously high intensity (for pressures < 1 torr) of a few lines in the $B^2\Sigma^+ \rightarrow X^2\Sigma^+$ violet bands is due primarily to the direct formation of perturbed $B^2\Sigma^+$ levels containing some admixture of $A^2\Pi$ character. At higher pressures the intensities associated with these perturbed $B^2\Sigma^+$ lines is further enhanced. Recently Lavolée and Tramer[13, 49] have observed collision-induced intersystem crossing from perturbed $A^1\Pi$ levels of CO by utilization of synchrotron excitation. These diatomic examples should provide a more quantitative test of the theoretical principles which are also applicable (with some appended summations over coupled states) to larger molecules like glyoxal.

A. Collision Dynamics

The coupled mixed pair of states reduce from (3.1) and (3.2) to

$$\hat{S} = \cos\frac{\theta}{2} S + \sin\frac{\theta}{2} T \tag{3.3a}$$

and

$$\hat{T} = -\sin\frac{\theta}{2} S + \cos\frac{\theta}{2} T \tag{3.3b}$$

where for $|\theta| < \pi/2$ S is the molecular eigenstate of primarily S parentage, while $\hat{T}$ is primarily triplet (T). The angle θ is determined by the intramolecular singlet–triplet coupling, v_{ST}, and the zeroth order singlet and triplet energies, $E_0(S)$ and $E_0(T)$, respectively,

$$\theta = \arctan\{2v_{ST}[E_0(S) - E_0(T)]^{-1}\} \tag{3.3c}$$

The perturbed energy levels are at $E_0(S) + \Delta_S$ and $E_0(T) + \Delta_T$ where Δ_S and Δ_T are readily written in terms of v_{ST}, $E_0(S)$ and $E_0(T)$ when required.

The interaction between the molecule and a collision partner, the perturber, is governed by a potential energy V, which generally depends on the electronic coordinates, the vibrational, rotational, etc. coordinates of the molecule and the perturber. In order to illustrate the essential physical mechanism for the collision-induced intersystem crossing, we dispense with the myriad of quantum numbers necessary to specify the sublevels in the zeroth order basis. The perturber is assumed to remain in its electronic ground state; we do not include the possibility for electronic energy transfer from the excited molecule. Thus, the only electronic states, which enter, are S and T and the perturber ground state. Omitting explicit reference to the latter implies that V has portions that are diagonal in the zeroth order pure-spin states S and T, V_{SS}, and V_{TT}, respectively, as well as the spin-flipping terms V_{ST} and V_{TS}, which must arise from spin–orbit and exchange interactions. (Note that the case of doublet–quartet state collision-induced intersystem crossing follows in an identical fashion. Here the words doublet and quartet are substituted for singlet and triplet, respectively.)

The experimental data on glyoxal, propynal, and other systems indicate that the spin-flipping V_{ST} and V_{TS} terms are negligible for collision-induced intersystem crossing. However, for collision-induced internal conversion T and S now have the same spin multiplicity, so direct electrostatic coupling terms contribute to V_{ST} and V_{TS}. We have previously suggested that in this situation V_{ST} and V_{TS} must be retained.[21, 24] Subsequently, Bondybey and Miller[50] have measured collision-induced internal conversion rates in the isoelectronic molecules CO^+ and CN involving the $X^2\Sigma^+$ and $A^1\Pi$ states. Their experiments demonstrate a collision-induced cascading $X \rightarrow A \rightarrow X \rightarrow A \ldots$ down successive interspersed vibrational levels of the X and A states. The collision rates demonstrate the importance of the interelectronic coupling terms V_{AX} and V_{XA}. A considerable magnitude for these couplings can be understood from the fact that, when even a perturbing atom is situated in a noncolinear position, the nuclear symmetry is no longer $C_{\infty h}$. Hence, the Π and Σ^+ symmetry designations are invalid and strong mixing is expected of A and X components into the two new triatomic electronic states and their potential energy surfaces for the interacting molecule–perturber system. This behavior can probably be readily illustrated semi-empirically by use of the diatomics in molecule method.[51] Here we concentrate on the collisions wherein the spin multiplicity is changed, so the couplings V_{ST} and V_{TS} can be neglected (or treated as small perturbations.)

The coupling functions V_{SS} and V_{TT} still depend on the molecular vibrational and rotational degrees of freedom as well as the relative molecule–perturber separation, $\mathbf{R}$. Since the experiments imply that the physical origin of the collision-induced intersystem crossing resides in long-range attractive interactions, we may adopt a semiclassical approximation where the quantum-mechanical variables for the relative translation is replaced by a classical trajectory, $\mathbf{R}(t)$, for the relative molecule–perturber motion. The internal dynamics is then influenced by the time-dependent interactions $V_{SS}[\mathbf{R}(t)]$ and $V_{TT}[\mathbf{R}(t)]$, which are still functions of molecular rotational and vibrational variables. For simplicity and for illustrative purposes we consider only the pair of coupled levels S and T and a pure triplet level T', which represents the molecular state after the collision. Note T' may differ in rotational and/or vibrational quantum numbers from T. (Cases where T' is part of another mixed pair, etc. are readily treated analogously.[22])

The time-dependent Schrödinger equation for the effect of the collision dynamics on the probability amplitudes for being in S, T, and T' is most conveniently represented in the mixed state basis (3.3), as this makes all coupling matrix elements vanish in the limit of infinite molecule–perturber separations. Within the mixed basis set, the matrix elements of V can be worked out to give

$$\langle\hat{S}|V|\hat{S}\rangle=\cos^2\frac{\theta}{2}V_{SS}[\mathbf{R}(t)]+\sin^2\frac{\theta}{2}V_{TT}[\mathbf{R}(t)] \tag{3.4a}$$

$$\langle\hat{T}|V|\hat{T}\rangle=\sin^2\frac{\theta}{2}V_{SS}(\mathbf{R}(t))+\cos^2\frac{\theta}{2}V_{TT}(\mathbf{R}(t)) \tag{3.4b}$$

$$\langle\hat{S}|V|\hat{T}\rangle=\tfrac{1}{2}\sin\theta[V_{SS}(\mathbf{R}(t))-V_{TT}(\mathbf{R}(t))] \tag{3.4c}$$

$$\langle\hat{S}|V|T'\rangle=\sin\frac{\theta}{2}V_{TT'}(\mathbf{R}(t)) \tag{3.4d}$$

$$\langle\hat{T}|V|T'\rangle=\cos\frac{\theta}{2}V_{TT'}(\mathbf{R}(t)) \tag{3.4e}$$

Note that the coupling between the initial "singlet" level and the final triplet in (3.4d) arises from perturber-induced triplet–triplet couplings $V_{TT'}[\mathbf{R}(t)]$ multiplied by the mixing coefficient $\sin(\theta/2)$. In the absence of singlet–triplet mixing, the terms $V_{TT'}[\mathbf{R}(t)]$ are the couplings responsible for collision-induced vibrational and rotational transitions (and elastic collisions) within the triplet manifold. It is thus clear how small S–T mixings, small $\sin(\theta/2)$, can enable the mixed singlet, $\hat{S}$, to "borrow" collisional couplings for triplet collisional processes to drive the collision-induced intersystem-crossing process. It is instructive to note the analogy with the much simpler phenomenon of intensity borrowing in vibronically induced radiative transitions.

When $\hat{S}$ and $\hat{T}$ can be spectrally resolved and when their splitting satisfies

$$|E_0(S)+\Delta_S-E_0(T)-\Delta_T|\gg\hbar/\tau_{\text{excitation}} \tag{3.5}$$

where $\tau_{\text{excitation}}$ is the duration of the excitation process—say laser radiation—then excitation can be taken to initially produce a pure molecular eigenstate. We consider the case where the initial state is $\hat{S}$, but it follows similarly if $\hat{T}$ is the initial state and some S' or T' is the final one, etc. When (3.5) is violated, coherent excitation of $\hat{S}$ and $\hat{T}$ is, in principle, possible, and it might be interesting to study this situation in level anti-crossing spectroscopy where the splitting on the left in (3.5) is reduced to a minimum by magnetic (or electric) field tuning of the zeroth order levels of S and/or T.[54]

Since the molecule is taken to be in $\hat{S}$ initially, as $t \to -\infty$, the wavefunction at subsequent times, t, during the collision is given by

$$\Psi(t) = \sum_{n=\hat{S},\hat{T},T'} a_n(t) \exp\left[-(i/\hbar)\int^t E_n(t')\,dt'\right] |n\rangle \tag{3.6}$$

where the initial condition translates into

$$a_n(-\infty) = \delta_{n\hat{S}} \tag{3.7}$$

and the energies $E_n(t')$ are the full diagonal matrix elements,

$$E_{\hat{S}}(t') = E_0(S) + \Delta_S + \cos^2\frac{\theta}{2} V_{SS}[\mathbf{R}(t')] + \sin^2\frac{\theta}{2} V_{TT}(\mathbf{R}(t')) \tag{3.8a}$$

$$E_{\hat{T}}(t') = E_0(T) + \Delta_T + \sin^2\frac{\theta}{2} V_{SS}[\mathbf{R}(t')] + \cos^2\frac{\theta}{2} V_{TT}[\mathbf{R}(t')] \tag{3.8b}$$

$$E_{T'}(t') = E_0(T') + V_{T'T'}(\mathbf{R}(t')) \tag{3.8c}$$

The three-level molecular Schrödinger equation becomes

$$\begin{aligned} i\hbar\frac{\partial}{\partial t} a_{\hat{S}}(t) = {} & a_{\hat{T}}(t)\exp\left\{-\left(\frac{i}{\hbar}\right)\int^t [E_{\hat{T}}(t') - E_{\hat{S}}(t')]\,dt'\right\} \\ & \times \frac{1}{2}\sin\theta[V_{SS}[\mathbf{R}(t)] - V_{TT}[\mathbf{R}(t)]] \\ & + a_{T'}(t)\exp\left\{-\left(\frac{i}{\hbar}\right)\int^t [E_{T'}(t') - E_{\hat{S}}(t')]\,dt'\right\}\sin\left(\frac{\theta}{2}\right)V_{TT'}[\mathbf{R}(t)] \end{aligned} \tag{3.9a}$$

$$\begin{aligned} i\hbar\frac{\partial}{\partial t} a_{\hat{T}}(t) = {} & a_{\hat{S}}(t)\exp\left\{-\left(\frac{i}{\hbar}\right)\int^t [E_{\hat{S}}(t') - E_{\hat{T}}(t')]\,dt'\right\} \\ & \times \frac{1}{2}\sin\theta[V_{SS}(\mathbf{R}(t) - V_{TT}(\mathbf{R}(t))] \\ & + a_{T'}(t)\exp\left\{-\left(\frac{i}{\hbar}\right)\int^t [E_{T'}(t') - E_{\hat{T}}(t')]\,dt'\right\}\cos\left(\frac{\theta}{2}\right)V_{TT'}(\mathbf{R}(t)) \end{aligned} \tag{3.9b}$$

$$\begin{aligned} i\hbar\frac{\partial}{\partial t} a_{T'}(t) = {} & a_{\hat{S}}(t)\exp\left\{-\left(\frac{i}{\hbar}\right)\int^t [E_{\hat{S}}(t') - E_{T'}(t')]\,dt'\right\}\sin\left(\frac{\theta}{2}\right)V_{T'T}[\mathbf{R}(t)] \\ & + a_{\hat{T}}(t)\exp\left\{-\left(\frac{i}{\hbar}\right)\int^t [E_{\hat{T}}(t') - E_{T'}(t')]\,dt'\right\}\cos\left(\frac{\theta}{2}\right)V_{T'T}[\mathbf{R}(t)] \end{aligned} \tag{3.9c}$$

B. Types of Collisional Events

For sufficiently small mixing, small θ, and the initial conditions (3.7), (3.9) may be approximated by first-order time-dependent perturbation theory. In this case the approximate solution of (3.9c) is

$$a_{T'}^{(1)}(t) \simeq (i\hbar)^{-1} \sin\frac{\theta}{2} \int_{-\infty}^{t} dt' V_{TT'}[R(t')]$$
$$\times \exp\left\{ -\left(\frac{i}{\hbar}\right) \int^{t'} [E_{T'}(t'') - E_{\hat{S}}(t'')]\, dt'' \right\} \tag{3.10}$$

The $\hat{S} \to T'$ transition probability, $P_{\hat{S} \to T'}$ is then obtained as

$$P_{\hat{S} \to T'} \simeq |a_{T'}^{(1)}(+\infty)|^2 \tag{3.11}$$

while the evaluation of the total transition probability requires a summation over all possible final levels T' and an averaging over all relative velocities and impact parameters. It suffices here to consider the elemental transition probability (3.11). A necessary condition for the validity of first-order time-dependent perturbation theory is

$$|a_{T'}^{(1)}(+\infty)|^2 \ll 1 \tag{3.12}$$

When (3.12) is violated, various exponentiated forms of perturbation theory can be employed.[61] However, (3.11) serves to illustrate the salient physical features. Situations in which considerable $\hat{S} \leftrightarrow \hat{T}$ transition amplitudes occur during the collision can be treated by accounting for a variation of the effective mixing coefficient $\sin^2[\theta(t)/2]$ during the collision. This is discussed in Section III.D where it is shown how (3.10) emerges as a special limiting case with $\sin^2(\theta/2)$ essentially an effective collision-dependent mixing coefficient.

Consider first a state T' which has an energy far from $E_{\hat{S}}$ in the sense that

$$|E_0(T') - E_0(T)| \gg |E_0(S) + \Delta_S - E_0(T) + \cos^2\frac{\theta}{2} V_{SS}[\mathbf{R}(t'')]$$
$$+ \sin^2\frac{\theta}{2} V_{TT}(\mathbf{R}(t''))| \qquad \text{all } t'' \tag{3.13}$$

so the integral in the exponential factor in (3.10) may be replaced by $(E_{T'} - E_T)t'$. Condition (3.13) merely states that the absolute magnitude of the collisional shift of energy of $\hat{S}$ during the collision, plus the small

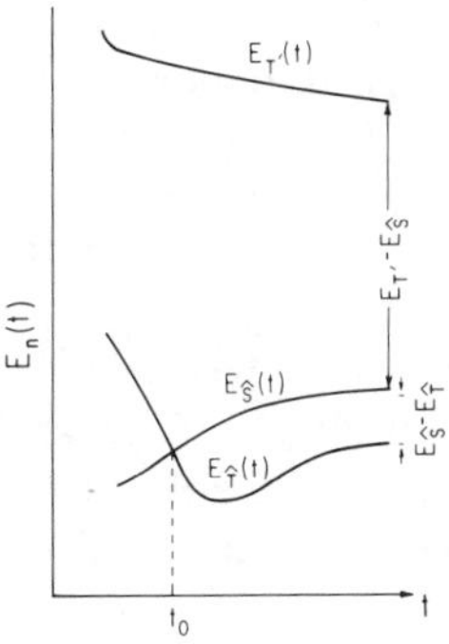

Fig. 6. Schematic representation of the collisional dependence of individual state energy levels. The time variation of $E_{T'}(t)-E_{\hat{S}}(t)$ during a collision is small compared to their asymptotic separation, condition (3.13). The pair $E_{\hat{S}}(t)$ and $E_{\hat{T}}(t)$ obey condition (3.15) and produce a point of stationary phase, Eq. (3.16), at t_0.

splitting $E_0(S)-E_0(T)+\Delta_S$ be small compared to the zeroth order $T \to T'$ energy transfer, $|E_0(T')-E_0(T)|$. This case is illustrated by curves $E(T)$ and $E(T')$ in Fig. 6. Using (3.13) converts (3.10) and (3.11) into

$$P_{\hat{S}\to T'} \simeq \hbar^{-2}\sin^2\frac{\theta}{2}\left|\int_{-\infty}^{\infty} dt\, V_{TT'}[\mathbf{R}(t)] \times \exp\{-(i/\hbar)[E_0(T')-E_0(T)]t\}\right|^2 \quad \text{if (3.13) satisfied} \tag{3.14}$$

which is simply $\sin^2(\theta/2)$ multiplied by the transition probability for a collision-induced $T \to T'$ transition between zeroth order pure-spin triplet levels (calculated, of course, by first-order perturbation theory.) The analogy with intensity borrowing in radiative transitions is perhaps made clearer by (3.14).

When the collisional fluctuations (3.4a) in the S-state energy can become comparable or larger than the energy transferred in the collision, $|E_0(S)+\Delta_S-E_0(T')|$, that is, when

$$|E_0(S)+\Delta_S-E_0(T')| \lesssim \left|\cos^2\frac{\theta}{2}V_{SS}[\mathbf{R}(t'')]+\sin^2\frac{\theta}{2}V_{TT}[\mathbf{R}(t'')]\right| \quad \text{some } t'' \tag{3.15}$$

the exponential factor in (3.10) must be retained as is.[22] Note that this produces a dependence of the exponential term on $\sin^2(\theta/2)$. Hence the condition (3.15) implies the contribution of collisional events in which the dependence of the transition probability, the cross-section, and the rate constant are no longer simply proportional to $\sin^2(\theta/2)$.

The condition (3.15) implies that the integral in the exponent of (3.10) is dominated by those times, t_0, for which the phase is stationary. This stationarity condition is satisfied for (3.10) when

$$E_{T'}(t_0) = E_{\hat{S}}(t_0) \tag{3.16}$$

that is, when the collision-dependent energies for the two states coincide. This case is illustrated by curves $E_T(t)$ and $E_S(t)$ in Fig. 6. Chu and Dahler assume that a surface crossing mechanism is the dominant one for collision-induced intersystem crossing.[52] Here it emerges as a special, albeit important, contribution.

If the instantaneous collisional energies cross to produce a stationary phase point in (3.10), this occurs twice during a collision, once as the molecule and perturber approach each other and once as they recede. As the transition can occur either on the inward or outward portion of the trajectory, the transition probability is given by the expression[53]

$$P_{\hat{S}\to T'} \simeq 2|a_{T'}^{(1)}(+\infty)|^2\{1-|a_{T'}^{(1)}(+\infty)|^2\} \tag{3.17}$$

Writing (3.14) as $\sin^2(\theta/2)\ f[\sin^2(\theta/2)]$ implies that (3.17) exhibits a nonlinear dependence on $\sin^2(\theta/2)$,

$$P_{\hat{S}\to T'} \simeq 2\sin^2\left(\frac{\theta}{2}\right)f\left[\sin^2\left(\frac{\theta}{2}\right)\right]\left\{1-\sin^2\left(\frac{\theta}{2}\right)f\left[\sin^2\left(\frac{\theta}{2}\right)\right]\right\} \tag{3.18}$$

from the double pass curve crossing nature of the collision in addition to the dependence imparted by the $f[\sin^2(\theta/2)]$ factor.

As an illustration of the kinds of f that may occur, consider a simple model where the long-range behavior of the potentials $V_{SS}[\mathbf{R}(t)]$ and $V_{TT}[\mathbf{R}(t)]$ arise from van der Waals interactions and vary as $A_{S,T}[\mathbf{R}(t)]^{-6}$. Given the existence of a stationary phase point, (3.16), the stationary phase approximation to (3.10) yields the general result

$$|a_{T'}^{(1)}(+\infty)|^2 \propto \sin^2\frac{\theta}{2}|V_{TT}'[\mathbf{R}(t_0)]|^2\left|\left(\frac{\partial}{\partial t}[E_{T'}(t)-E_{\hat{S}}(t)]\right)_{t=t_0}\right|^{-2} \tag{3.19}$$

which in the current case can be reduced to[22]

$$|a_{T'}^{(1)}(+\infty)|^2 \propto \frac{\sin^2(\theta/2)}{[\cos(\theta/2)]^{10/3}}\frac{A_T^2}{|A_S-A_T|^{4/3}|E_0(S)+\Delta_S-E_0(T')|^{1/3}} \tag{3.20}$$

Equations (3.20) and (3.17) show how transitions between states with asymptotic energy differences, which are smaller than collisional perturbations on these energy differences, yield contributions to collision-induced intersystem-crossing rates that are not just linearly dependent on the fractional mixing, $\sin^2(\theta/2)$. Another nonlinearity is generated by the introduction of collisionally dependent values, $\sin^2[\theta(t)/2]$ as described in Section III.D.

Experiments by Tramer and co-workers[13, 49] on collision-induced intersystem crossing from the $A^1\Pi$ state in CO display a nonlinear dependence on $\sin^2(\theta/2)$. Diatomic molecules like CO with monatomic perturbers present excellent cases for testing the general theory of collision-induced intersystem crossing as it is, in principle, possible in these circumstances to calculate the potential energy surfaces and to describe the collision dynamics to a reasonable degree of accuracy. Furthermore, an analysis of the perturbations in the spectrum can provide empirical values of $\sin^2(\theta/2)$. Because the CO experiments employ a synchrotron radiation source, the incident radiation excites a whole vibronic band with a distribution of $\sin^2(\theta_J/2)$ for the different rotational sublevels. Nevertheless, the experiments suffice to exhibit a nonlinear variation of the collision-induced intersystem-crossing rates with the average of $\sin^2(\theta_J/2)$ for the vibronic band. More experimental work on the atom–diatomic case is desirable, as information gathered for this simplest situation is used to provide a generalized conceptual framework for explaining the phenomenon in larger systems where a complete theoretical description is not currently feasible.

Larger molecules provide an increased density of states. Consequently, there may be a large set of singlet and triplet levels that satisfy conditions like (3.15). Hence, during a collision there are many energies, $\{E_n(t)\}$, which cross at sets of times $\{t_j\}$ where transition probabilities are enhanced. Multiple curve crossing approaches could be invoked if the crossings are well enough separated and if the kinetic energy is high enough, but the net effect of this process is to appear to produce a quasistatistical distribution among this set of crossing levels, despite the fact that the collision is a direct one that does not proceed through a long-lived collision intermediate.

C. Collision-Induced Intersystem Crossing in Large Molecules

It is hopefully clear by now that even the simplest diatomic molecule–atomic collision partner limit presents us with a number of possible types of contributing collisions to the collision-induced intersystem-crossing rates. It should be emphasized that a truly quantitative treatment would have to even include any hyperfine energies as the mixing angles (3.3c) can be greatly affected by zeroth order hyperfine shifts that are comparable to

$2|v_{ST}|$. Thus an idealized experiment would involve the preparation of a single pure or mixed rovibronic level and the observation of the state-to-state collision-induced intersystem-crossing rates. In larger molecules the larger level density produces a higher density of mixed levels and more terms possibly contributing to the mixed states (2.6). Otherwise, the basic phenomena and types of collisional events are the same. There is, however, an increased likelihood of multiple crossing of the nearby $\{E_n(t)\}$ curves as discussed above.

For a given initial level, $\hat{S}$, it is necessary to sum over all possible contributing final levels, T_l', which may be reached in the collision event. Likewise, if more than one rovibronic level is excited initially, it is necessary to add the contributions from all of the mixed levels which are prepared by the exciting light. The collision-induced intersystem crossing cross-section is of the form

$$\sigma \simeq \langle \sin^2 \frac{\theta}{2} \rangle n_M n_T \sigma_{\text{rot}} \tag{3.21}$$

Here $\langle \sin^2(\theta/2) \rangle$ denotes an average mixing coefficient for the initially excited mixed rovibronic levels, n_M is the average number of mixed singlets which are initially prepared, n_T is the average number of final triplet levels, $T_{l'}'$, (rotational, vibrational, spin, and hyperfine), which are accessible as final levels after the collision, and σ_{rot} is comparable to pure S or T electronic state rotational relaxation cross-sections within a single vibronic level or to pure T-state vibrational relaxation cross-sections (from one vibronic level to another), which should be comparable in magnitude. The $\langle \sin^2(\theta/2) \rangle$ dependence arises, strictly speaking, from collisions, satisfying conditions like (3.13), whereas the ones obeying conditions like (3.15) yield more general variations with mixing coefficients, but the overall effect is written symbolically as $\langle \sin^2(\theta/2) \rangle$. If each mixed singlet contains a number of different triplets, $\sin^2(\theta/2)$ stands for $\Sigma_i \sin^2(\theta_i/2)$, where the sum runs over the different triplet components of a given singlet. Alternatively, we may write this as $n_M \langle \sin^2(\theta_i/2) \rangle$, the form given in Ref. 21 where the n_T was inadvertently omitted. When many mixed initial levels contribute, the simple form of (3.21) is adequate for an initial semiquantitative estimate. The number of final levels, n_T, may be estimated as $k_B T$ times the total density of triplet vibronic sublevels. For instance, in glyoxal if we use the extreme lower limits of one mixed singlet per vibronic level, $n_M = 1$, and n_T calculated from only the vibronic density of states as,[54]

$$n_T \sim (1\text{ cm})(200\text{ cm}^{-1})$$

then values of $\sigma_{rot} \approx 100$ Å^2 and $\sigma \approx 10$ Å^2 yield the estimate from (3.21) of

$$\langle \sin^2 \frac{\theta}{2} \rangle \sim 0.5 \times 10^{-3} \tag{3.22}$$

which is small enough that the singlet rotational spectrum is not expected to be noticeably perturbed. The estimates (3.21) and (3.22) for large molecules like glyoxal are rather crude. Nevertheless, they do indicate how very small mixings can have substantial effects on collision-induced intersystem-crossing transition rates. Recent experiments by Tramer and co-workers[54] on double resonance in the magnetic field effect on collision-induced intersystem-crossing rates in glyoxal imply mixing coefficients, $\sin^2(\theta/2)$ in the range of $10^{-2}-10^{-3}$, thereby substantiating the utility of simple approximations like (3.21) for large molecules. Further experimental work will aid in stimulating refinements in the theory.

When $\hat{S}$ and $\hat{T}$ refer to two mixed vibrational levels of the same electronic state, the zero-pressure splittings of levels of $\hat{S}$ from those of $\hat{T}$ can be considerable, for instance tens of wavenumbers. Thus any substantial nonresonant process must have an additional term on the right in (3.22) to account for energetically unfavorable circumstances. Despite the differences, the experimental evidence to date[55] points to very rapid collisional transitions between Fermi resonance-mixed vibrational levels in accord with the above theoretical discussions.

D. Magnetic Field Effects on Collision-Induced Intersystem-Crossing Rates

Because the collision-induced intersystem crossing is induced by small spin-nonconserving mixings arising from near degeneracies between singlet and triplet sublevels, it is natural to investigate the effects of external fields that can shift the energies of the zeroth order levels and hence the degree of singlet–triplet mixing. To date, magnetic field effects have been considered,[54, 56, 57] but electric field studies for molecules with permanent electric dipole moments and preferably first-order Stark effects should provide a similar type of probe. Because of the experimental orientation, we discuss the magnetic case as the electric field situation can be treated by a straightforward generalization.

The effects of a magnetic field on the zero-pressure limit energies is well understood. The zeroth order pure singlet and triplet states have degeneracies lifted by the field (the former if there is unquenched orbital angular momentum) and display first-order Zeeman effects at low field strengths with second and higher order terms entering as the field strength is increased. Thus the field alters the energies of the zeroth order levels, $E_0(S)$ and $E_0(T)$, in a field (and quantum sublevel) dependent fashion. Since the magnetic field does not alter the intramolecular coupling between pure spin zeroth order singlets and triplets, it is clear that the magnetic field alters the mixing coefficients (3.3c) and, consequently, the mixed states (3.3a, b) of the isolated molecule. From (2.7) it follows that this can lead to a magnetic field dependence of individual sublevel lifetimes, a trivial field effect, which need not be discussed further.

It should be noted that the $\Delta J=0, \pm 1$ selection rules for the Zeeman coupling matrix elements imply that the field can induce singlet–triplet mixing between states that cannot be coupled in field free cases.[20, 24] Likewise, strong mixing may occur among a group of levels rather than just two as in (3.3). The treatment of this case involves a straightforward generalization, so we concentrate on the simplest two-level example to illustrate the basic physical effects.

Singlet–triplet spin–orbit coupling matrix elements are subject to certain symmetry selection rules. Thus a given singlet sublevel may often be coupled to only one of the closely spaced triplet spin sublevels. The magnetic field serves to introduce couplings between the three zeroth order spin levels, thereby enabling the singlet level to mix with all three spin sublevels. This mechanism has been invoked by Stannard[58] to explain magnetic field effects on collision-free lifetimes. Unfortunately, his mechanism involved irreversible intramolecular intersystem crossing for molecules that correspond to the small or too-many level small molecule limits for which this irreversible decay is impossible. Nevertheless, it now appears that Stannard's spin decoupling mechanism can explain the low magnetic field effect on collision-induced intersystem-crossing rates.[54]

Experiments on SO_2 and glyoxal exhibit a magnetic field effect[54, 56, 57] with a cross-section $\sigma(\mathcal{H})$ that varies with field, $\mathcal{H}$, in a fashion which differs markedly from the traditional $\mathcal{H}^2$-dependence observed in ordinary magnetic quenching experiments.[59] In SO_2 and glyoxal $\sigma(\mathcal{H})/\sigma(\mathcal{H}=0)$ displays a universal curve for a given molecule which is independent of the perturber. $\sigma(\mathcal{H})/\sigma(\mathcal{H}=0)$ saturates, becoming independent (or rather weakly dependent) of field at the low field strengths of 1 kG. The ratio $\sigma(\mathcal{H}\sim 1\,\text{kG})/\sigma(0)\simeq 1.4$ provides the "high field" magnetic enhancement of the collision-induced intersystem-crossing rates.

The perturber independence of $\sigma(\mathcal{H})/\sigma(\mathcal{H}=0)$ follows rather simply from (3.21).[24] The average mixing coefficients, $\langle \sin^2[\theta(\mathcal{H})/2]\rangle$, and number of mixed levels, $n_M(\mathcal{H})$, naturally vary with the field $\mathcal{H}$. However, the number of accessible triplet vibronic levels, n_T, and the typical collision rates, σ_{rot}, which are perturber dependent, are unchanged by the introduction of the magnetic field. Hence, we have the ratio

$$\frac{\sigma(\mathcal{H})}{\sigma(\mathcal{H}=0)}=\frac{\langle\sin^2[\theta(\mathcal{H})/2]\rangle n_M(\mathcal{H})}{\langle\sin^2[\theta(0)/2]\rangle n_M(0)} \tag{3.23}$$

which is manifestly perturber independent, but field dependent. Note, of course, that $\langle\sin^2(\theta/2)\rangle$ implies an average over all contributing levels with an inclusion of the type of nonlinear dependences on mixing coefficients that arise in Section III.B.

Experiments by Tramar and co-workers[54] on glyoxal indicate that the triplet spin splittings in glyoxal are small, and Tric has shown[54] that the spin-decoupling model can account for the observed magnetic field dependence of $\sigma(\mathcal{H})/\sigma(0)$. An alternative proposal has been advanced by Selzle et al.,[60] who invoke a collision complex model wherein intramolecular intersystem crossing occurs in the long-lived collision complex. However, such a statistical model cannot describe the resonances subsequently observed in the very high field experiments (up to 10^5 G) of Tramer and co-workers.[54] These resonances can be broad and of the order of ~7% of $\sigma(0)$ or they can arise as a group of weaker, narrow, closely spaced resonances. This resonance behavior arises from a mechanism that we previously discussed in detail[24] for a simple diatomic molecule example where the triplet spin splittings are assumed to be reasonably large. Nevertheless, the high field can introduce level crossings (or anticrossings) between the zeroth order (mixed) levels $\hat{S}$ and $\hat{T}$. Hence, near a particular field strength, $\mathcal{H}_0$, for a given pair of sublevels S and T, the mixing coefficients $\sin[\theta(\mathcal{H}_0)/2]\approx(2)^{-1/2}$, so the states (3.3) approach 50–50 mixing. Given the zero-field mixings $\langle\sin^2[\theta(0)/2]\rangle\approx10^{-2}$–$10^{-3}$ in glyoxal, it is clear that the anticrossing pair can yield a huge, resonant-type enhancement of the collision-induced intersystem-crossing rates. Because the experiments excite and therefore average over a number of mixed levels, the observed rates are averages over one pair of anticrossing levels near $\mathcal{H}_0$ with a number of others which still have small mixing coefficients. Consequently, the overall magnitude of the effect is smaller than would be observed if the particular anticrossing sublevels were the only ones to be excited. Note that this sublevel S involves a particular M_N, so the ideal individually excited level experiments would be impossible except in small molecules at high fields where the individual M_N sublevels can be resolved.

A full theoretical description of the resonances in the magnetic field dependence requires at minimum the consideration of the three states $\hat{S}$, $\hat{T}$, and T' in (3.9). Because of the large mixings, $\sin[\theta(\mathcal{H}_0)/2]=(2)^{-1/2}$, the simple perturbation treatment is likely inadequate; however, exponentiated perturbation treatments should suffice.[61] Since the zeroth order anticrossing levels have their energies, $E_S(t)$ and $E_T(t)$, varying during the collision, the effective mixing coefficient can be viewed as fluctuating during the collision, so its value may depart from the crossing one of $(2)^{-1/2}$ throughout a substantial portion of the collision. As a simple approximation designed to display some of the basic physics and not to be fully quantitative, let us consider a simple adiabatic model in which the time-dependent mixing coefficient,

$$\theta(\mathcal{H},t)=\arctan\{2v_{ST}[E_S(\mathcal{H},t)-E_T(\mathcal{H},t)]^{-1}\} \tag{3.24}$$

is slowly varying with time, $\partial\theta(\mathcal{H},t)/\partial t \simeq 0$. Hence we may consider the adiabatic basis states,

$$\hat{S}(t)=\cos\left[\frac{\theta(\mathcal{H},t)}{2}\right]S+\sin\left[\frac{\theta(\mathcal{H},t)}{2}\right]T \tag{3.25a}$$

$$\hat{T}(t)=-\sin\left[\frac{\theta(\mathcal{H},t)}{2}\right]S+\cos\left[\frac{\theta(\mathcal{H},t)}{2}\right]T \tag{3.25b}$$

in the theory of Section III.B. For simplicity, we specialize to the large energy transfer limit of (3.13) where the time variation of the energy levels in the exponentials can be safely ignored. Hence, the $\hat{S}\to T'$ transition probability becomes

$$P_{\hat{S}\to T'}\simeq\hbar^{-2}\left|\int_{-\infty}^{\infty}\sin^2\left[\frac{\theta(\mathcal{H},t)}{2}\right]V_{TT'}[\mathbf{R}(t)] \times\exp\left\{-\left(\frac{i}{\hbar}\right)[E_0(T')-E_0(T)]t\right\}\right|^2 \tag{3.26}$$

Note that the variation of $\sin^2[\theta(\mathcal{H}_0,t)/2]$ during the collision produces a considerably smaller value for (3.26) than given by the simple estimate provided by (3.14) with the free molecule limit of $\sin^2[\theta(\mathcal{H}_0,t=\pm\infty)/2]=\frac{1}{2}$. For a single isolated anticrossing in the adiabatic limit, the time variation of $\sin^2[\theta(\mathcal{H},t)/2]$ affects both the magnitude and width of the resonant part of the collision induced intersystem crossing rate. Since $\sin^2[\theta(\mathcal{H}_0,\ t=\pm\infty)/2]=\frac{1}{2}$, experiments may also excite $\hat{T}$, so $P_{\hat{T}\to T'}$ must also be included. It differs from (3.26) by the appearance of $\cos^2$ instead of the $\sin^2$. Cases violating the simplifying case of (3.13) are readily treated just yielding more lengthy formulas. A quantitative treatment of the resonances within the three-level model requires the retention of the

$\frac{\partial}{\partial t}\theta(\mathcal{H},t)$ terms;[62] however, the adiabatic perturbation method of Light[63] should be adequate here.

It should be noted that (3.26) (or the more general nonadiabatic treatment) can be utilized even in the field free situation, $\mathcal{H}=0$. Then (3.24) describes the effects of the fluctuations in the zeroth order energies, during the collision, in producing fluctuations in $\theta(t)$. When the collisional perturbations negligibly alter the magnitude of θ, that is, when

$$|E_0(S)+\Delta_S-E_0(T)-\Delta_T|\gg|V_{SS}[\mathbf{R}(t)]-V_{TT}[\mathbf{R}(t)]| \qquad (3.27)$$

then (3.26) for $\mathcal{H}=0$ reduces to the simple limit of (3.14). On the other hand, when condition (3.27) is violated but $\theta(t)$ of (3.24) is slowly varying during the collision, the $\sin[\theta(t)/2]$ factor in the integrand of (3.26) may approximately be removed from the integrand and be evaluated as $\sin[\theta(\bar{t})/2]$ with $\bar{t}$ a time during the collision where $V_{TT}[R(t')]$ is appreciable.

IV. PRESSURE DEPENDENCE OF COLLISION-INDUCED INTERSYSTEM CROSSING

As noted in the Introduction, for the too-many level small-molecule limit the density of triplet vibronic levels considerably exceeds that for the singlet. Hence, once a collision transports a molecule from $\hat{S}$ to T' or $\hat{T}'$, the likelihood of collisional return to the singlet manifold becomes negligible. (Of course, in diatomic molecules with singlet and triplet vibronic densities of comparable magnitude, the collisional $T'\rightarrow\hat{S}''$ or $\hat{T}'\rightarrow S''$ back transfers cannot be neglected in general.) Thus, we can anticipate that the collision-induced intersystem-crossing rates in intermediate case molecules exhibit a Stern–Volmer quenching law,

$$k_{S\rightarrow T}=k_0+k_1P \qquad (4.1)$$

with P the pressure of the quencher and k_0 and k_1 constants. This expectation ignores the slight nonlinearities that could be induced by variations of the collision-induced intersystem-crossing rates with the initial different $\hat{S}$ that are rapidly collisionally accessible to each other. A similar type of nonlinearity occurs in the pressure dependence of fluorescence[64–66] and triplet yields[67] because of the variation of nonradiative decay rates with vibrational energy and the fact that collisions alter this vibrational energy. However, at high enough pressures these sets of $\hat{S}$ levels are collisionally equilibrated, and the Stern–Volmer quenching kinetics should ensue.

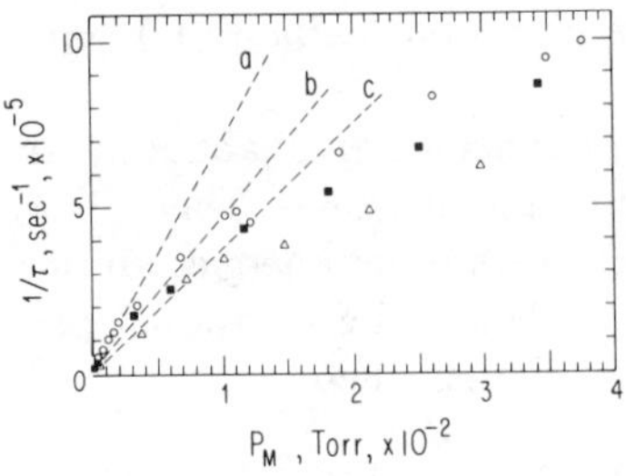

Fig. 7. Decay rate versus pressure for the $^3B_1$ state of SO_2 taken from Su et al.[14c] The quenchers are CO_2 (circles), CO (squares) and N_2 (triangles). The lines *a*, *b*, and *c* are the linear Stern–Volmer kinetics extrapolated from the low-pressure region for CO_2, CO, and N_2, respectively.

The theory of collision-induced intersystem crossing[20–22] yields Stern–Volmer kinetics only in the *limit of low enough pressures*. The remarkable feature of the theory[20, 36] is that is predicts *very non-Stern–Volmer* quenching kinetics, *especially in the high-pressure limit.* The quenching rate is predicted to reach a high-pressure limiting value, *independent of pressure and perturbing molecule*. (This is for the case where V_{ST} couplings are negligible; otherwise V_{ST} can produce an additional Stern–Volmer type contribution.) This type of behavior has been observed, subsequent to the theoretical predictions, by Strickler and Rudolph[14a, b] and by Su et al.[14c] in the $^3B_1$ state of SO_2 and by Weisshaar et al.[23] in the S_1 state of formaldehyde. The data on SO_2 are given in Figs. 7 and 8. Parmenter[68] has observed the onset of this pressure saturation effect in glyoxal, and this work should provide fundamental information on the shapes of the pressure saturation curves.

The $SO_2(^3B_1)$ state had previously been the source of considerable conceptual difficulties. Phosphorescence quantum yields were reported which were smaller than unity. The density of lower SO_2 rovibronic levels and typical radiative decay rates places the $SO_2(^3B_1)$ level squarely in the small molecule limit, so quantum yields of less than unity violate the laws of quantum mechanics. $SO_2(^3B_1)$ also displayed an enhanced higher pressure bimolecular reactivity with CO and with ethylene in conflict with all reasonable expectations. The prediction[20] and observation[14] of pressure saturation effects then made it clear that the previous experiments were based on higher pressure measurements where nonlinear Stern–Volmer kinetics were appearing. Notice in Fig. 1. that this nonlinear region is already apparent at pressures of 10 mtorr! Given the pressure saturation effect, the low-pressure data yield a phosphorescence quantum yield of unity and no need for enhanced reactivities of SO_2 at higher pressures.[69] Formaldehyde is a more complicated case because photochemical decomposition may proceed from the lowest vibronic levels of S_1 if the threshold is below these levels, so quantum yields of less than unity are theoretically

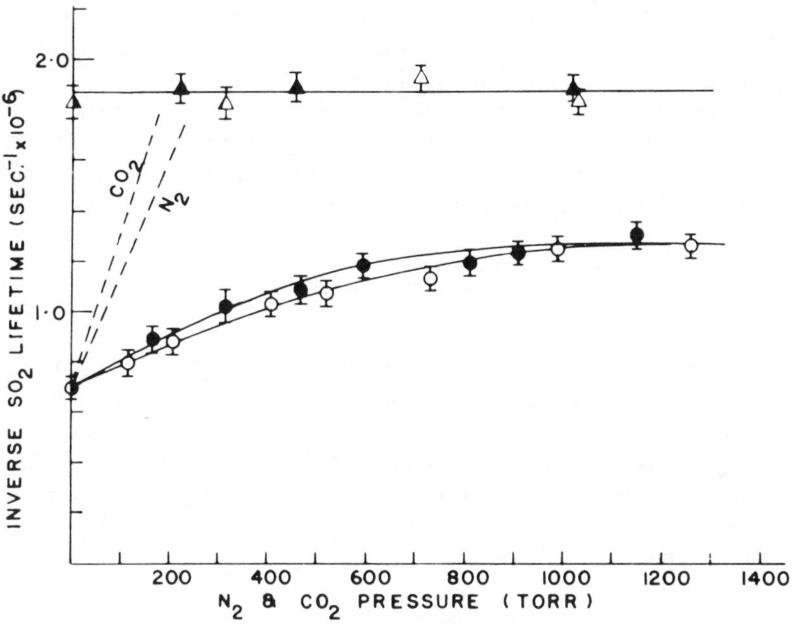

Fig. 8. Higher pressure data of Strickler and Rudolph[14b] for the decay rate of the SO_2 ($^3B_1$) state as a function of the pressure of N_2 (open circles) and CO_2 (closed circles). Dashed lines give the linear Stern–Volmer kinetics extrapolated from the low-pressure region.

possible from this photochemical channel. The experiments indicate the additional importance of collision-induced, internal conversion,[23] but the role of direct $V_{S_1S_0}$ collisional coupling is not yet known.

The full theoretical description of the pressure saturation of collision-induced intersystem crossing is quite involved.[20] The intermediate pressure region has yet to be treated satisfactorily. Hence we consider here a simple qualitative rationale for its occurrence. Consider an individual singlet vibronic manifold, S, which is coupled to a set of triplet vibronic levels, $\{T_l\}$, where ρ_l is small enough to place the molecule in the too-many level small molecule limit, the intermediate case. Zeroth order $\{T_l^0\}$ harmonic levels are badly mixed through extensive anharmonic couplings among nearly degenerate vibronic levels. Thus, it is expected that $T_l - T_{l'}$ vibrational relaxation rates are large between nearly isoenergetic pure-spin basis states, the rates being of order of k_{rot}, a rotational relaxation rate within a single vibronic level. When the collision rate, k_{rot}, becomes comparable to spacings between the $\{T_l\}$ sublevels on single-collision time scales, the $\{T_l\}$ levels act as if they are a continuous manifold of collision broadened states. This collisionally induced continuous $\{T_l\}$ state density of states can then drive irreversible electronic relaxation of the S state much as in the purely intramolecular large-molecule limit.

When the $\{T_l\}$ states are collisionally broadened, the individual S-state decay is given by (2.17) with ρ_{lj} a collisionally broadened state density. However, the collisions imply a collision-induced distribution of S sublevels, p_{si}, so (2.17) is converted to the average,[36]

$$\bar{\Delta}_s = \frac{2\pi}{\hbar} \sum_{i,j} p_{si} |v_{si,lj}|^2 \rho_{lj} \tag{4.2}$$

When the collision-induced electronic relaxation is much slower than k_{rot}, the distribution p_{si} is a thermalized Boltzmann one, but at lower pressures the detailed kinetics must be simulated to derive p_{si}. Nevertheless, the high-pressure limit of (4.2) is pressure independent as the pressure-broadened state density, ρ_{lj}, suffices to enable the replacement of the summation over j in (4.2) by an integration. For example, if ρ_{lj} were modeled by simple Lorentzians,

$$\rho_{lj} = \frac{\pi \gamma_{lj}}{(E - E_{lj})^2 + (\gamma_{lj}/2)^2} \tag{4.3}$$

with the binary collision pressure dependence,

$$\gamma_{lj} \propto P \tag{4.4}$$

when the spacing, ϵ_{lj}, between T-levels satisfies

$$\frac{\gamma_{lj}}{\epsilon_{lj}} \gg 1 \tag{4.5}$$

the large-molecule limit ensues, and this statistical limit value of (4.2) becomes independent of the $\{\gamma_{lj}\}$ and, hence, of the pressure. The actual situation[20] is somewhat more complicated by virtue of the fact that collisional transitions among the nearly degenerate $\{T_l\}$ levels can appear, but (4.2)–(4.5) provides an adequate qualitative description. The condition for the emergence of nonlinear Stern–Volmer kinetics is that the pressure, P_0, satisfy[20]

$$\hbar k_{rot} P_0 \sim \epsilon_{lj} \tag{4.6}$$

Between the pressures of P_0 and those where the quenching rate tends toward its high-pressure limit, the fluorescence decay of the prepared (mixed) S-state need not be a single exponential decay[70]—the multistate "kinetics" must, in general, be analyzed.

V. DISCUSSION

It is quite remarkable how small singlet–triplet couplings and mixing coefficients can wholly alter the collision-induced intersystem-crossing phenomenon from that which is anticipated on the basis of standard quenching theory. The situation is reminiscent of the small singlet–triplet couplings that enable phosphorescence to proceed. In the latter case the couplings are off-resonance in character, involving mixing between electronic states with considerable energy differences (e.g., 1 eV) between their origins. On the other hand, the collision-induced intersystem crossing is promoted by resonant coupling between accidentally nearly degenerate pairs (or sets) of singlet and triplet levels.

The fortuitous near-degeneracy of the levels poses severe difficulties to a complete quantitative description of the process in the too-many level small-molecule limit, and even in the small-molecule limit itself. The interaction potentials are poorly known. It is still unclear why collisional processes proceed on a much more rapid time scale in electronically excited states than in ground electronic states.[5] The intuitive explanation, that the excited states are larger, is insufficient. An analysis of the excited potential energy surfaces should prove enlightening in this regard.

Hyperfine couplings produce small shifts of individual sublevels, but these small shifts may greatly affect the singlet–triplet mixing coefficients and hence the microscopic dynamics of the collision-induced intersystem-crossing process. Thus these small energy shifts will probably have to be incorporated in the treatment of the simplest molecular collision-induced intersystem-crossing process involving a diatomic molecule with a rare-gas perturbing atom. The use of magnetic and electric fields provides a tool for shifting around these individual sublevels, but the quantitative analysis of the very high field Zeeman or Stark effects is rather difficult.

We have, therefore, focused attention on explaining the general physical manifestations of the collision-induced intersystem crossing in order to calibrate our physical intuition and to stimulate further experimental studies. Thus we consider the highly simplified two- and three-level model cases where an understanding of the dependence on mixing coefficients, intramolecular couplings, perturber interactions, etc. can be deduced in a simple, yet physically meaningful, fashion.

Many facets of collision-induced intersystem crossing, as well as other collision phenomena of electronically excited molecules, remain to be studied both experimentally and theoretically. The nature and rates of these processes have a strong influence on the characteristics of many laser chemistry processes. The rapidity of collision-induced intersystem crossing and of vibrational and rotational relaxation rates of highly excited molecules vitiates some claims of collision-free experiments, but it introduces additional interesting collision-dependent aspects of laser chemistry to be explored.

Acknowledgments

I am grateful to the experimentalists in the area, C. W. Lineberger, S. Nagakura, A. Tramer, S. J. Strickler, T. G. Calvert, and J. Weisshaar, who have communicated their results to me prior to publication and have stimulated this work. This research is supported, in part, by National Science Foundation Grant CHE80-23456.

References

1. K. F. Freed, *Topics Appl. Phys.*, **15**, 23 (1976).
2. K. F. Freed, *Acct. Chem. Res.*, **11**, 74 (1978).
3. J. Jortner and S. Mukamel, in *The World of Quantum Chemistry*, R. Daudel and B. Pullman, eds., Reidel, Boston, Mass., 1974.
4. P. Avouris, W. M. Gelbart, and M. A. El-Sayed, *Chem. Rev.*, **77**, 793 (1977).
5. S. A. Rice, this volume.
6. L. G. Anderson, C. S. Parmenter, H. M. Poland, and J. D. Rau, *Chem. Phys. Lett.*, **8**, 232 (1971); L. G. Anderson, C. S. Parmenter, and H. M. Poland, *Chem. Phys.*, **1**, 401 (1973).
7. T. W. Eder and R. W. Carr, *J. Chem. Phys.*, **53**, 2258 (1970).
8. R. A. Beyer and W. C. Lineberger, *J. Chem. Phys.*, **62**, 4024 (1975).
9. P. F. Zittel and W. C. Lineberger, *J. Chem. Phys.*, **66**, 2972 (1977).
10. J. T. Yardley, *J. Chem. Phys.*, **56**, 6192 (1972).
11. C. A. Thayer and J. T. Yardley, *J. Chem. Phys.*, **57**, 3992 (1972).
12. K. G. Spears and M. El-Maguch, *Chem. Phys.*, **24**, 65 (1977).
13. M. Lavollée and A. Tramar, *Chem. Phys. Lett.*, **47**, 523 (1977).
14. (a) R. N. Rudolph and S. J. Strickler, *J. Am. Chem. Soc.*, **99**, 3871 (1977); (b) S. J. Strickler and R. N. Rudolph, *J. Am. Chem. Soc.*, **100**, 3326 (1978). (c) F. Su, F. B. Wampler, J. W. Bottenheim, D. L. Thorsell, J. G. Calvert, and E. K. Damon, *Chem. Phys. Lett.*, **51**, 150 (1977).
15. A. Frad, F. Lahmani, A. Tramer, and C. Tric, *J. Chem. Phys.*, **60**, 4419 (1974).
16. R. Naaman, D. M. Lubman, and R. N. Zare, *Chem. Phys.*, **32**, 17 (1978).
17. R. van der Werf and J. Kommandeur, *Chem. Phys.*, **16**, 125 (1976); R. van der Werf, E. Schutten, and J. Kommandeur, *Chem. Phys.*, **16**, 151 (1976).
18. W. M. Gelbart and K. F. Freed, *Chem. Phys. Lett.*, **18**, 470 (1973).
19. M. Seaver, Ph.D. thesis, Indiana, 1978; C. S. Parmenter and M. Seaver, *J. Chem. Phys.*, **70**, 5458 (1979); H. M. Lin, M. Seaver, K. Y. Tang, A. E. W. Knight, and C. S. Parmenter, *J. Chem. Phys.*, **70**, 5442 (1979).
20. K. F. Freed, *J. Chem. Phys.*, **64**, 1604 (1976).
21. K. F. Freed, *Chem. Phys. Lett.*, **37**, 47 (1976).
22. K. F. Freed and C. Tric, *Chem. Phys.*, **33**, 249 (1978).
23. J. C. Weisshaar, A. P. Baronavski, A. Cabello, and C. B. Moore, *J. Chem. Phys.*, **69**, 4720 (1978).

24. K. F. Freed, *Adv. Chem. Phys.*, **42**, 207 (1980).
25. G. W. Robinson, *Excited States*, **1**, 1 (1974).
26. G. B. Kistiakowsky and C. S. Parmenter, *J. Chem. Phys.*, **42**, 2942 (1965).
27. E. M. Anderson and G. B. Kistiakowsky, *J. Chem. Phys.*, **48**, 4787 (1968); C. S. Parmenter and A. H. White, *J. Chem. Phys.*, **50**, 1631 (1969).
28. B. K. Selinger and W. R. Ware, *J. Chem. Phys.*, **53**, 3160 (1970); C. S. Parmenter and M. W. Schuyler, *Chem. Phys. Lett.*, **6**, 339 (1970).
29. K. G. Spears and S. A. Rice, *J. Chem. Phys.*, **55**, 5561 (1971); A. S. Abramson, K. G. Spears, and S. A. Rice, *J. Chem. Phys.*, **56**, 2291 (1972); C. Guttman and S. A. Rice, *J. Chem. Phys.*, **61**, 651 (1974).
30. K. F. Freed, *Chem. Phys. Lett.*, **42**, 600 (1976).
31. K. F. Freed, *J. Chem. Phys.*, **45**, 4214 (1966).
32. R. E. Smalley, B. L. Ramakrishna, D. H. Levy, and L. Wharton, *J. Chem. Phys.*, **61**, 4363 (1974).
33. F. Paech, R. Schmiedl, and W. Demtröder, *J. Chem. Phys.*, **63**, 4369 (1975).
34. P. Wannier, P. M. Rentzepis, and J. Jortner, *Chem. Phys. Lett.*, **10**, 102, 193 (1971); G. E. Busch, P. M. Rentzepis, and J. Jortner, *Chem. Phys. Lett.*, **11**, 437 (1971); *J. Chem. Phys.*, **56**, 361 (1972); D. Zevenhuijzen and R. van der Werf, *Chem. Phys.*, **26**, 279 (1977).
35. K. F. Freed and J. Jortner, *J. Chem. Phys.*, **50**, 2916 (1969).
36. K. F. Freed, *J. Chem. Phys.*, **52**, 1345 (1970).
37. K. F. Freed, *Topics Curr. Chem.*, **31**, 105 (1972).
38. M. Bixon, Y. Dothan, and J. Jortner, *Mol. Phys.*, **17**, 109 (1969).
39. C. Tric, *Chem. Phys. Lett.*, **21**, 83 (1973); *J. Chem. Phys.*, **55**, 4303 (1971).
40. F. Lahmani, A. Tramer, and C. Tric, *J. Chem. Phys.*, **60**, 443 (1974).
41. W. M. Gelbart, D. F. Heller, and M. L. Elert, *Chem. Phys.*, **7**, 116 (1975).
42. J. Chaiken, T. Benson, M. Gurnick, and J. D. McDonald, *Chem. Phys. Lett.*, **61**, 195 (1979).
43. A. Villaeys and K. F. Freed, *Chem. Phys.*, **13**, 271 (1976).
44. R. L. Swofford, M. E. Long, and A. C. Albrecht, *J. Chem. Phys.*, **65**, 179 (1976); R. G. Bray and M. J. Berry *J. Chem. Phys.*, **71**, 4909 (1980).
45. S. Mukamel and J. Jortner, *J. Chem. Phys.*, **65**, 5204 (1976).
46. P. J. Robinson and K. A. Holbrook, *Unimolecular Reactions*, Wiley, New York, 1972.
47. K. F. Freed, *Faraday Disc.*, **67**, 231 (1979).
48. D. W. Pratt and H. P. Broida, *J. Chem. Phys.*, **50**, 2181 (1969); H. E. Radford and H. P. Broida, *J. Chem. Phys.*, **38**, 644 (1963).
49. D. Grimbert, M. Lavollée, A. Nitzan, and A. Tramer, *Chem. Phys. Lett.*, **57**, 45 (1978).
50. D. H. Katayama, T. A. Miller and V. E. Bondybey, *J. Chem. Phys.*, **71**, 1662 (1979).
51. J. C. Tully, *Mod. Theor. Chem.*, **7**, 173 (1977).
52. M. Y. Chu and J. S. Dahler, *Mol. Phys.*, **27**, 1045 (1974); K. C. Kuhlander and J. S. Dahler, *J. Phys. Chem.*, **80**, 2881 (1976); *Chem. Phys. Lett.*, **41**, 125 (1976).
53. L. D. Landau and E. M. Lifshitz, *Quantum Mechanics*, Pergamon, New York, 1969.
54. A. Tramer (private communication); M. Lombardi, R. Jost, C. Michel, and A. Tramer, *Chem. Phys.*, **46**, 273 (1980).
55. E. Weitz and G. Flynn, *Ann. Rev. Phys. Chem.*, **25**, 275 (1974).
56. A. Matsuzaki and S. Nagakura, *Chem. Phys. Lett.*, **37**, 204 (1976); *J. Luminesc.*, **12-13**, 787 (1976); *Z. Phys. Chem.*, **101**, 283 (1976).
57. H. G. Kuttner, H. D. Selzle, and E. W. Schlag, *Chem. Phys. Lett.*, **48**, 207 (1977).
58. P. R. Stannard, *J. Chem. Phys.*, **68**, 3932 (1978).
59. J. H. van Vleck, *Phys. Rev.*, **40**, 544 (1932).
60. H. L. Selzle, S. H. Lin, and E. W. Schlag, *Chem. Phys. Lett.*, **62**, 230 (1979).
61. J. C. Light, *Meth. Comp. Phys.*, **10**, 111 (1971).

62. K. F. Freed (unpublished work).
63. J. C. Light, *J. Chem. Phys.*, **66**, 5241 (1977).
64. K. F. Freed and D. F. Heller, *J. Chem. Phys.*, **61**, 3942 (1974).
65. G. S. Beddard, G. R. Fleming, O. L. J. Gijzeman, and G. Porter, *Proc. Roy. Soc. (London) A*, **340**, 519 (1974).
66. R. G. Brown, M. G. Rockley, and D. Phillips, *Chem. Phys.*, **7**, 41 (1975); K. G. Spears, *Chem. Phys. Lett.*, **54**, 139 (1978); R. P. Steer, M. D. Swords, and D. Phillips, *Chem. Phys.*, **34**, 95 (1978).
67. K. H. Fung and K. F. Freed, *Chem. Phys.*, **14**, 13 (1976).
68. C. S. Parmenter (private communication).
69. F. Su, J. W. Bottenheim, D. L. Thorsell, J. G. Calvert, and E. K. Damon, *Chem. Phys. Lett.*, **49**, 305 (1977); F. Su and J. G. Calvert, *Chem. Phys. Lett.*, **52**, 572 (1977).
70. S. Mukamel, *Chem. Phys. Lett.*, **60**, 310 (1979); S. Mukamel and K. F. Freed (unpublished work).
71. This effect has been observed in the $A^1\Pi$ state of CO by M. Lavollée and A. Tramer, *Chem. Phys.*, **45**, 45 (1979).
72. J. B. Hopkins, D. E. Powers, and R. E. Smalley, *J. Chem. Phys.*, **72**, 5039 (1980); J. B. Hopkins, D. E. Powers, S. Mukamel and R. E. Smalley, *J. Chem. Phys.*, **72**, 5049 (1980).
73. K. F. Freed and A. Nitzan, *J. Chem. Phys.*, **73**, 4765 (1980).

COLLISIONAL EFFECTS IN ELECTRONIC RELAXATION

A. TRAMER

Laboratoire de Photophysique Moleculaire CNRS, Université Paris-Sud 91 405 Orsay, France

A. NITZAN

University of Tel-Aviv, Department of Chemistry Tel-Aviv, Israel

CONTENTS

In the past few years some attention has been given to collisional effects on electronic relaxation of gas-phase molecules. By now, a substantial amount of experimental results and some theoretical studies are available.

The aim of this chapter is to give a brief summary of the existing experimental data concerning these effects. These data are treated on the basis of a simple model describing the time evolution of the molecular system collisionally coupled to the translational energy continua. From this point of view our approach is complementary to that adopted in the review article by Freed,[1] where attention is focused more on calculating the cross-section for a collisionally induced transition.

In Section I, we outline principal problems and give a phenomenological description and classification of the collisional effects in some typical molecular systems.

In Section II, we describe briefly the primary collisional effects, vibrational and rotational relaxation and dephasing processes, and discuss their influence on the time evolution of an electronically excited molecular system.

Section III is devoted to a review and analysis of experimental data. Special attention will be given to the problems of the reversibility of the electronic relaxation and to the dependence of the electronic relaxation rates on the intramolecular parameters and on the properties of the collision partner.

Finally, in Section IV we treat some specific problems such as the electronic relaxation in van der Waals complexes, magnetic-field effects, and some finer details of relaxation paths.

Among electronic relaxation processes, the collision-induced intersystem crossing (singlet–triplet transitions in closed shell molecules) have been the most extensively studied. In addition, the intramolecular and intermolecular coupling mechanisms may be more easily separated in this case than in the case of the collision-induced internal conversion. For this reason, our attention will be focused at the problems of collision-induced transitions between different spin manifolds.

I. ESSENTIAL FEATURES OF THE COLLISION-INDUCED ELECTRONIC RELAXATION

A. General Remarks

The efficient quenching of the atomic and molecular fluorescence by collisions has been observed in early studies of the luminescence of gaseous compounds (for a review of early works see Ref. 2). In a large number of cases these processes have been explained by the electronic-to-vibrational energy transfer, charge transfer or excited-complex (excimer or exciplex) formation. There remains, however, an important class of collisional processes corresponding to the essentially intramolecular relaxation induced (or assisted) by collisions with chemically inert partners. In such cases, only a negligible part of the excited-system energy is transferred to the collision partner or transformed into the translational energy, its major part remains in the initially excited molecule but is redistributed in a different way between its internal degrees of freedom.

One can distinguish between two fundamental types of collision-induced processes:

1. Collisions that do not change the molecular electronic states. Such collisions only lead to relaxation and redistribution of the molecular vibrational and rotational energy.
2. Electronic relaxation: transition from a low (e.g., vibrationless) vibronic level of a higher electronic state to higher vibronic levels of the lower electronic state. Part of the electronic energy of the molecule is then transformed into vibrational energy. Such a transition from the initially excited singlet level to a quan-isoenergetic level of the triplet state is usually referred as the collision–induced intersystem crossing.

The first systematic investigations of collision-induced intersystem crossing have been carried out for a triatomic molecule with "anomalously" long decay time[3]—sulfur dioxide.[4–7] The fluorescence of SO_2, excited in the 260–335 nm spectral region, is efficiently quenched by collisions with ground-state SO_2 molecules and many other chemically inert quenchers with a simultaneous induction of the thermally equilibrated phosphorescence from low vibronic levels of the first excited triplet state. The rate of this transition is of the order of the gas-kinetic collision rate, indicating the high efficiency of such collisions. As a consequence of the high efficiency of the electronic relaxation, vibrational relaxation within the singlet manifold is practically absent.[5]

The collisional fluorescence quenching and phosphorescence induction processes have been later observed for a large number of small and medium-size molecules. The interpretation of these results was however rather confusing: collision-induced intersystem crossing considered as a transition from the pure singlet to the pure triplet state is in apparent contradiction with the Wigner rule of spin conservation,[8] at least in the case of light collision partners that cannot affect the intramolecular spin–orbit interaction.

The essential step in understanding collisionally induced intersystem crossing processes has been made by Gelbart and Freed.[9] In small molecules, the optical excitation does not prepare pure spin states but the quasistationary states resulting from the intramolecular singlet–triplet coupling. In a simple two-level model

$$|\sigma\rangle=\alpha|s\rangle+\beta|l\rangle \qquad \lambda=-\beta|s\rangle+\alpha|l\rangle \qquad (\alpha^2+\beta^2=1)$$

where $|s\rangle$ and $|l\rangle$ are the pure singlet and triplet states, respectively and where (except for the accidental s–l resonance) $\alpha^2 \gg \beta^2$. This idea has been developed by Freed[10,11] who showed that in the limiting case of very weak perturbations (treated by first-order time-dependent perturbation theory) the probability of the collision-induced intersystem crossing (and the corresponding cross-section σ_{ISC}) is proportional to the mixing coefficient β^2 characterizing the initially prepared σ state and goes to zero when $\beta^2 \to 0$.

The last conclusion has been confirmed by the experimental study of collision induced transitions between singlet and triplet manifolds of the simplest atomic system—the helium atom. It has been shown[12,13] that collision may transfer helium atoms from singlet to triplet states but only by a very specific reaction path: collisional relaxation within the singlet manifold populates high n^1F states—nearly degenerate and significantly mixed with the corresponding n^3F states; the $n^1F \rightleftharpoons n^3F$ transitions are allowed ($\beta^2 \neq 0$) and further collisional relaxation within the triplet manifold transfers the atoms to pure triplet levels.

The experimental evidence indicating the absence of transitions between pure spin states must be kept in mind in each treatment of collision-induced transitions carried out in the basis of such states. We have to assume that collisions cannot directly induce transitions between $\{|s\rangle\}$ and $\{|l\rangle\}$ states but only the relaxation within the $\{|s\rangle\}$ and the $\{|l\rangle\}$ manifolds, which in turn affects the populations of the $|s\rangle$ and the $|l\rangle$ states.*

We propose in the following a simple model describing the time evolution of an electronically excited molecule perturbed by collisions, where the role of the latter is limited to "primary" collisional processes: vibrational and rotational relaxation within each spin manifold and collisional dephasing. We follow here the way indicated by Voltz[15] and Mukamel[16] in their studies of collision-induced intersystem crossing and by Derouard et al.,[17,18] who applied a similar treatment to the specific case of the singlet–triplet anticrossing in strong magnetic fields. This model can account for the apparent wide variety of collisional effects on the fluorescence decay and quantum yield in different classes of molecules. The actually available experimental material is rich enough and makes it possible to distinguish between several types of behavior of different molecules and to relate them to the character of the interstate coupling in these species.

*This assumption is not strictly valid in the case of heavy collision partners, which may influence the intramolecular spin-orbit coupling by the heavy atom effect.[14] Collisions may also couple two electronic states belonging to the same spin manifold by the breakdown of orbital symmetry.

B. Phenomenological Description of Collisional Effects

The excited molecular systems may be roughly divided into a few groups characterized by the level-coupling schemes and the nature of the initially prepared state. Each of them shows a specific form of the fluorescence decay under collision-free conditions.[19, 20] The collisional effects in each group of molecules are also quite similar.

(a) "Large" molecules belonging to the "statistical limit" are well described in terms of a single radiative level $|s\rangle$ coupled to the $\{|l\rangle\}$ quasicontinuum. The fluorescence decay of the isolated molecule is exponential with decay time shorter than radiative lifetime (fluorescence yield smaller than unity). This decay is practically insensitive to collisional effects, when corrected for the vibrational and rotational relaxation within the $\{|s\rangle\}$ manifold. For benzene, the constant fluorescence yield in the wide pressure range has been first evidenced in the classical work of Kistiakovsky and Parmenter.[21] A similar behavior has been observed in the case of naphthalene, where the fluorescence decay time, as well as the triplet growth time and quantum yield do not depend on the pressure of the added inert gas.[22, 23]

(b) A group of medium-size molecules may be described in terms of a strong coupling between a single radiative $|s\rangle$ state and a number of discrete $|l\rangle$ states giving origin to a band of quasistationary states sharing the $|s\rangle$ oscillator strength. This is the strong-coupling sparse intermediate case of the theory of radiationless transitions. The excitation of individual quasistationary states results in a quasiexponential, anomalously long decay with lifetimes longer than the radiative lifetime. Collisions with inert partners induce a very efficient quenching in this case: the quenching constant k_q is equal to or higher than k_{coll}—the gas-kinetic collision rate in the hard sphere approximation. The decay remains exponential in a wide pressure range, the fluorescence decay time τ and fluorescence yield Q_f may be described by the Stern–Volmer law:

$$\frac{1}{\tau(p)}=\frac{1}{\tau(0)}+k_qP \qquad \frac{Q_f(0)}{Q_f(p)}=l+k_q\tau(0)P \tag{1}$$

SO_2,[5–7] benzophenone,[24–26] and benzoquinone[27] belongs to this group.

(c) The coherent excitation of a band of quasistationary states resulting from the s–l coupling leads to a quasibiexponential fluorescence decay: its first component corresponds to the dephasing of the initially prepared coherent superposition of states, while the latter one is due to the independent (incoherent) decay of individual states with anomalously long decay time as in the previous case. This is the coherently excited strong-coupling dense intermediate case of the theory of radiationless transitions. The first component of the decay is practically unaffected by collisions, while the second one is efficiently quenched, $k_q \gtrsim k_{coll}$. At sufficiently high gas pressures the first component prevails and the excited system shows an exponential decay as in the statistical case. Obviously, the pressure dependence of the fluorescence yield shows a strong deviation from the Stern–Volmer law.[28,29] Such a behavior has been observed in the case of pyrazine,[29] pyrimidine,[30] quinoxaline,[31,32] and biacetyl.[33]

(d) In a number of small polyatomic and diatomic molecules, the singlet–triplet (s–l) coupling may be treated in terms of the weak-coupling limit where efficient coupling exsists mostly between pairs of $|s\rangle$ and $|l\rangle$ states. The resulting $|\sigma\rangle$ and $|\lambda\rangle$ states have (except in the case if s–l quasiresonance) strongly predominant s- and l-character, respectively (see above). The optical excitation prepares essentially quasistationary states which, under collision-free conditions, decay exponentially with lifetime $\tau_\sigma \simeq \tau_s$. Collisions quench the fluorescence and the Stern–Volmer law may be applied, but the quenching constant is in general much lower than in previous cases (about 10% of k_{coll}) and depends strongly on the properties of the collision partner. Glyoxal[34,35] and propynal[36] as well as a number of diatomic molecules have to be classified in this group.

(e) In a few cases of diatomic molecules in excited states: the $A^1\pi$ state of CO[37,38] or the $B^3\pi_g$ state of N_2[39], the fluorescence decay which is quasiexponential under collision-free conditions is strongly modified by collisions. At sufficiently high gas pressures it may be approximated by a biexponential process with τ_1 shorter and τ_2 longer than the collision-free lifetime τ_σ. The pressure dependence of the decay and of the fluorescence quantum yield cannot be described by the Stern–Volmer law using a single quenching constant k_q.

We should keep in mind that the population of the $|l\rangle$ (triplet) levels isoenergetic with the initially excited $|s\rangle$ state may be also monitored, either by transient absorption or by resonant emmision. In the cases (a)–(c), the l-state population builds up due to the intramolecular processes at a pressure-independent rate.[22,23,32,40] Collisions lead to vibrational relaxation within the triplet manifold that populates the lower triplet levels and induces the (thermally equilibrated) phosphorescence emission. In the cases (d) and (e), the initial population of the $\{|l\rangle\}$ levels is close to zero and it grows with a pressure-dependent rate as shown by Slanger[41] in his study of the pressure dependence of the $d^3\Delta \to a^3\pi$ emission of CO under the optical excitation of the $A^1\pi$ state. One may conclude that in the first

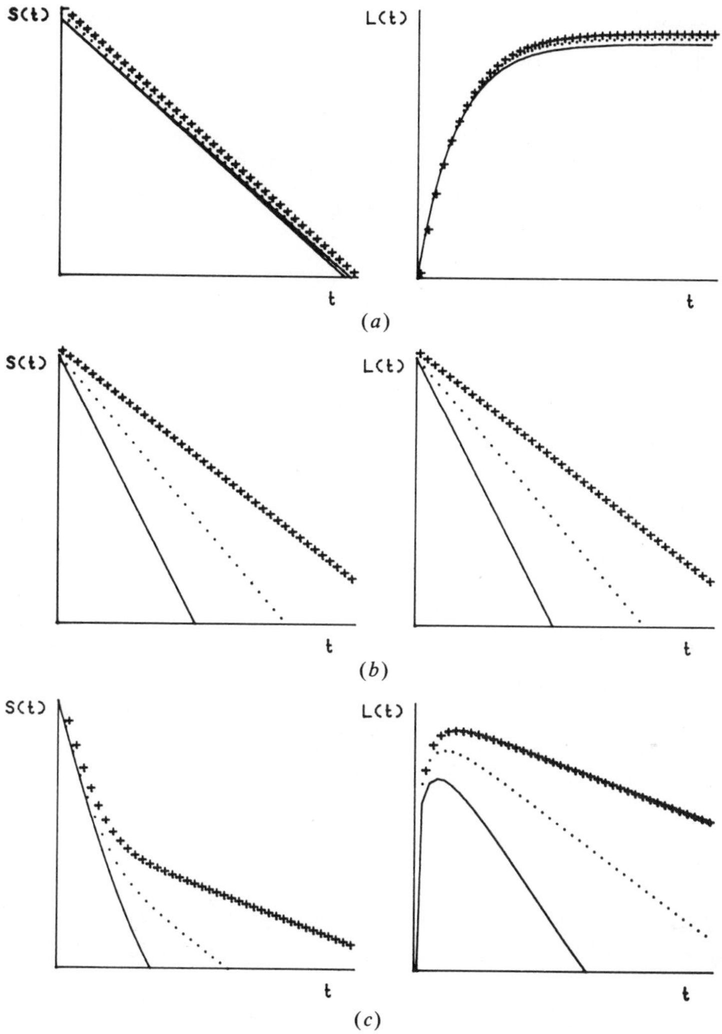

Fig. 1 Schematic representation of the pressure effect on the singlet ($S(t)=\log\langle s|\psi(t)\rangle^2$) and triplet ($L(t)=\log\langle 1|\psi(t)\rangle^2$) content of the excited molecular state. Crosses, collision-free conditions, points and solid line, increasing inert-gas pressure. (a) statistical-limit; (b) strong-coupling case (incoherent excitation); (c) strong coupling case (coherent excitation); (d) weak-coupling case (small polyatomics); (e) weak-coupling case (CO, N_2).

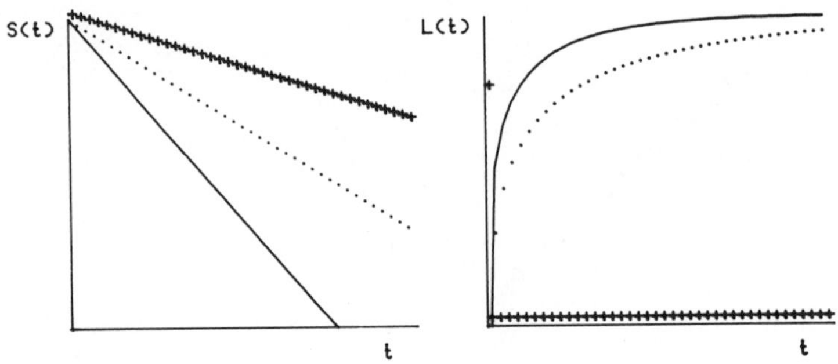

(*d*)

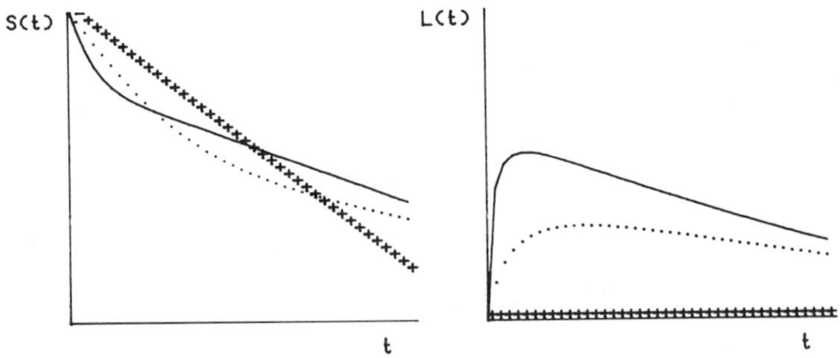

(*e*)

Fig. 1 (*Continued*)

case collisions maintain the inherent triplet character of the excited molecular state, while in the latter one they really induce the singlet–triplet transitions.

Pressure effects on the time dependence of the $|s\rangle$ and $|l\rangle$ states populations are schematically represented in Fig. 1.

II. MODEL TREATMENT OF THE COLLISION-INDUCED ELECTRONIC RELAXATION

A. Outline

In the absence of external perturbations, the time evolution of an excited, collision-free molecule is described in terms of its density matrix ρ

by the following set of equations:

$$\dot{\rho}_{ss}=i\sum_{l}(v_{ls}\rho_{sl}-v_{sl}\rho_{ls})$$

$$\dot{\rho}_{ll}=i\sum_{s}(v_{sl}\rho_{ls}-v_{ls}\rho_{sl})$$

$$\dot{\rho}_{ss'}=-i\epsilon_{ss'}\rho_{ss'}+i\sum_{l}(v_{ls'}\rho_{sl}-v_{sl}\rho_{ls'})$$

$$\dot{\rho}_{ll'}=-i\epsilon_{ll'}\rho_{ll'}+i\sum_{s}(v_{sl}\rho_{ls}-v_{ls}\rho_{sl'})$$

$$\dot{\rho}_{sl}=-i\epsilon_{sl}\mathrm{P}_{sl}+i\sum_{s'}v_{s'l}\rho_{ss'}-i\sum_{l'}v_{sl'}\rho_{l'l} \tag{2}$$

This evolution will be modified in the presence of collisions by what was previously defined as "primary" processes: vibrational and rotational relaxation within the $|s\rangle$ and $\{|l\rangle\}$ manifolds, dephasing of initially prepared coherent states and (in principle) coherence-transfer effects. The basis of zero-order states (pure-spin states in the case of intersystem crossing) has been chosen for this discussion, because it provides clear separation of intramolecular and intermolecular effects: the $|s\rangle$ and $|l\rangle$ states are mutually coupled by the intramolecular (spin–orbit) coupling v_{sl}, while collisions may uniquely couple $|s\rangle$ and $|s'\rangle$ ($|l\rangle$ and $|l'\rangle$) states within each manifold.

Without attempting a general discussion of dephasing and of vibrational and rotational relaxation processes it seems useful to give here a short summary of the essential features of "primary" collision effects.

B. "Primary" Collision Effects

1. *Rotational Relaxation*

It is well known that rotational relaxation in the ground and excited electronic states is highly efficient, and its rate is of the same order (or higher) as the gas-kinetic collision rate in the hard-sphere approximation.[42] This means that collisions with large impact parameters b involving the attractive part of the intermolecular potential are at least as important as head-on collisions. This assumption has been directly confirmed in cross-beam experiments[43] by the angular distribution of rotationally relaxed (or excited) molecules. In addition, a good correlation between cross-sections of rotational relaxation σ_{rot} and the depth of the potential energy well for a large series of colliders confirms the role of attractive interactions.[44]

Satisfactory correlations have been equally obtained between σ_{rot} and the strength of van der Waals forces expressed by such parameters of the collider as its dipole moment, polarizibility and ionization potential.[45]

Rotational relaxation (at least for nonresonant interactions) does not show any general propensity rules.[42] Differential cross-sections for $J \to J'$ rotational relaxation show only a slight dependence on ΔJ and ΔE, as long as ΔE does not exceed kT. The efficient transition into a large number of rotational levels in a single collision has been observed in many molecules such as OH,[46] I_2,[47] NO,[48] glyoxal,[49, 50] and benzene.[54]

In most cases, the rotation-level spacing is small as compared to kT. Therefore, rotational relaxation may be considered as a reversible process leading to a rapid establishment of the Boltzmann equilibrium between populations of rotational levels within a given vibronic state. The equilibrium (Boltzmann) density matrix is diagonal, i.e., all phase information is lost during the rotational (collision-induced) relaxation. The effect may be described by the supplementary terms to (2)

$$\dot{\rho}_{ll} = \sum_{l \neq l} (k_{ll'}\rho_{l'l'} - k_{l'l}\rho_{ll})$$

$$\dot{\rho}_{ll'} = -\frac{1}{2}\left(\sum_{l'' \neq l} k_{l''l} + \sum_{l'' \neq l'} k_{l''l'} \right)\rho_{ll'} \tag{3}$$

$k_{ll'}$ and $k_{l'l}$ are related to each other by the detailed balance requirement:

$$\frac{k_{ll'}}{k_{l'l}} = \frac{g_l}{g_{l'}}\, exp\left(\frac{\epsilon_{l'l}}{kT}\right) \qquad (\epsilon_{l'l} = \epsilon_{l'} - \epsilon_l) \tag{4}$$

where g_l and $g_{l'}$ represent degeneracies of $|l\rangle$ and $|l'\rangle$ levels and $\epsilon_{ll'}$—their energy spacing. Similar equations apply within the $\{|s\rangle\}$ manifold. We assume in the following that $k_{ll'}$, $k_{l'l}$, $k_{ss'}$ and $k_{s's}$ are proportional to the perturbing gas pressure.

2. *Vibrational Relaxation*

The rate of collisionally induced vibrational relaxation varies within extremely wide limits: for diatomic molecules with high vibrational frequencies, particularly in the ground electronic state, the vibrational relaxation rate constant k_{vib} is of the order of 1 s^{-1} $torr^{-1}$ (e.g., CO+He)[52], while in the A state of the CF_2 radical k_{vib} varies between $1 \cdot 10^4$ and $6 \cdot 10^4$ for $v=1$ level and between 2.10^5 and $1 \cdot 2 \cdot 10^6$ for $v=6$ level for different collision partners.[53] k_{vib} attains the gas-kinetic collision rate for relatively large polyatomic molecules in excited electronic states ($k_{vib} \simeq 10^7$ s^{-1} $torr^{-1}$ for 6^1 level of the B_{2u} state of benzene with hexane or heptane as collision partners).[54]

It seems well established that the vibrational relaxation in small molecules is correctly described by the SSH model[55] assuming the dominant role of repulsive interactions. The mechanisms of the vibrational relaxation in medium-size molecules is certainly more complex and not completely elucidated. Recent studies carried out in the low-pressure range (one collision per lifetime) indicate that collisions may lead to highly selective transitions to some specific vibronic states; these "propensity rules" cannot be explained by simple consideration of ΔE (energy) and Δv (vibrational quantum number) changes (as in the SSH model) but are related to some dynamical properties of vibrational modes[54,56,57]

It should be noted that even if the attractive part of the intermolecular potential plays a nonnegligible role in the vibrational relaxation, there is no apparent correlation between rotational and vibrational relaxation rates. As may be seen from the foregoing discussion, $k_{\text{rot}}/k_{\text{vib}}$ ratios may vary by many orders of magnitude.

In small molecules and also at the bottom of the vibrational manifold of larger molecules, the vibrational levels are sparsely spaced (vibrational energy spacing larger than kT), and vibrational relaxation is relatively slow and may be considered as irreversible. If we consider the interaction between the vibrationless level of the $\{|s\rangle\}$ manifold and between higher vibronic levels of the $\{|l\rangle\}$ manifold, the vibrational relaxation in the $\{|s\rangle\}$ manifold may be neglected. Thus the $|s\rangle$ level decays only radiatively with the rate constant $k_s = k_s^r$, while the decay rate of the $|l\rangle$ levels is given by the rate constant $k_l = k_l^r + k_l^{vr}$, k_l^{vr} (proportional to the gas pressure) being the rate of depopulation by vibrational relaxation within the $\{|l\rangle\}$ manifold. k_s^r and k_l^r are radiative relaxation rates (k_l^r is assumed to vanish when $|l\rangle$ belong to the lowest triplet state of the molecule). The radiative and vibrational irreversible decay may be accounted for by:

$$
\begin{aligned}
&\dot{\rho}_{ss} = -k_s^r \rho_{ss} \qquad \dot{\rho}_{ll} = -k_l \rho_{ll} \\
&\dot{\rho}_{ll'} = -\frac{1}{2}(k_l + k_{l'})\rho_{ll'} \simeq -k_l \rho_{ll'} \\
&\dot{\rho}_{ss'} = -\frac{1}{2}(k_s^r + k_{s'}^r)\rho_{ss'} \simeq k_s^r \rho_{ss'} \\
&\dot{\rho}_{sl} = -\frac{1}{2}(k_s^r + k_l)\rho_{sl}
\end{aligned} \tag{5}
$$

It should be noted that vibrational relaxation is taken here into account in a highly simplified way—a damping process that takes place in addition to radiative damping. This is sufficient for our purpose because, as stated above, vibrational relaxation is essentially irreversible and because we are interested in the population change only within the originally excited energy regime. Different situations may require that we treat vibrational relaxation similarly to rotational relaxation as described above.

3. *Dephasing Effects*

In addition to actual transitions between energy levels, collisions may result in a phase change in the molecular state without a change in the state itself. This occurs when an adiabatic energy change takes place during the collision, and leads to a phase relaxation in the molecular state. This dephasing effect may be accounted for by introducing damping of the off-diagonal elements in the density matrix with the corresponding rates γ_{sl}, $\gamma_{ll'}$, and $\gamma_{ss'}$ taken to be proportional to the gas pressure:

$$\dot{\rho}_{sl} = -\gamma_{sl}\rho_{sl} \qquad \dot{\rho}_{ll'} = -\gamma_{ll'}\rho_{ll'} \qquad \rho_{ss'} = -\gamma_{ss'}\rho_{ss'} \tag{6}$$

Dephasing plays a fundamental role in the homogeneous broadening of energy levels in the case of atoms where the population relaxation is relatively inefficient. Its importance for molecules has not been sufficiently elucidated: one can argue that with a high probability of rotational relaxation the importance of pure dephasing collisions will be smaller. Nevertheless, recent studies using the optical transient techniques suggest that the role of pure dephasing is nonnegligible: the cross-sections for dephasing collisions for iodine B^3O_u+ state are about twice as large as those for rotational relaxation[58] in this system.

C. Time Evolution of the Collisionally Perturbed Molecule

The time evolution of the molecule under the combined effect of collisions and of the intramolecular interactions is described by the set of equations[59,60]

$$\dot{\rho}_{ss} = i\sum_{l}(v_{ls}\rho_{sl} - v_{sl}\rho_{ls}) - k_s^r\rho_{ss} + \sum_{s'\neq s}(k_{ss'}\rho_{s's'} - k_{s's}\rho_{ss})$$

$$\dot{\rho}_{ll} = -i\sum_{s}(v_{sl}\rho_{ls} - v_{ls}\rho_{sl}) - k_l\rho_{ll} + \sum_{l'}(k_{ll'}\rho_{l'l'} - k_{l'l}\rho_{ll}) \tag{7a}$$

for the diagonal elements of the molecular density matrix and

$$\dot{\rho}_{ss'} = -i\epsilon_{ss'}\rho_{ss'} + i\sum_{l}(v_{ls'}\rho_{sl} - v_{sl}\rho_{ls'}) - \bar{\gamma}_{ss'}\rho_{ss'}$$

$$\dot{\rho}_{ll'} = -i\epsilon_{ll'}\rho_{ll'} + i\sum_{s}(v_{sl'}\rho_{ls} - v_{ls}\rho_{sl'}) - \bar{\gamma}_{ll'}\rho_{ll'}$$

$$\dot{\rho}_{sl} = -i\epsilon_{sl}\rho_{sl} + i\sum_{s'}v_{s'l}\rho_{ss'} - i\sum_{l'}v_{sl'}\rho_{l'l} - \gamma_{sl}\rho_{sl} \tag{7b}$$

for the nondiagonal density matrix elements.

In these equations, the rates $\bar{\gamma}$ include contributions from the rotational (reversible) relaxation, the vibrational and the radiative (irreversible) relaxations and from dephasing (proper T_2) processes:

$$\bar{\gamma}_{ss'}=\frac{1}{2}\left(\sum_{s''\neq s}k_{s''s}+\sum_{s''\neq s'}k_{s''s'}\right)+k_s^r+\gamma_{ss'}$$

$$\bar{\gamma}_{ll'}=\frac{1}{2}\left(\sum_{l''\neq l}k_{l''l}+\sum_{l''\neq l'}k_{l''l'}+k_l+k_{l'}\right)+\gamma_{ll'}$$

$$\bar{\gamma}_{sl}=\frac{1}{2}\left(\sum_{s'\neq s}k_{s's}+\sum_{l'\neq l}k_{l'l}+k_s^r+k_l\right)+\gamma_{sl} \tag{8}$$

In addition to relaxation and dephasing, the coupling to the thermal bath may shift the position of energy levels. In (7b) it is assumed that the energy differences ϵ_{sl} etc. already include this shift.

It should be noted that several approximations have to be introduced in order to get (7)–(8). The general expression for the time evolution of the molecular density matrix may be written in the form[61–63]

$$\dot{\rho}=-i[H_0+V,\rho]+\int_0^t R(t-\tau)\rho(\tau)\,d\tau \tag{9}$$

where H_0 is the zero-order molecular hamiltonian, V is the intramolecular $(\{|s\rangle\}-\{|l\rangle\})$ coupling and where the tetradic operator R accounts for the thermal relaxation effects. Equation (7) involves first the Markoffian approximation, neglecting memory effects in (9). In the Markoffian limit, we have:

$$\dot{\rho}_{ij}=-i\epsilon_{ij}\rho_{ij}+i\sum_k(v_{kj}\rho_{ik}-v_{ik}\rho_{kj})+\sum_{lm}R_{ijlm}\rho_{lm} \tag{10}$$

Equations (7) are of this form with explicit expressions for the elements of the tetradic R. Further approximations are involved at this point. It is common[59,60] to disregard elements r_{ijlm} for which $|\epsilon_{lm}-\epsilon_{ij}|>|R|$. In this spirit, elements of R that couple diagonal and off-diagonal elements of the density matrix are neglected in (7). So are R elements which couple coherence (off-diagonal) terms of ρ to other such terms. This approximation is valid as long as the pressure broadening is much smaller than the average level spacing. We shall use (7) also for higher pressure as a qualitative tool for estimating collisional effects. Note that in the high-pressure limit (pressure broadening much larger than the average level spacing) the pressure effect saturates, and the molecule behaves as a "large" (statistical-limit) molecule. Equation (7) correctly describes this limit (see below), so that the approximation involved here is expected to affect only details of the relaxation behavior in the intermediate pressure range.

In what follows, we shall consider a few simple models based on (7).

D. Small-Molecule Weak-Coupling Case

As a prototype model for the decay of an excited molecule in presence of collisions we first consider the system schematically represented in Fig. 2: Two pure-spin states $|svj\rangle$ and $|lv'J\rangle$ (s and l now denote the electronic states while v, v' and J, J' stand for the vibrational and rotational quantum numbers, respectively below we often drop the specification of the vibrational levels) are intramolecularly coupled by the matrix element v_{sl} while the intramolecular coupling between all other levels belonging to the $\{|s\rangle\}$ and $\{|l\rangle\}$ manifolds is disregarded. This picture corresponds to a simplified description of a strongly localized intramolecular perturbation between two sets of rovibronic levels $\{|s, J\rangle\}$ and $\{|l, J'\rangle\}$, where a nonnegligible mixing occurs only between a single pair of levels $|s, J\rangle$ and $|l, J\rangle$ (with energies ϵ_{sJ} and ϵ_{lJ}) and gives rise to a pair of quasistationary states $|\sigma, J\rangle$ and $|\lambda, J\rangle$:

$$|\sigma, J\rangle = \alpha|s, J\rangle + \beta|l, J\rangle$$

$$|\lambda, J\rangle = -\beta|s, J\rangle + \alpha|l, J\rangle \tag{11}$$

$$\text{where } \alpha = \cos(\phi/2), \beta = \sin(\phi/2), tg\phi = 2v_{sl}/(\epsilon_{sJ} - \epsilon_{lJ}) \tag{12}$$

For all other states with $J' \neq J$, we have:

$$|\sigma, J'\rangle \equiv |s, J'\rangle \qquad |\lambda, J'\rangle \equiv |l, J'\rangle \tag{13}$$

A further simplification consists in replacing the whole set of $\{|s, J'\rangle\}$ levels ($J' \neq J$) by a single state $|u\rangle$, and the set of $\{|l, J'\rangle\}$ levels ($J \neq J$)

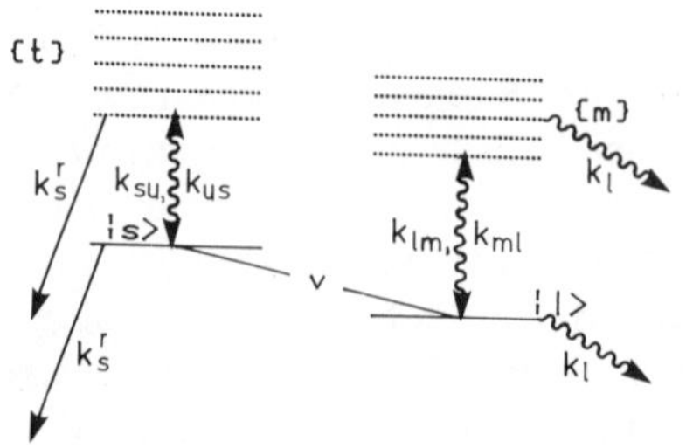

Fig. 2 Four-level model. Energy levels and transition rates.

by a single state $|m\rangle$. For this four-level model, (7) reduces to the form:

$$\begin{aligned}
\dot{\rho}_{ss} &= i(V_{ls}\rho_{sl} - V_{sl}\rho_{ls}) - k_s^r\rho_{ss} \\
\dot{\rho}_{ll} &= i(V_{sl}\rho_{ls} - V_{ls}\rho_{sl}) - k_l\rho_{ll} + (k_{lm}\rho_{mm} - k_{ml}\rho_{ll}) \\
\dot{\rho}_{mm} &= -k_m\rho_{mm} + (k_{ml}\rho_{ll} - k_{lm}\rho_{mm}) \\
\dot{\rho}_{uu} &= -k_s^r\rho_{ss} + (k_{us}\rho_{ss} - k_{su}\rho_{uu}) \\
\dot{\rho}_{sl} &= -i\epsilon_{sl}\rho_{sl} + iV_{sl}(\rho_{ss} - \rho_{ll}) - \bar{\gamma}_{sl}\rho_{sl} \\
\dot{\rho}_{lm} &= -i\epsilon_{lm}\rho_{lm} - \bar{\gamma}_{lm}\rho_{lm} \\
\dot{\rho} &= -i\epsilon_{su}\rho_{su} - \bar{\gamma}_{su}\rho_{su}
\end{aligned} \tag{14}$$

where:

$$\begin{aligned}
\bar{\gamma}_{sl} &= \gamma_{sl} + \tfrac{1}{2}(k_{ml} + k_{us}) + \tfrac{1}{2}(k_s^r + k_l) \\
\bar{\gamma}_{lm} &= \gamma_{lm} + \tfrac{1}{2}(k_{lm} + k_{ml}) + \tfrac{1}{2}(k_l + k_m) \\
\bar{\gamma}_{su} &= \gamma_{su} + \tfrac{1}{2}(k_{us} + k_{su}) + k_s^r
\end{aligned}$$

Equation (14) describes the time evolution of the intramolecularly coupled pair of levels following optical excitation, where collisional coupling to other levels is accounted for by considering two additional representative levels. We expect that this picture may qualitatively account for the more important features of the collisionally induced electronic relaxation.

The existence of the intramolecular coupling between the levels $|s\rangle$ and $|l\rangle$ implies that under selective excitation the initially excited state of the molecule is the stationary state of the isolated molecule $|\sigma, J\rangle$. This corresponds to the initial form of the density matrix in the s, l, u, m basis:

$$\rho(t=0) = \begin{bmatrix} \alpha^2 & \alpha\beta & 0 & 0 \\ \alpha\beta & \beta^2 & 0 & 0 \\ 0 & 0 & 0 & 0 \\ 0 & 0 & 0 & 0 \end{bmatrix} \tag{15}$$

Equation (7) implies that with this initial condition the off-diagonal elements of ρ involving the m and u levels will be zero for all times t:

$$\rho_{sm}(t) = \rho_{lm}(t) = \rho_{su}(t) = \rho_{lu}(t) = 0 \tag{16}$$

We will not try to solve (14) exactly but only to discuss approximate solution under different limiting conditions. We shall also present some results of computer simulations for physically reasonable sets of rate constants k and γ. We first estimate separately the influence of collisional dephasing and of rotational and vibrational relaxation on the time evolution of the excited system.

Since in most real experiments the population of the radiative states is monitored by the intensity of the (spectrally unresolved) fluorescence (emitted from the $|s\rangle$ and $|u\rangle$ levels) we are essentially interested in the overall population $\rho_{ss}+\rho_{uu}$. Similarly the overall population of the optically nonactive levels $\rho_{ll}+\rho_{mm}$ (which may be monitored by transient absorption or by emission within the triplet manifold) is of interest.

1. Pure Dephasing Only

To consider the effect of pure dephasing we take $k_{lm}=k_{ml}=k_{st}=k_{ts}=0$, $k_s^r\neq 0$, $\gamma_{sl}\neq 0$. If we suppose that dephasing is fast relative to the intramolecular evolution: $\gamma_{sl}>v_{sl}$, we may approximate ρ_{sl} by its instantaneous stationary value taking $\dot{\rho}_{sl}=0$ whereupon

$$\rho_{sl}=\frac{iv_{sl}(\rho_{ss}-\rho_{ll})}{i\epsilon_{sl}+\bar{\gamma}_{sl}} \tag{17}$$

This leads to a set of rate equations for ρ_{ss} and ρ_{ll} (obviously in this case $\rho_{mm}(t)=\rho_{uu}(t)=0$ for all times t)

$$\begin{aligned}\dot{\rho}_{ss}&=-K_{ls}\rho_{ss}-k_s^r\rho_{ss}+K_{sl}\rho_{ll}\\ \dot{\rho}_{ll}&=K_{ls}\rho_{ss}-K_{sl}\rho_{ll}\end{aligned} \tag{18}$$

where

$$K_{sl}=K_{ls}=\frac{2\bar{\gamma}_{sl}v_{sl}^2}{\epsilon_{sl}^2+\bar{\gamma}_{sl}^2} \tag{19}$$

and where

$$\bar{\gamma}_{sl}=\frac{k_s^r}{2}+\gamma_{sl}$$

We see that in this case, following the initial dephasing, the subsequent evolution of the system may be correctly described by rate equations involving only the populations of the zero-order states and formally identical to usual equations of chemical kinetics. It is, however, interesting to note that the transition rate K_{sl} is *not* proportional to the perturber pressure as it was assumed for the rates k_1^{vr}, k_{lm}, and k_{su}, but depends on the gas pressure in a more complex way. In the limiting case of weak intramolecular and intermolecular perturbations (weak relative to the level spacing, $\epsilon_{sl}^2 \gg v_{sl}^2, \bar{\gamma}_{sl}^2$) we have:

$$K_{sl} = K_{ls} = 2\beta^2\bar{\gamma}_{sl}$$

K_{sl} is thus proportional to β^2, i.e., to the l-state content in the initially prepared quasistationary state. This conclusion, which readily follows from (15) and (19), is the same as deduced by Freed[11] from first-order perturbation treatment of collision-induced electronic transitions. In the other extreme limit, $\bar{\gamma}_{sl} \gg \epsilon_{sl}$, the rates $K_{sl} = K_{ls}$ become inversely proportional to $\bar{\gamma}_{sl}$. This results from the dilution of the coupling over the pressure-broadened l level (similar effects have been discussed with respect to line broadening[64,65]). However, in this limit the broadened levels may overlap and a model based on two levels only breaks down (see below).

2. *Dephasing + Rotational Relaxation*

In all real systems the effects of dephasing and of rotational relaxation are superimposed. The overall behavior is similar to that described in the previous case, except that now the populations of the $|m\rangle$ and $|u\rangle$ levels do not vanish. After the rapid decay of coherences, the evolution of the system is governed by the rate equations:

$$\begin{aligned}
\dot{\rho}_{ss} &= -(K_{ls} + k_s^t)\rho_{ss} + K_{sl}\rho_{ll} + k_{su}\rho_{uu} \\
\dot{\rho}_{ll} &= K_{ls}\rho_{ss} - (K_{sl} + k_{ml})\rho_{ll} + k_{lm}\rho_{mm} \\
\dot{\rho}_{mm} &= k_{ml}\rho_{ll} - k_{lm}\rho_{mm} \\
\dot{\rho}_{uu} &= k_{us}\rho_{ss} - k_{su}\rho_{uu}
\end{aligned} \tag{20}$$

Here $K_{sl} = K_{ls}$ are again given by (19) with $\bar{\gamma}_{sl}$ defined by (11) (including the rotational relaxation contribution). Again these rate constants are proportional (in the very weak mixing limit) to the β^2 mixing coefficient of the initially excited state and show a linear dependence on $\bar{\gamma}_{sl}$ as long as $\bar{\gamma}_{sl} \ll \epsilon_{sl}$.

Collisions destroy the coherences associated with the initially prepared state and induce transitions between rotational levels. This results in redistribution of population between all interacting levels but the overall population (the trace of the density matrix) is not affected and decays only radiatively with the rate $k_s^r(\rho_{ss} + \rho_{uu})$. Collisions, thus, modify the form of the excited-state decay but the overall fluorescence yield remains equal to one unless the levels s, u, and/or l, m are subjected to additional nonradiative decay processes (e.g., in the case of predissociation).

3. *Effects of Vibrational Relaxation*

Vibrational relaxation opens an additional decay channel due to irreversible transitions from the $|l\rangle$ and $|m\rangle$ levels to lower-lying vibronic states of the $\{|l\rangle\}$ manifold. The overall set of rate equation describing the system after the initial dephasing is

$$\dot{\rho}_{ss} = -(K_{ls} + k_s^r)\rho_{ss} + K_{sl}\rho_{ll} + k_{su}\rho_{uu}$$

$$\dot{\rho}_{ll} = K_{ls}\rho_{ss} - (K_{sl} + k_{ml} + k_l)\rho_{ll} + k_{lm}\rho_{mm}$$

$$\dot{\rho}_{mm} = k_{ml}\rho_{ll} - (k_{lm} + k_m)\rho_{mm}$$

$$\dot{\rho}_{uu} = k_{us}\rho_{ss} - (k_{su} + k_s^r)\rho_{uu} \tag{21}$$

The rates k_l and k_m correspond to any nonradiative decay channel, which couple to the levels $|l\rangle$ and $|m\rangle$. It should be kept in mind, however, that reverse collisionally induced electronic transitions may follow vibrational relaxation in the $\{|l\rangle\}$ manifold, thus leading to further emission. This effect must be taken into account when the $|s\rangle$ level is not the lowest one in the $\{|s\rangle\}$ manifold (see Section III.B).

4. *Discussion*

Equation (21) may be easily solved yielding a time evolution characterized by biexponential decay. The first (fast) component of the decay corresponds to the initial equilibration of populations between the optically active and the optically nonactive levels. The second (slow) component corresponds to the radiative decay of the thermally equilibrated system.

In addition to solving (21) we have performed a numerical integration of (14) with the initial condition given by (15), thus avoiding the fast dephasing approximation. Some results of this calculation for a reasonable choice of parameters are presented in Fig. 3. It is seen that the decay is indeed biexponential (except for very brief periods immediately following the exciting pulse).

These calculations were carried out for various sets of parameters $\epsilon_{sl}, V_{sl}, k_i, k_{ij}$ and γ_{sl}. In addition, the number of relevant levels in the $\{|s\rangle\}$ and $\{|l\rangle\}$ manifolds was accounted for by assigning to the levels $|m\rangle$ and $|u\rangle$ statistical weights N and N', respectively, thus considering them as sets of N (N') degenerated levels.

As may be expected the rate associated with the fast decay component increases with β^2 (i.e., with v_{sl} for constant ϵ_{sl} values) and with $\bar{\gamma}_{sl}$ (provided that $\bar{\gamma}_{sl} \leqslant \epsilon_{sl}$). The rate associated with the slow (equilibrated)

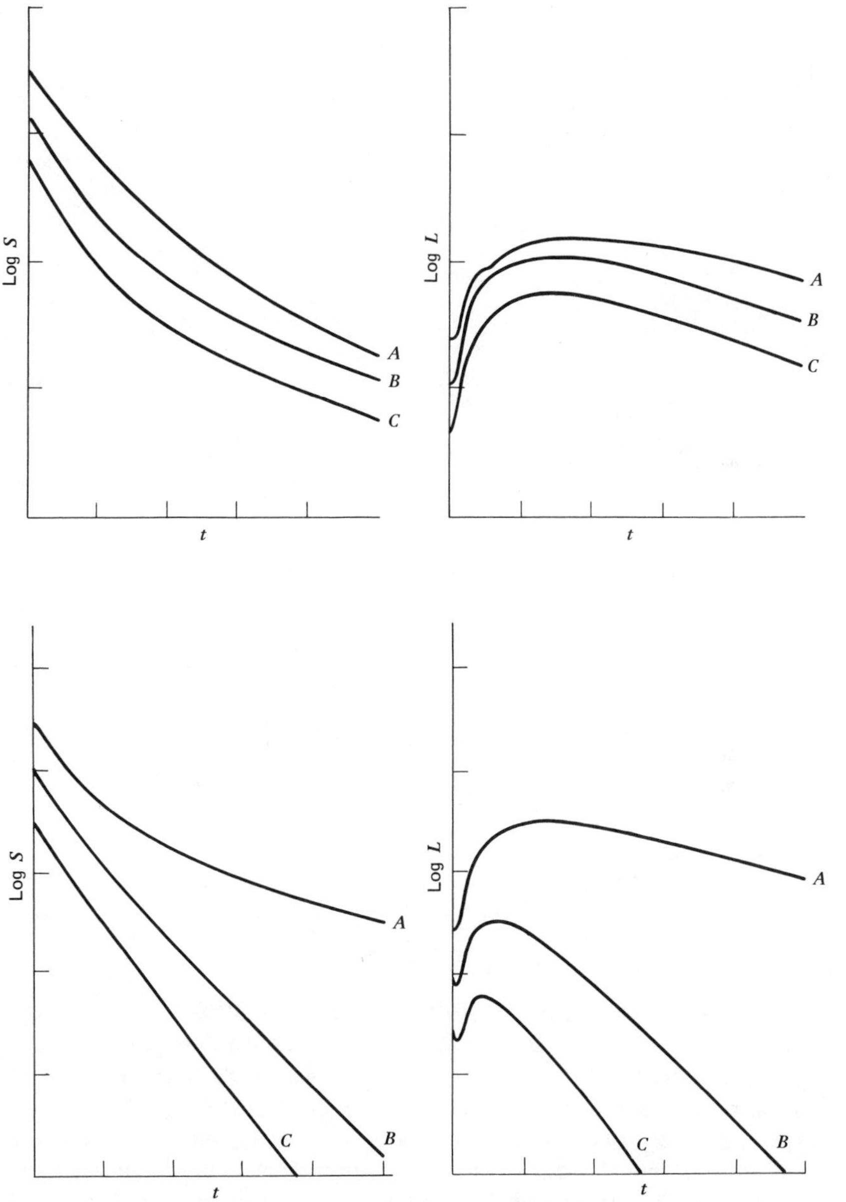

Fig. 3. Computer simulations of the time dependence of $S=\rho_{ss}+\rho_{uu}$ and $L=\rho_{ll}+\rho_{mm}$ for the four-level model. $\epsilon_{sl}=10$, $v_{sl}=3$, $k_s^r=0.5$, $N=N'=1$ and (top) $k_l=0$, $\bar{\gamma}_{sl}=3$ (A), 6 (B), 9 (C) (center) $\bar{\gamma}_{sl}=6$, $k_l=0$ (A), 1 (B) and 2 (C) (bottom $k_l=0$, $\bar{\gamma}_{sl}=6$. Initial excitation of the σ state (A) and of the λ state (B).

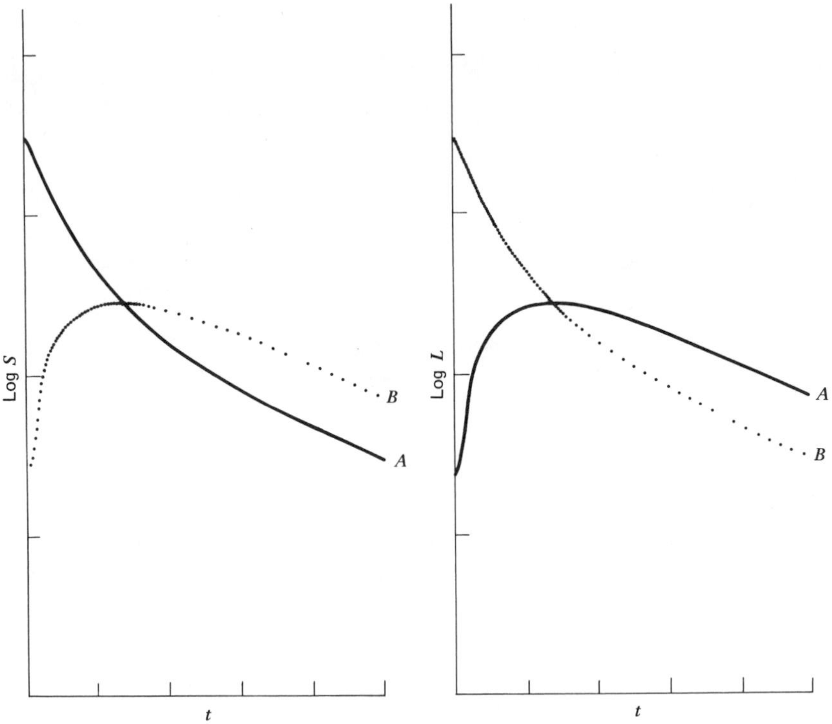

Fig. 3 (*Continued*)

decay component is given by

$$k_{\rm eq}=\frac{k_s^r(\rho_{ss}^{\rm eq}+\rho_{uu}^{\rm eq})+k_l(\rho_{ll}^{\rm eq}+\rho_{mm}^{\rm eq})}{Tr(\rho)}$$

where $\rho^{\rm eq}$ is the equilibrium (Boltzmann) density matrix. This is the total decay rate. The corresponding radiative decay rate (photon emission rate) $k_{\rm eq}^r$ is given by a similar expression with k_l replaced by k_l^r. Note that if (as is usually the case) $k_l=0$, and since $(\rho_{ll}^{\rm eq}+\rho_{mm}^{\rm eq})/(\rho_{ss}^{\rm eq}+\rho_{uu}^{\rm eq}\simeq(N+1)/(N'+1)\gg 1$, the long time component may be very slow, even undetectable, in the absence of vibrational relaxation. However, since vibrational relaxation rates vary within very wide limits we may expect to encounter both extreme limits corresponding to $K_{sl}\gg k_l$ and $K_{sl}\ll k_l$, as well as intermediate situations. In the first case the decay will be biexponential; in the latter one the equilibrium between the s,u and $l\ ,m$ level populations cannot be established because of the rapid nonradiative decay of the l and m levels. $\rho_{ss}+\rho_{uu}$ will thus decay exponentially with the rate $k_s^r+K_{ls}$. In the language of the chemical kinetics the first case corresponds to reversible and the latter one to irreversible $\{|s\rangle\}\to\{|l\rangle\}$ electronic relaxation.

Finally we note that if the initially excited state is the state $|\lambda\rangle$ with the predominant l-character, and if the vibrational relaxation within the l manifold is inefficient, the time dependence of ρ_{ss} and ρ_{ll} will be reversed relative to the former case: ρ_{ss} (and the fluorescence intensity) initially increases until the extablishment of the thermal equilibrium and then decays with the previously defined rate k_{eq}.

E. Intermediate-Size Molecules

The simple models discussed so far are appropriate for the small-molecule case. In larger molecules corresponding to the intermediate level structure cases many levels in the $\{|l\rangle\}$ manifold couple to the initial $|s\rangle$ level. The main new feature arising in this case can be seen by considering single $|s\rangle$ level coupled by the intramolecular interaction to a group of $|l\rangle$ levels. The relevant equations from the set (7) are:

$$\dot{\rho}_{ss}=i\sum_{l}(v_{ls}\rho_{sl}-v_{sl}\rho_{ls})-k_s^r\rho_{ss}$$

$$\dot{\rho}_{sl}=i\epsilon_{sl}\rho_{sl}+iv_{sl}\rho_{ss}-i\sum_{l'}v_{sl'}\rho_{l'l}-\bar{\gamma}_{sl}\rho_{sl} \tag{22}$$

with the corresponding equation for ρ_{ls}. To simplify the calculation, we again assume that on the time scale of interest $\dot{\rho}_{sl}=\dot{\rho}_{ls}=0$. We also disregard all terms of the form $\sum_{l'}v_{sl'}\rho_{l'l}$ in the equation for ρ_{sl} and for ρ_{ls}. This approximation is justified because when many levels l participate, $\rho_{l'l}$ is expected to be small at all times; also $v_{sl'}$ is expected to be an irregularly varying function of the index l', therefore we expect that:

$$\Big|\sum_{l'}v_{sl'}\rho_{l'l}\Big|\sim|v_{sl'}\rho_{l'l}|\ll|v_{sl}\rho_{ss}|$$

with this approximation we obtain (following the initial dephasing):

$$\dot{\rho}_{ss}=-\left(2\sum_{l}\frac{v_{sl}^2\bar{\gamma}_{sl}}{\epsilon_{sl}^2+\bar{\gamma}_{sl}^2}+k_s^r\right)\rho_{ss} \tag{23}$$

so that the level s decays exponentially with a nonradiative component given by K_{sl}. If $\langle\epsilon_{sl}^2\rangle\gg\langle\bar{\gamma}_{sl}^2\rangle$ only one term in this sum is important and we regain the small-molecule limit. When $\bar{\gamma}_{sl}$ (which is essentially proportional to the pressure) becomes of the same order of magnitude as ϵ_{sl} the pressure dependence saturates and for higher pressures the relaxation will be pressure independent.

Equation (23) describes the relaxation of fluorescence only after the initial dephasing. It is useful at this point to recall the corresponding isolated molecule case. In that case, when many $|l\rangle$ levels lie close enough to the level $|s\rangle$ (strong-coupling case of the isolated intermediate case molecule) so that the initial excitation results in a coherent superposition of exact molecular levels, the initial decay corresponds to the rapid intramolecular dephasing. In the isolated molecule, this initial prompt decay is followed by a long tail characterized by the diluted radiative lifetime of exact molecular levels. Equation (23) describes the finite pressure case corresponding to this long tail. The slow component is represented by the term $k_s^r\ \rho_{ss}$, which is small because of the population distribution over all strongly coupled $|s\rangle$ and $\{|l\rangle\}$ levels. In addition there exists a nonradiative decay component proportional to the pressure (because γ_{sl} is proportional to the pressure). As the pressure is increased, the initial dephasing time (dependent on the total width of the band of $\{|l\rangle\}$ levels effectively coupled to $|s\rangle$ and not on their widths) is practically unaffected. In contrast to this, the long time decay rate is enhanced by the pressure-induced component in (23) until, when the broadened levels overlap, the two decay components coalesce and the fluorescence decays according to a simple exponential law with a rate given by the golden rule:

$$2\sum_l \frac{|v_{sl}|^2\bar{\gamma}_{sl}}{\epsilon_{sl}^2+\bar{\gamma}_{sl}^2} \simeq 2\rho_l \int dx \frac{|v_{sl}|^2\bar{\gamma}_{sl}}{x^2+\bar{\gamma}_{sl}^2} = 2\pi |v_{sl}|^2 \rho_l$$

where ρ_l is the l-level density and we assume: $\bar{\gamma}_{sl}^2 \gtrsim \epsilon_{sl}^2$. As stated above, the pressure dependence disappears and the molecule behaves as a "large" (statistical-limit) molecule.

This discussion also indicates the expected behavior of larger molecules. Obviously, when the isolated molecule shows the statistical-limit behavior we do not expect any important modification of its decay character in presence of collisions with chemically inert molecules.

F. Some Specific Comments

(a) It has been assumed so far that collisions do not modify any intramolecular parameters. In the specific case of singlet–triplet transition this is essentially equivalent to the assumption that the spin–orbit coupling is the same in the collisional complex as in the isolated molecule. This approximation is reasonable as long as the collision partners do not contain heavy atoms. In experiments involving heavy collision partners such as Kr or Xe, the "external heavy-atom effect" may play a nonnegligible role. Similarly, when the s–l coupling is forbidden in an isolated molecule by orbital symmetry, collision perturbation (lower symmetry of the collisional complex) may induce coupling between the two electronic states. This seems to be the case of the $B^2\pi_g$ and $W^3\Delta_u$ isoenergetic states, of N_2 where the coupling is strictly forbidden by the $g \nleftrightarrow u$ Kronig rule. Collision-induced transitions between those states have been observed[(39)] and must be due to the breakdown of the $g \nleftrightarrow u$ rule in the collisional complex.

(b) The treatment presented here is based on the assumption that the duration of a collision is much shorter than all other time scales of interest, i.e., $t_{\text{coll}}\ V_{sl} \ll 1$ and $t_{\text{coll}}\Gamma \ll 1$ where Γ represents any of the k and γ parameters appearing in (7). If this conditions are not satisfied (e.g., if collision result in the formation of a long-time collisional complex) we may expect substantial deviations from the predicted time behavior.

As a matter of fact, recent observations of the vibrational, rotational, and electronic relaxation induced by low-energy collisions in supersonic jets involving a long interaction times between collision partners show a drastic increase of the cross-sections with respect to the room temperature values.[65a–e]

(c) As mentioned above, the s–l transitions (in absence of the vibrational relaxation) will affect the fluorescence yield only in the case when the $\{|l\rangle\}$ manifold is subjected to nonradiative decay. When the $\{|l\rangle\}$ manifold corresponds to a dissociative electronic state, we are in the specific case of the collision-induced (or enhanced) predissociation. Such a behavior has been observed for the BO_u^+ state of I_2, where collisions induce transitions to the isoenergetic $A\ 1u$ dissociative state (see below).

III. REVIEW AND ANALYSIS OF EXPERIMENTAL DATA

We will apply the model treatment defined in the preceding section to the analysis of experimental data and to the discussion of particular problems concerning the mechanism of collision-induced electronic relaxation processes. We will try to find in the actually available experimental material the answers to some—practically important—questions:

1. In what kind of molecular system the electronic relaxation may be considered as reversible? What is the influence of the reversible relaxation on the deactivation paths?

2. Is it possible to correlate the cross-sections for collision-induced s–l transitions with the intramolecular s–l coupling strength? Do the strongly mixed states play a specific role of "gates" between s and l manifolds?
3. What is dependence of $s \rightarrow l$ cross-sections on the properties of the collision partner?

We will center our attention at the most interesting case of small molecules and at the most extensively studied process of collision-induced intersystem crossing.

A. Reversibility of Collision-Induced Electronic Relaxation

As shown before (Section II. D.4) the electronic relaxation becomes irreversible when the l-level density is sufficiently large and/or when the vibrational relaxation within the $\{l\}$ manifold is much more rapid than the electronic relaxation:

$$\frac{\rho_l}{\rho_s} \gg 1 \quad \text{and/or} \quad k_l > K_{ls}$$

This means that even in relatively small polyatomic molecules the electronic relaxation is practically irreversible (the fluorescence of collisionally perturbed molecules exhibits a purely exponential decay with the pressure dependence of the lifetime and yield given by Stern–Volmer law [cf. (1)]. Such a behavior has been observed in the well-known case of intersystem crossing in glyoxal[34,35] and may be easily explained by the properties of the molecule. The ratio of triplet and singlet level densities in the vicinity of the ^1A_u 0° level may be roughly estimated as being of the order of: $\rho_l/\rho_s \cong 100$.[66] The rate of collision-induced $s \rightarrow l$ crossing for ground-state glyoxal as collision partner is $K_{ls} = 2.10^6$ sec^{-1} torr^{-1}. The vibrational relaxation rate within the triplet manifold is not known but is certainly not very different from that of relaxation in the singlet manifold: $k_s \cong 10^7$ sec^{-1} torr^{-1}.[35] Hence, the probability of the reverse crossing for the molecule transferred to the triplet level isoenergetic with the singlet 0° state is of the order 10^{-3}; the transition must be thus considered as practically irreversible.

Reversible electronic relaxation has been found in smaller molecules. In methylene (for which collision-induced intersystem crossing has been extensively studied for many years[67,68]) the reversible character of singlet–triplet transitions has been evidenced in a study of pressure effect on the CH_2 radical prepared either in the singlet state (by photochemical decomposition of diazirine) or in the triplet state (by photofragmentation of diazirine due to the triplet–triplet energy transfer). It has been shown that for a sufficiently high pressure of inert gas (N_2, SF_6) one attains an equilibrium $^1CH_2/^3CH_2$ concentration ratio, independent of the initially prepared state.[69] The nonexponential (quasibiexponential) decay of the $B^3\pi_g$ state of N_2 in presence of collisions with Ar has been equally explained by reversible, collision-induced transitions between closely spaced $B^3\pi_g$ and $w^3\Delta_u$ states.[39]

A direct proof of the reversibility of singlet–triplet transition has been deduced from time-resolved studies of the fluorescence observed under a selective excitation of $\{\sigma\}$ and $\{\lambda\}$ rovibronic states resulting from the perturbation between $A^1\pi$ ($v=0$) and $e^3\Sigma^-$ ($v=1$) vibronic states in CO. In the former case, the fluorescence shows a biexponential decay, while in the latter one it exhibits an initial induction followed by the decay (Fig. 4; see also Fig. 3 bottom). The ratio of rate constants for the $s\rightarrow l$ and $l\rightarrow s$ processes (determined by a simple kinetic treatment) is in a good agreement with the expected thermal distribution after a complete equilibration of the $(\{s\}+\{l\})$ system.[37,70]

Similar nonexponential decays induced by collisions with inert partners have been observed by Erman et al.[71–73] for CO, CN, and CS excited by the electron impact. For CN (prepared by electron-impact fragmentation of the parent compound with a high yield of vibrationnally "hot" ground-state molecules) the authors observe an increase of the fluorescence ($A^2\pi\rightarrow X^2\Sigma^+$) in presence of collisions. Such an unusual effect may be explained by collisionally induced inverse relaxation from strongly populated and long-lived high vibrational levels of the ground state to quasiresonant low levels of the $A^2\pi$ radiative state.[72] This explanation was confirmed by a time-resolved study of the CN fluorescence from selectively excited levels of the A state: the biexponential decay with a long component longer than the radiative lifetime of the A state is clearly due to reversible A X collision-induced transitions.[73a]

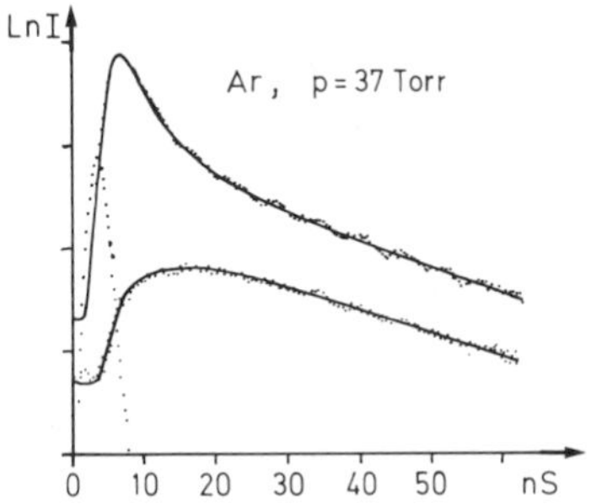

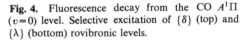

Fig. 4. Fluorescence decay from the CO $A^1\Pi$ ($v=0$) level. Selective excitation of $\{\delta\}$ (top) and $\{\lambda\}$ (bottom) rovibronic levels.

Complex and not entirely elucidated effects of the inert-gas pressure have been equally observed in the case of fluorescence from the second excited state ($B^2\Sigma^+$) of the CN radical in an early work by Luk and Bersohn.[74]

B. Vibrational Relaxation by Reversible Electronic Relaxation

As shown above, the electronic relaxation in small molecules may be more efficient than the vibrational relaxation within the same electronic state. If, moreover, one of interacting electronic states is nonradiative, that is, if the lifetime of molecules transferred to the $|l\rangle$ state is much longer than that of the $|s\rangle$ state, new specific deactivation channels may play an important role. The usual path of vibrational relaxation within the $\{s\}$ manifold: collision-induced transitions from initially excited vth to $(v-1)$th vibronic level may be less efficient than a many-step process involving

1. Electronic relaxation from the $|s, v\rangle$ state to resonant $\{l\}$ levels,
2. Vibrational relaxation within the $\{l\}$ manifold favored by the long lifetime of excited species.
3. The inverse electronic relaxation from the l manifold to lower lying $|s, v-1\rangle, \ldots, |s, v-n\rangle$ states.

Such a mechanism has been proposed by Slanger[41] and observed for the first time by Bondybey et al.[75,76] in the case of diatomic molecules inbedded in rare-gas matrices. In their subsequent work,[77] similar effects have been found for collisional processes in the gas phase. The vibrational relaxation of CO^+ excited to the $v=3$ and $v=2$ levels of the $A^2\pi$ state induced by collisions with He is more efficient by 4–5 orders of magnitude than in the ground-state CO+He system.[52] Moreover, the form of the fluorescence decay from the $v=1$ level observed under $v=2$ excitation cannot be fitted if a direct $v=2 \rightarrow v=1$ relaxation path is assumed: the induction time of the relaxed emission being much longer than the decay of the resonance fluorescence.

A similar behavior has been observed under the optical excitation of the $v=1$ level of the CO $A^1\pi$ state perturbed by the isoenergetic level of the triplet $d^3\Delta$ state. Again, a large delay of the emission from the $v=0$ level with respect to the $v=1$ level decay must be considered as due to the transit through long-lived triplet states. Since this delay may be fitted by the vibrational relaxation rate within the triplet manifold (directly measured in independent experiments), a three-step relaxation path:

$$A^1\pi(v=\,) \rightarrow d^3\Delta(v=5) \rightarrow d^3\Delta(v=4) \rightarrow A^1\pi(v=0)$$

has been proposed in this case.[70]

The vibrational relaxation occurring apparently in a single electronic state but due in fact to the reversible internal conversion (or intersystem crossing) may play an important role in diatomic molecules exhibiting a strong collisional coupling between radiative and nonradiative (long-lived) states. It may be responsible in part for surpisingly high rates of the (apparent) vibrational relaxation in excited states of diatomics: for the CO $A^1\pi$ state + Ar, the apparent rate constant for the vibrational relaxation in the singlet manifold amounts to $3.10^6\ s^{-1}\ torr^{-1}$ when a direct mechanism is assumed,[78] while in fact the process occurs in the triplet manifold with much lower rate of $2.10^5\ s^{-1}\ torr^{-1}$.[70]

The importance of this mechanism is certainly reduced in polyatomic molecules, where direct vibrational relaxation is efficient and the inverse electronic relaxation is less probable.

C. Dependence of Collisional Relaxation Rates on Intramolecular Parameters

As shown in Section II, the rate constant of the collision-induced electronic relaxation (K_{ls} or K_{sl}) depends on intramolecular parameters (s–l coupling constant—v_{sl} and s–l level spacing ϵ_{sl}) as well, as on the strength of the external, collisional perturbation (expressed in (19) by the overall dephasing rate $-\bar{\gamma}_{sl}$). The separation of both factors is not always self-evident. Nevertheless, for the simplicity sake we will first discuss the relation between the amount of the intramolecular s–l coupling (mixing) by assuming constant value of the intermolecular interaction.

The experimental material being still very limited, it seems preferable to start by a concise review of proposed theoretical models and consider the sparse experimental data as a mean to check their validity in a few specific cases.

1. *Theory*

As already mentioned, the first and essential step has been made by Gelbart and Freed,[9] who showed that a nonzero mixing between $|s\rangle$ and $|l\rangle$ states in isolated molecules is a necessary condition for collisional $|s\rangle \rightarrow |l\rangle$ transitions. Therefrom, the linear dependence of collisional cross-sections on the mixing coefficient β^2 has been deduced from the theory involving the first-order time-dependent perturbation method.[10,11] This approximation is valid for very weak external perturbations and/or very high kinetic energies of colliding particles (very short duration of the collisional perturbation, τ_{coll}). For a number of molecular systems these conditions are obviously not fulfilled and a more general treatment must be applied.

The case of the intersystem crossing in methylene[67,68] has been extensively discussed. The absence of detectable singlet–triplet radiative transitions may be considered as a direct proof of the weakness of the s–l mixing, while the nonradiative transitions are easily induced by collisions. Dahler et al.[79–81] adopted in the theoretical study of the methylene case an approach different from that of Freed and Gelbart. They consider as the essential mechanism of transitions the s–l level crossing due to different shifts of $|s\rangle$ and $|l\rangle$ levels in the collisional complex. In the vicinity of the crossing point, the s–l mixing is obviously much stronger than in the isolated molecule and the nonadiabatic transitions would populate the triplet state. Such a treatment, successful for the collisional relaxation within atomic multiplets (see, e.g., Ref. [82]), where all other (widely spaced) states may be neglected, is not so well adapted to molecules: in the collisional complex, a large number of close-lying rovibronic levels becomes strongly mixed. The physical meaning of two individual states that cross at the well-defined intermolecular distance is not evident.

This problem has been discussed in a recent work by Freed and Tric,[83] where strong collisional interactions involving level-crossing effects ("perturbed mixing collisions") are considered together with long range interactions ("simple mixing collision") for which the first-order perturbation approximation is valid (see also Ref.[1]). The important point is the dependence on the β^2 parameter: for "simple mixing collisions" we have again the collisional cross-section σ_{isc} proportional to β^2 but in "perturbed mixing collisions" σ_{isc} attains the maximum value for relatively low values of β^2, the average cross-section shows thus a kind of saturation.

Similar conclusions have been deduced from the exact treatment of a simple molecular model: collisionally perturbed two-level system composed of a mixed state $|\sigma, J\rangle$ and a pure triplet state $|l, J'\rangle$ proposed by Grimbert et al.[38] By assuming, as in the Freed work,[11] the collisional coupling between $|\sigma\rangle$ and $|l\rangle$ equal to βV (where V is the collisional coupling between rotational levels in $\{s\}$ and $\{l\}$ manifolds) one obtains (for different types of the intermolecular potential) the analytical expressions for the $\sigma_{isc}=f(\beta^2)$. A new feature is a prediction of the temperature dependence of σ_{isc}.[84] The most important result concerns the dependence of $\sigma_{isc}=f(\beta^2)$ on the strength of external perturbations: for weak pertubers σ_{isc} is proportional to β^2 in the wide range, for stronger ones, the initial rapid increase of σ_{isc} is followed by much steeper dependence (which may be approximated by the linear dependence of σ_{isc} on $\ln \beta^2$). At last, for still stronger perturbations σ_{isc} attains rapidly an almost constant value of the order of gas-kinetic cross-sections, independent of the s–l mixing coefficient.

This conclusion will be useful in the analysis of experimental data.

2. *Experimental Results*

As mentioned above (Sec. I. B. 2,3), the long-lived fluorescence of molecules belonging to the strong-coupling intermediate case is efficiently quenched in collisions with inert partners (and still more efficiently self-quenched). The quenching rates are closely to the gas-kinetic collision rate in the case of biacetyl[33] and benzoquinone[31] and exceed them in the case of pyrimidine self-quenching ($k_q = 10^8\ s^{-1}\ torr^{-1}$)[(30)] and methylglyoxal self-quenching ($k_q = 8.10^7\ s^{-1}\ torr^{-1}$)[(86)], while for pyrazine–SF_6 collisions one obtains by combining Stern–Volmer parameters[29] and collision-free decay times[85]: $k_q = 6.10^7\ s^{-1}\ torr^{-1}$.

Such a high quenching efficiency fits well into the Gelbart–Freed picture: the long-lived fluorescence is emitted from essentially triplet ($\beta^2 \cong 1$) with a very slight singlet admixture; the relaxation to closely lying pure triplet levels will thus occur at a rate practically equal to the relaxation rate within the triplet manifold[87]

In the small molecules belonging to the weak-coupling limit, the initially prepared state has a predominant singlet character ($\beta^2 \ll 1$). As expected, the fluorescence quenching cross-sections are, in general, smaller than in the previous case and vary in wide limits. Unfortunately, the correlation between intramolecular parameters and quenching efficiencies cannot be done by comparing the behavior of different molecules: the intermolecular potential (the strength of the collisional perturbation) obviously depends not only on the collider but also on the excited species. The only way to check this correlation consists in the comparison of relaxation from different vibronic (rovibronic) levels of the same molecule, provided the *s–l* mixing coefficients for these levels have been previously determined.

To our knowledge, the only study of this kind has been carried out for CO $A^1\pi$ state, where the average mixing coefficients $\bar{\beta}_v^2$ characterizing vibronic levels $-v$ have been calculated from singlet–triplet coupling constants determined by the analysis of spectroscopic perturbations in the $A^1\pi$–$X^1\Sigma^+$ absorption spectra.[88,89] The results obtained under selective excitation of $v = 0 \div 8$ levels show a good correlation with $\bar{\beta}^2$: as expected from the treatment of Grimbert et al.[38]: σ_{isc} vary strongly in the case of He used as collision partner (from 0.5 Å^2 for $\bar{\beta}^2 = 10^{-3}$ to 3.5 Å^2 for $\bar{\beta}^2 = 0.1$), while for Ar this variation is strongly reduced (σ_{isc} changes from 10 to 25 Å^2) and for Kr the gas-kinetic quenching-cross-section is practically the same for all levels.

Recent, preliminary data on the fluorescence quenching from single rotational levels of CS $A^1\pi$ ($v = 0$) state of CS by O_2 and CO probably due to the intersystem crossing as in the case of CO[90] show little dependence of σ_{isc} on the mixing coefficients (k_q for O_2 increases from 6×10^6 to 8.4×10^6

sec^{-1} when β^2 varies from 0.09 to 0.19. This is not surprising in view of very high β^2 values and of strongly perturbing colliders used in this work (quenching rates closely to gas-kinetic collision rates). In NH_2 $^2A_1$ state, the electronic quenching rate is nearly the same for different vibronic levels[91] but the absence of spectroscopic information makes any discussion difficult. At last, in the case of the glyoxal self-quenching the rates of the intersystem crossing (about 0.1 of the rotational relaxation rates) do not vary in the significant way for low-lying ($\Delta E_{vib} \leqslant 1000$ cm^{-1}) vibronic levels of the ^1A_u state,[35,57] but for all these states the singlet–triplet perturbations are too weak to be detected by the high-resolution spectroscopy.

The collision-assisted predissociation in iodine B^3O_u+ state merits a detailed discussion. It is well known that B state is weakly coupled to the dissociative A $1u$ state by rotational and hyperfine-structure terms in the molecular Hamiltonian. The natural predissociation rate strongly depends on the vibrational quantum number (pronounced maxima for $v=5$ and $v=25$, a minimum for $v=15$), this dependence being due to a variation of the Franck–Condon factor.[92–94] The predissociation rate is enhanced by collisions. In absence of a detailed theoretical treatment of the collision-assisted I_2 predissociation, one can suppose that the asymmetric perturbation (breakdown of the orbital symmetry) in the collisional complex affects electronic and rotational wavefunctions but does not change the nuclear geometry.

As a matter of fact, the collisional cross-sections exhibit a clear dependence on the Franck–Condon factor. Their variation is relatively weak in the case of the very efficient self-quenching (σ varying between 55 and 90 Å^2)[94] and much more pronounced in the case of light atoms.[95–98] For He σ varies from 0.3 Å^2 for $v=15$ to 1.8 Å^2 for $v=25$ and to 1.34 Å^2 for $v=43$ (an intermediate value of the Franck–Condon factor).

We can conclude by repeating our initial remark: the amount of experimental data is still not sufficient and the further work is necessary. Moreover, in most actual studies carried out in the gas phase only the average cross-sections may be determined and this necessitates the averaging over the statistical distribution of velocities and of impact parameters in all theoretical treatments. From this point of view, crossbeam experiments, where the dependence of quenching cross-sections on the relative velocity of colliding particles (and even on the impact parameters, if the angular distribution of products is studies) may be determined, would be of highest interest.

D. Role of Strongly Mixed States (Problem of "Gates")

The collision-induced electronic relaxation is, in general, slow as compared to the rotational relaxation within a single vibronic state. This implies that even under selective excitation of one rotational level in the $\{s\}$ manifold the thermal equilibrium between rotational levels populations is rapidly attained. The electronic relaxation will thus involve a large number of initial $\{s, J\}$ and final $\{l, J'\}$ levels (excepted for the case of anomalously efficient electronic relaxation[99]). Determination of differential cross sections for all $|s, J\rangle \to |s', J'\rangle$ and $|s, J\rangle \to |l', J'\rangle$ transitions would necessitate a selective excitation of a single rovibronic level $|s, J\rangle$ and selective monitoring of populations of all final levels in the single-collision conditions.

In real experiments using broadband fluorescence detection, the overall population of the vibronic level belonging to the $\{s\}$ manifold is monitored; in order to deduce therefrom the values of individual relaxation rates for $|s, J\rangle$ levels, we are usually obliged to make some simplifying assumptions. One can either assume that the overall relaxation rate is a weighted average of transition rates from all rotational levels or suppose that transitions occur uniquely between a limited number of strongly mixed levels playing the role of "gates" between $\{s\}$ and $\{l\}$ manifolds. The choice of the suitable approximation depends obviously on the character of the interstate coupling in the given molecular (or atomic) system.

As already mentioned (Section I.A), in the helium atom a significant singlet–triplet mixing occurs only for the quasidegenerate n^1F and n^3F states, which act effectively as "gates" between two manifolds of (practically) pure spin states.[8,12] A similar behavior would be expected in molecules when the coupling constant v_{sl} is much smaller than the average spacing between $|s, J\rangle$ and $|l, J\rangle$ states. A significant mixing takes place only in exceptional cases of an accidental $|s, J\rangle - |l, J\rangle$ degeneracy. A good example of such behavior is the $A^1\Sigma$ state of the BaO molecule. In the chemiluminescence spectrum of BaO (produced in the $Ba + N_2O$ reaction), the $P(46)$ and $R(44)$ lines of the (1,1) and (1,2) bands in the $A^1\Sigma$–$X^1\Sigma$ transition show anomalously high intensities under low perturbing-gas pressures ($P_{Ar} = 0.2$ torr). The radiative $A^1\Sigma$ state is populated by collision-induced transitions from the nonradiative $a^3\pi$ state. The $A^1\Sigma$–$a^3\pi$ coupling is extremely weak and a significant perturbation occurs only for the $A^1\Sigma(v=1, J=45)$ level. The anomalous intensity of emission is obviously due to its selective population in collision-induced transitions. As may be easily shown, such an effect is expected only when the transition rate $|\sigma, J\rangle \rightleftharpoons |\lambda, J\rangle$ between "gate" levels is of the same order of magnitude as the rotational relaxation rates within $\{s\}$ and $\{l\}$ manifolds and when the transition probabilities to other levels are vanishingly small.[100,101]

In the molecules with a stronger s–l coupling, the spectral perturbations are not so "localized" and mixing coefficients are nonnegligible for a large number of rotational levels. The transitions would thus occur between many $|\sigma\rangle$ and $|\lambda\rangle$ levels. In the case of CO narrow-band excitation to the

$A^1\pi$ rotational levels, the collisionally induced $d^3\Delta$–$a^3\pi$ emission is composed of a large number of rotational lines, even in the low-pressure limit. It suggests a nonselective population of $\{l, J\}$ states in a good agreement with a relatively strong $A^1\pi$–$d^3\Delta$ coupling ($v_{sl} \cong 10\ \text{cm}^{-1}$). On the other hand, a good correlation between cross sections for collisional depopulation of a given vibronic level –v of the A state and average values of mixing coefficents for this level –β_v^2 suggests that transitions take place from (almost) all initially populated levels.

In polyatomic molecules belonging to the weak-coupling case (glyoxal, propynal) the density of rovibronic triplet levels $\{l, J'\}$, which may be collisionally coupled to the initially excited $|s, J\rangle$ state, is already so high that strong deviations from the average σ_{isc} value cannot be expected. As a matter of fact, the phosphorescence excitation spectra of glyoxal recorded under low gas pressure do not show any features corresponding to an anomalously high (or low) transition probabilities from individual rotational levels of the ^1A_u state to the ^3A_u state.[35]

E. Dependence of Collisional Relaxation Rates on the Intermolecular Potential

Systematic measurements of the fluorescence quenching rates as a function of the collision partner have been carried out for a number of diatomic and small polyatomic molecules: glyoxal,[35] propynal,[36] I_2,[95–98] SO_2[6] and OH.[102] In the first two cases, the quenching is due to the collision-induced intersystem crossing, in the third one, to the collision-assisted transition into a dissociative continuum. In the case of SO_2, both: $S_1 \rightsquigarrow T_1$ and $S_1 \rightsquigarrow S_0$ processes seem to be important, while for OH the final state has not been identified.

In all cases, a reasonable agreement has been found between relaxation rates and the strength of attractive van der Waals interactions described by the semiempirical formulas in terms of ionization potentials, polarizibilities and dipole moments of colliding molecules.[36,103]

More recently, Parmenter et al.[44,104] proposed the correlation between collisional cross-sections (for a wide variety of relaxation processes) in $A^* + M$ collisions and the potential energy well ϵ_{A^*M}. The form of this dependence is given by:

$$\sigma_M = C \exp\left[\frac{\epsilon_{A^*M}}{kT}\right] \tag{27}$$

If supposed that ϵ_{A^*M} may be approximated by a geometric mean between

$\epsilon_{A^*A^*}$ and ϵ_{MM} (known for a number of small molecules) one obtains:

$$\sigma_M = C \exp\left[\delta\left(\frac{\epsilon_{MM}}{k}\right)^{1/2}\right] \tag{28}$$

where $\delta = (\epsilon_{A^*A^*}/kT^2)^{1/2}$ is constant for a given excited species.

This formula may be directly checked, and excellent correlations are found for a large number of literature data as long as the rotational and electronic relaxation is considered.[44] The model fails for the vibrational relaxation of small molecules, this failure being obviously due to an important role of repulsive interactions in the case of "strong" collisions necessary for the induction of vibrational transitions.

The important conclusion from the Parmenter work is a close relation between rotational and electronic relaxation, both of them depending mainly on attractive part of the intermolecular potential. This may be clearly seen in the case of glyoxal, where the cross-sections for the collision-induced intersystem crossing[35] and for the rotational relaxation[105] are correlated. We consider this result as a strong argument in favor of the model discussed in Section II and based on the assumption that the electronic relaxation is induced by long-range interaction through a reversible rotational relaxation and dephasing processes.

In the further discussion[104] the authors assume that the parameter δ deduced from experimental results using (28) gives directly the depth of the potential well for A^*-A^* interactions $\epsilon_{A^*A^*}$. This statement is controversial: experimental data for I_2[95–98] show a strong dependence of δ on the vibrational level in the B^3O_u+ state.[104] The same is true for the $A^1\pi$ state of CO. This difference can be hardly explained by the dependence of the intermolecular forces on the vibrational quantum number.

From our point of view, δ depends not only on the $\epsilon_{A^*A^*}$ potential but also on the intramolecular parameters of the A^* excited species, relevant for a particular relaxation process. As discussed before [Section II.D.1; (19)] we expect more complex relations between rotational and electronic relaxation rates than a simple proportionality supposed in (28). As a matter of fact neither in glyoxal nor in iodine can both processes be described by the same set of parameters.

As previously we will insist on the necessity of further experimental studies. The most significant results would be obtained by a simultaneous determination of rotational, vibrational, and electronic relaxation rates from a well-defined excited states of model molecules as a function of the collision partner.

IV. NEW PROBLEMS AND PERSPECTIVES

In the last section we emphasized very often the scarcity of available experimental data and a need of more detailed studies in a direct line of previous works. It seems, nevertheless, that an essential progress may be attained only by the extension of the research field to new techniques and to new problems. In the following sections we attempt to outline—in a very subjective way—some of the most promising perspectives. Since only the first steps have been made and the number of reported experimental results is extremely limited, we will focus our attention on expected rather than at already accomplished works.

A. Electronic Relaxation in van der Waals Complexes

Recent developments of the supersonic-nozzle technique opened a new field: the study of dynamics of electronically excited van der Waals complexes.[106] This problem is closely related to that of collisions involving electronically excited molecules. The analogy clearly appears when the system composed of an excited molecule A* and of a perturber M is described in the reference of its center of mass. The electronic excitation of the free molecule (followed by a collision) corresponds to the electronic transition between ground- and excited-state dissociative continua (Fig. 5).

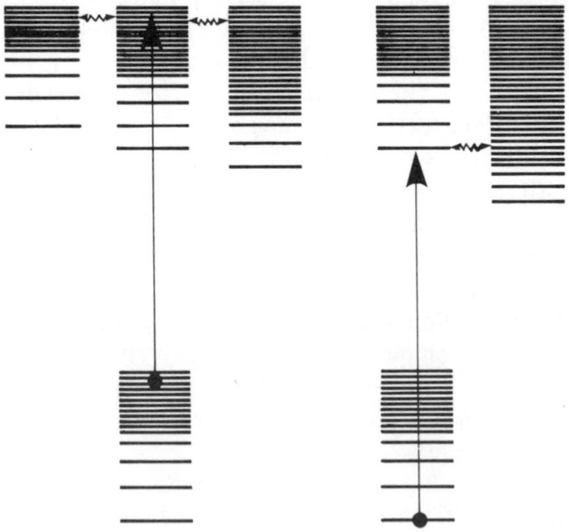

Fig. 5. Schematic representation of optical transitions and coupling patterns in the M··· A system. Excitation of the free A molecule (at left) and of the van der Waals complex (at right).

Collision-induced $|s\rangle \rightarrow |l\rangle$ transitions in the excited molecule M will result from the coupling between dissociative continua of the $|s\rangle$ and $|l\rangle$ states of the M ...A* complex. This coupling strength depends on the relative velocities of colliders, impact parameters. orientation of molecular axes, etc. The theoretical treatment of the problem necessitates thus an averaging over the statistical distribution of all relevant parameters.

The direct excitation of the complex formed in the ground electronic state prepares the same system in its bound state $|s\rangle$ isoenergetic with dissociative continua of lower $\{s'\}$ and $\{l\}$ states. We are dealing here with a much simpler problem of the coupling between a discrete state with a well-defined geometry and a limited number of continua. Experimental[107,108] and theoretical[109,110] studies of the I_2. He complex excited to higher vibronic levels of the $I_2B^3O_u+$ state showed a highly selective character of the coupling: the bound state of the I_2. He complex corresponding to v vibrational quanta of the I$\cdots$I vibration is strongly coupled to the dissociative continuum of the neighbor v-1 vibrational state and very weakly to all other states. The excited complex decays thus almost exclusively through the vibrational predissociation:

$$I_2(v)\cdot He \rightarrow I_2(v-1)+He$$

We may expect that the electronic predissociation of the excited complex will be also highly selective. If the initially excited $|s\rangle$ state is the vibrationless level of the $\{s\}$ electronic manifold, the coupling to lower lying vibronic levels of the $\{l\}$ manifold will induce the transition:

$$A^*(s)\cdot He \rightarrow A^*(l)+He$$

The identification of final (l) states populated in the dissociation process (relatively easy in absence of collisional relaxation in molecular beams) would be of higher interest for the understanding of coupling mechanisms.

On the other hand, if the initial excitation prepares one of higher vibronic states in the $\{s\}$ manifold, this state will be coupled to the lower $\{s'\}$ levels as well, as to the $\{l\}$ levels. The vibrational relaxation within the $\{s\}$ manifold will thus compete with the electronic $|s\rangle \rightarrow \{l\}$ relaxation process.

First significant results have been obtained for the $I_2\cdot$He complex in the $I_2B^3O_u+$ state. As mentioned above, all vibronic levels with $v \geqslant 1$ are vibrationally predissociated; the predissociation is evidenced by the emission of the I_2 $(v-1)$ fluorescence under a selective excitation of the $I_2\cdot$He(v) state. However, no fluorescence could be detected when the vibrationless level of the complex (vibrationally nonpredissociated) was excited. This effect has been tentatively explained by the electronic predissociation (slow as compared to the vibrational predissociation of higher levels, but much more rapid than the radiative decay rate) transferring the molecule to the dissociative A state[111]

$$I^*_2(B)\cdot He \rightarrow I^*_2(A) + He \rightarrow I + I + He$$

The analogy with the collision-assisted predissociation of I_2 B state is in fact very close.

This result can be probably generalized for a large class of molecules. It is interesting to note that in the fluorescence-excitation spectra of helium supersonic jets containing "large" (statistical-limit) molecules the lines of van der Waals complexes appear with a high intensity,[112] while these bands are absent or very weak in the case of "small" or "intermediate-case" molecules such as chromyl chloride,[113] biacetyl,[114,115] and glyoxal.[116] All molecules belonging to the latter group exhibit a very efficient fluorescence quenching in collisions, one can thus suppose that the absence of fluorescence from excited complexes is also due to the induction of electronic relaxation.

The case of glyoxal complexes with He and Ar has been studied in more detail.[117] In the fluorescence excitation spectrum, no lines corresponding to the $C_2H_2O_2\cdot X$ complexes could be detected in the vicinity of the ${}^1A_u - {}^1A_g 0_0^0$ transition, while they appear with a non-negligible intensity in transitions to higher vibronic levels of the ${}^1A_u$ state. In contrast to it, the phosphorescence excitation spectrum in the 0_0^0 spectral region is composed mainly of bands corresponding to $C_2H_2O_2\cdot X$ complexes. As could be expected, the 0^0 level of the complex lying below the 0^0 level of the free molecule yields on dissociation free molecules in high vibronic levels of the phosphorescent ${}^3A_u$ state but not in the fluorescent ${}^1A_u$ state. From higher vibronic levels of the complex ${}^1A_u$ state both deactivation channels are opened; the observed branching ratio between ${}^1A_u$ and ${}^3A_u$ products depends on relative probabilities of transitions to strongly coupled but sparse lower ${}^1A_u$ levels and to weakly coupled but dense ${}^3A_u$ ones.

B. Electronic Relaxation of Photofragmentation Products

The photodissociation of van der Waals complexes may be considered as prototype of more complex processes occurring during the photofragmentation of a polyatomic molecule: $ABC + h\nu \rightarrow (ABC)^{**} \rightarrow AB^* + C$. As in the previous case, we expect a close correlation between initially prepared state of the parent molecule and the energy distribution in AB* and C fragments. However, after the breakdown of the B—C, the fragments are still closely together and their interaction during the separation ("half-collision") process may induce the relaxation in fragments. Observed populations would be modified with respect to the initial ones.

This problem has been discussed in recent theoretical[118,119] and experimental[120] studies of photodissociation phenomena, but the experimental material is still very poor. We can give as an example the case of CH_3ONO photodissociation by the far-ultraviolet radiation ($\lambda = 1100 \div 1300$Å) yielding the NO fragment in the $A^2\Sigma^+(v=0\div3)$, $C^2\pi(v=0)$ and $D^2\Sigma^+(v=0)$ states. The relative population of the $A(v=3)$ level is anomalously high and, since the efficient $C(v=0)\rightarrow A(v=3)$ and $D(v=0)$ collisional relaxation is known, the anomaly may be tentatively explained by the electronic relaxation during the dissociative "half-collision."[121]

As in the case of van der Waals complexes, the relation between the dissociation process:

$$(AB^{**}--C)\rightarrow AB^{*}+C$$

and the collisional relaxation:

$$AB^{**}+C--\rightarrow AB^{*}+C$$

is very close. One can hope that with further developments in laser and crossed-beam techniques, a comparative study of both processes will be possible at least for simple molecular systems (e.g., HCN photodissociation and H+CN* collisions).

C. Magnetic-Field Effects

There exists, in the past few years an increasing interest in the influence of external (magnetic and electric) fields on the dynamics of excited molecular states. This interest is not surprising if we are reminded of the role played by this kind of studies in the development of the atomic physics. We will limit our discussion to the phenomena related to the collisional electronic relaxation; application of magnetic fields in the studies of predissociation[93] and of dephasing processes in isolated molecules[122] will not be treated here.

As far, as collisionally perturbed molecular systems are concerned one can distinguish two kinds of phenomena which seem not to be directly related:

1. Level-anticrossing effects taking place in relatively strong magnetic fields.
2. So-called "magnetic fluorescence quenching" observed in weak magnetic fields.

The mechanism of the level-anticrossing is relatively well understood, while the nature of the weak-field effect is still subject to controversions.

1. *Level-Anticrossing and Electronic Relaxation*

The level-crossing and anticrossing phenomena have been observed for the first time for hyperfine components of atomic energy levels.[123] The level-anticrossing spectroscopy was then applied to the study of very weak singlet–triplet (or doublet–quadruplet) coupling in atoms[11, 12] and in diatomic molecules (see, e.g., Refs. 124 and 125 and references therein).

We consider the case of $|s\rangle$ and $|l\rangle$ levels so weakly coupled that their mixing is vanishingly small ($v_{sl} \ll \epsilon_{sl}$). The effect of the magnetic field on the spin–orbit coupling (i.e., on the v_{sl} constant) can be usually neglected, the magnetic-field effect consists simply in the Zeeman shift of the $m_s \neq 0$ component of the paramagnetic (e.g., triplet) state, which may induce the crossing of the $|s\rangle$ and $|l\rangle$ levels. In the vicinity of the crossing point ($\epsilon_{sl}(H) \to 0$) both states are strongly mixed and give origin to a pair of $|\sigma\rangle$ and $|\lambda\rangle$ states sharing radiative and nonradiative widths of $|s\rangle$ and $|l\rangle$ levels. This implies the change of the fluorescence intensity from the $|s\rangle$ state populated by optical excitation or by electric discharge, when the magnetic field is scanned across the crossing point (optically detected level-anticrossing signal). The major part of the early work was carried out in collision-free conditions and the decay of excited species was essentially radiative. In this case, the v_{sl} coupling constants can be directly deduced from the widths of anticrossing signals. It has been, however, observed in course of this work[17, 18] that anticrossing signals are substantially modified (broadened in most cases) by collisions with inert partners (ground-state He atoms in the case of the n^1D–n^3D anticrossing in He). This influence may be explained by two closely related effects:

1. A rapid decay of coherence between initially prepared states due to pure dephasing and reversible relaxation between rotational levels and/or magnetic sublevels.
2. A nonlinear dependence of the transition probabilities between σ and λ states on their mixing coefficients α^2 and β^2.

The level-anticrossing techinque offers thus a unique possibility for detailed studies of electronic relaxation, because intramolecular parameters which determine the relaxation rate (the spacing of zero order $|s\rangle$ and $|l\rangle$ levels and of quasi-stationary $|\sigma\rangle$ and $|\lambda\rangle$ states, the mixing coefficients α^2 and β^2) vary in a well-defined way with the strength of the magnetic field.

This technique has been recently applied to glyoxal, a classical example of a polyatomic molecule corresponding to the weak-coupling limit. Under a selective excitation of single rotational levels of the vibrationless ^1A_u 0^0 state (gas pressure corresponding to a few collisions per lifetime), each anticrossing event induces a drop of the fluorescence intensity and an enhancement of phosphorescence. Because of the $\Delta J=0$ selection rule, the anticrossing spectra are entirely different for each rotational level. They contain a large amount of purely spectroscopic information concerning the density and distribution of $\{l, J\}$ triplet states which may couple to the $|s, J\rangle$ state, the spin-orbit and spin-rotation coupling constants, etc.[126] Still more important, from our point of view, is the possiblility to deduce from the signal shapes for different gas pressures the detailed form of the dependence of quenching cross-sections σ_{isc} on the level spacing and mixing coefficients.[127]

2. *Fluorescence Quenching in Weak Magnetic Fields*

This effect has been recently reported for several small molecules but detailed studies have been carried out only for glyoxal,[128–133] while for CS[134] and NO_2[135,136] a more general discussion seems to be still premature in view of the limited amount of data and of the extreme complexity of their spectra.

The characteristic features of the weak-field effect in glyoxal may be summarized in the following way:

1. In the collision-free conditions the fluorescence lifetime and yield are unchanged (initially reported data suggesting a shortening of the collision free lifetime are not confirmed in further works[130–133]). On the other hand, cross-sections for the collision-induced intersystem crossing are increased by about 40% for all collision partners,
2. The effect is practically independent of the excitation wavelength in a wide frequency range ($\Delta E_{\mathrm{vib}} = 0 \div 1000$ cm^{-1}).[132] No significant difference was found, either, when different rotational levels have been selectively pumped.[133]
3. For all exciting wavelengths, the effect is saturated in the magnetic field of the order of 800 G.

The contrast with previously discussed level-anticrossing effects is striking. Moreover, the triplet-level densities (of the order of 1/cm^{-1}) and singlet–triplet coupling constants ($v_{sl} \leqslant 10^{-2}$ cm^{-1}) deduced from level-anticrossing experiments[126] show that in the magnetic fields $H \leqslant 1$ kG (corresponding to Zeeman shifts of $\sim$0.1 cm^{-1}) a significant modification of mixing coefficients is highly improbable. These arguments seem to be strong enough to rule out the explanation of the weak-field effect by the level-anticrossing mechanism.[1,129]

Another possible explanation, proposed by Stannard[137] and by Küttner et al.[132] consists in the magnetic mixing of the triplet sublevels (T_x, T_y, T_z). Since in absence of rotation and of external fields, the singlet state is coupled to one of the triplet components,[138] their mixing will increase the density of final states and the efficiency of collisional transitions. It has been shown[133] that rotation (except for very high K values) cannot entirely mix T_x, T_y and T_z states, while this mixing is induced by the magnetic field and is complete (the field effect saturates) when the spin–rotation coupling is broken (the molecular analog of the Paschen–Bak effect). From direct observation of Zeeman effect in the ^3A_u–^1A_g absorption spectrum[133] corroborated by the value of the spin–rotation coupling constant deduced from level-anticrossing experiments[126] the magnetic-field strength necessary for induction of the Paschen–Bak effect would be of the order of 500 G.

If this interpretation of the weak-field effect is correct (its confirmation in the case of other molecules would be necessary), application of weak magnetic fields in the study of collisional electronic relaxation opens new, interesting possibilities: without affecting the structure of $\{s\}$ levels, the weak magnetic field modifies the character of the $\{l\}$ manifold.

V. FINAL CONCLUSIONS

The analysis of actually available experimental data shows that the simple model proposed in Section II accounts for the fundamental properties of the collision-induced electronic relaxation. Let us summarize some of the most important points:

1. The reversible character of electronic relaxation (detectable for obvious reasons only in the case of diatomic molecules) is now a well-established experimental fact.
2. The efficiency of collisions for induction of $|s\rangle \to |l\rangle$ transitions is closely related to the s–l coupling strength in the isolated molecule, even if the details of this relation are still not completely elucidated.
3. The electronic and rotational relaxation are closely related: in both cases the efficiency of collisions may be correlated to the attractive part of the intermolecular potential.

We emphasized the scarcity and dispersion of experimental data and a need in further studies extended to new molecular systems and applying modern techniques (highly selective optical excitation, crossed-beam experiments, external field effects).

We discussed also the relation between collisional relaxation and processes occuring in "half-collisions" (dissociation of molecules and of van der Waals complexes). We expect a rapid development of work in this area.

At last, it must be kept in mind that the electronic collisional relaxation represents a specific case of collisional processes involving electronically excited species such as electronic-energy transfer and reactive collisions.

Acknowledgments

One of the authors (A.T.) is highly indebted to Drs. J. A. Beswick, F. Lahmani, B. Soep, and R. Voltz for many exciting discussions and interesting suggestions, A. N. thanks Dr. Don Heller and Allied Chemical Corporation for their hospitality during part of the period when the manuscript

References

1. K. F. Freed, in *Potential Energy Surfaces*, K. Lawley, ed. Wiley, London, 1980.
2. P. Pringsheim, *Fluorescence and Phosphorescence*, Wiley Interscience, New York, 1949.
3. A. E. Douglas, *J. Chem. Phys.* **45**, 1007 (1966).
4. S. J. Strickler and D. B. Howell, *J. Chem. Phys*, **49**, 1947 (1968).
5. H. D. Mettee, *J. Chem. Phys.* **49**, 1784 (1968).
6. H. D. Mettee, *J. Phys. Chem.*, **73**, 1071 (1969).
7. L. E. Brus and J. R. McDonald, *J. Chem. Phys.*, **61**, 97 (1974).
8. For discussion see R. M. St. John, and R. G. Fowler, *Phys. Rev.*, **122**, 1816 (1961).
9. W. M. Gelbart and K. F. Freed, *Chem. Phys. Lett.*, **18**, 470 (1973).
10. K. F. Freed, *Chem. Phys. Lett.*, **37**, 47 (1976).
11. K. F. Freed, *J. Chem. Phys.*, **64**, 1604 (1976).
12. T. A. Miller, R. S. Freud, F. Tsai, T. J. Cook, and B. R. Zegarskei, *Phys. Rev.*, **9A**, 2474 (1974).
13. J. Derouard, R. Jost, M. Lombardi, T. A. Miller, and R. S. Freund, *Phys. Rev.*, **14A**, 1025 (1976).
14. see e.g. S. P. Mc Glynn, T. Azumi, and M. Kinoshita, *Molecular Spectroscopy of the Triplet State*, Princice-Hall, Englewood Cliffs, N.J. 1969.
15. R. Voltz (unpublished).
16. S. Mukamel, *Chem. Phys. Lett.*, **60**, 310 (1979).
17. J. Derouard, R. Jost, and M. Lombardi, *J. Phys. Lett.*, **37**, L135 (1976).
18. J. Derouard, 3rd cycle thesis, Grenoble, 1976.
19. A. Nitzan, J. Jortner, and P. Rentzepis, *Proc. Roy. Soc.* **A327**, 367 (1972).
20. A. Tramer and R. Voltz, in *Excited States* Vol. 4, E. C. Lim, ed., Academic Press, New York, 1979.
21. G. B. Kistiakovsky and C. S. Parmenter, *J. Chem. Phys.*, **42**, 2942 (1965).
22. B. Soep, C. Michel, A. Tramer, and L. Lindqvist, *Chem. Phys.*, **2**, 293 (1973).
23. H. Schröder, H. J. Neusser, and E. W. Schlag, *Chem. Phys. Lett.*, **54**, 4 (1978).
24. G. K. Busch, P. M. Rentzepis, and J. Jortner, *J. Chem. Phys.*, **56**, 361 (1972).
25. R. M. Hochstrasser and J. E. Wessel, *Chem. Phys. Lett.*, **19**, 156 (1973).
26. D. Zevenhuijzen and R. van der Werf, *Chem. Phys.*, **26**, 279 (1977).
27. L. E. Brus and J. R. Mc Donald, *J. Chem. Phys.*, **58**, 4223 (1973).
28. A. Nitzan, J. Jortner, J. Kommandeur, and E. Drent, *Chem. Phys. Lett.*, **9**, 273 (1971).
29. A. Frad, F. Lahmani, A. Tramer, and C. Tric, *J. Chem. Phys.*, **60**, 4419 (1974).
30. K. G. Spears and M. El-Manguch, *Chem. Phys.*, **24**, 65 (1977).

31. J. R. Mc Donald and L. E. Brus, *J. Chem. Phys.*,**61**, 3895 (1974).
32. B. Soep and A. Tramer, *Chem. Phys.*, **7**, 52 (1975).
33. R. van der Werf, E. Schutten, and J. Kommandeur, *Chem. Phys.*; **16**, 125 (1976).
34. L. G. Anderson, C. S. Parmenter, and H. Poland, *Chem. Phys.*, **1**, 401 (1973).
35. R. A. Beyer and W. C. Lineberger, *J. Chem. Phys.*, **62**, 4026 (1975).
36 C. A. Thayer and J. T. Yardley, *J. Chem. Phys.* **61**, 2487 (1974).
37. M. Lavollée and A. Tramer, *Chem. Phys. Lett.*, **47**, 523 (1977).
38. D. Grimbert, M. Lavollée, A. Nitzan, and A. Tramer, *Chem. Phys. Lett.*, **57**, 45 (1978).
39. R. F. Heidner, D. G. Sutton, and S. N. Suchard, *Chem. Phys. Lett.*, **37**, 243 (1976).
40. Y. Hirata and I. Tanaka, *Chem. Phys.*, **25**, 381 (1977).
41. T. Slanger and G. Black, *J. Chem. Phys.*, **63**, 969 (1975).
42. T. Oka, *Adv. Atom. Mol. Phys.*, **9**, 127 (1973).
43. J. P. Toennies, *Ann. Rev. Phys. Chem.*, **27**, 225 (1976).
44. H. M. Lin, M. Seaver, K. Y. Tang, A. E. W. Knight, and C. S. Parmenter, *J. Chem. Phys.*, **70**, 5442 (1979).
45. C. A. Thayer and J. T. Yardley, *J. Chem. Phys.*, **57**, 3992 (1972).
46. R. K. Lengel and D. R. Crosley, *J. Chem. Phys.*, **67**, 2086 (1977).
47. J. I. Steinfeld and W. Klemperer, *J. Chem. Phys.*, **42**, 3475 (1965).
48. H. P. Broida and T. Carrington, *J. Chem. Phys.*, **38**, 136 (1963).
49. B. F. Rordorf, A. E. W. Knight, and C. S. Parmenter, *Chem. Phys.*, **27**, 11 (1978).
50. H. M. Ten Brink, J. Langelaar, and R. P. H. Rettschnick, *Chem. Phys. Lett.*, **62**, 263 (1900).
51. C. S. Parmenter (private communication).
52. D. J. Miller and R. C. Millikan, *J. Chem. Phys.*, **53**, 3384 (1970).
53. D. L. Atkins, D. S. King, and J. C. Stephenson, *Chem. Phys. Lett.*, **65**, 257 (1979).
54. C. S. Parmenter and K. Y. Tang, *Chem. Phys.* **27**, 127 (1978).
55. R. N. Schwartz, Z. I. Slavsky, and K. F. Herzfeld, *J. Chem. Phys.*, **20**, 1591 (1952).
56. D. A. Chernoff and S. A. Rice, *J. Chem. Phys.*, **70**, 2521 (1979).
57. H. M. Ten Brink, Thesis, Amsterdam, 1979.
58. R. G. Brewer and S. S. Kano, in *Laser-Induced Processes in Molecules*, K. L. Kompa and S. D. Smith, eds., Springer, Berlin, 1979.
59. C. P. Slichter, *Principle of Magnetic Resonance*, Harper and Row, New York, 1963, Chap. 5.
60. C. Cohen-Tannoudji, *Frontiers in Laser Spectroscopy*, Les-Houches 1975, R. Balian, S. Haroche, and S. Liberman, eds. North-Holland, Amsterdam, 1977.
61. R. Zwanzig, *Lectures in Theoretical Physics III*, Wiley-Interscience, New York 1961, p. 106.
62. R. Zwanzig, *Physica*, **30**, 1109 (1964).
63. H. Mori, *Prog. Theor. Phys.*, **33**, 423 (1965).
64. G. W. Robinson and R. P. Frosch, *J. Chem. Phys.*, **37**, 1962 (1962); **38**, 1187 (1963).
65. A. Nitzan and J. Jortner, *Theor. Chim. Acta*, **29**, 97 (1973).
65a. J. Tusa, M. Sulkes and S. A. Rice, *J. Chem. Phys.*, **70**, 3136 (1979).
65b. J. Tusa, M. Sulkes and S. A. Rice, *J. Chem. Phys.*, (in press).
65c. T. D. Russell, B. M. Dekoven, J. A. Blazy and D. H. Levy, *J. Chem. Phys.*, **72**, 3001 (1980).
65d. C. Jouvet and B. Soep, *J. Chem. Phys.*, (in press).
65e. B. Soep (to be published).
66. C. Michel and A. Tramer, *Chem. Phys.*, **42**, 315 (1979).
67. R. F. W. Bader and J. I. Generosa, *Can. J. Chem.*, **43**, 1631 (1965).
68. T. W. Eder and R. W. Carr, *J. Chem. Phys.*, **53**, 2258 (1971).

69. F. Lahmani, *J. Phys. Chem.*, **80**, 2623 (1976).
70. M. Lavollée and A. Tramer, *Chem. Phys.*, **45**, 45 (1979).
71. N. Duric, T. A. Carlson, P. Erman, and M. Larsson, *Z. Physik.*, **A287**, 123 (1978).
72. N. Duric, P. Erman and M. Larsson, *Phys. Scripta*, **18**, 39 (1978).
73. T. A. Carlson, J. Copley, N. Duric, P. Erman, and M. Larsson, *Chem. Phys.* **42**, 81 (1979).
73a. D. H. Katayama, T. A. Miller, and V. E. Bondybey, *J. Chem. Phys.*, **71**, 1662 (1979).
74. C. K. Luk and R. Bersohn, *J. Chem. Phys.*, **58**, 2153 (1973).
75. V. E. Bondybey and L. E. Brus, *J. Chem. Phys.*, **63**, 794 (1975).
76. V. E. Bondybey, *J. Chem. Phys.* **66**, 995 (1977).
77. V. E. Bondybey and T. A. Miller, *J. Chem. Phys.*, **69**, 3597 (1978).
78. E. H. Fink and F. J. Comes, *Chem. Phys. Lett.*, **25**, 190 (1974).
79. M. Y. Chu and J. S. Dahler, *Mol. Phys.*, **27**, 1045 (1974).
80. K. C. Kuhlander and J. S. Dahler, *J. Phys. Chem.*, **80**, 2881 (1976).
81. K. C. Kuhlander and J. S. Dahler, *Chem. Phys. Lett.*, **41**, 125 (1976).
82. E. E. Nikitin, *Adv. Chem. Phys.*, **28**, 317 (1975).
83. K. F. Freed and C. Tric, *Chem. Phys.* **33**, 249 (1978).
84. D. Grimbert and A. Nitzan (unpublished).
85. A. E. W. Knight and C. S. Parmenter, *Chem. Phys.*, **15**, 85 (1976).
86. R. van der Werf, E. Schutten, and J. Kommandeur, *Chem. Phys.*, **16**, 151 (1976).
87. F. Lahmani, A. Tramer, and C. Tric, *J. Chem. Phys.*, **60**, 4431 (1974).
88. R. W. Field, B. G. Wick, J. D. Simmons, and S. G. Tilford, *J. Mol. Spect.*, **44**, 383 (1972).
89. R. W. Field, M. Lavollée, R. Lopez-Delgado, and A. Tramer (to be published).
90. A. J. Hynes and J. H. Brophy, *Chem. Phys. Lett.*, **63**, 93 (1979).
91. J. B. Halpern, G. Hancock, M. Lenzi, and K. H. Welge, *J. Chem. Phys.*, **63**, 4808 (1975).
92. J. Tellinghuisen, *J. Chem. Phys.*, **57**, 2397 (1972); **58**, 2821 (1973).
93. J. Vigué, M. Broyer and J. C. Lehmann, *J. Chem. Phys.*, **62**, 4941 (1975).
94. K. Sakurai, G. Capelle, and H. P. Broida, *J. Chem. Phys.*, **54**, 1220 (1971).
95. R. L. Brown and W. Klemperer, *J. Chem. Phys.*, **41**, 3072 (1964).
96. J. I. Steinfeld and W. Klemperer, *J. Chem. Phys.*, **42**, 3475 (1965).
97. R. B. Kurzel and J. I. Steinfeld, *J. Chem. Phys.*, **53**, 3293 (1970).
98. J. I. Steinfeld and A. N. Schweid, *J. Chem. Phys.*, **53**, 3304 (1970).
99. L. A. Melton and K. C. Yin, *J. Chem. Phys.*, **62**, 2860 (1974).
100. C. R. Jones and H. P. Broida, *J. Chem. Phys.*, **60**, 4369 (1974).
101. R. W. Field, C. R. Jones, and H. P. Broida, *J. Chem. Phys.*, **60**, 4377 (1974).
102. M. Kaneko, Y. Mosi, and I. Tanaka, *J. Chem. Phys.*, **48**, 4468 (1968).
103. J. E. Selwyn and J. I. Steinfeld, *Chem. Phys. Lett.*, **4**, 217 (1969).
104. C. S. Parmenter and M. Seaver, J. Chem. Phys. **70**, 5458 (1979).
105. D. W. Lindle, C. S. Parmenter, and B. F. Rordorf (to be published) cited in Ref. 44.
106. see e.g. D. H. Levy, L. Wharton, and R. E. Smalley, in *Chemical and Biochemical Applications of Lasers*, Vol. 2, C. B. Moore, ed., Academic Press, New York, 1977.
107. M. S. Kim, R. E. Smalley, L. Wharton, and D. H. Levy, *J. Chem. Phys.*, **65**, 1216 (1976).
108, K. E. Johnson, L. Wharton, and D. H. Levy, *J. Chem. Phys.*, **69**, 2719 (1978).
109. J. A. Beswick and J. Jortner, *J. Chem. Phys.*, **68**, 2279 (1978).
110. J. A. Beswick and J. Jortner, *J. Chem. Phys.*, **68**, 2525 (1978).
111. R. E. Smalley, L. Wharton, and D. H. Levy, *Chem. Phys. Lett.*, **51**, 392 (1977).

112. R. E. Smalley, L. Wharton, D. H. Levy, and D. W. Chandler, *J. Chem. Phys.*, **68**, 2487 (1978).
113. J. A. Blazy and D. H. Levy, *J. Chem. Phys.* **69**, 2901 (1978).
114. J. Chaiken, T. Benson, M. Gurnick, and J. D. Mc Donald, *Chem. Phys. Lett.*, **61**, 195 (1979).
115. R. Campargue and B. Soep, *Chem. Phys. Lett.*, **64**, 469 (1979).
116. B. Soep and A. Tramer, *Chem. Phys. Lett.*, **64**, 465 (1979).
117. C. Jouvet and B. Soep, *Chem. Phys.* (submitted).
118. S. Mukamel and J. Jortner, *J. Chem. Phy.*, **65**, 3735 (1976).
119. Y. B. Band and K. F. Freed, *J. Chem. Phys.*, **67**, 1462 (1977).
120. G. A. West and M. J. Berry, *Chem. Phys. Lett.*, **56**, 423 (1978).
121. F. Lahmani, C. Lardeux, and D. Solgadi, (to be published).
122. H. G. Weber, P. J. Brucat, and R. N. Zare, *Chem. Phys. Lett.*, **60**, 179 (1979).
123. T. G. Eck, L. L. Foldy, and H. Wieder, *Phys. Rev. Lett.*, **10**, 239 (1963).
124. T. A. Miller and R. S. Freund, *J. Mol. Spectr.* **63**, 193 (1976).
125. R. Jost and M. Lombardi, *J. Phys.*, **39**, **C1**. 26 (1978).
126. R. Jost, M. Lombardi, C. Michel, and A. Tramer, *Chem. Phys.* **46**, 273 (1980).
127. R. Jost, M. Lombardi, C. Michel, and A. Tramer, (work in course)
128. A. Matsuzaki and S. Nagakura, *Chem. Phys. Lett.*, **37**, 204 (1976).
129. A. Matsuzaki and S. Nagakura, *Helv. Chim. Acta* **61**, 675 (1978).
130. H. G. Küttner, H. L. Selzle, and E. W. Schlag, *Chem. Phys. Lett.*, **48**, 207 (1977).
131. H. G. Küttner, H. L. Selzle, and E. W. Schlag, *Chem. Phys.*, **28**, 1 (1978).
132. H. G. Küttner, H. L. Selzle, and E. W. Schlag, *Israel J. Chem.* **16**, 264 (1977).
133. C. Michel and C. Tric, *Chem. Phys.* (in press)
134. A. Matsuzaki and S. Nagakura, *Bull. Chem. Soc. Jap.*, **49**, 359 (1976).
135. S. Butler and D. H. Levy, *J. Chem. Phys.*, **62**, 815 (1975).
136. S. Butler, C. Kahler, and D. H. Levy, *J. Chem. Phys.*, **66**, 3538 (1977).
137. P. R. Stannard, *J. Chem. Phys.*, **68**, 3932 (1978).
138. I. Y. Chan and K. R. Walton, *Mol. Phys.*, **34**, 65 (1977).

ELECTRONIC TO VIBRATIONAL ENERGY TRANSFER FROM EXCITED HALOGEN ATOMS

PAUL L. HOUSTON*

Department of Chemistry, Cornell University, Ithaca, New York 14853

CONTENTS

*Alfred P. Sloan Fellow.

I. INTRODUCTION

Electronic to vibrational (E→V) energy transfer is an area of research that is currently receiving extensive investigation. From the practical point of view this field is important to an understanding of the chemistry of the upper atmosphere, to the development of new laser systems that employ either E→V or V→E pumping mechanisms, and to the prediction of the direction and efficiency of photochemical reactions. From the theoretical point of view electronic to vibrational energy transfer provides one of the simplest examples of a nonadiabatic collision process. As a consequence, this field forms a testing ground for the evaluation of theories that treat the nonadiabatic coupling between two or more electronic surfaces.

Several previous articles have reviewed the importance of E→V transfer in the quenching of electronically excited species.[1-11] While most earlier studies have inferred the role of vibrational modes from the correlation between the rate of quenching and the vibrational frequency of the quencher, more recent studies have focused on direct observation of the vibrational excitation imparted to the quenching molecule. Many of these direct experiments have been reviewed by Lemont and Flynn,[11] who covered the period from 1962 to 1976 and discussed experiments on a wide variety of electronically excited species. In this chapter I will concentrate on direct observations of electronic to vibrational energy transfer from the excited spin–orbit states of $Br(4^2P_{1/2})$ and $I(5^2P_{1/2})$. Experiments performed between 1970 and 1979 are summarized. Although no effort has been made to provide a comprehensive review of the theoretical literature, three of the numerous theoretical interpretations and their relation to the experimental data are discussed in Section V.

The spin–orbit properties of the stable halogen atoms are listed in Table I. Although the ${}^2P_{1/2}$–${}^2P_{3/2}$ energy differences for F and Cl are comparable to lower frequency vibrational separations in many molecules, no observation of E→V transfer from these atoms has yet been reported. For Br and I, on the other hand, the spin–orbit separations are large enough so that E→V transfer might populate more than one quantum, even for high-frequency vibrations. The shorter radiative lifetimes of $Br({}^2P_{1/2})$ and $I({}^2P_{1/2})$ also facilitate the observation of their transient concentrations by fluorescence detection, as discussed below.

TABLE I
Spin–Orbit Properties in Halogen Atoms

	$E(^2P_{1/2}-^2P_{3/2})$		Rad. lifetime
Atom	(cm^{-1})	(kcal/mole)	(s)
F	404	1.15	830.
Cl	881	2.52	83.
Br	3685	10.53	1.1
I	7603	21.72	0.13

Before embarking on our review of E→V transfer from the excited halogens, it is appropriate to ask what information it is that we wish to obtain. We consider the transfer of energy from an excited halogen atom $X^*(*={}^2P_{1/2})$ to an acceptor molecule A. Perhaps the most basic quantity of interest is the total rate constant k^E with which A quenches X*. For a molecule with only one vibrational mode this constant will be the sum of the individual rate constants leading to the population of A in the ith vibrational level; i.e., $k^E=\Sigma k_i$, where the summation runs from $i=0$ to $i=v_{max}$, and k_i is the rate constant for the microscopic process $X^*+A(v=0)\rightarrow X+A(v=i)$. The quantity v_{max} is the maximum vibrational level that may be populated by the transfer consistent with conservation of energy. Ideally, of course, we would like to determine each of the constants k_i and its dependence on temperature. Most often, however, we must settle for less. In practice it is usually possible to obtain k^E and its temperature dependence. While it is sometimes possible to obtain the individual values of k_i, it more frequently occurs that only the linear combination Σik_i is available from experiment. It should be noted that the quantity $\Sigma ik_i/\Sigma k_i =\Sigma ik_i/k^E$ is just the average number of vibrational quanta excited per deactivation of X* by A. We will denote this average by the letters Q.E. While this review will often incorporate experiments that provide measurements of k^E or its temperature dependence, it will focus only on those experiments that provide direct evidence that Q.E.$\neq$0.

Donovan, Husain, and Stephenson[12] appear to have been the first to observe vibrational excitation following electronic energy transfer from I* or Br*. In their experiment transfer from Br* to HBr was monitored by ultraviolet absorption following flash photolysis of HBr. Leone and Wodarczyk[13] subsequently introduced an improved experimental technique for examination of this process. Their technique is based on generation of the excited halogen by laser flash photolysis and observation of the time-dependent concentration changes by fluorescence detection of the spin–orbit transition in the halogen or the vibrational transition in the quenching molecule. This method, which has now provided direct measurements of E→V transfer from Br* and I* to a wide variety of quenching molecules, will be discussed in Section II. Section III will outline the relationship between the observed fluorescence signals and the desired rate constants. The experimental results will be summarized in Section IV and discussed in Section V.

II. EXPERIMENTAL TECHNIQUES

A. Rate Constant Determinations: The Apparatus

The experimental technique that has provided the vast majority of results on E→V transfer from excited halogens is that introduced in 1974 by Leone and Wodarczyk.[13] In brief, X* is produced by a photolysis source, and the concentrations of X* and of the vibrationally excited collision partner are monitored by observing their time-dependent infrared fluorescence. In our laboratory the apparatus has taken the form depicted in Fig. 1. It consists basically of a source and sample cell, a detector system, and an electronic system for signal enhancement and analysis.

A pulsed laser source is used to photodissociate a parent compound. The dissociation creates a transient concentration of X* in a time short compared with the energy-transfer rates of interest. Typical parent molecules, dissociation wavelengths, and laser sources are listed in Table II. Although

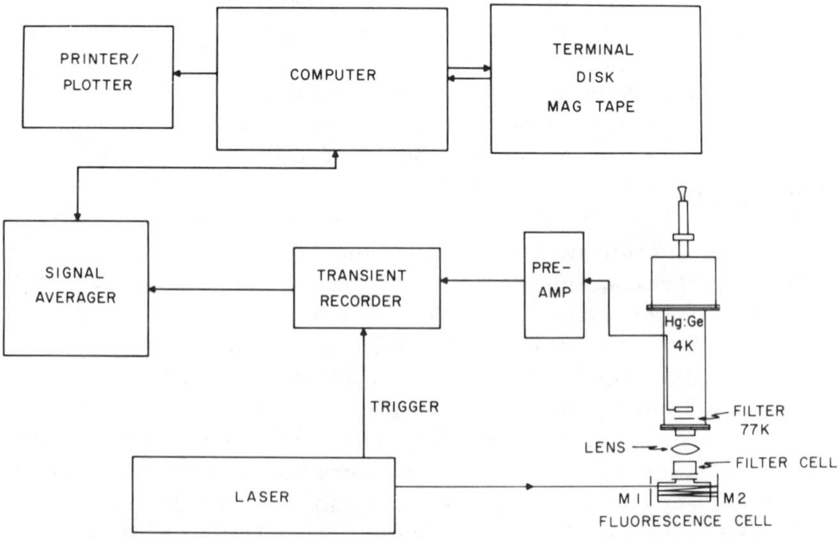

Fig. 1 Schematic diagram of apparatus for determining E→V rate constants. (From Ref. 22.)

TABLE II
Parent Molecules for Production of X*

X*	Parent molecule	Dissociation wavelength (nm)	Source	Ref.
Br*	Br_2	473	2×(Nd:YAG)	13,15,25
	Br_2	480–500	Dye	17,27,31,36, 38,40,41
I*	I_2	490	Dye	22,23,50
	CF_3I	266	4×(Nd:YAG)	42,43

the laser beam is directed by mirrors through the sample cell several times, only a small fraction of the parent molecule is dissociated per pulse under normal operating conditions. In addition to containing the parent compound, the fluorescence cell also contains a fixed pressure of buffer gas such as argon and a variable pressure of the quenching gas. The buffer gas is included to ensure that the X* translational temperature remains close to the temperature of the cell wall. Its partial pressure should be high enough so that the X* atom undergoes several collisions with the buffer before transferring its energy to the quenching gas. However, an exceedingly high buffer pressure or an inappropriate choice of buffer might lead to excessive quenching of X* by the buffer itself. Fortunately, the rare gases are very inefficient at relaxing either I* or Br*, so that the inclusion of even several hundred torr of these species causes only a small amount of deactivation while ensuring thermal equilibrium. The constraints are more serious on the variable pressure of the quenching gas. This pressure must be large enough so as to greatly exceed the transient pressure of X*. Otherwise, deviations from pseudo-first-order kinetics would complicate the kinetic analysis. On the other hand, the pressure of the quenching gas must be low enough so that E→V transfer is slow compared to either the laser pulse duration or the characteristic time response of the detection system. A further constraint is that the pressure of the quenching molecules should not be so high that the vibrational fluorescence is trapped by the unexcited molecules near the walls of the cell. In practice, the pressure of quenching gas can usually be varied over about an order of magnitude. Pressures are measured with a capacitance manometer.

The detection system typically consists of a lens to focus the fluorescence, an interference filter to isolate the transition of interest, and an infrared detector element cooled in a dewar. A filter cell filled with a quenching gas is sometimes interposed between the fluorescence cell and the detector to selectively block the A(1)→A(0) fluorescence while passing, for example, the A(2)→A(1) fluorescence. Although any one of several different detector elements can be employed, there is a great advantage in choosing a detector element whose noise equivalent power is limited only by the thermal background. For such a detector a substantial improvement in signal-to-noise can be achieved by cooling the interference filter and by limiting the detector's field of view to the region of the sample cell. The details of this enhancement are provided in any standard text on infrared detection.[14] For infrared detectors, as for any electrical transducer, there is a tradeoff between the signal-to-noise ratio and the time response. Typical time responses for the detectors used to measure E→V transfer are in the range of 0.2–1.0 μs. However, the time response of the detection electronics is often chosen to be longer to minimize the observed noise on more slowly varying fluorescence signals. For some measurements to be described below it is necessary to determine the wavelength response of the detector. A practical method for obtaining this response curve has been presented elsewhere.[15,16]

The signal recovery system consists of a preamplifier, a transient recorder, a signal averager, and a computer. Following each pulse of the laser the transient recorder digitizes the waveform received from the detector. The numerical data representing the signal are then added to the memory of the signal averager. The recorder is subsequently armed to accept the information from the next laser pulse, and the cycle is repeated. Typical averaging systems of this type can run at repetition rates up to 50 Hz while providing a 2000-point representation of the waveform. After a sufficient number of repetitions of the experiment has been performed so that the accumulated signal has an acceptable signal-to-noise ratio, the averaged signal is sent to a computer for analysis, storage, and plotting.

The signal-to-noise ratio obtained by this technique is ultimately proportional to the product of the number of X* molecules in the field of view of the detector and the Einstein A coefficient of the emitter. In our apparatus signal-to-noise ratios exceeding 30 can be obtained when this product is on the order of 2×10^{14} emitter s^{-1}.

The relationship between the fluorescence signals and the rate constants of interest is the subject of Section III.

B. Stimulated Emission Following E→V Transfer

Although the apparatus described in the previous section is capable of providing accurate determinations of the extent and rate of vibrational excitation, the selective nature of E→V transfer is more easily investigated by observation of stimulated emission between vibrational levels of the acceptor molecule. Peterson, Wittig, and Leone were the first to observe such emission following E→V transfer.[17] The apparatus they employed consisted of a laser tube, an optical cavity, a monochromator, and a detector. Several improvements[18-20] on the original design have led to the system described below.

The E→V transfer takes place in a laser device similar to that illustrated in Fig. 2. The device consists of three concentric tubes. A gas mixture containing Br_2, the acceptor molecule, and perhaps some inert gas is contained in the center tube. This tube is surrounded by two annular sections which contain, in succession, a flowing dye solution and a xenon flashlamp. When the flashlamp is fired, the dye solution absorbs the light and emits strong fluorescence in the blue–green region. This blue–green light dissociates the Br_2 in the gas mixture to yield Br* + Br. Finally, E→V transfer from Br* to the acceptor molecule selectively populates the upper laser level.

A block diagram of the entire apparatus is shown in Fig. 3. The stimulated emission is extracted by an optical cavity formed from a grating and a partially transmitting mirror. A monochromator placed between the laser output mirror and the detector is used to identify the lasing transition.

In addition to its usefulness as a laser device, the apparatus described above offers one major advantage to the investigator of E→V transfer. The emission signal is strong enough so that its exact wavelength can be resolved with the monochromator. Consider, for example, the case of E→V transfer from Br* to CO_2. Although the apparatus of Section IIA could readily detect the *spontaneous* emission of 4.3 μm radiation following E→V transfer, it is unlikely that the signal would be strong enough to distinguish spectrally between the possible fluorescing levels. The (10°1), (02°1), and (00°1) levels might all be possibilities. On the other hand, detection of the

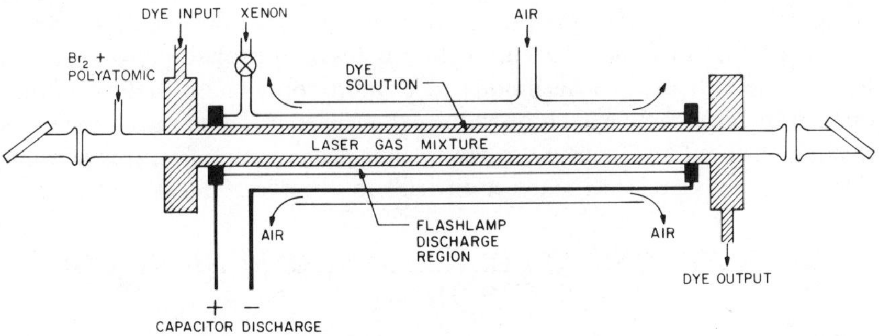

Fig. 2. Cross-sectional view of flashlamp and photolysis tube for production of stimulated emission following E→V transfer. (From Ref. 20.)

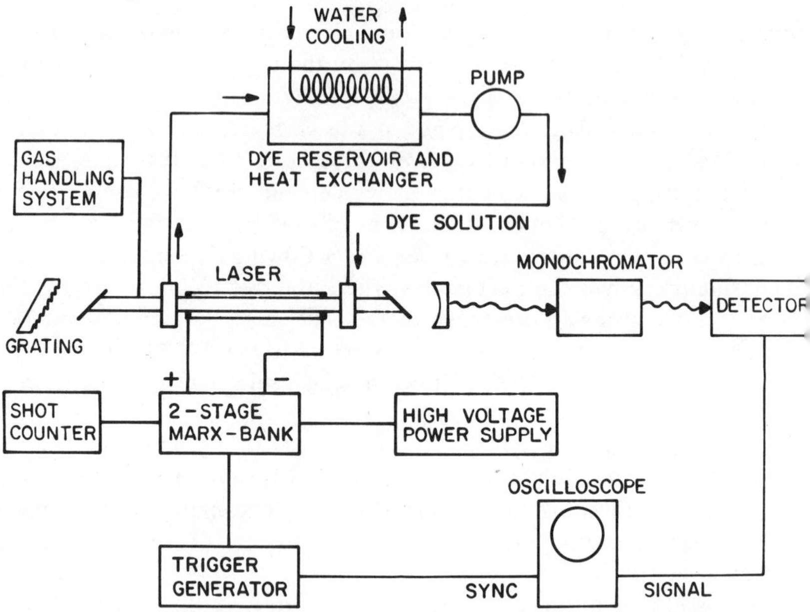

Fig. 3. Block diagram of experimental apparatus for identifying stimulated emission following E→V transfer. (From Ref. 20.)

stronger *stimulated* emission in the apparatus of Fig. 3 showed conclusively that the major emission was from the $CO_2(10^{\circ}1)$ level.[20] A more detailed discussion of this example is given in Section IV.A.7. The stimulated emission technique is most useful in determining the states which are selectively populated by E→V transfer.

Unfortunately, there is one major limitation to this technique. For lasing to occur, the rate of vibrational pumping by the E→V process must compete effectively with the rate of vibrational redistribution. Since vibrational redistribution often occurs with a collisional probability of $\sim 10^{-2}$, it is possible to use this technique to examine only those molecules that quench the electronic excitation with high efficiency. Since most molecules quench I* about ten times more slowly than they quench Br*, it is perhaps this limitation which has made it difficult to observe I* E→V lasers.

III. RATE CONSTANT DETERMINATIONS: THE KINETIC SCHEME

Although it is impractical to present and solve a kinetic scheme for the most general case of E→V energy transfer, the simple scheme presented below will illustrate the connection between the rate constants of interest and the form of the fluorescence signal observed in the apparatus of Section II.A. We consider energy transfer from a halogen X* to a diatomic molecule A whose vibrational mode may be populated only up to $v=1$ by the transfer. The important steps to our understanding of the E→V process are

$$X^* + A \xrightarrow{k_1} X + A(1) \tag{1}$$

and

$$X^* + A \xrightarrow{k_0} X + A(0) \tag{2}$$

The object of our study is to learn the total rate of X* deactivation by A, namely, (k_0+k_1), as well as the individual rate constants k_0 and k_1. However, other deactivation mechanisms will also be present. For example, the excited X* atom will be quenched by its molecular precursor, typically X_2:

$$X^* + X_2 \xrightarrow{k_{X_2}} X + X_2 \tag{3}$$

There will also be V→R, T mechanisms by which A will lose its vibrational excitation. We denote such loss mechanisms by the equation

$$A(1) + M \xrightarrow{k_V} A + M \tag{4}$$

where M may be any collision partner that causes energy to disappear from the observed vibrational level. It often occurs that the most efficient collision partner in process (4) is A itself. We have assumed in this scheme that radiative decay and diffusion of excited molecules to the wall are slow compared to the collisional deactivation processes.

The differential equations corresponding to processes (1)–(4) may be solved to yield a general solution for the concentrations of X* and A(1) subject to the initial conditions $X^* = X_0^*$ and $A(1)=0$. The solution is

$$X^* + X_0^* \exp\{-[(k_1+k_0)P_A + k_{X_2}P_{X_2}]t\} \tag{5}$$

$$A(1) = X_0^* \frac{k_1 P_A}{[(k_1+k_0)P_A + k_{X_2}P_{X_2} - k_V P_A]} \times \left(\exp\{-[k_V P_A]t\} - \exp\{-[(k_1+k_0)P_A + k_{X_2}P_{X_2}]t\}\right) \tag{6}$$

It is easily seen from (5) that the quantity $[(k_1+k_0)P_A+k_{X_2}P_{X_2}]$ is the decay rate of X* emission. The decay rate of A(1), on the other hand, depends on whether $k_V P_A$ is greater than or smaller than the decay rate of X*. We consider these two cases separately.

Case I: $[(k_1+k_0)P_A+k_{X_2}P_{X_2}]>k_V P_A$. For this case we find that the preexponential constant in (6) is positive and that the A(1) fluorescence signal rises with the rate of the X* decay, namely $[(k_1+k_0)P_A+k_{X_2}P_{X_2}]$. Consequently, a plot of either the decay rate of X* fluorescence or the rise rate of A(1) fluorescence vs. pressure of A gives a line whose slope is (k_1+k_0). The decay of the A(1) signal takes place at the rate $k_V P_A$, so that a plot of this rate vs. P_A gives a line whose slope is k_V. Finally k_{X_2} can be obtained, for example, from a plot of the X* fluorescence decay rate vs. P_{X_2}.

Case II: $k_V P_A>[(k_1+k_0)P_A+k_{X_2}P_{X_2}]$. For this case we find that the preexponential constant in (6) is negative and that the identities of the rise and decay rates for the A(1) fluorescence are reversed. The decay of the A(1) fluorescence is now equal to the decay of the X* fluorescence, while the rise of the A(1) fluorescence is equal to $k_V P_A$. The constant k_{X_2} may still be obtained from the decay of X* fluorescence as a function of P_{X_2}.

For either of the above cases the sum (k_1+k_0) may be obtained from each of two sources. This redundancy offers a valuable check as to whether the assumptions of the kinetic scheme are valid. However, it is not true that the time dependence of either the A(1) or X* fluorescence gives a separate value for k_0 or k_1. This information must be obtained from the ratio of the two signal amplitudes.

The value of k_1 is obtained by taking the ratio of the amplitude of the A(1) fluorescence to the amplitude of the X* fluorescence. As seen from (5) and (6), this ratio is simply the absolute value of $\{k_1 P_A/[(k_1+k_0)P_A+k_{X_2}P_{X_2}-k_V P_A]\}$. Since the sum (k_1+k_0) and the rate constants k_{X_2} and k_V are known from the kinetic measurements, the measured amplitude ratio may be used to determine k_1. For our purposes it will be most convenient to express the results in terms of the average number of quanta excited per X* deactivation by A. In general this quantity is Q.E.$=\Sigma i k_i/\Sigma k_i$, where k_i is the rate constant that populates A(i). For the case at hand the average number of quanta excited is simply $k_1/(k_0+k_1)$.

It is advisable to point out the experimental pitfalls in the determination of k_1 by the above method. Before it can be set equal to the ratio of preexponential factors in (5) and (6), the measured amplitude must first be corrected for (a) any variation in detector sensitivity to A(1) vs X* fluorescence, (b) any difference in transmission for the infrared interference filters used to isolate the A(1) vs X* fluorescence, (c) the difference in radiative lifetimes of the emitters, and (d) possible self-absorption of the A(1) fluorescence. The problems listed in (a) and (b) above can be overcome by measuring the detectivity and the filter transmission, respectively. The correction for radiative lifetimes is more troublesome. In particular, it is extremely difficult to measure accurately the radiative lifetime of either I* or Br*. The calculated values of Garstang[21] have been used in most reports. Unfortunately, it sometimes occurs that there is also a conflict over the radiative lifetime of the acceptor molecule, A. Thus, while measurements of the amplitude ratio may be accurate to a few percent, it is unlikely that any measurement of Q.E. based on the A(1)/X* fluorescence ratio is accurate to better than $\pm 20\%$. The final problem listed in (d) above is of most concern for strong absorbers such as CO_2 or HF. Although a relatively simple method can be used to correct for self-absorption in the cases of weaker absorbers,[22] this effect introduces a further uncertainty into most measurements of Q.E.

Generalization of this simple kinetic scheme to the situation arising when E→V transfer from X* can populate levels up to A(2) is straightforward. Detailed solutions for examples corresponding to Cases I and II have been presented in Ref. 22 and 23, respectively. An additional complication is that near-resonant V→V energy transfer of the type

$$A(2)+A(0) \overset{k_{VV}}{\rightarrow} 2A(1) \tag{7}$$

often takes place much more rapidly than E→V transfer or V→V,T deactivation of A(1). In the absence of any vibrational deactivation, the fraction of A(2) product molecules would be simply $k_2/(k_0+k_1+k_2)$, while the fraction of A(1) product molecules would be $k_1/(k_0+k_1+k_2)$. If process (7) is rapid compared to E→V transfer and deactivation of A(1), then the fraction of A(1) molecules becomes $(k_1+2k_2)/(k_0+k_1+k_2)$. The factor of 2 in the numerator is due to the fact that deactivation of A(2) by process (7) yields 2 molecules A(1). Measurements of the A(1) fluorescence then proceed as in the simple case above. Cases I and II still apply, except that one obtains the E→V rate constant equal to $(k_0+k_1+k_2)$. The ratio of the fluorescence amplitudes now gives the quantity k_1+2k_2, which is just the numerator of Q.E.$=(k_1+2k_2)/(k_0+k_1+k_2)$.

In the case where $v=2$ is the highest vibrational level populated by the E→V transfer, an alternate technique may be used to solve for k_2 directly. This technique involves blocking the A(1)→A(0) fluorescence with a filter cell containing a high pressure of A. The signal with gas in the filter cell is compared with the I* signal to obtain the ratio $k_2P_A/[(k_0+k_1+k_2)P_A+k_{X_2}P_{X_2}-k_{VV}P_A]$. The constant k_{VV} in the denominator can be obtained from the time dependence of the A(2)→A(1) fluorescence, so that the amplitude ratio can be used to determine k_2. The details of this technique are reported elsewhere.[23]

Generalization of the kinetic scheme to cases where E→V transfer from X* to A populates vibrational levels higher than $v=2$ becomes increasingly complicated. Under certain conditions, however, the interpretation of the fluorescence signals remains relatively simple, and the amplitude ratio of A(1) to X* fluorescence still yields the numerator of Q.E. $=\Sigma ik_i/\Sigma k_i$. The conditions under which these simple relations hold are outlined elsewhere.[16,22]

Finally, it is important to note that in polyatomic molecules, where more than one mode is accessible, it is theoretically possible to measure a value of Q.E. for each mode. We will still define the numerator of Q.E. as Σik_i where the index runs over all levels of the particular mode of interest. The denominator, however, will be the total deactivation rate k^E which is the sum of individual rate constants to all possible levels of all possible modes. With these generalizations Q.E. for a particular mode still corresponds to the number of quanta in that mode excited per deactivation of X*.

We close this section with an experimental example[22] that illustrates how the various rate constants of interest are extracted from the data. Figure 4 displays an I* fluorescence signal which might have been obtained from a mixture of argon, I_2, and NO. The dots represent the digitized data, while the solid line is a nonlinear least squares fit of a single exponential decay to the data. When the reciprocal time constant for this decay is plotted against the NO pressure, one obtains the upper straight line displayed in Fig. 5. (The other lines in this plot are obtained for I* deactivation by HBr and HCl.) By comparison to (5) and after generalization to the case of NO where $v_{max}=4$, it is found that the slope of the τ^{-1} vs. P_{NO} plot is simply $(k_0+k_1+k_2+k_3+k_4)=k^E$, the total deactivation rate of I* by NO. The numerical value of k^E in this case is $(3.88\pm 0.32)\times 10^3$ s^{-1}torr^{-1}.

Figure 6 displays the fluorescence obtained when the interference filter is changed to allow observation of NO vibrational fluorescence near 5.3 μm. The dots are again the digitized data, while the solid line is a fit to the data of a function of the form of (6). Vibrational self-relaxation of NO is known[24] to take place at a rate slower than the rate of I* deactivation, so that Case I of the kinetic scheme is expected to hold. This expectation can be confirmed by measuring the rise rate of the NO fluorescence as a function of NO pressure. Equations analogous to (5) and (6) predict for

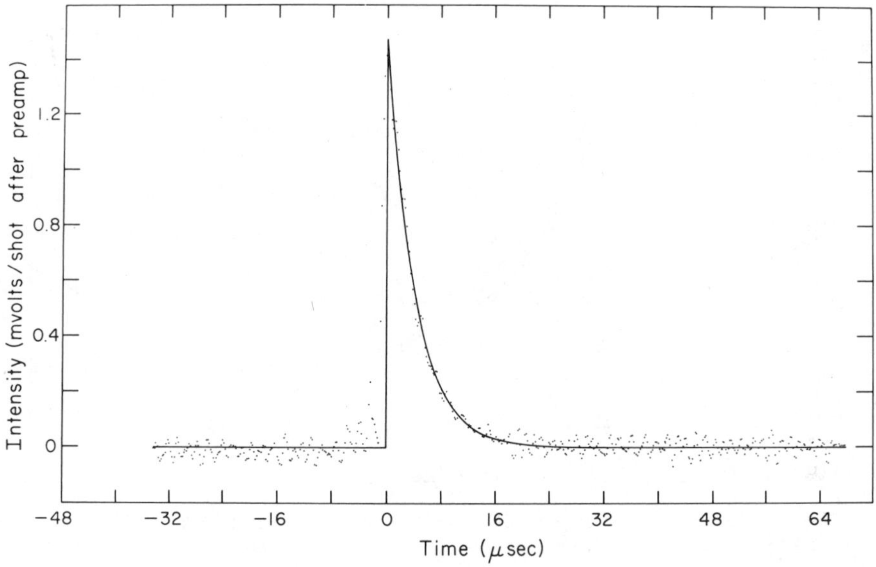

Fig. 4. Typical I* fluorescence signal obtained at 1.3 μm. The dots represent the digitized data, while the solid line is a fit to the data of a single exponential decay.

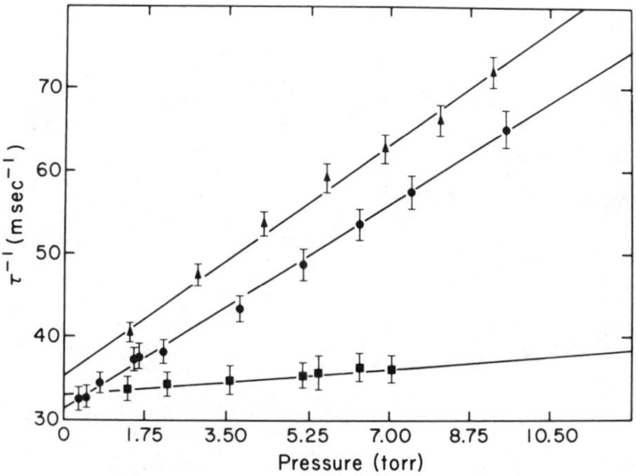

Fig. 5. Variation of τ^{-1} for I* fluorescence decay vs pressure of HCl (squares), HBr (circles), and NO (triangles). (From Ref. 22.)

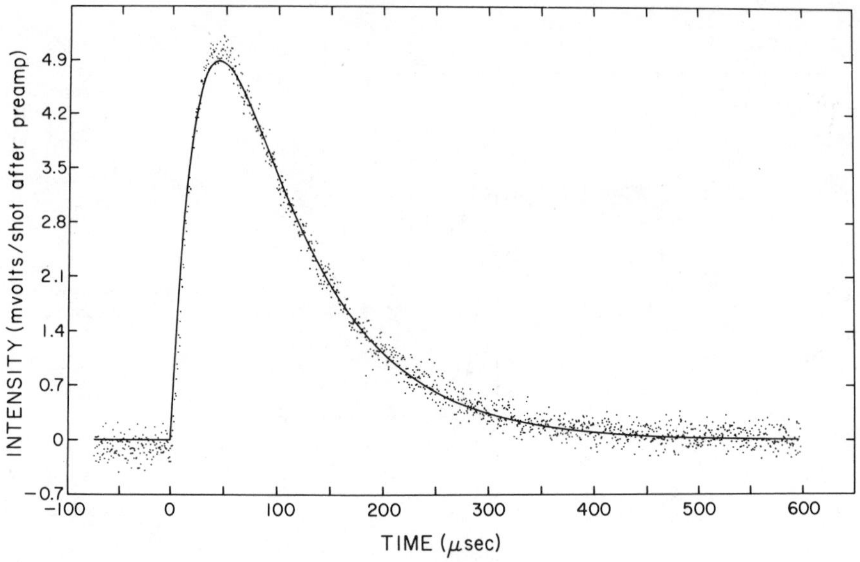

Fig. 6. NO fluorescence signal (dots) following laser photolysis of a mixture of 0.030 torr of I_2 and 4.31 torr of NO. The signal is the average of 4096 laser shots. (From Ref. 22.)

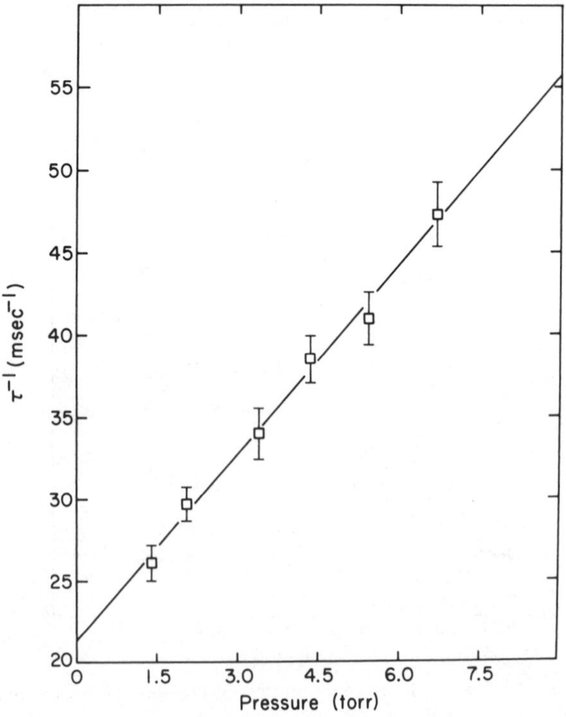

Fig. 7. Variation of τ^{-1} for rise of NO vibrational fluorescence with pressure of NO. (From Ref. 22.)

Case I that the NO fluorescence should rise with the same rate constant observed for the I* decay. Figure 7 displays the reciprocal rise time for NO fluorescence as a function of NO pressure. From the slope of the linear plot one obtains a rate constant of $(3.84 \pm 0.34) \times 10^3$ $sec^{-1} torr^{-1}$, in excellent agreement with the result obtained for the I* fluorescence decay.

Having determined $k^E = \Sigma k_i$, the remaining task is to try to learn the extent of vibrational excitation in the NO. When the NO/I* fluorescence amplitude ratios are (1) measured as a function of NO pressure; (2) corrected for detector response, filter transmission, and radiative lifetime; and (3) extrapolated to zero NO pressure to correct for any NO self-absorption, a ratio of 3.4 ± 0.7 is obtained. Generalization of the kinetic scheme to the case where $v_{max} = 4$ shows that this fluorescence ratio should be equal to $\Sigma i k_i / \Sigma k_i$, the average number of NO quanta excited per deactivation of I*. Thus, we find that Q.E. (NO) $= 3.4 \pm 0.7$.

IV. EXPERIMENTAL RESULTS

A. E→V Transfer from $Br(^2P_{1/2})$

1. *Hydrogen Fluoride*

Wodarczyk and Sackett[25] have measured the room temperature quenching constant for HF deactivation of Br* by observing both Br* and HF fluorescence in an apparatus similar to that shown in Fig. 1. Electronic deactivation proceeds rapidly enough so that Case I of the kinetic scheme was observed. A comparison of the HF and Br* fluorescence intensities showed that nearly every deactivating collision resulted in the population of HF($v = 1$). The value of the deactivation rate constant was found to be $(1.1 \pm 0.2) \times 10^6$ $s^{-1} torr^{-1}$.

Quigley and Wolga[26] observed the reverse process, HF(1) + Br→HF(0) + Br*, by exciting HF in a flowing mixture of HF, Br, Br_2, and argon. The observed double exponential decay for the HF fluorescence was thought to be caused by the fast equilibrium HF(1) + Br⇄HR(0) + Br* followed by the slower vibrational deactivation HF(1) + M→HF(0) + M. A value of $(1.0 \pm 0.5) \times 10^6$ $s^{-1} torr^{-1}$ was obtained for the V→E process. Although this value is roughly a factor of 2 different from that calculated by application of microscopic reversibility to the E→V process,[25] the discrepancy is admissable by the error limits for these two different experimental techniques.

2. *Hydrogen Chloride*

Vibrational excitation of HCl following its deactivation of Br* was first observed by Leone and Wodarczyk,[13] who measured a total deactivation rate of 8.6×10^{-12} cm^3 molec^{-1} s^{-1}. By comparing the fluorescence intensities of Br* and HCl, these authors determined that 96% of the quenching collisions yielded HCl($v=1$).

More recently, Reisler and Wittig[27] have examined the temperature dependence of the quenching cross-section. In the temperature range from 296–608 K the cross-section was found to be nearly constant. Because of (a) the high magnitude of the cross-section, (b) its temperature invariance, and (c) the lack of resonance in this system, the authors suggested that the E→V transfer was caused by a curve crossing rather than by a long-range attractive interaction. A detailed consideration of these theories will be deferred until Section V.

3. *Hydrogen Bromide*

Donovan, Husain, and Stephenson[12] were the first to observe E→V transfer from Br* to HBr. The vibrational excitation was monitored by ultraviolet absorption following flash photolysis of HBr.

Leone and Wodarczyk[13] subsequently examined this E→V transfer using an apparatus similar to that in Fig. 1. A rate constant of 1.4×10^{-12} cm^3 molec^{-1} s^{-1} was obtained in reasonable agreement with previous work.[28]. In comparing the HBr and Br* fluorescence intensities, Leone and Wodarczyk used a lifetime of 104 ms for HBr[29] and obtained an average number of excited quanta equal to 0.51. However, if a lifetime of 239 ms is assumed,[30] the value of Q.E. would become just slightly larger than unity.

4. *Hydrogen*

Grimley and Houston[31] have indirectly monitored electronic to vibrational energy transfer between Br* and three isotopic forms of hydrogen by observing subsequent vibrational to vibrational energy transfer between hydrogen and CO. The energy flow pathways for the H_2 and HD systems are shown in Fig. 8. From the decay of Br* fluorescence at 2.71 μm, they found the total deactivation rates of Br* by H_2, HD, and D_2 to be 8.7×10^4, 2.1×10^5, and 2.2×10^4 s^{-1}torr^{-1}, respectively. These rate constants are substantially lower than those of a previous determination.[32]

No CO fluorescence was observed when hydrogen was omitted from the cell. For hydrogen/CO mixtures, however, a comparison of the amplitude of the CO fluorescence to the amplitude of the Br* fluorescence indicated that the lower limit to the number of vibrational quanta excited per Br* deactivation is 0.66 ± 0.11 for H_2, 0.78 ± 0.22 for HD, and 0.094 ± 0.033 for D_2.

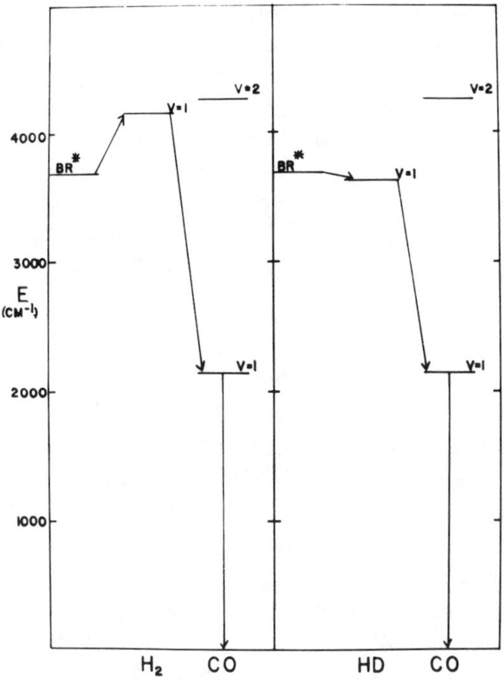

Fig. 8. Energy transfer scheme used for examination of E→V transfer between Br* and hydrogen. The E→V transfer is followed by V→V transfer from hydrogen to CO. CO ($v=1$) fluorescence is measured to determine the amount of energy transferred.

5. Carbon Monoxide

Observation of E→V transfer from Br* to CO has been reported by Lin and Shortridge,[33] who probed the excited vibrational levels of CO by monitoring their absorption of cw CO laser. Br* was created by flash photolysis of CF_3Br at $\lambda < 200$ nm. CO($v=1$) was observed to appear in ~10 μs for a mixture of 0.42 torr of CF_3Br, 2.08 torr of CO, and 2.5 torr of argon. No absorption from higher vibrational levels was observed.

In retrospect, there is some question as to whether the observed excitation was caused by E→V transfer. While CO($v=1$) appeared with a rate constant of ca. 5×10^4 $s^{-1}torr^{-1}$, the deactivation of Br* by CO is known to proceed with a rate constant of only 350 $s^{-1}torr^{-1}$,[31] and vibrational relaxation of CO($v=1$) by CO takes place at a rate of less than 0.02 $s^{-1}torr^{-1}$.[34] Case I of the kinetic scheme should apply, and the appearance time of CO($v=1$) should be governed by the 350 $s^{-1}torr^{-1}$ rate of quenching of Br*. Since the observed CO($v=1$) appears with a much faster rate constant, it is possible that the excitation is caused by some other effect. Neither Leone[15] nor Grimley and Houston[31] were able to observe any CO vibrational fluorescence following dissociation of Br_2 in an apparatus similar to that of Fig. 1. While the cw CO laser probing technique used by Lin and Shortridge might be more sensitive than the fluorescence technique, it is also possible that the flashlamp used by these authors to dissociate CF_3Br could create other excited species, which then might transfer their energy to CO. Vibrationally excited CF_3, CF_3Br, C_2F_6, and/or more highly excited electronic states of Br than Br(4 $^2P_{1/2}$) are possibilities.

6. *Nitric Oxide*

Although lasing has been observed[19] to occur on the $2\rightarrow1$ transition in NO as a result of E→V transfer from Br*, no systematic study of energy transfer in this system has been published. An unpublished report[35] places the deactivation rate constant at 2.0×10^{-12} cm^3 molec^{-1}s^{-1}. The fact that the $2\rightarrow1$ transition is at least partially inverted suggests that Q.E. is probably greater than 1.5 for this system. Because of the $Br_2+2NO\rightleftharpoons2$ NOBr equilibrium, an alternative to Br_2 as a parent molecule will be needed to measure accurately the quenching rates and the value of Q.E. for this system.

7. *Carbon Dioxide*

Energy transfer between Br* and CO_2 has been the most carefully studied of all E→V systems. Following the initial observation of this transfer by Leone,[15] the total quenching rate and the fraction of collisions that result in population of the ν_3 mode were measured in an apparatus similar to that of Fig. 1.[17,36] The total quenching rate was found to be 1.5×10^{-11} cm^3 molec^{-1}s^{-1}, while the average number of asymmetric stretching quanta excited per deactivation was determined to be Q.E.$(\nu_3)=$ 0.40. In addition, stimulated CO_2 emission at 10.6 μm was observed following flash photolysis of a Br_2/CO_2 mixture in the apparatus of Fig. 3.[17] Subsequent investigation of this laser system revealed that stimulated emission could also be obtained at 4.3 and 14.1 μm.[18] Observation of these latter transitions suggested that the (10°1) and perhaps the (02°1) levels of CO_2 received a large fraction of the Br* energy. Experiments incorporating a grating into the laser cavity established firmly that the (10°1) state was the state responsible for the lasing transitions.[20] Although it is possible that other vibrational levels are also populated by the transfer, the laser experiments clearly confirm that CO_2(10°1) is selectively excited by E→V transfer from Br*, and that this state is one of the major products.

The temperature dependence of the total quenching rate of Br* by CO_2 was measured by Reisler and Wittig.[27] In the range 296–608 K the quenching cross-section falls from about 3.2 Å^2 to about 2.2 Å^2. This result was taken as a confirmation that long-range attractive forces were responsible for the transfer, in much the same way as they are for V→V transfer.[37] Subsequent comparison of the quenching of Br* by $^{12}CO_2$ vs $^{13}CO_2$ revealed that the cross-sections for both quenching processes fell with temperature.[38] It was also demonstrated that the fraction of deactivations resulting in population of the ν_3 mode was independent of temperature and slightly higher for $^{13}CO_2$ (Q.E. = 0.50) than for $^{12}CO_2$ (Q.E. = 0.40). However, while an approximate calculation based on multipolar expansion of the interaction potential[37] predicted the observed negative temperature dependence for $^{12}CO_2$, it indicated that a postive temperature dependence should be observed for $^{13}CO_2$, in contrast to the experimental observation. It is not clear, therefore, whether the long-range attractive quadrupole–dipole term of the interaction potential is the major factor in this E→V transfer.

8. *Nitrous Oxide*

Energy transfer from Br* to N_2O was reported to take place with a rate constant of 2.6×10^{-12} cm^3 molec^{-1}s^{-1} and to excite vibrational fluorescence from both the symmetric and asymmetric stretches.[17] Weak lasing was reported on the (001)–(100) transition.

9. *Carbon Disulfide and Carbon Oxysulfide*

Hariri and Wittig[36] have investigated E→V transfer from Br* to CS_2 and OCS and found the quenching rates to be 1.1×10^{-12} and 1.4×10^{-12} cm^3 molec^{-1} s^{-1}, respectively. The latter rate is some 20% higher than that found in a previous measurement.[39] The fraction of collisions resulting in excitation of one quanta of ν_3 vibration was found to be 0.2 for OCS and 0.4 for CS_2.[36]

10. *Water*

Peterson, Braverman, and Wittig[19] have observed laser action on the (020)→(010) transition in H_2O following flash photolysis of $Br_2/He/H_2O$ mixtures in an apparatus similar to that of Fig. 3. A subsequent study of the Br*/H_2O transfer system[40] revealed that 100% of the deactivating collisions lead to excitation of either H_2O(100) or H_2O(001). The total quenching constant was found to be 6.2×10^{-11} cm^3 molec^{-1} s^{-1}. An energy-level diagram for this molecule is given in Fig. 9.

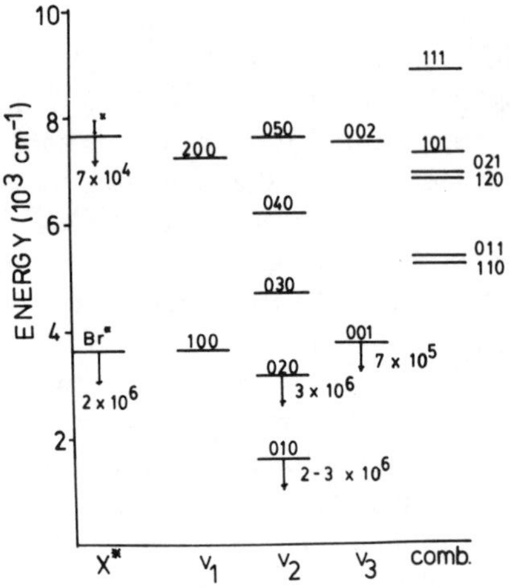

Fig. 9. Energy level diagram fro H_2O.

11. *Hydrogen Cyanide*

Stimulated emission on the (001)–(010) and (001)–(100) transitions of HCN has been observed following flash photolysis of Br_2/HCN/He mixtures in an apparatus similar to that of Fig. 3.[17] A detailed examination of the E→V transfer from Br* to HCN has found the total quenching rate to be 2.0×10^{-11} cm^3 molec^{-1} s^{-1}.[41] By comparing the HCN fluorescence intensity to the HCl intensity obtained under similar Br* quenching conditions, it was concluded that 88% of the quenching collisions result in population of the ν_3 mode of HCN.

B. E→V Transfer from I($^2P_{1/2}$)

1. *Hydrogen Fluoride*

Coombe and Pritt[42,43] have investigated E→V transfer from I* to HF in an apparatus similar to that shown in Fig. 1. A transient concentration of I* was generated by photolysis of CF_3I at 266 nm with a quadrupled YAG laser. Since HF(1) and HF(2) are relaxed faster by collisions than they are populated by E→V transfer, Case II of the kinetic scheme is observed. A total quenching rate of 8.7×10^4 s^{-1}torr^{-1} was observed. The authors determined the fraction of quenching which led to HF(2) by making a detailed comparison of the HF fluorescence signals obtained with and without HF in a filter cell held between the fluorescence cell and the detector. Roughly 35% of the total quenching led to HF(2) while $5 \pm 4\%$ led to HF(1). Thus, the average number of quanta excited is Q.E. $= \Sigma i k_i / \Sigma k_i = 0.75$. This determination was based on the HF lifetimes reported by Merideth and Smith[44] and on the I* lifetime reported by Comes and Pionteck.[45] Use of the I* lifetime calculated by Garstang[21] gives Q.E. = 0.91.

2. *Hydrogen Chloride*

Transfer of energy from I* to HCl was examined by Grimley and Houston[22] using the apparatus in Fig. 1. The total quenching rate was measured to be $(4.46 \pm 0.77) \times 10^2\ s^{-1}\ torr^{-1}$ from the decay of I* fluorescence at 1.3 μm. This value was in good agreement with the results of two previous investigations[46,47] but was twice as large as that found by Pritt and Coombe.[42] Direct observation of HCl fluorescence demonstrated the importance of the E→V mechanism for this system. A comparison of the HCl and I* fluorescence amplitudes indicated that Q.E. was 1.7 ± 0.5. The maximum possible quantum level which can be populated consistent with conservation of energy is $v_{max} = 2$.

3. *Hydrogen Bromide*

Vibrational excitation of HBr following its deactivation of I* was first observed by Grimley and Houston,[22] who measured k^E both from the decay of I* fluorescence and from the rise of HBr fluorescence. Figure 10 shows a plot of the reciprocal time constant for the HBr appearance as a

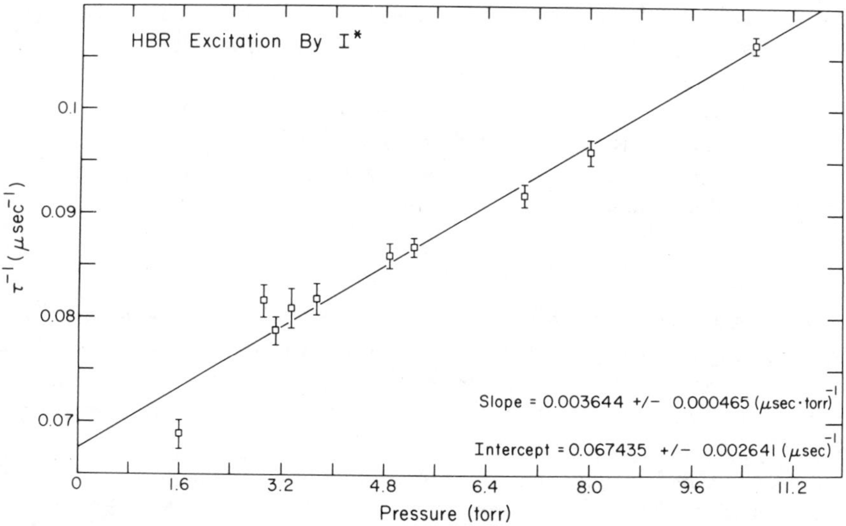

Fig. 10. Variation of τ^{-1} for rise of HBr vibrational fluorescence with pressure of HBr. (From Ref. 22.)

function of HBr pressure. The slope of the linear least squares fit gives a value of $k^E = (3.64 \pm 0.46) \times 10^3\ s^{-1} torr^{-1}$, where the uncertainty is equal to 2 standard deviations. This value agreed with that obtained for the I* fluorescence decay to within experimental error, but both values were some 30% below the average of three previously reported values.[42,47,48] The possibility of small amounts of Br_2 impurity in the HBr was suggested as a source of the discrepancy.

Comparison of the amplitudes of the HBr and I* fluorescence suggested that the average number of HBr quanta excited was 2.3 ± 0.5. This determination implies that at least $(76 \pm 17)\%$ of the deactivation is caused by E→V transfer but is based on the assumption of a 239 ms radiative lifetime for HBr.[30] If the lower value [29] of 104 ms were adopted, the value of Q.E. would drop to 1.00, and a minimum of 33% of the deactivation would be caused by the E→V process.

Wisenfeld and Wolk[48] measured the amount of I produced from the nonreactive deactivation of I* by HBr and found it to be negligible. Their results imply that $(96 \pm 12)\%$ of the deactivation is a result of the reaction to form HI + Br. This conclusion, which is in contradiction to the E→V results, is supported by the fact that the I*/HBr quenching constant is much higher than one would predict from a plot of energy discrepancy vs normalized E→V rate.[42] On the other hand, the isotope effect for HBr vs DBr, $\sigma_H/\sigma_D \geqslant 2.6$, is larger than one might expect for a reactive channel and is more consistent with that expected from resonance considerations in E→V transfer.[47] In addition, the agreement between the observed negative temperature dependence for the quenching rate and that calculated from a resonant E→V theory[49] lends further support to the argument that E→V transfers plays a significant role in the deactivation. Additional work on the I*/HBr system is badly needed to resolve this controversy concerning reactive vs. nonreactive quenching. A direct observation of the reaction products HI and Br would be particularly useful.

4. *Hydrogen*

Deactivation of I* by three isotopic forms of hydrogen was monitored by observing the decay of I* emission and by inferring the E→V transfer between I* and hydrogen from observation of subsequent V→V transfer between hydrogen and CO_2.[50] HD relaxes I* 2.5 times more efficiently than H_2 and over 200 times more efficiently than D_2. The observed rate constants are in fair agreement with those obtained in other studies.[46,51–54]

Detection of the hydrogen excitation was accomplished by the scheme shown in Fig. 11. If I* transfers some fraction of its electronic energy to the hydrogen vibrations, it is likely that much, if not all, of that excitation

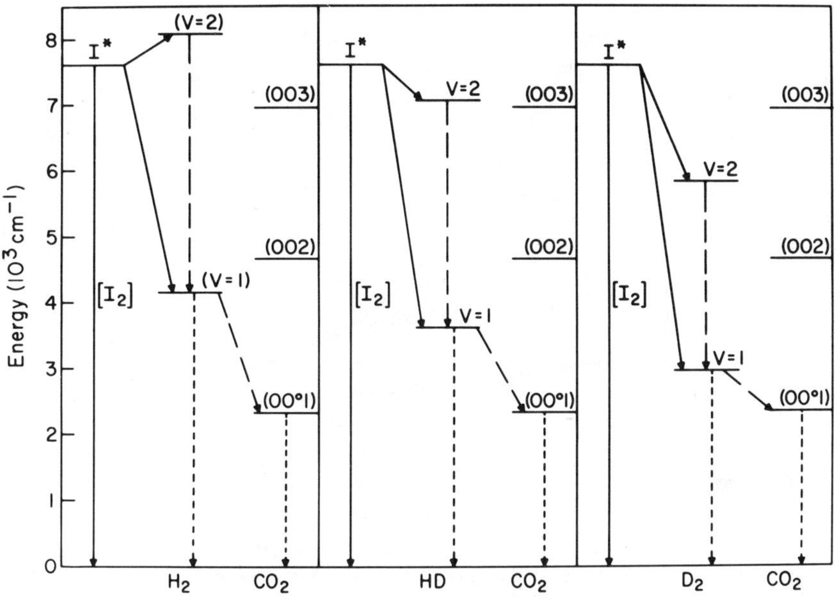

Fig. 11. Energy transfer scheme used for examination of E→V transfer between I* and hydrogen. The E→V transfer is followed by V→V transfer from hydrogen to CO_2. CO_2(001) fluorescence is measured to determine the amount of energy transferred (From ref. 50).

will be subsequently transferred to the asymmetric stretching mode of CO_2. Observation of CO_2 fluorescence at 4.3 μm would indicate the presence of E→V transfer between I* and hydrogen. CO_2 was chosen as a probe molecule because it quenches I* with a rate constant of only 4.1 $s^{-1}torr^{-1}$.[54,55] Thus, CO_2 does not itself become vibrationally excited by E→V transfer to any appreciable extent.

The influence of E→V transfer is clearly shown in Fig. 12, which displays the CO_2 fluorescence following dissociation of I_2 in mixtures of hydrogen and CO_2. In the absence of hydrogen, virtually no CO_2 fluorescence is observed. However, when 1.6 torr of H_2 is added to ~1 torr of CO_2, the smaller of the two signals is observed. The larger of the two signals is obtained if the H_2 is replaced by HD.

A quantitative analysis of the ratio of the I* and CO_2 fluorescence amplitudes provided a lower limit estimate of Q.E. for each of the I*/hydrogen systems. Only a lower limit is obtained because of approximations in solving the kinetic scheme and because of possible underestimation of self-absorption of the CO_2 fluorescence. The minimum number of

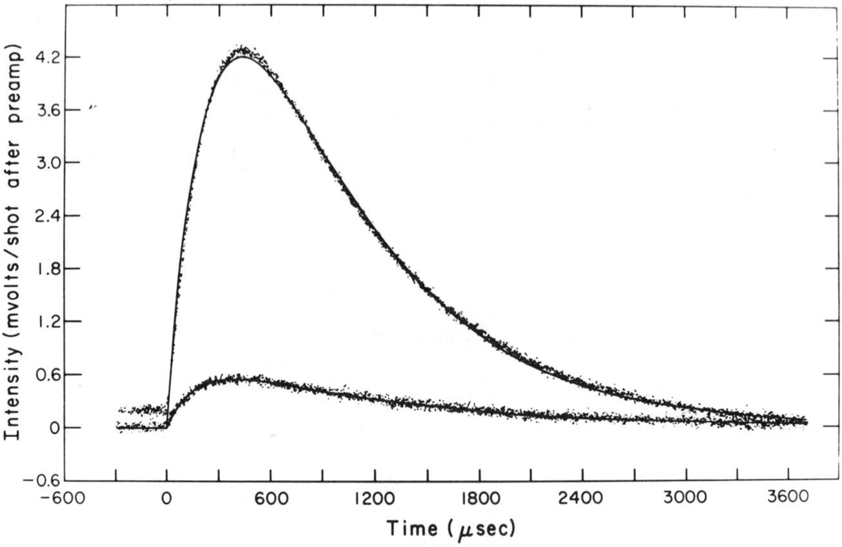

Fig. 12. CO_2 fluorescence following dissociation of I_2 in mixtures of hydrogen and CO_2. Larger signal is for HD while smaller signal is for H_2. Virtually no fluorescence was seen from CO_2 if hydrogen was excluded from the mixture.

quanta excited by the E→V transfer was found to be 0.61 ± 0.12 for HD, 0.51 ± 0.10 for H_2, and 0.10 ± 0.03 for D_2.

The temperature dependences of the total quenching constants have been measured by Deakin and Husain[56] and by Burde and McFarlane.[53] The rate constants were found to increase with increasing temperature for H_2 and HD, but to decrease with increasing temperature for D_2. Butcher et al.[52] have interpreted these temperature variations in terms of near resonant energy transfer of the type

$$I^* + H_2(v'', J'') \rightarrow I + H_2(v', J') + \Delta E$$

where ΔE is the amount of energy that must be converted into translation. Their analysis predicts the correct qualitative dependence on temperature for quenching of I* by these hydrogens.

5. *Nitric Oxide*

As described in Section III, the total quenching rate of I* by NO has been determined both from the decay of I* fluorescence and from the rise of NO fluorescence.[22] NO($v = 1$) was relaxed slowly under the experimental conditions, so that Case I of the kinetic scheme was obtained. Although the measured quenching rate was found to be an order of magnitude less than that reported previously,[46] the constants obtained from the I* and NO fluorescence signals agreed to within the experimental uncertainty.

A comparison of the I* and NO fluorescence amplitudes led to the conclusion that the average number of NO vibrational quanta excited per deactivation of I* is 3.4 ± 0.7. The maximum value of Q.E. consistent with conservation of energy is 4.

A long-lived intermediate complex in conjunction with nonadiabatic curve crossing was suggested as a possible mechanism for the deactivation of I* by NO. If the NO is considered as a pseudo halogen atom in its ground electronic state, then combination of $I(^2P_{1/2})$ with NO might lead to a bound "diatom" similar, for example, to IBr in its $^3\Pi_0$ state. Possible potential curves for such an interaction are shown in Fig. 13. It is well-known that halogens and interhalogens in the $^3\Pi_0$ state are predissociated by nonadiabatic transitions to potential curves such as $^1\Pi_1$, which correlate to ground-state atoms. In the case of I*+NO the proposed (INO)* complex would be predissociated to potential curves yielding $I+NO(v=i)$.

6. *Carbon Monoxide*

In a method similar to that described in Section IV.A.5, Lin and Shortridge have observed vibrational excitation in CO following flash photolysis of CF_3I/CO mixtures. As in the case of Br* transfer to CO,

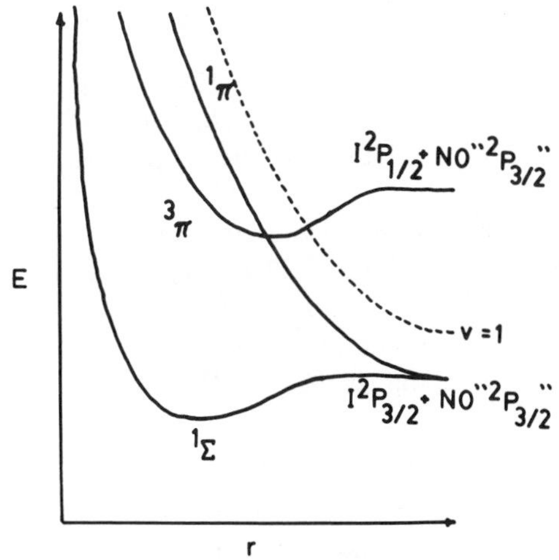

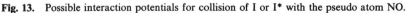

Fig. 13. Possible interaction potentials for collision of I or I* with the pseudo atom NO.

there is some question as to whether the excitation of CO observed following photolysis is actually due to transfer from I*. Deactivation of I* by CO is known to take place at a rate of only 39 $s^{-1} torr^{-1}$,[54,58] while vibrational relaxation of CO occurs with a rate constant of less than 0.2 $s^{-1} torr^{-1}$.[34] The CO observed by Lin and Shortridge appears with a rate constant of roughly 1×10^5 $s^{-1} torr^{-1}$. Furthermore, using the apparatus of Fig. 1, Houston[57] has failed to observe I*-CO energy transfer for conditions under which transfer to other molecules gave large fluorescence signals. As discussed in Section IV.A.5, it is possible that CO is excited by transfer from some species other than I*.

7. *Water*

The only polyatomic molecule for which energy transfer from I* has been adequately examined is water. Grimley and Houston[23] have measured as a function of P_{H_2O} and P_{D_2O} the rates for (a) the decay of I* fluorescence, (b) the rise and decay of total H_2O stretching fluorescence, and (c) the rise and decay of the (002)→(001) stretching fluorescence. The I* fluorescence decay rate was also measured as a function of P_{HDO}. Since the symmetric and asymmetric stretching levels are strongly coupled by collisions, no attempt was made to distinguish between them in this experiment. These levels are relaxed to unobserved bending levels more rapidly than they are populated by E→V transfer, so that Case II of the kinetic scheme was observed to apply. An energy level diagram for H_2O has been presented in Fig. 9.

H_2O and HDO relax I* roughly 50 times more efficiently than D_2O. The rate constants obtained were in fair agreement with those reported elsewhere.[47,55,58] Donovan et al.[47] have discussed the large isotope effect. They concluded from their analysis that E→V transfer from I* to H_2O takes place by a long-range resonant mechanism and that the quenching cross-section is a strong function of the change in vibrational quanta number ΔV.

Analysis of the relative fluorescence amplitudes by Grimley and Houston[23] revealed that the quenching of I* by H_2O and HDO is, in fact, due entirely to an electronic to vibrational energy-transfer mechanism that produces two quanta of stretching vibration in the water molecule. A lower excitation limit of 1 stretching quanta for D_2O was estimated.

8. *Other Systems*

Electronic to vibrational energy transfer from I* has been directly observed to populate the vibrational modes of NH_3, C_2H_2, and HCN.[59] Although complete studies of these systems have not been performed, it seems clear that a substantial fraction of the I* electronic energy is transferred to the vibrational modes of these molecules.

C. Summary of Results and Their Correlations

A complete summary of the known data on E→V transfer from Br* and I* is given in Tables III and IV. For each quenching gas the total quenching rate and the average number of vibrational quanta excited per deactivation are listed. For polyatomic molecules the column labeled Q.E. also provides the observed vibrational mode in parenthesis. The values ΔE and V are listed for the process

$$X^* + A \rightarrow X + A(V) + \Delta E$$

where V is the most resonant vibrational level that accepts the electronic energy, and ΔE is the amount of energy that must be taken up by rotation and translation.

Figure 14 shows that there is good correlation between the total quenching rate, the absolute value of the energy defect ΔE, and the total change of vibrational quantum number ΔV. The solid points of this figure are for

TABLE III
Summary of E→V Transfer From Br*

Molecule	Quenching rate (298 K) $k^E = \Sigma k_i$ (cm^3 molec^{-1} s^{-1})	Quanta excited $\Sigma i k_i / \Sigma k_i$ Q.E. (mode)	ΔE^a (cm^{-1})	Vibrational^a state V	Ref.
HF	3.4×10^{-11}	1.00	−276	1	25
HCl	8.6×10^{-12}	0.95	789	1	13,27
HBr	1.4×10^{-12}	1.00^b	1126	1	13
H_2	2.7×10^{-12}	⩾0.60	−465	1	31
HD	6.4×10^{-12}	⩾0.78	40	1	31
D_2	6.8×10^{-13}	⩾0.094	1335	1	31
NO	2.0×10^{-12}	$\geqslant 1.5^c$	39	2	17,35
$^{12}CO_2$	1.5×10^{-11}	$0.40(\nu_3)$	−30	(101)	17,36,38
$^{13}CO_2$	7.1×10^{-12}	$0.50(\nu_3)$	52	(101)	38
N_2O	2.6×10^{-12}	—	100	(101)	17
COS	1.4×10^{-12}	$0.2(\nu_3)$	−100	(201)	36
CS_2	1.1×10^{-12}	$0.3(\nu_3)$	0	(102)	36
H_2O	6.2×10^{-11}	$1.0(\nu_1$ or $\nu_3)$	0	(100) or (001)	40
HCN	2.0×10^{-11}	$0.88(\nu_3)$	374	(001)	41

[a] ΔE for process $Br^* + A \rightarrow Br + A(v) + \Delta E$.
[b] Assuming a 239 ms radiative lifetime for HBr.[30]
[c] Suggested minimum based on published observation of lasing on 2→1 transition.

TABLE IV
Summary of E→T Transfer from I*

Molecule	Quenching rate (298 K) $k^E = \Sigma k_i$ (cm^3 molec^{-1} s^{-1})	Quanta excited $\Sigma i k_i / \Sigma k_i$ Q.E. (mode)	ΔE^a (cm^{-1})	Vibrational[a] state V	Ref.
HF	2.7×10^{-12}	0.91^b	−136	2	43
HCl	1.4×10^{-14}	1.7	1940	2	22
HBr	1.1×10^{-13}	2.3^c	198	3	22
H_2	1.1×10^{-13}	$\geqslant 0.51$	−500	2	50
HD	3.0×10^{-13}	$\geqslant 0.61$	500	2	50
D_2	1.4×10^{-15}	$\geqslant 0.10$	1750	2	50
NO	1.2×10^{-13}	3.4	267	4	22
H_2O	2.2×10^{-12}	$2(\nu_1$ or $\nu_3)$	30	(200), (101), or (002)	23
HDO	1.9×10^{-12}	$2(\nu_3)$	30	(002)	23
D_2O	3.7×10^{-14}	$\gtrsim 1$ $(\nu_1$ or $\nu_3)^d$	2077	(200), (101), or (002)	23

[a] ΔE for process $I^* + A \rightarrow I + A(v) + \Delta E$.
[b] Corrected value based on calculated I* lifetime of 130 msec.[21]
[c] Assuming a 239 ms. radiative lifetime for HBr (see text).
[d] Estimated lower limit.

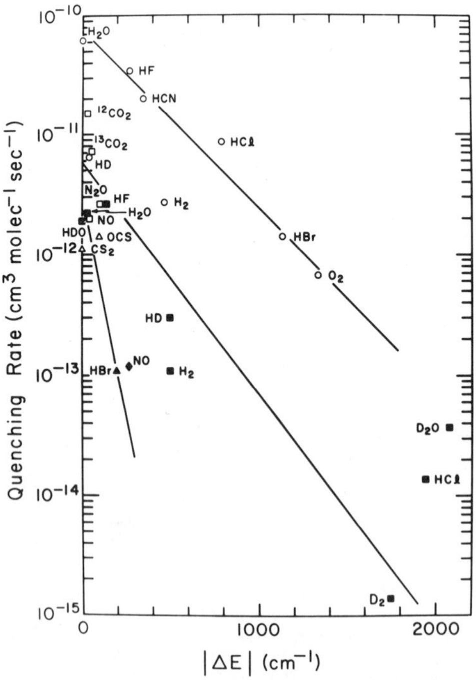

Fig. 14. Plot showing the correlation between quenching rates of Br* (open figures) or I* (solid figures) and the energy defect $|\Delta E|$. The circles are for $\Delta V = 1$, the squares are for $\Delta V = 2$, the triangles are for $\Delta V = 3$, and the diamond for $\Delta V = 4$.

deactivation of I*, while the open points are for deactivation of Br*. The circles represent cases where the change in vibrational quantum number is 1, the squares represent a change of 2, and the triangles represent a change of 3. The one example where $\Delta V=4$ is plotted as a diamond.

The correlation of total quenching rate with $|\Delta E|$ is strongest for Br* and $\Delta V=1$, as shown in the upper line of the figure. The quenching rate drops by roughly an order of magnitude for each increase of 650 cm^{-1} in $|\Delta E|$. Molecules in which the vibrational quantum numbers change by two have consistently lower quenching rates than those in which $\Delta V=1$. Again, there is a correlation between quenching rate and $|\Delta E|$, although the scatter about the line of the figure is rather pronounced. Molecules with $\Delta V=3$ have still lower quenching rates which seem to be even more sensitive to ΔE, as shown by the lowest of the three lines in the figure. We conclude from these observations that the total quenching rate is a strong function of ΔE and ΔV and that the higher ΔV becomes the more sensitive the quenching rate is to ΔE.

There is no strong correlation between the amount of energy transferred and either ΔV or ΔE. In fact, for the molecules in Tables III and IV, comparison of Q.E.(ν_i) with $\Delta V(\nu_i)$ shows that the ratio Q.E.(ν_i)/$\Delta V(\nu_i)$ does not fall below about 20%, except for the possible cases of I* and Br* deactivation by D_2. The number of vibrational quanta excited is always substantial when compared to the number of quanta necessary for acceptance of all the energy in a resonant fashion. The quantity Q.E./ΔV does not correlate well with ΔE; HBr and HCl accept as high a fraction of energy as H_2O even though ΔE varies from 0–1126 cm^{-1} for these molecules. Nor does the ratio of Q.E./ΔV seem to depend on ΔV; NO accepts almost as high a fraction of energy from I* as does HCl from Br* even though $\Delta V=4$ in the former case and $\Delta V=1$ in the latter. We conclude, therefore, that while ΔE and ΔV have a strong effect on the quenching rate, they have virtually no effect on the fraction of energy which is transferred.

V. DISCUSSION

The summary of results provided in the previous section has demonstrated the importance of the energy defect, ΔE, and the change in vibrational quantum number, ΔV, in determining the rate of total deactivation for excited halogen atoms. The observation that these two factors make very little difference in determining the *amount* of energy transferred implies that, even for large ΔE and ΔV, the rate of any E→T,R process must be slower than the rate of the E→V process. Although the question of why this E→V process should dominate the X* deactivation has been addressed by several theoretical studies,[60–71] the discussion in this section will concentrate on a comparison of the experimental results with the predictions of only three theoretical approaches to E→V transfer. Each approach correctly predicts that the quenching rate will be a strong function of both ΔE and ΔV, but the three approaches emphasize different aspects of the interaction potential. The aim of the current discussion will be to decide which aspects of that potential are most responsible for inducing E→V transfer.

A. Quantum-Mechanical Calculations

One approach to a theoretical understanding of E→V transfer is to perform rigorous quantum-mechanical trajectory calculations on each system of interest. By such an approach one might hope to obtain exact results for an assumed interaction potential and to determine by comparison with experiment which parts of that potential are important to the E→V process. Unfortunately, given the current limitations of computer technology, many approximations must be introduced in order to make the quantum calculations tractable. While the quantum approach can provide information on part of the overall problem, it cannot yet take all possible effects into account.

In the spirit of this approach, Zimmerman and George[65] have calculated the probabilities for E→V transfer between Br* and I* and the three isotopic forms of hydrogen: H_2, HD, and D_2. While concentrating on the effect of resonance, they have restricted their study to collinear configurations and have chosen the potential energy surfaces to be two diatomics-in-molecules surfaces separated by the asymtopic spin–orbit energy. Although this approach ignores the effects of rotation and long-range attraction, it might be expected to give a qualitative picture of both the effect of resonance on the quenching rate and the extent of vibrational excitation.

The results of the calculation are summarized as follows. For Br* the theory predicts that HD will have the largest quenching cross-section, followed by H_2 and then D_2. This trend, which might be expected from resonance considerations, is in qualitative agreement with the experimental results of Section IV.A.4. Furthermore, the theory predicts that the main quenching contribution for all three species is from collisions which leave the hydrogen in its first excited vibrational level. This finding is again in good agreement with experiment[31] where the minimum average number of quanta excited was found to be 0.60 for H_2 and 0.78 for HD. It is less clear experimentally that the quenching of Br* by D_2 is dominated by the E→V channel.

The results of the calculation for I* were also in good qualitative agreement with experiment. The order of quenching efficiencies was correctly predicted to favor HD over H_2 and H_2 over D_2. The theory also predicted that the total quenching of I* by these molecules should be much slower than the rate for quenching of Br*, in agreement with experimental results. Finally, it was found that the probability for the $\Delta V=2$ transition in hydrogen was higher than for the $\Delta V=1$ or $\Delta V=0$ transition. Although the experimental results do not directly confirm this prediction, they are not inconsistent with it.[50]

In summary, the quantum-mechanical calculations predict the experimental observations that the quenching is dominated by the E→V channels and that the quenching rate is a sensitive function of ΔE and ΔV. That such a theory, which includes no attractive contribution to the atom–molecule potential, correctly predicts these qualitative trends suggests that attractive forces do not play a strong role in the quenching. The main inadequacies of the quantum calculation at this stage are (a) its expense, and (b) the current impracticality of including rotational degrees of freedom in an exact way. Although a semiempirical method for incorporating rotation has been presented,[65] the results are not in as good quantitative agreement with experiment as those of the collinear model.

B. Long-Range Attractive Forces

A second theoretical approach to understanding E→V transfer is based on the notion that long-range attractive forces might couple the electronic and vibrational degrees of freedom. Ewing[62] has developed a theory for such coupling that is similar to the theory for V→V transfer proposed earlier by Sharma and Brau.[37] First order perturbation theory is used in this approach to calculate the probability for the E→V transition:

$$P=\hbar^{-2}\left|\int_{-\infty}^{+\infty}\langle i|V|f\rangle e^{i\omega t}dt\right|^2 \tag{8}$$

where ΔE, the energy discrepancy, is equal to $\hbar\omega$, and $V=V(t)$ is the interaction potential. In the theory of Ewing, V is taken to be an angle-averaged multipole expansion. For E→V transitions between X* and H_2, for example, the highest order nonvanishing term in the expansion would be the quadrupole–quadrupole interaction,

$$V(t)=\left(\frac{14}{5}\right)^{1/2}\frac{Q^{\mathrm{X}}Q^{\mathrm{M}}}{R(t)^5} \tag{9}$$

where Q^X is the quadrupole moment of the atom and Q^M is the quadrupole moment of the molecule. After assumption of straight line trajectories one obtains the result[62]

$$P(\omega, v, b) = \left(\frac{14}{20}\right)\pi\hbar^{-2}|Q_{if}^X|^2|Q_{if}^M|^2 \frac{\left[(\omega b/v)^2 K_2(\omega b/v)\right]^2}{\Gamma(5/2)v^2 b^8} \tag{10}$$

where K_2 is a second-order modified Bessel function, b is the impact parameter, and v is the velocity. The quantities Q_{if}^X and Q_{if}^M are the matrix elements for the atomic and molecular transitions, respectively. Although the equation above should still be integrated over impact parameters and velocities before being compared to experimental results, the qualitative trends in the E→V probability are clear. For large values of $\omega b/v$, $K_2(\omega b/v) \to \exp[-\omega b/v]$, so the theory correctly predicts the exponential falloff in the quenching rate with ΔE. In addition, since the probability is proportional to $|Q_{if}^M|^2 = |\langle v_f|Q|v_i\rangle|^2$, the rate should decrease rapidly with increasing $\Delta V = v_f - v_i$.

French and Lawley[71] have recently performed a more detailed calculation based on the quadrupole–quadrupole attractive forces. Improvements over the simple theory above included the use of calculated rather than straight line trajectories and the inclusion of rotational degrees of freedom. For the case of I* deactivation by H_2, HD, and D_2 they found that while the theory correctly predicts the magnitudes of the rate constants when ΔE is close to zero, it does not do as well for large deviations from resonance. In particular the k_{H_2}/k_{HD} ratio is off by more than two orders of magnitude. They conclude that other effects besides the quadrupole–quadrupole interaction must be responsible for the E→V process when ΔE is not near to zero and suggest that short-range coupling might be involved.

Pritt and Coombe[42] and Donovan, Fotakis, and Golde[47] have extended the ideas behind the Ewing theory to cover the case of quadrupole–dipole attractive forces, which vary as R^{-4}. This is exactly the type of force for which Sharma and Brau[37] originally performed the perturbation theory calculation. For small values of $\omega b/v$ the cross-section averaged over temperature and impact parameter is given by[37]

$$I(T) = \frac{3\pi^2\mu|\mu_{if}^M|^2|Q_{if}^X|^2}{64\hbar^2\sigma^2 kT} \tag{11}$$

where σ is the hard-sphere collision diameter, μ is the reduced mass, and μ_{if}^M is the dipole matrix element between the vibrational levels of the molecule. This matrix element μ_{if}^{M} predicts the observed decrease in cross-section with increase in ΔV. For large $(\omega b/v)$ the deactivation probability falls off as $\exp(-\omega b/v)$, in agreement with the experimental results. It is interesting to note that $I(T)$ has a pronounced negative temperature dependence, in accord with the experimental results on Br^*/CO_2 (Section IV.A.7), and I^*/HBr (Section IV.B.3).

Pritt and Coombe[42] have normalized the rate constant for hydrogen halide deactivation of I* and Br* by dividing each observed constant by the product $|\mu_{if}^{M}|^2|Q_{if}^{A}|^2$. When plotted against the energy defect ΔE, the logarithm of the normalized rates for the I*/HF, Br*/HF, Br*/HCl, Br*/HBr, and I*/HCl systems all fall near to the same line. This observation offers evidence that all of these systems are dominated by a similar E→V mechanism. The systems I*/HBr and I*/HI give normalized rate constants which are much larger than expected. Pritt and Coombe attribute this deviation to the possibility of reactive channels. The controversy over the extent of reaction vs E→V transfer in the I*/HBr system has been previously discussed in Section IV.B.3.

Donovan, and Fotakis, and Golde[47] have analyzed the quenching rates of I* by hydrogen and deuterium halides and by H_2O and D_2O in terms of the dipole–quadrupole attractive forces. Although they conclude that the long-range mechanism can adequately account for the quenching by H_2O, D_2O, HBr, and DBr, they point out that for HCl and DCl there are no resonant transitions with $\Delta E < 200$ cm^{-1} and small values of ΔJ. It is unlikely that the approximations of the Sharma–Brau theory are valid for such large energy discrepancies.

In summary, although E→V theories based exclusively on long-range attractive forces correctly predict the qualitative dependence of the quenching rate constant on ΔE and ΔV, the occurrence of E→V transfer at very large ΔE suggests that the long-range forces may not be the only mechanism by which E→V transfer takes place. This suggestion is supported by the comparison between the quantum calculations and the long-range multipole expansion calculations. Both predict the observed qualitative dependence of the quenching rate on ΔE and ΔV, but the former theory neglects long-range attraction, while the latter theory includes nothing else.

C. Curve-Crossing Mechanisms

The third of the theoretical approaches we will examine was introduced by Nikitin[61] and is based on the idea that E→V transfer involves a crossing between the zero-order potential curve correlating to X* + A(0) and that correlating to X + A(v). For the interaction between a halogen, X, and a

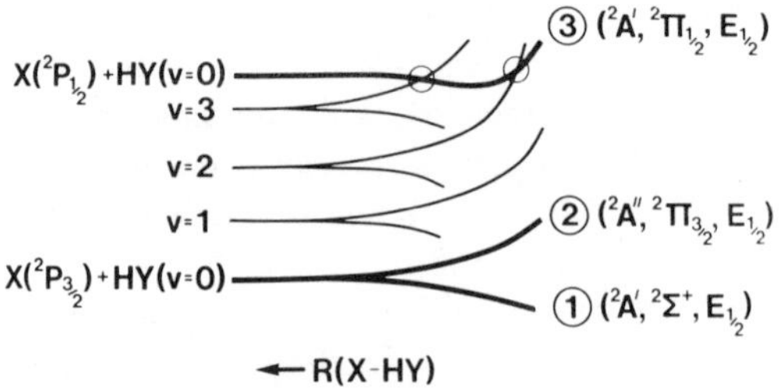

Fig. 15. Diagramatic representation of a section through the potential energy surfaces for the X+HY system. (From Ref. 47.)

hydrogen halide, HY, the relevant curves are shown in Fig. 15. The potential curves corresponding to $X+HY(v=i)$ are obtained approximately by shifting the $X+HY(v=0)$ potential curve up by an energy $ih\nu$. If the repulsive part of the X+HY curve is stronger than that of the X*+HY curve, then the ground and excited curves will cross. Reactants move along the upper curve and reach the point of intersection where they undergo a nonadiabatic transition to the lower curve. This transition corresponds to the deactivation of the atom and excitation of the hydrogen halide.

The details of the nonadiabatic transition can be handled in a variety of ways.[60,61,63,64,66,72–74] The simplest approach is to use the Landau–Zener formula for the probability of crossing.[61,72–74] For a single crossing the probability of a transition from the zero-order curve corresponding to X*+HY(0) to that corresponding to $X+HY(v)$ is given by

$$P_{\mathrm{EV}} = 1 - \exp\left(- \frac{2\pi |V_{if}|^2}{\Delta F \hbar v} \right) \tag{12}$$

where V_{if} is the matrix element of the interaction and $\Delta F = \delta/\delta R\,(V_{ii} - V_{ff})|_{R=R_c}$ is a measure of how fast the two surfaces approach one another. A useful estimate for ΔF is $\Delta F \cong \Delta E \alpha$, where ΔE is the asymptotic energy defect and α^{-1} is the length over which the coupling between zero-order potential curves is strong.

It can be easily verified that (12) predicts the main qualitative features of E→V transfer. First, the matrix element V_{if} will include the normal vibrational contribution, which falls off rapidly as ΔV increases. In the limit of small V_{if}, P_{EV} will be linearly proportional to V_{if}. Second, since ΔF is proportional to the energy defect ΔE, (12) predicts that when ΔE is not too large P_{EV} falls exponentially as ΔE increases. Finally, it should be noted that in the limit of low probabilities and under the assumption that the crossing is thermally accessible, (12) predicts that $\ln(P_{EV})$ will have roughly a $T^{-1/2}$ temperature dependence.

Although the Landau–Zener theory predicts all the experimental trends observed for E→V transfer, and, although more sophisticated techniques for handling the nonadiabatic transition are available,[60,63,64,66] no calculations have been performed as yet for deactivation of I* or Br* by this mechanism. The major difficulty is that the theory is very sensitive to the details of the potential curves in the region of interaction. In general, these "curves" will be multidimensional surfaces. On the other hand, if our object is to learn about the potential from comparison of theory and experiment, it is just this sensitivity that we should demand from theory. It seems likely, therefore, that future theoretical work in this area will be very beneficial to our understanding of E→V transfer.

D. Summary

The three theories of E→V transfer considered above all predict a strong dependence of the E→V quenching rate on ΔE and ΔV. While the quantum calculations include no attractive forces, the long-range theory is based entirely on the attractive part of the potential. The influence of attractive forces in the curve-crossing mechanism is more subtle. Although the potential crossing occurs at short range and does not explicitly depend on attractive forces, it seems unlikely from Fig. 15 that a crossing would occur for thermally accessible energies if the upper curve were not more attractive than the lower. Since all the theoretical approaches are consistent with the qualitative trends in the experimental data, it seems that the question of long-range vs short-range forces cannot be answered at this time. While more accurate theories and experiments might someday resolve this question, it is likely that both effects play an important role.

VI. CONCLUSION

Experimental studies of electronic to vibrational energy transfer from excited halogen atoms have been reviewed and summarized. The major trends of the experimental data are as follows: (1) the E→V channel often dominates the deactivation process, (2) the quenching rate falls rapidly with an increase in the amount of energy ΔE which must be taken up by rotation and translation, (3) the quenching rate decreases rapidly as the change in number of vibrational quanta ΔV increases, and (4) the *amount* of energy transfer is high and independent of ΔV and ΔE. These trends have been discussed in terms of three theoretical approaches to E→V transfer: quantum-mechanical calculations, a long-range attractive mechanism, and a curve-crossing mechanism. All three theoretical approaches predict the observed qualitative trends in the experimental data, despite the fact that the three approaches are based on different assumptions concerning which parts of the interaction potential are important. It seems unlikely at this point that a distinction between the effectiveness of long-range vs short-range forces can be made based on available theories and experiments.

Acknowledgments

It is a pleasure to acknowledge the collaboration of Dr. A. J. Grimley on the experimental work in our laboratory which led to Ref. 16, 22, 23, 31, and 50. This work has been sponsored in part by the National Science Foundation under grant number CHE-76-21991 and in part by the Air Force Office of Scientific Research under grant number AFOSR-78-3513. I would like to thank the organizing committee of the International Conference on Laser Chemistry for inviting me to participate.

References

1. J. C. Polanyi, *J. Quant. Spect. Radiat. Transf.*, **3**, 471 (1963).
2. A. B. Callear, *Ann. Rep. Prog. Chem.*, **61**, 48 (1964).
3. J. C. Polanyi, *Appl. Opt. Suppl.*, **2**, 109 (1965).
4. A. B. Callear, *Appl. Opt. Suppl.*, **2**, 145 (1965).
5. A. B. Callear, *Photochemistry and Reaction Kinetics*, P. G. Ashmore, F. S. Dainton, T. M. Sugden, eds., Cambridge University Press, London, 1967, Chap. 1.
6. A. B. Callear and J. D. Lambert, *Comp. Chem. Kinetics*, Vol. 3, C. H. Bamford and C. F. H. Tipper, eds., Elsevier, New York, 1969, Chap. 4.
7. R. B. Cundall, *Transfer and Storage of Energy by Molecules*, Vol. 1, G. M. Bennett and A. M. North, eds., Wiley-Interscience, New York, 1969, Chap. 1.
8. R. J. Donovan and D. Husain, *Chem. Rev.*, **70**, 489 (1970).
9. R. J. Donovan and D. Husain, *Adv. Photochem.*, **8**, 1 (1971).
10. J. Wm. McGowan, ed., *The Excited State in Chemical Physics*, Wiley-Interscience, New York, 1975.
11. S. Lemont and G. W. Flynn, *Ann. Rev. Chem. Phys.*, **28**, 261 (1977).
12. R. J. Donovan, D. Husain, and C. D. Stephenson, *J. Chem. Soc. Faraday Trans.*, **66**, 2148 (1970).
13. S. R. Leone and F. J. Wodarczyk, *J. Chem. Phys.*, **60**, 314 (1974).
14. R. D. Hudson, *Infrared System Engineering*, Wiley, New York, 1969.
15. S. R. Leone, Ph.D. Dissertation, University of California, Berkeley, 1974.
16. A. J. Grimley, Ph.D. Dissertation, Cornell University, 1979.
17. A. B. Peterson, C. Wittig, and S. R. Leone, *Appl. Phys. Lett.*, **27**, 305 (1975).
18. A. B. Peterson, C. Wittig, and S. R. Leone, *J. Appl. Phys.*, **47**, 1051 (1976).
19. A. B. Peterson, L. W. Braverman, and C. Wittig, *J. Appl. Phys.*, **48**, 230 (1977).
20. A. B. Peterson, and C. Wittig, *J. Appl. Phys.*, **48**, 3665 (1977).
21. R. H. Garstang, *J. Res. Nat. Bur. Std.*, **A68**, 61 (1964).
22. A. J. Grimley and P. L. Houston, *J. Chem. Phys.*, **68**, 3366 (1978).
23. A. J. Grimley and P. L. Houston, *J. Chem. Phys.*, **69**, 2339 (1978).

24. J. C. Stephenson, *J. Chem. Phys.*, **59**, 1523 (1973).
25. F. J. Wodarczyk and P. B. Sackett, *Chem. Phys.*, **12**, 51 (1976).
26. G. P. Quigley and G. J. Wolga, *J. Chem. Phys.*, **62**, 4560 (1975).
27. H. Reisler and C. Wittig, *J. Chem. Phys.*, **68**, 3308 (1978).
28. R. J. Donovan and D. Husain, *Trans. Faraday Soc.*, **62**, 2643 (1966).
29. L. A. Gribov and V. N. Smirnov, *Sov. Phys. Usp.*, **4**, 910 (1962).
30. B. S. Rao, *J. Phys.*, **B4**, 791 (1971).
31. A. J. Grimley and P. L. Houston, *J. Chem. Phys.*, **70**, 5184 (1979).
32. J. R. Wiesenfeld and G. L. Wolk, *J. Chem. Phys.*, **1976**, 1506.
33. M. C. Lin and R. G. Shortridge, *Chem. Phys. Lett.*, **29**, 42 (1974).
34. J. C. Stephenson, *Appl. Phys. Lett.*, **22**, 576 (1973).
35. S. Lemont and G. W. Flynn, unpublished work quoted in Ref. 11.
36. A. Hariri and C. Wittig, *J. Chem. Phys.*, **67**, 4454 (1977).
37. R. D. Sharma and C. A. Brau, *Phys. Rev. Lett.*, **19**, 1273 (1967); *J. Chem. Phys.*, **50**, 924 (1969).
38. H. Reisler and C. Wittig, *J. Chem. Phys.*, **69**, 3729 (1978).
39. S. Lemont, Ph.D Dissertation, Columbia University, New York, 1976.
40. A. Hariri and C. Wittig, *J. Chem. Phys.*, **68**, 2109 (1978).
41. A. Hariri and A. B. Peterson, and C. Wittig, *J. Chem. Phys.*, **65**, 1872 (1976).
42. A. T. Pritt, Jr. and R. D. Coombe, *J. Chem. Phys.*, **65**, 2096 (1976).
43. R. D. Coombe and A. T. Pritt, Jr., *J. Chem. Phys.*, **66**, 5214 (1977).
44. R. E. Meridith and F. G. Smith, *J. Quant. Spect. Radiat. Trans.*, **13**, 89 (1973).
45. F. J. Comes and S. Pionteck, *Chem. Phys. Lett.*, **42**, 558 (1976).
46. J. J. Deakin and D. Husain, *J.C.S. Faraday Trans. II*, **68**, 41 (1972).
47. R. J. Donovan, C. Fotakis, and M. F. Golde, *J.C.S. Faraday Trans. II*, **72**, 2055 (1976).
48. J. R. Wiesenfeld and G. L. Wolk, *J. Chem. Phys.*, **67**, 509 (1977).
49. C. Fotakis and R. J. Donovan, *Chem. Phys. Lett.*, **54**, 91 (1978).
50. A. J. Grimley and P. L. Houston, *J. Chem. Phys.*, **70**, 4724 (1979).
51. R. J. Donovan and D. Husain, *Trans. Faraday Soc.*, **62**, 1050 (1966).
52. R. J. Butcher, R. J. Donovan, and R. H. Strain, *J.C.S. Faraday Trans. II*, **70**, 1837 (1974).
53. D. H. Burde and R. A. McFarlane, to be published.
54. D. Husain and J. R. Wiesenfeld, *Trans. Faraday Soc.*, **63**, 1349 (1967).
55. D. H. Burde and R. A. McFarlane, *J. Chem. Phys.*, **64**, 1850 (1976).
56. J. J. Deakin and D. Husain, *J.C.S. Faraday Trans. II*, **68**, 1603 (1972).
57. P. L. Houston, unpublished data.
58. R. J. Donovan and D. Husain, *J.C.S. Faraday Trans.*, **62**, 2073 (1966).
59. A. T. Young and P. L. Houston, unpublished data.
60. J. C. Tully, Dynamics of molecular collisions, in *Modern Theoretical Chemistry*, Vol. 2, W. H. Miller ed., Plenum, New York, 1976.
61. E. E. Nikitin, *Opt. Spectrosk.* **9**, (1960); A. Bjerre and E. E. Nikitin, *Chem. Phys. Lett.*, **1** 179 (1967); E. E. Nikitin, *Theory of Elementary Atomic and Molecular Processes in Gases*, Oxford University Press, New York, 1974.
62. J. J. Ewing, *Chem. Phys. Lett.*, **29**, 50 (1974).
63. J. C. Tully and R. K. Preston, *J. Chem. Phys.*, **55**, 562 (1971).
64. W. H. Miller and T. F. George, *J. Chem. Phys.*, **56**, 5637 (1972).
65. I. H. Zimmerman and T. F. George, *J. Chem. Phys.*, **61**, 2468 (1974); *J.C.S. Faraday Trans. II*, **71**, 2030 (1975); *Chem. Phys.*, **7**, 323 (1975).
66. W. H. Miller, *J. Chem. Phys.*, **68**, 4431 (1978).
67. G. Karl, P. Kruus, and J. C. Polanyi, *J. Chem. Phys.*, **46**, 224 (1967).

68. P. G. Dickens, J. W. Linnett, and O. Sovers, *Disc. Faraday Soc.*, **33**, 52 (1962).
69. E. R. Fisher and E. Bauer, *J. Chem. Phys.*, **57**, 1966 (1972); E. Bauer, E. R. Fisher and F. R. Gilmore, *J. Chem. Phys.*, **51**, 4173 (1969); E. R. Fisher and G. K. Smith, *Chem. Phys. Lett.*, **13**, 448 (1972).
70. R. D. Levine and R. B. Bernstein, *Chem. Phys. Lett.*, **15**, 1 (1972); A. D. Wilson and R. D. Levine, *Mol. Phys.*, **27**, 1197 (1974).
71. N. P. D. French and K. P. Lawley, *Chem. Phys.*, **22**, 105 (1977).
72. L. D. Landau, *Phys. Z. Sowjet. URSS*, **2**, 46 (1932).
73. C. Zener, *Proc. Roy. Soc. (London)*, **A137**, 696 (1932).
74. E. C. G. Stuckelberg, *Helv. Phys. Acta*, **5**, 369 (1932).

Section 5

STUDIES IN CONDENSED MEDIA

COHERENT OPTICAL TRANSIENT STUDIES OF DEPHASING AND RELAXATION IN ELECTRONIC TRANSITIONS OF LARGE MOLECULES IN THE CONDENSED PHASE

DOUWE A. WIERSMA

Picosecond Laser and Spectroscopy Laboratory of the Department of Physical Chemistry, State University, Nijenborgh 16, 9747 AG Groningen, The Netherlands

CONTENTS

I. INTRODUCTION

Since the pioneering work by Davydov[1] and McClure[2] on the optical spectra of molecules in pure and mixed crystals, spectroscopists have been actively engaged in the assignment of electronically excited states and analysis of complex molecular spectra. As a result, our understanding of the effects of vibronic coupling,[3] vibronic mixing[4] and spin–orbit coupling[5] on the spectrum of a particular electronic transition and radiationless processes in the excited state[6] has grown considerably. In this era, scientific interest was mainly focused on the intensities and positions of spectral lines.

It is well known, however, that the width of a spectral line,[7] at least in principle, yields information on the dephasing dynamics of the optical transition. Spectral lineshapes of purely electronic transitions in solids unfortunately are seldom determined by dynamic interactions, but, at least at low temperature, quite often by the effects of strain. The observed, named inhomogeneous linewidth is therefore of little interest. In case of vibronic transitions, however, the effect of vibrational relaxation on the lineshape may exceed the inhomogeneous linebroadening. Even so, classical spectroscopy quite often fails to elucidate the nature and strength of the perturbing forces on the optical (homogeneous) lineshape.

It was only after the invention of the laser that new possibilities for performing experiments arose whereby the intrinsic (homogeneous) optical lineshape could be studied. After the first successful generation and detection of a photon echo in ruby, by Abella, Kurnit, and Hartmann,[8] there has been a growing interest in the exploitation of this phenomenon to obtain new information on the dephasing of optical transitions. Next to the photon echo, line-narrowing (or hole-burning),[9] and optical free induction decay (OFID)[10] have proven to be extremely useful techniques to study the intrinsic (homogeneous) optical lineshape. With the advancement in construction of pulsed and cw dye lasers,[11] *any* optical transition can now be excited and studied. This has led to usage of coherent optical phenomena to the study of optical transitions in molecular mixed crystals.[12] With the recent success of extending the optical free induction[13] and photon-echo[14] decay measurements into the picosecond domain these techniques have reached their full potential. Recently quite a number of new laser-induced phenomena as resonance Rayleigh scattering,[15] polarization,[16] and four-wave mixing spectroscopy[17] have been shown also to yield information on the optical lineshape. Coherent anti-Stokes Raman–scattering[18, 19] and gain[20] measurements have been employed to study the vibrational dephasing dynamics of vibrations in the electronic ground state. These latter techniques will not be dealt with in this chapter, but we will concentrate on the progress in the field that has been made using photon-echo, free induction decay, or hole-burning techniques. Each of these coherence probe techniques has its own merits and pitfalls, and a critical evaluation seems appropriate. As the coherent transient phenomena themselves are basically understood,[21, 22] we will focus our interest in this report not on the phenomena but on the results and in particular on the results obtained in the study of optical dephasing in molecular mixed crystals.

II. OPTICAL COHERENCE

In the field of molecular spectroscopy the word coherence has different meanings. Energy transport in molecular solids is said to be coherent,[23, 24] when the phenomenon is best described in terms of Bloch states which in a one-dimensional lattice are of the form:

$$\psi^f(\mathbf{k}) = N^{-1/2} \sum_{n\alpha}^{N} \exp i(\mathbf{k}\cdot\mathbf{r}_{n\alpha})\phi^f_{n\alpha} \tag{1a}$$

where

$$\phi^f_{n\alpha} = A(\phi^f_{n\alpha}) \prod_{\substack{m\beta \\ \neq n\alpha}} \phi^0_{m\beta} \tag{1b}$$

with $\phi^f_{n\alpha}$ representing an excited state f at site $n\alpha$. $\mathbf{k}$ represents the wave vector, $\mathbf{r}_{n\alpha}$ the lattice vector from the origin to site $n\alpha$ and A the antisymmetrization operator working between molecules. Exciton coherence in this case is thus defined with respect to the on-site Frenkel[25] states and coherent exciton motion just means that the intermolecular exchange dominates the electron–phonon coupling. In the other extreme where the energy transport is best described as a stochastic process, of hopping between sites, we say that the energy migration is incoherent or diffusive.

The same kind of terminology has been used to discuss the optical[26] and magnetic resonance[27] spectra of the naphthalene dimer (miniexciton). For a recent review on the optical and magnetic resonance spectra of dimers the paper by Burland and Zewail is recommended.[28]

The word coherence has also been used in the description of radiationless processes (RP) in molecules in the gasphase. In the so-called molecular eigenstate basis set-description of RP, the initial state prepared after flash excitation can be written as a superposition of quasistationary states:

$$\psi(t=0)=\sum_n A_n|n\rangle \tag{2a}$$

This state develops in time as follows:

$$\psi(t)=\sum_n A_n \exp\left(\frac{-iE_n t}{\hbar}\right)|n\rangle \tag{2b}$$

It may be shown[29] that the time-resolved emission from such a system is given by

$$I(t)=\Gamma_s\Big(\sum_{m\neq n}\sum A_n A_m^* \alpha_n \alpha_m^* \exp(-i(\epsilon_n-\epsilon_m)t_e-\{(\gamma_n+\gamma_m)/2\}t)$$
$$+\sum_n |A_n|^2|\alpha_n|^2\exp(-\gamma_n t)\Big) \tag{2c}$$

where Γ_s is the emission rate constant: $\epsilon_i=E_i/\hbar$ and γ_i is the Wigner–Weisskopf damping rate constant of level i. The first term in (2c) is called the coherent contribution to the molecular radiative decay. It is due to contructive interference between the radiating molecular eigenstates. This emission only persists for a time Δ^*, where Δ^* represents the spread in E_i. In a sparse level structure this interference effect will lead to quantum beats[30] which have recently also been observed in a large molecule.[31] Note that coherence in this case defines a microscopic state. The second term in (2c) is called the incoherent decay component, which in this description is due to independently decaying states.

In this paper we discuss *optical coherence* phenomena which are based on the presence of a macroscopic optical polarization. Such a macroscopic optical polarization (P) may be created, when we excite by using a laser an ensemble of molecules into a superposition of phased states. For a two-level system the superposition state can be described as

$$\psi(t)=c_1(t)\exp\left(\frac{-iE_1 t}{\hbar}\right)|1\rangle+c_2(t)\exp\left(\frac{-iE_2 t}{\hbar}\right)|2\rangle \tag{3a}$$

The existence of a macroscopic oscillating dipole moment implies that

$$P=\langle\Psi(t)|er|\Psi(t)\rangle=c_1^*(t)c_2(t)\exp[-i(E_2-E_1)t/\hbar]\mu_{12}+c.c.$$
$$=Tr(\rho\mu) \tag{3b}$$

where the brackets $\langle\ \rangle$ stand for ensemble average, ρ is the density matrix and μ the transition moment operator. As μ has only off-diagonal elements, the existence of P implies ρ_{12} to be nonzero. During the existence of P the emission of all molecules in the ensemble is correlated, which implies that the emission intensity is proportional to:

$$I_N^c \alpha \left|\left(\sum_i^N \langle\mu_{12}^i\rangle\right)^2\right| \cong N\sum_i |\langle\mu_{12}^i\rangle|^2 \tag{4}$$

where N is the number of molecules in the ensemble and μ_{12}^i the two-level transition moment at site i.

The *coherent* emission intensity of an ensemble of N molecules is therefore N times stronger than the *incoherent* emission. This result is due to the $N(N-1)$ cross-terms in the expansion as first shown by Dicke.[32] A closer look[8] at this enhanced spontaneous emission shows that the coherent emission is also highly *directional*, in fact in a sample of macroscopic size the constructive interference effects only occur in the direction of the exciting laser beam.

III. OPTICAL COHERENCE EFFECTS

A. Photon Echo

An extremely useful description of coherence effects in two-level systems may be given by representing the effect of the laser field on the ensemble of molecules in the form of the well-known optical analog of the Bloch[33] equations in the rotating frame[34]:

$$\frac{d\langle P^R\rangle}{dt}=\langle P^R\rangle\times E_{\text{eff}}-\frac{(\langle P_x^R\rangle\hat{x}+\langle P_y^R\rangle\hat{y})}{T_2}-\frac{(\langle P_z^R\rangle-\langle P_0^R\rangle)\hat{z}}{T_1} \tag{5}$$

with

$$\langle P^R\rangle=\text{Tr}\,\rho P^R;\ E_{\text{eff}}=\begin{bmatrix}\gamma E_0\\ 0\\ \Delta\Omega j\end{bmatrix}\qquad \text{where } \gamma=\sqrt{2}\cdot\frac{\mu_{12}}{\hbar}$$

$\Delta\Omega j = \Omega_j - \Omega$ where Ω is the frequency of the applied laser field $E_x = \sim 2E_0 \cos \Omega t$ and Ω_j the optical frequency of the molecule excited. T_1 and T_2 are the longitudinal and transverse optical relaxation constants; also

$$\frac{1}{T_2} = \frac{1}{T_2^*} + \frac{1}{2T_1} \tag{6}$$

in the Markoff approximation, where T_2^* is the pure dephasing time. We further note at this point that the transverse optical relaxation time is related to the width (FWHM) of the homogeneous lineshape (Δv_h) as follows

$$\Delta v_h = (\pi T_2)^{-1} \tag{7}$$

In Fig. 1 we show a schematic representation of the precession that the macroscopic polarization (P^R) undergoes in the rotating frame. For a strong resonant field ($\gamma E_0 \gg \Delta\Omega_j$) the effect of a pulse on P^R can be easily figured out by inspection. When P^R is rotated over $\pi/2$ by a laser pulse we say that the applied laser pulse was a $\pi/2$ pulse, etc.

Feynman, Vernon, and Hellwarth[35] were the first ones who showed that *any* two-level system interacting with a field may be represented in the form of a torque equation. This assures the formal equivalence of different type of echo effects. Compared to nuclear spin echoes first observed and interpreted by Hahn[34] the photon echo distinguishes itself by its spatial properties, which are due to the effect that the exciting wavelength is much smaller than the crystal size. The relaxation terms in (5), describing the interaction of a molecule with its surroundings, are added in a phenomenological way. The microscopic interpretation of these constants must come

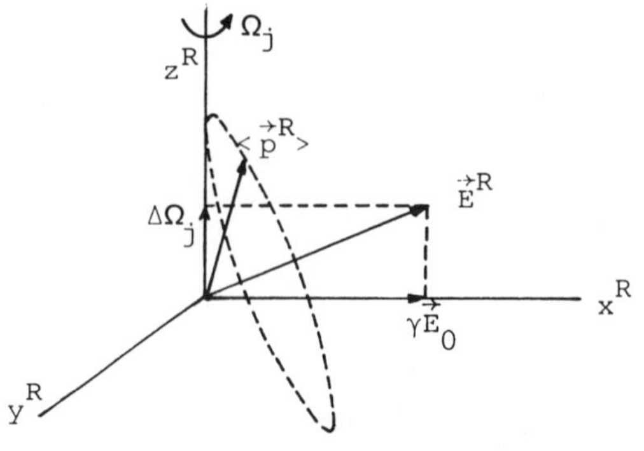

Fig. 1. Precession of the macroscopic polarization (P) in the rotating frame.

from a model Hamiltonian describing the interaction of the molecule studied, with the heat bath. This is the subject of Section IX. We also wish to emphasize that only in a (pseudo) two-level system the relaxation may be described by only two constants T_1 and T_2. In a multilevel system the observed echo decay is not necessarily exponential but may become oscillatory and the dephasing time is not a well-defined quantity.

When all molecules of an ensemble have identical optical resonance frequencies (as in a beam), the decay of the coherent optical emission is only determined by processes that affect the phase of the wavefunction of the excited molecules. In solids as in gases, quite often an appreciable spread in resonance frequencies is observed caused by either crystal inhomogeneities or the Doppler effect. The observed optical lineshape, named *inhomogeneous*, is in this case due to the sum of the lineshapes of all oscillators with varying frequencies. If the spread in frequencies is Gaussian, the observed lineshape is Gaussian and the situation may be sketched as shown in Fig. 2. The lineshape of a collection of identical oscillators is named the *homogeneous* lineshape and contains information on optical dephasing processes. Imagine now what would happen if we were to excite, with a very short laser pulse, such an inhomogeneous distribution of optical oscillators. As the initial phase of all oscillators is identical, we expect constructive interference in the emission whereby the emission should follow an N^2 intensity dependence and be highly directional. This free induction decay (OFID) will be determined by the Fourier transform of the inhomogeneous distribution as follows:

$$\text{OFID}\ \alpha \int_{\infty}^{\infty} \exp\left[\frac{-(\omega-\omega_0)^2 T_2^{i2}}{2}\right] \exp[i(\omega-\omega_0)t]\, d\omega = \alpha \exp\left(\frac{-t^2}{2T_2^{i2}}\right) \qquad (8)$$

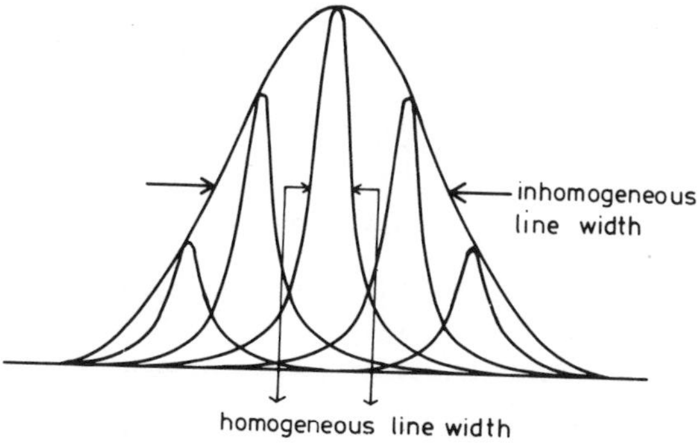

Fig. 2. Illustration of the difference between homogeneous (microscopic) and inhomogeneous (macroscopic) line width.

where ω_0 is the center frequency and $2(2\ \ln 2)^{1/2}T_2^i$ (in rad s^{-1}) is the full width at half-maximum (FWHM) of the Gaussian inhomogeneous distribution.

From (8) we see that the decay of the OFID is completely determined in this case by the spread in resonance frequencies up to the point where the incoherent decay takes over. It is well known[34] that a second laser pulse effectively induces time reversal, which implies that, at twice the exciting pulse separation, rephasing of the *macroscopic oscillating* dipole moment occurs, which results in the emission of a pulse (photon echo) riding on top of the incoherent emission. With short-pulse excitation the echo pulse shape is composed of two free induction decays back to back as indicated in Fig. 3. The decay of the photon echo with increasing pulse separation is, next to spontaneous emission and radiationless relaxation, only due to pure dephasing processes, which interrupt the phase of the participating optical oscillators. Abella et al.[8] have shown that for two pulses having wave vectors $\mathbf{k}_1$ and $\mathbf{k}_2$ the unrelaxed photon echo intensity is given by

$$I(k)=\tfrac{1}{4}N^2I_0(k)|\{\exp i(\mathbf{k}_e+\mathbf{k}_1-2\mathbf{k}_2)\cdot r\}_{\mathrm{av}}|^2 \tag{9}$$

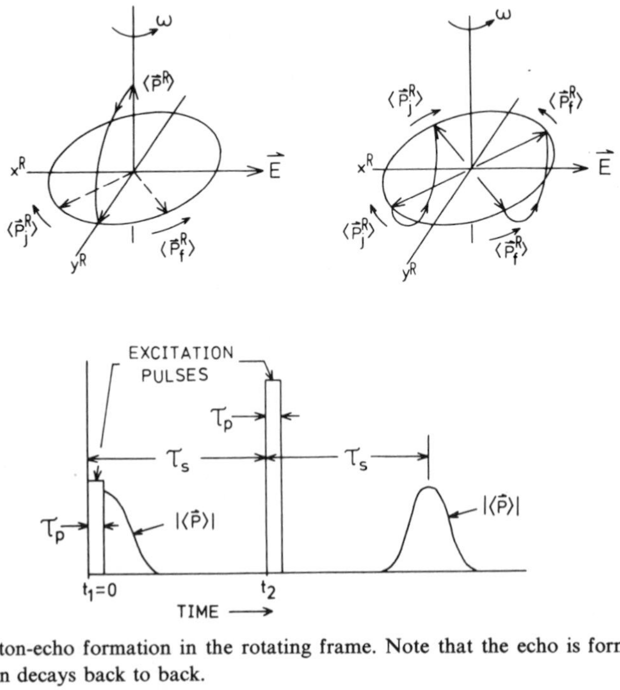

Fig. 3. Photon-echo formation in the rotating frame. Note that the echo is formed as two free induction decays back to back.

where $\mathbf{k}_e$ is the wave vector of the echo, the averaging is done over all crystal sites, and $I_0(k)$ is the incoherent emission intensity of an isolated molecule. For collinear excitation pulses and the molecules packed in a cubic array, $I(k)$ is calculated, in the limit of a large Fresnel number, to be[8]

$$I(k)=\frac{1}{4}I_0(k)\frac{N^2\lambda^2 l}{A\epsilon l} \tag{10}$$

where A is the excited cross-section area, ϵ the dielectric constant of the crystal, λ the wavelength of the exciting radiation, and l the crystal thickness. Abella et al.[8] remarked that the coherence volume factor $(\lambda^2 l/\epsilon)$ is very important in the sense that its omission would predict an extremely short lifetime of the excited ensemble, thus precluding photon-echo observation. Another important point of practical interest is that according to (9) the photon-echo intensity peaks when $\mathbf{k}_e=2\mathbf{k}_2-\mathbf{k}_1$, which enables spatial separation of the echo from the excitation pulses.

B. Optical Free Induction Decay

In a photon-echo experiment the presence of inhomogeneous broadening is crucial. In the free induction decay method for probing of optical dephasing processes, on the contrary, inhomogeneous broadening is a nuisance. In an OFID experiment, the optical transition is excited, with a cw laser whose linewidth is narrow compared to the homogeneous linewidth, until steady state is reached. Hereafter the laser excitation is suddenly stopped and the free induction decay of the excited molecules follows. In practice this OFID is monitored by looking at the heterodyne beat between the OFID and the laser at a different frequency. For an ensemble of inhomogeneously broadened two-level systems Hopf et al.[36] calculated for the decay of the ensemble average of $\rho_{12}(t)$ in the rotating frame:

$$\langle\tilde{\rho}_{12}(t)\rangle\propto\chi\left[1-\left(1+\chi^2T_2T_1\right)^{-1/2}\right]\exp\left\{-\frac{1}{T_2}\left[1+\left(1+\chi^2T_1T_2\right)^{1/2}\right]\right\} \tag{11}$$

with the definitions

$$\tilde{\rho}_{12}(t)=\rho_{12}\exp\left[-i(\Omega t-kz)\right] \qquad \chi=\frac{\mu_{12}E_0}{\hbar}$$

where μ_{12} is the transition dipole and E_0 the field strength of the applied laser field $E_x(z,t)=E_0\cos(\Omega t-kz)$, which propagates in the z-direction while being polarized along x. Finally T_1 and T_2 are the earlier defined longitudinal and transverse optical relaxation constants of the optical Bloch equations. Equation (11) clearly shows that the OFID decay is only determined by purely dephasing processes in the limit where the power-broadening term $(\chi^2 T_1 T_2)$ is negligible.

C. Optical Nutation

There is another optical coherence effect, optical nutation,[37] that is quite useful especially in measuring the transition dipole. With an intense laser pulse the system is driven from the ground state to the upper state (stimulated absorption) and then back to the ground state (stimulated emission), the cycle repeating until the pulse ends. The transmitted laser beam being alternately absorbed and emitted, will therefore display oscillations at the Rabi frequency, which is being determined by the laser field strength and the transition dipole. The damping of the beats again is being determined by the dephasing dynamics.

D. Hole-Burning

There is another way of measuring T_2 of an optical transition which is not based on a coherent optical effect but also employs a coherence property of the laser, namely, its monochromaticity. In the hole-burning[38] technique, as in the OFID method, a cw dye laser is used to create a hole in the absorption spectrum. When this hole is transient, its width, being determined in the low-intensity limit by $2(\pi T_2)^{-1}$, may be probed by side-band modulation as first demonstrated by Szabo on ruby.[39] When the hole is permanent, as is the case in photochemical hole-burning, the width may be easily measured by means of a narrow-band excitation spectrum as first performed by de Vries and Wiersma on dimethyl *s*-tetrazine in durene.[40]

Prior to continuing with a more detailed examination of these coherence effects and discussing the results obtained, it seems important to briefly reflect on the relationship between the density matrix and the optical homogeneous lineshape and consider the question which state initially is excited in a large molecule.

IV. COHERENT EFFECTS AND OPTICAL LINESHAPE

The time evolution of any system can be described by the density matrix equation

$$\frac{d}{dt}\sigma = \frac{-i}{\hbar}[H^f, \sigma] \tag{12}$$

where σ is the density matrix and H^f the full Hamiltonian of the complete system. It is often convenient to decompose H^f into parts such that we can focus our interest on the dynamics of a subensemble of the complete system. We therefore expand H^f as follows:

$$H^f = H_m + V_{mb} + H_b + H_r + H_{mr} \tag{13}$$

whereby H_m is the Hamiltonian of the subensemble (e.g.) molecule, H_b the hamiltonian of the remaining degrees of freedom (bath), V_{mb} the interaction between molecule and bath, H_r the radiation field Hamiltonian and H_{mr} the interaction between radiation field and molecule. Note that the Hamiltonian in (13) already is an approximation of the exact one in the sense that the interaction of the radiation field with the bath has been neglected. There is a second important remark to be made on the decomposed form of H^f in (13). In general there is no unique way of partitioning the complete Hamiltonian in a part belonging to the molecule and the bath. This ambiguity, which is of fundamental importance in the discussion on relaxation and dephasing, is further looked at and discussed in greater detail in Section IX. What is quite often done, in a phenomenological approach of the description of coherent optical effects, is to concentrate on the dynamics of the subensemble, while taking care of the various interactions through the introduction of ad hoc constants describing feeding and decay. While this procedure is obviously not very satisfactory and the limitations of this approach not transparent, it has shown to be extremely useful. The time evolution of the molecular density matrix ρ, the so-called reduced density matrix, in this approach is then described by the following phenomenological equation:

$$\frac{d}{dt}\rho = \frac{-i}{\hbar}[H, \rho] - \tfrac{1}{2}\{K, \rho\} + F \tag{14}$$

hereby is $H = H_m + H_{mr}$, K the decay matrix which incorporates the effect of radiative and radiationless decay, and F the feeding matrix. Quite often in the optical domain, H_{mr} is approximated by the interaction of the radiation field with the molecular transition dipole (μ_{ij}) between the initial (i) and final (j) state; $H_{mr} = -\boldsymbol{\mu}_{ij} \cdot \mathbf{E}_0 \cos(\Omega t - kz)$. So far we have dealt only with a "homogeneous" molecule density matrix equation, while one should realize that only information on the ensemble averaged density matrix ($\langle \rho \rangle$) can be obtained. In fact, experimentally we obtain only information on $\langle \rho \rangle$ through measurement of the transient signal electric field during or after excitation of the sample. We therefore must calculate the ensemble-averaged polarization

$$\langle P(z,t) \rangle = \mathrm{Tr}(\mu \rho(t)) \tag{15}$$

which acts as a source term in the Maxwell equation to generate the transient signal field. It has been shown[41] that for optically thin samples, Maxwell's equation may be solved by using the "slowly varying envelope approximation." This results in the following linearized wave equation, which connects the signal field with the induced polarization:

$$\left(\frac{\partial}{\partial z}+\frac{1}{\eta c}\frac{\partial}{\partial t}\right)\langle E_s(z,t)\rangle=\frac{-\Omega}{2\epsilon_0 c}\langle P(t)\rangle \tag{16}$$

This may be even further simplified by realizing that the spatial variation of $\langle E_s(z,t)\rangle$ in most experiments, whereby sample lengths of $\gtrsim 1$ mm are used, greatly exceeds the temporal variation in $\langle E_s(z,t)\rangle$. In this case we obtain the simple result that

$$\langle E_s(z,t)\rangle \cong \frac{\Omega\eta l}{2c\epsilon_0}\langle P(t)\rangle \tag{17}$$

where η is the index of refraction of the sample, l the length, c the speed of the light, and ϵ_0 the vacuum permittivity constant. From the foregoing discussion it is clear that only the ensemble-averaged molecular density matrix needs to be calculated to describe coherent optical phenomena in weakly absorbing samples. We also note that (15) shows that for a calculation of $\langle P(t)\rangle$ only the off-diagonal elements of the density matrix need to be known. Therefore effectively the signal field for a two-level system is given by $E_s(z,t)\cong-(\Omega\eta l/2c\epsilon_0)\langle\rho_{12}(t)\rangle$. In optically thick samples (17) does not hold and the transient signals can only be described by simultaneous solution of the Schrödinger and Maxwell equation. The interesting effect of self-induced transparency, which results of this coupling, has been observed in solids and gases and described in great detail by McCall and Hahn[42] and Slusher and Gibbs.[43] This effect, however, is outside the scope of this chapter and will not further be dealt with.

We further wish to emphasize the formal relationship that exists between the optical lineshape and the elements of the molecular density matrix. The general expression for the absorption lineshape is [44]:

$$I(\omega)\alpha\int_{-\infty}^{\infty}dt\,e^{-i\omega t}\langle\langle\mu(0)\mu(t)\rangle\rangle \tag{18}$$

where ω is the photon frequency, μ the ensemble's transition dipole and $\langle\langle\ \rangle\rangle$ denotes averaging over the initial ensemble states (of the molecule and bath). In this lineshape formula the exciting field is absent, so that saturation effects are not incorporated. We may further write, assuming a two-level system,[45]

$$\langle\langle\mu(0)\mu(t)\rangle\rangle = \langle\mu_{if}(0)\mu_{fi}(t)\rangle = \mu_{if}(0)\langle\mu_{fi}(t)\rangle$$
$$= \rho_{fi}(t)|\mu_{if}|^2 \tag{19}$$

where we use the fact that $\mu_{if}(0)$ is independent of the bath and therefore commutes with the trace over the bath. We thus have, combining (18) and (19):

$$I(\omega)\alpha\int_{-\infty}^{\infty} dt\, e^{i\omega t}\rho_{fi}(t) = 2\,\mathrm{Re}\int_{0}^{\infty} dt\, e^{i\omega t}\rho_{fi}(t) \tag{20}$$

where $\rho_{fi}(t)$ is the solution of (14) with the initial condition $\rho_{fi}(0)=1$. We have thus seen that the transient signal field and the optical lineshape are both determined by the off-diagonal elements of the density matrix. This ascertains that measurement of the optical coherent transients can yield information on the dephasing dynamics of optical transitions.

Another problem of immediate interest to the observation of coherent transients in condensed organic molecules is the question, which electronic state initially is prepared by optical excitation? This question, which is the central one in the theory of radiationless processes, has been studied in great detail and the interested reader is referred to a recent review on the subject by Jortner and Mukamel.[46] For optical excitations in solids it seems safe to assume that Born–Oppenheimer singlet and triplet states are the initially prepared states.

In the Born–Oppenheimer approximation the total vibronic wave function of a molecule may be written as a product function:

$$\psi_{im}(q,Q) = \varphi_i(q,Q)\chi_m(Q) \tag{21}$$

where $\varphi_i(q,Q)$ is the electronic and $\chi_m(Q)$ the vibrational wavefunction. In this approach the total energy may be given as a sum $E_{im}(q,Q) = E_i(q,Q) + E_m(Q)$, which enables description of the optical excitations in the form of energy diagrams. A most convenient diagram is obtained for the equilibrium configuration $Q=Q_0$ of the different states and in Fig. 4 we show such a so-called Jablonski diagram of the lower electronic states of a typical organic molecule. The coupling between the lowest excited singlet and triplet state, induced by the spin–orbit interaction in the Born–Oppenheimer description, has been examined in great detail.[47] This coupling, which in the solid phase leads to decay of the singlet state into the triplet state, is named intersystem crossing. Another process that depletes the excited singlet state without emission, is that where the excited

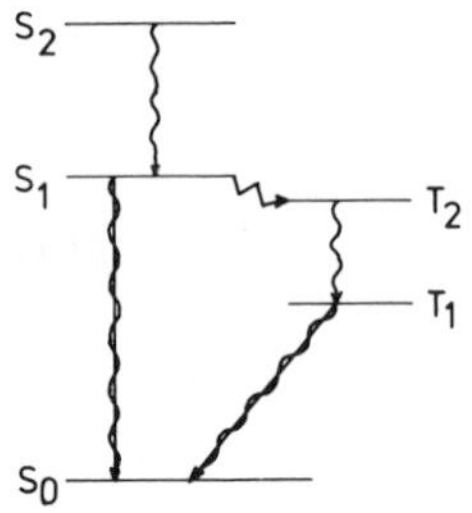

Fig. 4. Jablonski diagram for a typical (large) organic molecule. The straight arrows indicate radiative decay, while the wavy arrows stand for radiationless decay.

singlet state decays into the ground state. This radiationless process is called internal conversion. These processes together are responsible for the fact that the emission quantum yield of a molecule is often far less than 1. For our purpose it is sufficient to know that these processes may be accounted for in the solid by phenomenological constants in the decay and feeding matrix.

V. HOLE-BURNING IN OPTICAL TRANSITIONS

A. Introduction

In recent years high-resolution photochemical hole-burning[40] has shown to be an important technique to study homogeneous lineshapes of molecules in the condensed phase. Especially the unique possibility of measuring simultaneously the low-temperature linewidth and shift[48] has added much to our understanding of optical dephasing processes. So far however, the relationship between the observed hole-width and optical dephasing constants was not firmly established.

De Vries and Wiersma[40] used in their analysis of the deconvoluted hole-width (W) the following relationship $W=2.\ (\pi T_2)^{-1}$, where T_2 is the optical dephasing constant in the optical analog of the famous Bloch equations.[33] Such a relationship may indeed be derived as Sargent et al.[49] showed when the molecule is treated as a statistically closed two-level system and if the hole-depth is negligible. Voelker et al.[50] showed how a graphical relationship between the hole-width and depth could be made, which was used to extract the homogeneous linewidth from the observed hole-width. This procedure seemed satisfactory as at low temperature, the hole-width so determined was very close, though not identical, to the one calculated from the upper state fluorescence lifetime. The remaining difference was assigned to "some minor temperature independent dephasing mechanism." In a previous section we have shown that, especially in a quasi steady state situation, the upper state may not be considered isolated from the triplet state. It therefore seems necessary to investigate optical hole-burning starting from a microscopic model that incorporates all the important decay channels.

In the following we will examine transient hole-burning in a multilevel system using a density matrix calculation. It will be shown that with an appropriate extrapolation procedure from the observed hole-width not only optical T_2 can be measured but also the transition dipole and intramolecular relaxation constants may be obtained. We will then proceed by using a kinetic model to examine the effect of an irreversible decay channel (photochemistry) on the hole-width. A detailed account of this work will be published elsewhere.[51]

B. A Density Matrix Description of Hole-Burning in a Three-Level System

For a three-level system as depicted in Fig. 5 and with the exciting laser field $E_x = E_0\cos(\omega t - kz)$ (near)-resonant with the $\langle 2|\leftrightarrow|1\rangle$ optical transition the following set of equations may be derived[52]

$$\begin{aligned}
\dot{\rho}_{11} &= k_{21}\rho_{22} + k_{31}\rho_{33} + \tfrac{1}{2}i\chi(\tilde{\rho}_{21} - \tilde{\rho}_{12}) \\
\dot{\rho}_{22} &= -K_2\rho_{22} + \tfrac{1}{2}i\chi(\tilde{\rho}_{12} - \tilde{\rho}_{21}) \\
\dot{\rho}_{33} &= k_{23}\rho_{22} - k_{31}\rho_{33} \\
\dot{\tilde{\rho}}_{12} &= (-\Gamma + i\Delta)\tilde{\rho}_{12} + \tfrac{1}{2}i\chi(\rho_{22} - \rho_{11}) \\
\rho_{11} + \rho_{22} + \rho_{23} &= 1
\end{aligned} \tag{22}$$

with the following definitions:

$$\begin{aligned}
\Gamma &= T_2^{-1} \\
K_2 &= k_{21} + k_{23} \\
\Delta &= \omega_{\text{sat}} - \bar{\omega}_0 \\
\chi &= \frac{\mu_{12}E_0}{\hbar} \quad \text{the Rabi frequency and} \\
\tilde{\rho}_{12} &= \rho_{12}e^{-i(\omega t - kz)}
\end{aligned} \tag{23}$$

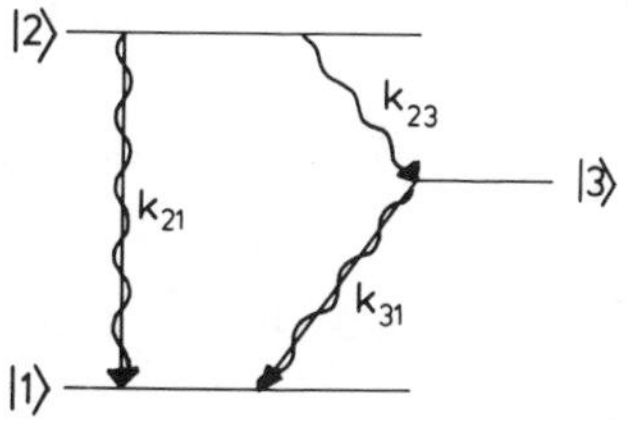

Fig 5. Typical pseudo three-level system for an organic molecule. The spin-degeneracy of the triplet state may quite often be ignored, in case only one level in the pumping cycle plays a role.

In the steady-state limit, obtained by putting $\dot{\rho}_{11}=\dot{\rho}_{22}=\dot{\rho}_{33}=\dot{\rho}_{12}=0$, these equations can easily be solved. After some algebraic manipulations we obtain for the density matrix elements $(\rho_{11}-\rho_{22})$, which determine the hole absorption:

$$\rho_{11}-\rho_{22}=\frac{\Gamma^2+\Delta^2}{\Gamma^2+\Delta^2+(\chi^2\Gamma/2K_2)(2+A)} \qquad \text{where } A=\frac{k_{23}}{k_{31}} \tag{24}$$

We note here that in the case where all triplet spin sublevels are incorporated (24) remains identical except for A, which becomes

$$\sum_i \frac{k_{23}^i}{k_{31}i}$$

The probe absorption coefficient $\alpha(\nu)$ can now be calculated by incorporating the (Gaussian) inhomogeneous distribution and convoluting with the homogeneous lineshape. In the limit of low optical density $\alpha(\nu)l \ll 1$, where l is the crystal length, $\alpha(\Delta_p)$ is found to be proportional to

$$\alpha(\Delta_p)(:)W(\omega_p)\left\{1-\frac{\hat{K}^2}{\Gamma'}\cdot\frac{\Gamma+\Gamma'}{(\Gamma+\Gamma')^2+\Delta_p^2}\right\} \tag{25}$$

where $\Delta_p=\omega_p-\omega_{\text{sat}}$

$$\Gamma=T_2^{-1}; \quad \Gamma'=\Gamma(1+\hat{K}^2T_2^2)^{1/2}$$

where $\hat{K}^2=(\chi^2\Gamma/2K_2)(2+A)$ and $W(\omega_p)$ represents the value of the inhomogeneous distribution at frequency ω_p.

From expression (25) we derive for the *observed* hole-width

$$\Delta v_{\text{FWHM}}^{\text{hole}}=\frac{1}{(\pi T_2)}\left[1+\left(1+\frac{\chi^2T_2}{2K_2}\left(2+\frac{k_{23}}{k_{31}}\right)\right)^{1/2}\right] \tag{26}$$

This expression shows that only in the limit of zero saturation intensity $(E\rightarrow 0)$ the observed hole-width equals twice the homogeneous line width

$$\Delta v_{\text{FWHM}}^{\text{hole}}=\frac{2}{\pi T_2}=2\Delta v_{\text{hom}}.$$

Is is also interesting to note that there is a simple relationship between the observed hole-width and the decay constant of the optical free induction decay:

$$\Delta\omega_{\mathrm{FWHM}}^{\mathrm{hole}} = \frac{2}{\tau_{\mathrm{OFID}}}$$

(See also Section VII.B.) Expression (26) further shows that also in optical hole-burning experiments the triplet state forms a bottleneck in the optical pumping cycle and leads to an increased hole width. We finally note that from the observed hole-width as a function of saturation power not only optical T_2 may be obtained but also the transition dipole.

C. A Kinetic Description of Photochemical Hole-Burning

In a photochemical hole-burning experiment molecules are removed from the optical pumping cycle, which implies that the trace of the density matrix is no longer constant. We have approached this problem by using a kinetic theory, which is warranted in the regime where $t \gg T_2$. In fact we have shown that a kinetic treatment of the steady hole-burning case dealt with in the previous section leads to an identical result as obtained from the density matrix. Especially for the solution of the transient populations in the non steady-state regime a kinetic approach is preferred as it leads to a less complicated set of equations that need to be solved. We will direct our attention to a molecule that photoconverts in the excited singlet state, photoconversion in the triplet state, however, leads to the same result. The model scheme is depicted in Fig. 6. The kinetic equations for this set of levels will be

$$\begin{aligned}\dot{N}_1 &= w(N_2 - N_1) + k_{21}N_2 + k_{31}N_3\\ \dot{N}_2 &= w(N_1 - N_2) - k_{21}N_2 - k_{23}N_2 - k_d N_2\\ \dot{N}_3 &= k_{23}N_2 - k_{31}N_3\end{aligned} \tag{27}$$

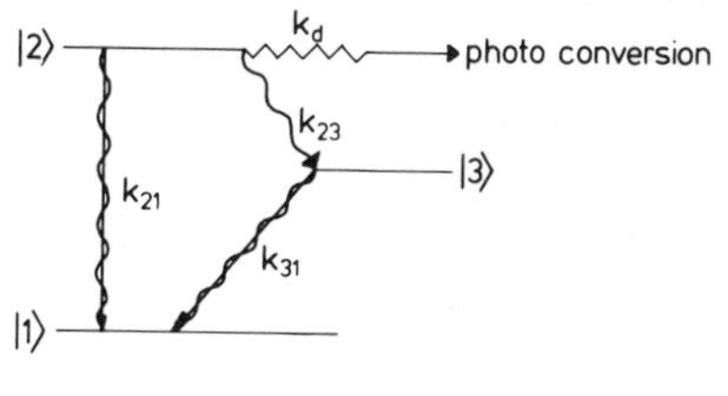

Fig. 6. Model scheme of a molecule which undergoes photoconversion or dissociation from the excited singlet state.

with w the pumping rate

$$w(\Delta)=\frac{T_2}{2}\left(\frac{\mu E}{\hbar}\right)^2\frac{1}{T_2\Delta^2+1} \tag{28}$$

$$\Delta=\omega_{\text{burn}}-\bar{\omega}_0$$

and initial conditions $N_1(0)=N$, $N_2(0)=N_3(0)=0$. This set of equations was solved by Laplace transformation and as the hole is probed after the burning laser is off, we are interested in the groundstate population $N_1(t)$ for $(t-t_{\text{off}})\gg$ all lifetimes in the system. The following relation is obtained at a time long after burning (t_{ab})

$$\frac{N_1}{N}(t_{ab},\Delta)=\left\{\frac{N_1}{N}(t_{\text{off}},\Delta)+(1-\phi_{\text{diss}})\frac{N_2}{N}(t_{\text{off}},\Delta)+\frac{N_3}{N}(t_{\text{off}},\Delta)\right\} \tag{29}$$

where $N_i(t_{\text{off}},\Delta)$ are complex functions of the pumping rate and all decay constants. The probe absorption coefficient again is calculated by incorporating the effect of inhomogeneous broadening and probe excitation. We find in the limit of very weak probe beam absorption

$$\alpha_p(:)\frac{1}{\pi T_2}\int_{-\infty}^{\infty}\frac{1}{\left(w_p-\bar{w}_0\right)^2+\dfrac{1}{T_2^2}}\frac{N_1}{N^c}(t_{ab},\Delta)d\Delta \tag{30}$$

where N for reason of convenience is taken at the inhomogeneous center frequency. This integral was calculated numerically for every probe beam frequency and for different burning times. We have performed computer calculations specifically for porphin whereof all relevant constants are known[53] and hole-widths as a function of burning time were reported.[50] Figure 7 shows some results for typical burning times. We have also included the results obtained from a steady-state theory of a two- and three-level system.

We finally note that computer analysis of the hole width as a function of burning power shows that the functional form of the power dependence of the photochemical hole is identical to the one obtained for steady-state hole-burning. This assures that an appropriate extrapolation procedure will enable us to determine the optical homogeneous lineshape from a photochemical hole-burning experiment. In the following section we will review the high-resolution photochemical hole-burning experiments that have been performed so far and emphasize the most important results obtained.

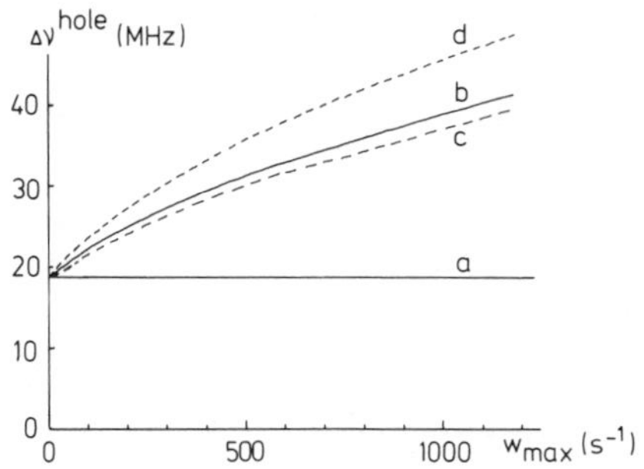

Fig. 7. Hole-width versus on-resonance pumping rate w_{max} for a two-level system (a), a closed three-level system (b) and a partially open system with burning times of 0.1 s (c) and 1 s (d), using the decay parameter values for porphin. Note the influence of the triplet bottleneck [(b) compared to (a)] and the influence of the burning time relative to the triplet decay time [(c) and (d) compared to (b)].

D. Results

The first high-resolution photochemical hole-burning experiment, which exposed the homogeneous linewidth of an electronic transition in an organic molecule at low temperature, was reported by de Vries and Wiersma in a study of the system dimethyl-*s*-tetrazine (DMST) in durene.[40] These experiments were similar in nature but not in execution to the frequency line-narrowing experiments performed by Szabo[38] on ruby. They were also related to preceding (low-resolution) hole-burning observations in solid solutions of organic molecules by Kharmalov, Personov, and Bykovskaya[54] and by Gorokovskii, Kaarli, and Rebane.[55] In the hole-burning experiment of de Vries and Wiersma the fact that DMST is photochemically unstable[56] was exploited. The idea of the experiment was that within the inhomogeneous distribution of the optical absorption only those molecules would photodissociate whose homogeneous linewidths overlapped with the exciting narrow-band (~30 MHz) laser. The low-temperature absorption and emission spectra of DMST in durene are given in Fig. 8. We note here that the observed inhomogeneous line width of the purely electronic transition is exceedingly sharp for organic mixed crystals; more surprisingly, we even found that the inhomogeneous linewidth of one of the naturally abundant isotopes was only 3 GHz! The hole burning experiment showed as displayed in Fig. 9a, b, that the hole in the origin lineshape was only 120 MHz from which a homogeneous linewidth of 55 ± 10 MHz was calculated. The experiment was later redone[57] with a frequency-stabilized cw dye-laser whereby the hole-width was extrapolated to zero laser burning time. In this case a hole-width of ~25 MHz was obtained from which was concluded that at low temperature the homogeneous linewidth $[(\pi T_2)^{-1}]$ is only determined by the upper state 6 ns decay constant. The identity

$$\frac{1}{T_2} = \frac{1}{T_2^*} + \frac{1}{2T_1}$$

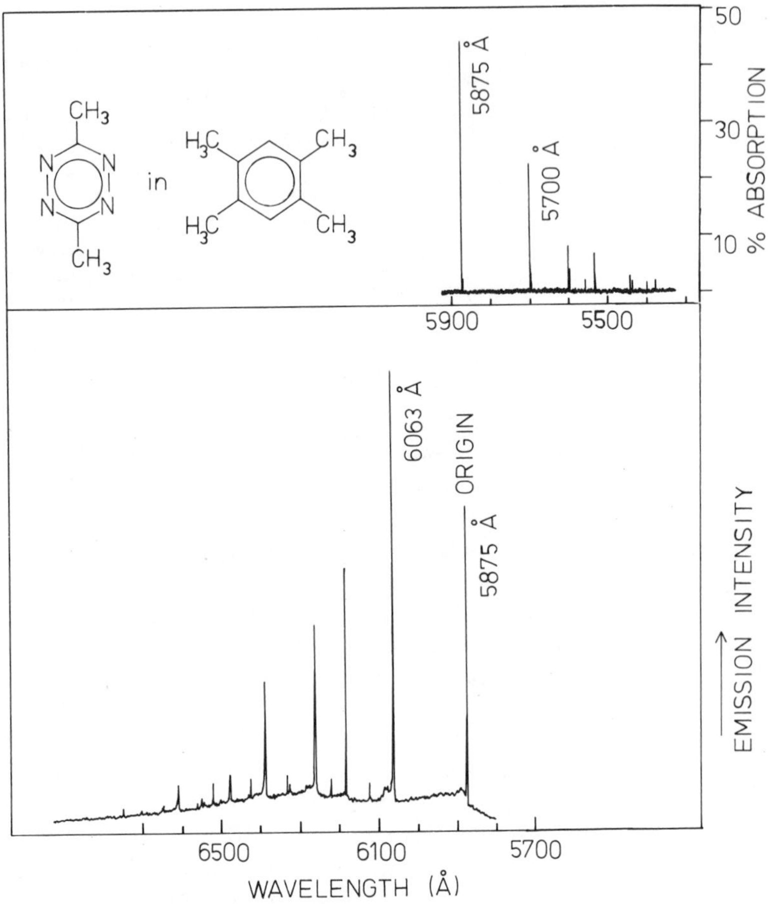

Fig. 8. Absorption and laser-induced (at 5700 Å) emission spectra of DMST in durene at 2 K. The intensity distribution in emission is not realistic because of inevitable decomposition of DMST during excitation. (After Ref. 40.)

then implies the absence of pure dephasing (T_2^*) processes at low temperature for the purely electronic transition. De Vries and Wiersma showed that this fact could also be exploited in a measurement of ultrashort fluorescent lifetimes of photochemically unstable molecules. From a photochemical hole-burning experiment of *s*-tetrazine in benzene[57] they calculated a fluorescence lifetime of $455 \pm {}^{75}_{55}$ ps in good agreement with an earlier direct measurement by Hochstrasser et al.[58]

Another type of hole-burning experiment on DMST was done whereby a vibronic transition was excited. In this case, as Fig. 9c shows, *no* hole was

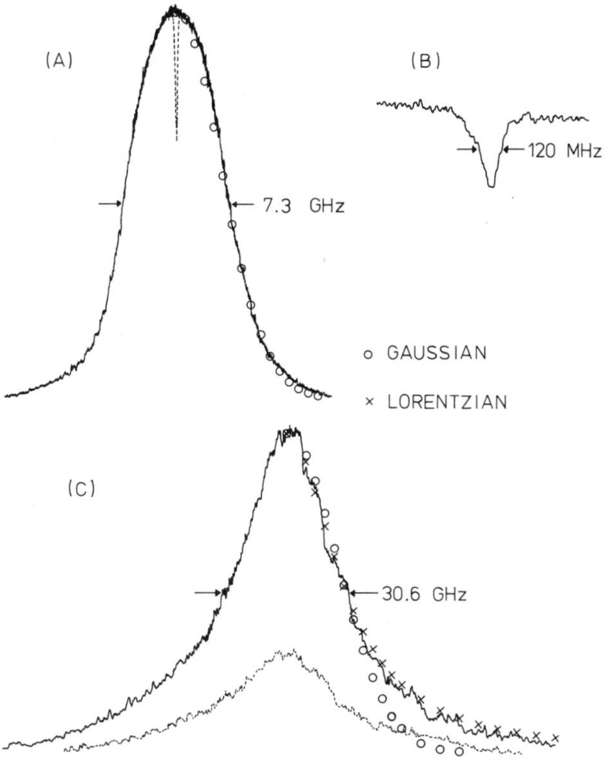

Fig. 9. cw-dye-laser-induced excitation (absorption) spectra of (a) the origin at 5875 A and (c) the vibronic transition at 5700 Å of DMST in durene at 2 K. The cw dye laser was slowly scanned (20 MHz/s) through the absorption regions with instantaneous bandwidth (~ resolution) of <20 MHz. The dotted spectra were obtained after 1 s laser burning (effective bandwidth 30 MHz) in the origin and 5 min burning (effective bandwidth 100 MHz) in the vibronic line (at the top). (b) High-resolution scan of the hole burned in the zero-phonon line. (After Ref. 40.)

burned in the transition, which implies that the vibronic transition studied is predominantly homogeneously broadened. The observed homogeneous linewidth of 28.8 GHz was ascribed to vibrational relaxation, which implies a vibrational lifetime of $\gtrsim$5.5 ps. This number was in sharp contrast with an earlier report on a vibrational lifetime of 200 ps in pentacene in *p*-terphenyl.[59] As is known now this number was based on a misinterpretation of the observed signal and the real lifetime of that mode is ca 2 ps.[60]

We note here that this experiment does not exclude the possibility of a pure dephasing contribution to the observed line width. This point is revisited in Section VI.C. Voelker et al.[50] showed that the homogeneous linewidth of the $S_1 \leftarrow S_0$ transition of free-base porphin may be also be determined by photochemical hole-burning. They found that at low temperature the hole-width was almost determined by the upper state fluorescence lifetime of 17 ns. This is a particular interesting case as the hole-burning is due here to photoisomerization of the molecule of one tautomer of the free base into the other by the shift of two hydrogen atoms.[61] This photointerconversion process is reversible, which has led to an interesting proposal of using this system as an optical information storage system. In a second paper Voelker, Macfarlane, and van der Waals[48] demonstrated that photochemical hole-burning can also be used to study the temperature dependence of the linewidth and line shift, from

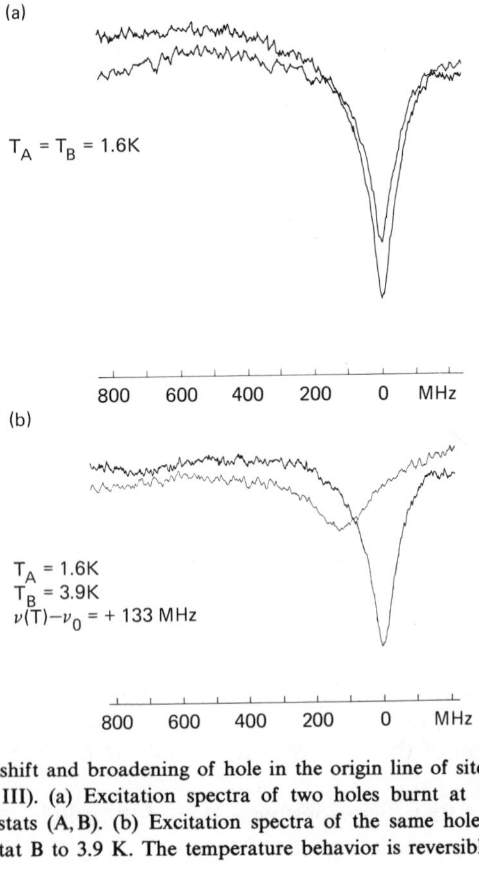

Fig. 10. Frequency shift and broadening of hole in the origin line of site B_1 of porphin in *n*-octane (see Table III). (a) Excitation spectra of two holes burnt at 1.6 K in identical samples in two cryostats (A, B). (b) Excitation spectra of the same holes after raising the temperature of cryostat B to 3.9 K. The temperature behavior is reversible. (After Ref. 48.)

which unique information on the optical dephasing processes was obtained. A picture of the effect of temperature on the width and shift of the hole is given in Fig. 10. The important result was that for the so-called *B*-sites of porphin in *n*-octane both the linewidths and shifts were exponentially activated with increasing temperature. The observed activation energy of ca 15 cm^{-1} strongly suggested that phonons were responsible for the observed optical dephasing effects. A similar conclusion was also drawn in a related photochemical hole-burning experiment on H_2-phtalocyanine in solid tetradecane by Gorokovski and Rebane.[62]

It seems worth mentioning here that Hesselink and Wiersma[63] recently also have been successful in measuring the homogeneous linewidth of the $S_1 \leftarrow S_0$ transition of the *A*-sites of Zn-porphin in *n*-octane. Also in this case over an extended temperature range, as shown in Fig. 11, an exponential activation of the linewidth is observed. The measured activation energy of 31 ± 2 cm^{-1} of these *A*-sites is in close agreement with the estimated[48] ~30 cm^{-1} activation energy for the *A*-sites of the free-base porphin in *n*-octane.

We will defer discussion of the proposed mechanism for optical dephasing to Section IX. In a recent paper by Voelker and Macfarlane[64] photochemical hole-burning is used to monitor (supposedly) the vibrational relaxation times of over a dozen excited state modes in porphin. A large variation in vibrational lifetimes (~1–50 ps) was measured whereby no systematic variation as a function of vibrational frequency was found. This result is graphically depicted in Fig. 12. It is not yet clear whether the

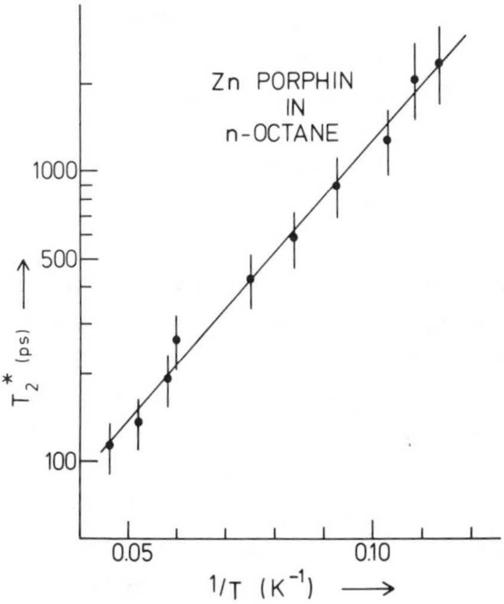

Fig. 11. Plot of T_2^* vs $1/T$ for the origin of the $S_1 \leftarrow S_0$ absorption of Zn-porphin in *n*-octane as measured in a picosecond photon echo experiment. Note that the echo decay time is $\frac{1}{2}T_2$, where $1/T_2 = 1/T_2^* + 1/2T_1$.

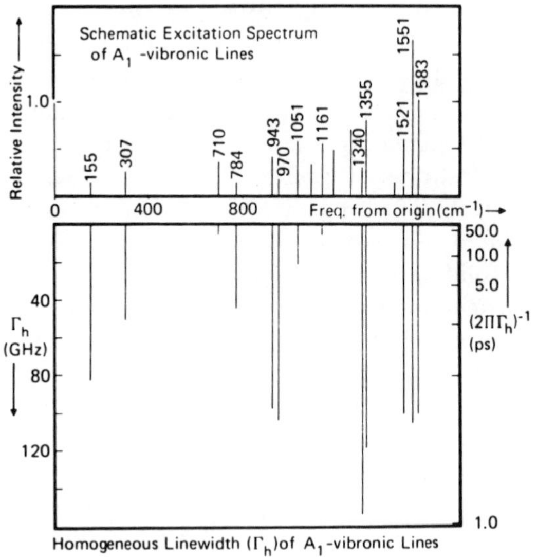

Fig. 12. Upper part: Schematic excitation spectrum of A_1 vibronic lines relative to the origin at 16,331 cm^{-1}. Lower part: homogeneous width of the same lines from hole-burning measurements. (After Ref. 64.)

observed differences in vibration lifetimes are due to a variation in electron–phonon coupling or that the intrinsic molecular relaxation is responsible for the observed effects. We wish to mention that also hole-burning of nonphotochemical origin has been reported in organics in glasses.[65] While this is an interesting effect, it is outside the scope of this paper. Also optical hole-burning at room temperature in solution of organics has been reported[66] but will not further be discussed here.

VI. PHOTON ECHOS IN ORGANIC MIXED CRYSTALS

A. Introduction

Soon after the report of de Vries and Wiersma[40] that at low temperature the coherence lifetime in $S_1 \leftarrow S_0$ electronic transitions of organics could approach the fluorescence lifetime, Aartsma and Wiersma[12] succeeded in detecting the first photon echo in an organic solid. The experiment was done at the $S_1 \leftarrow S_0$ transition of pyrene in biphenyl and is a beautiful illustration of the possible difference between optical T_1 and T_2. In Fig. 13 we reproduce the absorption spectrum of pyrene in the origin region, which may be identified as a "T_1-spectrum," and a photon-echo excitation spectrum, which may be called a "T_2-spectrum." The different lines in the

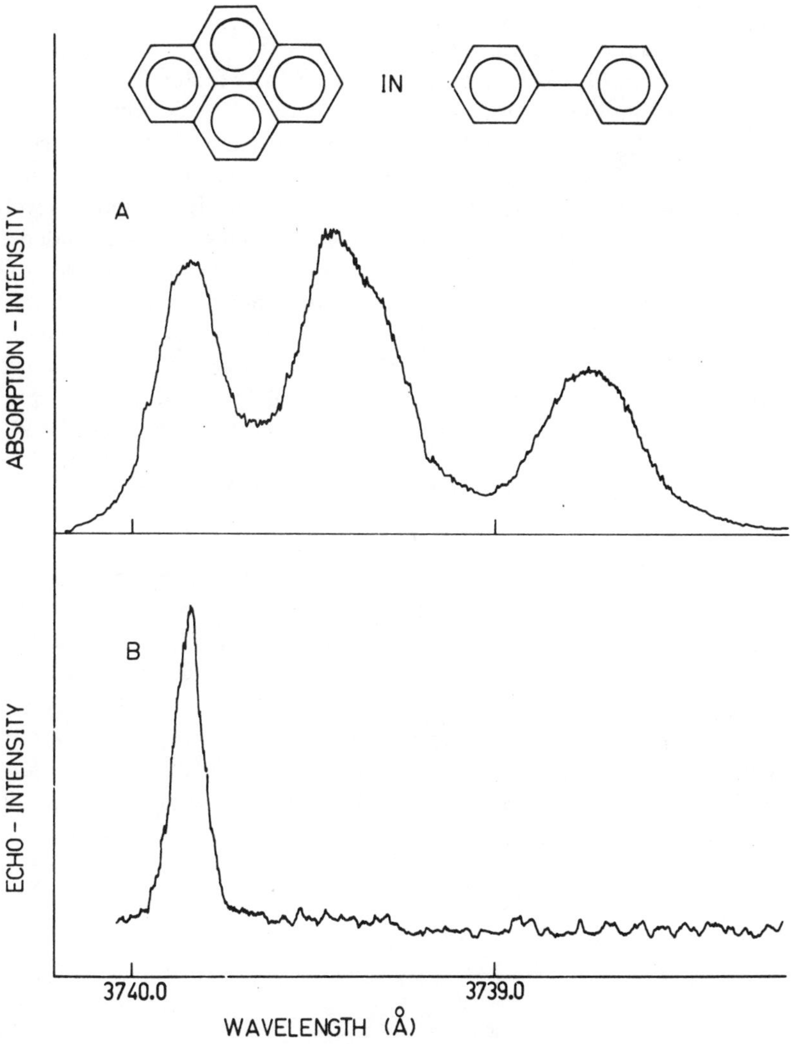

Fig. 13. (a) Absorption spectrum and (b) photon-echo excitation spectrum of the origin of the $^1B_{2u} \leftarrow {}^1A_{1g}$ electronic transition of pyrene in biphenyl at 2 K. The laser bandwidth in these excitation experiments was ca 0.05 Å. (After Ref. 12.)

origin region may be assigned to different sites of pyrene in biphenyl. The photon-echo excitation spectrum clearly shows that the optical dephasing times (T_2) of the different sites differ at least by a factor 3. The fluorescence lifetimes of these sites, however, were found to be identical (75 ± 2 ns). The photon-echo excitation spectrum further shows that even for one site, optical T_2 may vary throughout the inhomogeneous distribution. For further details we refer to Ref. 12. Since this first report, several photon-echo studies followed on tetracene and pentacene,[67] naphthalene,[68] and triphenyl methyl.[69] In the remainder of this section we will only concentrate on pentacene, as on this molecule not only photon-echo, but also optical free induction decay[70, 71] and optical nutation[71] experiments have been done. The main reason for the popularity of pentacene, the ruby of organic crystals, is undoubtedly due to the fact that its absorption spectrum lies in the spectral region covered by the rhodamine 6G laser.

B. Pentacene in Mixed Crystals

1. *Absorption Spectrum*

In Figs. 14 and 15 we show the low-temperature absorption spectra of pentacene in *p*-terphenyl and naphthalene. The lines of interest for this work are marked by arrows. The absorption spectrum is typical for a strongly allowed transition. Meyling and Wiersma showed[72] that this transition is short-axis polarized in agreement with theoretical predictions. Note that in *p*-terphenyl, pentacene exhibits four different sites[72] whereby the energetically higher sites behave distinctly different from the lower ones. The relevant data characterizing the electronic origins of these sites are gathered in Table I.[73]

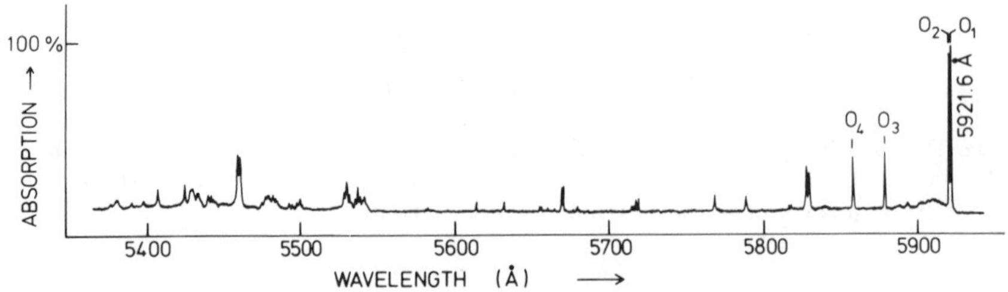

Fig. 14. Low-temperature (at 1.5 K) *b*-polarized fluorescence excitation spectrum of pentacene in p-terphenyl. Note that the absorption of the 0_3, 0_4 sites in this picture is artificially reduced by the detection setup. In a straight absorption experiment all four sites absorb equally.

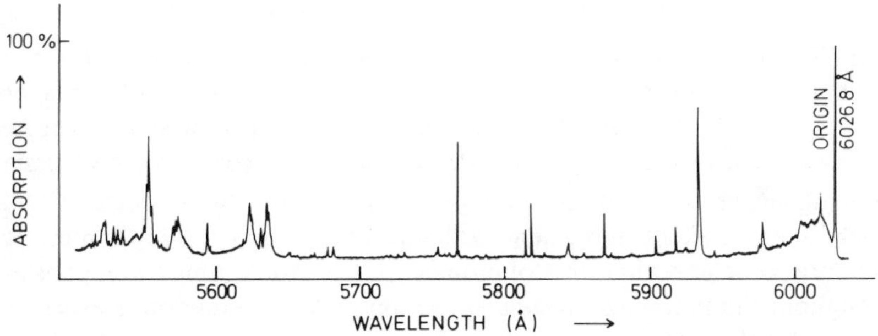

Fig. 15. Low-temperature (1.5 K) *b*-polarized fluorescence excitation spectrum of pentacene in naphthalene.

TABLE I
Origin Wavelengths and Lifetimes at 1.5 K of Pentacene in *p*-Terphenyl and Naphthalene

Host	Wavelength (λ_{air}, Å)	Lifetime (ns)
p-Terphenyl	5921.6 (O_1)	23.5 ± 1
	5920.1 (O_2)	23.5 ± 1
	5878.7 (O_3)	9.5 ± 1
	5858.4 (O_4)	10.0 ± 1.5
Naphthalene	6026.8	19.5 ± 1

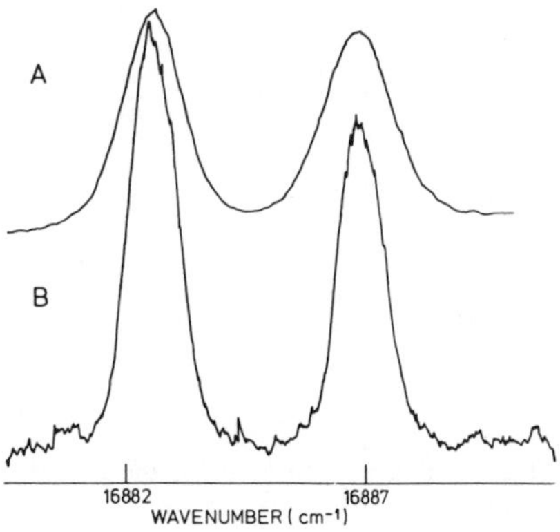

Fig. 16. Fluorescence excitation spectrum (A) and photon-echo excitation spectrum (B) of the origin of the $^1B_{2u} \leftarrow {}^1A_{1g}$ transition of pentacene in *p*-terphenyl at 1.3 K.

2. *Photon-Echo Excitation Spectrum*

In Fig. 16 we reproduce a high-resolution absorption and photon-echo excitation spectrum of the energetically lower origins $(0_1, 0_2)$ of pentacene in *p*-terphenyl.[67] The photon echo spectrum clearly shows the expected narrowing due to the N^2 dependence of the photon echo. For a Gaussian inhomogeneous distribution we expect a narrowing by a factor of $\sqrt{2}$ which is in excellent agreement with the observation for both sites. We therefore conclude that the photon-echo relaxation times are constant throughout the inhomogeneous distribution. This information is especially relevant for the picosecond photon echo experiments to be discussed later.

3. *Low-Temperature Photon-Echo Relaxation*

In the section on photochemical hole-burning it has become clear that in all photoexcitation experiments the role of the triplet state needs to be investigated. In Section IV we have presented a density matrix equation which is able to describe coherent optical transients effects in a system of coupled states. This equation is[73]

$$\rho(t) = -\frac{i}{\hbar}[H, \rho] - \tfrac{1}{2}\{K, \rho\} + F \tag{14}$$

For pentacene the appropriate level scheme is given in Fig. 17 from which we calculate for the decay matrix

$$K = \begin{bmatrix} 0 & 0 & 0 \\ 0 & K_2 & 0 \\ 0 & 0 & k_{31} \end{bmatrix} \text{ and for } F = \begin{bmatrix} k_{21}\,\rho_{22} \quad +k_{31}\,\rho_{33} & 0 & 0 \\ 0 & 0 & 0 \\ 0 & 0 & k_{23}\rho_{22} \end{bmatrix}$$

where $K_2 = k_{21} + k_{23}$. In the absence of an exciting field and pure dephasing processes we then find:

$$\dot{\rho}_{12} = -\tfrac{1}{2}K_2\rho_{12} \quad \text{or} \quad \rho_{12}(t) = \rho_{12}(0)e^{-1/2K_2 t}$$

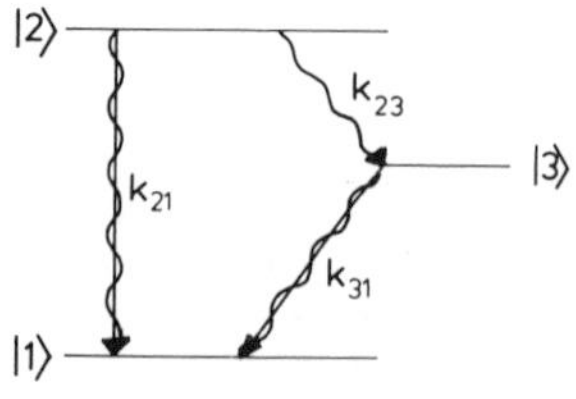

Fig. 17. Level scheme for pentacene used for the description of the photon-echo decay.

The important point to note here is that the photon echo does not suffer from the triplet state bottleneck in the same sense as hole-burning or optical free induction decay. It is only the repetition rate in the photon-echo experiment that is limited by the triplet state decay rate. Morsink et al.[74] showed that at low temperature (~2 K) the photon-echo lifetime of dilute PTC-h_{14} and PTC-d_{14} in p-terphenyl crystals is identical to the fluorescence lifetime. This implies that at this temperature pure dephasing processes are absent. The homogeneous linewidths are therefore 5.9 MHz (PTC-d_{14}) and 7.1 MHz (PTC-h_{14}) at this temperature, which is almost a factor of 10^4 less than the inhomogeneous width.

4. *Stimulated Photon Echo*

Using three optical excitation pulses a so-called stimulated photon echo (3PSE) may be generated at a time after the third pulse which is identical to the splitting between the first two excitation pulses. The phase-match

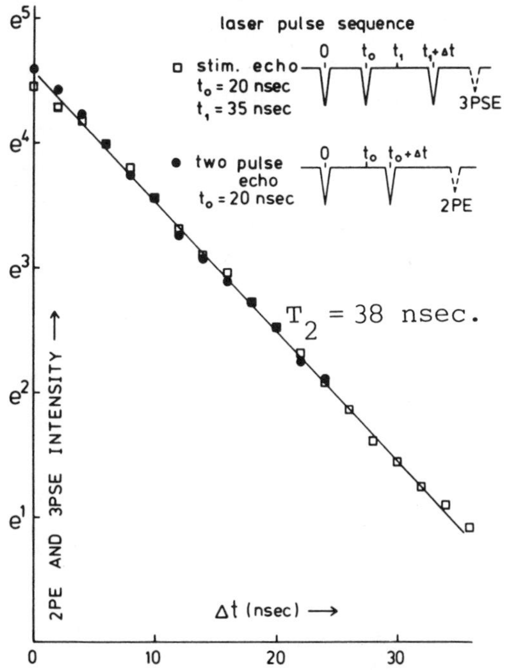

Fig. 18. Results of a two-pulse and a three-pulse photon-echo measurement at low temperature in a mixed crystal of pentacene in p-terphenyl. Note that this mixed crystal is not dilute in the sense that $T_2 = 2T_1$ at low temperature (1.5 K).

condition for this echo is $\mathbf{k}_e = \mathbf{k}_3 + \mathbf{k}_2 - \mathbf{k}_1$ which shows that the 3PS echo may be looked at as a two–pulse echo, whereby half of the second pulse is delayed. The decay of this echo is known to be influenced only by T_1-type processes.[33]

In Fig. 18 we exhibit the results of both a two-pulse and a three-pulse echo measurement at low temperature on pentacene in *p*-terphenyl. The important point to note is that the echo decay times are identical but shorter than the fluorescence lifetime for this more concentrated crystal. The implication is that at higher concentration, optical dephasing is also caused by energy-transfer processes in this system and that the process is irreversible (T_1-type).

The photon-echo method seems to be eminently suitable to provide additional information on energy transfer in mixed crystals and particular on the question to what extent energy back-transfer is important.

5. *Temperature Dependence of the Photon-Echo Intensity*

Aartsma and Wiersma[67] were the first to report on the temperature dependence of the photon echo of pentacene in *p*-terphenyl. These initial experiments were performed using a nanosecond pulsed dye laser and measuring the echo intensity as a function of temperature for a fixed time separation of the exciting pulses. Figure 19 shows the latest result using this method for the O_1-site of pentacene in *p*-terphenyl. In a separate experiment it was ascertained that the fluorescence lifetime of 23.5 ns remained constant up to 110 K.[73] This change in echo intensity as a function of temperature is thus a manifestation of a *pure* (T_2^*) dephasing contribution to the echo lifetime. Experimentally it was found that an

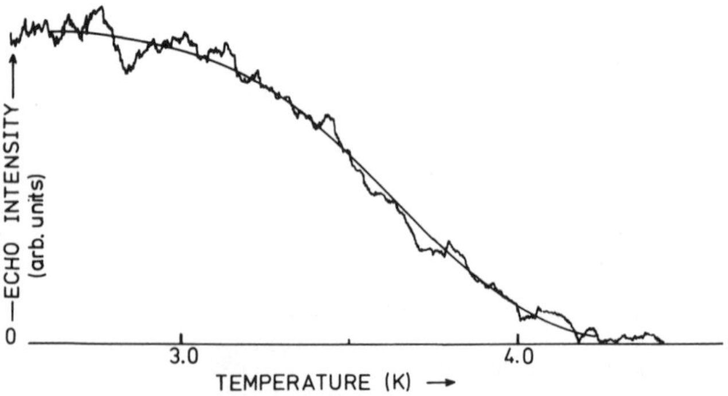

Fig. 19. Temperature dependence of the echo intensity for the 0_1-site of pentacene-h_{14} in *p*-terphenyl. The laser pulse separation was 25 ns. The smooth curve represents a fit with $\Delta E = 28 \pm 1$ cm^{-1} and $T_2^*(\infty) = 2 \pm 1$ ps.

excellent fit could be obtained of the experimental results by assuming that T_2^* is exponentially activated $T_2^* = T_2^*(\infty)\exp(\Delta E/kT)$. The homogeneous line width in this system at low temperature was therefore exponentially activated. This was explained by Aartsma and Wiersma by assuming that pseudolocal phonons were responsible for the optical dephasing. A similar behavior was established for tetracene in *p*-terphenyl,[67] porphin in *n*-octane,[50] H_2-pthalocyanine in tetradecane[62] and naphthalene in durene.[68] These photon echo measurements however remained unsatisfactory from two points of view. First, the temperature range in which measurements could be performed was very small and the important temperature range of ~4–12 K not covered. One should note that above 12 K, T_2^* can be obtained from linewidth measurements. Secondly no *direct* information on the echo lifetime was obtained, which resulted in great uncertainty, especially in $T_2^*(\infty)$. Also in the analysis it was *assumed* that the photon-echo relaxation at every temperature was exponential.

This unsatisfactory situation motivated Hesselink and Wiersma[14, 75] to attempt to generate and detect picosecond photon echos. Using a picosecond synchronously pumped dye laser system for excitation, and optical mixing as an echo-detection scheme they succeeded in measuring *directly* photon-echo relaxation times in the picosecond time domain. In Fig. 20 we show the results of such a picosecond photon-echo measurement on the

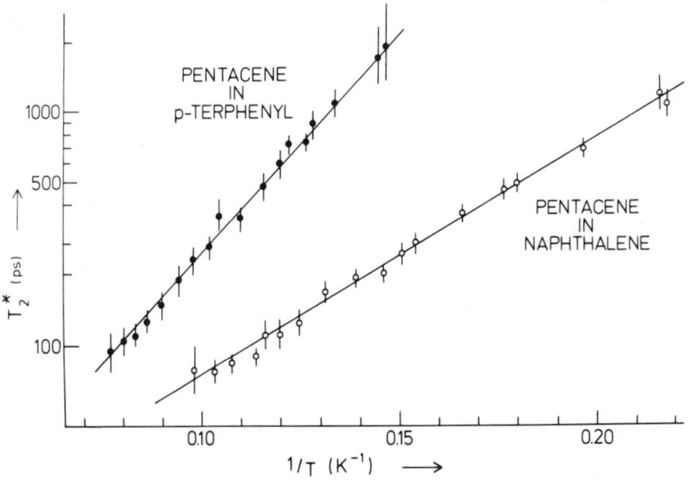

Fig. 20. Results of picosecond photon-echo decay measurements on the origin of pentacene in *p*-terphenyl and naphthalene as a function of temperature. Note that the echo decay term is $\frac{1}{2}T_2$.

electronic origin of pentacene in *p*-terphenyl and naphthalene. The surprising result is that up to ca. 15 K, the pure dephasing contribution to the echo relaxation is of the form $T_2^* = T_2^*(\infty)\exp(\Delta E/kT)$ in agreement with the earlier low-temperature results. It is also satisfactory to note that in the case of PTC in *p*-terphenyl, excellent agreement between the results obtained with a nanosecond and picosecond laser system exists. These picosecond results therefore confirm the suggestion of Aartsma and Wiersma[67] that photons of a restricted frequency range play a dominant role in the optical relaxation processes.

6. *Line-Shift Measurements*

One of the great advantages of photochemical hole-burning over coherent optical techniques in obtaining information on optical relaxation is that low-temperature line-shift measurements can be done as shown by Voelker et al.[48] in a study of porphin in *n*-octane. We have therefore tried to develop another technique that would permit low-temperature line-shift measurements in photochemically stable systems. Preliminary temperature-modulation experiments were done whereby the absorption spectra for different temperatures were subtracted. Figure 21 shows a modulation

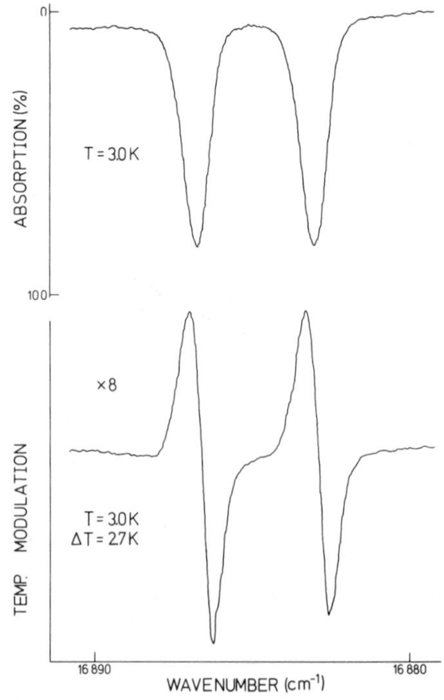

Fig. 21. Absorption (upper) and temperature modulation (lower) spectrum of pentacene in *p*-terphenyl. The temperature jump was ca. 2.7°.

spectrum for a temperature jump of ca. 2.7 K. The shift in this experiment in only 0.06 cm^{-1}. The temperature modulation was done by chopping the radiation of a tungsten lamp. This could probably be better done by using the temperature-jump technique developed by Francis and co-workers.[76] The great problem we encountered in this experiment is a reliable measurement of the temperature jump. We suggest one way of doing this would be to use an echo measurement in determining the effective temperature. It might be advantageous then to use a pulsed CO_2 laser to induce the temperature jumps. This method in principle should be capable of measuring line shifts of ca. 100 MHz. An important prerequisite, fulfilled for pentacene in *p*-terphenyl, is that the relaxation is independent of the position in the inhomogeneous distribution.

C. Vibronic Relaxation and Dephasing in Pentacene

1. Introduction

In the analysis of the results of photochemical hole-burning on vibronic transitions in dimethyl *s*-tetrazine,[40] *s*-tetrazine[57] and porphin[64] it is assumed that the homogeneous linewidth is exclusively determined by the vibrational lifetime, i.e., $T_2 = 2T_1$. The absence of pure dephasing contributions however was not rigorously ascertained. This motivated Hesselink and Wiersma[60] to attempt a photon-echo experiment on a vibronic transition. The selection of the vibration was made on basis of a lifetime measurement using a pump-probe technique. On basis of these measurements the system pentacene in naphthalene was chosen. In the following we will first show the results of pump-probe experiments which also contain some interesting new information on the spectroscopy of pentacene. We then continue by presenting the results of psec photon-echo experiments.

2. Picosecond Pump-Probe Experiments in Pentacene in Naphthalene

With a similar setup as used by Ippen et al.[77] for pump-probe experiments, except for an intensity stabilizer in both beams, we performed experiments on the electronic origin at 6027 Å and vibronic transitions at 5933 and 5767 Å. The results of these experiments are shown in Fig. 22. Except for minor details, the transient on the purely electronic transition is in agreement with our expectation that the singlet excited state is long lived (19.5 ns) on a picosecond time scale. The transient on the 261 cm^{-1} vibration confirms what was already known from the optical absorption spectrum, namely, that it is very short lived. From the near Lorentzian lineshape at low temperature we calculate a 3.3 ps relaxation time in

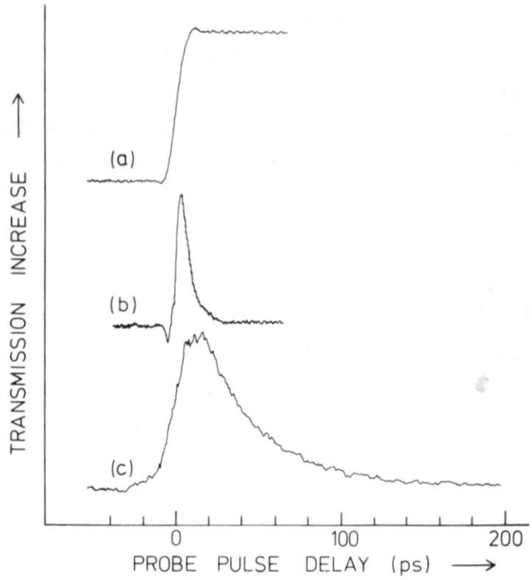

Fig. 22. Results of psec pump-probe experiments on the origin (a), 261 cm^{-1} vibronic band (b), and 748 cm^{-1} vibronic band (c) in the absorption spectrum of pentacene in naphthalene. The zero-time point was determined in a separate pump-probe optical mixing experiment. The autocorrelation of the excitation pulses was 4.3 ps (a), 19 ps (b), and 8.5 ps (c).

agreement with our previous report. Of particular interest for our aim is the transient observed for the 748 cm^{-1} vibration. Analysis of the transient shows that a lifetime of 32 ± 2 ps may be attributed to the vibrationally excited state. This lifetime is substantial longer than obtained for the same vibration of pentacene in *p*-terphenyl. In Table II the vibrational lifetimes as measured by the pump-probe technique are gathered. From this table it is clear that an obvious choice for the photon-echo measurement is the 748 cm^{-1} vibration of pentacene in naphthalene. There is another feature that we draw attention to at this point, namely, that for the vibronic transitions the probe-pulse absorption prior and "long" after the saturation pulse is

TABLE II
Excited State Vibrational Lifetimes of Pentacene in *p*-Terphenyl and Naphthalene at 1.5 K

Host	Wavelength (λ_{air}, Å)	Lifetime (ps)
p-Terphenyl	5788.0 (O_3)	2 ± 1
	5631.6 (O_3)	15 ± 4
Naphthalene	5933.8	2.8 ± 1
	5767.1	32 ± 2

identical. This is unexpected as the groundstate recovery is *not* identical to the vibrational lifetime. This fact can only be explained by assuming that there is a strong excited-state absorption at the probe wavelength. The transient absorption-lineshape of the electronic origin has a similar explanation. For a more detailed interpretation of the transients we refer to Ref. 60.

3. *Picosecond Photon Echo Measurements*

By combining the technique of echo-excitation under a small angle[8] with echo detection by optical mixing[14] we have been able to observe picosecond photon echos from the vibronic transition at 5767 Å of pentacene in naphthalene. The schematics of the setup are shown in Fig. 23. The echo-relaxation at 1.5 K is shown in Fig. 24 from which a low-temperature echo lifetime of 33 ± 1.5 ps is calculated. It is interesting to note that this echo lifetime is *identical* to the vibrational lifetime measured in the pump-probe experiment. We note that this is the first time that it has been shown that the homogeneous lineshape at low temperature of a vibronic transition is exclusively determined by vibrational relaxation, i.e., $T_2 = 2T_1$. This in fact seems to justify the assumption of Voelker and Macfarlane[64] that the photochemical vibronic hole-width in porphin is only determined by vibrational relaxation. We have also measured the temperature dependence of the dephasing time and the result of this measurement is shown in Fig. 25. Analysis of the data shows that, as in the case of the electronic origin, the temperature-dependent contribution to T_2 can be written as $T_2^*(T) = T_2^*(\infty)\exp(\Delta E/kT)$. For $T_2^*(\infty)$ we obtain 5.4 ± 1.5 ps and $\Delta E = 16.6 \pm 1.5$ cm^{-1}. While for the vibronic transition $T_2^*(\infty)$ is somewhat smaller, the activation energy is identical to what is observed for the purely electronic transition. In view of these results it seemed also interesting to probe the temperature dependence of the vibrational relaxation of this state. Much to our surprise we found, as also shown in Fig. 25, that the temperature dependence of the vibrational lifetime was *identical* to that of the echo lifetime, in other words $T_2 = 2T_1$ over the whole temperature range investigated. The very interesting conclusion must be drawn that vibrational relaxation into the vibrationless excited state proceeds via a narrow range of host phonons (presumably pseudolocalized). A different mechanism for vibrational relaxation was recently suggested by

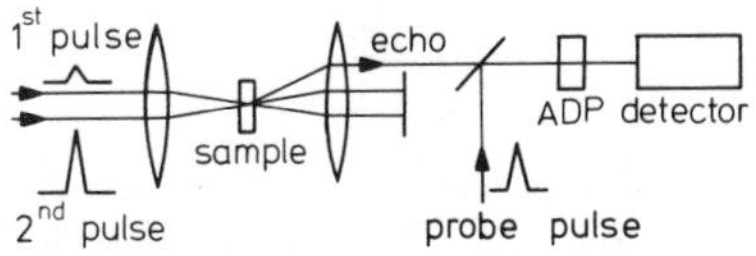

Fig. 23. Schematics of the photon echo setup used for study of the vibronic transitions. For details consult Ref. 14.

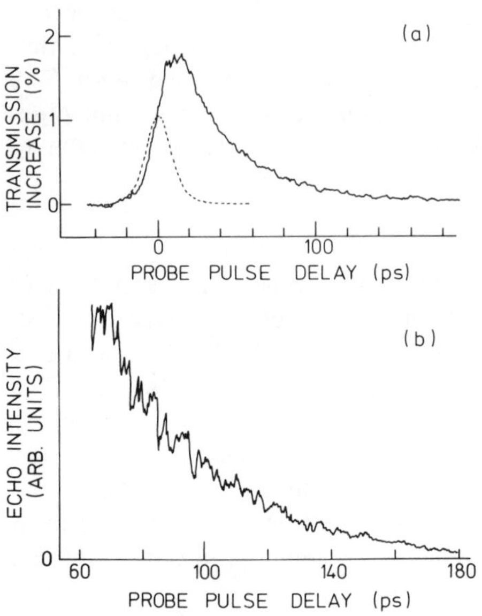

Fig. 24. The low-temperature transient absorption (a) and photon-echo relaxation (b) of the 5767 Å vibronic transition of pentacene in naphthalene at 1.3 K. The dashed line in (a) is the correlation between the pump and probe pulse. In (b) the wavelengh of the probe-mixing pulse was 5570 Å. (After Ref. 60.)

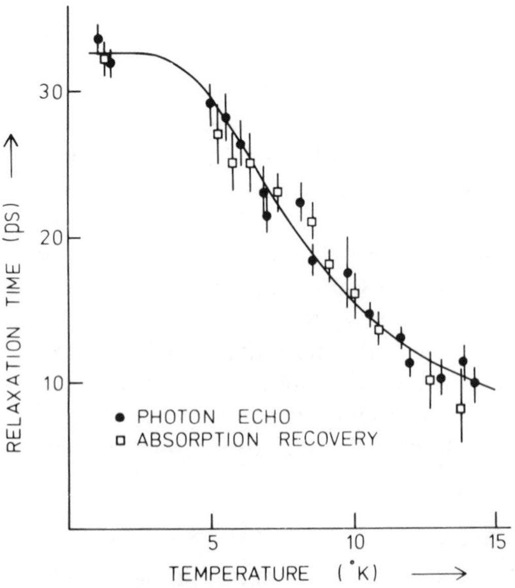

Fig. 25. Plot of the measured relaxation times as a function of temperature. Note that the echo decay time is $T_2/2$. Typical uncertainty in the temperature is ±0.1 K. (After Ref. 60.)

Hochstrasser and Nyi[78] in a study of vibrational energy redistribution after single vibronic level excitation in the $S_1 \leftarrow S_0$ transition of azulene in naphthalene. In this system phonon-induced relaxation *into* host levels was mainly held responsible for vibrational relaxation.

VII. OPTICAL NUTATION AND FREE INDUCTION DECAY

A. Introduction

In the last few years it has become clear that optical free induction decay as method to study optical dephasing processes is very attractive. While the original technique,[10] employing Stark switching of the molecular transition was limited to the study of polar molecules, the cw dye laser switching technique recently introduced by Brewer and Genack[79] can be used to study any optical transition. De Voe and Brewer[13] recently showed that the laser frequency switching technique may be extended to the 100 ps time domain, which makes it very useful for a study of fast optical dephasing phenomena. As of today the technique has been used to study dephasing of I_2 in the gas phase,[79] and in a molecular beam,[80] the system Pr^{3+}/LaF_3;[81] pentacene in *p*-terphenyl,[70, 71] ruby,[82] and very recently the duryl radical in durene.[83] Optical nutation[37] has been used to determine the transition dipole moment in ammonia[84] and for the $S_1 \leftarrow S_0$ transition of pentacene in *p*-terphenyl.[70]

B. Theory

In molecules, as noted by de Vries and Wiersma,[52] the application of free induction decay to study optical dephasing may be frustrated by the presence of an intermediate triplet state. The level scheme, which is representative for most molecules with an even number of electrons, is shown in Fig. 26. For an applied laser field $E_x = E_0\cos(\Omega t - k_z)$, that is resonant with the $\langle 2| \leftarrow |1\rangle$ transition, we may write, in the RWA approximation, the following steady-state density matrix equations, which describe the coherent decay after laser frequency switching[52]

$$\begin{aligned}
\dot{\rho}_{11} &= 0 = K_2\rho_{22} + \tfrac{1}{2}i\chi(\tilde{\rho}_{21} - \tilde{\rho}_{12}) \\
\dot{\rho}_{22} &= 0 = -K_2\rho_{22} + \tfrac{1}{2}i\chi(\tilde{\rho}_{12} - \tilde{\rho}_{21}) \qquad (32) \\
\dot{\tilde{\rho}}_{12} &= 0 = \left(-\frac{1}{T_2} + i\Delta\right)\tilde{\rho}_{12} + \tfrac{1}{2}i\chi(\rho_{22} - \rho_{11})
\end{aligned}$$

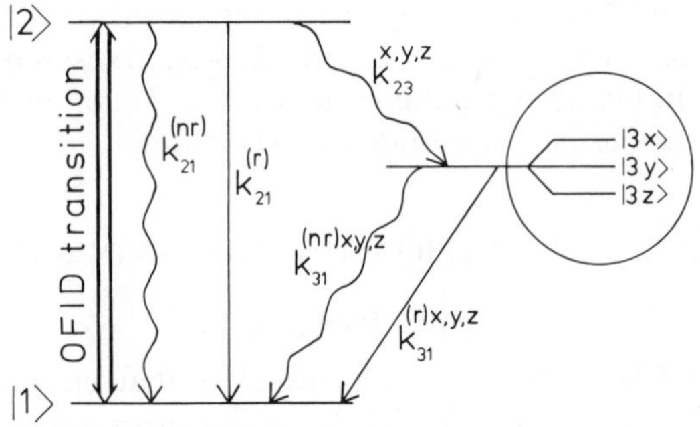

Fig. 26. Molecular model system. The meaning of the energy levels is discussed in the text. Straight and wavy arrows indicate radiative and radiationless transitions, respectively. The part within the circle has a greatly magnified (~10^4) energy scale.

where

$$K_2 = k_{21} + k_{23}^x + k_{23}^y + k_{23}^z$$

$$\chi = \text{the Rabi frequency.}$$

$$\tilde{\rho}_{12} = \rho_{12} e^{-i(\Omega t - kz)}$$

In addition we may write

$$\sum_i \rho_{ii} = 1 \text{ or } \rho_{11} = 1 - \rho_{22}(1 + A) \tag{33}$$

where

$$A = \sum_i k_{23}^i / k_{31}^i$$

De Vries and Wiersma showed, that using (32) and (33), the decay rate of the ensemble-averaged density matrix $\langle \rho_{12}(t) \rangle$, which determines the OFID decay rate is given by:

$$\tau_{\text{OFID}}^{-1} = \frac{1}{T_2}\left[1 + \left(1 + \frac{\mu_{12}^2 E_0^2}{\hbar^2} \frac{T_2}{2K_2}(2 + A)\right)^{1/2}\right] \tag{34}$$

We note that this expression is identical (except for the factor 2) to the transient hole-burning width (see Section V.B). The same problems that were encountered in the photochemical hole-burning are therefore present here. (We remark that the connection with the OFID decay rate for a two-level system is easily made by putting $A=0$ and $K_2=k_{21}$.[36]) The effect of the intermediate triplet state is then clearly recognized as an effective increase in off-resonance pumping in a pseudo-two-level system. The expression shows that measurement of T_2, in principle, still remains possible by extrapolating the OFID-decay rate to zero laser power. Experimentally, however, since the signal intensity is also proportional to the laser power,[36] we easily run into problems for a reliable T_2 measurement unless the following equation is obeyed[52]:

$$\frac{\mu_{12}^2 E_0^2}{\hbar^2}\frac{T_2}{2K_2}(2+A)\lesssim 8 \tag{35}$$

It turns out that for some systems, e.g., porphin in *n*-octane[50, 52] the exciting laser power must be chosen so low, to meet condition (35), that detection of the OFID signal is impossible. Also for the case of the lowest $S_1 \leftarrow S_0$ transition of naphthalene in durene[52] the factor A is so large, that detection of OFID is out of reach. As the photon-echo technique does not suffer from the triplet state bottleneck, this system can be studied using the photon echo, as shown by Aartsma and Wiersma.[68] For the energetically lowest sites of pentacene the situation is particularly favorable, as A is only 9[73] and OFID detection is easily possible.

C. Results

In Fig. 27 we reproduce a low-temperature optical free induction decay signal of pentacene-h_{14} in *p*-terphenyl. The insert in the figure shows that

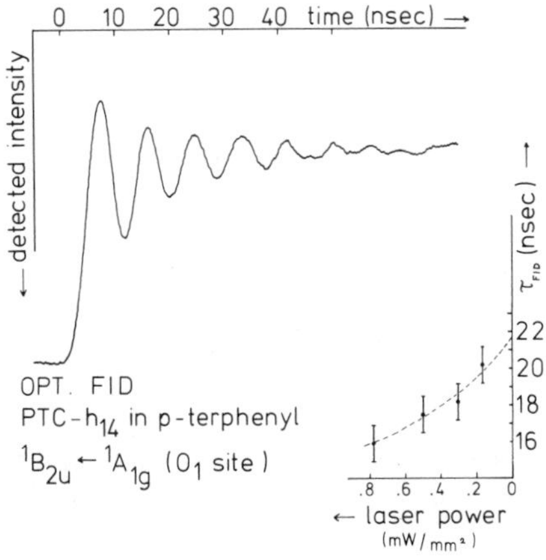

Fig. 27. Optical free induction decay signal of pentacene-h_{14} in *p*-terphenyl at 1.5 K. In the insert the power dependence of τ_{OFID} is shown. T_2 is obtained from a linear extrapolation for $E\rightarrow 0$, using least square fitting to the OFID formula written in the following form: $1/\tau^2 - 2/T_2\tau = (\mu/\hbar)^2 E^2 \times T_1/T_2$. The dotted curve is calculated using values for T_2 and $\mu^2 T_1$ obtained from the extrapolation. (After Ref. 70a).

extrapolation of the OFID decay rate leads to a value of $T_2=43.4\pm2$ ns, which is identical to $2T_1$. It is comforting to note that the measured optical dephasing constant is independent of the coherence bandwidth of the laser source in this pseudo-two-level system. Knowledge of the intersystem-crossing rates also enables determination of the $S_1 \leftarrow S_0$ electronic transition dipole. From a careful study of all factors, de Vries and Wiersma[73] obtained for $\mu_{12}=0.71\pm0.24$ D. Orlowski, Jones, and Zewail also reported the free induction decay of pentacene [70*b*]. In a recent detailed study of laser excitation experiments on pentacene, Orlowski and Zewail[71] also report OFID experiments from which is calculated that $T_2=45\pm2$ ns, which within the error limits is identical to twice the fluorescence lifetime (23.5 ± 1 ns) at low temperature.

Their experimental OFID curve together with computer fit is shown in Fig. 28. Orlowski and Zewail also report on the optical nutation in pentacene for which they derive the following expression for the intensity:

$$I(t)=\pi B_1^2\left\{I_0-2A\left(\frac{\chi}{t}\right)\operatorname{erf}\left(\frac{\chi t}{2}\right)J_1(\chi t)e^{-t/T_2}\right\} \tag{36}$$

where B_1 is the laser beam half-width $(1/e)$ of the intensity profile, A a

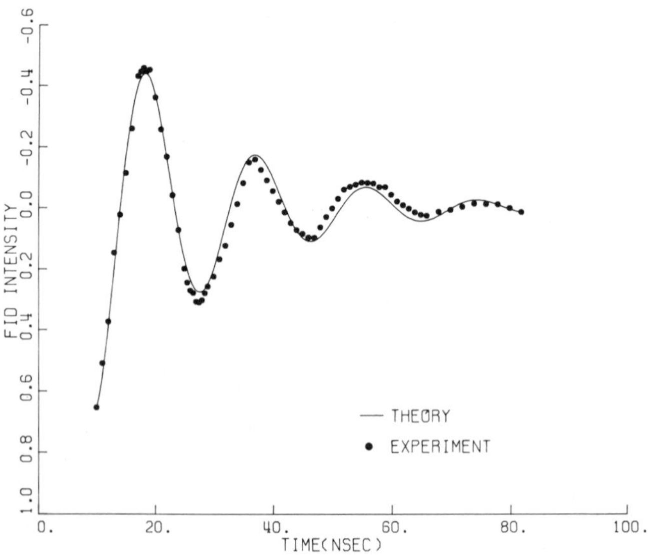

Fig. 28. The optical free induction decay observed in pentacene (0_1 origin) at 1.8 K and 400 μW of laser power. The computer fit provided a value of 45 ± 2 ns for T_2 from the observed transient decay. (After Ref. 71.)

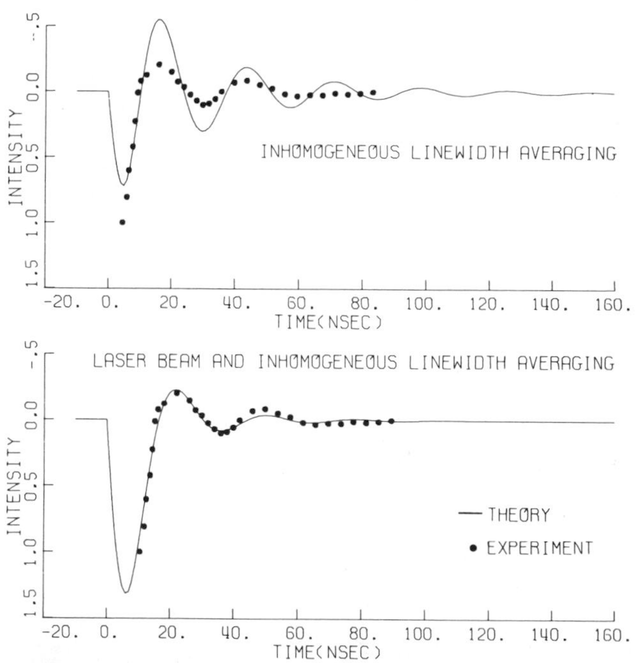

Fig. 29. Shown here are experimental data points and theoretical curves for the optical nutation in pentacene at 1.8 K. In each plot the vertical axis corresponds to relative absorption (arbitrary units) and $t=0$ is taken to be the switching time of the EO pulse. The top plot considers only inhomogeneous lineshape averaging. The bottom plot considers averaging over both the inhomogeneous line shape and the laser beam spatial profile. The fit is quite good in the latter case and provides values of 27.3 ± 1.3 ns for the nutation time and 45 ± 2 ns for T_2.

constant and $J_1(\chi t)$ the first-order Bessel function with argument χt. In Fig. 29 we show some calculated and experimental optical nutation curves as reported by Orlowski and Zewail. From such optical nutation spectra, Orlowski and Zewail also calculated the transition dipole for the lowest $S_1 \leftarrow S_0$ transition in pentacene and obtain $\mu = 0.7 \pm 0.1$ D, in excellent agreement with the value obtained from OFID.

VIII. SUMMARY OF THE RESULTS

It seems useful to summarize all optical relaxation data that have been obtained so far. In the preceding sections it was shown that for the temperature dependent part of T_2 the following phenomenological expression fitted all presently available data $T_2^*(T) = T_2^*(\infty)\exp(\Delta E/kT)$. For

TABLE III

Optical Data and Relaxation Parameters

Guest	Host	Transition	λ_{air}	Lifetime (nsec)	ΔE (cm^{-1})	$T_2(\infty)$ (ps)	Method	Ref.
Pentacene	*p*-Terphenyl	$1_{B_{2u}} \leftarrow 1_{A_{1g}}$;	5921.6	23.5 ± 1	20 ± 2	~20	ns photon echo	[a]
				23.5 ± 1	28 ± 1	2 ± 1	ns photon echo	[b]
				23.5 ± 1	30 ± 2	3 ± 1	ps photon echo	[c]
				24.5 ± 2	16.9 ± 1	10	ns nutation	[d]
				24.5 ± 2	15.1 ± 9	8	ns IRD	[d]
	Naphthalene		6026.8	19.5 ± 1	16.5 ± 1	7 ± 2	ps photon echo	[c]
Tetracene	*p*-Terphenyl	$1_{B_{2u}} \leftarrow 1_{A_{1g}}$	4931.1	19 ± 1	5.6 ± 0.5	70 ± 20	ns photon echo	[a]
	Anthracene		4941.3		6.4 ± 2	~32	ns photon echo	[b]
Naphthalene	Durene	$1_{B_{1u}} \leftarrow 1_{A_{1g}}$	3168.9	170 ± 5	17 ± 2	$>4<20$	ns photon echo	[e]
	Naphthalene-d_8		3170.5	107 ± 5	36 ± 3	$>2<10$	ns photon echo	[e]
H_2-tetra-4-*tert*-butyl pthalocyanine	Tetradecane	$S_1 \leftarrow S_0$	6930	5.2	36	6.6	hole-burning	[f]
		B_1-site	6135.7	17 ± 3	15 ± 3			
H_2-porphin	*n*-Octane	B_2-site	6145.5	17 ± 3	14.5 ± 3		hole-burning	[g]
		A_1-site	6123.3	17 ± 3	~30 cm^{-1}			
		A_2-site	6147.8	17 ± 3	~30 cm^{-1}			
Zn-porphin	*n*-Octane	$S_1 \leftarrow S_0$	5569.9	$2.1 \pm .15$	31 ± 2	15 ± 4	ps photon echo	[h]

[a] T. J. Aartsma and D. A. Wiersma, *Chem. Phys. Lett.*, **42**, 520 (1976).
[b] T. J. Aartsma, thesis, University of Groningen, 1978.
[c] W. H. Hesselink and D. A. Wiersma, *J. Chem. Phys.*, July 15th issue, (1980).
[d] T. E. Orlowski and A. H. Zewail, *J. Chem. Phys.*, **70**, 1390 (1979).
[e] T. J. Aartsma and D. A. Wiersma, *Chem. Phys. Lett.*, **54**, 415 (1978).
[f] A. A. Gorokhovski and L. A. Rebane, *Opt. Commun.*, **20**, 144 (1977).
[g] S. Voelker, R. M. Macfarlane, and J. H. van der Waals, *Chem. Phys. Lett.*, **53**, 8 (1978).
[h] W. H. Hesselink and D. A. Wiersma, unpublished results.

the *total* optical dephasing time (T_2) the following relation holds:

$$\frac{1}{T_2}=\frac{1}{2T_1}+\frac{1}{T_2^*}$$

where in the pseudo-two-level systems discussed, T_1 may be taken as the fluorescence lifetime (Section VI.B). The homogeneous linewidth is related to T_2 by the following expression:

$$\Delta\nu_{\text{FWHM}}=(\pi T_2)^{-1}$$

In Table III we have gathered all relevant optical data and relaxation parameters presently known.

IX. OPTICAL DEPHASING AND LINESHAPE

A. Introduction

In the previous section we have been confronted with the results of coherent transient experiments which showed that the low-temperature homogeneous lineshape of electronic transitions of impurity molecules is exponentially activated. Next to the optical domain such an exponential activation was observed in the microwave domain by van't Hof and Schmidt[85] and explained on the basis of the modified Bloch equations.

Recently quite a number of theories have been advanced to explain optical dephasing in electronic transitions of molecules in solids.

In this section we wish to briefly review the results obtained from the different theories. We are also interested in the assumptions made. A complete presentation of this work may be found in Ref. 86.

B. Reduced Density Matrix

The dynamics of any system are known if the following density matrix equation is solved:

$$\frac{d}{dt}\sigma=\frac{-i}{\hbar}[H,\sigma] \tag{37}$$

Hereby is σ the density matrix and H the full Hamiltonian of the system.

Except for simple problems where the number of degrees of freedom are limited, (37) cannot be solved and a reduced description is adopted, whereby the degrees of freedom are separated in those of the system (relevant ones) and those of the bath (irrelevant ones).

The full Hamiltonian is then partitioned according to

$$H = H_A + H_R + V_{AR} \tag{38}$$

where H_A describes the states of the system (atom or molecule), H_R the states of the bath (host), and V_{AR} contains the interaction between the system and the bath. The success of such a partitioning obviously crucially depends on the correlation between the variables of the bath and the system. When the characteristic time scale of the bath is much shorter than that of the system a reduced description may be expected to be quite successful.[87] The approach is then quite similar to the one made in the Born–Oppenheimer approximation where the electronic problem is solved first and then the nuclear motion is determined by the electronic potential. A sufficient condition for a reduced description to be useful is that the system and bath variables are weakly correlated.

There are several ways of proceeding from here to arrive at the master equation for the reduced density matrix. The Zwanzig projection operator technique in Liouville space[88] or the Kubo cumulant expansion[89] may be used and both methods have recently been applied[90, 45] to study optical dephasing in solids.

Prior to presenting some results of these approaches it seems useful to discuss the hamiltonian and partitioning thereof to describe optical dephasing.

C. The Mixed-Crystal Model Hamiltonian

Presently there is no consensus whether or not resonant phonons play a dominant role in optical dephasing phenomena. Indeed the only clear example of resonant phonon-induced dephasing is the case of tetracene in *p*-terphenyl.[67]

In this regard the situation for spin-dephasing in triplet excited states of molecules is more favorable. Schmidt and co-workers[91] have established that for a number of molecules the exponential activation of $T_2(T)$ in a microwave transition clearly could be related to pseudolocal phonons which were spectroscopically observed.

As shown in Table III, however, also in the optical domain exponential activation of $T_2(T)$ is the rule and an interpretation on basis of the presence of pseudolocal phonons is attractive.

This motivated de Bree and Wiersma,[86] following the theories presented for dephasing by van't Hof and Schmidt[85] and Harris,[92] to try and define the microscopic mixed-crystal Hamiltonian that would be suitable to emphasize the special role that resonant phonons may play in optical dephasing phenomena. For the simple case of an electronic transition interacting with one resonant phonon they arrive at the following Hamiltonian:

$$
\begin{aligned}
H &= H_{\mathrm{A}} + H_{\mathrm{R}} + V_{\mathrm{AR}} \\
H_{\mathrm{A}} &= \sum_f \left\{ \epsilon^f + \sum_\kappa V^f_{\kappa\kappa}\left(\bar{n}_\kappa + \tfrac{1}{2}\right) + \hbar(\Omega + \Delta\Omega^f)\left(B^+B + \tfrac{1}{2}\right) \right\} a_f^+ a_f \\
H_{\mathrm{R}} &= \sum_\kappa \hbar\omega_\kappa\left(b_\kappa^+ b_\kappa + \tfrac{1}{2}\right) + \frac{1}{3!} \sum_{\kappa\kappa'\kappa''} \{ U_{\kappa\kappa'\kappa''} b_\kappa b_{\kappa'}^+ b_{\kappa''}^+ + cc \} + \cdots \\
V_{\mathrm{AR}} &= V_a + V_{e-p} \\
V_a &= \frac{1}{3!} \sum_{\kappa\kappa'} \{ U_{\kappa\kappa'\lambda} b_\kappa b_{\kappa'}^+ B^+ + cc \} + \cdots \qquad (39) \\
V_{e-p} &= \sum_{f,\kappa} V_\kappa^f (b_\kappa + b_\kappa^+) a_f^+ a_f \\
&\quad + \sum_{f,\kappa} V^f_{\kappa\kappa}(b_\kappa^+ b_\kappa - \bar{n}_\kappa) a_f^+ a_f + \sum_{\substack{f,\kappa\kappa' \\ (\kappa\neq\kappa')}} V^f_{\kappa\kappa'} b_\kappa^+ b_{\kappa'} a_f^+ a_f \\
&\quad + \sum_{f,\kappa} V^f_{\kappa\lambda}(b_\kappa^+ B + b_\kappa B^+) a_f^+ a_f
\end{aligned}
$$

with

$$\Delta\Omega^f = \frac{V^f_{\lambda\lambda}}{\hbar}$$

and

$$\Delta\Omega^g = \frac{V_{g\lambda\lambda}}{\hbar} = 0$$

The index λ is used when the pseudolocalized phonon is involved. We note here that this Hamiltonian implies that the pseudolocal phonons are treated on equal footing with the electronic states of the molecule.

From (39) we extract the physical picture of the effective four-level system that will be studied.

H_{A} describes what will be called the subsystem or "molecule." Here only the vibrationless ground state (labeled as level 1) and the lowest excited vibrationless electronic state (level 2) will be considered in conjunction with states of the pseudolocalized mode with one excited vibrational quantum (levels 3 and 4)

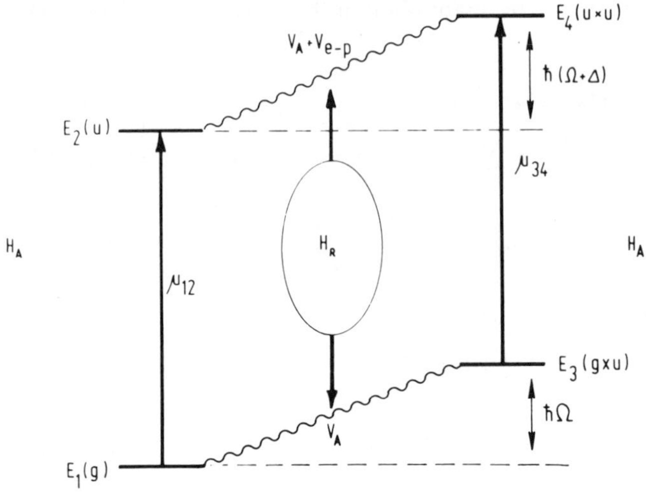

Fig. 30. Effective four-level system.

$$E_1 = \epsilon^0 + \tfrac{1}{2}h\Omega \qquad \text{ground state}$$

$$E_2 = \epsilon' + \sum_\kappa V'_{\kappa\kappa}\left(n_\kappa + \tfrac{1}{2}\right) + \tfrac{1}{2}h(\Omega+\Delta) \qquad \text{excited state}$$

$$E_3 = \epsilon^0 + \frac{3}{2}h\Omega \qquad \text{ground state plus resonant phonon}$$

$$E_4 = \epsilon' + \sum_\kappa V'_{\kappa\kappa}\left(n_\kappa + \tfrac{1}{2}\right) + \frac{3}{2}h(\Omega+\Delta) \qquad \text{excited state plus resonant phonon}$$

In Fig. 30 these four levels are depicted, where $\Delta\Omega^{f=2}$ represents the frequency-change of the pseudolocal phonon in going to the excited state. It is further assumed that only transitions from 1 to 2 and from 3 to 4 carry oscillator strength ($S_1 \leftarrow S_0$), with transition dipole moments of μ_{12} and μ_{34}, respectively. (Note that this choice implies that the pseudolocal phonon is of u-symmetry).

H_R represents the anharmonic band phonon reservoir, while V_{AR} is the molecule–bath interaction consisting of an anharmonic (V_a) and an electron–phonon coupling part (V_{e-p}). Here V_a accounts for the decay of the pseudolocal phonon into several band phonons.

The first three terms in V_{e-p} represent the electron–bandphonon coupling, while the last term is due to coupling of the band phonons with the pseudolocal phonon. For the pseudolocal phonon both diagonal and nondiagonal quadratic electron–phonon coupling is expected to be important. Note that by the above procedure only the nondiagonal part appears in V_{e-p}, as the diagonal contribution is already contained in H_A

by the frequency shift Δ. The scattering among the two lowest levels and the two highest levels, providing the through-bath interaction of the purely electronic transition with the hot pseudolocal phonon transition (Fig. 30) is, respectively, induced by V_a and $V_a + V_{e-p}$. Therefore, in general the scattering in the electronic ground state will differ from that in the excited state. Note further that all (BO) electron–phonon coupling terms neglected would also act on the excited state only.

The system–reservoir interaction operators are products of the type $V_{AR} = AR$, where A represents a "molecule" operator (a_f^+, a_f, B^+, B) and R a reservoir operator (b_κ^+, b_κ). Because of the BO approximation the electronic operators commute with the phonon operators. Moreover, the pseudolocalized phonon operators commute with those of the band phonons because of the boson commutation rules. This amounts to the statement that all molecule variables commute with the reservoir variables, $[A, R] = 0$.

D. Optical Redfield Theory

Following Cohen–Tannoudji[93] and using the Markoffian approximation we have derived[86] the equations of motion for the off-diagonal elements of the reduced density matrix, which determine the dephasing constant (T_2) and the optical lineshape. The following Redfield-like equations were obtained:

$$\dot{\rho}_{12} = -i(\omega_{12} + \Delta_{12/\hbar})\rho_{12} - \Gamma_{12}\rho_{12} + \Gamma_{34\to 12}\rho_{34} \tag{40a}$$

$$\dot{\rho}_{34} = -i(\omega_{34} + \Delta_{34/\hbar})\rho_{34} - \Gamma_{34}\rho_{34} + \Gamma_{12\to 34}\rho_{12} \tag{40b}$$

where $\Delta_{12} = \Delta_1 - \Delta_2$

$$\Delta_1 = P\sum_{k\alpha\beta}\sum p(\alpha)\frac{|\langle\alpha 1|V_{AR}|\beta k\rangle|^2}{E_{\alpha 1} - E_{\beta k}} \qquad \Delta_2 = P\sum_{k\alpha\beta}\sum p(\alpha)\frac{|\langle\alpha 2|V_{AR}|\beta k\rangle|^2}{E_{\alpha 2} - E_{\beta k}}$$

and a similar expression for $\Delta_{34} = \Delta_3 - \Delta_4$ obtained by interchanging the indices 1 and 2 by 3 and 4, respectively. P indicates that the principal part should be taken. These second-order shifts Δ_{ij} [though temperature-dependent through $p(\alpha)$] are assumed to be small and lead to a renormalization of the transition frequencies.

The decay (Γ_{ij}) and feeding ($\Gamma_{ij\to kl}$) constants take the following form:

$$\Gamma_{12}=\Gamma_{12}^{\mathrm{NA}}+\Gamma_{12}^{\mathrm{A}} \tag{41a}$$

$$\begin{aligned}\Gamma_{12}^{\mathrm{NA}}&=\tfrac{1}{2}(\Gamma_{1\to3}+\Gamma_{2\to4})\\&=\frac{\pi}{\hbar}\sum_{\alpha\beta}p(\alpha)|\langle\alpha1|V_{\mathrm{AR}}|\beta3\rangle|^2\delta(E_{\alpha1}-E_{\beta3})\\&\quad+\frac{\pi}{\hbar}\sum_{\alpha\beta}p(\alpha)|\langle\alpha2|V_{\mathrm{AR}}|\beta4\rangle|^2\delta(E_{\alpha2}-E_{\beta4})\end{aligned} \tag{41b}$$

$$\Gamma_{12}^{\mathrm{A}}=\frac{\pi}{\hbar}\sum_{\alpha}p(\alpha)\sum_{\beta}\delta(E_\beta-E_\alpha)|\langle\alpha1|V_{\mathrm{AR}}|\beta_1\rangle-\langle\alpha_2|V_{\mathrm{AR}}|\beta2\rangle|^2 \tag{41c}$$

$$\Gamma_{34\to12}=\frac{\pi}{\hbar}\sum_{\alpha\beta}p(\alpha)\langle\beta1|V_{\mathrm{AR}}|\alpha3\rangle\langle\alpha4|V_{\mathrm{AR}}|\beta2\rangle\{\delta(E_{\alpha3}-E_{\beta1})+\delta(E_{\alpha4}-E_{\beta2})\} \tag{41d}$$

and similar expressions for Γ_{34} and $\Gamma_{12\to34}$ obtained from the above by interchanging the indices 1, 2, by 3, 4.

Returning now to a discussion of the Redfield equations (40) we first note that, with our choice V_{AR} in (39), the only nondiagonal elements that are coupled to each other are ρ_{12} and ρ_{34}. Except for the driving-field terms these equations therefore have a form very similar to the modified Bloch equations postulated by McConnell.[94] The quantum–mechanical derivation of (40), however, leads to new insight and restrictions in the use of these equations in the optical domain (Section IV).

A first important conclusion to be made is that *coherence transfer is negligible*, unless the following relation holds:

$$\Delta=|\omega_{12}-\omega_{34}|\lesssim\Gamma_{34\to12} \tag{42}$$

This condition follows directly from the energy-conserving delta function in (41) and from the prerequisite that $\Gamma_{34\to12}=\Gamma^*_{43\to21}$. Therefore in our description, up to second order in the perturbation expansion, only coherence transfer among near-resonant states is allowed.

A second important conclusion to be made is that (41) show that coherence feeding and decay is due to distinctly different processes. In fact, the coherence decay itself is also due to two different types of effects.

The first one, the adiabatic contribution denoted by Γ_{12}^{A} and Γ_{34}^{A} induces adiabatic (secular) transitions in the system. Consultation of (39) shows that the quadratic electron–band phonon coupling, which leads to elastic (within the linewidth) Raman scattering processes, is responsible for the adiabatic contribution. A closer inspection reveals that this contribution is due to the *difference* of the independent *elastic* modulation of the initial and final state. These processes therefore can be classified as purely T_2-type processes. In the *Debeye* approximation for the band-phonon density of states they lead to the well-known T^7 and T^4 temperature dependence of the width and shift, respectively, of the ZPL[95–97] at low temperature.

The nonadiabatic (nonsecular) contributions $\Gamma_{12}^{\mathrm{NA}}$ and $\Gamma_{34}^{\mathrm{NA}}$ to the coherence decay are caused by inelastic T_1-type processes. Equation (41b) shows that these inelastic scattering processes are induced by anharmonicity (V_{A}) in the ground state and a combination of anharmonicity and electron–phonon coupling (V_{e-p}) in the excited state. Here V_{A} describes the decay (creation) of the pseudolocalized phonon into (from) two band phonons. The relevant part of V_{e-p} is in (39) the last term, which describes in the excited state the exchange of a pseudolocalized phonon with a band phonon. At low temperature ($kT \ll h\Omega$) we therefore expect the role of this last term to be negligible. Finally, note that $\Gamma_{ij}^{\mathrm{NA}}$ is the *sum* of two *inelastic* phonon scattering processes in the ground and excited state.

We now turn to the coherence feeding term $\Gamma_{ij\to kl}$. Equation (41d) shows that $\Gamma_{ij\to kl}$ is an *interference* term composed of the *inelastic* scattering amplitudes in the ground and excited state. Essential to the occurrence of coherence feeding in our model is therefore anharmonicity, as this provides a scattering mechanism that can act within both the electronic ground and excited state.

It is also interesting to note that if the scattering amplitude of one electronic state dominates that of the other, coherence transfer is negligible ($\Gamma_{34\to 12} \ll \Gamma_{12}^{\mathrm{NA}} \leqslant \Gamma_{12}$). In this case the optical Redfield relaxation equations (40a) and (40b) are no longer coupled and reduce to ordinary rate equations describing the incoherent decay.

In the other limit where anharmonicity dominates, coherence transfer to a near-resonant pseudolocalized phonon transition may be important.

E. Optical Lineshape

The optical lineshape under slow passage conditions may be directly derived from (40) through the relation

$$I(\omega) \propto \mathrm{Im}\left[(\rho_{12}+\rho_{34})\mu\right] \tag{43}$$

Neglecting the effects of radiative and radiationless decay, the low-temperature linewidth and shift for near-resonant states are calculated to be

$$[\pi T_2(T)]^{-1} = \left[\Gamma_{12} - \frac{\Gamma_{34}^{-1}|\Gamma_{34\to12}|^2 e^{-\hbar\Omega/kT}}{1+\Delta^2(\Gamma_{34})^{-2}}\right](\mathrm{s}^{-1})$$

$$\Delta\omega_{\mathrm{ex}}(T) = \frac{\Delta(\Gamma_{34})^{-2}|\Gamma_{34\to12}|^2 e^{-\hbar\Omega/kT}}{1+\Delta^2(\Gamma_{34})^{-2}}(\mathrm{rad\ s}^{-1}) \tag{44}$$

These equations reduce to the ones derived by van't Hof and Schmidt[85] and also Harris[92] when the scattering amplitudes in the ground and excited states are identical. We further note that in case of negligible exchange ($\Gamma_{12\to34} \ll \Gamma_{12}$) for both transitions also Lorentzian lineshapes are predicted with widths determined by Γ_{ij}. The temperature-dependent shift will now be induced by the acoustic phonons of the crystal.

In conclusion, we have shown that exponential activation of $T_2(T)$ may be understood on the basis of exchange effects.

F. Comparison with Other Lineshape Theories

Recently Jones and Zewail[90] have, using a Green function approach, also studied optical dephasing in solids. The Hamiltonian is partitioned as follows:

$$H = H_{\mathrm{M}} + H_{\mathrm{ph}} + V$$

where

$$H_{ph} = \sum_{qj} \hbar\omega_{qj}\left(a_{qj}^{+}a_{qj} + \tfrac{1}{2}\right)$$

$$V = V_1\epsilon + V_2\epsilon^2 + \tag{45}$$

with

$$\epsilon = i\sum_{qj}\left(\frac{\hbar(qj)^2}{2M\omega_{qj}}\right)^{1/2}(a_{qj} - a_{qj}^{+})$$

and H_{M} the impurity (molecular) Hamiltonian describing the harmonic phonons of wave vector q and mode index j. ϵ is the local strain tensor. An added restraint is that V_1 has no diagonal elements. An important difference with the mixed-crystal Hamiltonian we use is that in this case all phonons are treated on equal footing. This then leads to the following expression for $T_2'(T)$ at low temperature:

$$\frac{1}{T_2'} = \frac{9\times 6!\left(\frac{kT}{\hbar}\right)^7}{\rho_h^2 v_s^{10} 4\pi^3} |\langle f|V_2|f\rangle - \langle i|V_2|i\rangle|^2 \tag{46}$$

where the prime designates the purely dephasing contribution to T_2. ρ_h is the mass density of the host, v_s the sound velocity and i and f the initial and final electronic state. We note that this expression, in its temperature dependence is very similar to the McCumber–Sturge result.[95] An identical temperature dependence of a T_1-type contribution to T_2 is obtained, but the cross-section for this process is expected to be smaller.

Coupling of the electronic transition to the optical phonon branch (assumed to be dispersionless and of frequency ω_{op}) yields for

$$\frac{1}{T_2'(T)} \alpha \frac{e^{h\omega_{op/kT}}}{(e^{h\omega_{op/kT-1}})^2} \tag{47}$$

a result first derived by Abram[98] and later by Small.[99] At low temperature therefore this mechanism also leads to an exponential activation of T_2', namely, for $kT \ll \hbar\omega_{op}$

$$\frac{1}{T_2'} \alpha e^{-\omega_{op/kT}} \tag{48}$$

As discussed in Ref. 86 other approaches to the lineshape theory also lead to exponential activation of T_2.

We conclude this section by noting that, in contrast to a statement of Jones and Zewail in Ref. 90, the contribution of the optical phonon branch to optical dephasing may well be distinguished from that of exchange by studying the optical dephasing characteristics of several different guests in the same host crystal.

G. Conclusions

The pseudolocal phonon approach to dephasing, in the field of microwave and electronic spectroscopy rests on the assumption that insertion of a guest in a host crystal leads to formation of phonons which are substantially localized at the guest sites. There seems to be an urgent need for a model system where the pseudolocal phonons themselves are accessible for a direct relaxation study. Specifically the lifetime of this phonon and its temperature dependence are crucial quantities in a test of the dephasing theory proposed in Refs. 86 and 92. The mixed crystal system, tetracene in *p*-terphenyl seems most suitable, but the optical transition lies in a spectral region not accessible by the picosecond lasers at our disposal. We conclude that much work remains to be done to clarify the meaning of the observed dephasing characteristics of electronic transitions in mixed molecular crystals.

X. PHOTON ECHOES IN MULTILEVEL SYSTEMS

A. Introduction

Coherent optical measurements in multilevel systems may exhibit some interesting new features which are peculiar to such systems. It is well-known that in multilevel systems where optical branching occurs the emission may exhibit modulations (beats) which are due to interference between transition amplitudes. In optical coherence experiments the interesting new feature is that modulation effects may also be observed which are due to level-splittings in the electronic ground state. In fact the first photon-echo experiments were performed on a multilevel system, ruby,[8] where the electron spin in the ground and excited state of the chromium ion is strongly coupled to the nuclear spins of the neighboring Al-nuclei. To unravel this complexity many photon echo type experiments have been done by the group of Hartmann[100] and Abella and co-workers.[101] Photon-echo quantum beats were demonstrated and also the powerfulness of the so-called PENDOR (photon-echo ENDOR) technique was shown in the case of ruby.

In the field of molecular crystals Hesselink and Wiersma[69] showed that photon-echo studies of molecular radicals exhibited some new and exciting features. Very recently Burland et al.[83] showed that on such radicals (in that case the duryl radical) optical nutation and optical free induction

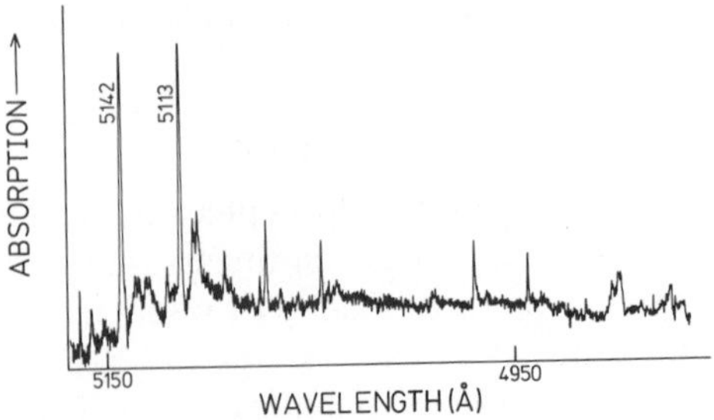

Fig. 31. Absorption spectrum of the lowest $^2D_1 \leftarrow {}^2D_0$ transition of triphenylmethyl in triphenylamine at 1.5 K.

decay experiments may be performed. In molecular radicals the multilevel nature of the electronic ground and excited state is due to hyperfine coupling and the interesting new thing is that for radicals isotope effects on the photon-echo decay may be studied.

B. Triphenylmethyl in Triphenylamine

1. *Optical Absorption Spectrum*

The optical absorption spectrum of triphenylmethyl (TPM) in triphenylamine (TPA), which has been investigated by Weissman,[102] is shown in Fig. 31. It seems quite certain that the lowest electronic transition in TPM is a $^2A_1 \leftarrow {}^2A_1$ transition (assuming local C_{3v} symmetry!) which is polarized parallel to the C_3-axis. We note here that the doubling of the spectrum is due to the occurrence of two sites in the crystal.

2. *Photon Echo Decay at Low Temperature*

In zero magnetic field the two-pulse photon-echo decay exhibits a beat which is shown in Fig. 32. Such beat patterns may be expected for radicals where the hyperfine splitting in the ground and excited state is different, in other words, the spin Hamiltonians in these states do not commute. The spectrum was not further analyzed but it was concluded that the beat was

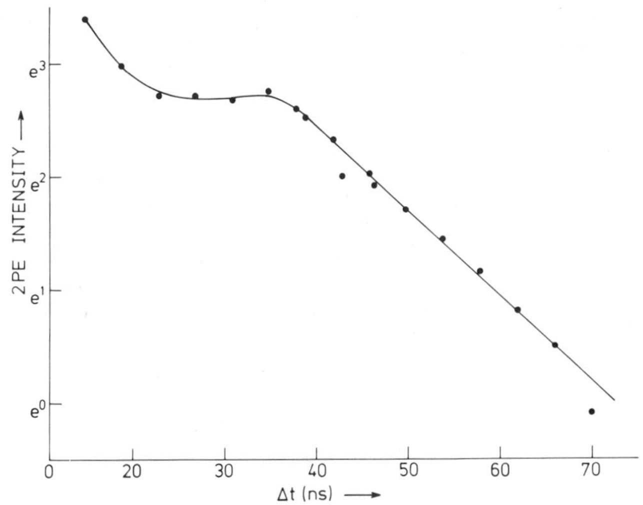

Fig. 32. Two-pulse echo decay of triphenylmethyl in triphenylamine at 1.8 K in zero magnetic field. Δt is the time separation between the exciting pulses.

possibly due to hyperfine coupling with orthohydrogen atoms of the benzene rings.

Morsink and Wiersma[103] recently showed that on deuteration the beat disappeared. This may be understood simply on basis of a threefold reduction of the magnetic moment of the deuteron compared to the proton. Also interesting and consistent with this interpretation was the observation that in a magnetic field the beat disappears. This is due to the fact that in a magnetic field all hyperfine levels are pure-spin states characterized by the quantum numbers $|m_S, m_I\rangle$ and therefore in the electronic transition, optical branching no longer occurs.

3. *Phonon-Induced Echo Relaxation*

The temperature dependence of the photon-echo intensity for several excitation pulse separations was also reported[69] and as an example we show in Fig. 33 a relaxation curve. In Ref. 69 it was reported that between 2.5 and 4.2 K the echo decay could be simulated by assuming that $T_2(T)$ was exponentially activated, but that below 2.5 K the relaxation behavior was inconsistent with this model. We have reexamined these echo relaxation curves and noted that an excellent fit of the echo decay could be obtained over the whole temperature range by assuming that $T_2(T)$ varies as CT^7. Such a temperature dependence of the linewidth is expected in the McCumber–Sturge model, where the electronic transition couples via Raman scattering process with the acoustic phonon branch assuming a Debeye density-of-states function. We note here that triphenyl methyl is isoelectronic with the host triphenylamine and the formation of pseudolocal phonons in this system therefore seems unlikely. We therefore expect

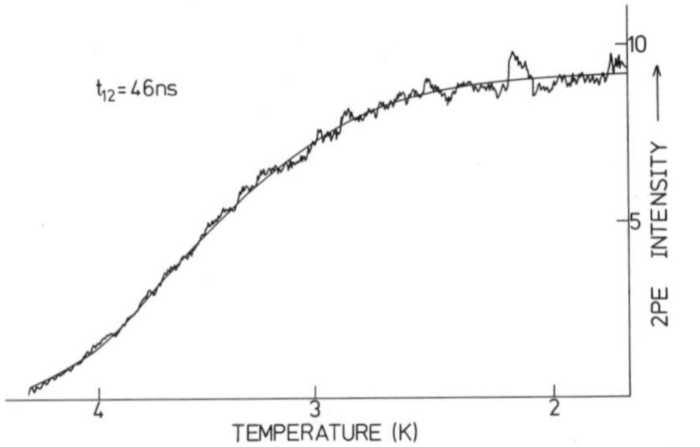

Fig. 33. Echo intensity as a function of temperature for triphenylmethyl in triphenylamine. The separation between the excitation pulses was 46 ns. The dotted curve is a fit whereby T_2 is assumed to be proportional to T^7.

indeed that TPM will in its relaxation behavior, probe the host phonon density of states function. We hasten to add that more experiments are necessary to ascertain the suggested temperature dependence of T_2. Especially accurate echo decay curves at every temperature in a magnetic field are mandatory.

4. *Stimulated Photon Echo*

The most interesting experiments on TPM turned out to be the three pulse or stimulated photon echo experiments. In Fig. 34 we show the results of these measurements,[104] which remained puzzling for a long time. First we concentrate on the decay in a high magnetic field. In this case an exponential echo decay is measured over more than four decades of intensity change. Hesselink and Wiersma explained this using the following Hamiltonian for the echo states.

$$H_i = \omega_0(1-\delta_{1i}) + \beta_e H\cdot g_i \cdot S + \sum_j I_j \cdot A_j^i \cdot S_j - \sum_j \beta_n H \cdot I_j \tag{49}$$

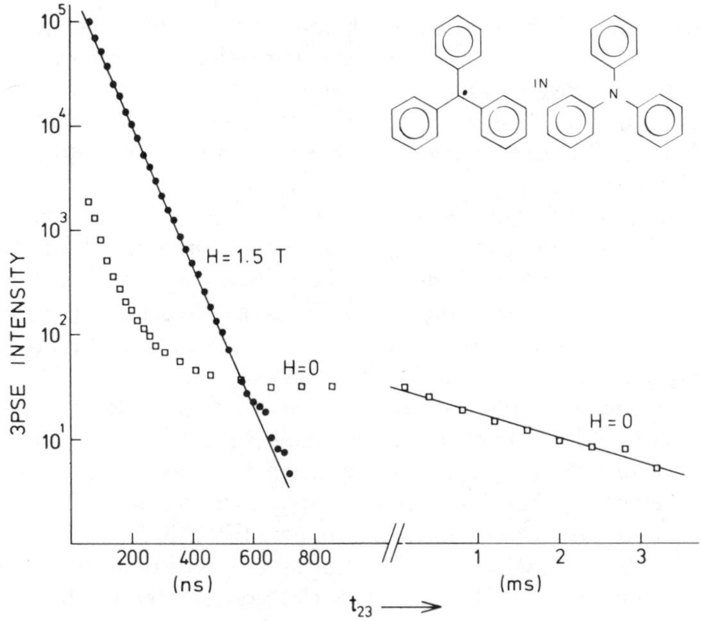

Fig. 34. Intensity of the stimulated photon echo of triphenylmethyl in triphenylamine at 1.5 K vs second-third pulse delay in zero field and in a magnetic field of 1.5*T*. The time separation between the first and second pulse was 40 ns.

where $i=1$ for the groundstate and $i=2$ for the excited state, while j sums over all hydrogen atoms in the molecule. ω_0 is the optical excitation frequency for a specific site, g_i the electron g tensor, and A_j^i the hyperfine tensor of nucleus j. It was further assumed that g_i and A_j^i were identical for all sites but that ω_0 varies from site to site, which gives use to the observed inhomogeneous optical linewidth. In a high magnetic field, $g\beta H \gg A_j(kl)$ the first-order energies and states may be written as:

$$E_i(m_S, m_I) = \omega_0(1-\delta_{1i}) + g_i(z)\beta_e H m_S + \sum_j A_j^i(z) m_S m_{I_j} - \sum_j \beta_n H m_{I_j} \tag{50a}$$

$$\phi_i(m_S, m_I) = |m_S, m_I\rangle \tag{50b}$$

where $m_I = |m_{I_1} m_{I_2} \cdots m_{I_n}\rangle$, $m_S, m_{I_j} = \pm\frac{1}{2}$, while $g_i(z)$ and $A_j^i(z)$ are the z-elements of the respective tensors in an axis system where H points along the z-axis. In a high magnetic field we then expect TPM to behave as a two-level system as the optical selection rule $\Delta m_S = \Delta m_{I_j} = 0$ is strictly obeyed. In this case the hyperfine splitting looks to the radiation field thus merely as an additional source of inhomogeneous broadening.

The decay rate of the stimulated photon echo in a two-level system is then expected to be given by $^2/T_1$ which in dilute crystals (no energy transfer) in the TPM system may be equated to $2/\tau_{fL}$ where τ_{fL} is the fluorescence lifetime. This is exactly what is observed experimentally for TPM where the lifetime of the 3PSE (65±2 ns) is identical to half the measured fluorescence lifetime (131±3 ns). In zero magnetic field the situation is far more complex. From the 2PE experiment we know that the electronic transition can no longer be described in terms of a two-level system and T_1 is no longer a well-defined quantity. Hesselink and Wiersma explained the initial nonexponential 3PSE decay as being due to cross-relaxation in the manifold of electron-nuclear spin states (spin-diffusion). This interpretation is supported by the recent finding of Morsink et al.[104] that in the perdeutero TPM radical the initial decay (up to 400 nsec) of the 3PSE in zero field is completely exponential, with a lifetime of half the fluorescence lifetime (160 ns), while a magnetic field no longer affects the 3PSE intensity. This shows that the spin diffusion processes, on this time scale, have been frozen out. We note here that this experiment shows that in TPM the intramolecular spin-diffusion processes dominate the echo decay behavior. This is an interesting fact to consider in the present discussion on the question whether or not an intramolecular heatbath can induce optical dephasing different from the (trivial) fluorescence decay. The most interesting feature of the 3PSE undoubtedly is the observation of a stimulated echo on a time scale *long* compared to the excited state lifetime. Only very recently[104] we have been able to understand this phenomenon. This photon echo is a new type of echo caused by probing of the optically induced nuclear spin polarization in the electronic ground state. This anomalous photon echo[105] is expected to occur in all systems where two or more levels in the electronic ground state have a transition amplitude to a common level in the excited state. This echo is related to other multilevel type echoes as the trilevel spin echo detected in Al,[106] the Raman echo,[107] observed in CdS[108] and the trilevel photon[109] and two-photon[110] echo generated in Na vapor.

In the following section we will present the results of calculations[111] on a model level system that gives a clear insight into this echo formation.

C. Anomalous Stimulated Photon Echo

In order to obtain physical insight into the anomalous 3PSE formation we have performed calculations on the simplest model system that has all the important features of the TPM radical. Figure 35 shows that we approximate the very complex level structure appropriate to TPM (there are 65,536 hyperfine levels in both the ground and excited state!) by a simple three-level model system. For this system we have calculated the

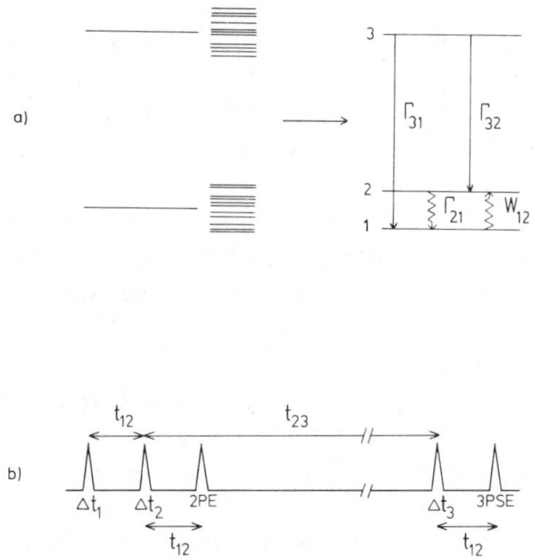

Fig. 35. (a) Model-level system for triphenylmethyl. Levels 1 and 2 are hyperfine levels of the electronic groundstate while level 3 is one of the hyperfine levels in the electronically excited state. (b) Optical pulse sequence used to generate the stimulated photon echo.

effect of a three-pulse sequence, shown in Fig. 35, on the polarization at the optical frequencies $\nu_{32}=(E_3-E_2/h)$ and ν_{31} at the time $t=2t_{12}+t_{23}+3\Delta t$. The following density matrix equation was solved[111]:

$$\rho(t)=\frac{1}{i\hbar}[H,\rho]-\tfrac{1}{2}\{\Gamma,\rho\}+F \tag{51}$$

whereby ρ is the homogeneous density matrix, H the Hamiltonian of the system including the radiation field, Γ the decay matrix, and F the feeding matrix. These matrices for our system take the following form:

$$\Gamma=\begin{pmatrix} W_{12} & 0 & 0 \\ 0 & \Gamma_{21} & 0 \\ 0 & 0 & \Gamma_{31}+\Gamma_{32} \end{pmatrix} \text{ and } F=\begin{pmatrix} \Gamma_{31}\rho_{33}+\Gamma_{21}\rho_{22} & 0 & 0 \\ 0 & \Gamma_{32}\rho_{33}+W_{12}\rho_{11} & 0 \\ 0 & 0 & 0 \end{pmatrix}$$

Γ_{32} and Γ_{31} are the optical relaxation rates (radiative and radiationless) of the upper level into the lower ones. Γ_{21} and W_{12} are the spin-relaxation rates among the hyperfine levels in the ground state. We note here that our calculations extend the work of Schenzle, Grossman, and Brewer[112] on a three-level system and that our results, for infinite relaxation times, for a two-pulse sequence are in agreement with theirs.

In our calculation we assume that $t_{23}\gg\Gamma_{32}^{-1}$, which implies that at time $t=t_{12}+2\Delta t+t_{23}$ the elements of the density matrix ρ_{13}, ρ_{31}, ρ_{23}, ρ_{32}, and ρ_{33} for all practical purposes equal zero. The calculations that followed the procedure suggested by Brewer[113] show that indeed at $t=2t_{12}+3\Delta t+t_{23}$ an optical echo may be expected whose decay, as a function of t_{23}, is only determined by relaxation among the lower levels. The physical picture for the anomalous 3PSE formation that emerges from the calculations is the following.

The first two excitation pulses at time $t=t_{12}+2\Delta t$ create within the inhomogeneous optical lineshape a frequency-dependent optical polarization. Such a polarization grating is known to yield the normal (two-pulse) photon echo. Simultaneously however at $t=t_{12}+2\Delta t$ the first two excitation pulses create also a *frequency-dependent spin polarization* in the electronic ground state. The important point to note here is that there exists a one-to-one correspondence between the optical- and spin polarization in any molecule of the ensemble. While the optical polarization decays fast (not slower than $\frac{1}{4}(\Gamma_{31}^{-1}+\Gamma_{32}^{-1})$) the spin polarization in the electronic groundstate decays slowly through spin–lattice relaxation. The anomalous 3PSE now is formed when a third excitation pulse comes along at a time $t=t_{12}+t_{23}+2\Delta t$ such that all optical polarization is lost, while significant spin polarization is still present. The third excitation pulse excites from this frequency-dependent spin polarization a frequency-dependent polarization at the optical frequency which mimics the situation present at $t=t_{12}+2\Delta t$. At time $t=2t_{12}+3\Delta t+t_{23}$ we therefore expect the formation of an optical echo, which we name the anomalous 3PSE. In the above trilevel system the anomalous 3PSE in the time t_{23} only decays through spin-lattice relaxation among the hyperfine levels in the groundstate. In a real multilevel system, cross-relaxation also destroys the original nuclear spin polarization. We therefore expect that in a system where there are only a few hyperfine levels, the echo lifetime would become much longer. This expectation is born out by our recent observation of anomalous 3PSE formation in the system Pr^{3+}/LaF_3.

D. Photon Echoes in the $^3P_0 \leftarrow {}^3H_4$ Transition of Pr^{3+} in LaF_3

The electronic states that participate in the photon echo formation in the $^3P_0 \leftarrow {}^3H_4$ transition of Pr^{3+} in LaF_3 are shown in Fig. 36. In this picture the coupling of the praseodimium spin with the fluor nuclei has not been taken into account. Optical branching is known to occur in this transition from the work of Hartmann and co-workers[114] but sofar no clear photon-echo quantum beat spectrum was reported.

We show in Fig. 37 a two-pulse echo spectrum, which clearly exhibits intense modulation at the $\langle \pm 1/2 | \leftarrow\rightarrow | \pm 3/2 \rangle$ ground state splitting of 8.47 MHz.[115]

In Fig. 38 we show a 3PS echo spectrum that exhibits modulations due to the excited state. For a detailed interpretation of these spectra we refer to Ref. 116. For our present purposes we are interested only in the 3PSE, at long times compared to the 47 μs fluorescence lifetime.

In zero magnetic field we indeed observed an anomalous 3PSE at seconds after the second pulse. These anomalous echoes had an intensity

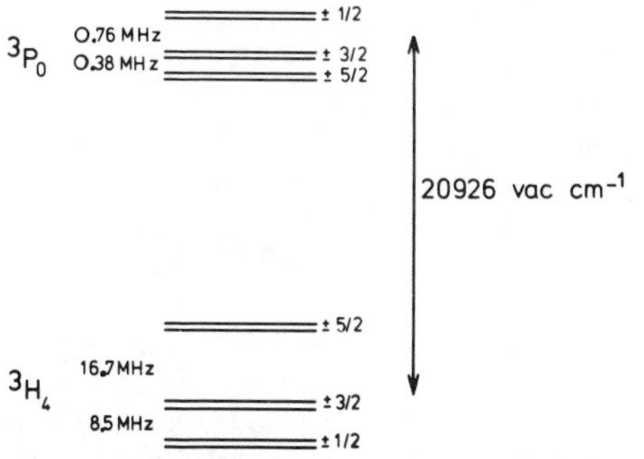

Fig. 36. Level structure in the ground and excited state of Pr^{3+} doped into LaF_3.

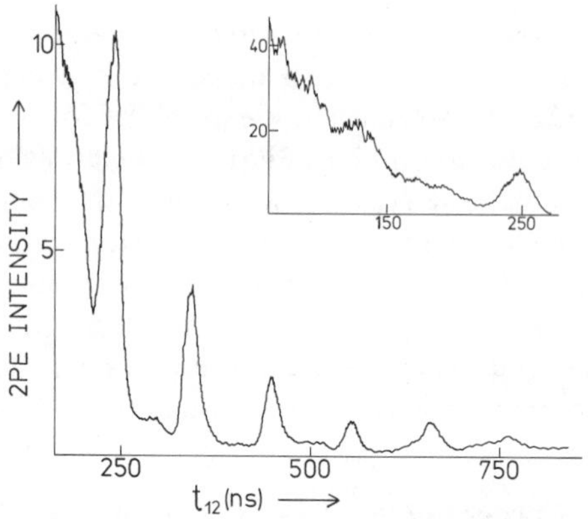

Fig. 37. Two-pulse echo intensity vs excitation pulse delay in the $^3P_0 \leftarrow {}^3H_4$ transition of Pr^{3+} in LaF_3 at 2 K.

of ca.15% of the 3PSE intensity at a probe delay of 0.1 μs. Even more surprisingly is the fact that in a magnetic field of 5 Tesla ($\perp C_3$) the echo could easily be picked up after 30 minutes! This is shown in Fig. 39 where the echo is shown immediately and at 30 min after the second pulse. The observed lengthening of the anomalous 3PSE lifetime is most likely due to suppression of the fluor spin-flip processes. A second effect of the field is that the system effectively reduces to the three-level system of Fig. 35. This means that in a high magnetic field the anomalous echo is formed in the $|-1/2\rangle, |-5/2\rangle$ m_I levels of the ground state coupled to either the $|-1/2\rangle$ or $|-5/2\rangle$ m_I state of the electronically excited state. For a more detailed interpretation of the Pr^{3+} echo phenomena we refer to Ref. 116.

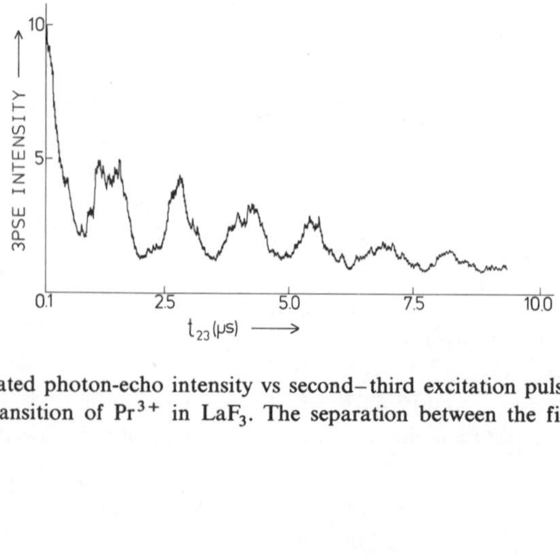

Fig. 38. Stimulated photon-echo intensity vs second–third excitation pulse delay at 2 K in the $^3P_0 \leftarrow {}^3H_4$ transition of Pr^{3+} in LaF_3. The separation between the first two excitation pulses is 50 ns.

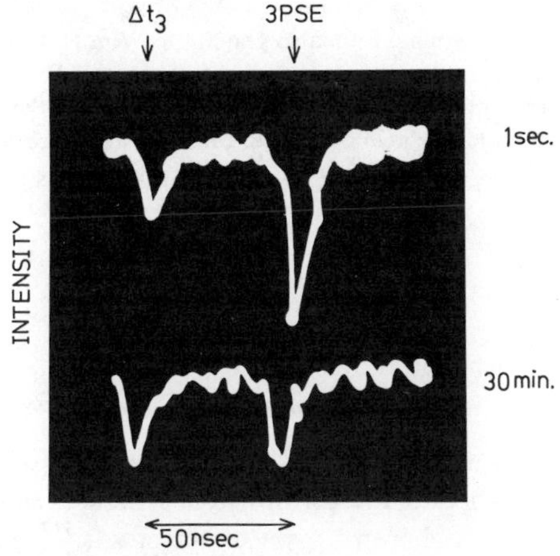

Fig. 39. Anomalous stimulated photon echo trace observed after the third probe pulse in a magnetic field of 4.8T at 2 K (the first pulse on the left is the probe pulse). The first two excitation pulses were separated by 50 ns.

E. Optical Free Induction Decay and Optical Nutation in a Multilevel System

In a recent paper by Burland, Carmona, and Cuellar[83] the first OFID measurements on an organic radical, duryl in a durene host crystal, were reported. At low temperature T_2 was found to be 212 ± 78 ns, which is substantial shorter than half the fluorescence lifetime (530 ns). The authors concluded that in this system also the nuclear spin-flip processes determine the OFID decay. We note here that in these multilevel systems the coherence bandwidth of the exciting laser will have a profound effect on the coherent emission.

In case the effective laser linewidth is less than the hyperfine splitting(s) excitation will prepare a two-level system. The effect of spin-flips on the coherence in this system will then manifest itself as a T_1-type process. No beats are expected in the decay of the optical free induction. With broadband excitation that spans some of the hyperfine splittings spin-flips will be monitored as T_2-type processes and quantum beats are expected in the photon-echo intensity vs probe delay. Burland et al. also demonstrated the feasibility of optical nutation in this system from which in principle, as from the OFID, the transition dipole could be calculated.

Acknowledgments

I am greatly indebted to my co-workers for their help in preparing this progress report. In particular I wish to acknowledge the help of Harmen de Vries who did most of the theoretical work on photochemical hole burning. I am also grateful to Wim H. Hesselink and Jos. B. W. Morsink for permission to reproduce some results here prior to publication. The permission of Dr. S. Voelker and co-workers and Prof. A. H. Zewail and co-workers to reproduce some figures in the text is also gratefully acknowledged. Finally I am indebted to the Netherlands Foundation for Chemical Research (S.O.N.) for generous financial support of the experiments described in this review.

Appendix

Since this paper was written several new developments have taken place. It was recently shown that the pump–probe measurements described in section VI.C. are novel coherent–type experiments.[117] The interpretation of the observations described in section VI.C.3. is therefore incorrect. The temperature dependence of vibrational relaxation therefore remains to be measured.

Furthermore, Fayer and co–workers have recently reported[118] detailed results of the concentration dependence of the photon echo decay in the mixed crystal of pentacene in naphthalene. At low guest concentration they find that at 1.4K $T_2 \neq 2\ \tau_{fl}$, whereby τ_{fl} is the fluorescence lifetime. They interpret this as due to a photon assisted modulation of the Van der Waals interaction.

At higher guest concentrations they observe a shortening of the echo decay time, which they attribute to a photon induced modulation of the guest–guest dipole–dipole interaction.

We question this interpretation mainly on basis of a quantitative calculation of this effect in the system naphthalene in durene.[119] The main point of our criticism is that Fayer et al. included in their calculation nearest neighbor interactions which, due to large static effects, should be excluded.

We believe that direct dipole-dipole coupling among the guest molecules, which leads to an "exciton"–type state in the mixed crystal, is responsible for the effect.[119] More work on these highly doped molecular crystals needs to be done.

References

1. A. S. Davydov, *Soviet Phys. JETP*, **18**, 210 (1948).
2. D. S. McClure, *J. Chem. Phys.*, **22**, 1668 (1954).
3. G. Herzberg and E. Teller, *Z. Phys. Chem.* (*Leipz.*), **21**, 410 (1933).
4. See, e.g., G. Orlandi and W. Siebrand, *J. Chem. Phys.*, **58**, 4513 (1973).
5. J. H. van der Waals and M. S. de Groot, in *The Triplet State*, A. B. Zahlan, ed., London, Cambridge University Press, 1967.
6. G. W. Robinson and R. P. Frosh, *J. Chem. Phys.*, **37**, 1962 (1962), **38**, 1187 (1963).

7. See. e.g., A. C. G. Mitchell and M. W. Zemansky, *Resonance Radiation and Excited Atoms*, Cambridge University Press, London, England, 1934.
8. N. A. Kurnit, I. D. Abella, and S. R. Hartmann, *Phys. Rev. Lett.*, **13**, 567 (1964); *Phys. Rev.*, **141**, 391 (1966).
9. A. Szabo, *Phys. Rev. Lett.*, **25**, 924 (1970); **27**, 323 (1971).
10. R. G. Brewer and R. L. Shoemaker, *Phys. Rev. Lett.*, **27**, 631 (1971).
11. *Ultrashort Light Pulses, Topics Appl. Phys.*, **18**, 16 (1977).
12. T. J. Aartsma and D. A. Wiersma, *Phys. Rev. Lett.*, **36**, 1360 (1978).
13. R. G. de Voe and R. G. Brewer, *Phys. Rev. Lett.*, **40**, 862 (1978).
14. W. H. Hesselink and D. A. Wiersma, *Chem. Phys. Lett.*, **56**, 227 (1978).
15. T. Yajima and H. Souma, *Phys. Rev. A*, **17**, 309 (1978).
16. J. J. Song, J. H. Lee, and M. D. Levenson, *Phys. Rev. A*, **17**, 1439 (1978).
17. P. F. Liao, N. P. Economu, and R. R. Freeman, *Phys. Rev. Lett.*, **39**, 1473 (1977).
18. D. von der Linde, A. Laubereau, and W. Kaiser, *Phys. Rev. Lett.*, **26**, 954 (1971).
19. A. Laubereau and W. Kaiser, *Rev. Mod. Phys.*, **50**, 607 (1978).
20. J. P. Heritage, C. P. Ausschnitt, and R. K. Jain, in *Picosecond Phenomena*, C. V. Shank, E. P. Ippen, and S. L. Shapiro, eds., Springer-Verlag, Berlin, 1978, p. 8.
21. S. R. Hartmann, *Sci. Am.*, **218**, 32 (1968).
22. R. G. Brewer, *Phys. Today*, **30**, 50 (1977).
23. A. H. Francis and C. B. Harris, *Chem. Phys. Lett.*, **9**, 181 (1971), *ibid.*, **9**, 188 (1971).
24. M. D. Fayer and C. B. Harris, *Phys. Rev. B*, **9**, 748 (1974).
25. J. Frenkel, *Phys. Rev.*, **37**, 17 (1931), *ibid.*, **37**, 1276 (1931).
26. J. P. Lemaistre, P. Pee, R. Lalanne, F. Dupuy, Ph. Kottis and H. Port, *Chem. Phys.*, **28**, 407 (1978).
27. B. J. Botter, C. J. Nonhof, J. Schmidt, and J. H. van der Waals, *Chem. Phys. Lett.*, **43**, 210 (1976).
28. D. M. Burland and A. H. Zewail, *Adv. Chem. Phys.*, **XL**, 369 (1979).
29. A. Frad, F. Lahmani, A. Tramer, and C. Tric, *J. Chem. Phys.*, **60**, 4419 (1974).
30. For a review see S. Haroche, in *High Resolution Laser Spectroscopy*, K. Shimoda, ed., Springer, Berlin, 1976.
31. J. Chaiken, T. Benson, M. Gurnick, and J. D. McDonald, *Chem. Phys. Lett.*, **61**, 195 (1979).
32. R. H. Dicke, *Phys. Rev.*, **93**, 99 (1954).
33. F. Bloch, *Phys. Rev.*, **70**, 460 (1946).
34. E. L. Hahn, *Phys. Rev.*, **77**, 297 (1950).
35. R. P. Feynman, F. L. Vernon, and R. W. Hellwarth, *J. Appl. Phys.*, **28**, 49 (1957).
36. F. A. Hopf, R. F. Shea, and M. O. Scully, *Phys. Rev. A*, **7**, 2105 (1973).
37. G. B. Hocker and C. L. Tang, *Phys. Rev. Lett.*, **21**, 591 (1968).
38. R. I. Personov, E. I. Al'shits, L. A. Bykovskaya, and B. M. Kharlamov, *Sov. Phys. JETP*, **38**, 912 (1974).
39. A. Szabo, *Phys. Rev. B*, **11**, 4512 (1975).
40. H. de Vries and D. A. Wiersma, *Phys. Rev. Lett.*, **36**, 91 (1976).
41. M. S. Feld, in *Frontiers in Laser Spectroscopy*, Vol. 1, Les Houches 1975, R. Balian, S. Haroche, and S. Liberman, eds., North-Holland, Amsterdam, 1977.
42. S. L. McCall and E. L. Hahn, *Phys. Rev.*, **183**, 457 (1969).
43. R. E. Slusher and H. M. Gibbs, *Phys. Rev. A*, **5**, 1634 (1972).
44. B. J. Berne and G. D. Harp, *Adv. Chem. Phys.*, **17** (1970).
45. S. Mukamel, *Chem. Phys.*, **37**, 33 (1979).
46. J. Jortner and S. Mukamel, in *The World of Quantum Chemistry*, R. Daudel and B. Pullman, eds., Reidel, Dordrecht-Holland, 1974, pp. 145–209.

47. For a recent interesting experiment confirming the theory in Ref. 5 see R. A. Schadee, C. J. Nonhof, J. Schmidt, and J. H. van der Waals, *Mol. Phys.*, **34**, 171 (1977).
48. S. Voelker, R. M. Macfarlane, and J. H. van der Waals, *Chem. Phys. Lett.*, **53**, 8 (1978).
49. M. Sargent, P. E. Toschek, and H. G. Danielmeijer, *Appl. Phys.*, **11**, 55 (1976).
50. S. Voelker, R. M. Macfarlane, A. Z. Genack, H. P. Trommsdorff, and J. H. van der Waals, *J. Chem. Phys.*, **67**, 1759 (1977).
51. H. de Vries and D. A. Wiersma, **72**, 1851 (1980).
52. H. de Vries and D. A. Wiersma, *J. Chem. Phys.*, **69**, 897 (1978).
53. W. G. van Dorp, W. H. Schoemaker, M. Soma, and J. H. van der Waals, *Mol. Phys.*, **30**, 1701 (1975), W. G. van Dorp, thesis, University of Leiden, 1975.
54. B. M. Kharlamov, R. I. Personov, and L. A. Bykovokaya, *Opt. Commun.*, **12**, 191 (1974).
55. A. A. Gorokhovski, R. K. Kaarli, and L. A. Rebane, *Pis'ma Zh. Eksp. Teor. Fiz.*, **20**, 474 (1974) [*JETP Lett.*, **20**, 216 (1974)].
56. J. H. Meyling, R. P. van der Werf, and D. A. Wiersma, *Chem. Phys. Lett.*, **28**, 364 (1974).
57. H. de Vries and D. A. Wiersma, see footnote in *Chem. Phys. Lett.*, **51**, 565 (1977).
58. R. M. Hochstrasser, D. S. King, and A. C. Nelson, *Chem. Phys. Lett.*, **42**, 8 (1976).
59. A. P. Marchetti, W. C. McColgin, and J. H. Eberly, *Phys. Rev. Lett.*, **35**, 387 (1975).
60. W. H. Hesselink and D. A. Wiersma, *Chem. Phys. Lett.*, **65**, 300 (1979); *J. Chem. Phys.*, **73**, 648 (1980).
61. S. Voelker and J. H. van der Waals, *Mol. Phys.*, **32**, 1703 (1976).
62. A. A. Gorokhovski and L. A. Rebane, *Opt. Commun.*, **20**, 144 (1977).
63. W. H. Hesselink and D. A. Wiersma, unpublished result.
64. S. Voelker and R. M. Macfarlane, *Chem. Phys. Lett.*, **61**, 421 (1979).
65. J. M. Hayes and G. J. Small, *Chem. Phys.*, **27**, 151 (1978).
66. M. R. Topp and H. B. Lin, *Chem. Phys. Lett.*, **50**, 412 (1977).
67. T. J. Aartsma and D. A. Wiersma, *Chem. Phys. Lett.*, **42**, 520 (1976).
68. T. J. Aartsma and D. A. Wiersma, *Chem. Phys. Lett.*, **54**, 415 (1978).
69. W. H. Hesselink and D. A. Wiersma, *Chem. Phys. Lett.*, **50**, 51 (1977).
70. H. de Vries, P. de Bree and D. A. Wiersma, *Chem. Phys. Lett.*, **52**, 399 (1977); erratum, **53**, 418 (1978). T. E. Orlowski, K. E. Jones, and A. H. Zewail, *Chem. Phys. Lett.*, **54**, 197 (1978).
71. T. E. Orlowski and A. H. Zewail, *J. Chem. Phys.*, **70**, 1390 (1979).
72. J. H. Meyling and D. A. Wiersma, *Chem. Phys. Lett.*, **20**, 383 (1973).
73. H. de Vries and D. A. Wiersma, *J. Chem. Phys.*, **70**, 5807 (1979).
74. J. B. W. Morsink, T. J. Aartsma and D. A. Wiersma, *Chem. Phys. Lett.*, **49**, 34 (1977).
75. W. H. Hesselink and D. A. Wiersma, in *Picosecond Phenomena* C. V. Shank, E. P. Ippen, and S. L. Shapiro, eds., Springer-Verlag, Berlin, 1978, p. 192.
76. S. J. Hunter, H. Parker, and A. H. Francis, *J. Chem. Phys.*, **61**, 1390 (1974).
77. E. P. Ippen, C. V. Shank, and A. Bergman, *Chem. Phys. Lett.*, **38**, 611 (1976).
78. R. M. Hochstrasser and C. A. Nyi, *J. Chem. Phys.*, **70**, 1112 (1979).
79. R. G. Brewer and A. Z. Genack, *Phys. Rev. Lett.*, **36**, 959 (1976).
80. A. H. Zewail, T. E. Orlowski, R. R. Shah, and K. E. Jones, *Chem. Phys. Lett.*, **49**, 520 (1977).
81. A. Z. Genack, R. M. Macfarlane, and R. G. Brewer, *Phys. Rev. Lett.*, **37**, 1078 (1976).
82. A. Szabo and M. Kroll, *Opt. Lett.*, **2**, 10 (1978).
83. D. M. Burland, F. Carmona, and E. Cuellar, *Chem. Phys. Lett.*, **64**, 5 (1979).
84. R. L. Shoemaker and E. W. van Stryland, *J. Chem. Phys.*, **64**, 1733 (1976).
85. C. A. van't Hof and J. Schmidt, *Chem. Phys. Lett.*, **36**, 460 (1975); **42**, 73 (1976).

86. P. de Bree and D. A. Wiersma, *J. Chem. Phys.*, **70**, 790 (1979).
87. S. Mukamel, I. Oppenheim, and J. Ross, *Phys. Rev.*, **A17**, 1988 (1978).
88. R. Zwanzig, in *Lectures in Theoretical Physics*, Vol. III, W. E. Britten, ed., Wiley-Interscience, New York, 1961, p. 106.
89. R. Kubo, *J. Math. Phys.*, **4**, 174 (1963).
90. K. E. Jones and A. H. Zewail, in *Advances in Laser Chemistry*, Vol. 3, A. H. Zewail, ed., *Springer Series in Chem. Phys.*, Springer, Berlin, 1978.
91. C. A. van't Hof and J. Schmidt, *Chem. Phys. Lett.*, **36**, 460 (1975); *ibid.*, **42**, 73 (1976); P. J. F. Verbeek, C. A. van't Hof, and J. Schmidt, *Chem. Phys. Lett.*, **51**, 292 (1977).
92. C. B. Harris, *J. Chem. Phys.*, **67**, 5607 (1977).
93. C. Cohen-Tannoudji, in *Frontiers in Laser Spectroscopy*, Vol. I, Les Houches 1975, R. Balian, S. Haroche, and S. Liberman, eds., North-Holland, Amsterdam, 1977.
94. H. M. McConnell, *J. Chem. Phys.*, **28**, 430 (1958).
95. D. E. McCumber and M. D. Sturge, *J. Appl. Phys.*, **34**, 1682 (1963).
96. B. di Bartolo, *Optical Interactions in Solids*, Wiley, New York, 1968.
97. M. A. Krivoglaz, *Sov. Phys. Solid State*, **6**, 1340 (1964).
98. I. I. Abram, *Chem. Phys.*, **25**, 87 (1977).
99. G. Small, *Chem. Phys. Lett.*, **57**, 50 (1978).
100. P. F. Liao and S. R. Hartmann, *Phys. Rev. B.*, **8**, 69 (1973); P. F. Liao, P. Hu, R. Leigh, and S. R. Hartmann, *Phys. Rev. A*, **9**, 332 (1974).
101. L. Q. Lambert, A. Compaan, and I. D. Abella, *Phys. Rev. A*, **4**, 2022 (1971); L. Q. Lambert, *Phys. Rev. B*, **7**, 1834 (1973).
102. S. I. Weissman, *J. Chem. Phys.*, **22**, 155 (1954); **37**, 1886 (1962).
103. J. B. W. Morsink and D. A. Wiersma, unpublished result.
104. J. B. W. Morsink, W. H. Hesselink, and D. A. Wiersma, *Chem. Phys. Lett.*, **64**, 1 (1979).
105. M. Aihara and H. Inaba, *J. Phys. A*, **6**, 1709, 1725 (1973); *Opt. Commun.*, **8**, 280 (1973).
106. H. Hatanaka, T. Terao, and T. Hashi, *J. Phys. Soc. Japan Lett.*, **39**, 835 (1975).
107. S. R. Hartmann, *IEEE J. Quant. Electron.*, **4**, 802 (1968).
108. P. Hu, S. Geschwind, and T. M. Jedju, *Phys. Rev. Lett.*, **37**, 1357, 1773 (E) (1976).
109. T. Mossberg, A. Flusberg, R. Kachru, and S. R. Hartmann, *Phys. Rev. Lett.*, **39**, 1523 (1977).
110. A. Flusberg, T. Mossberg, R. Kachru, and S. R. Hartmann, *Phys. Rev. Lett.*, **41**, 305 (1978).
111. J. B. W. Morsink and D. A. Wiersma, in *Laser Spectroscopy IV*, *Proc. 4th Int. Conf., Rottach-Egern 1979*, H. Walther and K. W. Rothe, eds., *Springer Series in Optical Sciences*, Springer-Verlag, New York, Heidelberg, Berlin, 1979.
112. A. Schenzle, S. Grossman, and R. G. Brewer, *Phys. Rev.*, **A13**, 1891 (1976).
113. R. G. Brewer, in *Frontiers in Laser Spectroscopy*, Vol. I, Les Houches 1975, R. Balian, S. Haroche, and S. Liberman, eds., North-Holland, Amsterdam, 1977.
114. Y. C. Chen, K. P. Cheang, and S. R. Hartmann, *Opt. Commun.*, **26**, 269 (1978).
115. L. E. Erickson, *Opt. Commun.*, **21**, 147 (1977).
116. J. B. W. Morsink and D. A. Wiersma, *Chem. Phys. Lett.*, **65**, 105 (1979).
117. W. H. Hesselink and D. A. Wiersma, *Phys. Rev. Lett.*, **43**, 1991 (1979).
118. D. E. Cooper, R. W. Olson, and M. D. Fayer, *J. Chem. Phys.*, **72**, 2332 (1980).
119. J. B. W. Morsink, B. Kruizinga, and D. A. Wiersma, *Chem. Phys. Lett.*, **76**, 218 (1980).

VIBRATIONAL POPULATION RELAXATION IN LIQUIDS

DAVID W. OXTOBY*

The Department of Chemistry and The James Franck Institute, The University of Chicago, Chicago, Illinois 60637

CONTENTS

I. INTRODUCTION

Vibrational population relaxation plays a fundamental role in many physical and chemical processes in liquids. In liquid-state photochemical reactions there is often competition between relaxation and reaction and the observed reaction rates and yields can therefore be quite sensitive to solute–solvent interactions. There has also been considerable recent interest in the possible development of liquid state chemical lasers based on the long vibrational lifetimes observed in simple liquids. For both these prob-

*Alfred P. Sloan Foundation Fellow.

lems it is necessary to have more detailed knowledge of the rates and mechanisms of vibrational relaxation in liquids.

Experimental studies of vibrational relaxation have been carried out since the 1950s, using ultrasonic attenuation as a probe. These methods are limited in the types of molecules that are accessible to study, however, and the real expansion of the field did not occur until the development of sensitive laser techniques. These experiments have revealed a rich variety of possible behavior, showing vibrational lifetimes ranging from picoseconds (for molecules of moderate size) to seconds (for liquid nitrogen). In addition, the rates show a surprising sensitivity to details of the molecular potential energy surface and the solvent interactions: for example, the relaxation rate of C—H stretching modes has been observed to vary over two orders of magnitude depending on the molecule and solvent involved. Previous assumptions that vibrational relaxation would always occur within 10–1000 vibrational periods in polyatomic liquids have therefore been called into question by these new experimental data.

Theoretical studies have approached relaxation in liquids from several points of view. Some have applied gaslike models, which involve hard collisions between pairs of molecules essentially unaffected by the surrounding medium, while others are solidlike in that they treat a central molecule surrounded by a fixed cage of neighbors. Still others have investigated the role of liquid-state collective modes in relaxation. In spite of much progress, a large number of unanswered questions remain.

In this chapter we review experimental and theoretical studies of vibrational population relaxation in liquids. This review is complementary to our previous article in the same series,[1] which treated vibrational phase relaxation (dephasing) in liquids due to vibrationally elastic interactions. A number of reviews have appeared recently on related subjects: vibrational relaxation in solid matrices has been covered elsewhere,[2] and several reviews[3] have been devoted to experimental studies of picosecond time-scale relaxation processes in liquids. Diestler[4] has recently reviewed theoretical studies of vibrational relaxation in liquids and solids; the focus of the present article is rather different from that of Diestler.

In Section II we survey a number of theoretical approaches to vibrational relaxation in liquids, and in Section III we summarize experimental studies to date. Section IV concludes with some recent theoretical work on the nonexponential relaxation, which is predicted for systems in which vibrational relaxation rates are as fast as those of other degrees of freedom.

II. THEORIES FOR VIBRATIONAL RELAXATION

With the exception of direct computer simulation of vibrational relaxation via molecular dynamics (see Section II.H) most theories for relaxation in liquids fall into two basic classes: gaslike isolated binary collision models in which the liquid structure affects only the collision frequency, and weak coupling theories in which the coupling of the vibrational degrees of freedom to those of the bath (e.g., translations and rotations) is assumed to be weak so that perturbation theory may be used. In the next two sections we outline the basic approaches involved and in subsequent sections discuss a number of extensions and elaborations of these models.

A. Isolated Binary Collision Model

The isolated binary collision (IBC) model for vibrational relaxation in liquids was developed by Herzfeld, Litovitz, and Madigosky.[5,6] Its fundamental assumption is that the vibrational relaxation rate from level i to level $j(\tau_{ij})^{-1}$ is given by the product of two factors

$$(\tau_{ij})^{-1} = P_{ij}(\tau_c)^{-1} \tag{2.1}$$

where $(\tau_c)^{-1}$ is the collision frequency (τ_c is the mean time between collisions) and P_{ij} is the probability per collision that a transition from state i to state j will take place. P_{ij} depends on the temperature T but is independent of the density ρ, since it is assumed to be a purely binary property. (τ_c^{-1}), which depends on both ρ and T, is proportional to ρ at low densities, but this proportionality is modified in the liquid phase. This approach is therefore a gaslike model in that the dense liquid affects only the collision frequency and not the dynamics of relaxation. To proceed, one must separately obtain P_{ij} and τ_c.

In some cases P_{ij} may be obtained experimentally from low-density gas-phase determinations of vibrational relaxation rates (although sometimes it is necessary to extrapolate downward to liquid state temperatures). At low densities $(\tau_c)^{-1}$ can be obtained from gas kinetic theory collisional cross-sections once a radius has been chosen, while τ_{ij}^{-1} can be found experimentally. Their ratio gives P_{ij}, the transition probability per collision. Purely theoretical calculations of P_{ij} can be carried out in principle using quantum scattering theory, although this becomes difficult for molecules of even moderate size. A number of semiclassical procedures have been proposed; among the most popular is the SSH model,[7] which gives[8]

$$P_{k-l}^{i-j}(a,b) = \left\{P_0(1)P_0(2)\left[V^{i-j}(a)\right]^2\left[V^{k-l}(b)\right]^2\right\}F_{\text{tr}} \tag{2.2}$$

where

$$F_{\text{tr}} = \left(\frac{8\mu}{2kt}\right)\left(\frac{8\pi^3\mu\,\Delta E}{h^2}\right)^2 \exp\left(\frac{-\phi_0}{kT}\right)$$

$$\otimes \int_0^\infty dv \frac{v}{\alpha^4}\left(\frac{r_c}{r_0}\right)^2 \frac{\exp(q-q')}{(1-\exp(q-q'))^2} \exp\left(\frac{-\mu v^2}{2kT}\right) \tag{2.3}$$

Here $P_{k-l}^{i-j}(a,b)$ is the probability per collision that mode Q_a in one molecule will change its state from i to j, while Q_b in the same or a different molecule changes from k to l. We have introduced

$$q=\frac{4\pi^2\mu v}{\alpha h} \qquad q'=\frac{4\pi^2\mu v'}{\alpha L}$$

and

$$\Delta E=h\nu_a(i-j)+h\nu_b(k-l)$$

with μ the reduced mass and v and v' the initial and final relative velocities. α is the exponential repulsion parameter in the potential, ϕ_0 the well depth, r_c the distance of closest approach, and r_0 the separation at zero potential energy. The vibrational factors $V^{i-j}(a)$ and $V^{k-l}(b)$ are often calculated in the breathing sphere approximation,[8] which in effect averages the vibrational motion out over the surface of a sphere. Recently Miklavc and Fischer[8] have proposed a modification of this model to account more accurately for the angular dependence of the interaction.

The collision frequency τ_c^{-1} must be obtained from some model for the liquid. If an effective hard sphere radius for the molecules can be chosen, the collision rate can then be obtained from Enskog theory[9] or from molecular dynamics simulations[10] of the hard sphere fluid. An alternative that has been popular is to use a cell model for the liquid.[11] In its simplest form[5, 12] a molecule moves in a cell created by fixed neighbors located on a lattice. For a cubic lattice, the distance between neighbors is $\rho^{-1/3}$ where ρ is the liquid-state number density. The central molecule then moves a distance $2\rho^{-1/3}-2\sigma$ between collisions, if its effective diameter is σ. The time between collision is then

$$\tau_c=2\bar{v}^{-1}\left[\rho^{-1/3}-\sigma\right] \tag{2.4}$$

where $\bar{v}$, the average thermal velocity, is

$$\bar{v}=\left(\frac{8kT}{\pi m}\right)^{1/2} \tag{2.5}$$

Madigosky and Litovitz[6] subsequently proposed a more elaborate cell model in which the cell walls were permitted to move; this modification leads to the result

$$\tau_c=\bar{v}^{-1}\left[\rho^{-1/3}-\sigma\right] \tag{2.6}$$

which is smaller by a factor of 2 than (2.4). τ_c is very sensitive to the choice of σ, so the procedure frequently used is to plot $\tau_{ij}=P_{ij}^{-1}\tau_c$ vs $\rho^{-1/3}$ and obtain σ from the intercept. If the points lie on a straight line and σ is comparable to the values of the molecular diameter obtained from gas-phase virial and transport data, this is taken as confirmation of the validity of the binary collision model.

B. Perturbation Theory

A number of approaches to vibrational relaxation in liquids have been suggested which involve the assumption that the coupling of the vibrational degrees of freedom to those of the bath (translations and rotations) is weak. Different formalisms have been used: time-dependent perturbation theory, projection operators, and cumulant expansions. We will outline the derivation of the relaxation rate using time-dependent perturbation theory and then point out that the other approaches give the same result in the weak coupling limit.

Suppose the Hamiltonian has the form

$$\mathcal{H}=\mathcal{H}_0+\mathcal{H}_B+V \tag{2.7}$$

where $\mathcal{H}_0$ is the Hamiltonian for the vibrational degrees of freedom; its precise form will be specified later. Effects of the anharmonicity of the molecular vibrations are present in $\mathcal{H}_0$. $\mathcal{H}_B$ is the "bath" Hamiltonian, which includes the rotational and translational degrees of freedom. V, which couples system to bath, is treated through perturbation theory. Let the eigenstates of $\mathcal{H}_0$ be labeled by $i, j\ldots$ and those of $\mathcal{H}_B$ by $\alpha, \beta,\ldots$. The initial state of the bath is not known precisely, but is given by the thermal equilibrium distribution $P_\alpha=\exp(-\mathcal{H}_{B\alpha/kT})=\exp(-E_\alpha/kT)$, where P_α is the probability that the bath is in state α. The transition rate τ_{ij}^{-1} from level i to level j due to V is then given by time-dependent perturbation theory[13] (Fermi's Golden Rule) as

$$\tau_{ij}^{-1}=\frac{2\pi}{\hbar}\sum_\alpha\sum_\beta P_\alpha|V_{i\alpha,j\beta}|^2\delta(E_i-E_j+E_\alpha-E_\beta) \tag{2.8}$$

A Fourier transform of the delta function gives

$$\begin{aligned}\tau_{ij}^{-1}&=\frac{1}{\hbar^2}\int_{-\infty}^{\infty}dt\sum_\alpha\sum_\beta P_\alpha\exp\left[\frac{it}{\hbar}(E_i-E_j+E_\alpha-E_\beta)\right]\\&\otimes|V_{i\alpha,j\beta}|^2\\&=\frac{1}{\hbar^2}\int_{-\infty}^{\infty}dt\exp(i\omega_{ij}t)\sum_\alpha\sum_\beta P_\alpha\exp\left(\frac{it}{\hbar}E_\alpha\right)V_{i\alpha,j\beta}\\&\otimes\exp\left(-\frac{it}{\hbar}E_\beta\right)V_{j\beta,i\alpha}\end{aligned} \tag{2.9}$$

Rewriting this in the Heisenberg representation gives

$$\begin{aligned}\tau_{ij}^{-1} &= \frac{1}{\hbar^2}\int_{-\infty}^{\infty} dt \exp(i\omega_{ij}t)\left\langle \exp\left(\frac{\mathcal{H}_B t}{\hbar}\right) V_{ij} \exp\left(\frac{-i\mathcal{H}_B t}{\hbar}\right) V_{ji}\right\rangle \\ &= \frac{1}{\hbar^2}\int_{-\infty}^{\infty} dt \exp(i\omega_{ij}t)\langle V_{ij}(t)V_{ji}(0)\rangle \end{aligned} \tag{2.10}$$

This can be rewritten[14] in terms of the time symmetrized anticommutator:

$$\begin{aligned}\tau_{ij}^{-1} = &\frac{2\hbar^2}{1+\exp(-\beta\hbar\omega_{ij})}\int_{-\infty}^{\infty} dt \exp(i\omega_{ij}t) \\ &\otimes\left\langle \tfrac{1}{2}\left[V_{ij}(t), V_{ji}(0)\right]_+\right\rangle \end{aligned} \tag{2.11}$$

where

$$[A, B]_+ = AB + BA$$

The treatment up to this point has been fully quantum mechanical; V_{ij} is an operator in the bath degrees of freedom. For many calculations on liquids, however, one wants to treat these degrees of freedom (rotations and translations) classically; the question then arises of what is the best way to replace a quantum correlation function with a classical one. A classical autocorrelation function is an even function of the time, a property shared by the anticommutator in (2.11) but not by the one sided correlation function of (2.10). It thus appears that the best place to make a classical approximation is in (2.11); in addition, doing so gives

$$\tau_{ij}^{-1} = \exp(\beta\omega_{ij})\tau_{ji}^{-1} \tag{2.12}$$

which satisfies detailed balance. In contrast, a classical approximation in (2.10) would give equal rates for the upward and downward transitions. The "semiclassical" approximation which we propose is then

$$\tau_{ij}^{-1} = \frac{2\hbar^2}{1+\exp(-\beta\hbar\omega_{ij})}\int_{-\infty}^{\infty} dt \exp(i\omega_{ij}t)\langle V_{ij}^{\text{class}}(t)V_{ji}^{\text{class}}(0)\rangle \tag{2.13}$$

What we are suggesting is that the appropriate way to take a semiclassical limit is to express all quantum correlation functions in terms of the corresponding anticommutators, and then replace them by classical approximations. We are not aware that this problem has been discussed elsewhere in the literature.

There are other ways to derive (2.10) and (2.11) besides the direct time-dependent perturbation theory approach. Diestler and Wilson[15] considered both dephasing and population relaxation and showed that for a harmonic oscillator linearly coupled to a heat bath the two rates differed only by a factor of 2. They used Zwanzig–Mori[16] projection operator methods in which the degrees of freedom in $\mathcal{H}_0$ are treated exactly and those in V through perturbation theory. Oxtoby and Rice[17] and Knauss and Wilson[18] extended the projection operator approach to nonlinear system-bath coupling (which leads to an additional "pure dephasing" term), and Diestler[19] applied similar methods to relaxation in a two-level system. Nitzan and Silbey[20] applied cumulant expansion techniques to the relaxation of a harmonic oscillator and a two-level system linearly coupled to a heat bath and in the weak coupling limit obtained the same relaxation rate as is given by perturbation theory. We will discuss both projection operator and cumulant expansion techniques more extensively in Section IV where we consider non-Markovian effects on vibrational relaxation.

To do calculations with (2.13) it is necessary to choose some form for the coupling potential V. There are different mechanisms that can contribute: the coefficient C_6 in the dispersion interaction varies with vibrational coordinate Q and so couples vibrations to translations, and in the same way dipole and multipole moments varying with Q couple vibrations with rotations and translations. A very important contribution comes from the repulsive part of the potential, which is often approximated as a set of atom–atom interactions between the atoms of the interacting molecules. Whatever form is taken for V, it can be expanded in the normal modes $\{Q^a\}$ of the excited molecule:

$$V = V_0 + \sum_{\alpha} F_1^{\alpha} Q^{\alpha} + \frac{1}{2} \sum_{\alpha} \sum_{\beta} F_2^{\alpha\beta} Q^{\alpha} Q^{\beta} + \dots \tag{2.14}$$

where

$$F_1^{\alpha} \equiv \left. \frac{\partial V}{\partial Q^{\alpha}} \right|_{Q=0}$$

$$F_2^{\alpha\beta} \equiv \left. \frac{\partial^2 V}{\partial Q^{\alpha} \partial Q^{\beta}} \right|_{Q=0}$$

Only the first few terms in the series need be retained in general, since vibrational amplitudes are small. Each term in V then has a product form in which the first part depends only on bath degrees of freedom and the second only on vibrational degrees. Vibrational matrix elements of V are then straightforward to calculate from the isolated molecule potential surface, since they depend only on $(Q^\alpha)_{ij}$, $(Q^\alpha Q^\beta)_{ij}$ and so forth. Anharmonicity can be included in the calculation of these matrix elements. The vibrational relaxation rate then becomes

$$\tau_{ij}^{-1}=\frac{2\hbar^{-2}}{1+\exp(-\beta\hbar\omega_{ij})}\int_{-\infty}^{\infty}dt\exp(i\omega_{ij}t)$$
$$\times\left\langle\left[\sum_{\alpha}(Q^\alpha)_{ij}F_1^\alpha(t)+\frac{1}{2}\sum_{\alpha,\beta}(Q^\alpha Q^\beta)_{ij}F_2^{\alpha\beta}(t)\right]\right.$$
$$\left.\times\left[\sum_{\alpha}(Q^\alpha)_{ij}F_1^\alpha(0)+\frac{1}{2}\sum_{\alpha,\beta}(Q^\alpha Q^\beta)_{ij}F_2^{\alpha\beta}(0)\right]\right\rangle \tag{2.15}$$

If some classical model is then introduced for the bath functions $F_1^\alpha(t)$ and $F_2^{\alpha\beta}(t)$ the relaxation rate may be calculated. Alternatively, these functions may be simulated through molecular dynamics, as discussed in Section II.H.

C. Is the Binary Collision Model Valid?

Since the introduction of the isolated binary collision model there has been considerable controversy over its applicability. A number of theoretical papers have challenged its basic assumptions and proposed corrections due to collective effects (usually within the weak coupling approximation of Section II.B), while others have supported and extended the model. In this section, we outline the development of the controversy over the binary collision approximation, which is not resolved even today.

The first published criticism of the binary collision model was due to Fixman;[22] he retained the approximation that the relaxation rate is the product of a collision rate and a transition probability, but argued that the transition probability should be density dependent due to the interactions of the colliding pair with surrounding molecules. He took the force on the relaxing molecule to be the sum of the force from the neighbor with which it is undergoing a hard binary collision, and a random force $mA(t)$. This latter force was taken to be the random force of Brownian motion theory, with a delta-function time correlation:

$$\langle A(t_1)A(t_2)\rangle=\beta\left(\frac{2kT}{m}\right)\delta(t_1-t_2) \tag{2.16}$$

where β is a friction constant and m the reduced mass of the colliding pair. He then showed that high-frequency components of $A(t)$ could affect the transition probability in the liquid phase, in disagreement with the binary collision model. However, as Fixman himself admitted, the Brownian motion approximation for $A(t)$ leads to a spectrum $S_A(\omega)$, which is much too large at high frequencies; as a result, the effect of the three-body interactions may be exaggerated in Fixman's theory. It would be interesting to extend his work to a more accurate stochastic model for $A(t)$, to see whether deviations from the IBC model persist.

Zwanzig[23] questioned the fundamental assumption that the relaxation rate can be written as a product of a collision frequency and a transition probability. Using the weak coupling theory outlined in Section II.B, he showed that there are both two-body and three-body contributions to the relaxation rate, and argued that if the time between collisions was comparable to or shorter than the vibrational period (as it can be in the liquid phase) then the three-body collision terms neglected in the IBC model would be comparable to the two-body terms. He then carried out an approximate calculation in which the potential $V_{ij}(t)$ coupling states i and j was taken to be proportional to the force on the center of mass; he related this force autocorrealtion function to the momentum autocorrelation function, which he assumed to fall off exponentially with time. The resulting transition rate is proportional to ω^{-1} for large frequencies, an unphysically slow decay, which results from the assumption that the momentum autocorrelation function is exponential even at short times.

Herzfeld[24] responded to the criticisms of both Fixman and Zwanzig. He pointed out that Fixman's model exaggerated the high-frequency spectral density of the random force and argued that in a more realistic picture its spectral density would be small and have little effect. In response to Zwanzig's comments, he pointed out that the relaxation rate is strongly dependent on the relative velocities of the interacting molecules, so that only the relatively rare hard collisions are effective in vibrational relaxation. The times between such hard collisions are long, so that interference effects will be small and the IBC model valid. In a subsequent Comment,[25] Zwanzig agreed with Herzfeld about the internal consistency of the IBC model. However, recently Lukasik[26] has observed from quantum-mechanical calculations that in fact the collisions effective in vibrational relaxation of hydrogen at low temperatures are *not* much higher in energy than the average collision, calling into question the whole IBC interpretation.

Davis and Oppenheim[27] examined the validity of the IBC model from the point of view of quantum statistical mechanics using a generalized Wigner distribution function. They thus avoided the weak coupling approximation used by both Fixman and Zwanzig. They pointed out an inconsistency in Herzfeld's[24] argument, since if a molecule undergoes a hard collision, it will come out with high momentum and therefore undergo another hard collision very soon; thus hard collisions will occur in groups rather than independently. Nevertheless, the basic result obtained by Davis and Oppenheim supports the binary collision model under most experimental conditions and provides a microscopic expression for the collision frequency in terms of the pair distribution function at a critical distance R^*. However, their approach is based implicitly on assumptions equivalent to Herzfeld's on the importance of rare hard collisions for relaxation; as mentioned above, Lukasik[26] has challenged this assumption.

Shin and Keizer[28] considered a simple model for a triple collision, in which a diatomic molecule undergoes collinear collisions with two atoms, one from each side. The initial velocities of the two atoms were assumed to be uncorrelated, and the time between the two collisions was taken to be random (a Poisson distribution). The interference effect was explicitly calculated and shown to change the relaxation rate for a low frequency mode of CS_2 in He by as much as 30%; the effect was smaller for high frequency modes or heavier collision partners. The Keizer–Shin model probably underestimates the true interference effect for two reasons: first, more than two collisions can interfere with each other, and second, hard collisions will tend to be grouped together.

It seems clear that at very low frequencies interference effects will be important and that the IBC model will fail; similarly, it seems likely (though by no means proven) that at very high frequencies binary interactions will dominate (Oxtoby[29] has argued this viewpoint recently). Much of the controversy then reduces to the question of where the borderline frequency range lies beyond which the IBC model will be valid. This basic question has not yet been resolved.

D. Cell Models

A simple cell model for vibrational relaxation was proposed by Herzfeld[12] in 1952. In this model the potential on a molecule due to its neighbor is replaced by an averaged spherical cell potential. This potential will be a minimum at the center of the cell and can be expanded in (even) powers of R, the distance from the center of the cell. In considering vibrational relaxation, one must include not only the distance of the center of mass from the cell center, R_c, but also the vibrational amplitude x, where x is generally small compared to R_c. Then

$$R = R_c + bx \tag{2.17}$$

where b is a direction cosine. The term in the potential that is quadratic in R will lead only to a frequency shift in vibrational and lattice modes, while the quartic term will give a contribution to V proportional to xR_c^3, which corresponds to the decay of a vibrational quantum into three lattice modes. Herzfeld used the model to calculate the relaxation rate of benzene and found apparently good agreement with experiment; however, a numerical error was subsequently corrected,[5] and the resulting rate was 100 times too small.

Diestler[30] introduced a more detailed cell model for diatomic molecules in rare gas liquids. Effects of rotation were neglected and only collinear collisions were considered. The cell potentials acting on the solvent atoms were approximated by square wells and the resulting eigenstates calculated directly. Coupling to vibrational degrees of freedom occurs because the length of the square well depends on the vibrational state of the impurity. Perturbation theory was then used to calculate relaxation rates. This approach appears not to treat mass effects properly, however, since the relaxation rate of liquid hydrogen was predicted to be slower than that for liquid nitrogen, whereas subsequent experiments[31, 32] showed it to be faster by six orders of magnitude.

E. Hydrodynamic Models

Several investigators have studied collective effects on vibrational relaxation through the use of hydrodynamic models. Whereas in the binary collision approximation the vibrationally excited molecule is considered to interact with only a single neighbor, in the hydrodynamic approximation one considers essentially the opposite limit in which the neighboring molecules are "smoothed out" and replaced by a viscoelastic continuum. While such a "macroscopic" approximation might seem rather drastic on a molecular scale, it has in fact been successfully applied to calculations of translational[33] and rotational[34] diffusion constants (giving results accurate to 10–20%) and to the velocity autocorrelation function of liquid argon.[33] Recently Oxtoby[35] developed a hydrodynamic model for vibrational dephasing and showed that it gave good agreement with experimental results for vibrational motions of atoms heavier than hydrogen. In this section we discuss the application of hydrodynamic models to vibrational population relaxation.

The first such study was one by Keizer,[36] who considered a heteronuclear diatomic in a viscous continuum. He showed that if the two atomic masses differ there is a dynamic coupling between the center of mass motion and the vibrational coordinate. Applying a hydrodynamic model to the relaxation of the center of mass momentum, he gave an expression for the rate of energy transfer from the vibrational degrees of freedom. Keizer used the resulting equation to calculate relaxation rates of heteronuclear diatomics. However, he used friction coefficients independent of the frequency (i.e., ignoring viscoelastic effects), which is not valid at frequencies characteristic of vibrational degrees of freedom. In addition, his proposed mechanism does not contribute to relaxation of symmetric linear molecules.

Metiu, Oxtoby, and Freed[37] applied a viscoelastic hydrodynamic model to the relaxation of a harmonic oscillator linearly coupled to a heat bath (a model for which the population relaxation and dephasing times differ by only a factor of 2; see Section II.B). They considered the two atoms of the diatomic to be half-spheres in contact with a continuum and used frequency-dependent (viscoelastic) shear and bulk viscosities with slip boundary conditions. The result they derived for the population relaxation rate was

$$T_1^{-1} = \frac{\mathrm{Re}\tilde{\zeta}(\omega_v)}{m} \tag{2.18}$$

where ω_v is the vibrational frequency of the harmonic oscillator, m is the atomic mass, and $\tilde{\zeta}(\omega)$ is expressed in Ref. 37 in terms of frequency-dependent shear and bulk viscosities. They estimated these using a single relaxation time (Maxwell) approximation which, while useful at low frequencies (up to a few hundred wavenumbers), cannot be used at the vibrational frequencies of interest for most diatomics. A full evaluation of the validity of the theory of Metiu et al. must therefore await a more accurate calculation of the frequency-dependent shear and bulk viscosities.

Velsko and Oxtoby[38] have developed a general model for vibrational relaxation in polyatomic liquids and investigated the usefulness of a hydrodynamic approximation for the forces present. The potential acting on the molecule is taken to be atom additive:

$$V = \sum_{\alpha=1}^{n} V^{\alpha} \tag{2.19}$$

where V^{α} is the potential acting on the αth atom (out of n) due to the surrounding solvent. This potential is then expanded in vibrational normal coordinates $\{Q_a\}$

$$V = \sum_{\alpha} \sum_{a} \frac{\partial V^{\alpha}}{\partial Q_a} Q_a + \sum_{\alpha} \sum_{a} \sum_{b} \frac{\partial^2 V^{\alpha}}{\partial Q_a \partial Q_b} Q_a Q_b \tag{2.20}$$

The ij matrix element of this (coupling levels i and j) is

$$V_{ij} = \sum_{\alpha} \sum_{a} \frac{\partial V^{\alpha}}{\partial Q_a} (Q_a)_{ij} + \sum_{\alpha} \sum_{a} \sum_{b} \frac{\partial^2 V^{\alpha}}{Q_a Q_b} (Q_a Q_b)_{ij} \tag{2.21}$$

The matrix elements $(Q_a)_{ij}$ and $(Q_aQ_b)_{ij}$ were calculated using the full anharmonic gas-phase potential energy surface (while there will be changes in the potential surface in the liquid, they have been shown to be small in inert solvents)[39]. The derivatives of V are then expressed in terms of atomic coordinates:

$$\frac{\partial V^\alpha}{\partial Q_a} = \sum_p \frac{\partial V^\alpha}{\partial r_p}\frac{\partial r_p}{\partial Q_a}$$

$$\frac{\partial^2 V^\alpha}{\partial Q_a \partial Q_b} = \sum_p \frac{\partial V^\alpha}{\partial r_p}\frac{\partial^2 r_p}{\partial Q_a \partial Q_b} + \sum_{p,q}\frac{\partial^2 V}{\partial r_p \partial r_q}\left(\frac{\partial r_p}{\partial Q_a}\right)\left(\frac{\partial r_q}{\partial Q_b}\right) \tag{2.22}$$

where r_p and r_q are the $3n$-coordinates of the n atoms. Their derivatives with respect to normal coordinates can be obtained form the gas phase potential energy surface. $-\partial V^\alpha/\partial r_p$ is just the pth component of the force acting on atom α if r_p is one of the coordinates of atom α; otherwise it vanishes. We also assume that the second derivative of V with respect to r_p is simply L^{-1} times the first derivative, where L is a characteristic potential range. While this might seem a drastic approximation, it turns out that these terms contribute only about 10% to the calculated relaxation rates, so that this crude treatment does not change the final result by very much.

Inserting these expressions into (2.13) gives the relaxation rate from level i to level j in terms of properties of the gas phase potential function and the correlation functions of the forces on the atoms. These correlation functions could be obtained in a number of ways, including a binary collision model or a molecular dynamics simulation. Another possibility is to investigate the predictions of a hydrodynamic model in which the force correlation function is obtained from the frequency-dependent shear and longitudinal viscosities, which are assumed to have Maxwell relaxation behavior. This should be satisfactory for vibrational transition involving energy gaps of up to 100 wavenumbers. Velsko and Oxtoby[38] applied this model to the relaxation of OCS and CO_2 in liquid nitrogen and obtained reasonable agreement of the overall reaction rates with experiment.[40, 41] In addition, they could study the detailed pathways of vibrational energy transfer in these molecules.

F. Superposition Approximations

The weak coupling theory prediction for the vibrational relaxation rate between two levels i and j was shown in Section II.B to have the form

$$\tau_{ij}^{-1} = \hbar^2 \int_{-\infty}^{\infty} dt \exp(-i\omega_{ij}t)\langle V_{ij}(0)V_{ji}(t)\rangle \tag{2.23}$$

If V arises from pairwise interaction between molecules, it can be written

$$V_{ij}(t)=\sum_{\beta}V_{ij}^{\alpha\beta}(t) \tag{2.24}$$

where α labels the molecule undergoing the vibrational transition and β ranges over its neighbors. Then τ_{ij}^{-1} has the form

$$\tau_{ij}^{-1}=\hbar^{2}\int_{-\infty}^{\infty}dt\exp(-i\omega_{ij}t)\sum_{\beta}\sum_{\gamma}\langle V_{ij}^{\alpha\beta}(0)V_{ji}^{\alpha\gamma}(t)\rangle \tag{2.25}$$

It is clear that τ_{ij}^{-1} depends on the correlated dynamics of three particles. Hills[42] has suggested an interesting approach for calculating τ_{ij}^{-1} based on a time-dependent superposition approximation previously applied by Hills and Madden[43] to vibrational dephasing.

Hills assumes that the molecules can be treated as spheres, so that the matrix elements of V depend only on the radial distance between pairs of molecules:

$$V_{ij}^{\alpha\beta}(t)=C_{ij}R\left[r_{\alpha\beta}(t)\right] \tag{2.26}$$

where R is some function of the radial separation $r_{\alpha\beta}(t)$ and C_{ij} is a constant that depends on the choice of intermolecular interaction and on the vibrational coordinates of molecule α. The relaxation rate is then

$$\tau_{ij}^{-1}=|C_{ij}|^{2}\hbar^{-2}\int_{-\infty}^{\infty}dt\exp(-i\omega_{ij}t)\sum_{\beta}\sum_{\gamma}\langle R(r_{\alpha\beta}(0))R(r_{\alpha\gamma}(t))\rangle \tag{2.27}$$

Hills shows that the integral can be rewritten in terms of density fluctuations

$$\delta\rho(\mathbf{r},t)=\rho(\mathbf{r},t)-\rho_{e}$$

$$\tau_{ij}^{-1}=|C_{ij}|^{2}\hbar^{-2}\int_{-\infty}^{\infty}dt\exp(-i\omega_{ij}t)\int d\mathbf{r}\int d\mathbf{r}'R(r)R(r')$$
$$\langle\delta\rho(\mathbf{r}-\mathbf{r}_{\alpha}(0),0)\delta\rho(\mathbf{r}'-\mathbf{r}_{\alpha}(t),t)\rangle \tag{2.28}$$

Hills now assumes that the probe molecule α does not move during the characteristic time scale for the motion of the surrounding molecules. In this case the expression for τ_{ij}^{-1} can be rewritten

$$\tau_{ij}^{-1}=|C_{ij}|^{2}\hbar^{-2}\int_{-\infty}^{\infty}dt\exp(-i\omega_{ij}t)\int d\mathbf{r}\int d\mathbf{r}'R(r)R(r')$$
$$\langle\delta\rho(\mathbf{r},0)\delta\rho(\mathbf{r}',t)\rangle^{*} \tag{2.29}$$

where the coordinate system is now centered on molecule α. The star in (2.29) indicates that the ensemble average is carried out with the constraint that there is a molecule at the origin. In other words, $\langle\delta\rho(\mathbf{r},0)\delta\rho(\mathbf{r}',t)\rangle^*$ is *not* equal to the van Hove correlation function as assumed by Hills. To derive his result one must introduce several further approximations. Since $\langle\delta\rho(\mathbf{r},0)\delta\rho(\mathbf{r}',t)\rangle$ is the probability that there will be a density fluctuation at $\mathbf{r}$ at time zero and one at $\mathbf{r}'$ at time t given that there is a particle at the origin, it is in fact a time-dependent three-particle correlation function:

$$\langle\delta\rho(\mathbf{r},0)\delta\rho(\mathbf{r}',t)\rangle^* = \rho_e^{-1}\langle\delta\rho(\mathbf{r},0)\delta\rho(\mathbf{r}',t)\delta\rho(\mathbf{0},0\ or\ t)\rangle \tag{2.30}$$

If we now make a time-dependent superposition approximation for this three-particle correlation function we obtain

$$\langle\delta\rho(\mathbf{r},0)\delta\rho(\mathbf{r}',t)\rangle^* = \rho_e^{-2}\langle\delta\rho(\mathbf{r},0)\delta\rho(\mathbf{0},0)\rangle \otimes \rho_e^{-2}\langle\delta\rho(\mathbf{r}',t)\delta\rho(\mathbf{0},t)\rangle\langle\delta\rho(\mathbf{r},0)\delta\rho(\mathbf{r}',t)\rangle \tag{2.31}$$

If we further assume that the first two correlation functions involving the probe molecule α are simply step functions (zero for $r<a$, one for $r>a$) we obtain Hills' result:

$$\tau_{ij}^{-1} = |C_{ij}|^2\hbar^{-2}\int_{-\infty}^{\infty} dt \exp(-i\omega_{ij}t)\int_{V-a^3} d\mathbf{r}\int_{V-a^3} d\mathbf{r}' R(r)R(r') \langle\delta\rho(\mathbf{r},0)\delta\rho(\mathbf{r}',t)\rangle \tag{2.32}$$

However, it should be emphasized that rather drastic superposition approximations have to be made to obtain this result; these assumptions are not explicit in the original paper.[42] From this point Hills shows that τ_{ij}^{-1} can be expressed in terms of the van Hove correlation function $S(k,\omega_{ij})$:

$$\tau_{ij}^{-1} = |C_{ij}|^2\hbar^{-2}8\rho_e\int_0^{\infty} dk\, k^2 S(k,\omega_{ij})V^2(k) \tag{2.33}$$

where

$$V(k) = \int_a^{\infty} dr\, r^2 j_0(kr)R(r)$$

He then chooses an exponential form for $R(r)$ and shows that at high frequencies

$$\begin{aligned}\tau_{ij}^{-1} &\propto |C_{ij}^2|\hbar^{-2}\rho_e T^{1/2}\omega_{ij}^{-2}\\ &\propto \rho_e T^{1/2}\omega_{ij}^{-3}\end{aligned} \tag{2.34}$$

However, this cannot be rigorously correct at high frequencies. For a continuous potential,

$$\sum_{\beta}\sum_{\gamma}\left\langle \frac{d}{dt}R(r_{\alpha\beta}(t))\frac{d}{dt}R(r_{\alpha\gamma}(t))\right\rangle$$
$$=-\frac{1}{2\pi}\int_{-\infty}^{\infty}d\omega\,\omega^2\left[\int_{-\infty}^{\infty}dt\exp(-i\omega t)\sum_{\beta}\sum_{\gamma}\langle R(r_{\alpha\beta}(0))R(r_{\alpha\gamma}(t))\rangle\right] \tag{2.35}$$

is finite. Therefore the expression in brackets must fall off faster than ω^{-2} at high frequencies (in fact it must fall off faster than any power of ω in order for all of its moments to be finite). Hills conclusion, that this expression falls off as ω^{-2}, must therefore result from the approximations he has introduced.

While we have pointed out some difficulties with this approach, it is clearly a promising method that deserves further study. It would be very desirable if one could reduce many-body dynamics problems to two-body problems as Hills has done.

G. Green's Function Methods

Rice and co-workers[44, 45] have proposed a different approach to vibrational relaxation in liquids based on analogies with the theory of radiationless transitions. Kushick and Rice[44] considered the case in which an initially excited vibrational level $|1\rangle$ is coupled to a continuum $|2,\epsilon\rangle$ where $|2\rangle$ is a second vibrational state and $|\epsilon\rangle$ is a continuous variable representing solvent states whose translational energy has been increased by ϵ. The coupling is chosen, for mathematical convenience, to be the square root of a Lorentzian function of ϵ:

$$\langle 1|V|2,\epsilon\rangle = v_{12}(\epsilon)=\left\{\frac{b^2v^2}{\left[b^2+(\epsilon-\epsilon_0)^2\right]}\right\}^{1/2} \tag{2.36}$$

For this choice, an exact solution of the population evolution was possible and the result was expressed in terms of v, b, and the density of translational states ρ_0 which was assumed to be constant. In the limit $b\rightarrow\infty$, $v_{12}(\epsilon)=v$ (constant coupling) and the population of state $|1\rangle$ decreases exponentially, but for smaller values of b nonexponential decay is predicted.

Muthukumar and Rice[45] studied a model in which an initially populated state $|s\rangle$ is coupled to a manifold of states $|l\rangle$, which in turn are coupled to a continuum of states $|m\rangle$. The distribution of couplings ξ_l (ξ_l can represent V_{sl}, V_{lm}, or $V_{sl}\,V_{lm}$) was taken to be

$$P(\xi_l)=\exp\left[-\frac{1}{2y_0(0)}\xi_l^2-\frac{\eta}{2y_0(0)^2}f(\xi_l)\xi_l^3\right] \tag{2.37}$$

The first term is the random component, independent of l, while the second term represents a nonrandom coupling through the dependence of $f(\xi_l)$ on the energy of state l. $f(\xi_l)$ was chosen to be Lorentzian

$$f(\xi_l^j)=\frac{a_j^4}{b_j^2+(E_l-E_{0j})^2} \qquad j=1,2,3 \tag{2.38}$$

with different values of $a_j b_j$ and E_{0j} for the different choices of ξ_l^j ($\xi_l^1=V_{sl}$, $\xi_l^2=V_{lm}$, $\xi_l^3=V_{sl}V_{lm}$). This model was solved in the limit of small η, which corresponds to a small nonrandom coupling perturbing a larger random coupling and under the assumption of constant density of states ρ_l. The population decay rate found was

$$P_{ss}(t)=\exp[-\Gamma_s t][1+\eta^2(\delta+t\epsilon)+O(\eta^4)] \tag{2.39}$$

where Γ_s, δ, and ϵ are complicated functions of the parameters of the model. It can be seen that nonrandom, Lorentzian coupling leads to nonexponential population decay.

These Green's function approaches provide considerable insight into certain model problems, but suffer from several deficiencies. First, the initially excited state is actually a continuum because of the translational states which the system can occupy, so that relaxation really takes place between continua rather than between a discrete level and a continuum. Second, in general, not only will the coupling V_{sl} be a function of ϵ but in addition the density of states in the l manifold will vary with energy; the latter effect has not been considered. Finally, and most important, the Green's function models are soluble only for certain simple choices of $v(\epsilon)$ and the parameters that appear are difficult to correlate with properties of the liquid. It is more difficult to use knowledge about liquid state dynamics in the Green's function approaches than in time correlation function methods.

H. Molecular Dynamics Simulations

The method of computer simulation via molecular dynamics has been applied to classical atomic fluids for some twenty years[46]; a wealth of information about collective and single particle motion in liquids has been obtained. More recently, simulations of diatomic fluids such as liquid nitrogen[47] have allowed the study of rotational relaxation in the classical approximation. Vibrational phase and energy relaxation have not been accessible to computer simulation until quite recently. There are two reasons for this: first, the fundamentally quantum nature of vibrations even in liquids makes a purely classical simulation more questionable than in the case of rotations; and second, the vibrational period is often so much shorter than the vibrational relaxation time that a simulation of vibrating molecules would require an extremely short integration time step and a long total simulation time, leading to lengthy calculations.

There have nevertheless been two recent direct simulations of vibrational relaxation in liquids. Riehl and Diestler[48] studied a one-dimension model of a pure diatomic liquid arranged head to tail. They were forced to take an unphysically small vibrational frequency (near 10 cm^{-1}) to carry out the calculation. For this system they found vibrational lifetimes on the order of a few picoseconds, several times longer than the dephasing times which they also calculated. Nordholm and co-workers[49] simulated the vibrational relaxation of bromine in argon both in the dense gas and near liquid state densities. They used a Morse potential for the bromine, and Lennard–Jones potentials for argon–argon and argon–bromine atom interactions. The bromine molecules were given initial vibrational energies close to dissociation, so that their vibrational periods were rather long. Detailed information was extracted about such properties as the distribution of changes in (classical) vibrational energy over the collisions and the distribution of collision durations. In the liquid phase the relaxation rate increased faster than the density, and this was interpreted as a deviation from the IBC model; in fact, the collision frequency (obtained, for example, from Enskog theory) increases still more rapidly with density in the liquid phase so that their results show if anything a slightly *slower* relaxation rate than would be predicted from an IBC model.

Since the direct simulation of vibrational relaxation in condensed phases is clearly a difficult and lengthy procedure for molecules with realistic vibrational frequencies, Shugard et al.[50] proposed an alternative approach based on work of Adelman and Doll[51] and applied it to relaxation of diatomic impurities in solid matrices. The motion of atoms near the impurity was simulated directly and the effect of more distant atoms was taken into account through a stochastic force, which was constructed from the phonon spectrum of the solid. This method still requires that the relaxation time not be too long compared to the vibrational period (i.e., that the vibrational frequency not be too high) but the calculation is much faster than a full molecular dynamics simulation since only a few degrees of freedom are simulated directly. It should be possible and interesting to use methods of this type in the liquid phase, although in this case the stochastic properties of the random force will have to be guessed using physical intuition rather than directly calculated from the properties of the harmonic solid.

Oxtoby, Levesque, and Weis[52] proposed an alternative approach to the simulation of vibrational relaxation in liquids. It is particularly useful for studying vibrational dephasing and Raman and infrared lineshapes, since in these cases one only needs to make an adiabatic approximation (vibrational motion fast compared with solvent motion), and one can then calculate vibrational lineshapes in either weak or strong coupling situations, for molecules with high vibrational frequencies that could not be simulated directly. Oxtoby et al. applied the method to a study of dephasing in liquid nitrogen, using both a Lennard–Jones atom–atom potential[52] and a more accurate potential[53] due to Raich.[54] Applications to vibrational population relaxation can be carried out under somewhat more restrictive conditions than dephasing. If weak coupling theory is valid, then the results of Section II.B express the relaxation rate in terms of correlation functions of bath degrees of freedom (rotations and translations) which are straightforward to simulate as Oxtoby et al.[52] have shown. This method is, however, restricted to moderate energy differences $\hbar\omega_{ij}$, since calculation of Fourier transforms of correlation functions simulated by molecular dynamics becomes less accurate as the frequency increases.

I. Relation to Solid- and Gas-Phase Theories

In this section we consider briefly several theoretical approaches to vibrational relaxation in the solid phase[4] (especially for molecular impurities in rare-gas matrices) and in the low-temperature gas phase. We discuss the relationship of this theoretical work to liquid-state relaxation.

An early paper by Sun and Rice[55] considered the relaxation rate of a diatomic molecule in a one-dimensional monatomic chain. An analysis similar to the Slater[56] theory of unimolecular reaction was used to obtain the frequency of hard repulsive core–core collisions, and then (in the spirit of the IBC model) this was multiplied by the transition probability per collision from perturbation theory and averaged over the velocity distribution to obtain the population relaxation rate. This was apparently the first prediction that condensed-phase relaxation could occur on a time scale as long as seconds.

A number of theoretical studies have investigated multiphonon relaxation in solids. Nitzan and Jortner[57] considered a harmonic oscillator coupled to a harmonic lattice; the coupling potential was taken to be linear in the vibrational coordinate and of high order in phonon displacements.

Nitzan et al.[58] explored the temperature dependence of the multiphonon relaxation and predicted that the rates would follow an energy-gap law and fall off exponentially with vibrational frequency. Weissman et al.[59] subsequently showed that quadratic coupling contributed significantly to the relaxation rate and gave a weaker frequency dependence than the exponential energy-gap law. Diestler[60] considered the change in the lattice phonon potentials due to the vibrational state of the impurity molecule and used this coupling in perturbation theory to calculate the relaxation rate; Lin[61] proposed a generalized version of the same approach based on the breakdown of the adiabatic approximation resulting from weak coupling between vibrational and lattice modes.

Experiments by Legay[2] on hydrides and deuterides in solid matrices showed isotope effects opposite to the predictions of the energy-gap model: even though the hydrides had higher frequencies than the deuterides, their relaxation rate was faster. Legay proposed that the rotations of the impurity are in fact the principal accepting mode for vibrational energy. Recent theoretical work on relaxation in solids has focused on rotations and other local modes participating in vibrational relaxation. Freed and Metiu[62] showed that with a reasonable choice of potential coupling vibrations and rotations through the lattice results consistent with experimental data were obtained. Berkowitz and Gerber[63] considered both rotations and local translational accepting modes and carried out a quantitative calculation, which showed the dominance of rotations as the accepting mode for NH and ND in solid argon. Diestler et al.[64] considered planar rotational motion coupled to vibrational degrees of freedom and obtained closed form expressions in reasonable agreement with experiment. Knittel and Lin[65] investigated the effect of rotational barriers on vibration–rotation coupling.

One final solid phase calculation is a study by Shin[66] of diatomic relaxation in condensed phases, which includes ideas from binary collision, cell model, and multiphonon relaxation theories. He calculated the energy transferred both to oscillatory motion of the surrounding lattice atoms and to the oscillations of the molecule itself.

The question arises of to what extent these solid-phase calculations are applicable to the liquid state. Can the liquid be replaced by a nearly harmonic lattice? Is rotation important as an accepting mode in the liquid only for hydride rotations, or will it contribute for heavier atoms as well? There are a number of questions to be answered through both theoretical and experimental studies on the relationship between relaxation in the solid and liquid phases.

We close this section by mentioning briefly some results on vibrational relaxation in gases and their relationship to liquid state relaxation. A number of studies have shown gas-phase relaxation rates, which at low temperature begin to *increase* with decreasing temperature. Audibert et al.[67] suggest that this is a result of the importance of the attractive part of the potential for the relaxation. Ewing[68] suggested that relaxation occurs through van der Waals molecule formation at low temperatures which then undergo vibrational predissociation leaving the previously excited molecule in a lower vibrational state; he calculated predissociation lifetimes and found that this mechanism could contribute to the overall vibrational relaxation. The reason these results are important for liquid phase processes is that they show that attractive forces may play a significant role and that the hard collision IBC model may not apply. Further low-temperature gas-phase studies may help to elucidate this problem.

III. EXPERIMENTAL STUDIES OF VIBRATIONAL RELAXATION

A. Ultrasonic Absorption Studies

Much of the early experimental data on vibrational relaxation rates in liquids came from ultrasonic attenuation studies.[5] In the absence of relaxation, the ratio α/f^2, where α is the sound absorption coefficient and f the sound-wave frequency, remains constant as f is changed. What is instead observed for a number of liquids composed of moderate-sized molecules is that α/f^2 drops significantly as f is increased through the experimentally accessible range (from about 0.5 to 300 Mc). The reason for this frequency dependence is that the energy in internal degrees of freedom relaxes toward equilibrium at a finite rate; if the perturbation from the sound wave is of higher frequency, those degrees of freedom will be unable to respond and the sound absorption will be reduced. By studying the frequency dependence of α/f^2 one can obtain the relaxation time τ for the internal energy.

This method has certain important limitations. First of all, one obtains only the overall relaxation time τ and not the relaxation time of individual vibrational levels. In many cases there is a single low-frequency vibrational mode that dominates the relaxation; if not, however, one will obtain only an average relaxation rate for several vibrational states. A second limitation is the restricted frequency range available, which allows measurement of lifetimes only if they lie in the nanosecond time domain. Thus slowly relaxing liquids (such as O_2 or N_2) or rapidly relaxing ones (such as many larger molecules) cannot be studied.

The method can nonetheless be usefully applied to a variety of liquids of moderate sizes. The book by Herzfeld and Litovitz[5] lists relaxation times obtained for 10 liquids, and considerably more work has been done since that time. Liquid CS_2 was studied extensively by early workers and provided the strongest evidence for the validity of the isolated binary collision model.[6] The temperature and pressure dependence of vibrational relaxation has also been studied, and studies on liquids containing small concentrations of impurities[5] provided the first evidence for the large effect that impurities can have on vibrational relaxation rates, especially when those impurities have low-frequency modes that can speed the relaxation.

B. Laser Studies of Simple Liquids

In this section we summarize experimental work on relaxation of pure simple liquids (in particular, diatomic liquids) using laser techniques. The relaxation in these cases is too slow to be observed through ultrasonic attenuation.

Renner and Maier[69] excited the lowest vibrational states of O_2 and N_2 through stimulated Raman scattering and attempted to measure vibrational relaxation through an optical Schlieren method, in which vibration to translation energy transfer causes heating and a density change, which leads in turn to a change in the measured refractive index. However, they later showed[70] that the microsecond to millisecond relaxation times they observed were due not to vibrational relaxation but to heat conduction. Calaway and Ewing[71] measured a relaxation time of 1.5 s for liquid nitrogen using stimulated Raman excitation and spontaneous anti-Stokes Raman scattering as a probe. However this relaxation time was later shown by Brueck and Osgood[31] to be due to the presence of very small concentrations (parts per million) of impurities in the nitrogen. These authors studies nitrogen doped with small amounts of CO; they excited the N_2 vibration through optical excitation of the weak collision-induced N_2 absorption band, and detected the fluorescence at the CO vibrational frequency (185 cm^{-1} below that for N_2). By extrapolating to zero CO concentration they obtained an intrinsic relaxation time for N_2 of 56 s, which corresponds to the lifetime due to spontaneous quadrupolar emission. Thus, collisional relaxation of liquid N_2 takes place on a time scale longer than 100 s.

Legay-Sommaire and Legay[32] investigated vibrational relaxation in liquid CO using optical pumping and fluorescence detection. For less pure samples they observed impurity-mediated relaxation, while in purer samples they saw nonexponential decay due to self-absorption and vibrational diffusion. By fitting a diffusion constant D and a relaxation rate k to their data they found k to be near 150 s^{-1}, only weakly dependent on temperature.

Delelande and Gale[32] studied vibrational relaxation in pure normal hydrogen at a number of densities and temperatures between 25 and 40 K in the gas, liquid, and solid phases. This was an extension of the earlier higher temperature gas phase work of Audibert et al.[67] They used stimulated Raman excitation and optical Schlieren or collision-induced fluorescence detection. They observed that, at fixed temperature, the lowest density liquid relaxation rate fell exactly on the extrapolation of the linear density dependence from the gas phase. Since collision rates are not linear with density, this result indicates a small reduction of the relaxation rate from that predicted by an isolated binary collision model. However, the data in the liquid phase could be fit quite well to the two-parameter form

$$\tau=\frac{1}{A}\left(N^{-1/3}-\sigma_c\right) \tag{3.1}$$

predicted by the IBC theory with a cell model collision time (here $N^{-1/3}$ is the average intermolecular separation and σ_c the collision diameter). σ_c was found to be approximately independent of temperature and equal to 2.9 Å, near the Lennard–Jones parameter for hydrogen ($\sigma_{LJ}=2.93$ Å). However, the authors point out that quantum effects not considered in the IBC theory should be significant for liquid hydrogen. Chatelet et al.[73] studied vibrational relaxation of room-temperature hydrogen under pressure at densities of up to 700 amagats, using stimulated Raman excitation and Schlieren detection. A fit of the high-density data to the cell model prediction of (3.1) gave a σ_c of 3.25 Å, which they considered too large relative to σ_{LJ} to support the use of the binary collision model in this case.

C. Laser Studies of Simple Liquid Mixtures

In the past several years systematic studies of the rates of vibrational energy relaxation in simple liquid mixtures have been carried out primarily by two research groups: Ewing and co-workers[41, 74, 75] and Brueck and Osgood.[31, 40, 76] These authors have used their results to test the validity of isolated binary collision models.

Ewing's experiments on nitrogen doped with impurities combine stimulated Raman vibrational excitation with spontaneous anti-Stokes–Raman scattering for detection of the time evolution of the population of excited molecules. Calaway and Ewing[74] doped liquid N_2 with O_2, CO, or CH_4. Their kinetic scheme included direct V–T relaxation of N_2^* by N_2, V–V exchange between N_2^* and dopant M*, and V–T relaxation of M* by N_2 (M–M collisions are rare and can be neglected at the low dopant concentrations used). For the N_2–CH_4 system, the observed rate was dominated by irreversible V–V transfer from N_2^* to CH_4^* followed by rapid CH_4 relaxation. The CO vibrational level is closer to N_2 than those of CH_4, so in this case equilibrium was established rapidly and the subsequent relaxation occurred through radiative relaxation of CO. The N_2–O_2

system was more complex; to explain the observed dependence on O_2 concentration it was necessary to postulate a mechanism feeding back energy from O_2 to N_2 through doubly or triply excited O_2 molecules. The experimental results were shown to be not inconsistent with an IBC model, although comparison was hindered by a lack of low temperature gas phase relaxation data. Manzanares and Ewing[41] studied relaxation in N_2 doped with CO_2, CD_4, and N_2O. In the first two cases equilibrium was rapidly established between N_2^* and M^* and the observed rate was dominated by relaxation of M^* to lower levels due to collisions with N_2; in the case of N_2O, self-relaxation by other N_2O molecules was important. The CO_2 results were shown to be consistent in order of magnitude with an IBC model, but extrapolation of the gas-phase data from 300 to 77°C was necessary to make the comparison. In a subsequent paper Manzanares and Ewing[75] studied N_2 doped with H_2 or D_2. Theoretical calculations of gas-phase cross-sections suggested that V–T relaxation should be more important than V–R, T or V–V. The absolute results (and the success of the IBC model) could not be estimated because of the extreme sensitivity of the calculated cross sections to the range of the repulsive exponential potential chosen.

Brueck and Osgood[40] used optical pumping and (CO) fluorescence detection to study relaxation of liquid N_2 doped with CO and OCS. With only CO present they observed (just as did Calaway and Ewing[74]) rapid V–V equilibration between CO and N_2 and radiative relaxation of CO. With OCS present as well, the kinetic scheme was more complicated. Since no OCS fluorescence was observed, the collisional deactivation of OCS was assumed to be rapid. Fits of the observed CO relaxation (the rise time was too fast to be observed) as a function of CO and OCS concentration gave V–V transfer rates for N_2–CO, N_2–OCS, and CO–OCS. While the transfer rates were comparable to those observed in the gas phase, an analysis, using detailed balance, of the reverse rate of V–V transfer from CO to N_2 was shown to be linearly proportional to the liquid N_2 density, which is not consistent with a cell model binary collision description. Brueck et al.[76] studied the relaxation of the ν_3 mode (C–F stretch) of CH_3F in liquid O_2 and Ar. They used both infrared absorption saturation and infrared double resonance techniques to monitor the relaxation rate. Using a cell model for the collision rate they obtained a transition probability only a factor of ten smaller than that given by gas phase room temperature measurements, much less than one would predict on the basis of the usual Landau–Teller temperature dependence.

Gale and Delalande[77] carried out an extensive experimental study of liquid H_2 doped with HD and D_2 as well as pure D_2 liquid. They studied both the V–T relaxation rate of vibrationally excited HD* and D_2^* by H_2

and the V–V transfer rate from H_2^* to HD* and D_2^*. As in their earlier study on pure liquid H_2,[32] they found the V–T rates to follow the cell model prediction $\tau \propto (\rho^{-1/3} - \sigma)$ in the compressed liquid phase, with the same value for σ as found previously. The reduced mass of the collision pair was found to be the most important factor in the rate of relaxation (the larger the reduced mass the slower the rate), even more important than the energy gap. The V–V transitions were found to be faster and less dependent on temperature than the V–T. No large changes in V–T or V–V rates were seen across the liquid–solid transition, although deexcitation appeared to be slightly less efficient in the solid.

D. Picosecond Vibrational Relaxation Experiments

Since the field of picosecond spectroscopy has been reviewed extensively in recent years,[3] we will confine this section to a brief discussion of experimental studies of vibrational relaxation in liquids. Picosecond techniques allow the study of relaxation in larger molecules than can be studied by other methods.

The most extensive work in this field has been carried out by Laubereau, Kaiser, and co-workers. They studied the relaxation of the C–H stretching mode of CH_3CCl_3 and CH_3CH_2OH (ethanol) in the liquid phase,[78] using stimulated Raman scattering to excite the vibrations, and spontaneous anti-Stokes–Raman scattering to probe the subsequent relaxation. They found a relaxation time $\tau = 20 \pm 5$ ps for ethanol and $\tau = 5 \pm 1$ ps for trichloroethane. Alfano and Shapiro[79] studied liquid ethanol using similar techniques, and monitored anti-Stokes scattering at 1464 cm^{-1} as well as at the C–H stretching frequency of 2928 cm^{-1}. They measured the growth and decay of these daughter vibrations and suggested that the primary decay mechanism of the ν_1 C–H stretch was into a pair of ν_8 C–H bending modes at 1454 cm^{-1}. They pointed out that they could be observing both the ν_8 mode and the $\nu_1 - \nu_8$ difference band. The ν_1 relaxation time they obtained was $\tau = 28 \pm 1.4$ ps, Laubereau et al.[80] continued the study on ethanol by measuring the full anti-Stokes Raman spectrum near both 2928 cm^{-1} and 1464 cm^{-1} as a function of delay time. They observed very rapid equilibration (in 1 ps) among C–H stretching levels followed by slow decay. They interpreted their results as showing participation of the $2\nu_8$ mode in the rapid equilibration process with the C–H stretching modes; the time constants they observed were $\tau = 22 \pm 5$ ps for the relaxation rate of the ν_1 and $2\nu_8$ mixture to ν_8, and $\tau = 40$ for the subsequent relaxation of ν_8. In another study, Laubereau et al.[81] studied relaxation of the C–H stretch of CH_3CCl_3 in solution in both CCl_4 and CD_3OD. In the first case they interpreted the observed concentration dependence as showing that the relaxation was due to triple interactions between CH_3 CCl_3 molecules with the CCl_4 not taking part in relaxation; in the second case they suggested that excitation of the C–D stretch was a major component of the observed relaxation. They saw no evidence for intramolecular relaxation. One final study using the technique of stimulated Raman excitation and spontaneous Raman detection was the study by Monson et al.[82] of relaxation of the C–H stretching mode of a series of hydrocarbons. They showed that the vibrational relaxation rate was proportional to the fraction of carbon atoms which were in CH_3 groups, and suggested that relaxation occurs most efficiently through these groups, so that the overall rate will be proportional to the fraction of the time that the vibration spends on them.

Spanner, Laubereau, and Kaiser[83] introduced a different technique which employed tunable infrared pulsed excitation followed by anti-Stokes –Raman detection. Using this method they studied a dilute solution of ethanol in liquid CCl_4 and observed a double exponential decay; the first relaxation (on a 2-ps time scale) they interpreted as being due to rapid equilibration among the modes near 2900 cm^{-1} and the second relaxation ($\tau = 40$ ps) as due to relaxation to the ν_8 level. They also studied CH_3I in ethanol and observed rapid equilibration between ν_1 (symmetric C–H stretch) and ν_4 (asymmetric C–H stretch) in 1.5 ps; the relaxation from these levels was much faster than in ethanol: $\tau(\nu_1) = 3$ ps and $\tau(\nu_4) = 0.5$ ps. Laubereau et al.[84] applied the same technique to relaxation of $CHCl_3$, CH_2Cl_2, and CH_3CCl_3 in CCl_4 solution. Chloroform showed a rapid exponential relaxation ($\tau = 2.5$ ps), while in CH_2Cl_2 a rapid initial decay (interpreted as equilibration between symmetric and asymmetric stretches) was followed by a slow decay with $\tau = 40$ to lower lying states. The relaxation time of CH_3CCl_3 was found to be $\tau = 100 \pm 30$ ps at a mole fraction $x = 0.2$, showing again the extreme concentration dependence in this molecule observed earlier.[81]

Other experimental techniques have been used to study the very fast relaxation of dye molecules in solution. Ricard and Ducuing[85, 86] studied rhodamine molecules in various solvents and observed vibrational rates ranging from 1 to 4 ps for the first excited singlet state. Their experiment consisted of two pulses with a variable delay time between them; the first excites molecules into the excited state manifold and the second measures the time evolution of stimulated emission for different wavelengths. Ricard[86] found a correlation between fast internal conversion and vibrational relaxation rates. Laubereau et al.[87] found a relaxation time of 1.3 ± 0.3 ps for coumarin 6 in CCl_4. They used an infrared pulse to prepare a well-defined vibrational mode in the ground electronic state, and monitored the population evolution with a second pulse that excited the system to the lowest singlet excited state, followed by fluorescence detection.

Quantitative interpretation of vibrational relaxation rates in polyatomic liquids is difficult because of the many processes that can occur. Laubereau et al.[84] have made some qualitative and semiquantitative estimates of the dominant contributions to vibrational relaxation of C–H stretches for five molecules in the neat liquid and CCl_4 solution. They suggest that rotational coupling will be very important for CH_3I relaxation and moderately important for $CHCl_3$ and CH_2Cl_2, while resonant transfer of one quantum of a second harmonic to a near neighbor is very important for CH_3CCl_3 and moderately important for CH_2Cl_2 and CH_3CH_2OH. Fermi resonance plays a role in the relaxation of all five molecules, especially $CHCl_3$, CH_3CCl_3, and CH_3I, while Coriolis coupling is less important and contributes only to $CHCl_3$ and CH_3I. Their arguments are based on approximate gas phase calculations and on experimental evidence from CCl_4 dilution studies.

IV. NON-MARKOVIAN EFFECTS ON VIBRATIONAL RELAXATION

The theory outlined in Section II was based on the assumption that vibrational relaxation occurs on a time scale slow compared with the bath (translation and rotation) degrees of freedom. In this case, a Markov approximation (separation of time scales) can be made and the relaxation can be described through rate equations; the resulting population decay is given by an exponential or a sum of exponentials. This time scale separation assumption is certainly valid for the small molecules that require a nanosecond or longer to relax, but in the picosecond or subpicosecond domain which applies to larger molecules non-Markovian effects may be present. In this section we outline the results of some theoretical studies of non-Markovian (nonexponential) relaxation.

We have already mentioned in Section II.G that the Greens function approach predicts nonexponential decay when the coupling matrix element is allowed to vary with energy. Lin[88] used a very different approach to investigate the limitations of the master equation description. He showed, using a density matrix formalism, that memory effects appear if perturbation theory is extended to fourth order in the coupling between two vibrational levels.

Abbott and Oxtoby[89] have carried out exact quantum mechanical simulations of a two-level system strongly coupled to a heat bath in order to investigate non-Markovian effects. The hamiltonian has the form

$$\begin{aligned}\mathcal{H}=&\hbar\omega_0|1\rangle\langle 1|+D(t)|1\rangle\langle 1|\\&+V(t)\left[|1\rangle\langle 0|+|0\rangle\langle 1|\right]\end{aligned} \tag{4.1}$$

Here $|0\rangle$ and $|1\rangle$ are the eigenfunctions of the zero-order Hamiltonian and D and V give the diagonal and off-diagonal coupling to the bath; they are time dependent because the bath is approximated as a classical stochastic process. In most of the simulations D was set to zero to focus on the role played by V in population relaxation (D gives rise primarily to dephasing).

Two models for the stochastic process $V(t)$ were considered. The first is a Poisson process in which the bath jumps at random intervals (at an average rate $b/2$ jumps per second) between two states with $V = \pm V_0$. This model can roughly simulate a process in which a molecule undergoes hard collisions with its neighbors at random intervals. The second model is a Gaussian process which corresponds to a molecule interacting simultaneously with many neighbors (long-range interactions). For both stochastic processes the two-time correlation function is

$$\langle V(t)V(0)\rangle = V_0^2 \exp(-bt) \tag{4.2}$$

but the four-time and higher order correlation functions differ. The simulations were carried out by taking the two-level system to be initially in the excited state of the *full* Hamiltonian. The Hamiltonian is a stepwise constant function of the time (this fact is self-evident for the Poisson process, while for the Gaussian process the continuous evolution of $V(t)$ was approximated as a series of steps with the step length taken to be very small). The exact wave function $\psi(t)$ of the two level system was calculated at a series of times and its projection onto the zero-order states $|\langle 0|\psi(t)\rangle|^2$ and $|\langle 1|\psi(t)\rangle|^2$ was monitored. Several thousand runs were carried out and the results averaged to obtain good statistics.

In the weak coupling (or rapid modulation) limit the relaxation is predicted to be exponential with a rate constant obtained from (2.11) of

$$\tau^{-1} = \frac{2V_0^2 b/\hbar^2}{b^2 + \omega_0^2} \tag{4.3}$$

In this limit the Poisson and Gaussian simulations should give the same results. If V_0 is larger, or b smaller, then the simulations will differ both from each other and from the rate equation prediction.

The results of the simulation are shown in Figs. 1 and 2. In both cases the time scale is written in reduced units of ω_0^{-1}. In Fig. 1 the choices

$$\frac{V_0}{\hbar\omega_0} = 0.5$$

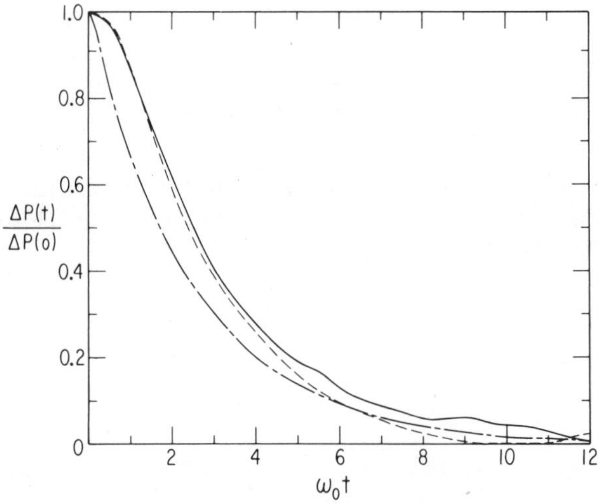

Fig. 1. Vibrational relaxation for $b/\omega_0 = 2.0$; $V_0/\hbar\omega_0 = 0.5$.—, Gaussian simulation;---, Poisson simulation;—-—, rate equations (Markov approximation).

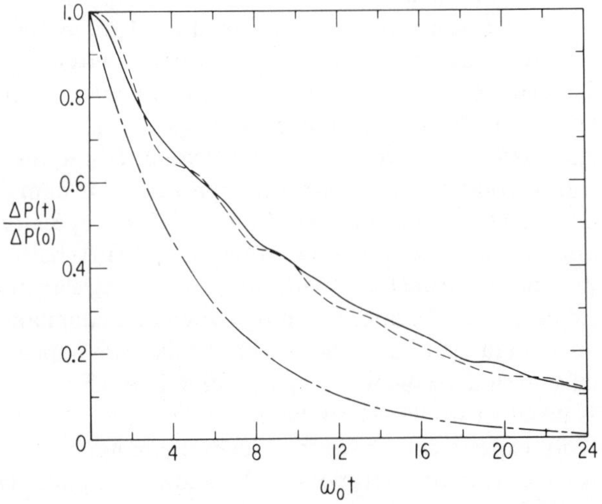

Fig. 2. Vibrational relaxation for $b/\omega_0 = 0.2$; $V/\hbar\omega_0 = 0.5$.—, Gaussian simulation;---, Poisson simulation;—-∞, rate equations (Markov approximation).

and

$$\frac{b}{\omega_0}=2.0$$

were made, which corresponds to relatively fast bath relaxation so that the rate equation description should be reasonably accurate. This is born out by Fig. 1. For Fig. 2, the parameters were

$$\frac{V_0}{\hbar\omega_0}=0.5$$

and

$$\frac{b}{\omega_0}=0.2$$

and non-Markovian effects are very much in evidence. The rate equation description no longer is accurate and the decay is nonexponential. The results from the Poisson and Gaussian simulations are surprisingly similar; this suggests that even in the strong coupling limit the relaxation is insensitive to correlation functions of $V(t)$ of higher order than second. There is a small but statistically significant difference between the two bath models: the Poisson process shows small oscillations (outside of the range of the error base estimated from statistical analysis of the data), while the Gaussian process does not. These oscillations persist even when diagonal fluctuations $D(t)$ are included as well as $V(t)$.

Abbott and Oxtoby[89] have investigated two different theoretical approaches for describing this behavior. An approach using cumulant expansion and truncation[20, 21] and one based on projection operators[16–18] have been generalized to allow for slow (non-Markovian) bath relaxation. Neither approach is satisfactory: both give too rapid relaxation and too large an amplitude of oscillation. It can be shown that cumulant truncation is rigorous for a Gaussian bath; the reason it fails here appears to be due to the initial correlations between system and bath. Further theoretical work on this problem would clearly be desirable.

To date, no experiments have provided unambiguous evidence for non-Markovian effects on vibrational relaxation in liquids. However, it seems likely that as time resolution and sensitivity are improved such effects will be observed.

Acknowledgments

I would like to thank Mr. S. Velsko and Dr. R. Abbott for helpful discussions. Acknowledgment is made to the donors of the Petroleum Research Fund of the American Chemical Society for partial support of this research (grant number ACS-PRF 10047-AC6).

References

1. D. W. Oxtoby, *Adv. Chem. Phys.*, **40**, 1 (1979).
2. F. Legay, in *Chemical and Biochemical Applications of Lasers*, Vol. 2, C. B. Moore, ed., Academic, New York, 1978.
3. A. Laubereau and W. Kaiser, *Rev. Mod. Phys.*, **50**, 607 (1978); *Ann. Rev. Phys. Chem.*, **26**, 83 (1975).
4. D. J. Diestler, *Adv. Chem. Phys.*, **42**, 305 (1980).
5. K. F. Herzfeld and T. A. Litovitz, *Absorption and Dispersion of Ultrasonic Waves*, Academic, New York, 1959.
6. W. M. Madigosky and T. A. Litovitz, *J. Chem. Phys.*, **34**, 489 (1961).
7. R. N. Schwartz, Z. I. Slawsky, and K. F. Herzfeld, *J. Chem. Phys.*, **20**, 1951 (1952).
8. A. Miklavc and S. Fischer, *Chem. Phys. Lett.*, **44**, 209 (1976).
9. S. Chapman and T. G. Cowling, *The Mathematical Theory of Non-uniform Gases*, Cambridge University Press, Cambridge, 1960.
10. B. J. Alder and T. E. Wainwright, *J. Chem. Phys.*, **33**, 1439 (1960).
11. H. Eyring and J. O. Hirschfelder, *J. Phys. Chem.*, **41**, 239 (1937).
12. K. F. Herzfeld, *J. Chem. Phys.*, **20**, 288 (1952).
13. L. I. Schiff, *Quantum Mechanics*, McGraw-Hill, New York, 1968.
14. B. J. Berne, in *Physcial Chemistry: An Advanced Treatise*, Vol. 8B, D. Henderson, ed., Academic, New York, 1971.
15. D. J. Diestler and R. S. Wilson, *J. Chem. Phys.*, **62**, 1572 (1975).
16. R. Zwanzig, *J. Chem. Phys.*, **33**, 1338 (1960); H. Mori, *Prog. Theor. Phys.*, **33**, 423 (1965).
17. D. W. Oxtoby and S. A. Rice, *Chem. Phys. Lett.*, **42**, 1 (1976).
18. D. C. Knauss and R. S. Wilson, *Chem. Phys.*, **19**, 341 (1977).
19. D. J. Diestler, *Chem. Phys. Lett.*, **39**, 39 (1976).
20. A. Nitzan and R. Silbey, *J. Chem. Phys.* **60**, 4070 (1974).
21. R. Kubo, *J. Math. Phys.*, **4**, 174 (1963).
22. M. Fixman, *J. Chem. Phys.*, **34**, 369 (1961).
23. R. Zwanzig, *J. Chem. Phys.*, **34**, 1931 (1961).
24. K. F. Herzfeld, *J. Chem. Phys.*, **36**, 3305 (1962).
25. R. Zwanzig, *J. Chem. Phys.*, **36**, 2227 (1962).
26. See C. Delalande and G. M. Gale, *Chem. Phys. Lett.*, **50**, 339 (1977), p. 342.
27. P. K. Davis and I. Oppenheim, *J. Chem. Phys.* **57**, 505 (1972).
28. H. K. Shin and J. Keizer, *Chem. Phys. Lett.*, **27**, 611 (1974).
29. D. W. Oxtoby, *Mol. Phys.* **34**, 987 (1977).
30. D. J. Diestler, *Chem. Phys.*, **7**, 349 (1975).
31. S. R. J. Brueck and R. M. Osgood, *Chem. Phys. Lett.*, **39**, 568 (1976).
32. C. Delalande and G. M. Gale, *Chem. Phys. Lett.*, **50**, 339 (1977).
33. R. Zwanzig and M. Bixon, *Phys. Rev.*, **A2**, 2005 (1970).
34. C. Hu and R. Zwanzig, *J. Chem. Phys.*, **60**, 4354 (1974); B. J. Berne and R. Pecora, *Dynamical Light Scattering*, Wiley, New York, 1976.
35. D. W. Oxtoby, *J. Chem. Phys.*, **70**, 2605 (1979).

36. J. Keizer, *J. Chem. Phys.*, **61**, 1717 (1974).
37. H. Metiu, D. W. Oxtoby, and K. F. Freed, *Phys. Rev.*, **A15**, 361 (1977).
38. S. Velsko and D. W. Oxtoby, *J. Chem. Phys.*, **72**, 2260 (1980).
39. See, for example, S. V. Ribnikar and O. S. Puzić, *Spectrochim. Acta*, **29A**, 307 (1973).
40. S. R. J. Brueck and R. M. Osgood, *J. Chem. Phys.*, **68**, 4941 (1978).
41. C. Manzanares and G. E. Ewing, *J. Chem. Phys.*, **69**, 1418 (1978).
42. B. P. Hills, *Mol. Phys.*, **35**, 1471 (1978).
43. B. P. Hills and P. A. Madden, *Mol. Phys.*, **35**, 807 (1978).
44. J. N. Kushick and S. A. Rice, *Chem. Phys. Lett.*, **52**, 208 (1977).
45. M. Muthukumar and S. A. Rice, *J. Chem. Phys.*, **69**, 1619 (1978).
46. See, for example, A. Rahman, *Phys. Rev.*, **136**, A405 (1964); L. Verlet, *Phys. Rev.*, **159**, 98 (1967).
47. J. Barojas, D. Levesque, and B. Quentrec, *Phys. Rev.*, **A7**, 1092 (1973).
48. J. P. Riehl and D. J. Diestler, *J. Chem. Phys.*, **64**, 2593 (1976).
49. D. L. Jolly, B. C. Freasier, and S. Nordholm, *Chem. Phys.*, **21**, 211 (1977); **23**, 135 (1977).
50. M. Shugard, J. C. Tully, and A. Nitzan, *J. Chem. Phys.*, **69**, 336 (1978); **69**, 2525 (1978).
51. S. A. Adelman and J. D. Doll, *J. Chem. Phys.*, **64**, 2375 (1976).
52. D. W. Oxtoby, D. Levesque, and J.-J. Weis, *J. Chem. Phys.*, **68**, 5528 (1978).
53. D. Levesque, and J.–J. Weis, and D. W. Oxtoby, *J. Chem. Phys.*, **72**, 2744 (1980).
54. J. C. Raich and N. S. Gillis, *J. Chem. Phys.*, **66**, 846 (1977).
55. H.-Y. Sun and S. A. Rice, *J. Chem. Phys.*, **42**, 3826 (1965).
56. N. B. Slater, *Theory of Unimolecular Reaction*, Cornell University Press, Ithaca, New York, 1959.
57. A. Nitzan and J. Jortner, *Mol. Phys.*, **25**, 713 (1973).
58. A. Nitzan, S. Mukamel, and J. Jortner, *J. Chem. Phys.*, **60**, 3929 (1974); **63**, 200 (1975).
59. Y. Weissman, A. Nitzan, and J. Jortner, *Chem. Phys.*, **26**, 413 (1977).
60. D. J. Diestler, *J. Chem. Phys.*, **60**, 2692 (1974).
61. S. H. Lin, *J. Chem. Phys.*, **65**, 1053 (1974).
62. H. Metiu and K. F. Freed, *Chem. Phys. Lett.*, **48**, 269 (1977).
63. M. Berkowitz and R. B. Gerber, *Chem. Phys. Lett.*, **49**, 260 (1977); **56**, 105 (1978); *Phys. Rev. Lett.*, **39**, 1000 (1977).
64. D. J. Diestler, E.-W. Knapp, and H. D. Ladouceur, *J. Chem. Phys.*, **68**, 4056 (1978).
65. D. Knittel and S. H. Lin, *Mol. Phys.*, **36**, 893 (1978).
66. H. K. Shin, *Chem. Phys. Lett.*, **60**, 155 (1978).
67. M. M. Audibert, C. Joffrin, and J. Ducuing, *Chem. Phys. Lett.*, **25**, 158 (1974).
68. G. Ewing, *Chem. Phys.*, **29**, 253 (1978).
69. G. Renner and M. Maier, *Chem. Phys. Lett.*, **28**, 614 (1974).
70. G. Renner and M. Maier, *Chem. Phys. Lett.*, **35**, 226 (1975).
71. W. F. Calaway and G. E. Ewing, *Chem. Phys. Lett.*, **30**, 485 (1975).
72. N. Legay-Sommaire and F. Legay, *Chem. Phys. Lett.*, **52**, 213 (1977).
73. M. Chatelet, G. Widenlocher, and B. Oksengorn, *C. R. Acad. Sci.*, **B287**, 195 (1978).
74. W. F. Calaway and G. E. Ewing, *J. Chem. Phys.*, **63**, 2842 (1975).
75. C. Manzanares and G. E. Ewing, *J. Chem. Phys.*, **69**, 2803 (1978).
76. S. R. J. Brueck, T. F. Deutsch, and R. M. Osgood, *Chem. Phys. Lett.*, **51**, 339 (1977).
77. G. M. Gale and C. Delalande, *Chem. Phys.*, **34**, 205 (1978).
78. A. Laubereau, D. von der Linde, and W. Kaiser, *Phys. Rev. Lett.*, **28**, 1162 (1972).
79. R. R. Alfano and S. L. Shapiro, *Phys. Rev. Lett.*, **29**, 1655 (1972).
80. A. Laubereau, G. Kehl, and W. Kaiser, *Opt. Commun.*, **11**, 74 (1974).
81. A. Laubereau, L. Kirschner, and W. Kaiser, *Opt. Commun.*, **9**, 182 (1973).

82. P. R. Monson, S. Patumtevapibal, K. J. Kaufmann, and G. W. Robinson, *Chem. Phys. Lett.*, **28**, 312 (1974).
83. K. Spanner, A. Laubereau, and W. Kaiser, *Chem. Phys. Lett.*, **44**, 88 (1976).
84. A. Laubereau, S. F. Fischer, K. Spanner, and W. Kaiser, *Chem. Phys.*, **31**, 335 (1978).
85. D. Ricard and J. Ducuing, *J. Chem. Phys.*, **62**, 3616 (1975).
86. D. Ricard, *J. Chem. Phys.*, **63**, 3841 (1975).
87. A. Laubereau, A. Seilmeier, and W. Kaiser, *Chem. Phys. Lett.*, **36**, 232 (1975).
88. S. H. Lin, *J. Chem. Phys.*, **61**, 3810 (1974).
89. R. J. Abbott and D. W. Oxtoby, *J. Chem. Phys.*, **72**, 3972 (1980).

EXPERIMENTAL STUDIES OF NONRADIATIVE PROCESSES IN LOW TEMPERATURE MATRICES

V. E. BONDYBEY

Bell Laboratories, Murray Hill, New Jersey 07974

CONTENTS

I. INTRODUCTION

While systematic studies of nonradiative relaxation in matrix-isolated molecules have only begun fairly recently, in the last few years there has been a considerable activity in this field and a substantial progress has occurred in our understanding of these processes.[1, 2]

At an early stage it has been recognized that relaxation of a small molecule in a periodic rare-gas lattice represents a prototype case of radiationless transition, which lends itself to rather straightforward and mathematically tractable modeling. Sun and Rice[3] have been the first to consider the simplest situation of an isolated diatomic (N_2) in an Ar lattice. Their semiclassical model, using realistic parameters predicted an $\approx 10^{-2}$ s lifetime for the vibrationally excited N_2, which was contrary to the generally accepted ideas of fast vibrational relaxation in condensed phases.

Several quantum-mechanical models were then developed[4–7] attempting to characterize the relaxation behavior of a two-level system coupled to the harmonic lattice. These theories, while differing in details of their treatment, agreed in predicting a very strong temperature dependence for the multiphonon relaxation process and formulating a so-called "energy-gap law," predicting an exponential decrease in the relaxation rates with increasing size of the guest vibrational frequency ω.

The first experimental observation of slow vibrational relaxation goes actually back to the early experiments of Vegard,[8] and several other examples were noted in the laboratories of Broida, Robinson, and other investigators.[9–13] Systematic studies of vibrational relaxation rates, however, only became possible with the advent of tunable dye lasers, permitting selective population of excited vibronic levels of the guest molecules.

Studies of radiationless transitions in matrix-isolated molecules represent a nice case of constructive interaction between theory and experiment. Most early experimental studies showed very poor agreement with the theoretical predictions. Thus in NH and OH, neither the expected temperature dependence, nor the energy-gap law predictions were fulfilled.[14, 15] Similarly, no steep temperature dependence of the relaxation rates was found in matrix-isolated CO.[16] The experimental studies, however, permitted to identify the reasons for the failure of the simple theories. This in turn led to development of new models, describing more adequately the experimental results.[17–19]

Conversely, with understanding of the reasons for the failure of the original simple models, it was possible to search experimentally for systems, which should fulfill better the assumptions and simplifications made in deriving the multiphonon relaxation theories. This indeed resulted recently in observations of multiphonon relaxation behaving in semi-quantitative agreement with the predictions of the original simple theories.[20] At present, the relaxation processes in matrices, and in particular the relaxation of diatomic molecules, promise to be among the best understood radiationless transitions in terms of quantitative agreement between experimental measurements and theoretical calculations.

In this chapter we will discuss some of the more recent developments in this field and review briefly the various types of nonradiative processes encountered in low-temperature solids.

II. MULTIPHONON VIBRATIONAL RELAXATION

Most of the earlier theoretical studies dealt with the simplest relaxation mechanism where the internal vibrational energy of the guest is dissipated directly into the delocalized and harmonic lattice phonons. The common results of these works[4–7] were, as we mentioned above, predictions of a strong temperature dependence for the relaxation and an exponential decrease in the rates with the size of the vibrational frequency. The former result has its origin in stimulated phonon emission; the conversion of vibrational energy into lattice phonons is greatly facilitated if some excited phonon states are thermally populated. The energy-gap law is due to the fact that the order of the multiphonon relaxation increases with the size of the vibrational quantum. A larger gap results in a higher order and hence less efficient process.

In spite of these unanimous predictions of the relaxation theories and the fact that also intuitively they appear to be correct, the experimental work failed to confirm them. In most systems where vibrational relaxation was studied, the strong temperature dependence of relaxation rates was not observed, and in a variety of molecules, in particular in small hydrides, the relaxation actually behaved in a direction opposite to the energy-gap law predictions.[14, 15] Explanation is to be found in the low efficiency of the true multiphonon relaxation process. The vibrational frequencies of light, strongly bound diatomic molecules are much higher than the frequencies of the rare-gas lattice, and multiphonon relaxation is thus a very high-order process. As a result, it is often masked by more efficient alternative relaxation pathways. In some cases, infrared radiation or energy transfer to impurities become important. In other instances, and in particular in light molecules, vibration→rotation transfer dominates the guest relaxation. We will discuss processes of this type briefly in the following sections.

The relative importance of the alternative relaxation mechanisms should clearly be reduced in molecules with lower vibrational frequencies. The effects of molecular rotation should also be reduced or eliminated in relatively heavy molecules with large moments of inertia and closely spaced rotational levels. With this in mind we have recently examined vibrational relaxation in the $B^3\Sigma_u^-$ state of matrix isolated S_2, which seemed to fulfill well the above requirements. Its vibrational frequency of $\approx$440 cm^{-1} is relatively low and should lead to rather efficient relaxation. Indeed, in solid Ne and in particular Ar, the vibrational decay is too fast to be studied with our experimental techniques, and mainly fully relaxed fluorescence is observed. The situation in the lighter hosts is further complicated by the presence of another electronic state in this energy range that perturbs the spectrum and participates in the relaxation process.[21] In solid Kr and in particular Xe, on the other hand, the relaxation processes slow down substantially, and extensive vibrationally unrelaxed $B^3\Sigma_u^-$ fluorescence can be observed. This was in the case of the Xe host already noted by Brewer and Brabson.[11] An advantage of carrying out this type of experiment in solid Xe matrix is the relatively wide range available for temperature-dependence studies.

Typical laser fluorescence spectra obtained by excitation of the $v=1$ $B^3\Sigma_u^-$ level are shown in Fig. 1. At 9 K in Fig. 1a only emission from $v=1$ is seen; apparently, vibrational relaxation is slow compared with the 16 ns radiative lifetime of the emitting state, and less than 0.5% of the $v=1$ molecules relax prior to reemission. The relaxation rate, however, increases with temperature, and at 28 K (Fig. 1c) more than 99% of the emission is

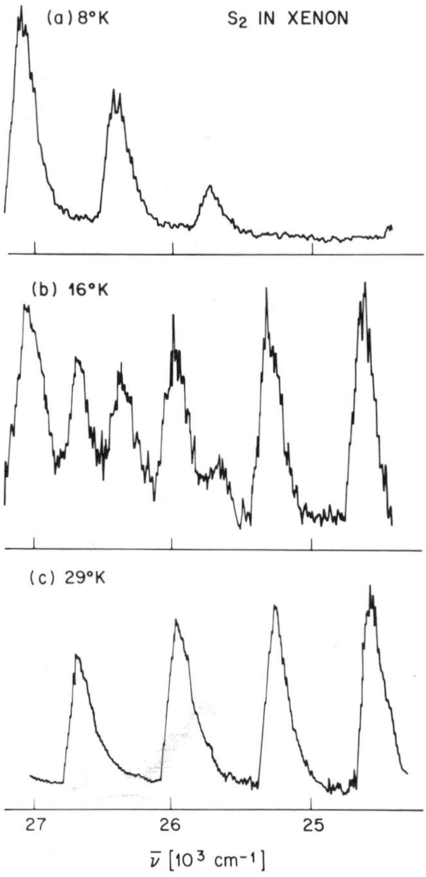

Fig. 1. Section of the $S_2\,B^3\Sigma_u^-$ emission spectrum is solid Xe following $v'=1$ excitation (a) 8 K; only the unrelaxed $v'=1$ progression is seen. (b) 16 K; both the $v'=1$ and $v'=0$ relaxed emission are seen. The increased noise is due to mild temperature fluctutation. (c) 29 K; only vibrationally relaxed emission is observed.

vibrationally relaxed. For a <20 K change in temperature, the relaxation rate changes by more than four orders of magnitude. It is encouraging that fitting the temperature-dependence data using expressions derived in the simple theoretical models yielded a realistic value of $\approx 40\ \text{cm}^{-1}$ for the accepting frequency.

This is clearly the dramatic temperature dependence of the vibrational relaxation rates predicted by theories of multiphonon relaxation. Apparently, in this case one deals with a true multiphonon relaxation process. It is reasonable to expect that also in many other heavier diatomics and in polyatomic molecules with small rotational constants or high barriers to free rotation the multiphonon relaxation mechanism will dominate.

Some time ago we have examined[22] the time-resolved emission from the $B^3\Sigma_u^-$ state of C_2^- and concluded that the rather efficient relaxation observed involves intersystem crossing into a nearby $a^4\Sigma_u^+$ state followed by vibrational relaxation within this state. This quartet state is characterized by a relatively low vibrational frequency $\omega_e = 1093$ cm^{-1}. The lifetimes of the intermediate $a^4\Sigma_u^+$ levels showed a rather strong temperature dependence and a conventional energy-gap law behavior, with increased relaxation rates in the heavier $^{13}C_2^-$ isotopic molecule. This time-resolved behavior at several temperatures is shown in Fig. 2. It seems quite likely that here also the relaxation involves a genuine multiphonon process.

Goodman and Brus[23] have recently excited the $B^2\Pi$ electronic state populated by intersystem crossing of matrix isolated NO using the 1933 Å ArF excimer laser, and observed phosphorescence from the lowest $a^4\Pi$ state. From the rise time on the phosphorescence they were able to deduce the rates of the $a^4\Pi$ vibrational relaxation. They found that the smaller

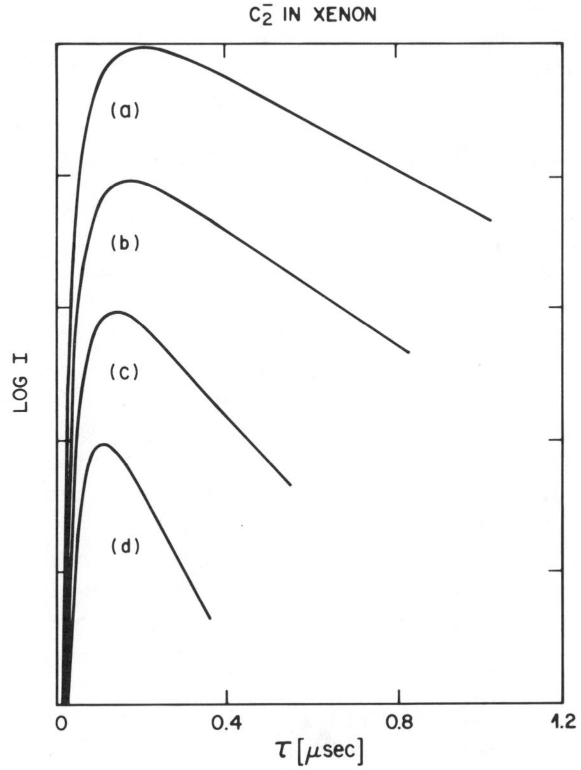

Fig. 2. Delayed $v'=1$ emission of C_2^- $B^2\Sigma_u^+$ in solid Xe following $v'=2$ excitation. Solid line represents best fit to the data assuming two intermediate lifetimes. (a) 4.2 K, 70 and 234 ns. (b) 20 K; 63 and 181 ns. (c) 32 K; 51 and 102 ns. (d) 42 K; 35 and 57 ns.

quanta in the $^{14}N^{18}O$ and $^{15}N^{16}O$ tend to relax faster than $^{14}N^{16}O$, as expected for multiphonon relaxation. The rates did, however, not strictly correlate with the reduced mass, and the expected temperature dependence was also not observed.

Even though we spoke in the preceding paragraphs about relaxation into the delocalized lattice phonons, the energy is clearly initially transferred into the motion of the nearest-neighbor host atoms, including the center of mass motion of the guest itself. Several theoretical models have recently exploited this idea. Nitzan *et al.*[24, 25] have developed a linear stochastic trajectory model considering a diatomic guest coupled to two nearest-neighbor atoms, which are in turn weakly coupled to the remaining lattice. A conceptually similar, quantum-mechanical model was formulated by Ladouceur and Diestler.[26] These studies have considered as a test case matrix-isolated Cl_2 molecules. Vibrational relaxation in Cl_2 is expected to be very fast, and its rates were not directly measured. It was, however, recently suggested[27] that a distinct broadening observed for high $B^3\Pi 0_u^+$ vibrational levels has its origin in a fast nonradiative relaxation process. Both theoretical models yielded rates in good qualitative agreement with the lifetimes and trends deduced from the spectroscopic data.

III. GUEST ROTATION AS THE ENERGY ACCEPTOR

The vibrational relaxation described in the preceding section assumes a more or less direct conversion of vibrational energy into the delocalized lattice phonons. Introduction of the impurity molecule will, however, modify the phonon structure in its neighborhood and introduce several degrees of freedom associated with the rotation and translation of the guest itself. In light molecules, particularly in the case of hydrides, the frequencies associated with these motions will be relatively high and, for the rotational modes, the spacing of the successive levels will increase rapidly with the degree of excitation. The vibration→rotation transfer will, under these circumstances, be a lower order process than the multiphonon relaxation and become important.

Several recent experimental studies in small hydride molecules have indeed demonstrated that molecular rotation is instrumental in accepting the energy. Thus the relaxation in the $A^3\Pi$ state of NH and ND was examined[14] using pulsed UV laser excitation and time-resolved fluorescence studies. The data showed that, contrary to the energy-gap law predictions, NH relaxes order of magnitude faster than ND. Very similar behavior was observed in the $A^2\Sigma^+$ state of OH in solid Ne.[15]

Relaxation rates in the NH $X^3\Sigma^-$ ground electronic state were also measured using an optical double resonance technique.[28] In these experiments an initial pump laser populates the $v''=1$ level and a second pulsed laser probes this population after a variable time delay. The absolute relaxation rates were a factor of $\approx$200 slower than in the excited electronic state, but the same trends could be observed. The ND lifetime was found to be $\approx$30 ms, possibly controlled by IR radiation; in NH, on the other hand, the $v''=1$ level relaxes several orders of magnitude faster, with $\approx$190 μs lifetime. All of these experiments found, furthermore, the rates to be only weakly temperature dependent.

While these observations would be hard to explain in terms of multiphonon relaxation, they are readily understood consequences of the vibration $\rightarrow$rotation transfer model. Isotopic substitution will change not only the vibrational frequency of the guest but also the spacing of the rotational levels, and will modify the localized phonon structure. Whereas the vibrational spacing is proportional to the square root of reduced mass, the rotational constant changes even faster, linearly in μ. As a consequence, the closest rotational level to $v=1$ is $J=13$ in NH and $J=16$ in ND. Thus, in spite of a smaller vibrational spacing, the deuteride relaxation is a higher order and hence less efficient process.

Similar trends were encountered in several other hydride molecules using a variety of experimental techniques. Moore and Wiesenfeld[29] have measured relaxation rates in the ground electronic state of HCl and DCl in solid Ar by exciting the guest using a tunable IR parametric oscillator and directly time resolving the infrared fluorescence. They found that at higher concentrations the decay is governed by energy transfer to impurities, possibly $(HCl)_2$ dimers. In more diluted samples, where the effects of the impurities are eliminated, HCl again relaxes 32 times faster than DCl.

Abouaf-Marguin *et al.*[30] have excited the lowest frequency C–F stretching modes of CH_3F and CD_3F using a pulsed CO_2 laser and monitored the ground-state recovery by means of a weak CW probe laser. They again find an order of magnitude slower relaxation in the deuteride. In both of these studies the results were rationalized by accepting the vibration$\rightarrow$ rotation relaxation mechanism.

Legay[1] carried the idea of rotational accepting modes further and suggested an empirical exponential relationship between the relaxation rate and the number of rotational levels needed to match the size of the energy gap,

$$J_m = \left(\frac{\omega}{B}\right)^{1/2}$$

and with this he correlated with some success the available experimental data. It should, of course be noted that perfectly free rotation can not accept vibrational energy, since the angular momentum has to be conserved. A deviation from spherical symmetry is needed to provide the vibration–rotation coupling. It is therefore clear that the relaxation rates will also be a sensitive function of the guest–host interaction potential and of the barrier to free rotation. Following the experimental demonstration of the importance of guest rotation, the localized translational and rotational modes were incorporated into the theoretical relaxation models.

Freed and Metiu[31, 32] have considered a simplified 2-*d* model of vibration–rotation coupling. Their model predicts the rates to decrease exponentially with $(\omega/B)^{1/2}$ providing theoretical basis for the empirical relationship of Legay, and to be only weakly dependent on temperature. Gerber and Berkowitz[33, 34] have developed a 3-*d* model of coupling to rotation. They use an experimentally determined Ar–HCl interaction potential to calculate the relaxation rates of NH and ND in solid Ar. The calculated relaxation rates are in an astonishingly good quantitative agreement with the experimental results. It would certainly be very interesting to extend this comparison of experimental and calculated rates to other rare-gas matrices. Such comparisons should provide a very sensitive test for the theoretical models.

IV. VIBRATIONAL ENERGY TRANSFER PROCESSES

In more concentrated samples, intermolecular vibrational energy transfer may occur and dominate the relaxation kinetics. Both in the classic studies of CO by Dubost and Charneau[16] and in the more recent work on HCl,[29] an extensive migration of vibrational energy among the guest molecules was detected. Two additional exothermic processes take place, reflecting the fact that entropy is unimportant in Boltzmann equilibria at 4 K. The vibrational energy flows into the heavier $^{13}C^{16}O$ and $^{12}C^{18}O$ isotopic species present in the samples in natural isotopic abundance, and processes of the type

$$CO(v=n)+CO(v=1)\rightarrow CO(v=n+1)+CO(v=0)$$

pump energy into high vibrational levels.

In highly diluted matrices the guest–guest interactions become unimportant, but vibrational energy transfer can occur unimoleculary between different vibrational modes of a polyatomic molecule. A few studies of relaxation in small polyatomic molecules were recently reported. They suggested that in addition to the energy gap considerations, the intramolecular intermode coupling elements will be of importance for the relaxation process.

In the excited $^3\Pi$ of the linear CNN radical[35, 36] the stretching vibrations relax via intramolecular conversion into the overtone of the bending mode. On the other hand in ClCF, the relaxation in the bending ν_2 manifold appears to occur independently[37] with mode-to-mode conversion being inefficient. Similarly studies of relaxation in the excited states of NCO suggest the importance of anharmonic intermode coupling.[38] Effects that enhance such intermode coupling, as for instance Fermi resonance, also facilitate nonradiative relaxation. One thus observes in NCO an extremely efficient relaxation between the components of Fermi polyads involving the ν_1 and ν_2 vibrational modes, and only the lowest member of each group is observed in emission. Relaxation between other levels is, on the other hand, slow compared with the upper state radiative lifetime.

The data available thus far do not appear to permit any generalization and more experimental studies and hopefully development of theoretical models will be needed in this area.

V. ELECTRONIC RELAXATION PROCESSES

Most of the original theoretical studies assumed that internal conversions and intersystem crossing between different electronic states will be in general too slow to compete with vibrational relaxation. Numerous recent experiments have, however, shown the nonradiative electronic relaxation to be a remarkably efficient process, which may in many cases dominate the small molecule relaxation.

A particularly clear-cut case of this behavior has been observed in the CN radical in solid Ne.[39, 40] The major relaxation pathway for excited vibrational levels of the low-lying $A^2\Pi$ electronic state involves crossing into the nearest available level of the $X^2\Sigma^+$ ground state followed by reverse crossing into the next lower lying $A^2\Pi$ level. The $A^2\Pi$ population reaches by this interelectronic cascade in a few microseconds the $v''=4$ level of the $X^2\Sigma^+$ ground state, which lies below the vibrationless level of $A^2\Pi$, and for which the electronic relaxation channel is no longer available. This level, as well as all the lower levels of the ground state relax several orders of magnitude more slowly. Their lifetimes are in the millisecond range and are probably controlled by infrared radiation.

This relative ease of electronic relaxation as compared with multiphonon vibrational relaxation was qualitatively interpreted using phonon Franck–Condon arguments. Change in the vibrational state of the guest molecule requires in general very little change in the equilibrium positions of the lattice atoms, and this results in poor Franck–Condon factors for multiphonon vibrational relaxation. Electronic transitions, on the other hand, are often accompanied by considerable changes in electron density distribution in the guest molecule, and this in turn requires rearrangement of the solvent. Such a resolvation then provides favorable Franck–Condon factors for conversion of the internal guest energy into the lattice phonons.

The rates of the individual steps of the $A^2\Pi$ nonradiative cascade were found to show qualitatively a steep dependence on the size of the energy gap. The CN system with its simple and well-understood spectroscopy is well suited for theoretical modeling of the interstate cascade relaxation mechanism and has indeed been the subject of several theoretical studies.

The simplest approach assuming that harmonic lattice phonons accept the energy, while predicting correctly the trends, failed to give good quantitative agreement. It predicted a much steeper energy gap dependence than was observed experimentally.[40] Much better agreement was obtained by Weissman, Nitzan, and Jortner[41] by considering quadratic effects and anharmonicity. A more recent study by Fletcher, Fujimura, and Lin[42] has established a very satisfactory numerical agreement of the calculated rates with experiment using a model with rotational localized modes accepting the energy.

Similar interelectronic relaxation processes, both spin-allowed and spin-forbidden have now been observed in a variety of other systems.[43, 44] For instance, such cascading relaxation was identified in C_2, C_2^-, and between the $C^3\Delta_u$ and $A^3\Sigma_u^+$ states of O_2. The available experimental evidence clearly suggests that interstate processes of this type should be rather commonplace in the relaxation of small matrix-isolated molecules.

It is interesting to note that the same relaxation mechanism was recently established for the collisional relaxation of the vibrational levels in the $A^2\Pi$ states of CO^+ and CN in room-temperature gas-phase experiments.[45, 46]

VI. HOST DEPENDENCE OF THE NONRADIATIVE RELAXATION

As we noted previously, an isolated small molecule cannot decay nonradiatively and the relaxation is therefore a sensitive function of the guest–host interaction potential. For this reason it is often instructive to compare the relaxation rates in different hosts. A particularly useful set of solvents for such comparison studies are the four rare-gas solids Ne, Ar, Kr, and Xe. While the pairwise interaction energy increases in this series from 60 cm^{-1} for two Ne atoms to $\approx$400 cm^{-1} in the case of Xe, this increase is paralleled by the increase in mass, and the Debye frequency remains for all the rare gases relatively constant near $\approx$65 cm^{-1}.

Where the magnitude of the guest–host interaction is the dominant factor, one could expect a steep increase in relaxation rates from Ne to the much more polarizable and strongly interacting Xe. Such trends have indeed been observed in most of those systems where the vibration→rotation transfer was established to be the rate controlling process. Thus in NH $A^3\Pi$ very little relaxation is observed in Ne matrix. The nonradiative decay is much more efficient in solid Ar, and in solid Kr the rates increase by more than another order of magnitude.[14] Even more dramatic increase in the relaxation rates is observed in OH $A^2\Sigma^+$, where the guest is found to hydrogen-bond rather strongly to the more polarizable heavier rare gases.[47]

A quite different behavior is encountered in those situations where rotation is believed to be unimportant and one deals with a true multiphonon relaxation. Thus in the $B^3\Sigma_u^-$ state of S_2 the relaxation is least efficient in solid Xe; the rates in the lighter hosts are faster by many orders of magnitude.[20] Also in the previously discussed case of C_2^- $a^4\Sigma_u^+$ relaxation the same trend was observed, with slowest relaxation in the Xe host. Allamandola *et al.*[49] have studied extensively the relaxation in the ground state of C_2^-. Fastest relaxation is again observed in solid Ne, with sharply decreasing rates in the heavier rare gases. It should, however, be noted that in this case it is believed that at least part of the C_2^- vibrational energy is being transferred to some as yet unidentified impurity.

Similar situation prevails in several polyatomic free radicals. Thus in the $A^2\Pi$ state of NCO, very little vibrational relaxation is observed on the time scale of the radiative lifetime, in solid Ar and Kr. In solid Ne, rather efficient relaxation takes place, permitting easy measurement of the individual relaxation rates. Also the excited electronic states of CNN and CF_2 exhibit most efficient relaxation in solid Ne.

While the principle of slower multiphonon relaxation in the heavier rare gases seems to be well established experimentally, its physical interpretation is not quite clear at this time. Classically, in a binary collision model a most efficient energy transfer will occur if the two collision partners are of comparable mass. Since most of the molecules discussed above are composed of first row elements, such a collisional model of relaxation would predict fastest relaxation in Ne, as observed. It is noteworthy that in the relatively heavy S_2 molecule, most efficient relaxation appears to occur in solid Ar; in Ne, while the relaxation is also fast, weak vibrationally unrelaxed emission has been detected.

Interesting in this context are also theoretical studies of vibrational predissociation in van der Waals complexes by Beswick and Jortner.[50] They conclude that the rates of vibrational predissociation of the rare gas atom-diatomic molecule complexes should be enhanced with decreasing mass of the rare-gas atom. If one could view relaxation of the guest molecule as a predissociation in a polyatomic van der Waals complex involving the guest and the nearest-neighbor rare-gas atoms, the observed trends would again be correctly predicted.

VII. SUMMARY

Matrix-isolated molecules display an astonishing variety in their relaxation behavior. In molecules characterized by small or intermediate vibrational spacing (less than ≈ 1000 cm^{-1}) and with larger moments of inertia or higher barriers to free rotation, one can observe behavior expected of multiphonon relaxation processes. One finds strong dependence of the rates on temperature and an exponential relationship between the rate and the size of the energy gap. In rare-gas hosts the relaxation shows a distinct trend toward increasing efficiency in the lighter hosts.

In light molecules and, in particular, in hydrides with large rotational constants, the molecular rotation is instrumental in accepting the energy. In this situation the temperature dependence of rates is absent and one observes an "inverse energy-gap law," with hydride molecules relaxing faster than the corresponding deuterides. The trend in different hosts is reversed in some cases, with most efficient relaxation in the heavier, more polarizable matrices. This may reflect the increased barrier to free rotation and hence increased vibration–rotation coupling.

Matrix-isolated molecules exhibit a surprising facility for interelectronic relaxation processes. Vibrational relaxation in excited electronic states is often dominated by interstate cascades involving other electronic states. The rates of the individual steps of such a cascade are modulated by the intramolecular Franck–Condon factors and exhibit qualitatively an exponential dependence on the size of the energy gap expected by multiphonon relaxation theories.

Finally, in strongly bound nonhydride molecules with large vibrational spacing (above ≈ 1500 cm^{-1}) both of the vibrational relaxation mechanisms described above may be very slow. Under these circumstances the guest relaxation is often controlled by alternative processes, for example, infrared radiation or intermolecular vibrational energy transfer.

Acknowledgment

The author wishes to acknowledge helpful discussions with P. M. Rentzepis and J. Jortner.

References

1. V. E. Bondybey and L. E. Brus, in *Adv. Chem. Phys.*, **41**, 269 (1980).
2. F. Legay, in *Chemical and Biological Applications of Lasers*, Vol. II, Academic, New York, 1977.
3. H. Y. Sun and S. A. Rice, *J. Chem. Phys.*, **42**, 3826 (1965).
4. A. Nitzan, S. Mukamel, and J. Jortner, *J. Chem. Phys.*, **60**, 3929 (1974).
5. A. Nitzan, S. Mukamel, and J. Jortner, *J. Chem. Phys.*, **63**, 200 (1975).
6. S. H. Lin, *J. Chem. Phys.*, **65**, 1053 (1976).

7. F. K. Fong, S. L. Naberhuis, and M. M. Miller, *J. Chem. Phys.*, **56**, 4020 (1972).
8. L. Vegard, *Z. Phys.*, **75**, 30 (1932).
9. C. M. Herzfeld and H. P. Broida, *Phys. Rev.*, **101**, 606 (1956).
10. D. S. Tinti and G. W. Robinson, *J. Chem. Phys.*, **49**, 3229 (1968).
11. L. E. Brewer and G. D. Brabson, *J. Chem. Phys.*, **44**, 3274 (1966).
12. V. E. Bondybey and J. W. Nibler, *J. Chem. Phys.*, **56**, 4719 (1972).
13. H. P. Broida and M. Peyron, *J. Chem. Phys.*, **32**, 1068 (1960).
14. L. E. Brus and V. E. Bondybey, *J. Chem. Phys.*, **63**, 786 (1975).
15. V. E. Bondybey and L. E. Brus, *J. Chem. Phys.*, **63**, 794 (1975).
16. H. Dubost and R. Charneau, *Chem. Phys.*, **12**, 407 (1976).
17. R. B. Gerber and M. Berkowitz, *Phys. Rev. Lett.*, **39**, 1000 (1977).
18. K. F. Freed and H. Metiu, *Chem. Phys. Lett.*, **48**, 262 (1977).
19. K. F. Freed, D. L. Yeager, and H. Metiu, *Chem. Phys. Lett.*, **49**, 19 (1977).
20. V. E. Bondybey and J. H. English, *J. Chem. Phys.*, **72**, 3113 (1980).
21. V. E. Bondybey and J. H. English, *J. Chem. Phys.*, **69**, 1865 (1978).
22. V. E. Bondybey and L. E. Brus, *J. Chem. Phys.*, **63**, 2223 (1975).
23. J. Goodman and L. E. Brus, *J. Chem. Phys.*, **69**, 1853 (1978).
24. A. Nitzan, M. Shugard, and J. C. Tully, *J. Chem. Phys.*, **69**, 2525 (1978).
25. M. Shugard, J. C. Tully, and A. Nitzan, *J. Chem. Phys.*, **69**, 336 (1978).
26. H. D. Ladouceur and D. J. Diestler, *J. Chem. Phys.*, **70**, 2620 (1979).
27. V. E. Bondybey and C. Fletcher, *J. Chem. Phys.*, **64**, 3615 (1976).
28. V. E. Bondybey, *J. Chem. Phys.*, **65**, 5138 (1976).
29. J. Wiesenfeld and C. B. Moore, *J. Chem. Phys.*, **70**, 930 (1979).
30. L. Abouaf-Marguin, B. Gauthier-Roy, and F. Legay, *Chem. Phys.*, **23**, 443 (1977).
31. K. F. Freed and H. Metiu, *Chem. Phys. Lett.*, **48**, 262 (1977).
32. K. F. Freed and D. L. Yeager, and H. Metiu, *Chem. Phys. Lett.*, **49**, 19 (1977).
33. R. B. Gerber and M. Berkowitz, *Phys. Rev. Lett.*, **39**, 1000 (1977).
34. R. B. Gerber and M. Berkowitz, *Chem. Phys. Lett.*, **49**, 260 (1977).
35. J. L. Wilkerson and W. A. Guillory, *J. Mol. Spectr.*, **66**, 188 (1977).
36. V. E. Bondybey and J. H. English, *J. Chem. Phys.*, **67**, 664 (1977).
37. V. E. Bondybey, *J. Chem. Phys.*, **66**, 4237 (1977).
38. V. E. Bondybey and J. H. English, *J. Chem. Phys.*, **67**, 2868 (1977).
39. V. E. Bondybey, *J. Chem. Phys.*, **66**, 995 (1977).
40. V. E. Bondybey and A. Nitzan, *Phys. Rev. Lett.*, **26**, 413 (1977).
41. I. Weissman, J. Jortner, and A. Nitzan, *Chem. Phys.*, **26**, 413 (1977).
42. D. Fletcher, Y. Fujimura, and S. H. Lin, *Chem. Phys. Lett.*, **57**, 400 (1978).
43. V. E. Bondybey, *J. Chem. Phys.*, **65**, 2296 (1976).
44. J. Goodman and L. E. Brus, *J. Chem. Phys.*, **67**, 1482 (1977).
45. V. E. Bondybey and T. A. Miller, *J. Chem. Phys.*, **69**, 3602 (1978).
46. D. H. Katayama, T. A. Miller, and V. E. Bondybey, *J. Chem. Phys.*, **71**, 1662 (1979).
47. J. Goodman and L. E. Brus, *J. Chem. Phys.*, **67**, 4858 (1977).
48. D. S. Tinti, *J. Chem. Phys.*, **48**, 1459 (1968).
49. L. J. Allamandola, A. M. Rohjantalab, J. W. Nibler, and T. Chappell, *J. Chem. Phys.*, **67**, 99 (1977).
50. J. A. Beswick and J. Jortner, *J. Chem. Phys..*, **68**, 2277 (1978).

PICOSECOND SPECTROSCOPY AND DYNAMICS OF ELECTRON RELAXATION PROCESSES IN LIQUIDS

GERALDINE A. KENNEY-WALLACE*

Department of Chemistry, University of Toronto, Toronto, Canada M5S 1A1

CONTENTS

Looking back on a decade of experimental and theoretical activity on the nature of excess electron states in liquids, we see a progressive transition from the early quantum-mechanical description of the electron as a polaron, dressed in the polarization field of the dielectric continuum, toward more molecular models. As the boundary of the dielectric continuum receded in order to permit varying geometries of short range, discrete liquid structure to provide an effective electron–molecule trapping

*Alfred P. Sloan Foundation Fellow.

potential, so the electron became dressed in a cluster of molecules embedded in the liquid. The rotational, vibrational, and configurational relaxations of the molecules comprising the host cluster must now become an intimate part of any description of the dynamics of the formation of the stable species (e_s^-) and of its binding energy in the ground and excited states. Although electron localization is by now a well-established phenomenon,[1–4] it is only recently that the details of the *dynamics* of electron relaxation processes have begun to emerge from picosecond absorption spectroscopy,[5] not only of electron localization in liquids[6–9] but in glasses,[10] semiconductors,[11] and crystals[12] too.

In this chapter we will highlight recent experimental data on the picosecond dynamics of electron localization and solvation in polar liquids and on the ultrafast radiationless transitions that accompany laser excitation of e_s^- in the same systems. The specific issues we address concern (1) the mechanism for electron localization in polar liquids, (2) the molecular description of the solvation process in forming the cluster, and (3) the dynamics of electron transfer following photodetachment of an electron from its cluster.

In Section I we briefly discuss the relationship between the theoretical parameters and experimental observables in these experiments in terms of the spectroscopy of electrons in liquids. Experimental techniques are considered in more detail in Section II, while the data from electron solvation in pure liquids are reviewed in Section III in the context of the molecular dynamics of the host liquid. Section IV presents current results on electron trapping in very dilute polar systems and leads to speculation on mechanisms of electron localization. In Section V the first direct observations of a photoselective, laser-induced electron-transfer process are presented, following which we summarize as yet unresolved issues and speculate on future directions in the laser spectroscopy of electron-relaxation processes.

I. SPECTROSCOPY OF ELECTRONS IN LIQUIDS

In order to introduce the necessary relationships between the electronic properties of e_s^- and the experimental observables, let us consider the following, admittedly simplified, case as illustrated in Fig. 1. When an electron is injected into a liquid following, for example, multiphoton photoionization of an impurity molecule in the liquid, the hot electron scatters through the conduction band of the fluid as it rapidly relaxes to the bottom of the band, V_0, also described as the mobility edge.[2, 3, 13] In fact, V_0 is the work function for injection of the electron into the liquid from vacuum and, if positive, represents a barrier to injection. For most

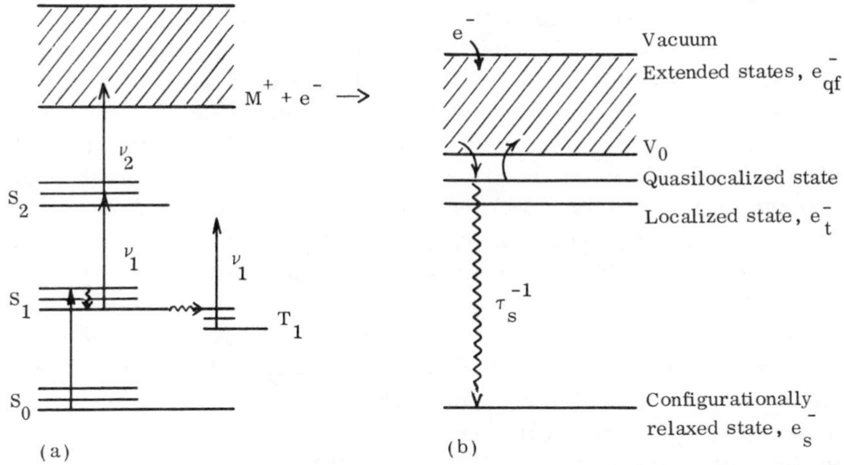

Fig. 1. Schematic of (a) multiphoton molecular photoionization, and (b) electron injection into liquid and subsequent transitions between extended and localized states. See text for discussion.

liquids, $V_0 < 0$.[2] At subpicosecond times in the conduction band, the electron is in an *extended state* and exhibits a high mobility (μ_e) as it undergoes single and multiple scattering in an unperturbed fluid. The scattering length, a, is governed by the details of the electron–atom scattering potential which, at least in rare gas liquids, is amenable to calculation. In more complex fluids, measured electron mobilities (μ_e) provide extremely useful data from which to attempt to extract the details of these scattering interactions, since $\mu_e \propto (\bar{a})^{1/2}$.[2, 13] At energies $\sim V_0$ the quasifree electron is now thermalized and has access to a tail of localized states immediately below the conduction band. The mobilities of electrons in some alkanes, for example, are indicative of these *quasilocalized states* in which the electron transport is partly that of a lower mobility, localized state and partly that of a high mobility extended state, since the electron can be promoted back into the conduction band by a thermal fluctuation in the liquid. Typically, μ_e in the rare gases is $\geqslant 10^2$ V^{-1} cm^2 s^{-1} but in the *n*-alkanes $0.1 < \mu < 10$ V^{-1} cm^2 s^{-1}.[2, 14]

Up to this point in time, $\sim 10^{-12}$ s, the electron–medium interactions have led to little, if any, medium reorganization, but at longer times the induced orientational polarization develops and contributes to the stabilization energy of the electron. We then observe the appearance of a solvated electron in a fully *configurationally relaxed ground state* (e_s^-) whose mobility[13] of 10^{-3} to 10^{-4} V^{-1} cm^2 s^{-1} is more reminiscent of an ion in a liquid. Although a sudden decrease in mobility is indeed a signature of electron localization and loss of carriers from the conduction band, the picosecond time scale of these processes in liquids is too fast to be resolved by conventional pulsed conductivity techniques and interpretations are complicated by the role of quasilocalized states, unless a model already exists to describe the electron transport in a two-level system. On the other hand, in picosecond laser spectroscopy we can directly probe the dynamics of the transition between the extended and localized states on a time scale comparable to the molecular reorientational motion in most liquids. Since localized electron states have high optical absorption cross-sections, $\sigma \leqslant 10^{-16}$ cm^2, they can be readily interrogated in a double-resonance experiment, which monitors the appearance of population in the ground state (e_s^-) following electron injection into the liquid via mechanisms discussed in Section II.

The optical absorption spectra of e_s^- are broad (typically fwhm$\sim$1 eV), structureless bands, which are skewed on the high-energy side of the spectrum and span the visible to near IR spectral region, as shown in Fig. 2.[4] While the breadth and intensity of the absorption band make optical detection of these species relatively simple, it is precisely these features that were for so long the intractable part of the theory of excess electronic states. The symbiotic relationship between the electron and its supporting fluid[5] first became apparent through experimental studies of e_s^- spectra in alcohols of diverse structural and dielectric properties, where it was clear that the position of E_{max}, and hence the binding and transition energies of e_s^-, were primarily a function of the short-range charge–dipole interactions and local structure in the liquid.[15] Similar conclusions concerning the dominance of short-range forces were reached in studies of electrons in supercritical water and subcritical ammonia vapor.[3, 16, 17] Thus the spectral grouping observed for electrons in a wide range of liquids can be rationalized and quantitatively predicted from a knowledge of the short-range forces and topology of the trapping site.[4, 15] As the focal points of theoretical models of electrons in fluids, the absorption maximum, linewidth and lineshape are now treated as the consequence of the strong coupling between the electronic motion and the vibrational and statistical configurational fluctuations of the molecules comprising the cluster to which the electron is confined.[18–21] These theories do not share identical conceptual bases but, as with the previous generalization of empirical semicontinuum[13, 22, 23] and cluster models,[24] each reproduces some features of the optical absorption spectra quite well, while falling short of agreement elsewhere. The limitations of space rather than interest prevent any summary or critique of the theories,[4, 25] but at the conclusion of this chapter we will conjecture on novel sources of spectroscopic data which will be required in the future as sensitive tests of such theories.

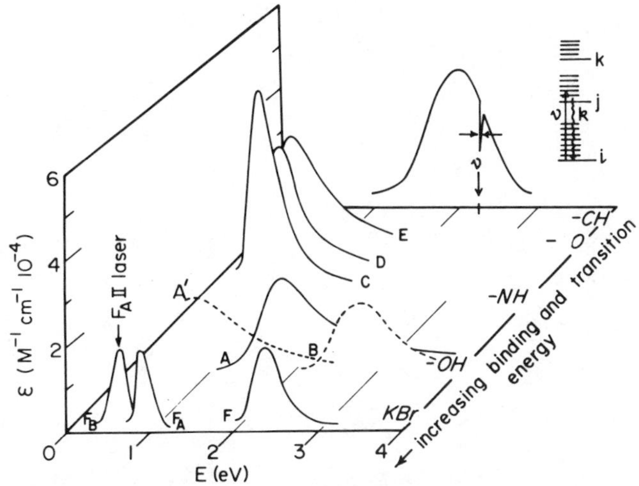

Fig. 2. Spectroscopy of electrons in fluids. Ensemble shows absorption maxima and coefficients in *n*-alcohols at picosecond (*A'*) and nanosecond (*A*) times, diols (*B*), amines (*C*), ethers (*D*), alkanes (*E*), and color center (*F*). The absorption (F_A) and stimulated (F_B) emission in KBr color center (F_AII) laser are also shown.

The absorption spectrum of e_s^- is indeed a fingerprint of its molecular environment, as Fig. 2 shows in ensemble form. For interest, the *F* center in KBr and its color center ($F_{\Pi}A$) laser[26] are shown too. More quantitative spectra are shown in Fig. 3, which displays for the first time the full visible and near-IR nanosecond absorption spectra of e_s^- in 1-octanol and 1-decanol in comparison to 2-octanol, 3-octanol, and 2-decanol.[27] The absorption spectra for e_s^- in the pair of 1-alcohols peak at $\lambda_{max}=650$ nm[15] and are clearly superimposable across the band, as are the spectra for 2-alcohols to which the former absorbances have been normalized at A_{max}. The λ_{max} for e_s^- in 3-octanol (unnormalized data) is similar to 2-octanol, $\lambda_{max}=800$ nm, but less than 900 nm observed in 4-heptanol,[15] reflecting the fact that the clustering of the dipolar molecules about the excess electron is impeded by the sterically awkward location of the −OH, leading to less stabilization energy from the equilibrium molecular structure of e_s^-. The dynamics of the clustering process should also reveal such effects.

The $E_{\lambda_{max}}$ of all e_s^- spectra in the series methanol to dodecanol fall within 1.80 ± 0.08 eV,[15] and their maximum absorption cross-sections lie within $1-2\times10^4$ M^{-1} cm^{-1}.[28] Therefore, it is reasonable to suppose that the final quantum state of e_s^- in each case experiences the same short-range electron–medium interactions and hence has the same binding energy. By recording the temporal evolution of these absorption bands in picosecond

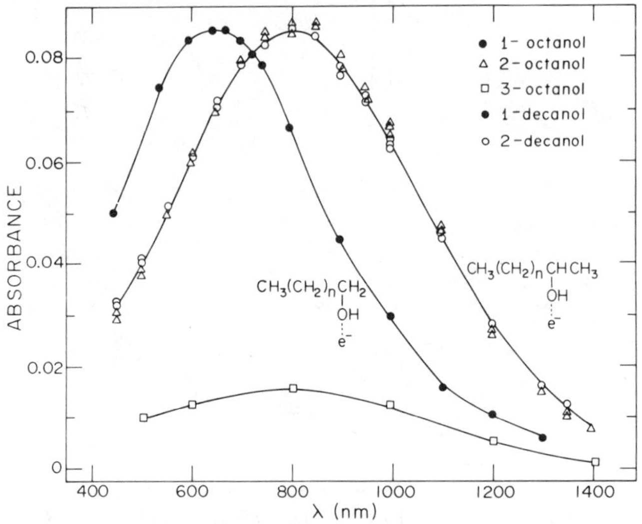

Fig. 3. Nanosecond absorption spectra of e_s^- in alcohols at 300 K: (●) 1-alcohols, (△) and (○) 2-alcohols, (□) 3-alcohols.

spectroscopy, we expect to differentiate these spectra via *dynamical* effects that will offer some insight on the mechanisms of electron localization and solvation, in alcohols in particular, and in polar liquids in general.

II. EXPERIMENTAL DESIGN

Electron relaxation processes have been studied with several double resonance or pump-probe techniques. Electron localization has been investigated in picosecond absorption spectroscopy (1) following photodetachment or photoionization of an ion or a molecule in the host liquid with the harmonics from mode-locked, solid-state lasers, and (2) through direct ionization of the liquid with a picosecond pulsed electron beam. With the exception of electrons generated in metal ammonia solutions, electrons in other liquids have a relatively short kinetic lifetime, typically 10^{-10} to 10^{-6} s, and thus it is not surprising that their discovery and subsequent investigations of their spectroscopic and kinetic properties have paralleled the rapid technological developments of the past decade in pulsed lasers, pulsed electron beams and fast time-scale measurements in general. We review below the different molecular mechanisms and techniques that are currently employed in studying the dynamics of these species.

A. Mechanisms for Electron Generation in Liquids

There are several different molecular mechanisms for the sudden generation of electrons in liquids, following which we may follow the picosecond relaxation processes involved in electron localization, solvation, and the radiationless transitions that occur following optical excitation of the stabilized species, e_s^-. The experimental designs discussed below are really complementary approaches to the problem of interest since no experiment shares an *identical* set of experimental variables with the others. Therefore the information obtained will not be redundant in terms of the theory of these electron relaxation processes.

Electrons can be generated via photoionization of an impurity molecule in the liquid under study,[29, 30] via photodetachment of an ion in a liquid,[7] or via direct ionization of that liquid.[6, 8] Laser-induced multiphoton photoionization (MPI) is employed to generate electrons from, for example, aromatic molecules whose ionization potentials (I) generally fall within the range 5–7 eV. In the gas phase, four types of highly excited states have been identified for these aromatic molecules, continuum ionized states, autoionizing states, Rydberg states and valence states, each of which exhibits different spectral features.[31] When a molecule is introduced into a liquid, these states acquire a less discrete character, and the continuum states are lowered in energy by an amount which corresponds to the sum of P^+, the high-frequency polarization energy of the positive ion and V_0, the work function for electron ejection into the conduction band of the liquid. The difference between

$$I_l = I_g + P^+ + V_0 \tag{1}$$

the gas phase (I_g) and liquid phase (I_l) ionization potential is such that former highly excited valence states may now be embedded in the continuum and behave as autoionizing states.[3, 31]

In such an experiment, therefore, we have the opportunity to study the mechanism of MPI, the nature and lifetimes of the intermediate states, and the competition between vibronic relaxation and excitation into the continuum, by varying the absorption steps (simultaneous or sequential) and polarization of the photons, as Fig. 1 shows.[29, 31, 32] Since electron trapping in all liquids proves to be exceedingly fast, the sudden appearance of a localized electron spectrum, will signify the onset of photoionization of the molecule in that liquid and the location of the conduction band. This quantitative information can then be used to refine models of excess electron states in liquids, since for most liquids V_0 is an empirically determined parameter.[2–4] Furthermore, by tuning the energy of the third photon through the conduction band edge we can determine the photoionization quantum yield as a function of incident photon energy, and thus measure the escape probability of electrons vs the geminate electron-hole recombination processes as a function of the excess kinetic energy acquired by the outgoing electron. These data are also extremely pertinent to the question of image potentials in the liquid, and to the probability of electron solvation in sites already structured by the presence of the parent solute.[33, 34]

MPI studies of electron relaxation processes can be used as well to probe the dynamics of the intramolecular relaxations of the guest molecule and its highly excited electronic states lying within thermal fluctuations of the continuum. For example, pyrene in alcohol can be photoionized with two photons at either 347 nm, the second harmonic of ruby, or 355 nm, the third harmonic from Nd glass.[29] Photoionization has been also seen with three photons at 530 nm, the second harmonic from a Nd glass or Nd:YAG laser.[29] Alternatively, by adding the third harmonic from a Nd:YAG to a UV dye laser pulse (obtained from frequeny mixing 1.06 μm with the tunable emission from a Rh6G dye laser, pumped with the 350 nm second harmonic, in a nonlinear crystal) we can explore the ionization edge with tunable UV pulses.[29]

In photodetachment studies we need not be concerned with the coulombic interaction that dominates the neutralization channel in MPI at times $< 10^{-10}$ s, but since the threshold for ionization of an anion is higher in the liquid than in the gas phase, we obtain once again information on the polarization energy of the ion in the liquid prior to photodetachment.[35, 36] The laser photodetachment mechanism has been used with inorganic ions, such as $Fe(CN)_6^{4-}$ [7, 37] or organic ions such as phenolate,[37, 38] supported in a wide range of aqueous or organic liquids.

The third class of mechanisms, involving either direct ionization of the liquid[6, 8] or electron ejection via field-emission,[39] has been used to study the behavior of quasifree, localized, and solvated electrons. In contrast to the *photoselectivity* of the previous two schemes, a cascade of events occurs when high-energy electrons impart energy to a liquid.[40] The resulting ions, excited states, and excess electrons provide a complex spectrum to unravel. However, the temporal evolution of each of the various species differs significantly and we are able to focus on the primary picosecond event, electron localization, with little interference.

B. Experimental Techniques

Picosecond absorption spectroscopy has been employed to follow the time-dependence of the optical absorption during the formation of e_s^-, while pulsed-laser saturation spectroscopy has been used to examine the

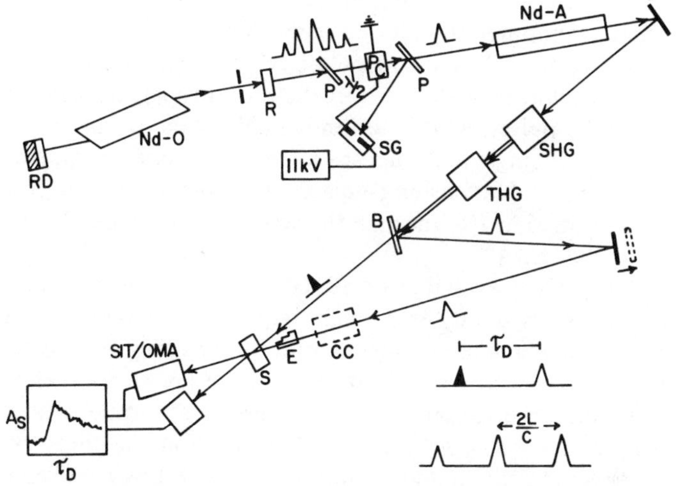

Fig. 4. Mode-locked Nd–glass oscillator amplifier system for studies of electron relaxation processes and optical Kerr effects in liquids. See text for details.

picosecond time history of the radiationless relaxations of e_s^-. In all of the techniques described below a pump pulse triggers electron ejection into the liquid and a probe pulse interrogates the absorption following a known delay time.

Figure 4 illustrates the neodymium glass oscillator–amplifier system which has been employed to study transient phenomena in systems where the time-evolution of strong absorptions and emissions can be recorded with an optical delay line and the echelon (E) technique.[41] In our system, a single picosecond pulse of typically 6 ps duration is extracted from the mode-locked pulse train via a spark gap (SG) triggered polarization gate (PP), amplified and used to generate harmonics of the fundamental 1.06 μm pulse through successive frequency-mixing in the nonlinear crystals (SHG, THG). The UV "pump" laser pulses (**Λ**) are then directed to a sample (S), wherein multiphoton ionization or photoexcitation can occur, while the "probe" pulses [Λ, at a single frequency or broadband from the picosecond continuum (CC)[42]] are directed along an optical delay line to interrogate the absorption (A_s) of the transients at a known time interval τ_D after the pump pulse. The temporal resolution of the measurement is derived from a precise knowledge of the difference in optical pathlengths travelled by the pump and probe pulses, taking group velocity dispersion into account.[41] A silicon vidicon tube (SIT) at the exit slits of a spectrometer and coupled to an optical multichannel analyzer (OMA), operating in a two-dimensional mode, records both kinetic and spectral data, which can be stored in a computer and plotted on an XY recorder.[43] In studies of the optical Kerr effect[43, 44] crossed polarizers are placed either side of the sample cell, collinear with the direction of propagation of the visible polarized probe pulses, while the fundamental 1.06 μm laser pulse enters the cell at a small angle and imposes a temporary orientational anisotropy across the sample. The ensuing polarization rotation of the probe beam and thus transmission $I(t)$ through the second, analyzing polarizer is again recorded on the OMA.[43]

While the powers available ($\leqslant 10$ GW) in these picosecond Nd laser pulses are necessary in any experiment requiring multiphoton excitation or sequential nonlinear interactions, or one investigating field-induced Kerr phenomena, the typical 10^{-2} Hz repetition rate presents a real obstacle to making reliable observations on *weak* transient absorptions or emissions. Therefore, for recording very weak e_s^- absorptions in dilute systems we must use direct ionization and excitation of a liquid by a pulsed e^--beam. The unique single-pulsed, stroboscopic LINAC facility[45] shown schematically in Fig. 5(a), delivers a 24 psec burst of reasonably monoenergetic, relativistic electrons, which pass through a cell of high-pressure xenon to generate a picosecond Cerenkov continuum for a probe pulse and then enter the sample cell (through which the liquid is flowing to avoid thermal effects) as the pump pulse. The trajectory, and hence arrival time of the e^--pulses, can be controlled through deflection about a 270° magnet, and the pulses are available at a repetition rate up to 60 Hz. Spectra and kinetics of transients are followed by recording the time-dependence absorption of the Čerenkov continuum or of a laser, pulsed synchronously with the electron beam, in conjunction with a variable optical delay line. The data acquisition system has a large dynamic range and presently permits very weak absorption or emission signals to be reliably recorded from 250–750 nm up to times as long as 3.8 ns following a single pump pulse.[6]

The third and possibly most flexible technique for the future, employs syncronously pumped, argon–ion jet stream dye lasers.[46–48] The principles of operation that have led to the new era of subpicosecond dye laser pulses have been discussed recently in detail, and we will only briefly outline them here.[46] Figure 5(b) illustrates an actively mode-locked (41 MHz) argon-ion laser, which synchronously pumps a jet stream dye laser in a folded three-mirror, extended cavity. With an average power of 900 mW and argon laser pulses of < 120 ps duration, we routinely obtain dye laser pulses of <2 ps and ca.nJ in energy, which are extremely stable over long periods of time.[49] As the autocorrelation traces show, the dye pulse shape is excellent and does not exhibit long tails.[49] This is a crucial point, since

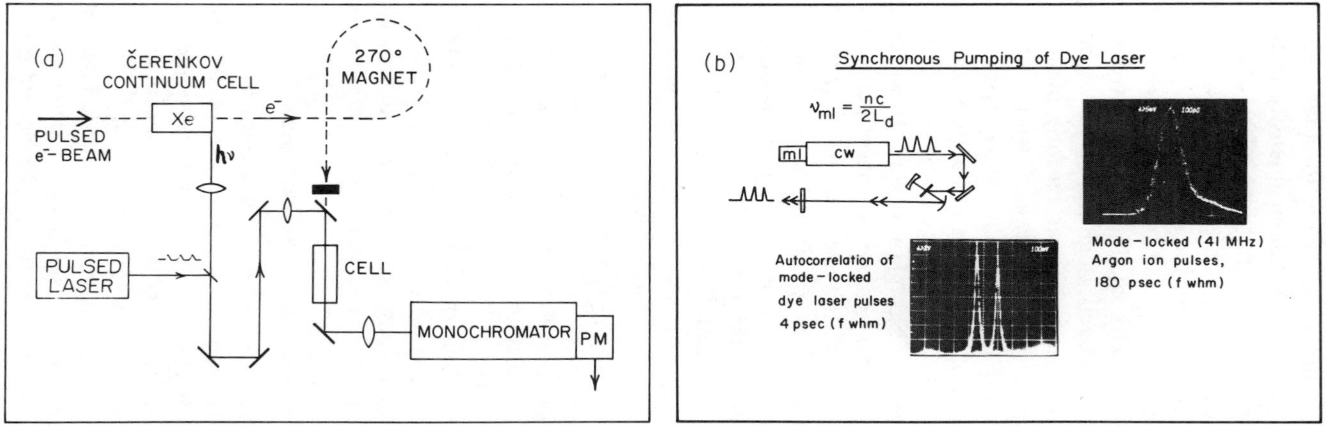

Fig. 5. High repetition rate pulsed excitation systems for picosecond absorption and emission studies. (a) Pulsed e^--beam with Ĉerenkov or laser probe pulses; (b) actively mode-locked, synchronously pumped argon ion jet stream dye laser. See text for further details.

the advantages of having ultrashort pulses to study chemical dynamics become questionable if there is substantial intensity in the wings of the pulse. The integrated intensity from the wings, often extending into many picoseconds, becomes a significant component of the excitation profile and masks the true temporal evolution of the signal. Thus, particular care must be shown in establishing the response function of the system, as subpicosecond data are the eventual goal. In order to investigate phenomena based on nonlinear optical effects, high-power pulses are required. We have amplified these dye laser pulses, as recently demonstrated for the *passive* mode-locked subpicosecond dye laser,[46] with a series of dye amplifiers transversely pumped (at 10 Hz) by the frequency-doubled output of a Nd-YAG laser. Full details of this design will appear elsewhere[49] but with an anticipated gain of 10^6 the laser pulses will be of GW powers and thus comparable to those from the Nd–glass system. However, the important additional advantage of achieving these ultrashort pulses not only at a 10 Hz repetition rate but also with a wavelength tunability that extends across the visible, achieved by direct pumping or resonant energy-transfer schemes,[50] is one that presently eludes the subpicosecond passive mode-locked dye lasers. Extending the repetition rate to over 100 Hz can be accomplished by using a pulsed XeCl laser (308 nm) as the amplifying system.[49]

Finally, electrons have been generated with nanosecond pulses from the KrF exciplex laser at 248 nm via photodetachment of OH^- and Cl^- in aqueous and methanol solutions,[51, 52] and could be via one-photon photoionization in liquids with the ArF laser at 193 nm, and via one and two-photon processes in aqueous naphthol and naphtholate with nanosecond pulsed nitrogen lasers at 337 nm.[53] With the improved fluxes and short pulses now available from gain modulation techniques applied to the TEA exciplex lasers, semiconductor lasers[54] and *F*-center lasers,[26] they may prove to be convenient sources for future studies of electron relaxation and transfer processes from the ultraviolet to infrared region.

C. Laser Saturation Spectroscopy

Saturation of an optical transition ("hole-burning") and subsequent analysis of the time and frequency-dependence of the recovery of the ground-state species has become a well-known technique for the study of the picosecond photophysics of radiationless transitions in stable molecules, transient species, and laser dyes in particular.[55–60]

Our present experimental arrangement for kinetic laser saturation spectroscopy with a *Q*-switched ruby laser has been described in detail previously[55] where it has been shown that the amplitude and temporal profile of the probe laser pulses contain the *picosecond time history* of the *n*-level system.[59] In these experiments, the e_s^- are generated via the two-photon, nanosecond laser-induced photoionization of pyrene in alcohol at 347 nm[29] and then subjected to a laser saturating pulse of up to 700 mJ per pulse at 694 nm (peak powers $\leqslant 35$ MW, or a laser fluence of J cm^{-2}) some 50 ns later. At a number density of typically 10^{15} cm^{-3}, the kinetic lifetime of e_s^- is ca. 10^{-6} s, and the photon/e_s^- ratios vary up to 10^3.

The depletion and recovery of the ground-state e_s^- absorption spectrum are monitored through time-dependent absorption or transmission measurements of the photobleached system.

Mode-locked Nd–glass[58] or ruby lasers have also been used to investigate hole-burning and intramolecular dynamics in molecules, as have intracavity dye laser techniques,[43] which operate on the principle of transferring the loss of intensity at the frequency of the "hole" in the spectrum into the enhanced gain of a dye laser, whose broadband output overlaps that frequency.

III. ELECTRON RELAXATION IN PURE LIQUIDS

A. Electrons in Alcohols

When electrons are injected into pure alcohols, $CH_3(CH_2)_nOH$ at 298 K, an instantaneous structureless infrared spectrum is observed whose maximum has not yet been determined but is probably >1500 nm[8, 61] (see A′ in Fig. 2). This absorption rapidly transforms into the characteristic visible spectrum of e_s^- over a picosecond period of time, which systematically increases the longer the chain length, n.[6, 8] When the picosecond kinetics of this visible absorption are monitored at wavelengths on the high-energy side of $\lambda_{max} \sim 650$ nm, the absorption signals are characterized by two components. The fast component represents a significant fraction of the total absorption signal and appears within the resolution time of the apparatus; if there is any induction period, current limits place this time at $\leqslant 2$ ps.[61] The second, slower component grows in exponentially after the pulse and it is the analysis of this part of the signal which we will now discuss. It is evident that subpicosecond resolution is required to determine whether or not any time-dependent signal can be observed at earlier times, across both the visible and near IR bands. It also appears, from two independent studies at 1.06 μm[61] and 1300 nm,[8] that the infrared spectrum decays quickly but at the same rate as the visible spectrum grows in. Furthermore, although the amplitude of the visible absorption varies with wavelength at a given time, if the absorption signals on the high energy side of λ_{max} are normalized, the dynamics are exactly the same.[6, 62] Therefore, it is convenient to refer to absorption data taken between 400 and 600 nm as representative of the full spectral changes and avoid the

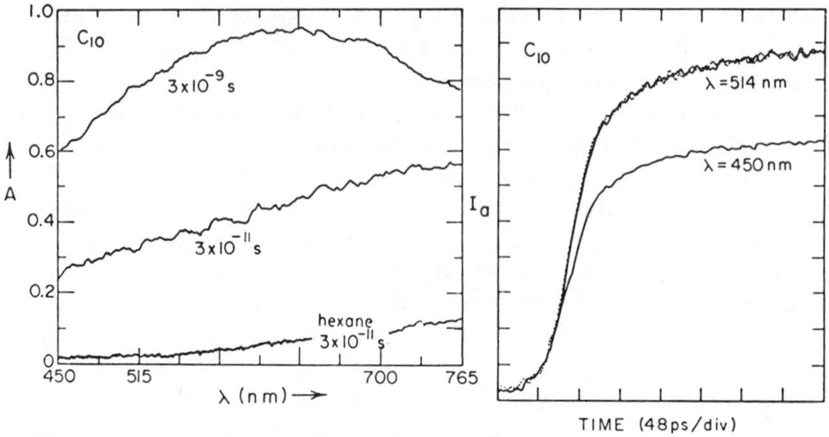

Fig. 6. Dynamics of electron localization in *n*-decanol (C_{10}) at 300 K. (a) Absorbance of e_s^- at 30 ps and 3 ns compared to tail of IR absorption in pure *n*-hexane. (b) The fast and slow component of the absorption signal shown here for $\lambda = 450$ nm, 514 nm in *n*-decanol. The dotted line is the 450 nm curve normalized to the 514 nm data.

spectral region where the observed signal is a convolution of both visible and IR absorptions which are simultaneously present. Figure 6 illustrates these features in the actual picosecond spectra and kinetics during electron trapping in 1-decanol.

We will now focus on the time-dependence of the slow component and its physical significance. Analysis of this part of the signal in all the alcohols reveals an exponential growth[6] whose reciprocal rate constant we have called τ_s and plotted in Table I, in conjunction with other properties of the liquids. The τ_s data are quoted to $\pm 10\%$. In alcohols where C_n, the number of carbon atoms, is greater than 6, the first-order signal tails off into an even slower component but this may in part be because e_s^- formed at early times are now undergoing chemical reaction at times $> 10^{-10}$ s.

When e_s^- dynamics are studied in ethoxyethanol, the λ_{max} having now shifted to ca. 800 nm,[15] the solvation time is significantly longer than that observed for ethanol. Similarly, the dynamics in 2-butanol and *t*-butanol monitored at 600 nm are given by $\tau_s = 38$ and 54 ps, where the λ_{max} are 800 and 1100 nm, respectively. In fact, the dynamics of e_s^- formation in pure 2-decanol appear identical to those in *t*-butanol, a fact we will investigate in more detail later in the paper. However, in 2-propanol, a mere alkyl group away from *t*-butanol, τ_s in 24 ps. What conclusions may we draw from Table I and these other apparently disparate examples[63] in branched alcohols? Two important correlations emerge. First, that the solvation

TABLE I
Comparison of Electron Solvation Times and Molecular Properties in Polar Liquids at 300K

ROH	C_1	C_2	C_3	C_4	C_5	C_8	C_{10}	H_2O	NH_3	$(CH_2OH)_2$
τ_s (ps)[a]	10	18	25	30	34	45	51	2–4,[b] <3[c]	<5[d]	<5[a]
τ_{rot} (ps)[e]	12	20	22	27	28	39	48			
η (cp)	0.55	1.10	2.00	2.60	3.35	8.95	14.1	0.89	0.6	19.9
$\bar{\rho}_d$ (10^{21} cm^{-3})	14.8	10.2	8.0	6.5	5.5	3.8	3.1	33.3	25.36	21.50

[a]ROH data, Refs. 6 and 63.
[b]Ref. 7.
[c]Ref. 8.
[d]Refs. 58 and 71.
[e]Ref. 64.

times bear a striking similarity to the rotational diffusion times of monomeric alcohols, τ_{rot},[6] as measured through dielectric dispersion techniques.[64] Second, that when all these τ_s data are plotted in terms of a hydrodynamic formalism, based on the Stokes–Einstein–Debye equations,[65, 66] the data fall on a straight line, as Fig. 7 shows. The rotational diffusion

$$\tau_s^{-1} = \alpha_{or}\eta^{-1}\frac{kT}{V} \tag{1}$$

times of these liquids also yield the same linear relationship,[65] as do the

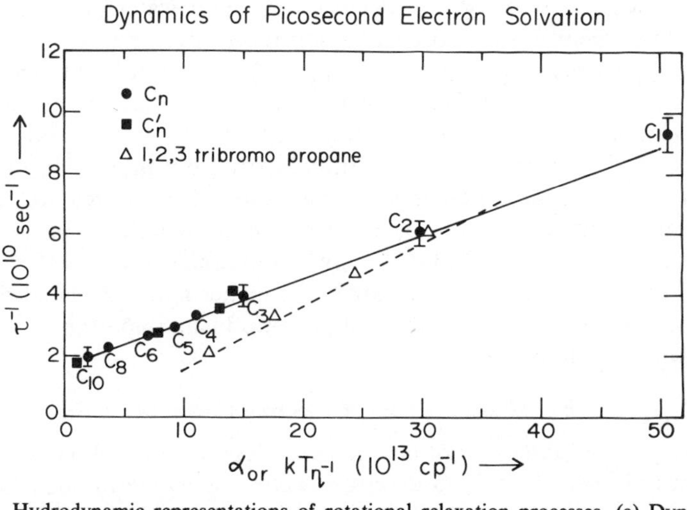

Fig. 7. Hydrodynamic representations of rotational relaxation processes. (a) Dynamics of electron solvation in (●) *n*-alcohols, (■) isomeric alcohols, (△) relaxation data from alkane.

dielectric relaxation data of 1, 2, 3-tribromopropane,[67] a liquid that does not support excess electron states. The clear implications from these two correlations is that in monitoring the dynamics of electron solvation we are observing the hydrodynamic response of the fluid to the sudden perturbation presented by the excess electron.[65] In other words, the observable picosecond dynamics of electron solvation are the intrinsic orientational dynamics of the liquid in which the molecular trap is being formed, and the spectral shifts correspond to a deepening of the potential well.[63]

In those cool alcohols studied, methanol to butanol, the final e_s^- absorption spectra are marginally blue shifted but the solvation times become longer and extend into nanoseconds.[8, 68, 69] In order to reexamine these data from this "hydrodynamic" perspective we calculated the Debye parameters, $\tau_{rot}T/\eta$ which should remain constant for a given liquid.[70] The results were fairly encouraging. In ethanol, for example, substitution of τ_s for τ_{rot}, where τ_s varied tenfold over an 85°C temperature range,[68] gave values from 5.69×10^3 at 298 K to 5.4×10^3 at 213 K, with a slight minimum at 253 K, 4.80×10^3.

Before analyzing the implications of these correlations, what other relaxation studies have been reported? Picosecond studies of electron solvation in ethylene glycol ($\eta=20$ cp) reveal that the visible absorption spectrum is fully developed within the time resolution of the apparatus, establishing $\tau_s \leqslant 5$ ps.[63] In liquid amines too, electron trapping is rapid and to date only limits can be placed on the solvation time, which for e_s^- in ammonia is $\leqslant 5$ ps at 223 K[71] and <5 ps in methylamine from 193 to 233 K.[58] Picosecond studies of electrons in liquid water also report extremely fast electron localization and solvation, placing this process less than 6 ps[56] between 2 and 4 ps[7] and <3 ps.[8] Reports of longer times[30, 37, 38] in aqueous systems may reflect strong electron-hole coulomb interactions, or temperature effects in these liquids, or the presence of absorbing transient species other than the electron. Apart from promoting ultrafast electron solvation, these liquids share one other common feature, a very high number density of dipoles, $\bar{\rho}_d$. For ethylene glycol (EG) $\bar{\rho}_d=2.15\times10^{22}$ cm^{-3} at 293 K, $\bar{\rho}(H_2O)$ is 3.33×10^{22} cm^{-3} and $\bar{\rho}(NH_3)$ is 2.5×10^{22} cm^{-3}. These $\bar{\rho}_d$ are at least an order of magnitude higher than calculated for the pure alcohols (see Table I), which will prove to be an important clue as to the role of local liquid structure.

B. Cluster Model of Electron Solvation

From the data and correlations in pure alcohols, the following model has been proposed[6, 62, 63, 65] to describe electron solvation in polar liquids. Further evidence to support and refine these ideas will be discussed in

Section IV. Trapping of electrons appears to occur in two stages, *localization* and then *solvation*. First, the quasifree electron initially scatters through the unperturbed liquid at subpicosecond times, inducing electronic polarization and sampling the statistical structure of the fluid, which is frozen-in on this time scale. Strong electron–medium scattering from phonons, density or configurational fluctuations may promote electron localization, and in these polar liquids the strongest scattering sites are identified as small molecular clusters arising from intermolecular hydrogen bonding. These local configurational fluctuations give rise to minima in the potential energy surface experienced by the quasifree electron but thermal fluctuations may promote the electron back into the continuum. The infrared absorption band observed at the earliest times in all liquids is assigned to this initially quasilocalized or shallowly localized electron state. Then, if the residence time in this state is greater than the vibrational and rotational relaxation time of the molecule, the potential well formed by the OH dipoles can deepen both through configurational relaxation of the initial trapping site, reminiscent of multiphonon processes in a lattice, and rotational alignment of adjacent molecules, which build up the size of the cluster.

In other words, the observed picosecond spectral shifts correspond to changes in the binding energy, hence the Franck–Condon factors involved in the electronic transitions, brought about by rotational diffusion of nearby molecules in the screened field around the newly formed cluster ion as it becomes solvated.[65] Whether or not this process can be described better theoretically as relaxation at constant charge or at constant field is still a matter for debate[40, 63] but it is clear that the observed time-dependence must be the net effect of several forces and counteracting torques. The latter can be envisaged as the internal friction which opposes the field-induced orientational polarization, and the former as the intermolecular forces that characterize the local liquid structure *prior to* electron localization. The short-range interactions must also minimize the internal H atom and dipole–dipole repulsions of the molecules defining the inner cavity of this cluster, which is now the final, configurationally relaxed, ground electron state of e_s^-.

We have pinpointed the essential processes contributing to the slow component in the time-dependent absorption signal as orientational relaxation of the liquid during electron solvation. But what other evidence is there to describe the fast, unresolved but significant IR absorption that is ascribed to *localization*? To what extent does the local liquid structure and density continue to play a quantitative role at earlier times? Is the electron self-trapped, as in a polaron model, during localization? It is to these questions that the experiments with very low number densities (ρ_d) of alcohol in alkanes were addressed.[62, 63] Mindful of the emerging role of the liquid, we first make a brief digression to analyze the exact implications of Fig. 7 in terms of current theories of liquid-state dynamics.

C. Interlude on Molecular Motion in Liquids

Hydrodynamic relationships, built on a *continuum* approach to the fluid, hide a multitude of microscopic correlations, as has been well documented in other areas.[70, 72–76] In any dynamical problem there must be both a microscopic or molecular contribution and a macroscopic, or many particle correlation, and hydrodynamics can only project out the latter. For all variables of the system, which fluctuate slowly in large volume elements and are conserved quantities, hydrodynamic treatments are relatively successful. Individual molecules, however, present a more chaotic picture. Molecules collide at times $\leqslant 10^{-12}$ s at liquid densities, move at relatively high velocities and thus are subject to rapid fluctuations in their local intermolecular interactions and angular correlations. Recently a number of important theoretical developments have occurred which mark the emergence of a microscopic description of the dynamical structure of simple fluids. By focusing on the short-range correlations and local liquid structure, significant deviations from hydrodynamic behavior have been predicted for rotational diffusion.[70, 72] To some extent this can be accounted for by adjusting the boundary conditions describing the coupling of the rotational and translational motions of the tagged molecule to those of its neighbors. When the tangential velocity of this molecule is zero, that is, a perfect coherence exists between the molecule and its nearest neighbors, the "stick" or hydrodynamic boundary conditions apply, and we arrive at the description of orientational relaxation based on the Stokes–Einstein–Debye relationships.[66] In the other limit described by "slip" boundary conditions, it is the details of the local mode–mode coupling which influences the magnitude of the torque experienced by the molecule.

Picosecond laser spectroscopy offers direct access to molecular vibrational (T_1) and phase (T_2) relaxation as well as orientational dynamics of molecules, τ_{rot}. In contrast to experiments on vibrational relaxation in liquids, all those on the reorientational process have been confined to large polyatomic molecules, particularly dye molecules in probe solvents[74, 76] for pragmatic reasons. Since the slip boundary conditions are also a sensitive function of the shape of the molecules and solute–solvent interactions, there is some uncertainty in deciding whether or not it is the local interaction in terms of the solvation volume, or the boundary condition, or both, that varies for a given molecule in a range of liquids such as the

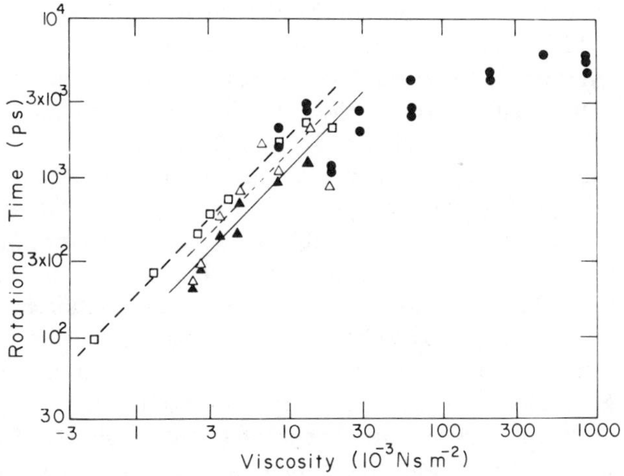

Fig. 8. Picosecond rotational relaxation times in alcohols from fluorescence depolarization studies[76] of (●) rhodamine 6G, (▲) eosin-*y*, (△) fluorescein in *n*-alcohols; (□) rhodamine 6G data.[77]

n-alcohols.[74, 76, 77] Let us now consider the graphs in Figs. 7 and 8 in this context.

The linearity of the plot in Fig. 7 suggests that e_s^- acquires a similar hydrodynamic volume (V) in each of these alcohols, but when an effective radius is calculated from V (assuming stick boundary conditions) from the slope, the result is clearly too small, $r = 1.1$ Å.[43] In fact, this is a typical result for rotating molecules where extensive slip is actually occurring.[72, 74] Thus a linear relationship between τ_{rot} and η is *not* sufficient evidence to ignore the molecular aspects of the liquid dynamics.

This point is further exemplified in Fig. 8 where the relaxation times of several xanthene dye molecules rotation in *n*-alcohols are plotted in the same hydrodynamic formalism.[76, 77] These data are from transient-induced dichorism and time-resolved fluorescence depolarization experiments. In the latter case, the anisotropy factor $r(t)$, which describes the intensities of fluorescence $I(t)$ parallel (||) and perpendicular ($\perp$) to the polarized exciting pulse, is related to the second Legendre polynomial, P_2,

$$r(t) = \left[I_{\|}(t) - I_{\perp}(t) \right] \left[I_{\|} + 2I_{\perp}(t) \right]^{-1} = \tfrac{2}{5} P_2 [e(\mathbf{0}) \cdot e(\mathbf{t})] \tag{3}$$

of the dipolar reorientation angle θ, made by unit vector $e(\mathbf{t})$ with z axis. For a rotating sphere and stick boundary conditions this reduced to the familiar rotational diffusion constant $D = kT/6V\eta$.[66] For molecules experiencing hindered rotation or rotating with a geometry that is other than spherically symmetric, then the final expression must include damping terms and a function describing the ratio of the axes about which the molecule tumbles.[74] From studies of $r(t)$ for fluorescent dye molecules in various liquids, information can be gleaned on the boundary conditions, the effective shape of the rotating species and hence on solvent attachment to the molecule in terms of its polarizability, charge and so forth. As Fig. 8 shows, there is at least a similar proportionality between τ_{rot} and η for three different dyes in the same liquid.[76, 77] However, the relationship fails in higher viscosity fluids, $\eta > 10$ cp (10^{-2} Nsm^{-2}), which in these experiments were varied mixtures of ethylene glycol and glycerol.[76] Stated simply, the rhodamine 6G molecule is rotating much faster than predicted by hydrodynamic considerations. The same conclusion applies to studies of aromatic molecules embedded in micellar environments.[76] A study of the temperature dependence for rotational relaxation of rhodamine 6G in glycerol yielded an activation energy for overcoming the rotational barrier of ca. 7 kJ mole^{-1}, while the data for the relaxation time of pure glycerol gave a value almost ten times larger.[76] These results not only point toward the dominance of liquid structure and short range correlations but also could explain the rapid electron solvation observed in diols as well. Both the electron and the rhodamine 6G appear to be located within a chelating site, that is a solvent cage arising from the multiple-hydrogen bonding interactions in these high ρ_d liquids. Little radial reorganization is required to adjust to the guest species, and changes to the macroscopic properties of the fluid (such as increases in η) scarcely influence the local solute–solvent internal frictional forces, hence the torque experienced by the rotating species remains fairly constant.

Finally, we should enquire as to whether or not it is reasonable to expect a *linear* response in describing the rotational diffusion of dipoles in the presence of a very strong local field, such as presented by the excess electron. The time window of the optical Kerr gate driven by a picosecond laser pulse depends on the relaxation of the molecules of the Kerr medium from an aligned orientation to an isotropic spatial distribution, once the applied optical field is switched off. For many liquids this relaxation time τ is the low field limit, namely, the Debye time. We might anticipate an asymmetry in the temporal response $\delta_n(t)$[44] of

$$\delta_n(t) = \frac{n_2}{\tau}\int_{-\infty}^{t} \bar{E}^2(t')\left[\exp\frac{-(t-t')}{\tau}\right]^{dt'}$$

the medium to the leading edge of the laser pulse (hence applied field, E)

in comparison to the switchoff time, since there is a finite time associated with the development of the full orientational polarizability associated with the nonlinear refractive index, n_2. This time will depend on the detailed interactions of the molecules with their neighbors and with the field. We have begun an experimental investigation on the time-dependence of the rise and decay transients of the electronic and nuclear components of the Kerr effect in liquids[43] in order to pursue the possibility of a nonlinear response or accelerated orientation of molecules rotating under the influence of a strong local laser field.

We conclude this section by stating that even though "hydrodynamic" behavior during electron *solvation* in pure liquids is observed, it does indeed mask the microscopic details of the solvation mechanism, although to no greater extent than is now realized in picosecond studies of the orientational dynamics of more classical molecules.

IV. ELECTRON LOCALIZATION IN DILUTE POLAR FLUIDS

Inspection of Table I reveals two further trends, implicit in the apparent linearity of the hydrodynamic plot in Fig. 5. The solvation time τ_s increases with increasing viscosity (η) and decreasing average number density ($\bar{\rho}_d$) of OH dipoles cm^{-3}. By studying the effects of lowering both the viscosity and density, the role of rotational and translational diffusion and short-range solvent structure should become more apparent.

A. Experimental Results

The following picosecond absorption data have been selected from an extensive recent study[62, 63] of electron localization in a series of alcohols diluted in a range of alkane of varying molecular structure and electron-scattering cross-sections or mobilities.[14] Before examining the data and implications in Figs. 9 and 10, we may summarize the results as follows. (1) For all pure alkanes and dilute alcohols in alkanes an *instantaneous*, weak, infrared absorption is still observed following electron injection into the system. (2) At a very low number density $\bar{\rho}_d$ of alcohol, this signal rapidly *disappears* within ca. 200 ps, while at a high value of $\bar{\rho}_d$ the absorption signal *grows in* over tens of picoseconds after the pump pulse and signals the onset of the familiar spectral shift for the IR to the visible, e_s^- absorption. (3) Furthermore, by knowing not only $\bar{\rho}_d$ but also the relative magnitude of the solute–solute and solute–solvent interactions prior to electron ejection, we can also predict the critical values of $\bar{\rho}_d$ that describe the transition from one type of kinetic behavior to the other. We will now discuss the quantitative details of these extremes of behavior and the

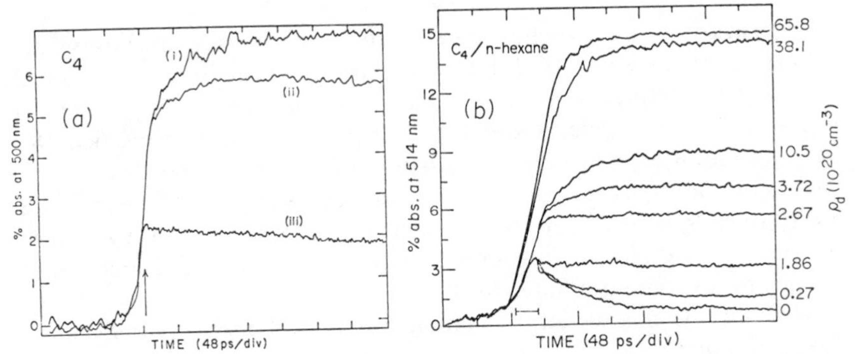

Fig. 9. Electron solvation in *n*-butanol as a function of alkane matrix. (a)$\bar{\rho}_d \sim 5 \times 10^{20}$ cm^{-3} in (i) *c*-hexane, (ii) *n*-hexane, (iii) isooctane. (b) Transient absorptions as a function of $\bar{\rho}_d$ of butanol in *n*-hexane.

kinetics and spectroscopy of the intermediate $\bar{\rho}_d$ region. It is apparent, however, that electron *localization* in the alkane matrix has occurred within a few picoseconds.

As Fig. 9(a) illustrates for *n*-butanol, the amplitude and time dependence of the absorption signal depend on the identity of the chosen alkane matrix. In trace (i), a mole fraction of alcohol $\chi \sim 0.1$ in cyclohexane, a signal whose profile is reminiscent of solvation signals in pure *n*-butanol is observed, but the solvation time is considerably longer, ~150 ps. The

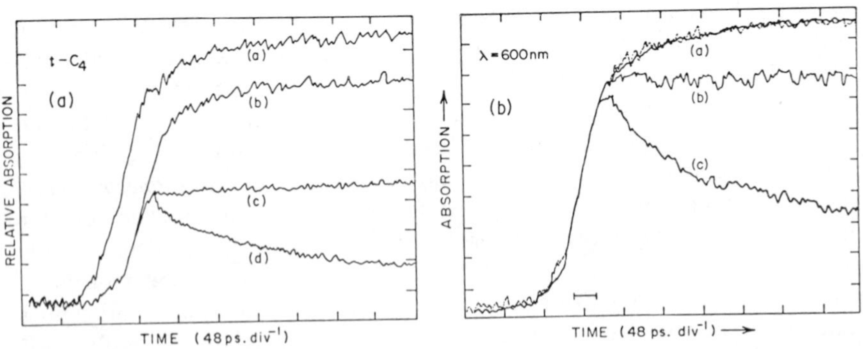

Fig. 10. Electron solvation in *t*-butanol and 2-decanol. (a) Absorption in pure *t*-butanol at 740 nm (trace *a*, displaced in time for clarity) and 600 nm (trace *b*); when normalized, *a* and *b* are superimposable. Upon dilution in *n*-hexane, trace *c* is recorded for $\bar{\rho}_d = 1.3 \times 10^{21}$ cm^{-3} and *d* for $\bar{\rho}_d = 1.5 \times 10^{20}$ cm^{-3}. (b) *a* normalized absorptions in pure 2-decanol and *t*-butanol (dotted) at 600 nm; *b* 2-decanol in *n*-hexane, $\bar{\rho}_d = 3.5 \times 10^{20}$ cm^{-3}; *c* 2-decanol in *n*-hexane, $\bar{\rho}_d = 3 \times 10^{19}$ cm^{-3}.

dynamics in *n*-hexane, trace (ii), appear somewhat faster but the curvature of the signal is less, while in isooctane, trace (iii), the absorption is notably weaker and decays after the pulse. The actual values of $\bar{\rho}_d$ do decrease slightly from 6×10^{20} to 4×10^{20} molecule cm^{-3} for cyclohexane to *n*-hexane and isooctane, respectively, but, since these values all fall into the high $\bar{\rho}_d$ regime (see below), the principal conclusion is that *different stages of solvation have been achieved in each case.*

Upon investigating electron localization in the *n*-butanol : *n*-hexane system over a wide range of $\bar{\rho}_d$, the data in Fig. 9(b) were obtained. In pure *n*-hexane ($\bar{\rho}_d = 0$) there is a fast, broad IR absorption signal, the tail of which extends into the visible and is shown decaying rapidly at 514 nm within about 150 ps. This we assign to a quasilocalized electron in *n*-hexane (e_t^-) whose ultimate fate is to undergo rapid ion recombination or capture by an impurity molecule or ion. Previously, e_t^- had been known to survive at 293 K over nanoseconds as a result of its relatively low electron mobility, $\mu_e = 0.09$ cm^2 V^{-1} s^{-1},[14] when the neutralization channel was suppressed. When a small number density of *n*-butanol is introduced, neither the kinetics nor the spectrum appear affected up to $\sim 10^{19}$ cm^{-3}. This appears to be a critical threshold, since by $\bar{\rho}_d = 2.7 \times 10^{19}$ cm^{-3} the decay is clearly slower, as Fig. 9(b) shows, and by $\bar{\rho}_d = 1.86 \times 10^{20}$ cm^{-3} the signal has flattened out. The amplitude of the signal and the IR spectrum, however, have not changed. Thereafter, further increases in $\bar{\rho}_d$ lead to an enhanced absorption signal and the spectrum begins to shift to the visible, after the pulse. At a density of $\bar{\rho}_d = 10^{21}$ cm^{-3} the dynamics of electron solvation clearly mimic those in pure fluids and the absorption signals, when normalized, are superimposable with those recorded for $\bar{\rho}_d = 3.8 \times 10^{21}$ cm^{-3} where $\tau_s = 51$ ps in comparison to 30 ps in pure *n*-butanol.[62] Figure 9(b) therefore presents a full profile of electron solvation in a dilute fluid, and its physical interpretation lies in the changing local liquid structure over this density regime.[62, 65] Before pursuing this discussion, we briefly indicate similar profiles in other dilute alcohols in alkanes, from which two further important conclusions are drawn. Figure 10 shows the kinetics of electron trapping and solvation in the *t*-butanol : *n*-hexane and 2-decanol : *n*-hexane systems. In the pure liquids the electron solvation times appear identical, $\tau_s = 54$ ps.[63] This is shown in trace (a) of Figure 10(b), where the normalized (dotted) absorption signal in *t*-butanol at 600 nm [originally trace (b) of Fig. 10(a)] has been superimposed on that from 2-decanol at 600 nm. The molecules differ drastically in shape, one being a long chain and the other a globular molecule, but the electron-trapping site about the OH bond differs only by a methyl group. However, upon dilution in a common solvent, *n*-hexane, each system displays quite different values of $\bar{\rho}_d$ at which transitions through the different stages of solvation are observed. In 2-decanol, the critical value of $\bar{\rho}_d$ is close to *n*-butanol namely 3×10^{19} cm^{-3} but in *t*-butanol $\bar{\rho}_d$ is very high, 1.5×10^{20} cm^{-3}. Similarly, the crossover to growth kinetics after the pulse does not occur in *t*-butanol until $\bar{\rho}_d > 3\times10^{21}$ cm^{-3} in comparison to 8×10^{20} cm^{-3} for 2-decanol.

The conclusions to be drawn from all of the above data direct our attention to the *short-range dynamical liquid structure*. First, even though electron solvation times are the same in two pure liquids, the microscopic details of the electron trapping and solvation steps may be distinctly different. Second, the critical density $\bar{\rho}_d$, above which the quasilocalized electrons decay in alkanes is affected by the alcohol molecules, is a function of both the alcohol and the alkane through their intermolecular interactions. Third, that a number density of alcohol molecules substantially below that of the pure alcohol, the system simulates the picosecond dynamics and spectroscopic changes associated with electron solvation in the pure liquid. Fourth, since all alcohols display over a range of $\bar{\rho}_d$ the three patterns of absorption *decay*, *invariant* absorption, and absorption *growth*, we therefore postulate that through electron localization we are sampling at least three density regimes which support different local liquid structures in these fluids. In order to search for and possibly identify these structures, independent investigations were made on the dynamics and viscosities of these same alcohol alkane systems, but in the absence of electrons.

B. Dynamical Structures in Low-Density Liquids

Pulsed NMR techniques provide a method of measuring rotational diffusion, segmental motion, internal motion, and torsional vibrations, all of which provide a temporal modulation at different frequencies of the magnetic fields associated with the molecule under study. We used ^{13}C Fourier transform NMR to measure the spin–lattice relaxation times (T_1, s) of each of the carbons in the alcohols and alcohol : alkane systems.[78] In the limit of the extreme narrowing approximations[79] and assuming the intramolecular dipole–dipole mechanisms dominate relaxation for hydrogen bearing carbon atoms, only rotational motions are considered and the reorientational correlation time (τ_c) of the molecule can be calculated from T_1. Any constraints on the motion of a segment of the alcohol molecule leading to a longer reorientational time will appear as a smaller T_1, for molecules whose correlation times are shorter than 10^{-8} s. In the case of the alcohols under study, τ_c vary but are $\leqslant 10^{-10}$ s. Thus lowering the viscosity or increasing the temperature should increase the T_1 values, whereas any form of multimer formation should have the opposite effect.

The formation of clusters in the alcohols is well documented, although the size and distribution of these clusters continue to receive much experimental and theoretical attention.[80]

Figure 11 shows the T_1 spin lattice relaxation times for *C(1), the atom attached to the —OH bond of 1-butanol, and *t*-butanol in deuterated cyclohexane. The inset shows T_1 times of *C(1) in the very low-density region for ^{13}C enriched 1-butanol. Note the dramatic increase in T_1 values at $\bar{\rho}_d < 10^{21}$ cm^{-3}. The steep dependence of T_1 on $\bar{\rho}_d$ slackens off about 5×10^{19} and reaches a value of 17.01 s at 6×10^{18} or $\chi_{ROH} = 10^{-3}$ in cyclohexane. In *n*-hexane the same T_1 relaxation time is already 16.23 at $\chi_{ROH} = 4\times10^{-3}$. The full details of this NMR and viscosity study have been given elsewhere[78b] but the pertinent conclusions are reiterated as follows. For all alcohols we have studied to date two extremes in behavior can be observed for the high- and low-density regimes. There is a substan-

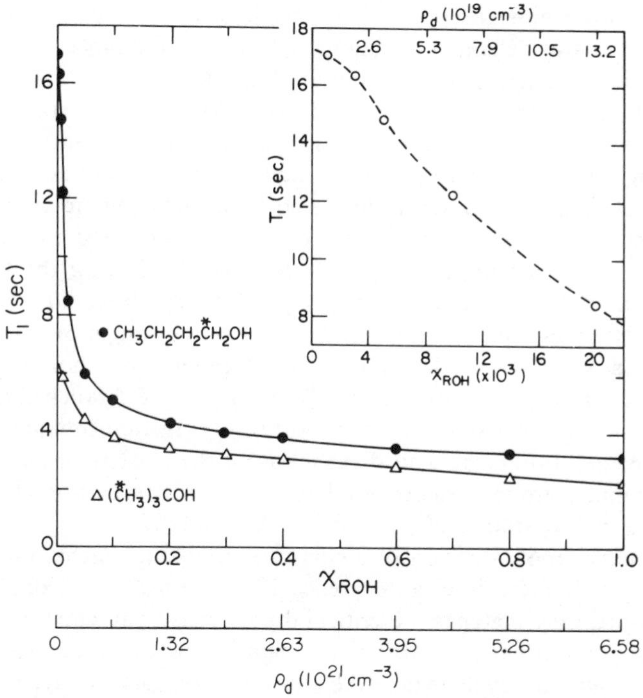

Fig. 11. Relaxation times and local liquid structure in dilute alcohols. Spin–lattice relaxation times (T_1) for (●) *n*-butanol and (△) *t*-butanol in *d*-cyclohexane, as function of χ_{ROH} and $\bar{\rho}_d$. Inset: (●) T_1 data for *n*-butanol in region of very low $\bar{\rho}_d$. See text for discussion.

tial drop in viscosity upon fivefold dilution in alkane to $\bar{\rho}_d \geqslant 10^{21}$ cm^{-3}, yet relatively small effects are observed on the T_1 times of the *C(1) atom.[78] On the other hand, notable increases in T_1 for *C(1) are seen as $\bar{\rho}_d$ drops below 5×10^{20} cm^{-3}, as shown in Fig. 11.

Comparison of the T_1 times of the C atom in the terminal methyl group with *C(1) for a *long* chain, 1-decanol, and a *short* chain, 1-butanol, shows an interesting effect in the pure liquids. The calculated effective correlation time for *C(1) in 1-decanol is ~37 ps, almost 10 times longer than the 4 ps calculated for C(10), the terminal methyl carbon. The small and invarient T_1 values for C(1) to C(6) are clear evidence too for a rigid backbone in this long chain molecule.[78] In 1-butanol, however, the effective correlation time for *C(1) is much faster, 6.5 ps and only a factor of two slower than the terminal C(4) measured at 3.5 ps. This implies that hydrogen bonding in 1-decanol is stronger and has a greater influence on the chain dynamics than does molecular association in 1-butanol. But by $\bar{\rho}_d = 3 \times 10^{20}$ cm^{-3} of 1-decanol in cyclohexane, the correlation time of *C(1) is already approaching 7 ps. From these and other data we infer that the viscosity of the liquid has two major components: strong hydrogen-bonding interactions and extensive alkyl chain interactions, which become more dominant as the alkyl chain lengthens. The initial drop in viscosity arises from a significant reduction in the intermolecular alkyl chain interactions, but since there is sufficient *internal* chain motion to contribute to the T_1 mechanisms even in entangled pure liquids, the T_1 values only show modest increases. However, the low alcohol density regime reveals relatively large T_1 changes in *C(1), which are clearly the consequence of removing the constraints on the *C(1) imposed by having the molecule anchored in the hydrogen bond. At $\bar{\rho}_d < 2 \times 10^{19}$ cm^{-3} it would appear that the alcohols are essentially monomeric. The exact $\bar{\rho}_d$ range over which the steep rise in T_1 is observed is similar for different 1-alcohols in a given alkane, but varies upon changing the alkane or going to branched alcohols. This is to be expected, since it is the local solute–solvent interactions that govern the equilibrium size and distribution of the clusters in these binary systems. Similar trends are reported for dielectric relaxation, ultrasonic relaxation, and IR studies of alcohol association.[80, 81]

In summary, upon addition of alcohol to an alkane matrix, the density region in which there is clear evidence for the onset of alcohol cluster formation precisely matches the critical dipole density at which changes in the pattern of the electron absorption signals are observed. As Fig. 11 shows, in *n*-butanol–cyclohexane the clustering appears to have reached a dynamic equilibrium about $\bar{\rho} \sim 10^{21}$ cm^{-3}, above which only minor changes in the effective correlation time of *C(1) atoms are observed. It should be recalled that it is this density at which the full picosecond spectral and kinetic changes in the formation of e_s^- mimic those of the pure 1-butanol. We will now return to the optical data and interpret the kinetics in the context of these results.

C. Mechanisms for Electron Solvation in Dilute Fluids

We proposed in discussing electron relaxation processes in pure fluids that liquid inhomogeneities such as density and configurational fluctuations, which provide an instantaneous structure on the subpicosecond time scale, played a major role in promoting rapid localization of the quasifree electron. The quantitative details of this primary step in the solvation sequence now begin to emerge from the data in very dilute alcohols. In pure alkane, the electron becomes quasilocalized, infrared absorbing, is still mobile and ultimately undergoes ion recombination. Not until a critical dipole density of alcohol molecules has been attained is there any discernible competition for the neutralization channel. This $\bar{\rho}_d$ threshold does not reflect a uniform density in the liquid, for it is only a parameter of the system, but it does acquire physical sense when the NMR data are analyzed. The critical density corresponds to the appearance of small alcohol clusters, dimers, and trimers in the alkane. These have higher cross-sections for electron capture than monomers, whose electron affinity is unfavorable.[36] Thus the mobile e_t^- can sample a certain volume of the fluid prior to recapture by a positive ion, and as the number of small clusters increases within this zone, capture by the relatively immobile alcohol clusters begins to modify the decay kinetics. Eventually neutralization is effectively blocked and the absorption following the pulse appears constant, over 500 ps at least. Further increases in $\bar{\rho}_d$ correspond to increases in both the size and distribution of clusters until these local liquid structures resemble the short range structure in a pure fluid. At this $\bar{\rho}_d$, the electrons can be localized and solvated in a molecular structure whose dynamical responses simulate those of the pure fluid. Typically $\bar{\rho}_d \sim 0.2\ \bar{\rho}_d$ (pure fluid). The alcohol clusters represent microphases in the liquid in which the electron ultimately becomes trapped. The transition in spectral and kinetic behavior of e_s^- over the range $5 \times 10^{18} < \bar{\rho}_d < 10^{20}$ cm^{-3} is truly a manifestation of the microscopic changes in the equilibrium liquid structure. Indeed, this behavior is somewhat similar to a phase transition as one moves from an isolated alcohol molecule in a matrix to a collective molecular response. There are echoes of the question, at what stage does a cluster of molecules behave like a liquid?[4]

We are still left, however, to speculate on the initial *localization* mechanism occurring at subpicosecond times. The critical dipole density above which the electron-scattering potential is modified by the presence of ROH appears to be similar for all *n*-alcohols in a given alkane, $\bar{\rho}_d \sim 10^{19}$ cm^{-3}. This leads us to propose that a density or configurational fluctuation of *minimum amplitude* with respect to kT is required before electron trapping takes place, which could be within the period of a vibration if strong electron–phonon coupling can occur. In high $\bar{\rho}_d$ systems, the frequency of desired fluctuations will be very high, hence we would predict ultrafast localization and solvation in NH_3, H_2O and diols, which is indeed observed. As the quasifree electron undergoes multiple scattering through the "statistically frozen lattice" of any liquid at times $\ll 10^{-12}$ s, it is accompanied by an induced electronic polarization of the lattice. Upon encountering a minimum amplitude fluctuation such that the residence time in that site is long enough for the liquid lattice to soften and distort to rearrange to accommodate the excess charge, the electron becomes trapped and solvated. Is this early IR stage a manifestation of polaron behavior? Is there a finite time associated with relaxation of a hot electron in the conduction band in liquids?

It is very interesting to note that in recent electron-scattering experiments in NH_3 vapor[82] a critical density for the onset of localization of 3×10^{19} cm^{-3} was reported, while localization of electrons in supercritical water is clearly well established at this $\bar{\rho}_d$.[3] Comparisons of electron localization studies in alcohols and alkanes via optical[63] and mobility[69] studies can be made, which show the expected mirror image relationship presented by decreasing μ_e and increasing optical absorption of e_s^- as a function of phase density.[83]

In order to treat these observations and hypotheses in a theoretical framework as successfully as the case of electron localization in helium,[13] we must first probe the dynamical properties of the IR absorptions in the subpicosecond regime. What perhaps is surprising and stimulating for future studies is the wealth of microscopic details that can be obtained on intermolecular interactions and electron transfer in liquids through picosecond spectroscopy, information of fundamental interest to chemical dynamics in the condensed phase. In this vein, we will conclude this chapter by an example of photoselective chemistry in electron transfer processes that occur following laser excitation of e_s^- in the cluster.

V. PICOSECOND LASER CHEMISTRY OF e_s^-

Although considerable theoretical effort has been directed toward developing a quantum-mechanical description of excess electron states,[13, 18–25] it is only recently through laser saturation spectroscopy (LSS) that direct experimental information has been reported on the nature of the excited states of the short-lived e_s^- and the degree of spectral homogeneity in the optical band. This information is vital for the construction of a theory in which the coupling of the electronic motion to the vibrational modes of the cluster molecules not only provides a width to the absorption band and a multiphonon mechanism for rapid vibronic relaxation but also appears as a basis for a microscopic treatment of electron transfer reactions in the context of radiationless transition theory.[18–20, 84] Photodetachment of an electron to participate in such reactions represents perhaps the simplest case of all to study, since the donor state collapses back into the solvent. We outline briefly here the first direct observations of a *laser-induced, intracluster electron transfer* for e_s^- in alcohols.[83]

Laser hole-burning experiments in the nanosecond time domain can reveal the picosecond time history of the relaxation times in an *n*-level system of the molecule if those relaxation times are fast (picosecond) compared to the duration of the laser saturating pulse (nanosecond).[55, 56] Under these circumstances, the levels of the system can be treated as photostationary states and the kinetic equations solved explicitly for the rate constant between any two specified levels. This leisurely approach to picosecond phenomena is equally valid in the picosecond time domain,

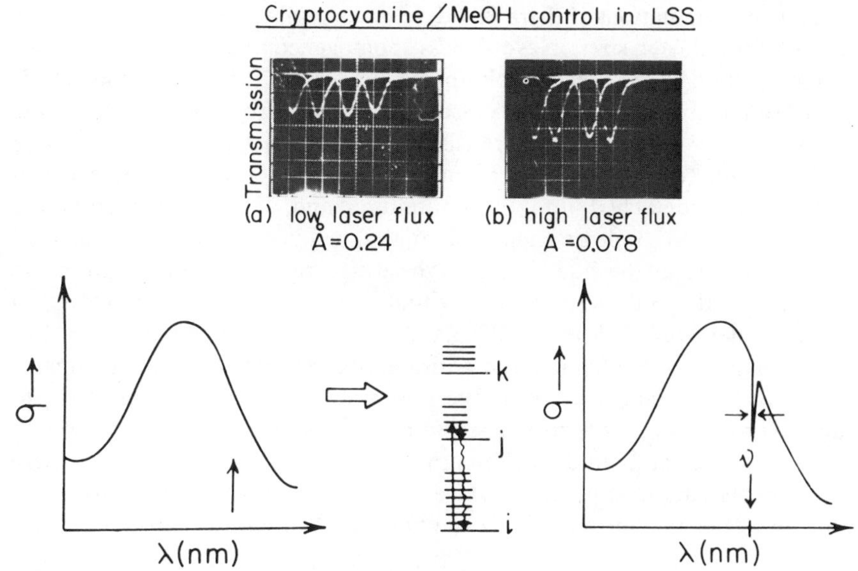

Fig. 12. Schematic for laser saturation spectroscopy (LSS), showing depletion and recovery in a two-level system. Oscilloscope traces show four transmitted laser pulses at (a) low and (b) high laser fluxes in the cryptocyanine methanol system.

where access to the femtosecond regime is now possible, and the precision of the measurement arises mainly from the accuracy with which the amplitude and shape of the laser pulse are known.[55–59] Figure 12 illustrates the general scheme for a hole-burning experiment and shows profiles of transmitted ruby laser pulses in cryptocyanine in methanol under (a) normal and (b) saturating conditions.[4] Note the substantial increase in amplitude in (b), which upon calculation[55] generates a relaxation time of 38 ps, in agreement with other measurements in the literature.[57]

A. Photophysics of Electrons in Liquids

Following laser optical excitation of e_s^- in alcohols, water and ammonia e_s^{-*} relaxes rapidly and nonradiatively to the ground state on a picosecond time scale, although it is conceivable any weak emission with a large Stokes shift would be reabsorbed by the solvent. We have shown through pulsed laser saturation spectroscopy that the absorption of e_s^- in alcohols at 298 K is homogeneously bleached across 1.5 eV width of the absorption band.[55] Although this is generally good evidence for homogeneous broadening, *on a time scale of* 10 *ps* we cannot neglect the possibility of some degree of inhomogeneity arising from high frequency vibrational fluctuations in the solvation structure, that is, the ground-state energy of e_s^-, and site-to-site fluctuations if there is a dynamic equilibrium between electrons in different cluster sizes. Nevertheless, homogeneous broadening has also been reported in picosecond photobleaching of e_s^- in amine liquids,[9, 58] and both these investigations present evidence which points toward the dominance of bound-bound transitions in the spectrum of e_s^- in these polar liquids. Undoubtedly at a high-incident photon energy the electron will be photodetached into the continuum and, if it escapes from the residual attractive hole or image potential, it will become rapidly resolvated at another site in the liquid. Under these circumstances, the nonradiative relaxation rate to the ground state should be comparable to the solvation time in that liquid. Whereas the data[7, 8, 56] for e_{aq}^- do not preclude this possibility because solvation and nonradiative recovery times are comparable to or faster than the laser probe pulses, leaving some uncertainty as to the deconvolution of the true times, at photon energies $\leqslant 1.8$ eV there is no such evidence for bound–continuum transitions in e_s^- in alcohols. There is, however, an unexpected feature of laser excitation of e_s^- which is unique to the alcohols and leads to photoselective laser chemistry.

B. Laser-Induced Electron Transfer

During high intensity laser saturation of e_s^- in alcohols, the experimental details of which are outlined in Section II and elsewhere,[83] we observed that following depletion of the ground state of e_s^-, the absorption signal did not fully recover. The *permanent* loss of absorbance followed the wavelength dependence of the e_s^- absorption cross-section and indicated that a highly reactive chemical channel opened up to e_s^{-*} at high laser intensities, in competition with vibronic relaxation.[55] Comparison with bleaching and recovery data of laser dyes in the same experiment indicated that electron loss via the chemical channel took place at times significantly faster than 10 ps, while chemical kinetic evidence ruled out mechanisms such as photodetachment, resolvation, or electron capture by an impurity molecule.[55, 83]

Figure 13 shows two kinetic traces of the formation and decay of e_s^- in methanol, (a) in the absence and (b) in the presence of a 20 ns ruby laser pulse. Saturating the e_s^- optical transition at 694 nm leads to a significant permanent loss of absorbers during the 0.4 J cm^{-2} laser pulse. Figure 14(a) shows the spectra of e_s^- in methanol and 2-propanol before and during bleaching at 694 nm, and it is clear that there is a uniform loss (ΔA) of intensity across the band. This loss, however, has a nonlinear dependence

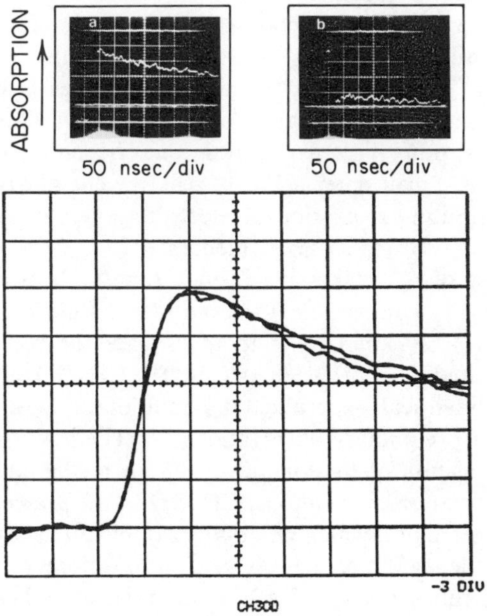

Fig. 13. Optical saturation of electrons in alcohols. Formation and decay of e_s^- in methanol (a) in the absence and (b) in the presence of a saturating laser pulse at 694 nm. The lower trace (20 ns div^{-1}) is a computer printout of data from e_s^- in CH_3OH and CH_3OD. See text for details.

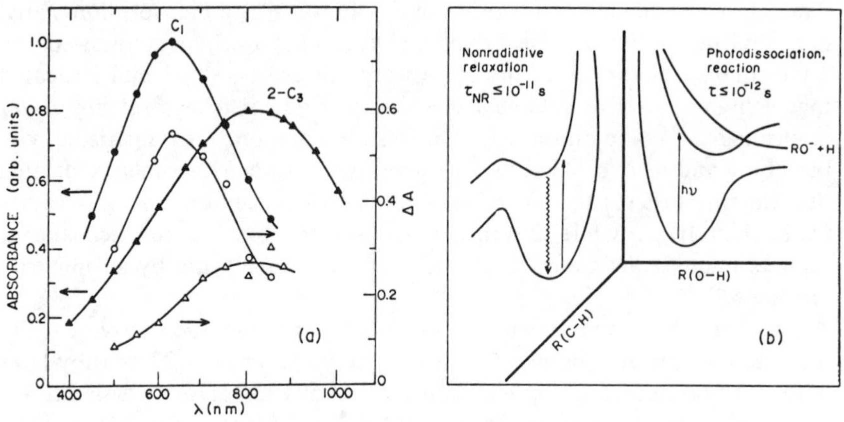

Fig. 14. (a) Spectra of e_s^- (●, ▲) before and (○, △) during saturation pulse at 694 nm; Δ A is loss of absorbance for methanol (C_1) and 2-propanol ($2\text{-}C_3$). (b) Relaxation versus reaction competition for $e_s^{-}*$ at high laser fluence.

on the laser fluence as illustrated in Fig. 15, which shows the fractional loss of absorbance (ϕ) recorded at 600 nm for the methanol system. From the above results[55] and earlier data,[85] it was speculated that reaction from $e_s^{-}*$ was occurring to give RO^- and H atoms at the high laser intensities, but since the slope in Fig. 15 decreases to a plateau, not all $e_s^{-}*$ appear to be lost through this route. The two central issues become: (1) the mechanism through which $e^{-}*$ can enter a highly reactive chemical channel, and (2) the other contributing relaxation or electron generating mechanisms that may come into play at higher laser fluences.

Since the loss of $e_s^{-}*$ takes place on a time scale faster than electron solvation, which is comparable to rotational diffusion in the liquid, we designed a series of experiments to investigate the role of vibration. A series of deuterated alcohols, such as CD_3OH, CH_3OD, and CH_3OH, were studied under identical experimental conditions and the following is a summary of our preliminary observations.[4, 83] The lower picture in Fig. 13 shows a superposition of the kinetics of e_s^- in methanol (CH_3OH) in the *absence* of a saturation pulse and in CH_3OD in the *presence* of a 20 ns, 0.3 J cm^{-2} ruby laser pulse, which enters the sample 50 ns after the beginning of the electron decay. Note there is very little difference between the two traces. Indeed, full recovery of e_s^- is observed after laser saturation. In contrast, significant and identical electron losses were observed for CH_3OH and CD_3OH subjected to the same saturation pulse, namely, $\phi_{600} = 0.30$. Table II summarizes these isotope effects seen in methanol and ethanol; the ϕ_λ are accurate to ± 0.05, and it can be seen that the magnitude of the

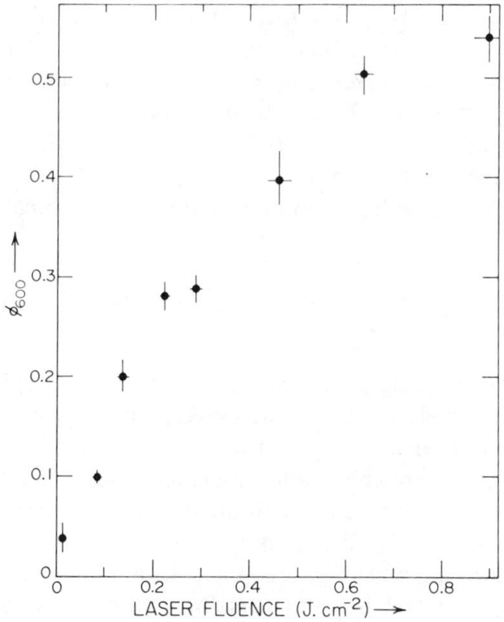

Fig. 15. Dependence of loss of electrons on laser fluence. See text for discussion.

isotope effect is also fluence-dependent. Deuteration in the —OH fragment of the molecule markedly alters the branching ratio between reaction and relaxation, and thus we identify the route to the chemical reaction as through a vibration of the OH bond. The fact that deuteration in the CD modes did not affect ϕ_λ despite the anticipated changes in density of states and Franck–Condon factors, leads us to conclude that, whereas multiphonon processes involving CH and OH vibrations must account for the rapid nonradiative relaxation normally shown by photoexcited e_s^{-*}, it is only the

TABLE II
Isotope Effects in Laser-induced Electron Transfer of e_s^- in ROH

Cluster	CH_3OH[b]	CH_3OD	CD_3OH	CH_3CH_2OH[c]	CH_3CH_2OD	EG	H_2O
ϕ_{694}^a	0.30	0.04	0.32	0.6	0.2	0.35	0
ϕ_{530}	0.17	0.06	0.16				0

[a] $\phi_{\lambda_1,\lambda_2}$ = fractional loss of absorbance following laser saturation at λ_1, λ_2 quoted to ± 0.05.
[b] methanol data, 0.35 J.cm^{-2}.
[c] ethanol data, 0.44 J.cm^{-2}.

OH vibration that leads to photochemistry, as Fig. 14(b) implies. The mixed ROD/ROH studies indicate that the molecule involved is a member of the inner group of molecules that comprise the cluster, since no isotope effect can be detected much below an equimolar mixture of the alcohols.[83]

These laser studies of isotope effects not only provide the first direct evidence that the mechanism leading to permanent photobleaching in liquid alcohols is via reaction (1) but also are the first direct manifestation of strong

$$e_s^- + h\nu \rightleftarrows e_s^{-*} \rightarrow RO^- + H \tag{1}$$

electron–phonon coupling in electrons in fluids. These couplings are a well known phenomenon elsewhere in condensed phase physics and in order to formulate a theoretical description, the vibration must be identified. The high frequency OH stretching and in-plane bending modes have been studied[86] in alcohol monomers and multimers and appear at ~3667 and ~1335 cm^{-1}, respectively, decreasing to ~2600 and ~864 cm^{-1} upon deuteration. However, recent laser optoacoustic studies of H_2O absorptions also show that OH has the stretching mode of a small but finite absorption at 694 nm, whereas D_2O does not.[87] To what extent are we directly exciting a local mode vibration or its harmonic at high laser intensities to promote the photodissociation and electron transfer? Future subpicosecond laser spectroscopy experiments will investigate the possibility of direct IR excitation of these vibrations to resonantly enhance the electron–phonon coupling and hence promote photodissociation and electron transfer, as well as exploring the fluence dependence of the isotope effect over a range of laser saturation wavelengths.

We can indeed claim that this is an example of *photoselective laser chemistry*. The competition between relaxation and reaction of photoexcited electrons in clusters represented in Fig. 14(b) is reminiscent of the competition in many laser-induced chemical processes, stimulated by the selective absorption of one or more photons, such as photodissociation, photoionization, isomerization, and so forth in polyatomic molecules, where the coupling of many vibrational modes provides energy randomization and relaxation on picosecond time scales.

We conclude this article on a note of optimistic speculation. Clearly the above results on solvated electrons establish the potential of ultrashort laser pulses to probe the fundamental details of the dynamics of electron transfer reactions, which will be the cornerstone for the development of microscopic theories of electron dynamics in the condensed phase. Electrons are ubiquitous species, and the practical reflection of this appears in research areas such as photosynthesis, dielectric breakdown, fast optical switches and solar energy cells, to name but a few. In fact, the radiationless relaxations of e_s^- may be employed as an ultrafast saturable absorber in mode-locking picosecond lasers. To provide new data for theories of electrons in fluids, it is imperative to explore new spectroscopic realms. In this spirit we are attempting to observe electrons in clusters via resonant Raman scattering, while polarization spectroscopy and magnetic circular dichroism[25] are potential tools for unraveling the electron–medium interactions. Clusters of known size can be made in the expansion from supersonic nozzles in molecular beam machines and attempts are under way to observe solvated ion species.[88] Clusters can be prepared by photoaggregation on low-temperature surfaces and their laser-induced photophysical properties studied *in situ*,[89] while the thermodynamics of nucleation[90] about ions and molecules in the atmosphere offers another active area of research pertinent to solvation phenomena in general. As we have continually emphasized in this chapter, the excess electron is a microscopic probe of the local structure and molecular dynamics of its host environment.

Acknowledgments

The author gratefully acknowledges the contributions to these studies from her students and collaborators whose work is cited in the references, particularly Dr. C. D. Jonah (ANL), and financial support from the Natural Sciences and Engineering Research Council of Canada, Research Corporation (U.S.A.), the Petroleum Research Fund, administered by the American Chemical Society, and the Connaught Foundation.

APPENDIX

Electrons residing in molecular clusters can be viewed as microscopic probes of both the local liquid structure and the molecular dynamics of liquids[4], and as such their transitory existence becomes a theoretical and experimental metaphor for one of the major fundamental and contemporary problems in chemical and molecular physics, that is, how to describe the transition between the microscopic and macroscopic realms of physical laws in the condensed phase. Since this chapter was completed in the Spring of 1979, several new and important observations have been made on the dynamics and structure of e_s^-, which, as a fundamental particle interacting with atoms and molecules in a fundamental way, serves to assist that transformation for electronic states in disordered systems. In a sense, *disorder* has become *order* on the subpicosecond time-scale, as we study events whose time duration is shorter than, or comparable to, the period during which the atoms or molecules retain some memory of the initial quantum state, or of the velocity or phase space correlations of the microscopic system. This approach anticipated the new wave of theoretical and experimental interest in developing microscopic theories of molecular dynamics and chemical dynamics, encouraged largely by the ability to photoselectively pump or probe molecular excitation or scattering with tunable lasers. Studies of photoionization and electron transport processes enjoy a natural place in this surge of interest in microscopic picosecond phenomena[91], as the following examples of recent work related to electron solvation and clusters amply illustrate.

The venerable debate over the role of long-range and short-range electron-medium interactions in electron localization and solvation mechanisms, the possible existence of pre-existing traps, and the degree of molecular reorientation involved in electron stabilization within the cluster, are all topics of discussion in this chapter. New Raman and picosecond optical absorption data on e_s^- in liquids, electron spin-echo magnetic resonance measurements on e_s^- in glasses, and molecular beam experiments with clusters have fulfilled many of our earlier predictions and still pose intriguing questions and speculations for the future. The initial trapping sites for electrons in alcohol : alkane systems have been identified for the first time[78b, 92] and the first subpicosecond laser spectroscopy observations reported for electron solvation in water[93]. In both studies it is clear that, as expected (see Section IV), the liquid structure offers pre-existing trapping sites for electrons. For example, in *n*-butanol and *t*-butanol diluted in cyclohexane and *n*-hexane, comparisons of ^{13}C NMR and picosecond optical absorption data in the very low $\bar{\rho}_d$ range (together with infrared and dielectric relaxation data) show that *n*-butanol and *dimers* and *t*-butanol *cyclic trimers* are the initial sites for alcohol traps. At higher $\bar{\rho}_d$, these mini-clusters build up to larger cluster sizes via molecular reorientation of adjacent alcohol molecules. Spin echo modulation measurements of e_s^- in ethanol glasses at 77K indicate[94] that the inner solvation shell comprises four molecules, whose *molecular* dipole appears oriented towards the electron, in contrast to aqueous glasses in which six H_2O molecules cluster about the electron with the OH *bond* dipole directed towards the electron. In water at 300K, the absorption at 615 nm of e_{aq}^-, formed from UV laser photodetachment of $Fe(CN)_6^{4-}$, is present within 0.3 ps[93], lowering the previous upper limit estimates of electron solvation in water by at least a factor of 10. Given the degree of three dimensional liquid structure in H_2O, and the high dipole density (see Section III), a minimum of dipole orientation about the electron would be required to reach a stable, cluster-like configuration. It now would be of great interest to combine the detailed studies on the intermolecular potentials[95] and structure of liquid water with the cluster-like M.O. calculations[24, 25], to see if the presence and dynamics of such traps can be projected from liquid water, while simultaneously explaining the close similarity of the e_{aq}^-

optical absorption spectrum to that of e_s^- in ice.[4, 10] We are pursuing similar subpicosecond experiments in the alcohols to explore the ideas outlined in Section IV.

If electrons are ejected into the liquid via *n*-photon absorption, such as illustrated for pyrene in Fig. 16, and the final state is S_n, a molecular autoionizing state of possibly finite lifetime, then experimental observations of electron solvation times may also include that lifetime. Thus, it is crucial, as in the observation of any laser-induced photophysical process, to observe the full spectral changes as a function of time. A corollary is as follows: if the liquid surrounding a negative ion is orientationally polarized with respect to the field of the ion, then any photodetachment process may result in ionization and electron trapping in an *already prepared* molecular site. Again, the time measured would be a reflection of the lack of work required for medium reorganization. Clearly these are points to be examined further, but recent picosecond data[96] on the photoelectron efficiency of 2, 6 toluidinyl naphthalene sulfonate ion (TNS) in water to give e_{aq}^- and in ethanol: water systems suggests that indeed this is the case. They propose that ultrafast electron transfer is facilitated by such a preformed configuration of water molecules about the TNS ion, but that it is abruptly impeded by the presence of a few ethanol molecules in the inner solvation shell.

Similar dramatic effects of selective solvation have been also seen in the laser-induced electron transfer (LIET) processes first discussed in Section V. More recent hole-burning experiments employing a Nd:YAG laser indicate that laser saturation at 530 nm also produces a significant, and

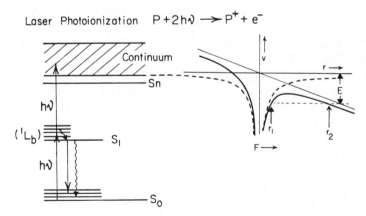

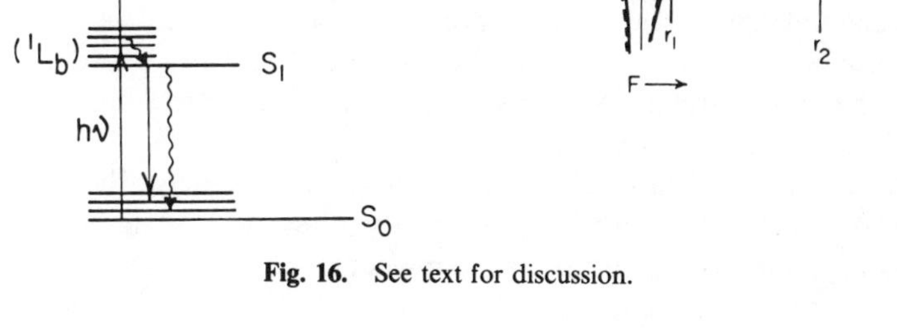

Fig. 16. See text for discussion.

isotopically sensitive electron transfer[92], as included in Table II. Coincidentally, the two wavelengths 694 nm and 530 nm probably overlap the 5th and 6th overtones of the methanol-OH stretching vibration and so we are examining the earlier notion of resonantly enhanced, photoselective laser dissociation and electron transfer in great detail. Ideally a resonant Raman scattering experiment would be most revealing here. The first evidence of deformation of a Raman band and lineshape due to e_s^- was recently reported employing coherent Raman ellipsometry (a polarization version of CARS) on solutions of e_s^- liquid hexamethyl phosphoramide (HMPA).[97] The $-CH$ vibration at 1486 cm^{-1} was perturbed, which would be expected if the $-CH_3$ groups comprise the inner part of the cluster as indeed inferred by the IR electronic spectrum (spectrum E in Fig. 1) and weak binding energy of e_s^- in HMPA. Similar experiments on electrons in alcohol clusters should reveal the $-OH$ perturbation and simultaneously identify the vibration involved in LIET. New laser/mass spectroscopy studies on methanol and water clusters ($2<n<6$) in supersonic nozzle beams show significant red shifts of the cluster-OH vibrational frequencies from the free monomer value,[98] and such data will undoubtedly assist in identifying probable cluster structures about e_s^-.

In experiments with crossed molecular and atomic beams, electron transfer to the methanol clusters was attempted via collision with a seeded Rb alkali atom beam, relying on the harpooning mechanism.[99] The distinction between such electron transfer[88, 99] and direct collisional electron bombardment[88b] of the cluster is that in the former harpooning mechanism the electron has negative energy with respect to the cluster while in the latter case it has a positive energy. Conclusions from the recent studies[99] (using mass spectrometry) were that electron transfer *did* show a higher electron attachment cross-section than electron bombardment, but that in most cases solvated anions were observed *i.e.* $(CH_3O^-)(CH_3OH)_n$ where $2 \leqslant n \leqslant 9$, rather than $(CH_3OH)_n^-$ or e_s^-. For $(CO_2)_n^-$ however, the case may be clearer.[88] Of course, if $(CH_3OH)_n^-$ were formed in a vibronically excited state, it seems clear in comparison to our LIET data,[83, 92] that dissociation and electron transfer to $CH_3O^-(CH_3OH)_{n-1}$ could take place at timescales $<10^{-13}s$. Since the collision time is $\sim 4\times 10^{-13}s$, much vibrational relaxation can occur, although reorientational motion of alcohol monomers ($10^{-11}s$) would be precluded in these cold clusters. Unless the latter are born with the correct "deep trap" configuration, electron *solvation* cannot occur.

Hydrated electrons have also been exploited as kinetic probes in model biological systems for many years. In the first tunable laser study of ionization potentials (I_p) of aromatic molecules in model membranes, we observed a significant reduction in the threshold I_p for pyrene in aqueous anionic sodium dodecyl sulphate micelles (SDS). The appearance of e^-_{aq}, and the pyrene positive ion, and simultaneous depletion of the electronically excited singlet pyrene $^1P^*$, were monitored as well-characterized, spectroscopic probes in a two-photon absorption, double laser pulse experiment.[100] The possibility of Stark-promoted ionization of high-lying Rydberg states of pyrene, whose excited states would be strongly perturbed by the high local field gradient (F) at the negatively charged micellar interface, was considered as a plausible explanation for this effect. Changes in the polarization energies (P^+) and V_o (see Section IIA) although present, are too small to account for the differences in I_p recorded on going from methanol to SDS. Figure 16 illustrates this hypothesis schematically, showing the distortion of the coulomb well in the Stark field and indicating the possibility of electron tunneling (from r_1 to r_2) out of high lying, excited states in the well. Further experiments are in progress to elucidate the picosecond dynamics of these processes.

However, as we noted earlier (in Section II), it is most important to have a full temporal and spatial profile of the picosecond laser pulses in order to correctly interpret the one or two photon excited molecular response functions (absorption, emission, Raman scattering) of the system under study. While autocorrelation techniques such as second harmonic generation (SHG) or two photon fluorescence have always been the route to such short pulse measurements,[5] a significant advance has recently been made in bringing these conventional autocorrelation measurements into the real-time domain.[49, 101]

The principle of these SHG correlation measurements is to split the intensity of the laser pulse (ω_1) into two replicas, each of which travel along an interferometer arm and are reflected back to be focused and superimposed in a nonlinear crystal. Only when the pulses overlap precisely in time and space is the SHG signal (at $\omega_2 = 2\omega_1$) observed. The intensity dependence of SHG is related to the second order, background free correlation function as follows:

$$I_{\omega_2}(t) \propto G_o^{(2)}(t) \propto \int_{t=0}^{\infty} I_{\omega_1}(t) I_{\omega_1}(t+\tau)\, dt$$

Conventionally these measurements were made by slowly scanning the optical delay (τ) in one interferometer arm to record $I_{\omega_2}(t)$. A "real-time" approach has been accomplished by introducing a *repetitively scanning*, optical reflector into the interferometer arm. Figure 17 illustrates our most recent development in which both interferometer arms end in corner cube reflectors, mounted on the cone of a speaker. The two speakers are each driven at 20 Hz (but are exactly 180° out-of-phase) by a sinewave generator (SWG), leading to an unprecedented dynamic range of 300 ps over

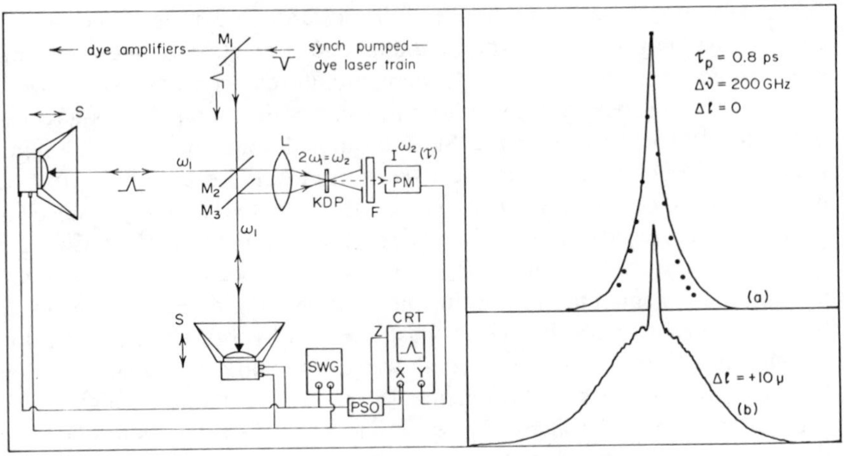

Fig. 17. See text for discussion.

which to make CW autocorrelation and cross-correlation laser measurements. The SHG signals are displayed continuously on an oscilloscope or can be recorded on an X–Y recorder.[49] Figure 17(a) and (b) illustrate the subpicosecond duration of the dye laser pulses ($\Delta\nu = 200$ GHz) and emphasize the importance of precise resonant cavity matching between the argon ion and dye laser, as predicted for synchronous pumping.[48] Whereas the dye laser pulse duration under optimum cavity conditions is 0.8 ps, when the dye laser cavity is 10 μm too long, partial mode-locking is seen as shown in (c). Alternatively, if the cavity is 5 μm too short the dye laser pulse totally collapses into "noise".[49] Thus our CW double-speaker autocorrelator acts as a picosecond sampling oscilliscope, and can be used to continuously monitor the quality and duration of the pulses during any experiment, as well as being used for optical mixing of unknown with reference laser signals.[92]

Amplified,[102] subpicosecond tunable laser pulses clearly present a powerful tool for future studies of ultrafast electron relaxation, and transfer processes in liquids and semiconductors as well as for molecular and chemical dynamics in liquids.

References

1. *Electrons in Fluids*, Proceedings International Conference, Banff, 1976, *Can. J. Chem.*, **55** (1977).
2. H. T. Davis and R. G. Brown, *Adv. Chem. Phys.*, **31**, 329 (1975).
3. J. Jortner and A. Gaathon, in Ref. 1, pp. 1801–1819.
4. G. A. Kenney-Wallace, *Acct. Chem. Res.*, **11**, 433 (1978).
5. C. V. Shank, E. P. Ippen, and S. L. Shapiro, eds; *Picosecond Phenomena*, Springer-Verlag, New York, 1978.

6. G. A. Kenney-Wallace and C. D. Jonah, *Chem. Phys. Lett.*, **39**, 596 (1976).
7. P. M. Rentzepis, R. P. Jones, and J. Jortner, *J. Chem. Phys.*, **59**, 766 (1973).
8. W. J. Chase and J. Hunt, *J. Phys. Chem.*, **79**, 2835 (1975).
9. D. Huppert and P. M. Rentzepis, *J. Chem. Phys.*, **64**, 919 (1976), and references therein.
10. L. Kevan, *Adv. Rad. Chem.*, **4**, 181 (1974).
11. B. Bosacchi, C. Y. Leung, and M. O. Scully, in Ref. 5, p. 244; D. von der Linde and R. Lambrich, in Ref. 5, p. 232; D. Adler and E. Yoffa, in Ref. 1, p. 1920.
12. R. T. Williams, J. N. Bradford, and W. L. Faust, *Phys. Rev B.*, **18**, 7038 (1978).
13. L. Kevan and B. Webster, eds., *Electron-Solvent Anion Solvent Interactions*, Elsevier, Amsterdam, 1976; see N. R. Kestner, p. 1.
14. W. Schmidt in Ref. 13, p. 213; J.-P. Dodelet and G. Freeman in Ref. 1, p. 2264.
15. R. R. Hentz and G. A. Kenney-Wallace, *J. Phys. Chem.*, **76**, 2931 (1972); **78**, 514 (1974).
16. A. Gaathon, G. Czapski, and J. Jortner, *J. Chem. Phys.*, **58**, 2648 (1973); A. Gaathon, Ph.D. Thesis, The Hebrew University of Jerusalem, 1974.
17. R. Olinger, S. Hahne, and U. Schindewolf, *Ber. Bunsenges. Phys. Chem.*, **76**, 349 (1972).
18. A. Banerjee and J. Simons, *J. Chem. Phys.*, **68**, 415 (1978).
19. K. Funabashi, I. Carmichael, and W. H. Hamill, *J. Chem. Phys.*, **69**, 2652 (1978), and references therein.
20. J. Ulstrup and J. Jortner, *J. Chem. Phys.*, **63**, 4358 (1975).
21. M. Cohen, in Ref. 1, p. 1906.
22. K. Fueki, D. F. Feng, and L. Kevan, *J. Am. Chem. Soc.*, **95**, 1398 (1973); and in Ref. 1, p. 1940.
23. N. R. Kestner, in Ref. 1, p. 1937.
24. M. Newton, *J. Phys. Chem.*, **79**, 2795 (1975).
25. For a contemporary critique, see B. C. Webster, *Ann. Rep. Chem.*, (UK), in press, 1980., and references therein.
26. L.F. Mollenauer, D. Bloom, and A. Del Gaudio, *Opt. Lett.*, **3**, 48 (1978).
27. G. E. Hall, G. A. Kenney-Wallace, and N. Klassen, to be published.
28. G. E. Hall and G. A. Kenney-Wallace, *Chem. Phys.*, **32**, 213 (1978).
29. G. E. Hall and G. A. Kenney-Wallace, *Chem. Phys.*, **28**, 205 (1978); G. E. Hall, Ph.D. Thesis, University of Toronto, 1980, and to be published.
30. Y. Yang, M. Crawford, M. J. McAuliffe and K. Eisenthal, *Chem. Phys. Lett.*, **74**, 160 (1980).
31. M. R. Robin, *Higher Excited States of Polyatomic Molecules*, Vols. 1 and 2, Academic Press, New York, 1974.
32. D. H. Park, J. O. Berg, and M. A. El-Sayed, in *Advances in Laser Chemistry*, Springer-Verlag, New York, 1978, p. 319; *Chem. Phys.* **42**, 379 (1979).
33. G. W. Robinson, R. J. Robbins, G. R. Fleming, J. Morris, A. Knight, and R. J. Morrison, *J. Am. Chem. Soc.*, **100**, 7145 (1978).
34. G. Dolivo and L. Kevan, *J. Chem. Phys.*, **70**, 5321 (1979).
35. U. Sowada and R. Holroyd, *J. Chem. Phys.*, **70**, 3586 (1979).
36. P. Engelking, G. B. Ellisen, and W. C. Lineberger, *J. Chem. Phys.*, **69**, 1826 (1978).
37. J.-C. Mialocq, J. Sutton and P. Goujon, *J. Chem. Phys.*, in press.
38. A. Matsuzaka, T. Kobyashi, C. D. Jonah, *J. Phys. Chem.*, **83**, 2554 (1979), and S. Nagakura, *J. Phys. Chem.*, **82**, 1201 (1978).
39. B. Halpern and R. Gomer, *J. Chem. Phys.*, **51**, 1048 (1969).
40. A. Mozumder, in Ref. 13, p. 139.
41. D. Huppert and P. M. Rentzepis, in *Molecular Energy Transfer*, R. Levine and J. Jortner, eds., Wiley, New York, 1975, p. 270.
42. R. Alfano and S. Shapiro, *Chem. Phys. Lett.*, **8**, 631 (1971).

43. B. A. Garetz and G. A. Kenney-Wallace, to be published; G. A. Kenney-Wallace, in Ref. 5, p. 208.
44. K. Sala and M. C. Richardson, *Phys. Rev. A*, **12**, 1036 (1975).
45. (a) C. D. Jonah, *Rev. Sci. Instr.*, **46**, 62 (1975); (b) G. Mavrogenes, C. D. Jonah, K. Schmidt, S. Gordon, G. Tripp, and L. W. Coleman, *Rev. Sci. Instr.*, **47**, 187 (1976).
46. E. P. Ippen and C. V. Shank, in Ref. 5, p. 103; E. P. Ippen, *Phil. Trans., Roy. Soc. (London)*, **A230** (1980).
47. D. J. Bradley, *J. Phys. Chem.*, **82**, 2259 (1978).
48. J. Heritage and R. R. Jain, *Appl. Phys. Lett.*, **32**, 101 (1978).
49. K. Sala, G. A. Kenney-Wallace and G. E. Hall, IEEE **QE16**, 990 (1980).
50. G. A. Kenney-Wallace, J. Flint, and S. C. Wallace, *Chem. Phys. Lett.*, **32**, 71 (1975).
51. S. C. Wallace, T. McKee, and B. P. Stoicheff, *Appl. Phys. Lett.*, **30**, 378 (1977).
52. G. A. Kenney-Wallace and T. McKee, unpublished data.
53. M. Ottolenghi, *Chem. Phys. Lett.*, **12**, 339 (1971).
54. P.-T. Ho, L. A. Glasser, E. P. Ippen, and H. A. Haus, in Ref. 5, p. 114; H. A. Haus, *Phil. Trans., Roy. Soc. (London)*, **A230** (1980).
55. G. A. Kenney-Wallace and K. Sarantidis, *Chem. Phys. Lett.*, **53**, 495 (1978).
56. G. A. Kenney-Wallace and D. C. Walker, *J. Chem. Phys.*, **55**, 447 (1971).
57. G. Mourou, B. Drouin, M. Bergeron, and M. Denariez-Roberge, *IEEE* **QE9**, 745 (1973).
58. D. Huppert, P. M. Rentzepis and W. S. Struve, *J. Phys. Chem.*, **79**, 2851 (1975).
59. L. Huff and L. G. deShazer, *J. Opt. Soc. Am.*, **60**, 157 (1970).
60. M. Topp and H.-B. Lin, *Chem. Phys. Lett.*, **50**, 412 (1977).
61. D. Huppert, G. A. Kenney-Wallace, and P. M. Rentzepis, submitted for publication.
62. G. A. Kenney-Wallace and C. D. Jonah, *Chem. Phys. Lett.*, **47**, 362 (1977).
63. G. A. Kenney-Wallace, B. A. Garetz, and C. D. Jonah, submitted for publication.
64. S. K. Garg and C. P. Smyth, *J. Phys. Chem.*, **64**, 1294 (1965).
65. G. A. Kenney-Wallace, *Phil. Trans., Roy. Soc. (London)*, **A230** (1980), *Chem. Phys. Lett.*, **43**, 529 (1976).
66. The Stokes–Einstein–Debye relationships are discussed in Ref. 70, 72 and 73.
67. S. K. Roy, K. S. Gupta, A. Ghatak, and A. Das, *J. Chem. Phys.*, **65**, 3593 (1976).
68. L. Gilles, M. R. Bono, and M. Schmidt, *Can. J. Chem.*, **55**, 2003 (1977).
69. J. H. Baxendale, *Can. J. Chem.*, **55**, 1996 (1977), and references therein.
70. See D. Kivelson in *Newer Aspects of Molecular Relaxation Processes, Faraday Symp.*, **11** (1977).
71. J. Belloni, M. Clerc, P. Goujon, and E. Saito, *J. Phys. Chem.*, **79**, 2848 (1975).
72. J. T. Hynes, R. Kapral, and M. Weinberg, *J. Chem. Phys.*, **69**, 2725 (1978).
73. B. J. Berne and G. D. Harp, *Adv. Chem. Phys.*, **17**, 63 (1970).
74. G. R. Fleming, A. E. Knight, J. Morris, R. J. Robbins, and G. W. Robinson, *Chem. Phys. Lett.*, **51**, 402 (1977), references therein; K. Spears and L. Cramer, *Chem. Phys.*, **30**, 1 (1978).
75. D. Oxtoby, this volume.
76. S. A. Rice and G. A. Kenney-Wallace, *Chem. Phys.*, **47**, 161 (1980).
77. T. Chuang and K. Eisenthal, *J. Chem. Phys.*, **57**, 5094 (1973); *Chem. Phys. Lett.*, **11**, 368 (1971).
78. R. Ling, G. A. Kenney-Wallace, and W. F. Reynolds, *Chem. Phys. Lett.*, **54**, 81 (1978); V. Gibbs, P. Dais, G. A. Kenney-Wallace, and W. F. Reynolds, *Chem. Phys.*, **47**, 407 (1980).
79. E. Breitmaier, K.-H. Spohn, and S. Berger, *Angew. Chem. Int. Ed. Engl.*, **14**, 144 (1975).
80. R. H. Stokes, *J. Chem. Soc., Faraday Trans. I*, **73**, 1140 (1977); F. Smith, *Austral. J. Chem.*, **30**, 23 (1977).

81. A. Djaranbakt, J. Lang, and R. Sana, *J. Phys. Chem.*, **81**, 2620 (1977), and references therein.
82. P. Krebs and M. Wantschik, *J. Phys. Chem.*, **84**, 1155 (1980).
83. G. A. Kenney-Wallace, G. E. Hall, L. A. Hunt, and K. Sarantidis, *J. Phys. Chem.*, **84**, 1145 (1980).
84. N. R. Krestner, *J. Phys. Chem.*, **84**, 1270 (1980), and references therein.
85. A. Bromberg and J. K. Thomas, *J. Chem. Phys.*, **63**, 2124 (1975).
86. A. J. Barnes and H. E. Hallam, *Trans. Faraday Soc.*, **66** 1920 (1970), and references therein.
87. A. C. Tam and K. Patel, *Nature*, Appl. Opt., **18**, 3348 (1979).
88. K. Bowen and D. R. Herschbach, to be published; K. Bowen, Ph.D. Thesis, Harvard University, 1978. C. Klots and R. N. Compton, *J. Chem. Phys.*, **69**, 1636 (1978).
89. J. Farrell, G. A. Kenney-Wallace, S. Mitchell, and G. A. Ozin, J. A. C. S., in press (1980).
90. A. W. Castleman, *Adv. Colloid Interface Sci.*, **10**, 73 (1979).
91. Picosecond Phenomena, Proceedings of 2nd International Conference 1980, R. M. Hochstrasser, W. Kaiser and C. V. Shank, eds. Springer-Verlag, New York, 1980.
92. G. A. Kenney-Wallace, L. A. Hunt and K. L. Sala, in Ref. 91, p. 203.
93. J. Weisenfeld and E. P. Ippen, *Chem. Phys. Lett.*, **73**, 47 (1980).
94. M. Narayana and L. Kevan, *J. Chem. Phys.*, **72**, 2891 (1980).
95. G. D. Carney, L. L. Sprandel, and C. W. Kern, *Adv. Chem. Phys.*, **37**, 305 (1978)
96. R. A. Auerbach, J. A. Synowiec and G. W. Robinson, in ref. 91, p. 215.
97. S. A. Akhamanov, L. S. Aslanyan, A. F. Bunkin, F. N. Gadzhiev, N. I. Koroteev and I. L. Shumai in Light Scattering in Solids, eds. J. Birman, H. Cummins and K. Rebane, Plenum Corp., New York, 1979.
98. H. S. Kwok, D. Krajnovich, M. Vernon, Y. R. Shen and Y. T. Lee, to be published.
99. E. Quitevis and D. Herschbach, to be published.
100. S. C. Wallace, G. E. Hall and G. A. Kenney-Wallace, *Chem. Phys.*, **49**, 279 (1980).
101. R. L. Fork and F. A. Beisser, *Appl. Opt.*, **17**, 3534 (1978).
102. A. Migus, C. V. Shank, E. P. Ippen and R. L. Fork, IEEE JQE in press, 1980.

STUDIES OF CHLOROPHYLL *IN VITRO*

K. J. KAUFMANN*

Department of Chemistry, University of Illinois, Urbana, Illinois 61801

and

M. R. WASIELEWSKI

Argonne National Laboratory, Chemistry Division, Argonne, Illinois 60439

CONTENTS

*Alfred P. Sloan Fellow.

I. INTRODUCTION

The absorption of photons by a green plant or a photosynthetic bacterium initiates a complex series of reactions. These reactions convert the energy of the photons into a charge separation.[1, 2] This charge separation is utilized to pump protons across the photosynthetic membrane. The energy

1

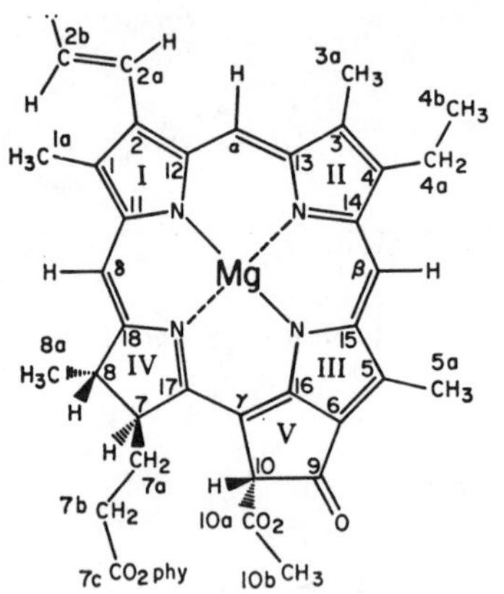

2

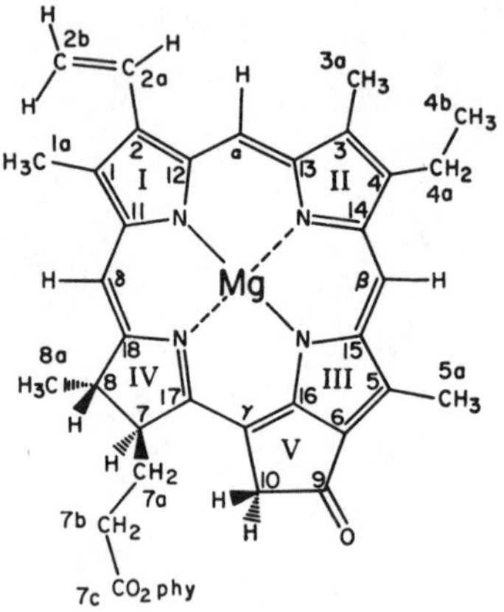

3

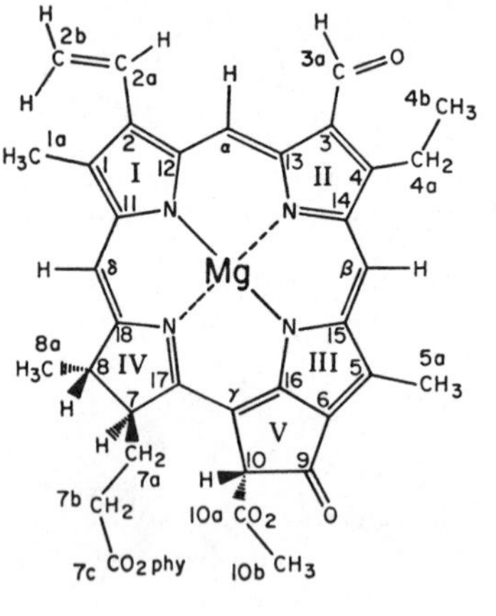

4

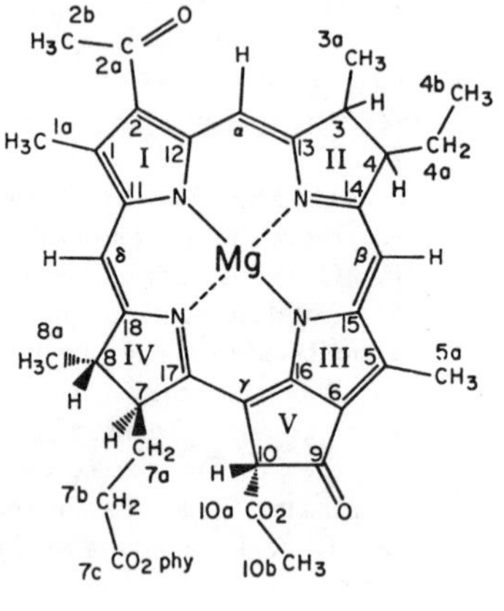

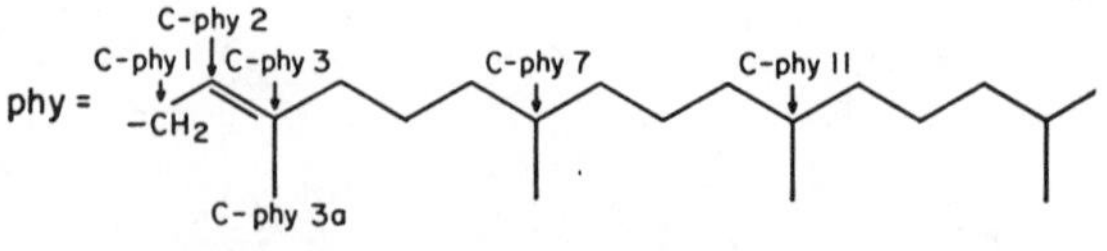

Fig. 1. Structure of the chlorophylls (1) chlorophyll *a*; (2) pyrochlorophyll *a*; (3) chlorophyll *b*; (4) bacteriochlorophyll *a*. The corresponding pheophytins are identical except that the central magnesium atom is replaced by two hydrogen atoms.

stored in the proton gradient is converted to ATP.[3] The charge separation is also utilized in the production of strong reducing agents. These reducing agents provide the energy for conversion of carbon dioxide into carbohydrates.[4] The substance most often associated with the photosynthetic process are the chlorophylls (Fig. 1). The two types of chlorophyll most often found in green plants are chlorophyll *a* and chlorophyll *b*. Despite their fame, the chlorophylls play a role only in the early steps of photosynthesis. This does not in any way diminish the importance of the chlorophylls in the photosynthetic process for without them, most biological solar energy conversion could not occur.

The photon flux striking a leaf is absorbed by an array of chlorophyll *a* and *b* molecules known as the antenna.[5, 6] The antenna chlorophyll is photochemically inactive. It serves merely to capture the incident radiation and transport it to a pigment–protein complex known as the reaction center.[6–8] Within this protein, a very specialized type of chlorophyll *a* converts the electronic excitation into a positive and negative charge. Since chlorophyll *a* functions both in the antenna and the reaction center, the environment must play a significant role in determining the function of each chlorophyll molecule.

Over the past thirty years, numerous workers have been studying the chlorophyll molecule and its derivatives *in vitro*. From these studies excellent models for the photoactive chlorophyll in green plants and photosynthetic bacteria have evolved. Most of the *in vitro* studies have involved comparisons between static properties of the chlorophylls *in vitro* and the chlorophylls *in vivo*; however, recently a number of groups have begun to probe the dynamics of some of the models for the photoactive chlorophyll found in the reaction center. Excellent reviews of primary events in photosynthetic bacteria and green plants can be found in Ref. 1, 2, and 9. Therefore, the details will only be briefly reviewed in the next section. The extensive work by Katz and co-workers on the *in vitro* properties of chlorophyll laid the foundation for all of the model systems that have been proposed so far. Since the early work of Katz's group has also been well reviewed in Ref. 10 and 11, it too will only be briefly discussed.

II. PRIMARY PROCESSES IN PHOTOSYNTHESIS

A. Green Plants

Green plants are capable of using the sun's energy to oxidize water to oxygen and to produce a strong enough reducing agent to incorporate carbon dioxide into carbohydrates.[2] To carry out these functions, plants have two separate photosystems (Fig. 2).[2, 9, 12] As can be seen in Fig. 2, the

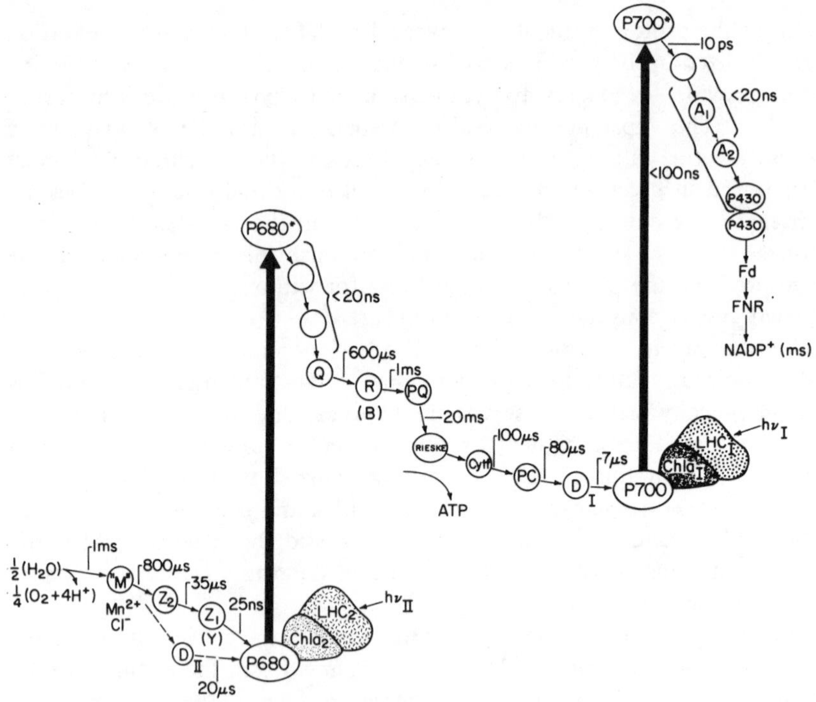

Fig. 2. Schematic diagram of photosystem I and II in green plants. This figure is taken from Govindjee, *Plant Biochem. J.*, SIRCAR Memorial. vol. 7, (1980).

photosynthetic pathway is extremely complex. It consists of two separate photochemical systems. Chlorophyll *a* is found in both systems. As can be seen from Fig. 2 the majority of the chemical reactions that take place do not involve the chlorophyll molecule in any way. These reactions occur unimpeded in the dark. Light is only required in the initial charge separation. Chlorophyll *a* plays a role during this step and the light harvesting that precedes it. The reactions involving chlorophyll *a* were designed by nature to quickly restore chlorophyll *a* to its native state. Thus the chlorophyll *a* molecule can be looked upon as a catalyst that converts photon energy into electrochemical energy. The major portion of the photosynthetic apparatus in green plants is designed to utilize the electrochemical potential generated by light interacting with chlorophyll into high-energy compounds such as ATP and carbohydrates.[3, 4] The photosystem that oxidizes water is known as photosystem II. The one that produces the reducing agent NADPH is called photosystem I. Although each photosystem can function independently a number of control loops exist that prevent one system from operating at a rate much faster than the other.[13] Each photosystem has it own reaction center protein. These proteins have not as yet been isolated. Despite this, a great deal of information is available on the reaction center of photosystem I.[2,9]

Oxidation, reduction, or excitation of many molecules in general, and chlorophyll *a* in particular, results in the bleaching of the red-most absorption band. When particles isolated from green plants, which have been enriched in photosystem I are excited with a pulse of light, photochemistry is initiated. In a very short time, perhaps in only a few picoseconds, the special photoactive chlorophyll *a* within the reaction center is oxidized.[14–16] It remains in this state for up to a few microseconds depending on the redox state of the system.[17] During this time it is possible to take a (light–dark) difference spectrum. Such a spectrum (Fig. 3) has a strong bleaching at about 700 nm.[18,19] Similar measurements can also be made in the steady state by shining a strong actinic light on the sample. Bleaching at 700 nm can also be observed by chemical oxidation of the photoactive chlorophyll. From these bleaching studies it is clear that the photoactive species in reaction centers of photosystem I must absorb at 700 nm.

Photooxidation of the reaction center not only produces an optical absorbance change, but results in an ESR signal as well.[20] This ESR signal has a kinetic behavior identical to the 700 nm bleaching and has therefore been assigned to the cation of the photoactive chlorophyll.[21] The ESR linewidth of the cation of the photoactive chlorophyll *a* is narrowed by

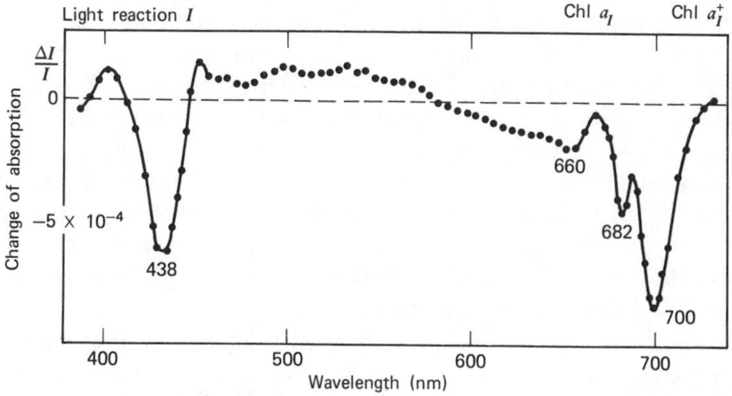

Fig. 3. Transient difference spectrum of photosystem I. The optical density changes observed can be ascribed almost entirely to the oxidized photoactive chlorophyll *a* of photosystem I. (From Ref. 19.)

$\sqrt{2}$ over the linewidth of the cation of monomeric chlorophyll *a in vitro*. This interesting anomaly was quite troublesome when the cation signal was first observed.[20] As we shall see in Section III the *in vivo* studies have shown that this narrowing of the ESR signal and the 700 nm absorbance have been crucial in determining that the photoactive species of photosystem I may be a dimer of chlorophyll *a*.

The primary events of photosystem I have not yet been elucidated. Preliminary results indicate that the photochemistry occurs on the picosecond time scale. There is also some evidence to indicate that the primary electron acceptor may be a chlorophyll complex.[16, 22, 23] Spectral changes associated with a number of intermediate acceptors have also been observed. Some of these acceptors have been identified as iron–sulfur proteins.[24] Photosystem II reaction centers have a photoactive form of chlorophyll that absorbs at 680 nm.[25] While very little else is known about the photoactive form in photosystem II, a great deal of information is available on the electron donors and acceptors of photosystem II.

B. Photosynthetic Bacteria

The biochemical architecture of photosynthetic bacteria is not as complex as that of green plants. For example, photosynthetic bacteria have only one photosystem, while green plants have two. The reaction center protein from several species of photosynthetic bacteria can be isolated from the photosynthetic membrane.[26] Reaction centers from the species *Rhodopseudomonas sphaeroides* have been extensively studied. Although minor details will change from one species to another, the important features are nearly identical. The reaction center protein has a molecular weight of about 70,000 daltons.[27] Within the reaction center protein extracted from the carotenoidless mutant strain R26 of the species *R. sphaeroides*, are found four molecules of bacteriochlorophyll *a*, two molecules of bacteriopheophytin *a*, one atom of nonheme iron, and, depending on the isolation procedure used, one or two molecules of ubiquinone. The absorption spectrum of the isolated reaction center has been well characterized.[26, 28] It is shown in Fig. 4. Based on *in vitro* absorption spectra, the bands at 870, 800, and 600 nm have been assigned to the bacteriochlorophyll *a* molecule. Bands at 760 and 530 nm have been attributed to the bacteriopheophytin *a*.

Photochemical oxidation results in small changes in the absorption spectrum of the reaction center with the exception of a large amount of bleaching in the red-most band at 870 nm.[28, 29] Accompanying this bleaching is an ESR signal, which is narrowed by $\sqrt{2}$ over the signal observed in the cation of monomeric bacteriochlorophyll *a*.[30, 31] Just as for

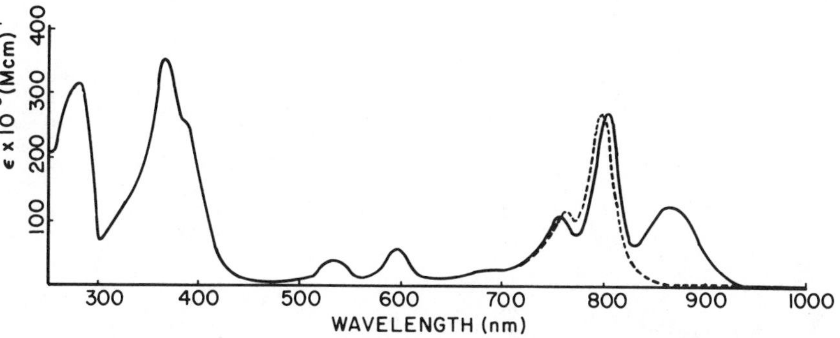

Fig. 4. The absorption spectrum of purified reaction centers from the photosynthetic bacteria *R. sphaeroides*. The dashed line shows the red portion of the absorption spectrum taken in the presence of actinic light. The light is strong enough to photooxidize all of the bacteriochlorophyll *a* that is photochemically active.

photosystem I, the narrowed ESR signal has been ascribed to a dimer of bacteriochlorophyll *a*. The quantum efficiency for the light-induced oxidation of the photoactive bacteriochlorophyll *a* dimer is unity.[32] Dynamic measurements of the primary events of bacterial photosynthesis demonstrate why the quantum efficiency is so high. After absorbing a photon the excited bacteriochlorophyll *a* dimer transfers an electron to a molecule of bacteriopheophytin *a* in only a few picoseconds.[33-36] It is not yet clear whether the bacteriopheophytin *a* is the first acceptor, or if a bacteriochlorophyll *a* molecule acts as an intermediate between the photoactive dimer and the bacteriopheophytin *a*.[37] The radiative lifetime of the dimer was estimated from the absorption spectrum to be about 20–30 ns.[38, 39] Both chlorophyll *a* and bacteriochlorophyll *a* have a singlet lifetimes *in vitro* of several nanoseconds.[40, 41] Thus the initial photochemical step is at least a thousand times faster than any competitive process. Once formed the bacteriopheophytin *a* anion quickly transfers an electron to one of the quinone molecules in only 150–250 ps.[33, 34] Rapid removal of the electron from the neighborhood of the dimer cation also maintains a high quantum efficiency. When photochemistry is blocked by chemical reduction of the quinone prior to the flash, the bacteriopheophytin *a* anion lasts for ~10 ns (at room temperature) prior to transferring the electron back to the bacteriochlorophyll *a* dimer to form the triplet and the ground state of the bacteriochlorophyll dimer.[42]

The primary events of bacterial photosynthesis are outlined in Fig. 5. The diagram is drawn such that the energy of each intermediate is represented by its position along the ordinate.[43] From this schematic

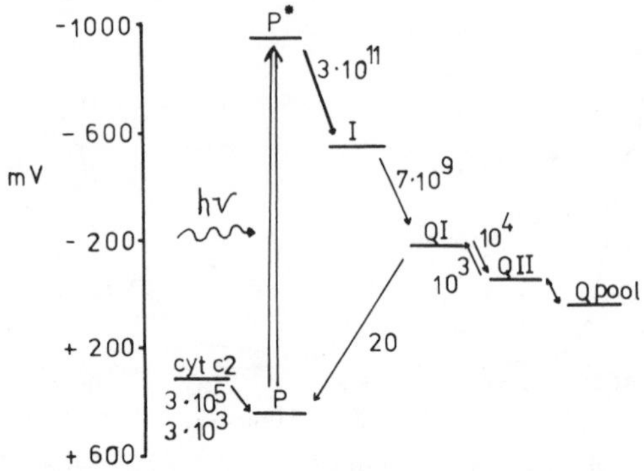

Fig. 5. Schematic of the primary events of bacterial photosynthesis. An approximate redox scale is shown on the left. The numbers above the arrows are the rate constants for the electron transfer reactions between the components. *P* is the photoactive bacteriochlorophyll *a*. I is the intermediate acceptor believed to be bacteriopheophytin *a*. The *Q*'s are ubiquinone molecules.

representation it is already clear how nature was able to design a system that has such a high quantum efficiency. For each step there is a loss of free energy large enough to assure a back-reaction much slower than the next reaction in the forward direction. The free-energy loss for each electron-transfer reaction is minimized to obtain the largest possible conversion of the photon energy into a charge separation. To a lesser degree the free energy drop is also minimized so that the forward reaction is not slowed by unfavorable Franck–Condon factors.[44, 45] Of course the distance between donors and acceptors is crucial to maintaining a high efficiency. Protein structure must also play a role. At present the contribution of the protein structure to the overall efficiency is not known.

The most critical step in the conversion of solar energy to chemical energy in both bacteria and green plants is the first one. Why nature chose chlorophyll *a* and bacteriochlorophyll *a* instead of some other substance to perform this crucial reaction has been the motivation for *in vitro* studies of chlorophyll.

III. CHLOROPHYLL *IN VITRO*

While biochemists were trying to isolate smaller and smaller functional units of the photosynthetic apparatus, a number of chemists were trying to understand the function of isolated chlorophyll *a* in solution. This work began over three decades ago. During the last thirty years a great deal of information has been learned about the photochemistry and the photophysics of chlorophyll *a in vitro*. From these studies have come a number of models that mimic *in vivo* chlorophyll *a*.[11] These will be discussed in the next section after a brief review of what is known about the chlorophyll molecule in solution.

The structure of chlorophyll *a* and bacteriochlorophyll *a* are found in Fig. 1. Heating of chlorophyll *a* in pyridine results in loss of the carbomethoxy group on ring V, producing pyrochlorophyll *a*. Pyrochlorophyll *a* is more stable in solution than chlorophyll *a*. The source of this stability is the removal of the β-keto ester. This group is well known to be sensitive to oxidation, especially in the presence of light.[46] Because of this stability pyrochlorophyll *a* is often used instead of chlorophyll *a* for *in vitro* studies.

A. Aggregation of Chlorophyll *a in Vitro*

In solution, the central magnesium atom prefers to be either five- or six-coordinate.[10, 47] In dry nonnucleophilic solvents, chlorophyll *a* will undergo self-aggregation to assure that the Mg atom is not coordinately unsaturated.[47] In a solvent such as carbon tetrachloride chlorophyll *a* forms only dimers even over a large concentration range.[48, 49] The fluorescence quantum yield from these dimers is considerably lower than that of monomeric chlorophyll *a*. The structure of these dimers is reasonably well understood based on the NMR and IR data.[47, 50, 51] The data reveal that there is an interaction between the keto carbonyl of one chlorophyll *a* molecule with the Mg of the second (Fig. 6). On the NMR time scale, there is a rapid equilibration between the two identical dimer structures.[50, 51] In hydrocarbon solvents such as hexane, the aggregate size grows with increasing concentration forming large oligomers.[49, 52] The interaction between macrocycles is identical to that of the dimers formed in CCl_4. Addition of nearly stoichiometric amounts of nucleophiles containing atoms such as O, S, or N breaks up the aggregate structure.[47] The coordination number of the Mg depends on the strength and the concentration of the nucleophile. Some nucleophiles such as dioxane contain more than one atom that can bind to the chlorophyll. In this case, oligomers of the following structure develop.[47, 53]

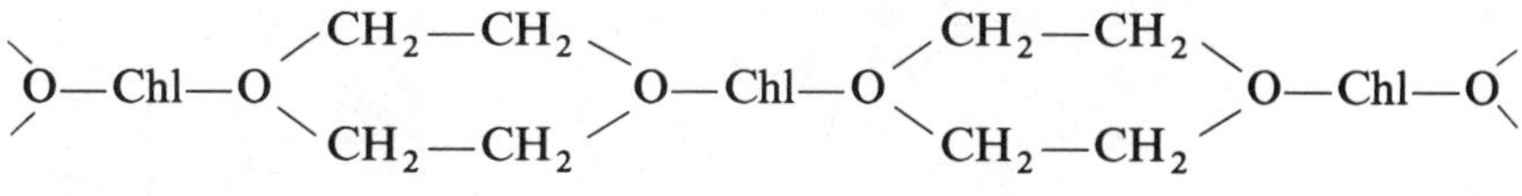

A bifunctional ligand such as ethanol or water contains both nucleophilic and electrophilic atoms.[54] Addition of the bifunctional ligands to chlorophyll *a* in hydrocarbon solvents results in the formation of a new type of

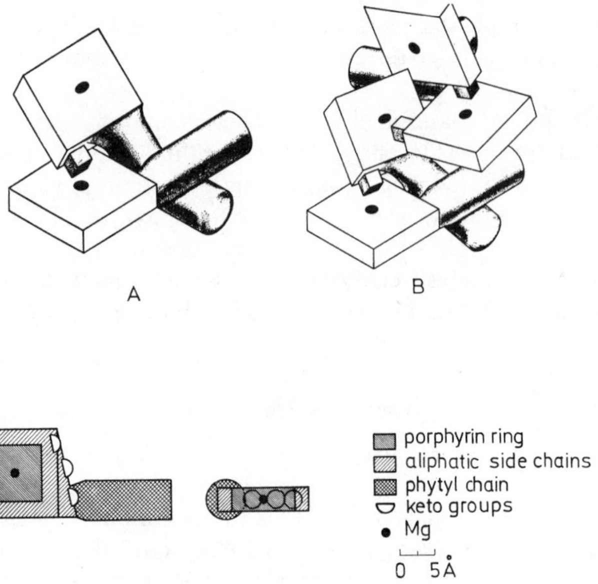

Fig. 6. Schematic structures of chlorophyll *a* species. (A) dimer; (B) tetramers. The dimensions of the Courtauld models are shown at the bottom of the figure. (From Ref. 10.)

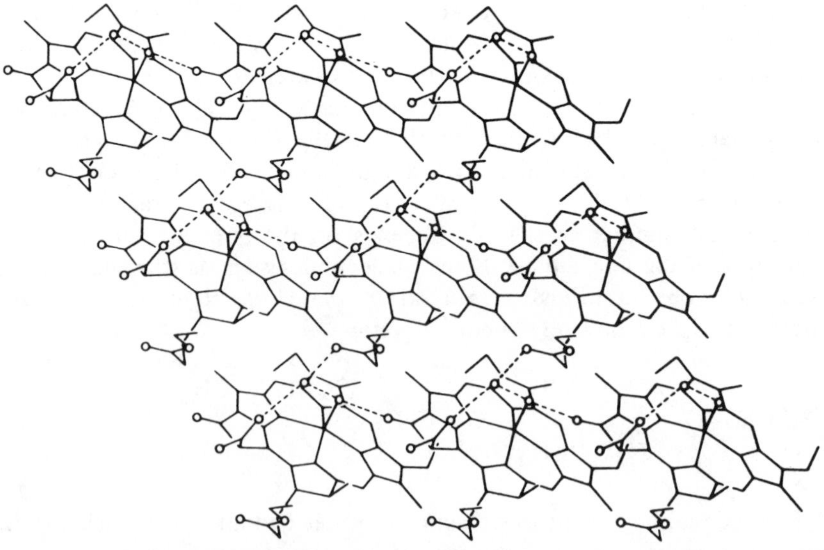

Fig. 7. One layer in the crystal structure of ethyl chlorophyllide $a \cdot 2H_2O$. Dashed lines represent hydrogen bonds. (From Ref. 55.)

oligomer. The oligomer is formed by the binding of the nucleophilic atom (e.g., O) of the bifunctional ligand to the Mg atom of one chlorophyll *a*, and the binding of the electrophile (almost always H) to the carbonyl oxygen on ring V of a second chlorophyll *a* molecule. When the long hydrocarbon chain, known as the phytyl chain, is replaced with an ethyl chain, the chlorophyll *a* oligomer that is formed can be grown into a crystal. The X-ray structure for this crystal has in fact been determined and is found in Fig. 7.[55]

B. Photochemistry of Chlorophyll *a* and Bacteriochlorophyll *a in Vitro*

Under solvation conditions where it is assured that chlorophyll exists only as a monomer the chlorophyll *a* macrocycle absorbs at about 665 nm while the related molecule bacteriochlorophyll *a* absorbs at 770 nm.[46, 56] For both molecules small solvent shifts can be measured, but they never exceed a few nanometers. Excitation of monomeric chlorophyll results in almost no electron transfer from the excited singlet state.[57] Instead, the addition of electron acceptors results in quenching of the chlorophyll *a* excited state. It has been postulated that during the collision a rapid electron transfer reaction between the singlet-excited chlorophyll and the acceptor takes place.[58, 59] Before the ions can become solvated and separated, a back-electron transfer between the reduced acceptor and the oxidized chlorophyll molecule takes place. Conservation of spin requires the back-transfer to return the chlorophyll to its ground state. If the ion pair lasts for several nanoseconds, triplet chlorophyll could also be formed via a hyperfine interaction.[60] No evidence for the formation of triplets has been found when high concentrations of quenchers are used to intercept the singlet excited state of chlorophyll *a*.[57] This is consistent with the belief that the ion pair has a very short lifetime, because the hyperfine interaction requires at least hundreds of picoseconds to produce a change in the spin state. When the electron acceptor has a large positive charge, electrostatic repulsion can separate the chlorophyll cation from the reduced acceptor. Even in this case, the efficiency is low. When chlorophyll *a* is in the triplet excited state, the forward transfer can again be fast, but the back reaction to the singlet ground state of chlorophyll is spin-forbidden.[61] Hyperfine interactions might induce some transfer back to the singlet ground state. Again the lifetime of the ion pair in the solvents studied is too short for the hyperfine mechanism to function.

The dimers formed in nucleophile free carbon tetrachloride absorb (λ_{max} 680 nm) to the red of monomeric chlorophyll.[62] Their photochemical activity has not been studied, since most electron acceptors are nucleophilic and would disaggregate the dimers. The larger oligomers formed in hydrocarbon solvents or through the addition of bifunctional ligands absorb very much farther in the red than does the monomer. For example, the water oligomer absorbs at 740 nm.[54] The water oligomer is the only one of the chlorophyll *a* species for which in the absence of any acceptor a photoreversible ESR signal has been detected.[63]

C. The ESR Linewidth of the Cation of Chlorophyll *a in Vivo* and *in Vitro*

The ESR linewidth of the cation of the water oligomer is a factor of ten or so narrower than that seen for the cation of monomeric chlorophyll *a*. It has been shown that the narrowed signal is due to a delocalization of the unpaired electron over a number of chlorophyll *a* molecules.[63] In fact, it has been shown that the linewidth is decreased in width by the factor $\sqrt{N}$ where N is the number of molecules that share the electron. This observation and the accompanying theoretical analysis of the mechanism for the line narrowing were crucial to the understanding of the ESR spectrum of the cation of photoactive species in photosystem I and in photosynthetic bacteria.[63] The narrowing by a factor of the $\sqrt{2}$ of the linewidth of the cations of chlorophyll *a* (from photosystem I) and bacteriochlorophyll *a* observed *in vivo* indicated that the oxidized form of the photoactive species is a dimer. A study of plants and bacteria containing deuterated chlorophylls, for which the hyperfine interaction is larger, also have an ESR

TABLE I
Observed and Calculated Linewidths (ΔH) of ESR Signal I in Plants and Bacteria[a]

	In Vivo ΔH (gauss)	
Organisms	Observed	Calculated
[^{1}H]*Syneccochocus lividus*	7.1±0.2	6.6±0.3
[^{2}H]*S. lividus*	2.95±0.1	2.7±0.1
[^{1}H]*Chlorella vulgaris*	7.0±0.2	6.6±0.3
[^{2}H]*C. vulgaris*	2.7±0.1	2.7±0.1
[^{1}H]*Scenedesmus obliquus*	7.1±0.2	6.6±0.8
[^{2}H]*S. obliquus*	2.7±0.1	2.7±0.1
[^{1}H]HP700	7.0±0.2	6.6±0.3
[^{1}H]*Rhodospirrilum rubrum*	9.1±0.5	9.1±0.4
[^{1}H]*R. rubrum*	9.2±0.6	9.1±0.4
[^{1}H]*R. rubrum*	9.5±0.5	9.1±0.4
[^{1}H]*R. rubrum*	4.0±0.5	3.8±0.1
[^{2}H]*R. rubrum*	4.2±0.3	3.8±0.1
[^{1}H]*Rhodopseudomonas sphaeroides*	9.6±0.2	9.1±0.4

[a]Taken from Ref. 63.

signal *in vivo* narrower by the $\sqrt{2}$ than that measured *in vitro*.[63] All of these results are summarized in Table I. ENDOR permits the observation of the hyperfine interaction at specific sites within the molecule. Therefore it is a much more sensitive probe of the electron distribution. ENDOR studies support the interpretation that the photoactive species is a dimer. The application of ENDOR also indicated that the dimer must be symmetrical.[64–67] Once it became clear that the photoactive species was a dimer; it was reasonable to assume that the red shift in the absorption of chlorophyll *a* from 665 nm *in vitro* to 700 nm in photosystem I and for bacteriochlorophyll *a* from 770 nm in solution to 870 nm in the reaction center, was due to an electronic interaction within the dimer. A number of dimeric models have been suggested over the past decade for the structure of the photoactive dimer in photosystem I. Although, these are reviewed in Ref. 11, they will be briefly discussed.

D. Models for the Photoactive Dimer

The main feature of all of the dimer models is that the planes of the two chlorophyll *a* molecules are parallel and that one or two O—H bridges are used to link the two macrocycles. The oxygen atom of the O—H bridge binds to the central Mg atom of one of the two chlorophyll *a* molecules, while the proton hydrogen bonds to one of the carbonyl oxygens of the second chlorophyll *a*. The only disagreement between the various models is the choice of the carbonyl group to which the proton bonds and the number of O—H bridges. Even though these variations are minor, they result in large structural differences. For example, the symmetry of the model is greatly affected by the number of O—H bridges and to a lesser degree by the choice of the carbonyl group. The plane-to-plane distance is influenced by the choice of carbonyl group to which the O—H bridge(s) is hydrogen bonded. The model shown in Fig. 8 uses the carbonyl on ring V and has an estimated planar separation of 0.36 nm.[68] The model in Fig. 9 utilizes the oxygen atom from the carbomethoxy as a site of attachment. It has a much larger separation of 0.57 nm. The distance between planes is of course crucial in determining the electronic interaction between the two macrocycles and therefore the size of the red shift in the absorption spectrum and the degree of electron delocalization. A number of theoretical calculations have been carried out,[68, 70–72] which demonstrate how one or the other structure is capable of reproducing the 700 nm absorbance seen *in vitro*. It is clear that for a large molecule such as chlorophyll *a*, semiempirical, let alone *ab initio*, calculations are not accurate enough to predict the subtle features in the electronic structure that would distinguish one model from the other as being the correct one. However, they may indicate what features of the proposed structures are important.

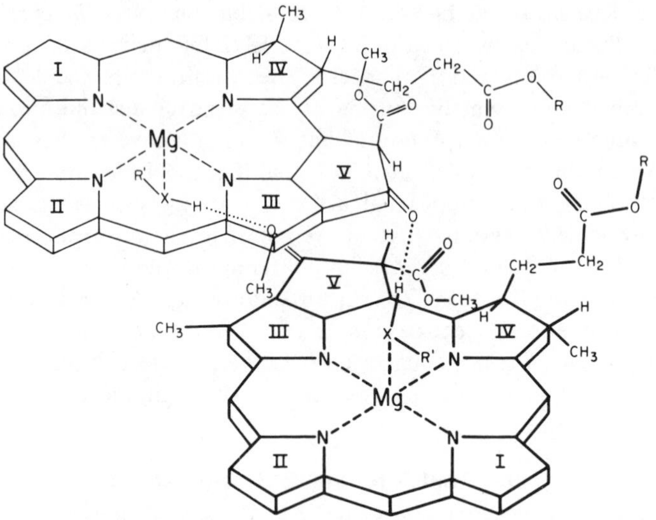

Fig. 8. Proposed structure for the photoactive chlorophyll *a* dimer in which the carbonyl oxygen on ring V is hydrogen bonded. *R* is the phytyl chain. *R'* can be a hydrogen atom or an alkyl group. It has been postulated that X could be 0, N, or S. (From Ref. 68.)

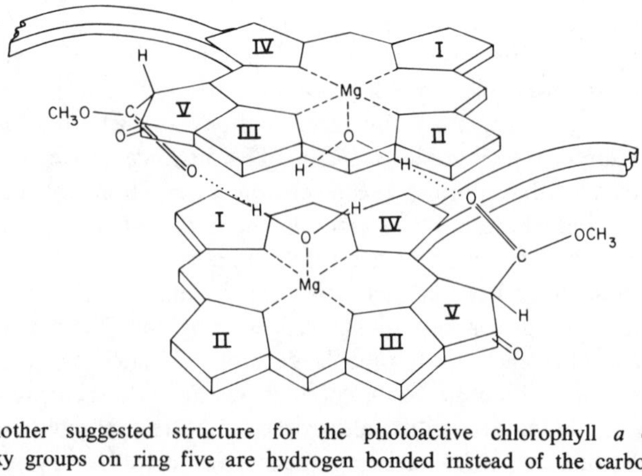

Fig. 9. Another suggested structure for the photoactive chlorophyll *a* dimer. The carbomethoxy groups on ring five are hydrogen bonded instead of the carbonyl groups. (From Ref. 69.)

E. Generation of *in Vitro* Dimers

A number of groups have constructed chlorophyll *a* or pyrochlorophyll *a* dimers *in vitro* to determine which of the proposed structures is correct. Since the entropy of dimerization is strongly negative and the enthalpy is probably negligible, the equilibrium constant for dimerization is small. By cooling chlorophyll *a* in the presence of an excess of water or ethanol, it is possible to overcome the entropy of dimerization and shift the equilibrium in favor of the dimer.[68, 73] Upon cooling a solution of chlorophyll *a* in a solvent such as toluene, the absorbance at 665 nm disappears to a large extent and a new absorbance around 702 nm is observed (Fig. 10). In the presence of molecular iodine, which has a redox potential below that of the chlorophyll *a* monomer, the absorbance at room temperature is unaffected.[73] As the dimer is formed by cooling, it becomes oxidized by the iodine. The oxidation of the dimer results in a bleaching of the red-most absorbance band (i.e., 702 nm). Thus upon cooling, not only is the 665 nm absorbance absent, but the 702 nm absorbance fails to be observed. In contrast, in solvents that prevent dimer formation, the absorbance of monomeric chlorophyll *a* was not affected by cooling in the presence of I_2. This demonstrates a third property of the dimer, namely, a

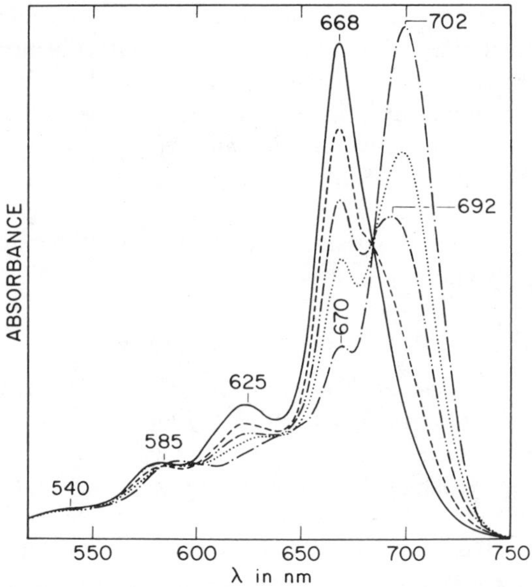

Fig. 10. Visible absorption spectrum of 0.094 M chlorophyll *a*, and 0.14 M ethanol in toluene at 298 K—, 273 K---, 247 K--···--, 224 K···, and 175 K-·-·. (From Ref. 68.)

lowering of its oxidation potential with respect to the potential of monomeric chlorophyll *a*. This is very important property for *in vivo* dimers, because it prevents oxidation of the antenna molecules by the photo-oxidized reaction center chlorophyll *a*. The smaller redox potential can easily be understood by a simple exciton picture. Exciton interaction between the two highest occupied orbitals on each chlorophyll macrocycle results in the formation of two new orbitals, one lower and the other raised in energy. The energy needed to remove an electron from the higher of the two new orbitals is decreased, thereby lowering the oxidation potential. The cation formed by oxidation with tetranitromethane at low temperatures had an ESR signal that was appropriately narrowed,[68] further supporting the contention that this species was a dimer. The two proposed structures for the low-temperature complex are those of Figs. 8 and 9.

In summary, the dimer formed by cooling chlorophyll *a* in the presence of ROH (R=H, CH_3CH_2) has three features that mimic those seen in photosystem I. The absorbance maximum is near 700 nm, the unpaired electron of the cation is delocalized over two macrocycles, and the oxidation potential is lower than that of the monomer. Despite their ability to mimic the *in vivo* dimer, these models have several drawbacks. Although the primary events of photosynthesis do take place at 4 K,[2] plants and bacteria usually operate in a very narrow temperature range near 290 K. Excited state dynamics such as internal conversion and intersystem crossing are very much affected by temperature.[74] Their influence on the singlet lifetime and photoactivity therefore cannot be determined with these complexes. More important, the ultimate test of any model is its photoreactivity. It is necessary (*in vitro*) for an electron acceptor to diffuse to within the Debye radius before an electron-transfer reaction can proceed. Thus solutions at 100 K where the viscosity is very high or the solvent already a glass, make it impossible to study the photochemical reactions of dimers formed by cooling.

F. Covalently Linked Dimers

The entropy of dimerization was overcome by covalently linking two molecules of pyrochlorophyll *a* together.[75] Later two chlorophyll *a* macrocycles were also successfully joined.[76] The covalently linked dimers of chlorophyll *a* or pyrochlorophyll *a* exist in one of three forms (Fig. 11). In the presence of nucleophiles the two molecules appear to be independent of one another. The absorbance and fluorescence spectra are identical to monomeric chlorophyll *a*. The only difference between monomeric chlorophyll *a* and the dimer in the presence of nucleophiles is a much smaller cross-section for stimulated emission. Under nucleophile-free conditions in

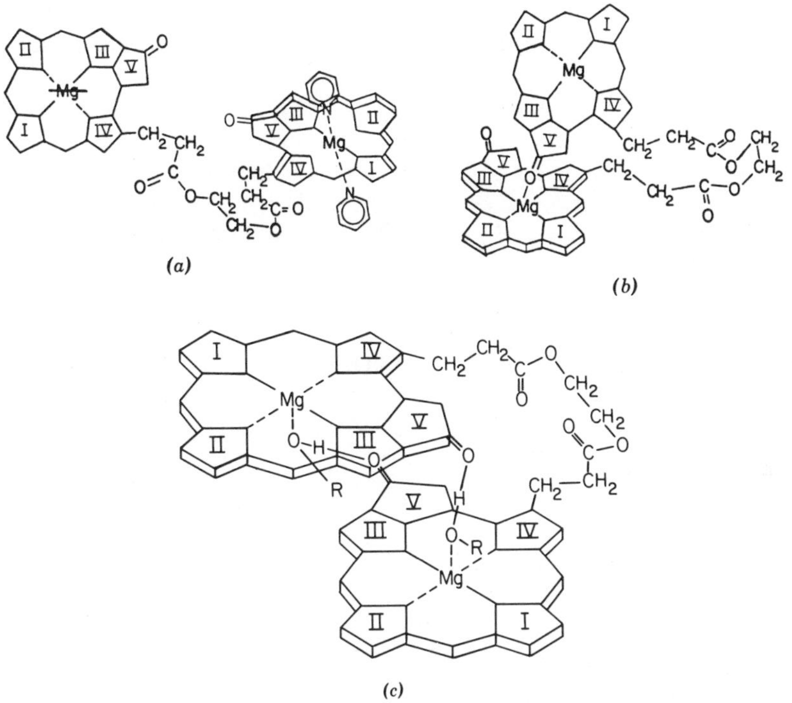

Fig. 11. The covalently linked pyrochlorophyll *a* dimer. The proposed structures are for three distinct solution conditions. (a) In neat pyridine the dimer is thought to be in its open form. (b) In nucleophile free solvents, the dimer is thought to be in its dry configuration. (c) In nucleophile free solvents to which a hundred fold excess of alcohol has been added, the dimer is in its folded configuration.

hydrocarbon solvents or in carbon tetrachloride, the two macrocycles interact with one another in the same way monomeric chlorophyll *a* does when it forms "dry" dimers (compare Figs. 6 and 11b). Addition of a bifunctional ligand (i.e., ROH) results in folding of the dimer to form a complex with a structure similar to that formed when chlorophyll *a* is cooled in the presence of water or ethanol (compare Figs. 8 and 11c). Both pyrochlorophyll *a* and chlorophyll *a* form the folded complex. However, 25% of the chlorophyll *a* dimers cannot be successfully folded at room temperature. This is because of partial epimerization about the C_{10} carbon which results in the formation of a stereoisomer of chlorophyll *a*. When two different stereoisomers are linked together they cannot be folded, since one of the two carbomethoxy groups interferes with the close approach of the two macrocycle planes. For this reason and also because pyrochlorophyll *a* is more photochemically stable, it has been the most often studied dimer.

Upon folding, the pyrochlorophyll *a* and the chlorophyll *a* dimer undergo a red shift in their absorption spectrum to 696 nm (Fig. 12). The cation formed by chemical oxidation appears to have its unshared electron delocalized over the two molecules.[76] The oxidation potential is reduced by about 100 mV from that of monomeric pyrochlorophyll *a*. Judging from the zero field splitting parameters, the triplet state of the dimer also appears to be delocalized over the two chromophores.[78] Such a triplet delocalization is found in reaction centers from photosynthetic bacteria.[79] Thus both the synthetic pyrochlorophyll *a* and chlorophyll *a* dimers in

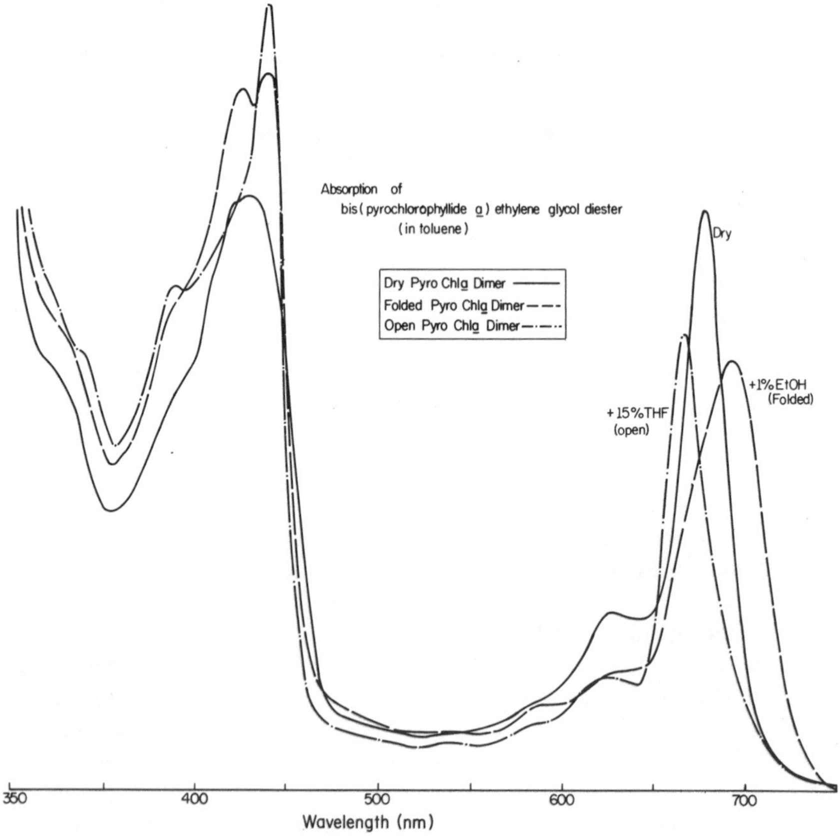

Fig. 12. Absorption spectrum of the pyrochlorophyll *a* dimer in its three configurations: Dry, —; Open-·-; Folded-----.

their folded configuration appear to have all of the static features associated with the photoactive chlorophyll *a* in photosystem I.

A covalent linkage between two bacteriochlorophyll *a* molecules results in a species which in the presence of the proper bifunctional ligand (e.g., ethanol) undergoes folding like the chlorophyll *a* dimer.[80] It does not reproduce the *in vivo* absorption spectrum, absorbing at 803 nm instead of 870 nm. Yet, it does have a delocalized cation and a delocalized triplet state.

All of the dimers discussed above suffer from the shortcoming that they require interaction with an external ligand such as ethanol. Without the alcohol they do not form the configuration that mimics the photoactive species. Binding of the ligand is controlled by an equilibrium. Thus a 10- to 100-fold molar excess of the folding agents such as alcohols or thiols are needed to attain complete conversion of the covalently linked dimers to the 700 nm absorbing configuration. Most solvents with a high dielectric constant have an unshared pair of electrons. This makes them nucleophilic. Since the solvent concentration is always in excess of 10 M, it can readily compete with the bifunctional ligands which are present at a concentration of 10^{-2} M or less. This competition results in the formation of dimer molecules which are principally in the open configuration (see Fig. 11a). Dissolving the dimers in solvents that are bifunctional ligands such as ethanol does not solve this problem because a second equilibrium reaction favors formation of hexacoordinate chlorophyll *a*. Chlorophyll *a* with 6 ligands (4 nitrogens and 2 external ligands) will not form the 700 nm absorbing species. This problem severely limits the solvation conditions under which the special dimer can be studied. To overcome this problem, the pyropheophytin *a* macrocycle was modified by oxidation of the vinyl group to an alcohol.[81] Following the oxidation, Mg could be reinserted into the molecule, the modified pyrochlorophyll *a* had an absorption maximum at 650 nm in contrast to pyrochlorophyll with a λ_{max} at 665 nm. The alcohol group provides a second site of attachment. A double esterification results in a structure shown in Fig. 13. The interplanar distance can

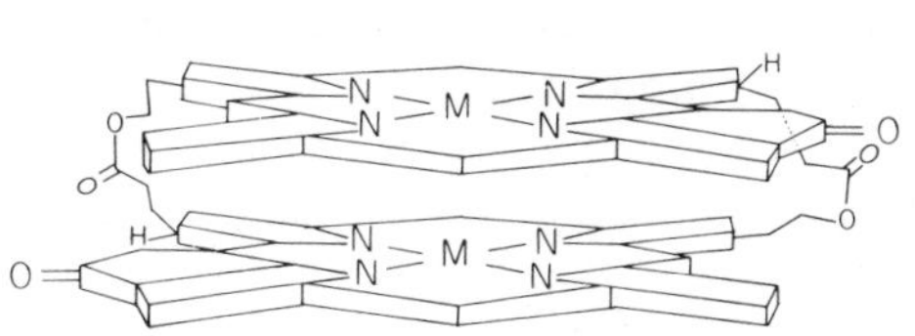

Fig. 13. Proposed structure for the pyrochlorophyll *a* cyclophane.

be adjusted by changing the size of the ester linkages.[82] It varies from 0.4 to 1.0 nm. Space-filling models suggest that the Q_y transition dipoles of the two molecules are on the average at 90° with respect to one another. This is in contrast to the Q_y transitions dipoles of the two macrocycles of the singly linked dimer (folded), which are nearly parallel.

The absorbance maximum of these cyclophanes is not shifted from that of the monomers. The only exception is that one dimer for which the macrocycles are believed to be within their van der Waals radii.[82] In this case the shift is only a few nanometers. The failure to observe a large red shift in the absorption spectrum of the cyclophanes with respect to the spectrum of the monomers is due to the orthogonal orientation of the transition dipoles. Their orientation inhibits any exciton interaction. The excited singlet state absorption spectrum of the Mg containing dimer is quite similar to that of the monomer. The triplet ESR also resembles the spectrum obtained from the isolated monomer. In contrast to the spectral observations the redox properties of the Mg containing dimer are very different from those of the isolated molecules. The oxidation potential is about 70 mV lower than that of chlorophyll *a*, while the reduction potential is about 150 mV more negative. The cation radical has a narrowed linewidth, indicating delocalization of the unshared electron. Thus it is possible for a pair geometry to exist, which can adequately account for both the lowered redox potential and the spin delocalization of the cation that is observed for the photoactive dimer *in vivo*, without giving rise to the optical spectra of the *in vivo* system. Perhaps the red shift in the absorption spectrum of the photoactive form is due to the influence of the protein environment and the presence of other chlorophyll molecules and not because of electronic interactions between the two chromophores. Further work both *in vivo* and *in vitro* are needed to determine the reason for the red shifts.

IV. TIME RESOLVED STUDIES OF CHLOROPHYLL *A* DIMERS

A. Photochemical Criteria for a Model System

Only tests of the static properties of the folded dimers have been used so far to verify that they are good models for the photoactive dimer found *in vitro*.[68, 73, 75, 76, 80] The most crucial test of course is a dynamic one, namely, whether or not the dimers undergo photochemistry in the singlet state and whether they are capable of stabilizing a charge separation. Monomeric chlorophyll *a* is capable of undergoing a stable and efficient photooxidation only when it is in its triplet level.[57, 59, 61] The dimers of pyrochlorophyll *a*, chlorophyll *a* and bacteriochlorophyll *a* undergo photochemical oxidation. This has been determined by observing a bleaching of the red-most absorbance band in the presence of electron acceptors.[68, 76, 80] Since all of the photochemical experiments were carried out with continuous illumination, or with submicrosecond time resolution, it is not clear whether the photochemical oxidation took place from the singlet or the triplet level. Also, the photochemical quantum yield has not been measured. In contrast, it has been well established that the primary events in photosynthetic bacteria occur within picoseconds. In photosystem I these events also take place rapidly, perhaps in only a few picoseconds.[33-37] In both systems photochemistry takes place from the singlet level. Also, the quantum yield in photosynthetic bacteria is unity,[32] in plants it is also believed to be close to 1.[83] Thus a high photochemical quantum yield and rapid electron transfer from the excited singlet level are important criteria that any model system must meet.

In searching for a suitable electron acceptor, it seems reasonable to just mimic what is known about the initial acceptors *in vivo*. In photosynthetic bacteria it has been well established that bacteriopheophytin *a* is one of the first electron acceptors.[33-37] In photosystem I of green plants a chlorophyll *a* dimer or monomer have been proposed as the acceptor.[22, 23, 92] Either chlorophyll *a* or pheophytin *a* would be excellent choices as electron acceptors. However, the singlet lifetime of the pyrochlorophyll *a* dimer in toluene is only 4 ns.[84] For a diffusion-controlled electron transfer reaction between the dimer and one of the *in vivo* acceptors to take place in a few nanoseconds would require a 10^{-1} to 10^{-2} molar concentration of pheophytin *a* or chlorophyll *a*. The molar extinction coefficient of these molecules is on the order of 40,000.[46] At a concentration of 10^{-2} M, the absorption of pheophytin *a* (or the chlorophyll *a* monomer) would be much too high. The solution of this dilemma is to link an electron acceptor such as pheophytin or chlorophyll to the dimer. Linking the dimer to an electron acceptor not only solves the diffusion problem, but also begins to mimic the photosynthetic reaction center.

B. Models for the Reaction Center

Such a reaction center model can be formed when a molecule of pyrochlorophyll *a* and one of pyrochlorophyll *b* are first connected with an ethylene glycol diester linkage.[85] The aldehyde linkage of the chlorophyll *b* (see Fig. 1) is reduced to an alcohol. A Mg or Zn atom is inserted into each ring. Then pyropheophorbide *a* is linked to the pyrochlorophyll *a* dimer through the previously prepared alcohol via an ester linkage. Addition of CH_3OH to nucleophile-free solutions of the trimer results in an absorption

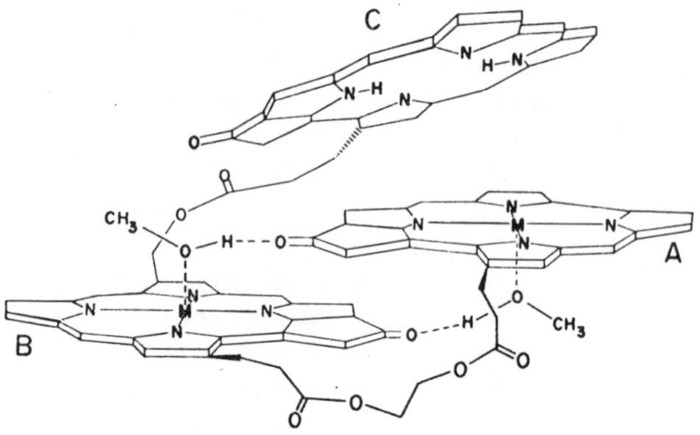

Fig. 14. Proposed average, solution structure of the trimer consisting of pyrochlorophyllide *a*, pyrochlorophyllide *b*, and pyropheophorbide *a* in the presence of hydroxylic solvents. (From ref. 85)

spectrum, which is a superposition of a dimer in the folded configuration, and a pheophytin *a* chromophore. Fluorescence is only observed from the dimeric portion of the complex, even when the pheophytin *a* is excited, indicating efficient energy transfer. The emission is quenched by a factor of 2 over that of a solution containing the folded dimer and free pheophytin *a* at the same concentration. The NMR spectrum of the trimer indicates interaction between the pheophytin *a* portion and the folded dimer part of the molecule. A schematic solution structure inferred from the NMR data is shown in Fig. 14.

Another type of reaction center model system is formed when three pyrochlorophyll *a* molecules are joined via a symmetric triester linkage.[82] Such a system is very interesting because of the proposed role of chlorophyll *a* as acceptor in photosystem I[16, 22, 27] and the possible role of bacteriochlorophyll *a* as an acceptor in photosynthetic bacteria.[37] The NMR spectrum of the pyrochlorophyll *a* trimer indicates that in the presence of a bifunctional ligand such as an alcohol, the three macrocycles take on a symmetric configuration. In contrast, optical absorption data indicate that the chromophores are not identical. A schematic representation of a structure that fits both the optical and the magnetic resonance data is shown in Fig. 15. Two of the macrocycles are in the proper configuration to produce a species absorbing at 700 nm. The third pyrochlorophyll appears free in solution, but is close enough to "switch" with one of the other two. At present, the photochemical properties of the trimer have not been explored. This molecule will provide us with a unique model for the study of photosynthesis and should provide a wealth of new information.

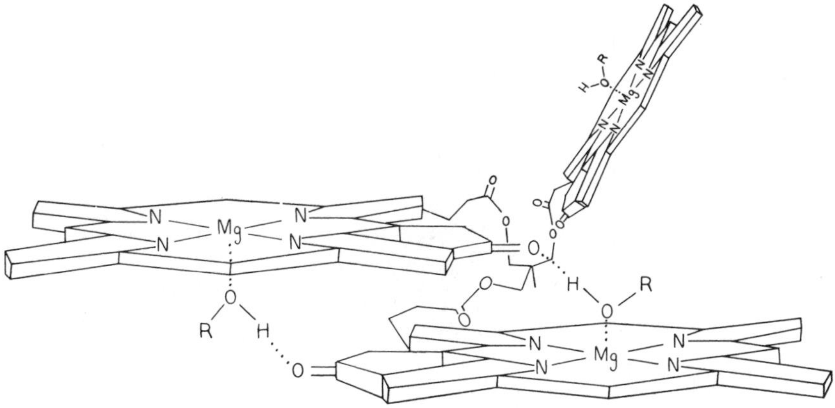

Fig. 15. Proposed solution structure of the pyrochlorophyllide *a* trimer.

A large number of alcohols ranging from ethanol to *p*-cresol can be utilized in the folding of the dimer.[87] By using an alcohol derivative of pheophytin *a*, pyropheophorbide *a* ethylene glycol monoester, to fold the dimer, it would be possible to link the pheophytin *a* chromophore to the dimer.[88] This procedure leads to the production of a complex whose proposed structure is shown in Fig. 16. This complex can only be formed when the solvent is completely free of all nucleophiles, especially water. This is probably due to the size of the pheophytin *a* macrocycle, since the binding of a smaller nucleophile to the pyrochlorophyll *a* would be favored. When the pheophytin *a* alcohol is added to a solution of dry dimer, a shoulder appears in the red-most absorbance band. This shoulder is due to the folding of the dimer by the alcohol. Coincident with the production of the shoulder, there is a quenching of the fluorescence from the dimer. The fluorescence was taken in a cell 2 mm × 10 mm in which the excitation is in the direction of 10 mm pathlength. Since the pheophytin *a* alcohol concentration is extremely high, its fluorescence is mainly quenched by self-absorption. The remainder appears to be quenched by energy

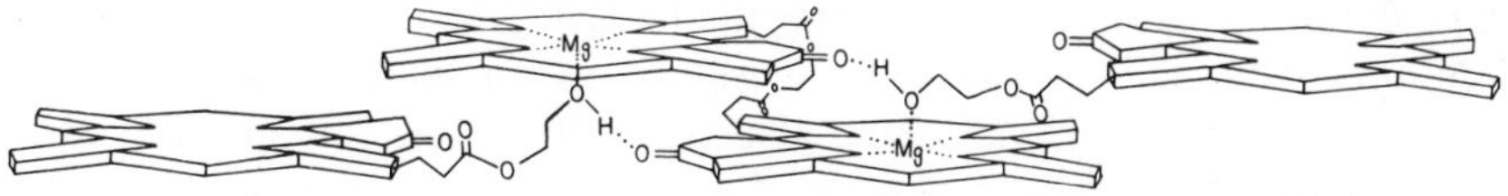

Fig. 16. 1a. Proposed structure of the complex formed through the interaction of the pyrochlorophyll *a*, dimer a, and pyropheophorbide *a* ethylene glycol monoester.

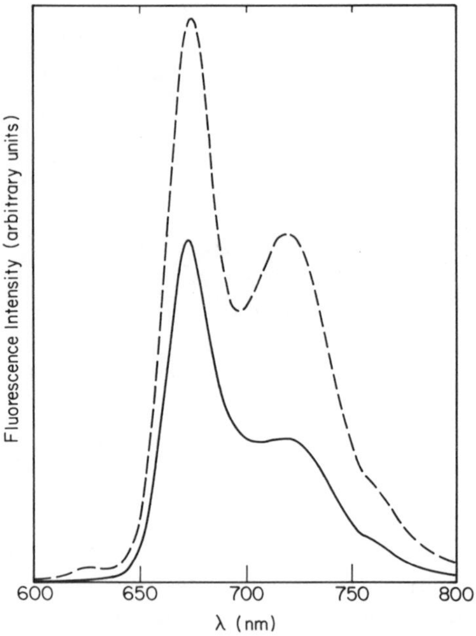

Fig. 17. Front face fluorescence from a toluene solution containing pyrochlorophyll *a* dimer and pyropheophorbide *a* ethylene glycol monoester—. The addition of ethanol disrupts the structure shown in Fig. 16 and results in an increase in emission---.

transfer to the dimer–pheophytin complex. Addition of ethanol disaggregates the complex producing free pheophytin *a* alcohol and the folded dimer. Front-face fluorescence detection does not suffer as much as from self-absorption. To minimize the inner filter effect for the front-face geometry, a 0.1 mm pathlength cell is used.[89] Because of the intense emission from the pheophytin *a*, the degree of fluorescence quenching does not seem as impressive. The front surface emission from the pheophytin *a* alcohol–pyrochlorophyll *a* dimer is shown in Fig. 17. Addition of ethanol to the solution destroys the complex. Equilibrium now favors folding of the dimer by the ethanol instead of by the pheophytin *a* alcohol. Therefore the pheophytin *a* macrocycle is no longer linked to the dimer, but is free in solution. As shown in Fig. 17 this results in an increase in the dimer emission at about 715 nm and also an increase in the pheophytin *a* fluorescence at 680 nm. Since the complex was formed with a twofold excess of pheophytin *a*, the 680 nm emission only grows by about a factor of 2.

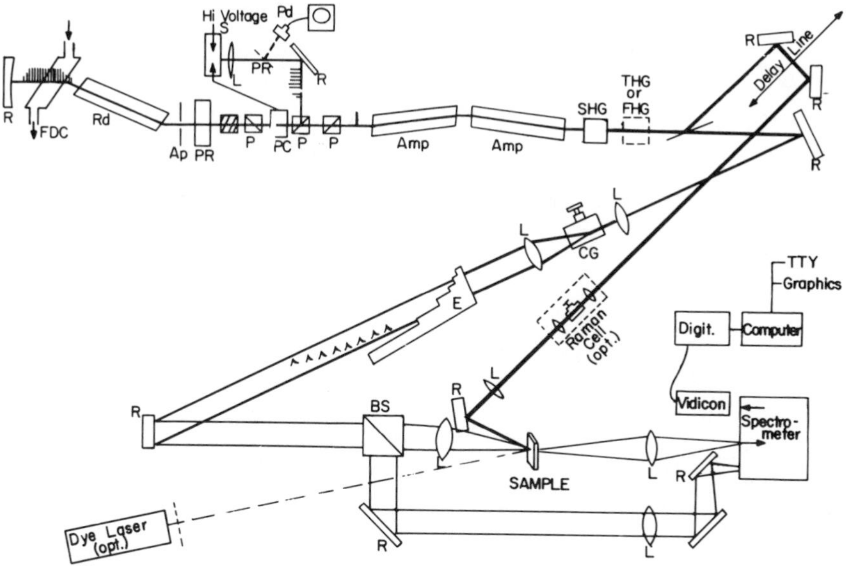

Fig. 18. Schematic of a picosecond absorption apparatus. R = reflector; FDC = flow in dye cell; Rd = neodymium laser rod; PR = partial reflector; P = polarizer; PC = Pockels cell; AMP = amplifier; SHG = second harmonic generator; L = lens; BS = beam splitter.

The observation of intense fluorescence quenching within the artificial reaction-center complex does not demonstrate that it performs efficient light-induced electron transfer, let alone stabilizes the charge separation. Time-resolved absorption has the capability of identifying the species present following excitation. From the magnitude of the fluorescence quenching, it is clear that any reactions that are taking place must be doing so on the picosecond time scale. A typical picosecond absorbance apparatus used to study the photochemistry of the dimer pheophytin complex is shown in Fig. 18. The details of this aparatus can be found in Ref. 90.

The excited state difference spectrum of the artificial reaction center (pheophytin *a*-pyrochlorophyll *a* dimer) is shown in Fig. 19. The salient features of this spectrum are a peak at 800 nm, bleaching at both 700 and 680 nm and a broadband absorbance between 650 and 450 nm. To interpret this spectrum one must compare it to spectra of the excited singlet states of the dimer and pheophytin, and the cation of the dimer and the anion of pheophytin (see Figs. 20 and 21).[82, 91]

The pheophytin *a* anion has a broadband absorbance between 700 and 900 nm, with a small peak near 790 nm. The cation of the pyrochlorophyll

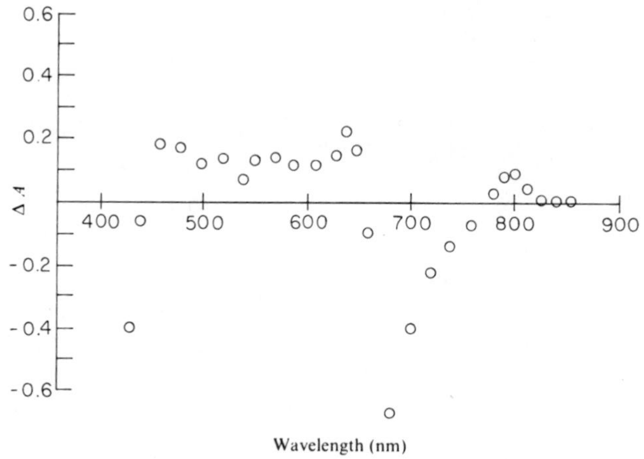

Fig. 19. Initial absorbance changes observed following 6 ps excitation with 530 nm light of toluene solution of 0.2 mM pyrochlorophyll a dimer and 0.8 mM pyropheophorbide a glycol monoester.

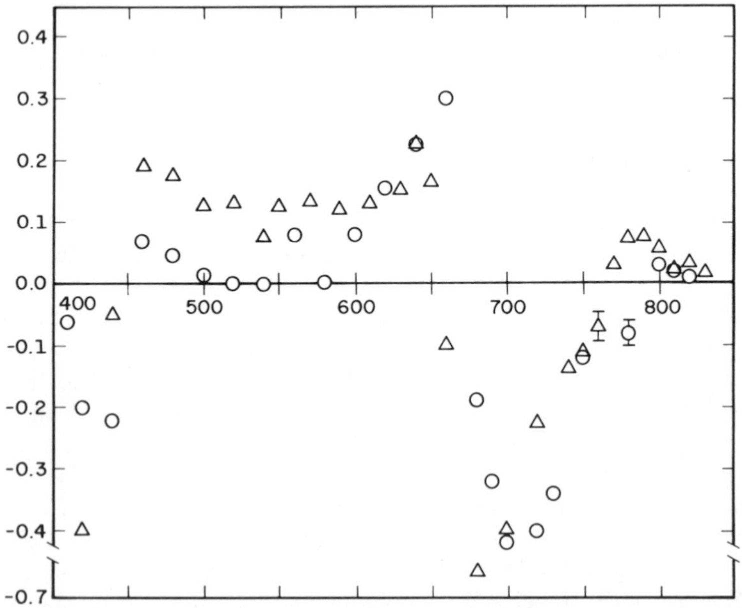

Fig. 20. Comparison of the excited state difference spectrum of the pyrochlorophyll *a* dimer in toluene (O) and the difference spectrum obtained upon exciting a dry toluene solution containing the pyrochlorophyll *a* dimer and pyropheophorbide *a* ethylene glycol monoester (Δ).

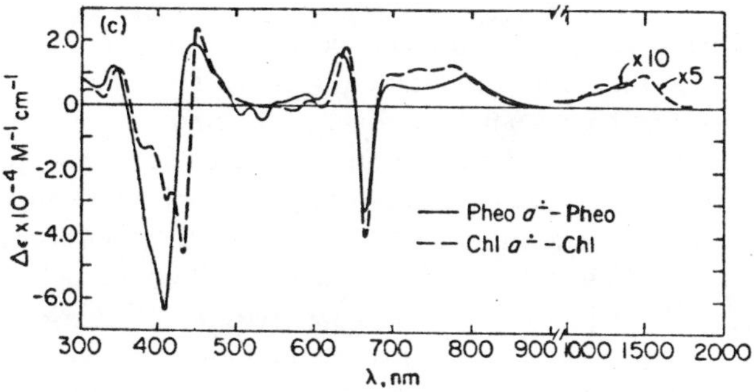

Fig. 21. Difference spectra of the pheophytin *a* anion, —; and the chlorophyll *a* anion,---. (From Ref. 91.)

a dimer[82] also has a broadband absorbance between 700 and 900 nm with a peak near 810 nm (Figure 22). The excited singlet state of pheophytin *a* has a nearly flat absorbance between 700 and 900 nm.[89] Thus the only two species which have actual peaks near 800 nm are the anion of pheophytin *a* and the cation of the dimer. Thus the small peak seen near 800 nm in Fig. 19 suggests that an anion of pheophytin *a* and a cation of the dimer are present. The bleaching both at 700 nm and 680 nm indicate that both the dimer and the pheophytin *a* part of the complex are being excited by the laser, but one cannot distinguish whether this is due to the singlet excited states of the dimer and pheophytin *a* or to the formation of a radical pair or a charge-transfer state. A broad absorbance in the visible part of the

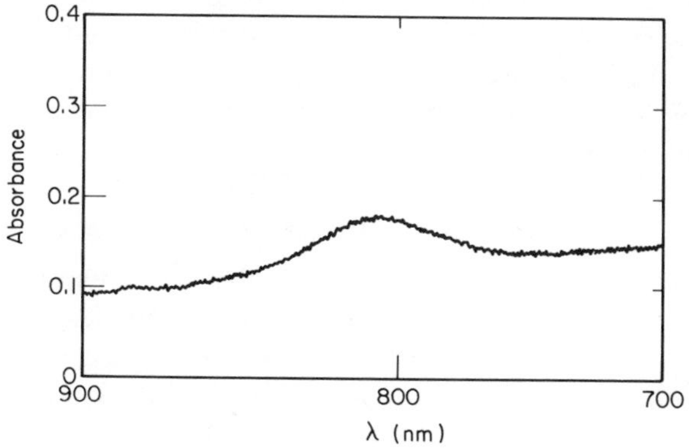

Fig. 22. The infrared absorption spectrum of the cation of the pyrochlorophyll *a* dimer.

spectrum around 500 nm is characteristic of both the excited states and the combination of the cation and anion spectra. However, at 540 nm the spectrum very much resembles a linear combination of the anion and cation signal, but does not look like the excited singlet of pheophytin *a*.

The absorbance changes throughout the region of observation disappear with an estimated decay time of 10–20 ns. This is much longer than the singlet lifetimes of pheophytin *a*[87] or the dimer,[84] and much shorter than the reported triplet lifetimes.[92] The combination of intense fluorescence quenching accompanied by a relatively long-lived difference spectrum are strong evidence for the existence of a separation of charges within the artificial reaction center following absorption of a photon. This separation of charges may result in two radical ions or be an excited state with a great deal of charge-transfer character. ESR measurements indicate that the state formed does not decay to a triplet, but returns to the ground state.[89] This too supports the idea that the complex undergoes a unique photochemistry that cannot be ascribed to isolated pheophytin *a* or the dimer.

The pheophytin alcohol–chlorophyll dimer complex reproduces faithfully most of the features associated with the primary reactions of photosynthetic bacteria. The absorption of a photon results quickly (< 10 ps) in the formation of a state that involves a cation of the photoactive dimer and an anion of pheophytin. The state is not fluorescent and decays in 10–20 ns when further photochemistry is blocked. These observations are in contrast to a number of complexes that have been recently synthesized by direct linkage of pyrochlorophyll[86] or pyropheophytin monomer to the pyrochlorophyll dimer.[85, 86] Although both these complexes are somewhat simpler to work with, since they can be prepared free of any other components, they show only a limited amount of fluorescence quenching. This is somewhat disappointing, but indicates how important the orientation of the dimer with respect to the electron acceptor must be to insure a rapid electron transfer reaction in the excited state. Once the structures of the several dimer–acceptor complexes have been determined, it is hoped that we will better understand the conditions necessary for effective electron-transfer reactions.

C. Electron-Transfer Reactions in Cyclophanes

The uncertainty in the structure of the pyrochlorophyll dimer–pheophytin complexes makes it difficult to identify the features that make the electron-transfer reaction between the excited singlet state of the photoactive and the initial electron acceptor so efficient. A number of research groups have taken the tack of using aromatic electron donors and acceptors in a well-defined geometry to study the electron-transfer reaction in general and to mimic the photosynthetic process in particular. The most thoroughly studied systems are of the type:

$$D—(CH_2)_n—A$$

where D is an electron donor such as dimethyl aniline and A is an electron acceptor such as anthracene or pyrene.[93–95] A tremendous amount of information on electron-transfer reactions has been learned from studying these systems. However, they lack one of the very important features of the photosynthetic apparatus, namely, that the electron donor (excited dimer) and the electron acceptor (probably a chlorophyll *a* or pheophytin *a* monomer) are at a fixed orientation and distance. This distance must play a crucial role in determining the rate of the forward reaction as well as any back-reaction.

A review of the theory of electron transfer reactions in solution and how it pertains to reactions of chlorophyll *a* can be found in Ref. 59. A brief summary of the salient features will be presented. During an electron transfer reaction in solution, the excited singlet donor (D*) and the acceptor (A) approach one another to form an encounter complex [(AD)*].

$$A + D^* \xrightarrow{k_1} (AD)^* \xrightarrow{k_2} [AD]^* \xrightarrow{k_3} (A^- \, D^+) \xrightarrow{k_4} A^+ + D^+$$
$$(AD)^* \xrightarrow{k_5} (A^- \, D^+)$$

The encounter complex can also be formed from A* and D. Within a few picoseconds the encounter complex undergoes vibrational relaxation to form an exciplex.[96] The solvent is reorganized about the exciplex in tens to hundreds of picoseconds to form an equilibrated exciplex, ([AD]*). At any time after formation of the encounter complex, electron transfer can proceed forming a charge-transfer complex [(A$^-$D$^+$)].[97, 98] The formation of the charge transfer complex usually manifests itself in the quenching of the fluorescence and the appearance of an absorption spectrum, which is a superposition of the anion spectrum of the acceptor and the cation spectrum of the donor.[99] The amount of charge-transfer character and the extent of the fluorescence quenching correlates with the redox properties of the donor and the acceptor as well as the solvent dielectric.[59, 97, 100–102] Rapid geminate recombination of the charge-transfer pair can lead to the donor and the acceptor being in the ground state, the triplet-excited state, or the singlet-excited state.[103–105] Geminate recombination will mostly take place at an encounter distance different than the one at which the original encounter complex is formed. Thus if the donor and the acceptor can be separated at a fixed distance during the reaction, it is possible to have a rapid forward electron-transfer reaction while slowing down the reactions that destroy the ion pair.[59] This situation cannot be attained in donor–acceptor complexes linked with only one hydrocarbon chain. However, it can be approached in cyclophane compounds in which two or more linkages are used.

Besides the doubly linked chlorophyll dimer[81] described on page 597 a number of porphyrin cyclophanes have been synthesized. One of these is *tetrameso*-[*p*, *p*′-(2-phenoxyethoxycarbonylphenyl)]-*strati*-bisporphyrin. This compound demonstrates exciton interaction in the form of red shifts with respect to the absorption and emission spectra of the monomer.[106] The zinc chelate spectrum of the diporphyrin also has a red shift in the absorption spectrum with respect to the monomer. The fluorescence of the zinc chelate is broadened and partially quenched when compared to the monomer. The authors suspected the quenching was due to electron transfer. The red shift and the fluorescence quenching are probably due to the forced stacking of the chromophores by the four covalent linkages. Another group has also synthesized a porphyrin dimer in which two covalent bonds force stacking of the macrocycles. However, this group has not reported any changes in the absorption spectrum between the monomer and the dimer. The NMR spectrum of the free base dimer and the one containing zinc are different.[107] This difference has been interpreted to be due to "parallel sliding of the two free-base porphyrins" and a "complete face-to-face configuration" for the zinc-containing complex. A series of covalently linked dimers has been synthesized in which the face-to-face separation is about 0.65 nm.[108] Visible absorption spectra and ESR spectra indicate that there is interaction between the two porphyrins. The motivation for synthesizing this series of dimers was to produce models for the enzymes that perform reduction of O_2 to H_2O, the reverse of the photosynthetic process.

A diporphyrin cyclophane (Fig. 23) has been synthesized in which the macrocycle planes are cofacial and separated by about 0.4 nm.[109, 110] It is possible to preferentially insert one or two metal atoms into these dimers. Thus it is possible to study three separate cyclophanes, containing two, one, or no magnesium atoms. The fluorescence emission of the dimers is much lower than that of the monomers with the dimetal compound having an emission three times larger than the other two. For the dimetal and the nonmetallated dimer the estimated energy of the ion pair is above the energy of the relaxed excited singlet state.[111] The time-resolved picosecond absorption spectrum of these compounds decay slowly from what appears

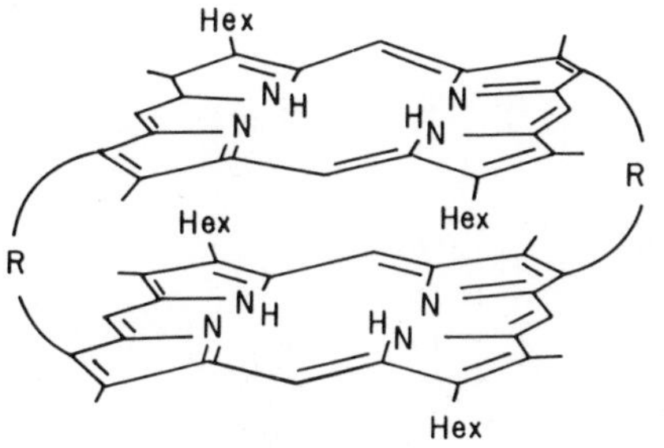

Fig. 23. Structure of a cofacial diporhyrin.

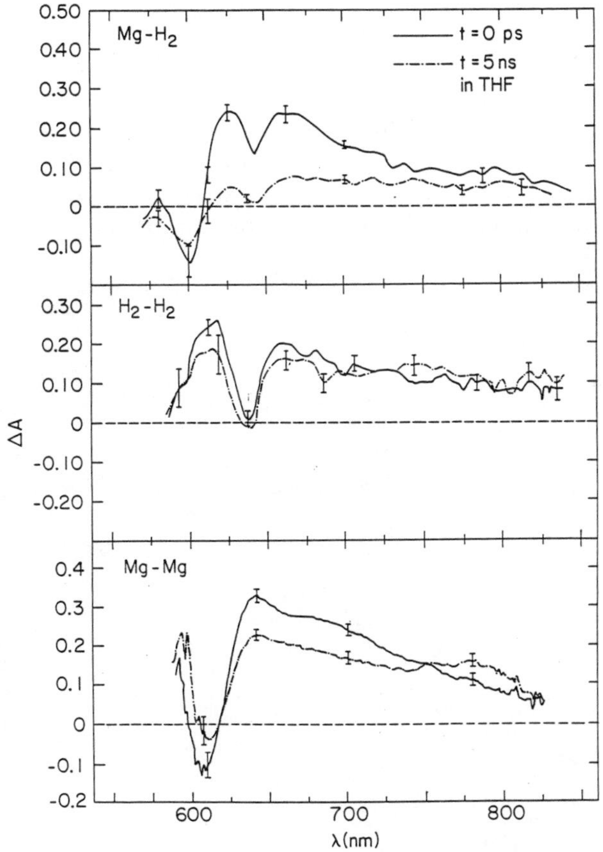

Fig. 24. Transient absorption spectra in THF of the cofacial diporphyrins shown in Fig. 23 with: one magnesium atom, no magnesium atoms, and two magnesium atoms. (From Ref. 111.)

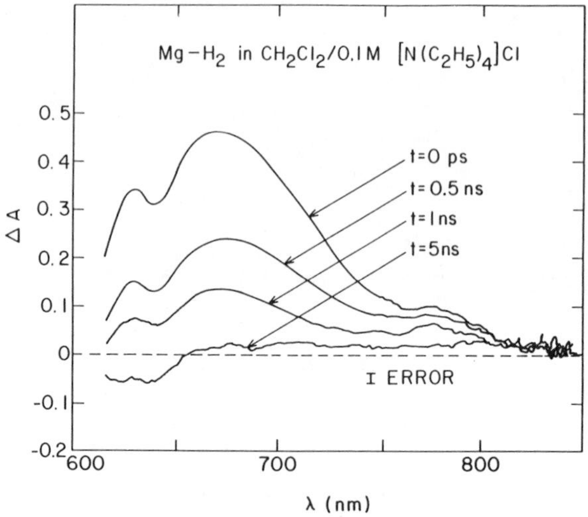

Fig. 25. Transient absorption spectra of a cofacial diporphyrin with only one magnesium atom (5.5×10^{-2}M) in CH_2Cl_2 and 0.1 M [$N(C_2H_5)4$] Cl. The transient decays with first-order kinetics and lifetime of 620 (± 20) ps. (From Ref. 111.)

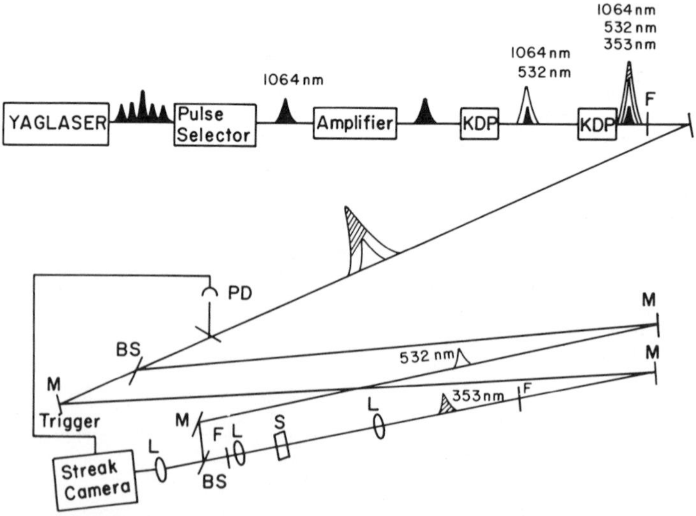

Fig. 26. Schematic of the picosecond fluorescence apparatus. F = filter; BS = beam splitter; L = lens; S = sample; M = mirror. For the chlorophyll studies the second KDP crystal was not used. All samples were excited with 530 nm light.

to be an excited singlet to an excited triplet (Fig. 25).[111] The compound with only one Mg atom has a difference spectrum in THF, which is believed to be that of a singlet state, Fig. 24 because in THF the estimated energy of the ion pair is above that of the singlet state. In CH_2Cl_2 the energy of the ions is below the singlet and a new excited-state spectrum appears, which is attributed to an anion of the metal free macrocycle and a cation of the Mg-containing portion of the dimer, Fig. 26. This absorbance decays in about 380 ps to a spectrum that resembles the triplet state. When $N(C_2H_5)_4Cl$ is added to a solution of the nonmetallated dimer in CH_2Cl_2, the ions are lowered in energy by about 170 mV due to complexing of the Cl^- with the Mg porphyrin.[112] The same radical ion spectrum can be seen (Fig. 25); however, it decays in 620 ps to the ground state instead of the triplet.[111] Further experiments need to be performed before it can be understood why the triplet level has not been populated. In any case, these experiments demonstrate that by fixing the donor–acceptor distance it is possible to obtain a rapid electron-transfer reaction and stabilize the ion pair for several hundred picoseconds.

D. Excited Singlet State Properties of Chlorophyll Dimers

As was presented in Section III, chlorophyll *a* adducts with ethanol have been prepared that successfully mimic the optical and ESR properties of photosystem I reaction center chlorophyll. These chlorophyll *a* special pair systems are assembled from the two monomer units by cooling a mixture of chlorophyll *a* in the presence of water or toluene to 100 K.[68, 73] The formation of the desired structure depends not only on the chlorophyll *a* concentration, but also on the mole ratio of chlorophyll *a* to nucleophiles in solution, the solvent, the rate of cooling, etc. There is reason to suppose that mixtures of various species of poorly defined structure are present in even the best of these preparations. The uncertainties in composition and structure and the experimental problems and restrictions imposed by working at low temperature in organic glasses have limited the information that can be derived from such model systems. The solution to this problem is to provide a mode of physical attachment between two chlorophyll molecules so that the magnitude of the entropy of dimerization is lowered.

Covalently linked dimers of both chlorophyll *a* and pyrochlorophyll *a* have been prepared, which mimic the spectroscopic and redox properties of P700.[75, 76] The two chlorophylls are joined in each case at their propionic acid side chains via an ethylene glycol diester linkage. The orientation of the chlorophyll macrocycles with respect to one another (Figs. 11 and 12) and consequently their electronic properties depend strongly on the solvent. The structure of most interest is the folded one (Fig. 11) because of its similarity to the photoactive dimer in photosystem I.

To study the singlet properties of the folded dimer, it is best to use an apparatus with subnanosecond time resolution. For time-resolved absorbance measurements, such an apparatus eliminates the ambiguities that arise when using a nanosecond laser due to the presence of triplet-excited states during the measurement. The picosecond absorbance apparatus has already been shown in Fig. 18. The fluorescence lifetime is also measured with a picosecond apparatus. The picosecond time resolution is required to eliminate the need for deconvolution of the data. To perform a picosecond emission experiment, it is necessary to use a streak camera. A technique has been recently developed through which it is possible to average a large number of streak camera experiments (Fig. 26).[113] Extensive averaging is needed to maintain a low-photon flux so as to insure that the data are not affected by nonlinear phenomena such as stimulated emission or two-photon excitation.

The fluoresecence lifetime of folded pyrochlorophyll *a* dimer in methylene chloride varied from 110 ps at room temperature to 4.6 ns at 200 K (Fig. 27).[84] However, the quantum yield for fluorescence of these samples remained relatively constant in the 290 to 200 K temperature range (Table II). The quantum yield data obtained at room temperature is in close

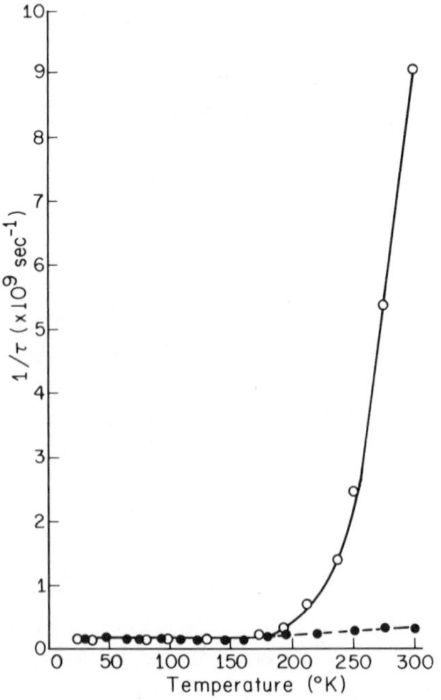

Fig. 27. Temperature dependence of the fluorescence lifetime of the pyrochlorophyll *a* dimer in toluene (●) and in methylene chloride (○).

TABLE II
Quantum Yield of 2×10^{-4} M Pyrochlorophyll *a* Dimer

Solvent	Temperature (K)	Quantum yield	Fluorescence lifetime (ns)
CCl_4	298	0.32	3.8
CH_2Cl_2	298	0.03	0.11
CH_2Cl_2	269	0.03	0.19
CH_2Cl_2	234	0.03	0.71
CH_2Cl_2	208	0.04	2.3
CH_2Cl_2	194	0.06	4.6
CH_2Cl_2	178	0.10	5.4

agreement with that measured previously at higher concentrations.[77] Below 200 K the lifetime lengthened slightly reaching a limit of 5.6 ns. No change in this lifetime could be detected down to 4 K. The measured fluorescence lifetime of the folded pyrochlorophyll *a* dimer was identical both for observation along the axis of the excitation beam and at 90° with respect to the excitation in all the solvents studied. Moreover, the lifetime was independent of concentration from 10^{-6} to 10^{-3} M. The pathlength of the cell was varied from 1 to 10 mm without affecting the fluorescence lifetime. In addition, observation of the fluorescence at polarizations parallel and perpendicular to the polarization of the excitation beam gave identical fluorescence lifetimes.

Additional information about the excited state dynamics of the dimer can be obtained from time-resolved absorption spectroscopy. The excited state difference spectrum of the folded pyrochlorophyll *a* dimer taken 6 ps after excitation is shown in Fig. 28. It shows a narrow band of bleaching at 700 nm. Such a bleaching of the red-most absorption band has also been seen in the excited singlet state of monomeric chlorophyll *a*.[57] An additional negative absorbance change is also seen at 440 nm. A small positive optical density change was observed at 460 nm. As the sample was cooled from 290 to 270 K the bleaching centered at 700 nm became somewhat broader and a new positive optical density change appeared at 660 nm (Fig. 28)

All light-induced absorption changes for folded pyrochlorophyll *a* dimer in CH_2Cl_2 decay with a lifetime of 110 ± 40 ps at room temperature. This decay rate is the same as the fluorescence lifetime of the folded pyrochlorophyll *a* dimer under identical conditions. When the temperature was lowered the lifetime of the excited state absorbance increased, while the magnitude of the initial bleaching at 700 nm remained unchanged. The 700 nm bleaching and the positive optical density change at 660 nm each

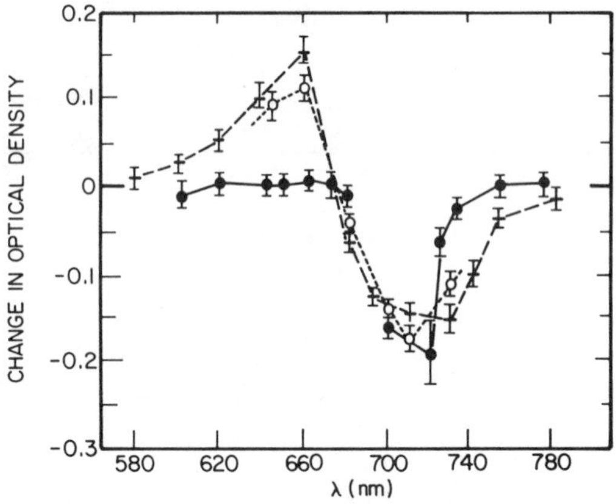

Fig. 28. Excited state difference spectrum taken 8 ps after excitation under the following conditions: carbon tetrachloride 298 k (-●-); methylene chloride 298 K (-+-); and methylene chloride 273 K (—o—).

possess a decay rate that was identical to the fluorescence lifetime of the dimer. In contrast to the measurements made with methylene chloride as the solvent, solutions of dimer in both carbon tetrachloride and toluene had an excited state difference spectrum that was independent of temperature. This spectrum resembles that seen in CH_2Cl_2 at temperatures below 270 K (Fig. 28).

The two most striking aspects of the data involve the strong dependence of the excited singlet state lifetime of the folded pyrochlorophyll *a* dimer on the solvent and the temperature. Before the strange temperature and solvent effects on the lifetimes can be ascribed to any new phenomena, other well-known mechanisms for shortening the fluorescence lifetime must be discounted. Rapid excited singlet-state decay rates have been observed in systems undergoing coherent processes such as stimulated emission. In these cases the fluorescence lifetime is strongly dependent on concentration, cell pathlength, and excitation intensity.[114] The shortening of a fluorescence lifetime due to stimulated emission using a relatively long pathlength cell is often more dramatic when the viewing angle with respect to the excitation beam is 0° rather 90°. This is due to the fact that the emitted photons sample more of the excited state population in the direction of the excitation than in the orthogonal direction. Moreover, in systems undergoing stimulated emission the decay of the excited singlet state is nonexponential. For example, these effects have been observed for 3,3′-diethyloxidicarbocynanine iodide (DODCI).[115] The decay rates for

DODCI measured using excited-state absorption as a probe depend strongly on concentration, pathlength, and excitation intensity. On the other hand, the experimental results show that neither changes in concentration, pathlength, viewing angle, or excitation intensity (over 2 orders of magnitude) effect the excited-state behavior of folded pyrochlorophyll *a* dimer in all the chlorocarbon solvents we examined.

Aggregation of the pyrochlorophyll *a* dimer molecules could also be responsible for the large variation of lifetime with solvent and temperature. The degree of aggregation of chlorophyll *a* is sensitive to concentration.[10, 47, 49, 54] The excited singlet-state lifetime of folded pyrochlorophyll *a* dimer in each chlorocarbon solvent used in this study does not depend on concentration in the range 10^{-6} to 10^{-3} M.

Energy transfer from the fluorescent dimer to a nonfluorescent aggregate would also shorten the singlet lifetime.[74] The rate of energy transfer is very sensitive to concentration; however, the lifetime at any given temperature appeared to be free of any concentration effects. Moreover, an estimate of the Förster transfer rate at 10^{-6} M for even the most favorable conditions indicated that transfer would occur at a much slower rate than the observed quenching.

The fluorescence lifteime of the folded pyrochlorophyll *a* dimer is independent of the orientation of the emission polarization with respect to the excitation polarization. Therefore, rotational reorientation does not influence the data. Fluorescence from chlorophyll *a* monomer is almost completely depolarized when 528 nm excitation is used. One would expect the polarization of the dimer to be similar. Thus, in any case the observed excited-state singlet lifetime should not be influenced by rotational reorientation.

Chlorinated hydrocarbons often undergo photochemical reactions. These usually involve free radical intermediates. The relative reactivity of the chlorinated hydrocarbons in these reactions is $CCl_4 > CHCl_3 > CH_2Cl_2$. The fluorescence lifetime of folded pyrochlorophyll *a* dimer in carbon tetrachloride, hexane, and toluene is identical. Therefore, it is unlikely that the CCl_4 is participating in a photochemical reaction with the pyrochlorophyll *a* dimer, since this would result in a change in the observed lifetime in CCl_4. Since the relative reactivity of $CHCl_3$ and CH_2Cl_2 is less than that of CCl_4, the likelihood of their participation in photochemical reactions leading to the observed reduction in lifetime in these solvents is even further reduced. Therefore, the usual explanations for extremely rapid fluorescence decay rates cannot be utilized to explain the data. It is necessary to examine the effect of temperature on the absorption and fluorescence data in both CCl_4 and CH_2Cl_2 to find a model that describes the experimental observations.

Changes in temperature alter three essential quantities for folded pyrochlorophyll *a* dimer in CH_2Cl_2. The fluorescence lifetime increases monotonically by a factor of 40 over a temperature range of 298 to 200 K. Over the same temperature range the quantum yield for fluorescence increases by only a factor of 2. The excited-state absorption data exhibit two temperature dependencies. As the temperature is changed the lifetime of the absorption changes closely follow the lifetime of the fluorescence changes. However, the size of the initial 660 nm absorption change following excitation has a different temperature dependence. In the range of 298 to 270 K the initial 660 nm absorption change attains its largest magnitude and remains essentially unchanged as the temperature is lowered further. This is in contrast to the dimer dissolved in CCl_4 for which the excited state absorption spectrum was essentially independent of temperature. It is necessary to explore several models for the excited singlet state of the pyrochlorophyll *a* dimer to determine which one can best explain the observation of a factor of 40 increase in the fluorescence lifetime as the temperature is lowered, while at the same time quantum yield remains constant. A single excited level would require that both the radiactive rate (k_{rad}) and the sum of the nonradiative (k_{nrad}) rates each change by a factor of 40 as the temperature is varied from 298 to 200 K. While such a change in the nonradiative rate has been previously seen,[116] there is no precedent to support a radiative rate change by such an amount. While the radiative rate could change somewhat with temperature due to changes in population of various vibronic levels, drastic changes in Franck–Condon factors from one vibrational level to another would then be necessary to explain a very large change in the radiative rate with temperature. Such a dramatic change in the Franck–Condon factors would almost certainly result in a change in the fluorescence spectrum. Only a slight temperature dependence of the emission spectrum is observed. It is also hard to imagine that the radiative rate would depend so strongly on temperature when the solvent was CH_2Cl_2 and be relatively temperature independent in CCl_4. Thus a single excited state cannot account for the observations.

In order to explain the data one needs at least a two-level scheme. These two states may reside on the same molecule or be representative of two distinct excited species in solution. Consider the case in which two distinct excited species exist in solution. In this context, the appearance of the 660 nm absorption change between 290 and 270 K would indicate a shift in the relative equilibrium concentration of the two species. If both species were fluorescent, the emission spectrum from each species would almost certainly be different. Thus one would expect that the luminescence spectrum would change with temperature. Moreover, unless the fluorescence lifetimes of the two species were identical the observed emission lifetime would exhibit a wavelength dependence. No changes in the fluorescence spectra were observed as the temperature was changed, in fact identical fluorescence lifetimes were measured throughout the fluorescence spectrum. Finally, the temperature dependence of the excited state difference spectrum indicates that the equilibrium should be shifted in favor of the 660 nm absorbing species at all temperatures lower than 270 K. Thus, below this temperature the system would again behave as a simple single-level system yet between 270 and 200 K the lifetime changes by more than a factor of 20. As we have argued above, such large changes in lifetime without parallel changes in quantum yield cannot be explained by a single-level scheme. Hence, two different ground-state conformations in equilibrium cannot give rise to the observed excited state behavior.

It is necessary to require that two levels are present in the same molecule. The observation of a 660 nm excited absorption for folded pyrochlorophyll *a* dimer dissolved in CCl_4 and in cold CH_2Cl_2 is indicative of only one excited state. Call this state S_1. Similarly call the excited state observed in CH_2Cl_2 at room temperature S_2. From the previous discussion it must be assumed that both states S_1 and S_2 must be excited states of the same molecule. In this view the effect of solvent is to shift the population of the excited pyrochlorophyll *a* dimer molecules from one excited state to a second excited state.

If an excited state equilibrium between S_1 and S_2 exists, the number of molecules in state S_1 as measured by the initial size of the 660 nm transient absorbance would depend on temperature. This is not the case. The excited-state absorption attributed to S_2 disappears in large part between 290 and 270 K (Fig. 28), yet the fluorescence lifetime is still strongly dependent on temperature. Therefore, a kinetic rather than thermodynamic relationship determines the relative population distribution between S_1 and S_2.

Given this relationship only one of the two states may be fluorescent. If both were fluorescent we would observe a change in the emission spectrum with temperature and a wavelength dependence for the fluorescence lifetime. These effects were not observed. This leaves only four possible kinetic relationships between S_1 and S_2.

1. $S_1 \rightarrow S_2$, and S_1 is fluorescent
2. $S_1 \rightarrow S_2$, and S_2 is fluorescent
3. $S_2 \rightarrow S_1$, and S_1 is fluorescent
4. $S_2 \rightarrow S_1$, and S_2 is fluorescent

The two possibilities, 1 and 4, may be quickly rejected. These situations would behave essentially as a single-level system because depletion of excited molecules from the fluorescent state is simply another nonradiative pathway.

Consider the data for the folded pyrochlorophyll *a* dimer in methylene chloride at room temperature. If relationship 3 is correct, then the lack of a 660 nm transient absorption at 298 K indicates that the lifetime of S_1 must be so short that it cannot be observed with picosecond resolution. This would imply that the sum of radiative and nonradiative decay rates out of S_1 must be much faster than the transfer rate (k_{21}). The fluorescence lifetime would then be determined by k_{21}. This yields a k_{21} of about $10^{10}\ s^{-1}$.

Since $k_{21} \ll k_{rad} + k_{nrad}$ and the fluorescence quantum yield, which is given by

$$\phi = \frac{k_{rad}}{k_{nr} + k_{rad}} \tag{1}$$

is about 0.03, the radiative rate (k_{rad}) would have to be much greater than $3 \times 10^{+8}\ s^{-1}$, which is much larger than one would expect.

We are left with relationship 2 as the only viable two-level scheme. This scheme is illustrated in Fig. 29. The rate equations for this scheme are

$$\frac{dN_1}{dt} = -(k_{10} + k_{12})N_1 \tag{2}$$

$$\frac{dN_2}{dt} = -(k_{20} + k_f)N_2 + k_{12}N_1 \tag{3}$$

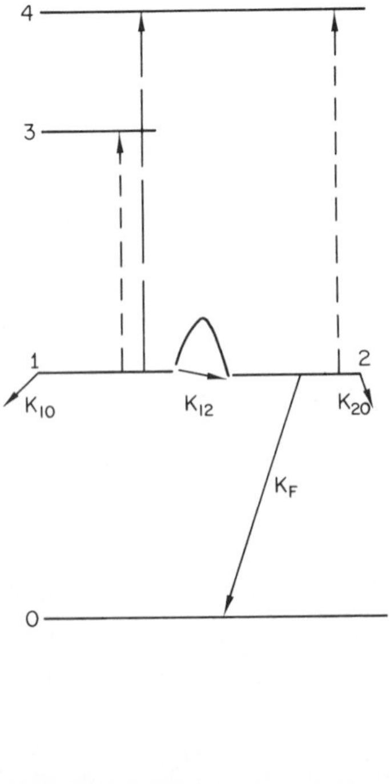

Fig. 29. Proposed two level scheme for the excited state dynamics of the pyrochlorophyll *a* dimer. K_{10} is a nonradiative decay rate. K_{12} is the rate of conversion from state 1 to state 2. K_{20} is a nonradiative decay rate, while K_f is a radiative decay rate.

where k_{12}, k_{10}, and k_{20} are rate constants for the processes illustrated in Fig. 30.

Their solution is given as

$$N_1(t)=N_1(0)\exp-(k_{10}+k_{12})t \tag{4}$$

$$N_2(t)=N_1(0)k_{12}(k_{10}+k_{12}-k_{20}-k_f)^{-1}\left[\exp-(k_{10}+k_{12})t - \exp-(k_{20}+k_f)t\right] \tag{5}$$

This assumes that the molecules are all initially in state 1. The quantum yield can also be readily calculated from the model and is given by

$$Q_y=\left(\frac{k_{12}}{k_{12}+k_{10}}\right)\left(\frac{k_f}{k_f+k_{20}}\right) \tag{6}$$

By choosing appropriate values of the rate constants these equations can be utilized to fit all the data presented for this system. Rate constant k_{12} is the rate that determines the presence of the 660 nm excited state absorption. For folded pyrochlorophyll *a* in CH_2Cl_2 at room temperature k_{12} must be faster than 10^{11} s^{-1} or, perhaps, S_2 is directly populated. Thus the population of molecules in S_1 is small even when observed 6 ps after excitation. As the temperature is lowered k_{12} decreases. The population is then observable 6 ps after excitation. This explains the results presented in Fig. 28.

Rate constant k_{12} is solvent dependent as well as temperature dependent. In carbon tetrachloride k_{12} is small. The measured fluorescence lifetime is a probe of $k_{10}+k_{12}$. Therefore the fluorescence lifetime is identical to the lifetime of S_1.

Equation (5) predicts that the appearance of the fluorescence from state 2 should not be instantaneous. When methylene chloride is used as a solvent the fluorescence does have a finite rise time. To show that the state 2 indeed has a finite rise time slower than the apparatus resolution rhodamine *b* fluorescence is compared to that from the pyrochlorophyll *a* dimer in carbon tetrachloride in Fig. 30. A finite rise time would not occur from different ground-state molecules in equilibrium. Thus this is additional evidence against two distinct ground-state molecules giving rise to observed phenomena.

The observations just described strongly suggest that dual excited states exist in the dimers of pyrochlorophyll *a*. The relative populations of these levels are controlled by kinetics and not by theormodynamics. This is a unique situation, which to the best of our knowledge has not been previously observed. The possibility exists that the switching between two

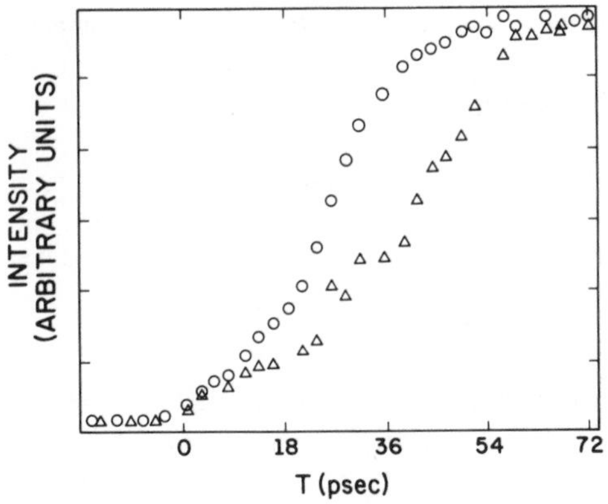

Fig. 30. The appearance of fluorescence of rhodamine *b* following 30 ps excitation at 530 nm (○), and the appearance of fluorescence from the pyrochlorophyll *a* dimer following excitation with 530 nm (△). Conditions were the same for the two experiments.

excited states may be a feature which allows the plant to store excitation energy for a brief period of time while waiting for an oxidized acceptor or a reduced donor to be formed. If the reaction center is unable to perform photochemistry, the molecule could then rapidly quench the excitation.

This sensitivity to the environment might explain the interesting results observed when *Rhodopseudomonas viridis* reaction centers were poised at a potential low enough to reduce the intermediate acceptor (Fig. 5) before being excited. Absorption features that correspond to the excited bacteriochlorophyll *a* special pair decayed in 20 ps.[36] Based on the lifetimes measured for the monomer in solution, the lifetime of the bacterial dimer should have been much longer. Also the dimer singlet decay rate must be much slower than the rate of oxidation of the dimer (~2 ps)[117] because the quantum yield for photochemistry approaches unity. The authors argued that a series of electron-transfer reactions between the reduced bacteriophoeophytin *a* and the singlet excited dimer, could lead to a quenching of the dimer excited state. An alternate explanation would be that the presence of the reduced intermediate acceptor alters the environment in a manner analogous to the change observed in going from CCl_4 to CH_2Cl_2 as the solvent.

The molecular mechanism responsible for the unusual dynamics of the pyrochlorophyll *a* dimers has not as yet been established. Several additional effects of the environment on the excited-state dynamics have been observed. The lifetime of the dimer depends smoothly on the mole fraction of CH_2Cl_2 in toluene (Fig. 31).[87] Thus a strong binding of the methylene chloride with the dimer either in the ground or the singlet excited state is

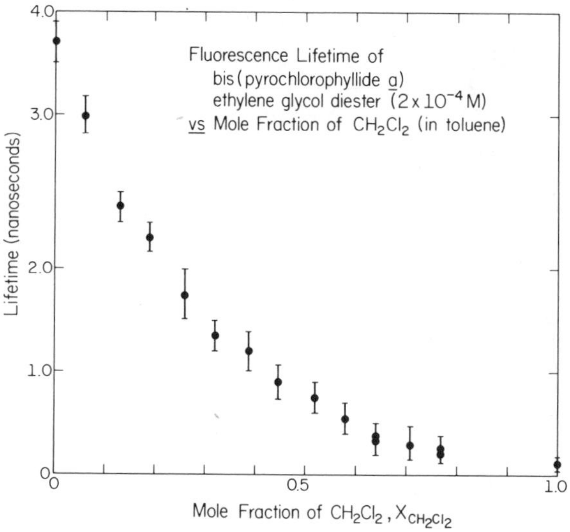

Fig. 31. Fluorescence lifetime of the pyrochlorophyll *a* dimer in toluene vs the mole fraction of methylene chloride.

not responsible for the unique behavior. In contrast when the pyrochorophyll *a* dimer is dissolved in dry toluene and CF_3CH_2OH is added to convert from the dry to the folded form, the fluorescence lifetime is decreased to about 600 psec. This decrease in the observed lifetime is commensurate with the observation of a red shift in both the absorption and the emission spectra. Measurements of the dependence of fluorescence lifetime on the amount of CF_3CH_2OH added is shown in Fig. 32. The data indicate that the binding of CF_3CH_2OH is needed to achieve a decrease in the singlet lifetime. The reasons for the shortening of the singlet lifetime of the fluorinated alcohol are also not clear. Is the effect due to the electron-withdrawing ability of the fluorine atoms or the higher acidity of the proton? This question was answered by folding the pyrochlorophyll dimer with *p*-cresol. This compound has a pK_A smaller than the trifluorinated ethanol, but is not as electron withdrawing. The singlet decay was not at all shortened, but was in fact somewhat longer than that observed when ethanol was used. Thus it appears that the removal of electron density from the dimer is producing the shortening of the singlet lifetime in pyrochlorophyll *a* dimers folded with CF_3CH_2OH. We suggest that removal of electron density from the macrocycle by methylene chloride is also responsible for the lowering of the barrier between the two excited states S_1 and S_2 resulting in a very fast decay time at 290 K.

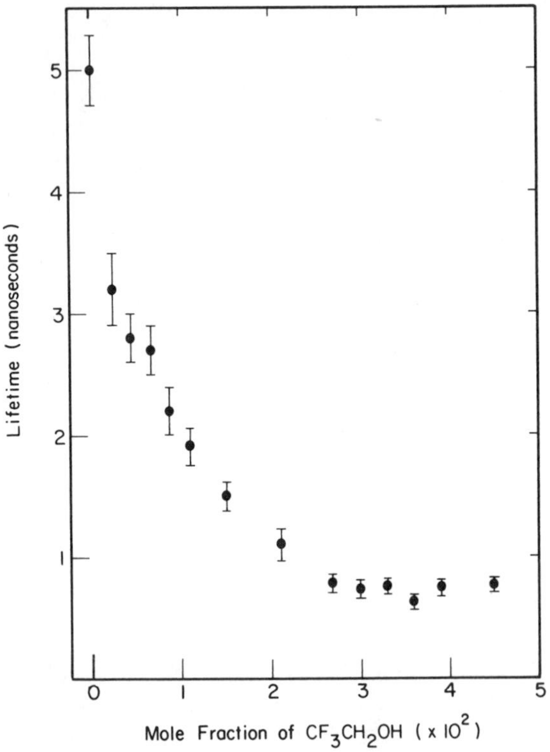

Fig. 32. The fluorescence lifetime of the dry pyrochlorophyll *a* dimer in toluene vs the mole fraction of CF_3CH_2OH added.

E. Triplet Dynamics of the Pyrochlorophyll *a* Dimer

Recently, the dynamics of the triplet states of pyrochlorophyll *a* dimer has been reported.[118] It was found that the triplet state of folded pyrochlorophyll *a* dimer opens into isolated triplet and ground-state molecules. The main evidence for this is the appearance of a positive optical density change near 660 nm and a decrease in optical density at 700 nm. The bleaching at 700 nm was interpreted to be caused by opening of the folded dimer. The "unpinned" monomer results in an increase in absorbance at 660 nm. In the presence of triplet quenchers and at low alcohol concentrations these workers were able to observe reformation of the folded dimer within a few microseconds. The appearance of a positive optical density change at 660 nm in Fig. 28 would be indicative of just an opening of the dimer in the singlet state. However, the rapid decay of the 700 nm bleaching indicates that the ground state of the folded dimer is being repopulated at the same rate the excited state is decaying. Thus, on the picosecond timescale the bleaching at 660 nm does not result from the opening of the folded dimer. Small changes in the excited-state geometry could destroy the electronic interaction leading to an excited-state difference spectrum which indicates the formation of two isolated molecules. However, during the triplet lifetime, the two macrocycles could diffuse apart resulting in the true dissociation observed in the microsecond experiments.[118]

V. CONCLUSION

A great deal has been learned by studying photosynthesis *in vivo* and in trying to reproduce its features *in vitro*. In the past the *in vitro* model systems have played a key role in understanding the *in vivo* structures. As our chemical expertise improves, we should be able to mimic the photosynthetic process in even greater detail. From these studies we not only hope to discover how plants carry out solar energy conversion, but perhaps to do it ourselves.

Acknowledgments

This work was supported by the National Science Foundation and the Department of Energy. K. J. Kaufmann thanks the Max Planck Institute for Biophysical Chemistry for their hospitality during the preparation of this manuscript.

References

1. R. K. Clayton and W. R. Sistrom, eds., *The Photosynthetic Bacteria*, Plenum Press, New York, 1978.
2. Govindjee, ed., *Bioenergetics of Photosynthesis*, Academic Press, New York, 1975.
3. P. Mitchell, *Biol. Rev. Cambridge Philos. Soc.*, **41**, 445 (1966).
4. M. D. Hatch and C. R. Slack, *Ann. Rev. Plant Physiol.*, **21**, 141 (1970).
5. J. P. Thornber, *Ann. Rev. Plant Physiol.*, **26**, 127 (1975).
6. R. Emerson and W. Arnold, *J. Gen. Physiol.*, **15**, 391 (1932); **16**, 191 (1932).
7. L. N. M. Duysens, Thesis, University of Utrecht, 1952.
8. B. Kok, *Biochim. Biophys. Acta*, **22**, 399 (1956).
9. *Primary Processes in Photosynthesis*, *Topics in Photosynthesis*, Vol. 2, J. Barber, ed., Elsevier, Amsterdam, 1977.
10. J. J. Katz and J. R. Norris, *Curr. Top. Bioeng.*, **5**, 41 (1973).
11. J. J. Katz, J. R. Norris, L. L. Shipman, M. C. Thurnauer, and M. R. Wasielewski, *Ann. Rev. Biophys. Bioeng.*, **7**, 393 (1978).
12. R. Emerson and C. M. Lewis, *Am. J. Bot.*, **30**, 165 (1943).
13. W. L. Butler and M. Kitajima, *Biochim, Biophys. Acta.*, **396**, 72 (1975).
14. J. M. Fenton, M. J. Pellin, Govindjee, and K. J. Kaufmann, *FEBS Lett.*, **100**, 1 (1979).
15. V. A. Shuvalov, B. Ke, and E. Dolan, *FEBS Lett.*, **100**, 5 (1979).
16. V. A. Shuvalov, A. V. Klevanik, A. V. Sharkov, P. G. Kryukov, and B. Ke. *FEBS Lett.*, **107**, 313 (1979).

17. B. Rumberg and H. T. Witt, *Z. Naturforsch.*, **19b**, 693 (1964).
18. B. Kok and G. Hoch, in *Light and Life*, W. D. McElroy and B. Glass, eds., John Hopkins Press, Baltimore, Md., 1961.
19. G. Döring, J. L. Bailey, W. Kreutz, J. Weikard, and H. T. Witt, *Naturwissenschaften*, **55**, 219 (1968).
20. B. Commoner, J. J. Heise, and J. Townsend, *Proc. Natl. Acad. Sci. USA*, **42**, 710 (1956).
21. H. Reinert and B. Kok, *Natl. Acad. Sci. Res. Council Publ.*, **1145**, 131 (1962).
22. K. Sauer, P. Mathis, S. Acker, and J. Van Best, *Biochim. Biophys. Acta*, **503**, 120 (1978).
23. V. A. Shuvalov, E. Dolan, and B. Ke, *Proc. Natl. Acad. Sci. USA*, **76**, 770 (1979).
24. R. Malkin and A. J. Bearden, *Biochim, Biophys. Acta*, **505**, 147 (1978).
25. G. Döring, H. H. Stiehl, and H. T. Witt, *Z. Naturforsch.*, **22b**, 639 (1967).
26. D. Reed and R. K. Clayton, *Biochem. Biophys. Res. Commun.*, **30**, 471 (1968).
27. M. Y. Okamura, L. A. Steiner, and G. Feher, *Biochemistry* **13**, 1394 (1974).
28. G. Feher, *Photochem. Photobiol.*, **14**, 373 (1971).
29. R. K. Clayton and R. T. Wang, *Meth. Enzymol.*, **23**, 696 (1971).
30. J. R. Bolton, R. K. Clayton, and D. W. Reed, *Photochem. Photobiol.*, **9**, 209 (1969).
31. J. D. McElroy, G. Feher, and D. C. Mauzerall, *Biochim. Biophys. Acta*, **172**, 180 (1969).
32. C. A. Wraight and R. K. Clayton, *Biochim, Biophys. Acta*, **333**, 246 (1974).
33. P. L. Dutton, R. C. Prince, D. M. Tiede, K. M. Petty, K. J. Kaufmann, T. L. Netzel, and P. M. Rentzepis, *Brookhaven Symp.*, **28**, 213 (1979).
34. M. G. Rockley, M. W. Windsor, R. J. Cogdell, and W. W. Parson, *Proc. Natl. Acad. Sci. USA*, **72**, 2251 (1975).
35. K. Peters, P. Avouris, and P. M. Rentzepis, *Biophys. J.*, **23**, 207 (1978).
36. D. Holton, M. W. Windsor, W. W. Parson, and J. P. Thornber, *Biochim. Biophys. Acta*, **501**, 112 (1978).
37. V. A. Shuvalov, A. V. Klevanik, A. V. Sharkov, J. A. Matveetz, and P. G. Krukov, *FEBS Lett.*, **91**, 135 (1978).
38. K. L. Zankel, D. W. Reed, and R. K. Clayton, *Proc. Natl. Acad. Sci. USA*, **61**, 1243 (1968).
39. L. Slooten, *Biochim. Biophys. Acta*, **256**, 452 (1972).
40. S. S. Brody and E. Rabinowitch, *Science*, **125**, 555 (1957).
41. L. A. Tumerman and A. B. Rubin, *Dokl. Akad. Nauk. SSSR*, **145**, 202 (1962).
42. W. W. Parson, R. K. Clayton, and R. J. Cogdell, *Biochim. Biophys. Acta*, **387**, 265 (1975).
43. R. C. Prince and P. L. Dutton, in *The Photosynthetic Bacteria*, R. K. Clayton and W. R. Sistrum, eds., Plenum Press, New York, 1978, Chap. 24.
44. J. J. Hopfield, *Proc. 29th Int. Cong. Soc. de Chimie Physique*, Elsevier, Amsterdam, 1977.
45. J. Jortner, *J. Chem. Phys.*, **64**, 4860 (1976).
46. L. P. Vernon and G. R. Seely, eds., *The Chlorophylls*, Academic Press, New York, 1966.
47. J. J. Katz, *Inorganic Biochemistry*, Vol. 2, G. L. Eichhorn, ed. Elsevier, Amsterdam, 1973, p. 1022.
48. J. J. Katz, G. L. Closs, F. C. Pennington, M. R. Thomas, and H. H. Strain, *J. Am. Chem. Soc.*, **85**, 3801 (1963).
49. K. Ballschmiter, K. Truesdell, and J. J. Katz, *Biochim. Biophys. Acta*, **184**, 604 (1969).
50. J. J. Katz, H. H. Strain, D. L. Leussing, and R. C. Dougherty, *J. Am. Chem. Soc.*, **90**, 784 (1968).
51. L. L. Shipman, T. R. Janson, G. J. Ray, and J. J. Katz, *Proc. Natl. Acad. Sci. USA*, **72**, 2873 (1975).

52. L. L. Shipman, T. M. Cotton, J. R. Norris, and J. J. Katz, *J. Am. Chem. Soc.*, **98**, 8222 (1976).
53. G. Sherman and E. Fujimori, *Arch. Biochem. Biophys.*, **130**, 624 (1969).
54. K. Ballschmiter, T. M. Cotton, H. H. Strain, and J. J. Katz, *Biochim. Biophys. Acta*, **180**, 347 (1969).
55. H. C. Chow, R. Serlin, and C. E. Strouse, *J. Am. Chem. Soc.*, **97**, 7230 (1975).
56. G. P. Gurnovich, A. N. Sevchesko, and K. N. Solovev, *Spectroscopy of Chlorophylls and Related Compounds*, U.S. Atomic Energy Commission Division of Technical Information AEC tr. 1799, 1971, Chap. 4.
57. D. Huppert, P. M. Rentzepis, and G. Tollin, *Biochim. Biophys. Acta*, **440**, 356 (1976).
58. M. Gouterman and D. Holten, *Photochem. Photobiol.*, **25**, 85 (1977).
59. G. R. Seely, *Photochem. Photobiol.*, **27**, 639 (1978).
60. H. J. Werner, Z. Schulten, and K. Schulten, *J. Chem. Phys.*, **7**, 646 (1977).
61. G. Tollin, F. Castelli, G. Cheddar, and F. Rizzuto, *Photochem. Photobiol.*, **29**, 147 (1979).
62. T. M. Cotton, A. D. Trifunac, K. Ballschmiter, and J. J. Katz, *Biochim. Biophys. Acta*, **368**, 181 (1974).
63. J. R. Norris, R. A. Uphaus, H. L. Crespi, and J. J. Katz, *Proc. Natl. Acad. Sci. USA*, **68**, 625 (1971).
64. H. Scheer, J. J. Katz, and J. R. Norris, *J. Am. Chem. Soc.*, **99**, 1372 (1977).
65. J. R. Norris, M. E. Druyan, and J. J. Katz, *J. Am. Chem. Soc.*, **95**, 1680 (1973).
66. J. R. Norris, H. Scheer, and J. J. Katz, *Ann. N.Y. Acad. Sci.*, **244**, 261 (1975).
67. G. Feher, A. J. Hoff, R. A. Isaacson, and L. C. Ackerson, *Ann. N.Y. Acad. Sci.*, **244**, 239 (1975).
68. L. L. Shipman, T. M. Cotton, J. R. Norris, and J. J. Katz, *Proc. Natl. Acad. Sci. USA*, **73**, 1791 (1976).
69. F. K. Fong, *Proc. Natl. Acad. Sci. USA*, **71**, 3692 (1974).
70. L. L. Shipman, J. R. Norris, and J. J. Katz, *J. Phys. Chem.*, **80**, 877 (1976).
71. F. K. Fong and W. A. Wassam, *J. Am. Chem. Soc.*, **99**, 2375 (1977).
72. A. Warshel, *J. Am. Chem. Soc.*, **101**, 745 (1979).
73. F. K. Fong and V. J. Koester, *Biochim. Biophys. Acta*, **423**, 52 (1976).
74. J. B. Birks, *Photophysics of Aromatic Molecules*, Wiley-Interscience, London, 1970.
75. S. G. Boxer and G. L. Closs, *J. Am. Chem. Soc.*, **98**, 5406 (1976).
76. M. R. Wasielewski, M. H. Studier, and J. J. Katz, *Proc. Natl. Acad. Sci. USA*, **73**, 4282 (1976).
77. J. C. Hindman, R. Kugel, M. R. Wasielewski, and J. J. Katz, *Proc. Natl. Acad. Sci. USA*, **75**, 2076 (1978).
78. J. R. Norris and M. R. Wasielewski, unpublished data.
79. P. L Dutton, J. S. Leigh, Jr., and D. W. Reed, *Biochim. Biophys. Acta*, **292**, 654 (1973).
80. M. R. Wasielewski, U. H. Smith, B. T. Cope, and J. J. Katz, *J. Am. Chem. Soc.*, **99**, 4172 (1977).
81. M. R. Wasielewski, W. A. Svec, and B. T. Cope, *J. Am. Chem. Soc.*, **100**, 1961 (1978).
82. M. R. Wasielewski, manuscript in preparation.
83. B. Ke, *Curr. Top. Bioenerg.*, **7**, 75 (1978).
84. M. J. Pellin, M. R. Wasielewski, and K. J. Kaufmann, *J. Am. Chem. Soc.*, **00**, 00 (1900).
85. S. G. Boxer and R. R. Bucks, *J. Am. Chem. Soc.*, **101**, 1883 (1979).
86. M. R. Wasielewski and G. L. Closs, manuscript in preparation.
87. M. J. Pellin, M. R. Wasielewski, and K. J. Kaufmann, unpublished results.
88. M. J. Pellin, K. J. Kaufmann, and M. R. Wasielewski, *Nature*, **278**, 54 (1979).
89. S. Litteken, M. R. Wasielewski, and K. J. Kaufmann, manuscript in preparation.

90. P. M. Rentzepis, *Science*, **202**, 174 (1978).
91. I. Fujita, M. S. Davis, and J. Fajer, *J. Am. Chem. Soc.*, **100**, 6280 (1978).
92. F. Castelli, G. Cheddar, F. Rizzuto, and G. Tollin, *Photochem. Photobiol.*, **29**, 153 (1979).
93. K. Gnädig and K. B. Eisenthal, *Chem. Phys. Lett.* **46**, 339 (1977).
94. N. Nakashima, M. Murakawa, and N. Mataga, *Bull. Chem. Soc. Japan*, **49**, 854 (1976).
95. J. Hinatsu, H. Masuhara, M. Mataga, Y. Sakata, and S. Misumi, *Bull. Chem. Soc. Japan*, **51**, 1032 (1978).
96. H. Leonhardt and A. Weller, *Ber. Buns. Phys. Chem.*, **67**, 791 (1963).
97. H. Knibbe, K. Röllig, F. P. Schäffer, and A. Weller, *J. Chem. Phys.*, **47**, 1184 (1967).
98. P. Froehlich and E. L. Wehry, in *Modern Fluorescence Spectroscopy*, Vol. 2, E. L. Wehry, ed., Plenum Press, New York, 1976.
99. R. Potashnik, C. R. Goldschmidt, M. Ottolenghi, and A. Weller, *J. Chem. Phys.*, **55**, 5344 (1971).
100. D. Rehm and A. Weller, *Israel J. Chem.*, **8**, 259 (1970).
101. M. Mataga, T. Okada, and N. Yamamoto, *Chem. Phys. Lett.*, **1**, 119 (1967).
102. A. Weller, in *Fast Reactions and Primary Processes in Chemical Kinetics*, S. Claeson, ed., Wiley, New York, 1967.
103. H. Schomburg, H. Staerk, and A. Weller, *Chem. Phys. Lett.*, **21**, 433 (1973), and **22**, 1 (1973).
104. H. J. Wasmer, H. Staerk, and A. Weller, *Chem. Phys. Lett.*, **68**, 2419 (1978).
105. N. Orbach and M. Ottolenghi, in *The Exciplex*, M. Gordon and W. R. Ware, eds., Academic Press, New York, 1975.
106. N. E. Kagan, D. Mauzerall, and R. B. Merrifield, *J. Am. Chem. Soc.*, **99**, 5484 (1977).
107. H. Ogoshi, H. Sugimoto, and Z.-i. Yoshida, *Tetrahedron Lett.*, **2**, 169 (1977).
108. J. P. Collman, C. M. Elliot, T. R. Halbert, and B. S. Tovrog, *Proc. Natl. Acad. Sci. USA*, **74**, 18 (1977).
109. C. K. Chang, *J. Heterocyclic Chem.*, **14**, 1285 (1977).
110. C. K. Chang, *Adv. Chem. Ser.*, **173**, 162 (1979).
111. T. L. Netzel, P. Kroger, C. K. Chang, I. Fujita, and J. Fajer, *Chem. Phys. Lett.*, **67**, 223 (1979).
112. M. S. Davis, A. Forman, L. K. Hanson, and J. Fajer, *Biophys. J.*, **25**, 148a (1979).
113. K. K. Smith, J. Y. Koo, G. B. Schuster, and K. J. Kaufmann, *Chem. Phys. Lett.*, **48**, 267 (1977).
114. G. R. Fleming, A. W. E. Knight, J. M. Morris, R. J. S. Morrison, and G. W. Robinson, *J. Am. Chem. Soc.*, **99**, 4306 (1977).
115. G. E. Busch and P. M. Rentzepis, *Science* **194**, 276 (1976).
116. K. K. Smith and K. J. Kaufmann, *J. Phys. Chem.*, **82**, 2286 (1978).
117. D. Holten, C. Hoganson, M. W. Windsor, C. C. Schenck, W. W. Parson, A. Migus, R. L. Fork and C. V. Shank, *Biochem. Biophys. Acta.*, in press.
118. N. Periasamy, H. Linschitz, G. L. Closs, and S. G. Boxer, *Proc. Natl. Acad. Sci. USA*, **75**, 2563 (1978).

PROTON TRANSFER: A PRIMARY PICOSECOND EVENT

P. M. RENTZEPIS AND P. F. BARBARA*

Bell Laboratories, 600 Mountain Avenue, Murray Hill, New Jersey 07974

CONTENTS

It is a widely accepted fact that electron transfer frequently plays a predominant role in many chemical and biological reactions. The transfer of proton[1] is also widely occurring in chemical processes even if it has not received as much attention as electron transfer. An all-inclusive treatise of proton and electron transfer is far beyond the scope of this chapter, therefore, it shall be restricted to the detection and measurement of proton translocation processes that occur in the picosecond regime. Specifically, we shall describe the initial events that involve proton transfer in such important and natural occurring processes as visual transduction.

The methodology of picosecond spectroscopy is widely known and need not be expounded upon even though variations of the original methods are being devised as the need for specific applications arises. A typical system may be divided into four components: (1) the pulse generating and amplifying system; (2) the timing and absorption continuum section, composed of a stage delay and/or echelon as the time-resolving element, and a cell containing a Raman scattering medium such as alcohol or water where the broad absorption continuum is generated; (3) the detection device usually consists of a spectrometer and vidicon for transient absorption studies and a monochromator/streak camera combination for recording transient emission spectra; in addition, picosecond Raman may be also measured, and, although this technique is not widely used yet, a consider-

*Department of Chemistry, University of Minnesota, Minneapolis, Minnesota.

able number of investigators are currently entering this field; (4) the data-processing component, which in most cases consists of an optical multichannel analyzer that feeds the data to a computer for data analysis and plotting of the relevant parameters.

Figure 1 shows a schematic diagram of the apparatus used in our laboratory, which incorporates all of these features.[2] Quite often the data are presented after analysis in the form of optical density change (ΔA) vs time in picoseconds (Fig. 2) for a selected region of the spectrum. It is

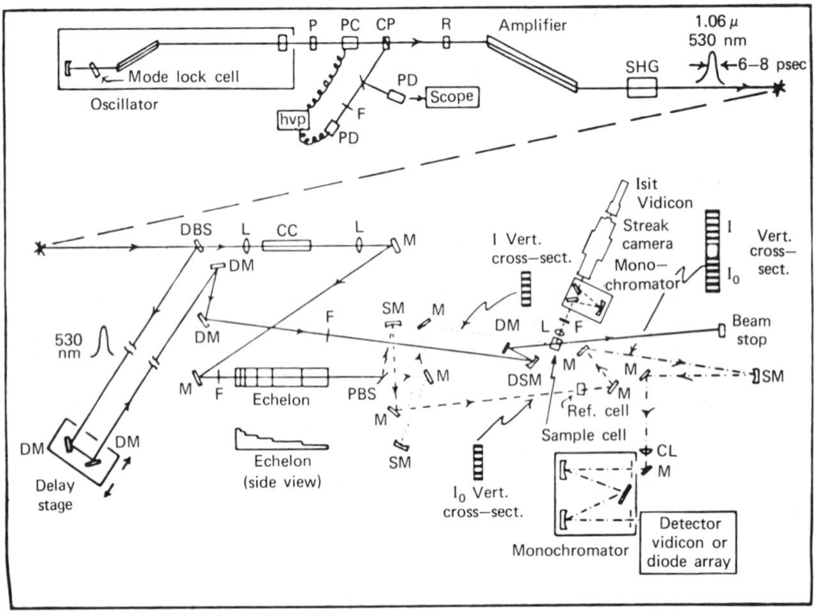

Fig. 1. *Picosecond experimental system*. The oscillator is a Nd^{3+}/glass or dye laser. Amplifier Nd^{3+}/YAG. The components along the optical path are P = polarizer, PC = Pockels cell, CP = crossed polarizer, PD = photodiode, HVP = high voltage pulse, R = rotator, SHG = second harmonic crystal, DBS = dielectric beam splitter, CC = continuum cell, DM = dielectric mirror, PBS = pelical beam splitter, SM = spherical mirror, L = lens and F = filters. For the train of 1060 nm pulses generated in the Nd^{3+}/glass oscillator, a single pulse is extracted by activating the Pockels cell. This pulse is amplified and sharpened in the Nd^{3+}/YAG rod and converted to 530 nm by the KDP crystal. A 530 nm pulse is used for excitation while the remaining 530 nm and 1060 nm are focused in the CC = continuum cell, generating a broad continuum. Passing the continuum through the echelon causes it to be dispersed into a set of pulses separated by a few picoseconds and split into reference I_0 and probing beam I by PBS. One pulse of the I beam and the excitation pulse are synchronized to arrive simultaneously by means of a delay state. After the cell, beams I and I_0 are imaged onto a slit of monochromator then onto a vidicon that detects the events and transforms the data to digital form for computer processing. The emission is detected by the streak camera and fed to computer via an ISIT vidicon. For more details, see *Science*, **202**, 174 (1978).

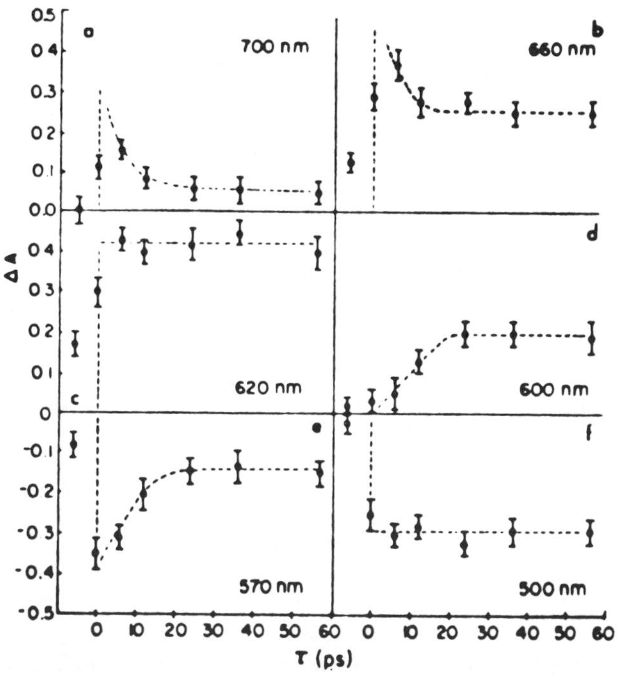

Fig. 2. Example of data display. In this case we present kinetic transients observed at various wavelengths after excitation of light-adapted bacteriorhodopsin at 530 nm, 20±5°C. The excitation pulse is 6 ps and has 2 mJ in total energy. (ā) 700 nm; (b) 660 nm; (c) 620 nm; (d) 600 nm; (e) 570 nm; (f) 500 nm. (From Ref. 2.)

trivially simple of course to display the spectrum at any selected period of time in the form of ΔA or I vs wavelength (λ) as in fact is the operating mode of flash photolysis, even though at a much slower time scale. Transient picosecond spectra recorded at a selected period of time have been in the literature for many a year. Figure 3 displays the transient spectrum of acridine 120 ps after excitation.[3] In fact Huppert et al.[4] several years ago demonstrated the feasibility of not only a transient spectrum at a fixed time but the simultaneous recording of time (ps) wavelength (λ) and ΔA (Fig. 4). The display of a transient absorption spectrum is conceptually simple and easy in practice to achieve by the use of a spectroscopic plate or preferably a vidicon placed across the exit of a spectrograph. The dynamic range of a vidicon at the present time at least, is not sufficient to accommodate the wide fluctuations in absorption intensity intrinsic to many chemical and biological molecules. This difficulty, which is accentuated by the intensity fluctuation of picosecond continua, results in an extremely difficult task, if not an unreliable one, when one records only the transient absorption species rather than utilize the double-beam method

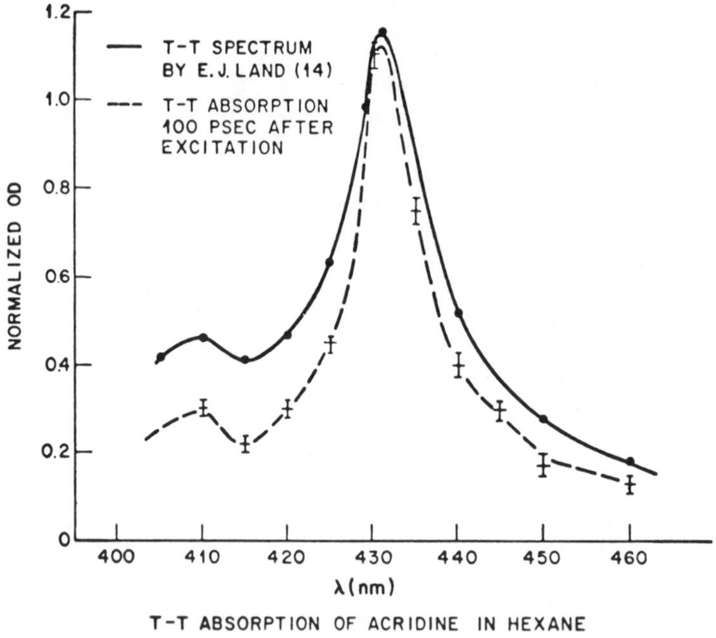

Fig. 3. Transient absorption spectrum of acridine recorded 100 ps after excitation.

developed by Netzel et al.[5] This method enables the simultaneous recording and comparison of a reference and interrogating time resolved spectra. Even with a double-beam method, one finds that the spectra that have widely fluctuating absorption band intensities must be corrected and preferably the spectrum taken in a sequence which accommodates a small range of intensities. The mechanics of obtaining spectra at a fixed interval of time after excitation is trivial once the time coincidence between the excitation pulse and interrogation is ascertained and the above conditions met. The improvement in sensitivity, reproducibility, and time resolution of picosecond spectroscopy is now almost entirely dependent on the availability of instruments such as streak cameras and vidicons rather than on inherent shortcoming of this field, if care is taken for the excitation pulse to be well characterized.

Even though the number of biological processes amenable to study by time-resolved spectroscopy is extremely large and range from hemoglobin[6] and vision[7] to photosynthesis,[8–10] this chapter shall be restricted to energy-transfer processes and specifically to the primary proton transfer in the visual pigments, the chromophore of vision, rhodopsin.

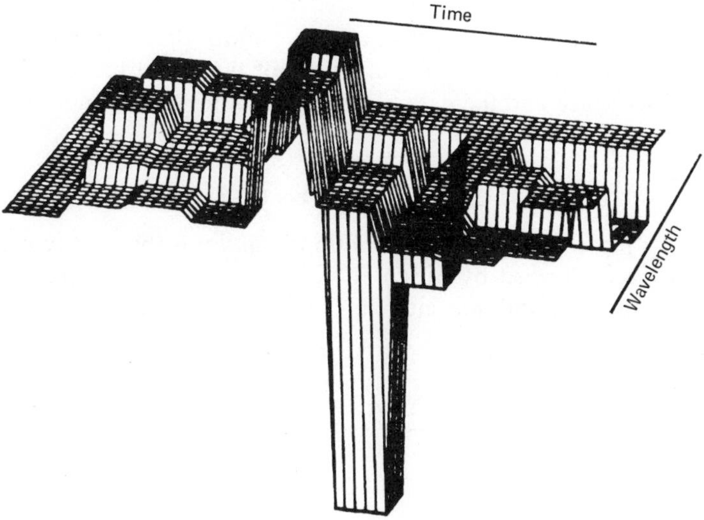

Fig. 4. Three-dimensional display of time (picoseconds) wavelength (nonometers), and absorbance change after excitation with a single 530 nm picosecond pulse. The wavelength range monitored is from 550 to 570 nm. The time, in 10-ps segments, covers 100 ps, and the absorbance changes are positive (upward) for absorbance and negative (downward) for bleaching. (From Ref. 3.)

I. PROTON TRANSFER IN RHODOPSIN

The sequence of events that lead to the visual transduction process has been altered lately, mainly as a result of the data obtained by picosecond spectroscopy, which suggested that proton translocation is a process that is of pivotal importance to the photoinduced visual process, if not the predominant process.

The first picosecond study on a biological molecule was centered on the kinetics of the primary intermediate, bathorhodopsin during the visual transduction process.[11] Low-temperature experiments had established the presence of an intermediate species referred to as bathorhodopsin, which seemed to be quite stable at low temperatures, 77 K[12] and 7 K.[13] However, because of its ultrafast formation time constant at room temperature, it could not be measured before the development of picosecond spectroscopy. It was established however that the first intermediate bathorhodopsin decayed, at room temperature, into a second intermediate lumirhodopsin then to two other species called metarhodopsin I and metarhodopsin II. Eventually, at some point, the chromophore isomerizes.[14–15] The intermediate at which the isomerization occurs is not known or at least unequivocally agreed upon. The recent picosecond experiments that we shall present have shown that a proton plays a predominant role during the initial events of the visual process; however, the exact sequence and the nature of the proton involved is not universally agreed upon.

II. PRIMARY PROCESSES IN RHODOPSIN

The first picosecond studies on rhodopsin and possibly of any biological system were performed by Busch et al.,[11] who studied the kinetics of the first photoinduced intermediate of rhodopsin, bathorhodopsin at room

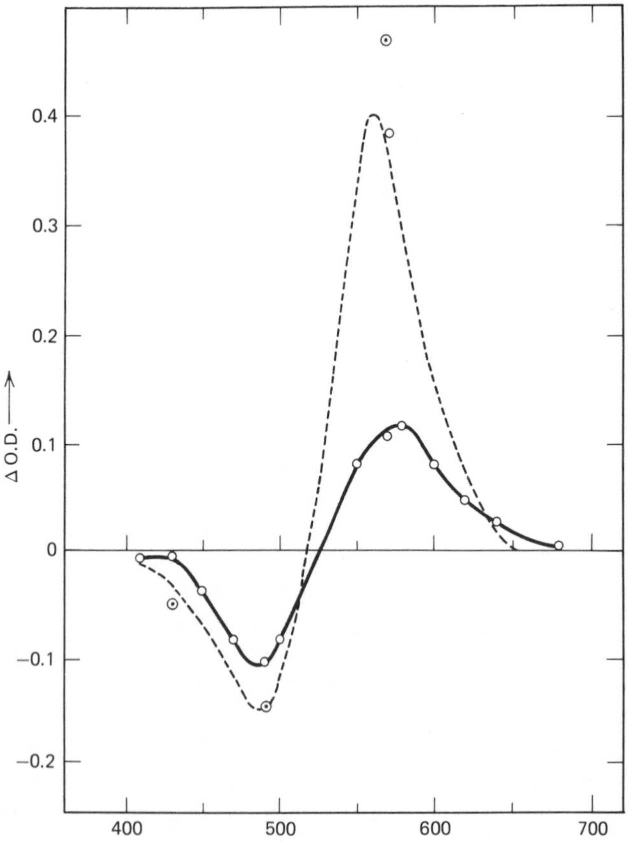

Fig. 5. Bathorhodopsin difference spectrum. 60 ps after excitation with a 0.5 mJ pulse at 530 nm. O, 293 K, O, 77 K, and O, 4 K; ··· photostationary difference spectrum at 77 K.

temperature in order to identify the nature of this species, and from its kinetics be able to clarify the sequence of events that generate the transients that lead to the visual recognition process. Some of the later transients forming were found to be isomers of the original rhodopsin, and therefore a number of workers assumed that the initial action of light causes isomerization of the retinal around 10–11 carbon position. To explain this supposition, various schemes for isomerization were proposed, including a bicycle model which was expected to be operative even at liquid helium temperatures.

The absorption spectrum of bathorhodopsin is known to have its maximum shifted to the red (λ_{max} 543 nm) of the normal dark-adapted rhodopsin (λ_{max} 500 nm). Thus after excitation of rhodopsin with a 530 nm pulse, a probe pulse at 561 nm, which is within the absorption band of prelumirhodopsin, will encounter and display an absorption. Utilizing this known fact, Busch et al.[11] excited rhodopsin at room temperature, with a 6 ps 530 nm pulse and measured the appearance of an absorption at 561 nm, which was interpreted as the formation of the first intermediate. The time constant of formation was found to be <6 ps, while the decay of the bathorhodopsin intermediate was measured to be $\sim$ 50 ns.

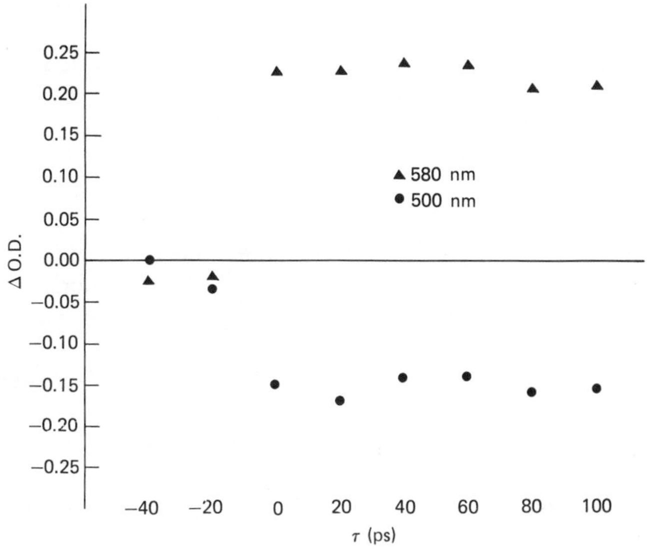

Fig. 6. Depletion rate of the rhodopsin band at 480 nm after excitation with the 530 nm pulse. Similar kinetics are observed over the entire spectrum of these two bands. Borth curves represent the average of five repeated runs.

These results provided the first experimental evidence for the existence of bathorhodopsin at physiological temperatures and also permitted conclusions to be drawn and models of the intermediate to be postulated that would be consistent with this time scale for formation. When later the complete difference spectra for bathorhodopsin at room temperature were obtained[16] in the time range of 6–300 ps (Fig. 5), it was possible to measure (Fig. 6) the formation kinetics of a band located between 530 and 680 nm that corresponds to bathorhodopsin having its maximum absorption at 580 nm and its isosbestic point at approximately 525 nm. These values correspond quite well with previously published data. Picosecond experiments performed by the same method revealed that rhodopsin also bleached with the same time constant as bathorhodopsin was formed, namely, <6 ps.

In order to resolve further the kinetics and elucidate the mechanisms, experiments were performed in the range of 300 to 4 K by Peters, Applebury, and Rentzepis.[17] These studies revealed that at temperatures as low as 77 K, the formation of bathorhodopsin as depicted by the appearance of 570 nm absorption band is still less than 6 ps and even at 4 K the rise time was only 36 ps. These results are shown in Fig. 7.

Such fast rates at these low temperatures are not expected if only the isomerization mechanism is assumed to be responsible for the formation of prelumirhodopsin at least at low temperatures. Another mechanism, therefore, becomes feasible because of the very fast and constant rate of bathorhodopsin formation at low temperatures. This mechanism is proton translocation. To decide if indeed isomerization at the chromophore skeleton or proton translocation is the phenomenon observed, the rhodopsin was immersed in D_2O and the normal procedure followed for substitution of all exchangeable protons with deuterium. Since it is well known that by the hydrogen on the retinal chromophore skeleton do not exchange, however, other rhodopsin protons including the proton near the Schiff base are exchanged for deuterium, the place where isomerization should occur is not altered. Therefore, if what we observe is isomerization, no noticeable effect in the rates of the formation between protonated and deuterium exchanged rhodopsin should be observed. However, a very strong effect is observed.

For the D-rhodopsin,* the time constants for formation of bathorhodopsin at the same low temperatures were found to be much larger than its protonated homolog. For example, at 40 K, the formation time constant

*D-rhodopsin designates the deuterated rhodopsin by D_2O exchange.

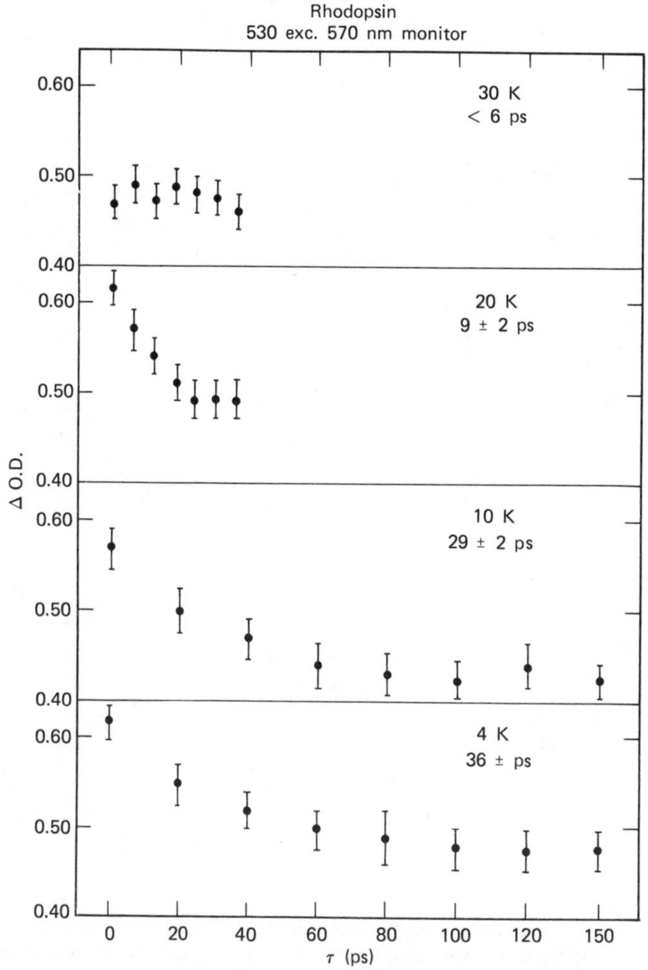

Fig. 7. Formation of bathorhodopsin at various temperatures monitored at 570 nm. Excitation of rhodopsin was with a 5-mJ, 530 nm, 6-ps pulse. The glass for low-temperature study was formed by mixing 1 part rhodopsin, solubilized in 0.3 M Ammonyx/0.01 M Hepes at pH 7.0, $A_{500} = 10.0$, with 2 parts distilled ethylene glycol. The variance is designated by the error bars.

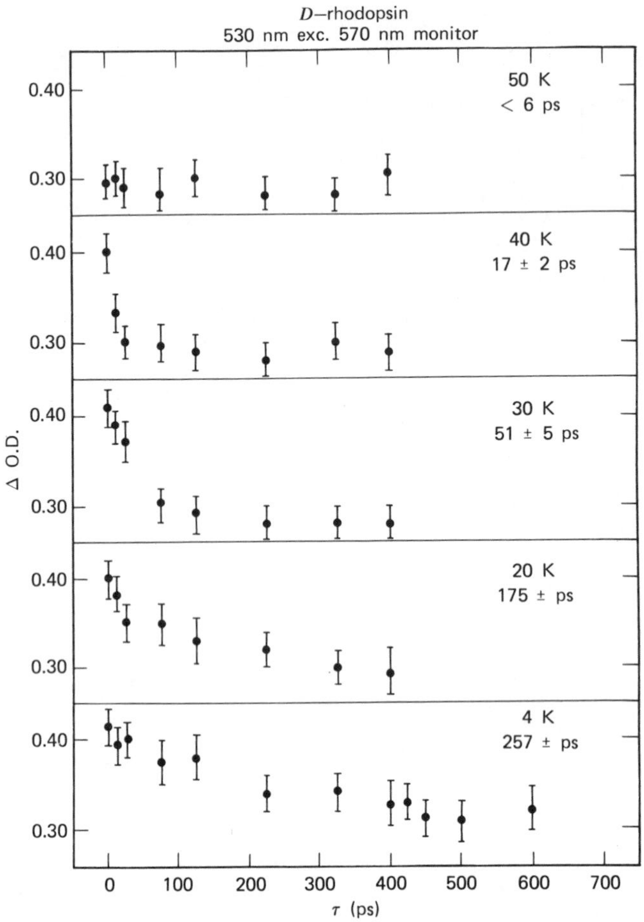

Fig. 8. The kinetics of formation of deuterium-exchanged bathorhodopsin under identical experimental conditions as for *H*-rhodopsin, except that the glass for low temperature study was formed by mixing *D*-rhodopsin with deuterated components. M Hepes at pH 7.0, $A_{500} = 7.0$, with 2 parts deuterium-exchanged glycol.

for *D*-bathorhodopsin was found to be 17 ps vs <6 ps for normal rhodopsin. As the temperature is lowered, the *D*-batho-formation time increases achieving the values of 51 ps at 30 K, and 256 ps at 4 K as shown in Fig. 8. Comparison of rhodopsin and *D*-rhodopsin rates of formation at various temperatures is given in Table I. Even a casual look at Table I makes evident the factor of 7 difference between these two rates.

TABLE I
Bathorhodopsin Formation

Temperature (°K)	Time Constant (ps) (H)	(D)
270	<6	<6
77	<6	<6
50	<6	<6
40	<6	<6
30	<6	51±5
20	9±2	175±15
10	29±2	200±20
4	36±2	257±28

III. PROTON TUNNELING

The photoinduced process, which converts rhodopsin into prelumirhodopsin presumably in the excited state, appears to be very efficient and no significant rhodopsin ground-state repopulation has been observed between the first 6 and 100 ps after excitation. This indicates that either that 30% of excited rhodopsin relaxes to the ground state quite faster than 6 ps. The process that is responsible for the formation of bathorhodopsin and possibly the initial events of vision could be better understood by considering the dependence of the formation rate on temperature and the difference of the rates between rhodopsin and *D*-rhodopsin at low temperatures. The plot of Fig. 9 in the form of an Arrhenium plot (ln K vs $1/T$) exhibits two prominent features. (1) The strong deviation from the linearity of a normal Arrhenius process, and (2) the fact that at low temperature K becomes temperature independent, and at the limit of $T \to 0$ K the rate remains finite.

Considering the effect of deuteration on the rate of formation of the first intermediate, it is seen that it is smaller than the protonated one by a factor of approximately 7. Because none of the protons in the retinal skeleton are exchangeable and in the absence of drastic changes of the protein structure by the D_2O, the observed isotope dependence suggests specific effects corresponding to proton transfer reactions. The non-linear Arrhenius plot and the temperature-independent rates at low temperature imply that a tunneling process is operative at the low-temperature limit. In addition to the temperature-independent rates, the magnitude of the isotope effects suggest that a proton transfer reaction is mainly responsible for the observed data and possibly for the formation of bathorhodopsin at low temperatures.

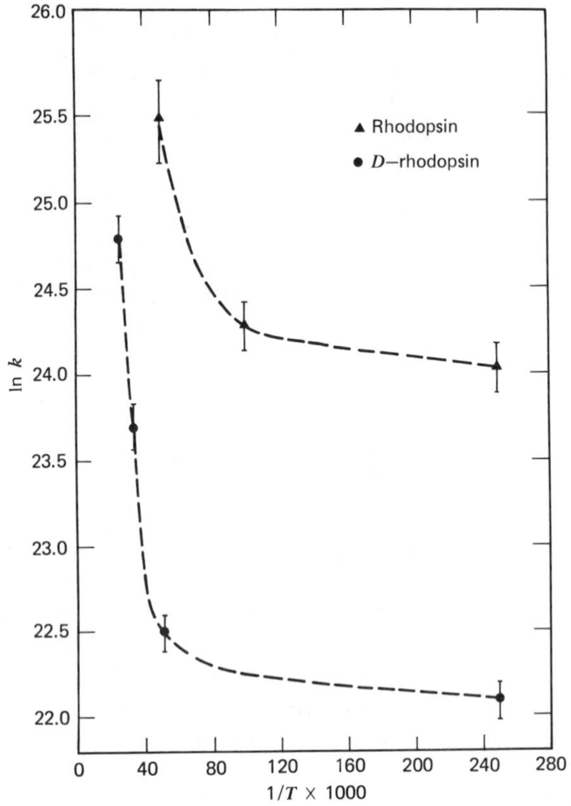

Fig. 9. Arrhenius plot for the data in Figs. 2 and 3 (ln k vs $1/T$ (K)$\times 10^3$.

There are several other results that also support the proton-transfer mechanism. The absorption spectrum of bathorhodopsin is red shifted with respect to the spectrum of rhodopsin itself, therefore, bathorhodopsin may be a more tightly protonated Schiff base than rhodopsin. Model studies by Sandorfy[18–19] show explicitly that translocating the proton toward the Schiff base nitrogen could account for the spectral red shift.

The mechanism for the formation of bathorhodopsin was assigned previously to a cis–trans isomerization, although no concrete proof exists for its support or against it. However, the extremely fast formation of prelumirhodopsin at 4 K and the isotope effect argue against the isomerization mechanism being the dominant process at least at these low temperatures. Regarding the mechanism for the proton–transfer process, the existing theories are not sufficiently developed to deal with tunneling in large molecules, neither are the present-day theories capable of providing a unique answer for the depth or width of the barrier or even predict the shape for the barrier. However, assuming that the expression

$$K_t = \nu_0 \exp \frac{-\pi^2 a_0 \kappa_0}{h} (2ME)^{1/2} \tag{1}$$

is to an extent applicable for the case of rhodopsin, we can calculate an activation energy or barrier height, E_a, and a width of the barrier or translocation distance a_0. Using the experimental rate constant of 2.8×10^{11} s^{-1} at 4 K, and the mass of the proton and deuterium, we can calculate an E_a value of 5.9 kJ (1.4 kcal). This value for E_a is obtained if we assume that ν_0 takes the value of the N—H band frequency of 1500 cm^{-1}, however, if we use the more reasonable value of 10^{13} for ν_0 we obtain a barrier of 3.1 kJ (740 cal), therefore, a barrier height slightly above room temperature kT. This implies that at room temperature the proton translocation could be practically an activated process proceeding without the need for invoking a tunneling process. Such a barrier height would explain the ultrafast formation time constant at room temperature and temperature-independent slower rates at 4 K. Employing the same arguments and expression (1) the width of the barrier a_0, through which the proton translocates is calculated to be ~0.5 Å.

IV. MODELS FOR PROTON TRANSFER

The low-temperature picosecond data have been found to be consistent with at least two mechanisms (presented schematically in Fig. 10) which can account for the formation of bathorhodopsin. (a) A concerted double hydrogen transfer leading to a retroretinal structure which is similar to a previously proposed model, and (b) a proton translocation to the Schiff base nitrogen atom, generating a carbonium ion.[20] The formation of a carbonium ion was exposed previously also by Mathies and Stryer.[21] Very recent experiments by Applebury et al.[22] on bacteriorhodopsin between 298 and 1.4 K indicate that bacteriorhodopsin behaves in a very similar manner as rhodopsin and at low temperatures proton tunneling and a deuterium isotope effect of $K_H/K_D = 2.4/1$ have been observed. This value, although smaller than the rhodopsin ratio, nevertheless strongly suggests proton translocation via tunneling at low temperatures with a barrier height of ~0.6 kJ (140 cal).

Several other mechanisms have been proposed for the formation of the first intermediate, the majority of which are strictly conceptual models

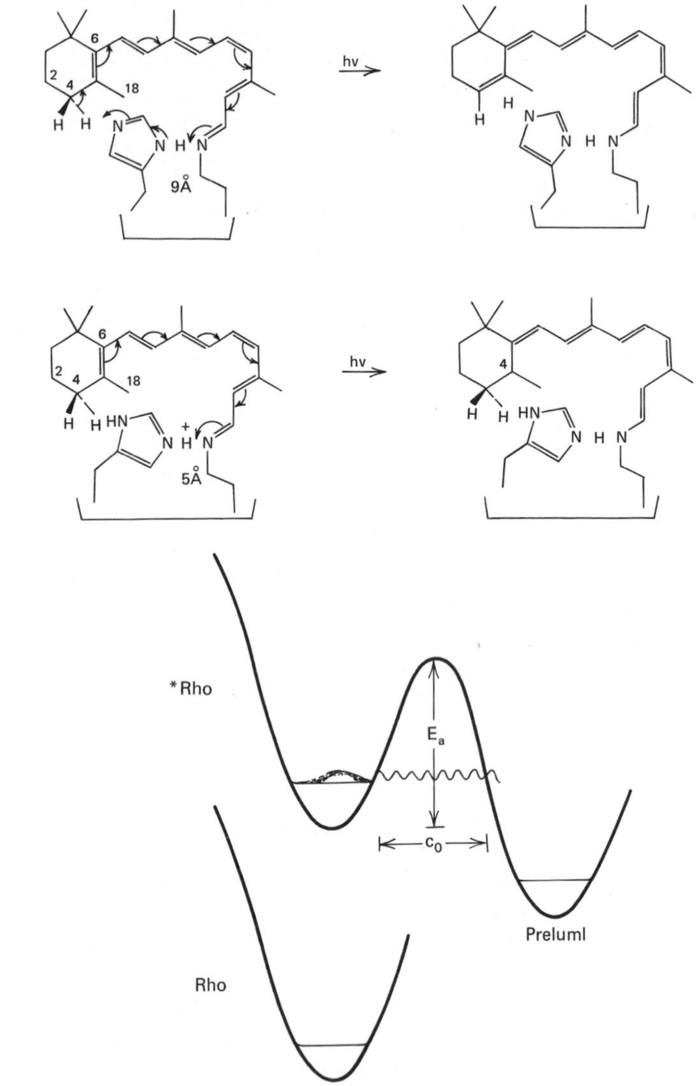

Fig. 10. Models proposed for proton translocation: (Model A) single proton translocation with carbonium ion formation. (Model B) concerted double proton translocation with retroretinal formation. (c) Tunneling potential energy barriers for formation of prelumirhodopsin. E_a represents the energy barrier and a_0 the width or translocation distance.

without strong experimental evidence; these we shall not discuss. There is an amount of experimental work that is based on resonance Raman and infrared experiments, which seem to be of relevance, even though two opposing views precipitate from these data. It is very important to be pointed out that in the resonance Raman studies even though the conclusions drawn refer to the nitrogen–hydrogen bond, of the protonated Schiff base nitrogen, yet the N—H resonance Raman spectrum has not been observed. The assumption is that the C=N band reflects on the integrity of the N—H bond. Similarly it is assumed that the weak band at 1655 cm^{-1} is due to N=C, even though the question has been raised that this might as well be an overtone or combination of other bands. Additional criticism on the Raman studies is voiced that even if this band is N=C, yet the shifts shape and intensity are not sufficiently resolved to provide unequivocal proof for any process. Finally resonance Raman data for the same bands taken in a flow cell have produced results that are interpreted as a proof that the Schiff base is equally protonated before and after excitation. For these results the previous questions hold; in addition they are in direct contrast to experiments on model compounds, which show that under the pH conditions of rhodopsin Schiff base retinals will be unprotonated. However, after excitation the Schiff base nitrogen becomes more electron-negative (basic) thus providing the force to attract a proton closer forming protonated Schiff base retinal.

Even though the translocation of a proton toward the Schiff base seems to be an extremely attractive model and one which accounts for the data, yet other models are also quite plausible. With regard to isomerization, one can say with certainty that it does take place; however, at what step is not clear at this time. Small nuclear deformation in contrast to cis-trans isomerization is a necessary prerequisite to proton tunneling, therefore, some small structural deformation does occur very fast and at an early stage of the visual transduction process. It is quite unlikely though that complete cis-trans isomerization—rotation around a double bond—is solely responsible for the initial events especially at low temperatures. It is more likely that as the retinal is excited the Schiff base nitrogen becomes basic resulting in electron transfer and the sequential double bond shift. This transforms the original double bonds, i.e.,$C_{11}=C_{12}$ to $C_{11}—C_{12}$ and allows for the rotation around the now $C_{11}—C_{12}$ single bond quite freely in addition to inducing a proton transfer. This model would account for both the novel picosecond proton transfer results and the apparent isomerization. Whether this model is correct, only unequivocal experimental data will provide the answer rather than more conceptual models.

References

1. P. Schuster, P. Wolschann, and K. Tortschunoff, *Dynamics of Proton Transfer in Solution*, *Chemical Relaxation in Molecular Biology*, I. Pecht and R. Rugler, eds., Springer-Verlag, New York, 1977, p. 107–174.
2. P. M. Rentzepis, *Science*, **202**, 174–182 (1978).
3. V. Sundstrom, E. C. Lim, and P. M. Rentzepis, *J. Chem. Phys.*, **66**, 4287–4293 (1977).
4. D. Huppert, K. D. Straub, and P. M. Rentzepis, *Proc. Natl. Acad. Sci. USA*, **74** 4139–4143 (1977).
5. T. L. Netzel and P. M. Rentzepis, *Chem. Phys. Lett.* **29**, 337–342 (1974).
6. R. H. Austin, K. Beeson, L. Eisenstein, H. Frauenfelder, and I. C. Gousalus, *Phys. Rev. Lett.* **32** 403–505 (1974).
7. T. Yoshiazwa and S. Moriuchi, *Biochemistry and Physiology of Visual Pigments*, H. Langer, ed., Springer-Verlag, New York, 1973, p. 69.
8. M. G. Rockley, M. W. Windsor, and W. W. Parsons, *Proc. Natl. Acad. Sci. USA*, **72** 2251 (1975).
9. G. Porter, J. A. Synowiec, and C. J. Tredwell, *Biochim. Biophys. Acta*, 454, 329 (1977)
10. K. J. Kaufmann, P. L. Dutton, T. L. Netzel, J. Leigh, and P. M. Rentzepis, *Science*, **188**, 1301 (1975).
11. T. Yoshizawa and G. Wald, *Nature*, **197**, 1274 (1963).
12. M. L. Applebury, D. Zuckerman, A. A. Lamola, and T. Jovin, *Biochemistry*, **13**, 3448 (1974).
13. T. Yoshizawa, in *Handbook of Sensory Physiology*, Vol. VII/1, H. Dartnall, ed., Springer-Verlag, New York, 1972, p. 146.
14. G. Wald, Nature, **219** (1968) 800.
15. T. Rosenfeld, B. Honig, M. Ottolenghi, M. Hurley, and T. G. Ebrey, *Pure Applied Chem.*, **49**, 341 (1977).
16 V. Sundstrom, P. M. Rentzepis, K. Peters, and M. L. Applebury, *Nature*, **267**, 645 (1977).
17. K. Peters, M. L. Applebury, and P. M. Rentzepis, *Proc. Natl. Acad. Sci. USA*, **74**, 3119–3123 (1977).
18. J. Favrot, J. M. Leclerq, R. Roberge, C. Sandorfy, and D. Vocelle, *Chem. Phys. Lett.*, **53**, 433 (1978).
19. J. Favrot, C. Sandorfy, and D. Vocelle, *Photochem. Photobiol.*, **28**, 271 (1978).
20. L. Salem, *Science*, **191**, 822 (1976).
21. R. Mathies and L. Stryer, *Proc. Natl. Acad. Sci. USA*, **73**, 2169 (1976).
22. M. L. Applebury, K. Peters, and P. M. Rentzepis, *Biophys. J.* **23**, 375–382 (1978).

LASER STUDIES OF PROTON TRANSFER

D. HUPPERT

Department of Chemistry, Tel Aviv University, Ramat Aviv, Israel

M. GUTMAN

Department of Biochemistry, Tel Aviv University, Ramat Aviv, Israel

and

K. J. KAUFMANN*

Department of Chemistry, University of Illinois, Urbana, Illinois

CONTENTS

I. INTRODUCTION

One of the simplest of all chemical reaction is the exchange of a proton. Unlike electron-transfer reactions, which involve mainly the exchange of charge between reactants, the transfer of a proton also results in the transport of mass. Reactions involving protons are ubiquitous because of their simplicity. They have been studied in a large number of systems

*Alfred P. Sloan Fellow.

ranging from the base-catalyzed decomposition of nitramide to the dissociation of water.[1,2] A great deal of our understanding of chemical equilibria, chemical kinetics, isotope effects, mass transport in solution, free-energy relationships, and chemical reactivity have developed from the study of proton-transfer reactions. The reactions involving protons are not only very common and instructive, but are found in a number of processes important to our everyday existence. Many commercial chemicals are prepared in part via acid or base-catalyzed reactions. Protons are involved in a host of enzymatic reactions, such as the synthesis of ATP in mitochondria and chloroplasts.[3] A chemiosmotic solar cell may someday be developed that can play a role both in solar energy conversion and storage. Nature has developed such a system in the form of halophilic bacteria, which convert a photon flux into a proton gradient.[4]

The rates of proton reactions have been shown to span at least 15 orders of magnitude.[5] A wealth of kinetic data has produced a good qualitative description of the proton-transfer process in the form of reaction mechanisms.[6] For reactions in solution there is a dearth of experimental data on the intimate details of proton-transfer reactions. Detailed theories of proton-transfer reactions are still developing. However, these are often based on information obtained from gas-phase reactions (i.e., molecular beam studies), which are free of solvent participation.[7] The transport of hydrogen ions through water and ice have been very important in understanding the structure and properties of these materials.[8–10] The structure of the solvated proton in water as well as in other solvents has been one of keen experimental and theoretical interest.[11]

NMR, rapid mixing, and perturbation techniques have been most useful in the study of hydrogen ion reactions and transport. Since stopped flow is limited to the study of relatively slow reactions, most of the information about rapid proton-transfer reactions such as those between oxygen or nitrogen atoms is due either to NMR or to perturbation techniques.[8,12]

The perturbation methods utilize a change in the equilibrium constant to monitor the kinetics. Equilibrium is often perturbed by a rapid change in temperature or pressure. For a number of reactions, including those involving protons, light can be used to quickly change the equilibrium constant.[13] The perturbation arises from the fact that the ground-state properties of a molecule are different than those of the excited state. Since molecules absorb a photon in 10^{-15} s, in principle the equilibrium can be perturbed on this time scale, and thereby initiate proton release or uptake with light. Under proper conditions such reactions need no longer be diffusion limited, and it is possible to observe features of the transfer process such as participation of the solvent and motions of the proton within or near the hydration sphere of the donor/acceptor.

II. EXCITED-STATE PROTON TRANSFER

It was observed nearly half a century ago that the wavelength of the fluorescence from 1-naphthylamine-4-sulfonate was dependent on the pH.[14] Förster correctly interpreted these data in terms of a change in the p*K* of the aromatic amine as a result of the excitation.[15] Even today a quantitative molecular picture of this phenomena has not yet been obtained due to the difficulty of accurately calculating the excited-state electronic distributions of such large aromatic molecules. A qualitative valence bond picture in which the charge is delocalized in the excited state can be used to account for the p*K* change.[16] Such a qualitative argument has been used to rationalize the observation that aromatic amines and alcohols become more acidic in the excited state, while aromatic acids and ketones become more basic.[16] In addition to the resonance arguments, configuration interaction with charge-transfer states has been used to explain why excited singlets of aromatic alcohols (ketones) are more acidic (basic) than the triplets, which are more acidic (basic) than ground-state alcohols (ketones).[17] Since triplet states of xanthone and benzephenone appear to be more basic than their singlet counterparts, such qualitative arguments are not always correct.[18]

Several methods are available for the determination of the excited singlet state p*K* (p*K**). The one that is most popular due to its straightforward and simple application is the Förster cycle.[19] A schematic energy-level diagram for the Förster cycle is shown in Fig. 1. If one assumes that ΔS is the same for the ground and the excited-state proton-transfer reactions, then it follows that

$$pK^* - pK = \frac{h\nu_A - h\nu_{AH}}{2.3RT} \tag{1}$$

where ν_{AH} and ν_{A} are the frequencies of the electronic transitions that connect the ground and excited states of the acid (AH) and its conjugate base (A). Since electronic transitions take place between Franck–Condon states, and the p*K** only has meaning for an equilibrated system, it is

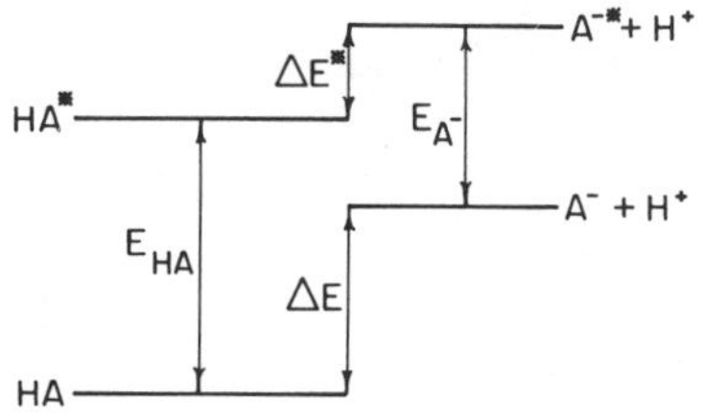

Fig. 1. Electronic energy levels for an acid and its anion used in the Förster cycle.

necessary to correct the ν_A and ν_{AH} for relaxation of the solvent around the excited state. Usually an average of the fluorescence and absorption spectra are used as an approximation of the difference in energy between the equilibrated forms of the molecule in the ground and excited states. For molecules that do not fluorescence such an averaging technique is not possible. For molecules that show poor mirror symmetry the averaging technique will result in errors in the estimated pK^*. Such an effect has been seen for anthroic acids, where rotation of the carboxyl group in the excited state affects relaxation of the solvent.[20,21]

Dynamic measurements of the transfer rate have been used as an alternative to the Förster cycle.[22] The proton-transfer reaction is represented by the following scheme

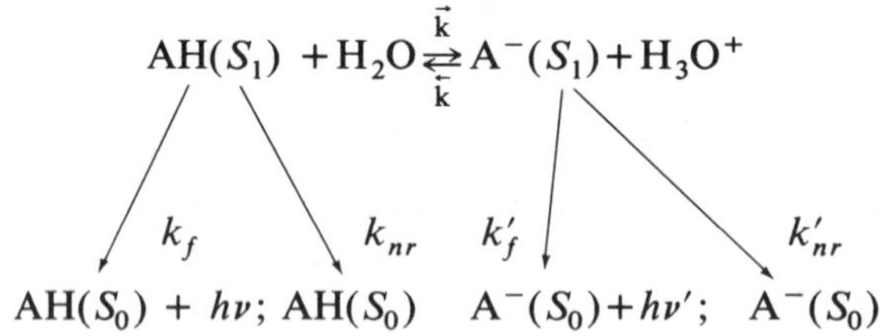

(I)

where $\vec{k}$ is the pseudo-first-rate constant for proton transfer to the solvent $\overleftarrow{k}$ is the bimolecular rate constant for the back-reaction; k_f and k_f' are radiative rate constants; k_{nr} and k_{nr}' the nonradiative rate constants.

For such a reaction scheme the rate equations are

$$-\frac{d[\mathrm{AH}(S_1)]}{dt}=(\vec{k}+k_f+k_{nr})[\mathrm{AH}(S_1)]-\overleftarrow{k}[\mathrm{H_3O^+}][A^-(S_1)] \tag{2}$$

$$-\frac{d[\mathrm{A^-}(S_1)]}{dt}=(k_f'+k_{nr}')[\mathrm{A^-}(S_1)]+\overleftarrow{k}[\mathrm{A^-}(S_1)][\mathrm{H_3O^+}]-\vec{k}[\mathrm{AH}(S_1)] \tag{3}$$

These equations have been intergrated and result in a complex dependence of the $[\mathrm{AH}(S_1)]$ and $[\mathrm{A^-}(S_1)]$ on the rate constants.[23] In the limit that the pH is high, but the ground-state population still consists of the conjugate acid, then the solutions to the differential equations become

$$[\mathrm{AH}(S_1)]=[\mathrm{AH}(S_1)]_0 e^{-t/\tau} \tag{4}$$

$$[\mathrm{A^-}(S_1)]=\frac{\vec{k}[\mathrm{AH}(S_1)]_0[e^{-t/\tau'}-e^{-t/\tau}]}{\vec{k}-\overleftarrow{k}[\mathrm{H_3O^+}]+k_f-k_f'+k_{nr}-k_{nr}'} \tag{5}$$

where τ is the lifetime of $AH(S_1)$ at $pK^* < pH < pK$; it is equal to $1/(\vec{k}+k_f+k_{nr})$ and $\tau' = 1/(k'_f+k'_{nr})$, which is the lifetime of $A^-(S_1)$ at high pH.

These equations have been applied to the study of the deprotonation reaction in 2-naphthol.[22,24,25] The apparatus used in such studies had a resolution comparable to the singlet decay rate of the protonated form, $1/\tau$. Hence deconvolution to obtain the transfer rate in 2-naphthol resulted in the incorrect conclusion that solvent relaxation results in a decrease in the transfer rate.[25] If the transfer rate is slower than the apparatus resolution, it is possible to obtain directly the rate of proton transfer from the fluorescence rise of the anion product. When the hydrogen ion concentration becomes large enough so that $\overleftarrow{k}[H_3O^+]$ is of the same order of magnitude as the forward reaction rate, $\vec{k}$, then the rise time of the fluorescence signal actually decreases. From the integrated rate equations, it is then possible to obtain $\overleftarrow{k}$.[22] Dynamic measurements of $\vec{k}$ and $\overleftarrow{k}$ lead directly to the excited state equilibrium constant and are the best way to determine the pK^*. The method is only useful for deprotonation reactions when the $\vec{k} > k_f + k_{nr}$ (i.e., when equilibrium is established during the excited state lifetime). From a practical point of view this is not too large a drawback because the pK^* is not very meaningful unless equilibrium is established in the excited state. Often measurement of $\vec{k}$ requires an apparatus capable of making picosecond measurements, since $1/(k_f+k_{nr})$ is almost always on the order of a nanosecond.

An alternative to the kinetic measurement of proton transfer is the use of the steady-state assumption.[13,27] This results in two equations for the relative quantum yields

$$\frac{\phi}{\phi_0} = \frac{1+\overleftarrow{k}[H_3O^+]/(k'_f+k'_{nr})}{1+\vec{k}/(k_f+k_{nr})+\overleftarrow{k}[H_3O^+]/(k'_f+k'_{nr})} \tag{6}$$

$$\frac{\phi'}{\phi'_0} = \frac{\vec{k}/(k_f+k_{nr})}{1+\vec{k}/(k_f+k_{nr})+\overleftarrow{k}[H_3O^+]/(k'_f+k'_{nr})} \tag{7}$$

where ϕ is the fluorescence quantum yield (QY) from the neutral form [AH] as a function of pH. ϕ_0 is the fluorescence QY from AH at a low enough pH to quench all anion fluorescence. ϕ' is the fluorescence QY from the anions as a function of pH, while ϕ'_0 is the fluorescence QY at a $pH > pK$. If the proton transfer reactions do not result in deactivation of the singlet state then $\phi/\phi_0 + \phi'/\phi'_0 = 1$. Once the rates $k_f + k_{nr}$ are known from the fluorescence lifetimes it is possible to determine both $\vec{k}$ and $\overleftarrow{k}$ and hence pK^*. In applying the steady-state methods, a change in the absorption spectrum with pH must be taken into account in order to obtain the proper QYs.[13] Corrections to the steady-state equations must also be made for the fraction of hydronium ions that are initially found in the reaction sphere, and which react immediately with the excited state.[13,23] From such steady-state measurements rate constants for the back-reaction $\overleftarrow{k}$ have been found to be within a factor of 2 of the diffusion rate for protons. Analogous equations can be derived for reactions in which pK^*>pK.

Despite the errors that can arise in both the Förster cycle and the steady-state fluorescence the pK^* obtained by both methods usually agree to within 1 or 2 pK units (see tables in Ref. 28). For triplet states application of either method is often difficult or even impossible. Instead it is necessary to use a kinetic approach. The amounts of conjugate acid and base in the triplet state are followed by time-resolved excited-state absorption spectroscopy.[29]

The Förster cycle is not capable of observing the details of the transfer reaction, since it is a thermodynamic determination of pK^* (and hence $\Delta G_R^0{}^*$). Thermodynamics provide no direct information about the kinetics. When used in conjunction with a model for proton transfer, such as the Marcus theory, thermodynamic quantities such as $\Delta G_R^0{}^*$ do provide some insight into the reaction rate.[7,30,31] The rates derived using the thermodynamic quantity $\Delta G_R^0{}^*$ when compared with the measured rate provide a test of the Marcus theory. Free energy relationships like Brønsted plots provide information on how chemical reactivity varies with chemical structure and on the nature of the transition state.[1,32,33,]

A comparison of $\overleftarrow{k}$ in the excited state with $\overleftarrow{k}$ in the ground state can be used to obtain some information about the protonation step. For example, the anion of 8-hydroxypyrene-1,3,6-trisulfonate (HPS) has a rate constant for reprotonation ($\overleftarrow{k}$) of 2×10^{11} liter mole^{-1}s^{-1} in the ground state, while the reaction rate in the excited state was a factor of 4 smaller.[34] It was argued that in the excited state the three SO_3^- groups of the molecule have nearly the same affinity

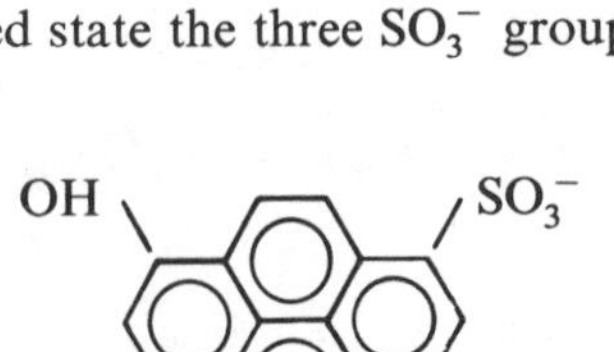

HPS

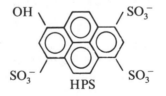

for the proton as does the oxygen. Thus a steric factor of 1/4 "slows" down the excited state reactions. In the ground state, the affinity of the oxygen is much larger than that of the three SO_3^- groups, and hence it immediately traps the proton within the reaction sphere.

In the general scheme for proton transfer

$$AH + B \underset{k_{-1}}{\overset{k_1}{\rightleftarrows}} AH\cdots B \underset{k_{-2}}{\overset{k_2}{\rightleftarrows}} A\cdots BH \underset{k_{-3}}{\overset{k_3}{\rightleftarrows}} A^- + BH^+ \qquad \text{(II)}$$

the reaction sphere is the volume at which the rate constants k_2 and k_{-2} are appreciable.[13,35] Outside this volume they are zero. In fact, for excited-state proton-transfer reactions k_2 and k_{-2} are believed to be so much larger than the other rate constants that the encounter complex is equilibrated. The radius of the reaction sphere in water is taken to be about 0.6–0.75 nm. This is based on the fact that a solvent shell two to three water molecules thick can transfer a proton to an acceptor. The rate constants for formation and decay of the encounter complex (from ions) are given by the theory of diffusion controlled reactions:[36,37]

$$k_{\text{diffusion}} = 4\pi a D N' \frac{\delta}{\exp(\delta) - 1} \exp\left(\delta \frac{\kappa a}{1 + \kappa a}\right)$$

$$\delta = \frac{z_1 z_2 e^2}{\varepsilon k T a} \qquad (8)$$

where z_1e and z_2e are the charges at the proton donor and acceptor, ε is the dielectric constant, D is the sum of the diffusion coefficients, N' is one thousandth of Avogadro's number and κ is the reciprocal radius of the ionic atmosphere. For a large number of proton-transfer reactions in the ground and excited state, it is possible to account for the data by an efficiency γ for the proton transfer within the encounter complex giving $\overleftarrow{k} = \gamma k_D$.[13,35] The efficiency γ consists of a probability for transfer, which can be calculated from the rate constants in scheme II. This probability must then be multiplied by the steric factors for the donor and the acceptor.[38] For HPS a steric factor of $\frac{1}{4}$ would be needed in the excited state for proton transfer to the anion, since the three sulfonate and the one oxygen have the same proton affinity. In the ground state the steric factor would be 1.[38]

A more sophisticated model has been worked out in which the reaction sequence is[39]

$$\underset{\substack{\text{Translational}\\\text{Diffusion}}}{AH + B \underset{k_t'}{\overset{k_t}{\rightleftarrows}} AH, B} \underset{\text{Rotation}}{\underset{k_r'}{\overset{k_r}{\rightleftarrows}}} \underset{\text{Reaction}}{AH\cdots B \overset{kx}{\rightarrow} A^- \cdots HB^+} \qquad \text{(III)}$$

Diffusion again controls the rate of encounter. However, after the encounter complex is formed, rotation within the encounter complex to the proper position is necessary for proton transfer to proceed. The model is quite similar to the earlier one, the only difference being the inclusion of the rotational correlation time. The steric factors are still included, but now are called the fraction of surface area, which are reactive. They are usually determined from analysis of the data. This model is very useful for "large" molecules in which orientational relaxation times are the same order as the lifetime of the encounter complex.

To summarize, the proton transfer reaction can be broken into three distinct parts: Diffusion of the reactants to within the radius of the ionic atmosphere; accelerated diffusion to within the encounter distance; and subsequent conversion of the encounter complex to products. For reactions in which the equilibrium is rapidly established within the encounter complex, the rate equations are dominated by the diffusion process. This results in the loss of information about the dynamics of the encounter complex. For such a reaction some information can be obtained about the ionic radius by varying the ionic strength and using an electrostatic theory (such as is done for Deby–Hückel activity coefficients) to calculate the effect of shielding by the ions.[13,36,38]

III. INTRAMOLECULAR PROTON TRANSFER

To understand the dynamics within the encounter complex it is necessary to utilize a system in which a stable complex is always present. Such a complex consists of a molecule with an intramolecular hydrogen bond. The reaction can be quickly initiated by changing the properties of the reactant. For example with UV excitation a large pK gradient can be "switched on" in 10^{-15} s. The proton, because of this perturbation, undergoes a translocation across the hydrogen bond. Changes in the molecular structure and the solvent during the proton rearrangement can be probed, since the artificial encounter complex is prepared instantly. Temporal information is no longer smeared out by the diffusion process. However, intramolecular systems are not a perfect model for the encounter complex. They do not mimic the rotational dynamics that may take place following a diffusion controlled encounter. Also, when electronic structure changes take place in the same molecule, the influence of the donor site and acceptor site on one another is much greater than the site interaction within an encounter complex. Despite these drawbacks much can be learned about the dynamics that take place within an encounter complex through the study of intramolecular proton-transfer reactions.

Some of the above shortcomings can be overcome by incorporating two functional groups in a molecule in such a way that an intramolecular hydrogen bond is not possible. Following absorption of a photon, one of the groups must still undergo a positive pK change, while the other realizes a negative pK change. In the case of 7-azaindole

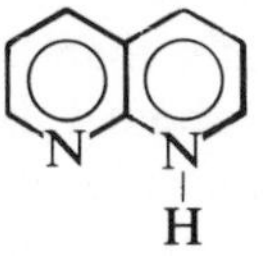

one nitrogen that is protonated in the ground state becomes more acidic in the excited state, while the second nitrogen becomes more basic. This results in a double proton transfer reaction occurring either within a symmetric dimer of 7-azaindole or between the chromophore and an alcohol.[40,41]

The temperature dependence of the proton transfer reaction within the 7-azaindole dimer can be monitored by observing the fluorescence spectrum as shown in Fig. 2. The violet component around 350 nm arises from the original dimer and the green 450 nm component originates from the tautomer formed by a rapid double proton translocation. The ratio of green to violet emission increases as the temperature is lowered. This is attributed to increased dimerization of the 7-azaindole. Below about 200 K the fluorescence ratio decreases again because of a barrier to proton motion. Below 77 K the ratio becomes temperature independent. From the temperature dependence a lower limit of 1.4 kcal/mole was estimated for

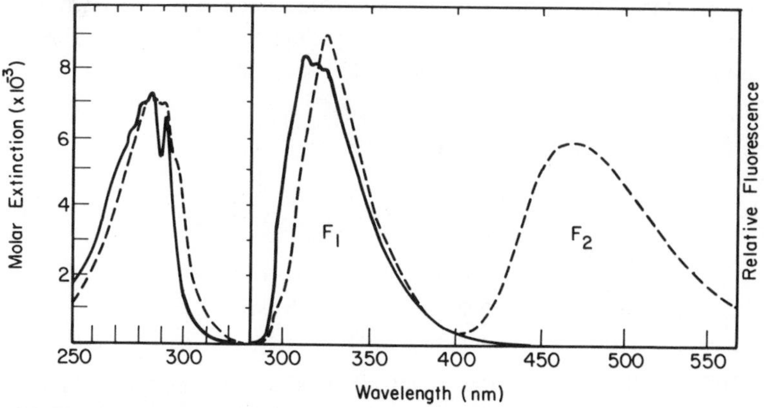

Fig. 2. Room-temperature absorption (left) and fluorescence (right) of 7-azaindole in 3 methyl pentane at 1.0×10^{-5} (—) and 1.0×10^{-2} M (---). The excitation wavelength was 285 nm. (From Ref. 40.)

this barrier. At 77 K the ratio of the proton to the deuteron transfer rate was found to be 2.9. The constant ratio of violet to green emission below 77 K and calculations of the size of the isotope effect indicate that proton translocation proceeds via a quantum-mechanical tunneling mechanism. Recent picosecond experiments found that the double proton transfer occurs within about 5 ps at 290 K.[42] At 77 K the rise of the tautomer fluorescence was biphasic. Greater than 80% of the emission appeared in under 5 ps. The remainder took about 1 nsec. The kinetics at 77 K suggest that there are two separate channels for the production of the tautomer: a rapid rate $>10^{11}$ s^{-1} for double proton translocation from the unrelaxed excited state and a slower rate that can be thermally induced over the barrier or via tunneling through it. The idea of a dual pathway is supported by the observation of a wavelength-dependent emission spectrum between 318 and 325 nm. Below 318 nm the spectrum is independent of wavelength. This work would indicate that the actual isotope ratio for motion through the barrier might be much larger than that observed in the earlier emission studies because a substantial part of the tautomer emission comes from the unrelaxed excited dimers. Their contribution to the emission should have only a weak isotope dependence. It will be necessary to measure both the proton and deuteron transfer rate over the entire temperature range to determine more accurately the height of the barrier as well as the size of the isotope effect for both the transfer through and over the barrier. The picosecond studies also suggest that the proton motion is faster than the vibrational deactivation.

If two functional groups are capable of forming an intramolecular hydrogen bond in the ground state, then absorption of a photon will result in the transfer of the proton across the intramolecular hydrogen bond. An example of such a molecule is methyl salicylate (MS). The absorption maximum of the red-most band of methyl salicylate

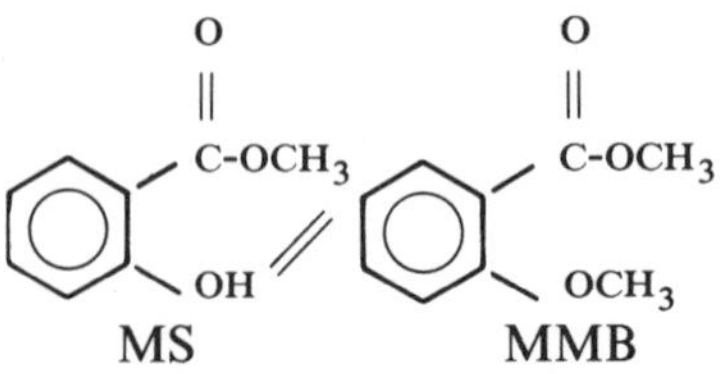

is at 308 nm, the fluorescence spectrum has two components. The major component at 450 nm, with a minor component that is found at 340 nm. The component at 450 is believed to be due to the zwitterion, which results from intramolecular proton transfer in the excited state. By analogy with the fluorescence spectrum of methyl-*o*-methoxy benzoate (MMB), the 340 nm component is assigned to the neutral form.[43] The 340 nm and 450 nm emissions were equally quenched by CS_2.[43] Thus it was concluded that the two forms of the excited molecule were in equilibrium. This assumption of equilibrium occurring during the singlet lifetime set of a lower limit of 10^9 s^{-1} on the rate of intramolecular proton transfer. As the temperature was lowered, the ratio of the 450–340 nm fluorescence increased. From the temperature dependence of the ratio a ΔH of about -1.0 kcal/mole between the neutral and the zwitterion was calculated. Since proton transfer occurred even at 4 K, it was felt that it proceeded via a tunneling mechanism.[44] The relative intensity of the two emissions changed when methyl salicylate was dissolved in acetonitrile instead of cyclohexane. This change was originally interpreted to be due to stabilization of the excited neutral form by the higher dielectric strength of acetonitrile. However, excitation spectra for the two emissions indicate the presence of two different ground-state forms.[45,46]

Naiboken has proposed a somewhat different mechanism for the dual fluorescence from methyl salicylate.[47] Absorption of a photon results in an excess of vibrational energy. This excess energy ruptures the intramolecular hydrogen bond. After relaxation takes place in the excited state, the hydrogen bond can then be reformed. Molecules which reform the hydrogen bond in the excited state emit at 450 nm, while those that do not reform the hydrogen bond will emit at 340 nm. The absorption maximum of methyl salicylate is found at 308 nm, while the absorption maximum for methyl-*o*-methoxybenzoate occurs at 293 nm. If dissociation of the hydrogen bond takes place during the absorption of a photon one might expect the absorption of methyl salicylate to be to the blue of the absorption of methyl-o-methoxybenzoate which does not have an intramolecular hydrogen bond. Calculations of the changes in the abortion of carbonyl compounds when they are placed in hydrogen-bonding solvents have been carried out.[48] They predict that n–π^* transitions are blue shifted in hydrogen-bonding solvents because of the dissociation of the hydrogen bond in the excited state. This effect is not seen when one compares the absorption spectrum of MS to MMB. Also, the difference in energy between the blue and ultraviolet emissions is much larger than that of a hydrogen bond. The separation between the maximum of the two fluorescence components corresponds to about 6600 cm^{-1}, while the hydrogen bond in methyl salicylate is estimated at about 2400 cm^{-1}.[49] Even if one assumes vastly different Franck–Condon factors for the species with and without a hydrogen bond, one cannot account for the large difference in the the energy of the two emissions. In contrast, it has been well documented that the protonation of carboxyl groups and the deprotonation of phenolic groups in the excited state lead to large changes in the emission spectrum.

In addition to its interesting photophysics methyl salicylate had been used to protect polymers against damage from ultraviolet radiation. It has since been replaced by a number of other molecules. However, many of these, such as *o*-hydroxybenzophenone, possess intramolecular hydrogen bonds in the ground state. In fact an empirical relationship exists between the strength of the intramolecular hydrogen bond in the ground state and the ability of the molecule to protect polymers from damage by solar radiation.[50] Therefore, motion of the proton in the excited state appears to play a role in the quenching of singlet and triplet excitation.

Rapid proton dynamics in the singlet excited state can often be studied by picosecond emission spectroscopy.[51] A typical apparatus used for such a study consists of a mode-locked neodymium phosphate glass laser. The laser produced a train of pulses each about 8 ps in duration (Fig. 3). A pulse selector extracted a single pulse which was amplified about 50–100 times. The laser light was frequency doubled to 528 nm with an angle tuned KDP crystal. It was then convert to 264 nm with a temperature tuned ADP crystal. A small portion of the light was reflected into a photodiode to trigger a streak camera. The green and ultraviolet components of the pulse were separated with a dichroic beam splitter. Residual

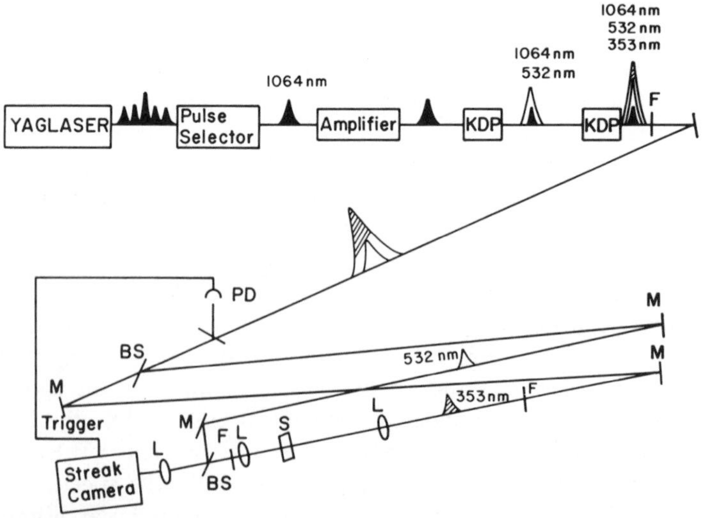

Fig. 3. Block diagram of picosecond fluorescence apparatus. F = filter, M = mirror, BS = beam splitter, S = sample, L = lens, and PD = trigger photodiode.

green light in the ultraviolet beam was filtered out with a polarizer, a second dichroic mirror, and a set of color filters. The green pulse traveled along a path parallel to the ultraviolet pulse. The pathlength of the green pulse was adjusted so that it arrived at the sample prior to the ultraviolet pulse. The 528 nm light passed through the sample without being absorbed and was imaged onto the slit of the streak camera. The ultraviolet beam excited the sample and the fluorescence was collected and also imaged onto the streak camera slit. The phosphor of the streak camera was viewed with a SIT vidicon camera. The output of the camera's video preamplifier was amplified and digitized by a multichannel analyzer. The digital data was transferred to a minicomputer for storage and analysis. The green prepulse always arrived at the photocathode a fixed period before the fluorescence of the sample. Therefore, it served as an accurate time mark in each streak record. By using this time marker the computer was capable of averaging a number of streak records to improve the signal-to-noise ratio.[51] The power of this averaging method is illustrated in Fig.4. In this figure the data from one laser shot are super imposed on the average of 15

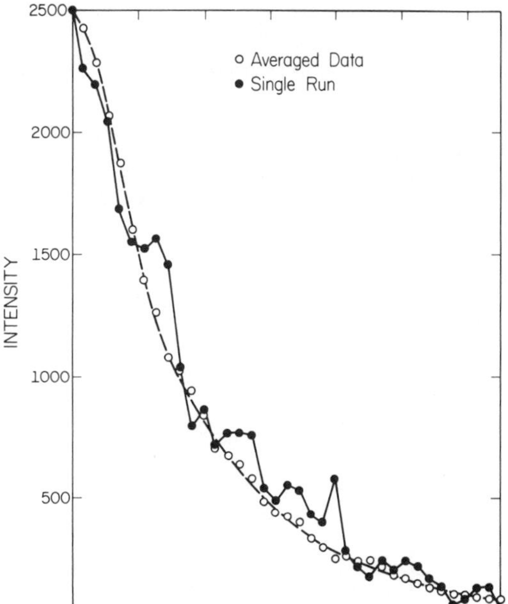

Fig. 4. Signal-to-noise improvement by averaging 15 laser shots. Open circles represent average of 15 shots while solid circles represent one of those 15 runs.

such measurements. One can see from Fig. 4 that data obtained from a single streak record were in good agreement with those made using the averaging procedure.

Variations of the above apparatus are used in a number of laboratories. Several experiments utilize a type II KDP crystal to generate the third harmonic at 353 nm instead of the fourth harmonic at 264 nm. To obtain a higher repetition rate at a sacrifice in temporal resolution, the glass laser can be replaced with a mode-locked neodymium YAG laser.

The formation of the excited zwitterion of methyl salicylate is accompanied by fluorescence at 450 nm. Therefore measurement of the rise of the fluorescence at this wavelength should give a lower limit for the rate of intramolecular proton transfer. The results of such an experiment are shown in Fig. 5. The recorded fluorescence signal is a convolution of the apparatus function with the fluorescence from the sample. In Fig. 5 the solid dots represent the apparatus function, while the open circles represent the actual data. The data are almost perfectly fit by the response of the streak camera. This response function is for light at 528 nm striking the photocathode. The wavelength-dependent response of the streak camera is

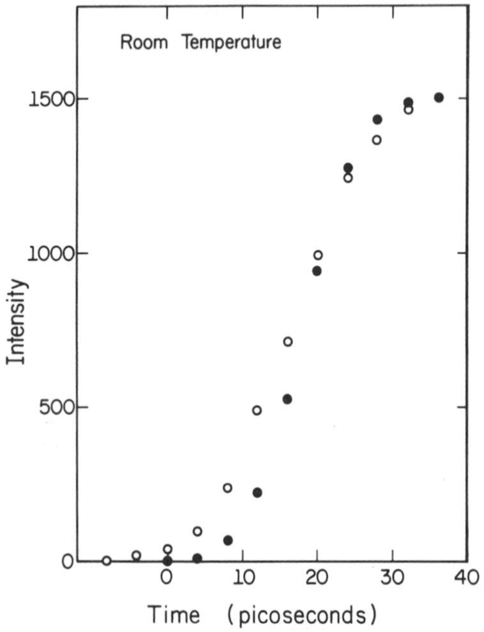

Fig. 5. Rise time of 450 nm methyl salicylate fluorescence at room temperature. Open circles represent averaged data. Solid dots represent apparatus function.

expected to broaden the apparatus function for emission at 450 nm.[53] This might account for the discrepancy between the data and the apparatus function. Deconvolution of the data indicates that the transfer rate must be greater than 10^{11} s^{-1}. The data did not change even as the temperature was lowered to 4 K. Replacing the proton with a deuteron and cooling to 4 K also did not alter the data. However, the 7-azaindole experiments have shown that the transfer rate could depend on the excitation energy. Using a tunable picosecond dye laser one might find a slower transfer rate then has been measured with 265 nm excitation.

The lifetime of the 450 nm fluorescence from methyl salicylate was measured over the range 40–353 K. The lifetime was nearly temperature independent between 40 and 160 K. Above 160 K the lifetime became shorter as the temperature was increased (see Table I and Fig. 6). Over a temperature range 253–333 K the relative quantum yield was found to decrease with increasing temperature. The change in the relative quantum yields of the 450 nm emission closely paralleled those observed in the lifetime data. Thus the radiative rate remains constant, while the nonradiative rate depends on the temperature. Using the fluorescence lifetime at 40 K as an estimate for a temperature-independent radiative rate it was possible to calculate the temperature-dependent nonradiative rate using the

TABLE I
Temperature Dependence of the Fluorescence Lifetime and Nonradiative Decay Rate for Methyl salicylate

Temp. (K)	τ (ps)	$k_{nr}(T)$, s^{-1}	Relative quantum yield[a] at 450 nm	at 340 nm
100	8300			
184	5020	7.87×10^{7}		
197	4650	9.45×10^{7}		
212	2250	3.24×10^{8}		
213	2092	3.58×10^{8}		
233	1189	7.21×10^{8}		
253	803	1.12×10^{9}		
273	470	2.01×10^{9}	2.02	1.15
296	280	3.45×10^{9}	1.00	1.00
303	239	4.06×10^{9}	0.88	1.20
313	218	4.47×10^{9}	0.63	1.20
323	176	5.56×10^{9}	0.48	1.13
333	150	6.55×10^{9}	0.35	1.03
343	124	7.94×10^{9}		
353	113	8.73×10^{9}		

[a] Normalized so that at this wavelength the quantum yield is 1 at 296 K.

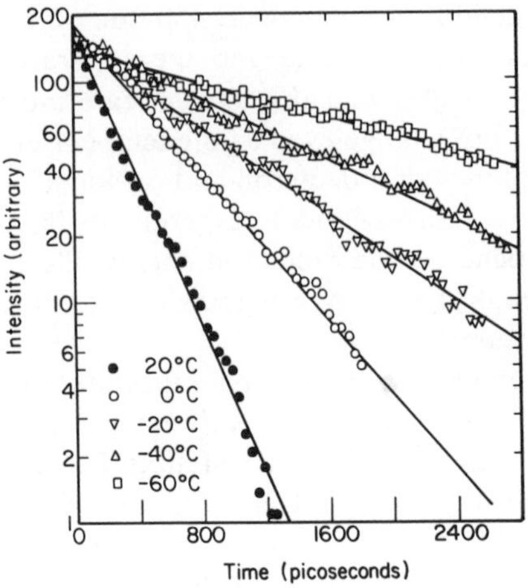

Fig. 6. Plots of the fluorescence lifetimes observed for the 450 nm component of methyl salicylate at various temperatures.

formula

$$\frac{1}{\tau_f} = k_r + k_{nr}(\tau) \tag{9}$$

where τ_f was the measured fluorescence lifetime, k_r was chosen to be 1.2×10^8 sec^{-1} from low temperature work, and k_{nr}(T) was the temperature dependent nonradiative decay rate. For three separate runs a plot of the ln k_{nr}(T) vs $1/T$ yields an activation energy of 3.7 kcal/mol (Fig. 7). Replacing hydroxyproton with a deuteron gave an indentical energy of activation. One can also use the relative quantum yields and an estimate of the radiative rate ($k_r \sim 1.2 \times 10^8$ s^{-1}) to calculate the temperature-dependent nonradiative decay rate from the equation:

$$k_{nr} = \frac{k_r}{\phi_f} - k_r \tag{10}$$

One then obtains an activation energy of about 4.7 ± 0.6 kcal/mol. Even though the temperature range used is much smaller and hence more prone to errors, this value is in good agreement with the activation energy calculated from the variation of fluorescence lifetime with temperature.

Methyl salicylate was also dissolved in heptylcyclohexane and octadecane. At 293 K, the former solvent has a viscosity about a factor of 4 larger than that of methylcyclohexane, while the latter is a solid. The lifetime of the 450 nm fluorescence was measured to be about 350 and 450 ps, respectively, in these two solvents. Presumably these changes represent differences in the nonradiative rate.

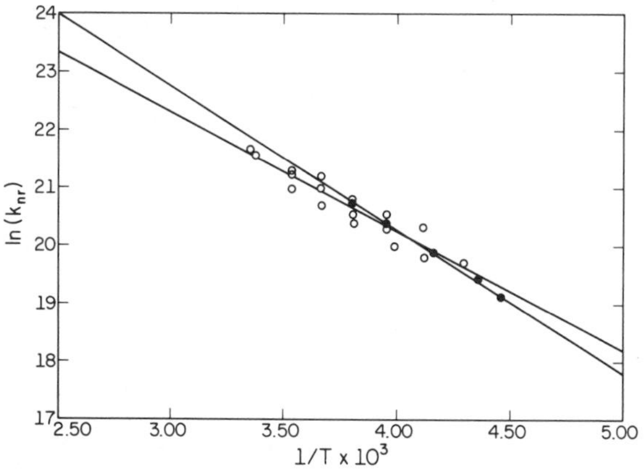

Fig. 7. Plot of ln k_{nr} vs $1/T$. Open circles represent data for three separate experiments with methyl salicylate. Solid dots represent one run with the hydroxy proton replaced with deuterium. Solid lines represent least-squares fit of the data for the normal and deuterated samples.

The steady state fluorescence of the 340 nm component did not change measurably in intensity from 253 to 333 K, while in contrast (see Table I) blue emission decreased by more than a factor of 6 as the temperature was increased. The excitation spectra for the 450 and 340 nm emission were slightly different as well. The amount of 340 nm emission from methyl salicylate in methylcyclohexane was too small to accurately measure with the streak camera. However, in acetonitrile the 340 nm light has a decay time of about 1 ns, while the 450 nm luminescence lasted only 100 ps. Similar differences in the lifetime for the blue and the red fluorescence components have been recently observed in a study of salicylanilide.[53]

The excited state species which emit at 340 and 450 nm are obviously not in equilibrium, since their fluorescence lifetimes differ by about a factor of 10 in acetonitrile. In addition, while the fluorescence intensity of the 450 nm emission increases in parallel with an increased fluorescence lifetime as a consequence of lowering the temperature, the 340 nm emission remains relatively constant. The rapid formation of the 450 nm emission and the very different lifetimes for the two fluorescence components would also indicate that two species are formed immediately after absorption of a photon. Perhaps they originate from different ground state molecules. This is suggested by the observation of differences in the excitation spectrum of the two emissions. The data suggest that in the ground state the location of the hydroxyproton with respect to the carbonyl oxygen must be different for molecules which will emit at 450 nm and those that will emit at 340 nm. The small change in the size of the 340 nm component with temperature suggests that the ground state equilibrium is not very sensitive to temperature. The hypothesis that ground-state equilibria are responsible for the dual fluorescence is strongly supported by the fluorescence observed from methyl 2,6-dihydroxybenzoate.[54]

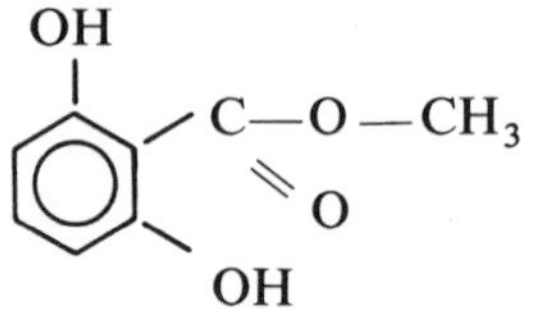

In nonhydrogen bonding solvents, this molecule was shown to emit only at 500 nm. Emission at 390 nm could only be obtained in alcohols. These results demonstrate that the ultraviolet fluorescence component is due to molecules in which the carbonyl oxygen is not hydrogen bonded to the hydroxyl group. Equilibrium cannot be established between the two tautomers in the excited state; only molecules which are in the correct ground state configuration will participate in a proton translocation. This proton translocation takes place in under 10 ps. Since neither a temperature dependence nor an isotope effect could be observed, the dynamic measurements do not elucidate the mechanism for the proton translocation. This question has been answered in part by the application of laser-induced fluorescence.

Methyl salicylate was placed in a rare gas matrix at 4.2 K.[54] A tuneable dye laser was used to excite the sample, while a monochromator was used to measure the emission spectrum. By tuning the dye laser it was possible to measure the excitation as well as the fluorescence spectrum of methyl salicylate. A partially resolved vibrational progression was observed both in the emission and in the excitation spectra. A fluorescence lifetime of 12 ns was measured at all of the emission wavelengths. As expected deuteration changed the vibrational progressions. It also lengthened the fluorescence lifetime. From the vibronic structure it was argued that a double minimum potential does not exist in the excited states. An estimate of 0.003 was made for the ratio of Franck–Condon factors for the vertical to nonvertical transitions shown in Fig. 8. This would imply that a false Franck–Condon origin 300 times larger than the true origin should be observed above it. This intense false origin was not observed. This observation rules out the existence of a double minimum potential. Instead a vertical Franck–Condon excitation must result in a large change in the geometry of the methyl salicylate. This was supported both by the rapid appearance of the 450 nm emission and the absence of a mirror symmetry

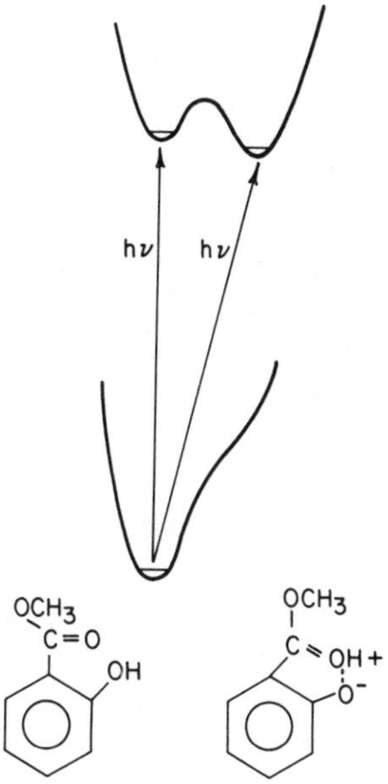

Fig. 8. Schematic of ground and excited state of methyl-salicylate. The excited state shown here has the postulated double minimum. The arrows show the vertical and nonvertical Franck-Condon transitions that are expected for an excited state with a double minimum potential. (From ref. 55).

for the Franck–Condon envelopes observed in the excitation and the fluorescence spectra.

In nonhydrogen bonding solvents, the triplet quantum yield for orthohydroxybenzophenone

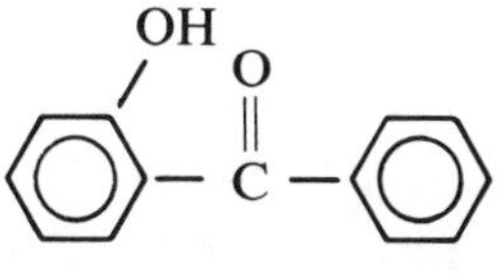

is known to be negligible. In hydrogen-bonding solvents the triplet yield increases.[56] This is in contrast to benzophenone and other hydroxybenzophenones in which the triplet yield is very large in all solvents. Since *O*-methoxybenzophenone also has a large yield for intersystem crossing, an excited-state proton-transfer reaction was believed to be responsible for the rapid quenching of either the singlet state in competition with intersystem crossing or by the sequence:[57]

$$S_0 \xrightarrow{h\nu} S_1(\text{neutral}) \xrightarrow{\text{H}^+\text{transfer}} S_1(\text{zwitterion}) \xrightarrow{isc} T_1(\text{zwitterion})$$

$$\xrightarrow{H^+\text{transfer}} T_1(\text{neutral}) \xrightarrow{ic} S_0(\text{neutral}) \qquad \text{(IV)}$$

To identify the pathway responsible for the deactivation, *O*-hydroxybenzophenone was excited with a 25 ps pulse at 355 mn.[58] The ground-state repopulation was found to occur within about 35 ps. The neutral triplet state is believed to absorb at 532 nm. During the 35 ps lifetime of the excited molecule, no absorption at 532 nm could be detected. Therefore it was concluded that rapid internal conversion to the ground state was responsible for the deactivation of the singlet excited state. Whether this rapid quenching proceeds through the neutral or the zwitterion has not been determined. When *O*-hydroxybenzophenone was dissolved in ethanol a fraction of the excited singlet returns directly to the ground state, while a long-lived absorption (1.5 ns) at 532 nm was observed. Energy transfer to 1-methylnaphthalene demonstrated that this long-lived absorption is due to the triplet state. The shortening of the triplet lifetime of *O*-hydroxybenzophenone with respect to the triplet lifetime of benzophenone was predicted in earlier work which estimated the triplet levels of the *o*-hydroxybenzophenone.[57]

Intramolecular proton transfer also occurs in the excited singlet state 6-(2-hydroxy-5-methylphenyl)-*s*-triazines.[59–61] The triazines that have been

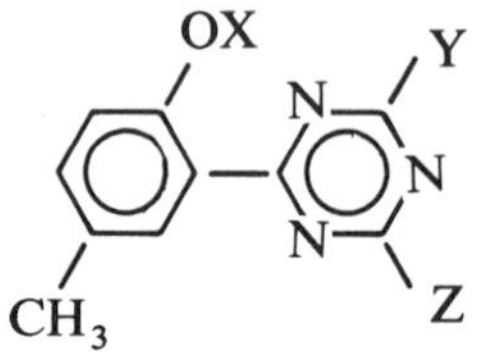

studied are summarized in Table II. For the triazines the singlet excited-state dynamics are summarized in Fig. 9. The emission and absorption spectra are shown in Fig. 10. The normal fluorescence is absent in the observed emission. This has been attributed to a very rapid decay of the enol via intramolecular proton transfer. The fluorescence lifetime (τ') and the fluorescence quantum yield (ϕ_f') depend on the temperature. While the temperature dependence as well as the magnitude of ϕ_f' and τ' of the triazines are sensitive to substitution at Y and Z, they are independent of the isotope at X (H or D). The change in the quantum yield and the lifetime do not parallel one another as the temperature is altered. This can be seen in Fig. 11 where the ratio ϕ_f'/τ' is plotted. This requires that either the radiative rate be temperature dependent or that a temperature-dependent nonradiative rate from the enol excited singlet state competes with the proton-transfer reaction. The temperature dependence has been interpreted as being due to the latter, namely, a change in the nonradiative

TABLE II
Samples of 6-(2-Hydroxy-5-methylphenyl)-*s*-triazines That Have Been Examined

Samples	X	Y	Z
$(OO)_h$	H	$(CH_3$	$(CH_3$
$(ON)_h$	H	OCH_3	$N(CH_3)_2$
$(ON)_d$	D	OCH_3	$N(CH_3)_2$
$(NN)_h$	H	$N(CH_3)_2$	$N(CH_3)_2$
$(NN)_d$	D	$N(CH_3)_2$	$N(CH_3)_2$

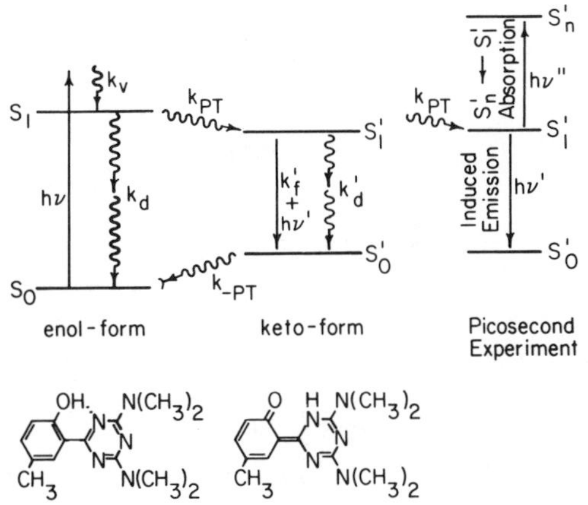

Fig. 9. A schematic energy state diagram for the intermolecular proton transfer in excited 2, 4-bis(dimethylamino)-6-)2-hydroxy-5-methylphenyl)-*s*-triazine. (From Ref. 59.)

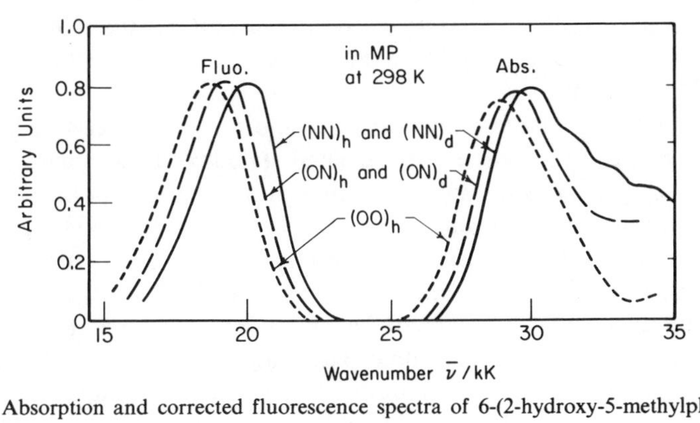

Fig. 10. Absorption and corrected fluorescence spectra of 6-(2-hydroxy-5-methylphenyl)-*s*-triazines. (From Ref. 60.)

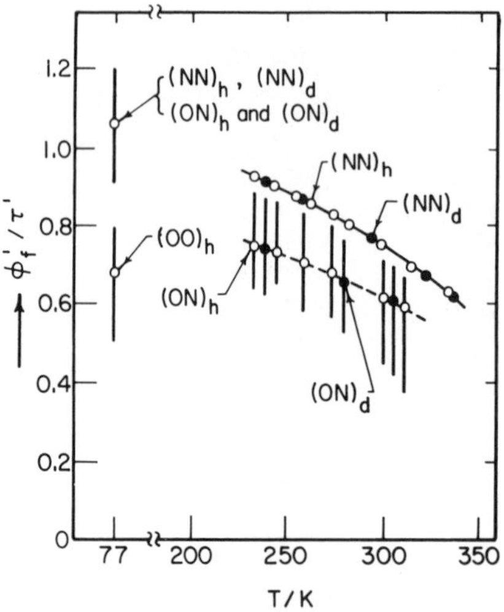

Fig. 11. Plots of Φ_f'/τ' vs T. (From Ref. 60.)

decay rate (k_d). From an analysis of the absolute fluorescence quantum yield and the fluorescence lifetime, the efficiency of proton transfer (γ) was found to be independent of whether a proton or deuteron was involved in the reaction. The transfer efficiency is given by:

$$\gamma = \frac{k_{pT}}{k_{pT} + k_d} \tag{11}$$

where k_{pT} is the proton transfer rate and k_d is the rate of deactivation of the excited state. γ was isotope independent: 0.57 for $(ON)_{H \text{ or } D}$ and 0.69 for $(NN)_{H \text{ or } D}$. Since γ is isotope independent, the proton transfer rate should also be independent of the isotope. An alternative explanation would require both k_{pT} and k_d to be dependent on the isotope, a most unlikely possibility. Although the efficiency of proton translocation in the excited state of triazine could be obtained by steady-state methods, the rate of intramolecular proton transfer could only be determined by measuring the rate of formation of the excited-state keto form. Instead of using picosecond emission spectroscopy to make this determination, excited-state absorption spectroscopy was used.[61] A single pulse at 694 nm was extracted from a train of pulses generated by a mode-locked ruby laser. The single pulse was amplified and then split in half. One of the halves was converted to light at 347 nm by second harmonic generation. This UV pulse was used to excite the sample. The remaining half of the 694 nm light was used to generate a quasicontinuum. The quasicontinuum served as a probe for time-resolved excited-state absorption measurements. Details of the apparatus used for this study can be found in Ref. 61, while a general review of the picosecond absorption technique can be found in Ref. 62. Since the keto tautomer is not present in the ground state, intramolecular proton transfer will result in a population inversion. This population inversion can be probed by observing an apparent increase in the transmission of light at 520 nm. The appearance of a new absorbance band in the near-ultraviolet can also be used to monitor the formation of the singlet excited keto tautomer. The dynamics were isotope independent, but depended on the substituents. Proton transfer rates of $1.2 \times 10^{10}\ s^{-1}$ and $1.8 \times 10^{10}\ s^{-1}$ were obtained for the (ON) and the (NN) derivatives when the proton-transfer efficiencies were used to extract a transfer rate (k_{pT}) from the observed rate of formation of the keto form. The absence of an isotope effect was used to argue that the energy of activation for the proton transfer reaction was close to zero and also that the reaction did not proceed via quantum-mechanical tunneling. Although this may be the case, further studies of the temperature dependence of k_{pT} are needed to support this conclusion.

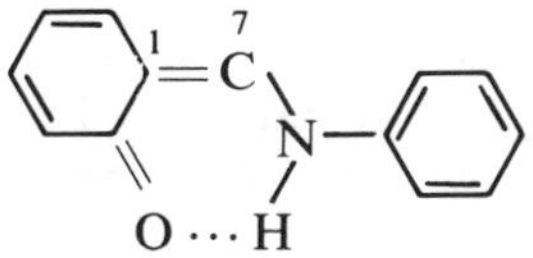

Light-induced intramolecular proton transfer has long been suspected to play an important role in the mechanism for photochromism in both crystals and solutions. One class of compounds that has been extensively studied is the *N*-salicylidene anilines.[62] The main evidence for invoking an excited-state proton-transfer step in the photochromic mechanism comes from the observation that the colored species exists as the keto amine tautomer (in the ground state) while the bleached form exists as an enol imine (in the ground state). Replacement of the labile hydrogen with a methyl group prevents the photochemical reaction that leads to the colored species. Picosecond transient spectroscopy was used to study *N*-salycylidene-*o*-toluidine (NST) and the related compound 2-(*o*-hydroxyphenyl)benzothiazole (HBT).[64]

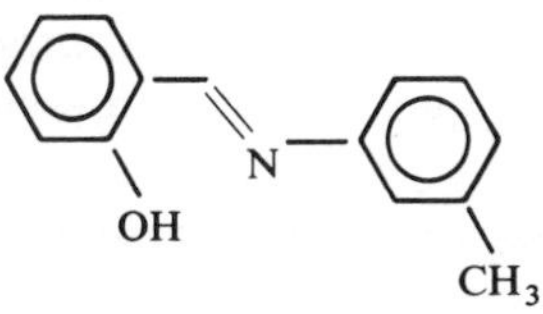

NST

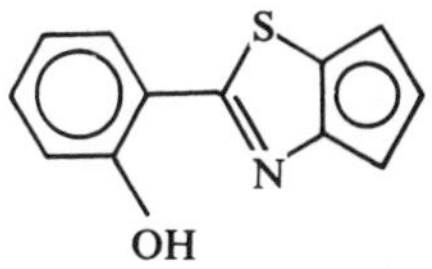

HBT

In a solution of isopentane: methylcyclohexane (1:1 volume ratio) the colored form of *N*-salycylidene-*o*-toluidene was generated in 84 ps, while in a solution of cyclohexanol the rate of appearance was much longer. This is believed to be due to the effect of viscosity on the rate of formation of the photochromic species since the viscosity of the cyclohexanol is more than an order of magnitude larger than the mixed hydrocarbon solvent. The viscosity dependence indicates that a large geometrical change must precede formation of the colored species. Since intramolecular proton transfer occurs very rapidly in methyl salicylate even in high viscosity solvents, this geometrical change cannot be due merely to excited-state proton transfer. Much more significant structural changes must take place to form the precursor to the ground state of the keto amine.

Evidence for such a precursor can be found in the excited-state absorption spectra of 2-(*o*-hydroxyphenyl)benzothiazole. The colored species is formed in 53 ps, while a transient absorbance near 385 nm was found to decay in exactly the same amount of time. The excited *cis*-keto amine is the logical choice for such an intermediate. However, its fluorescence lifetime is about 700 ps. Thus the structure of the intermediate that forms the colored species in 53 ps is not clear. Consequently, it is called X. It has been postulated from the picosecond data, the wavelength dependence of the quantum yield for production of the photochromic species, and the wavelength dependence of the fluorescence yield from the *cis*-keto amine that X is a vibrationally excited state of the *cis*-keto amine.[65] Such a long lifetime for vibrational excitation in large molecules in fluid solution is exceptional, since vibrational decay in the excited singlet manifold is believed usually to occur in under 10 ps. However, the use of a long vibrational decay time has also been invoked to interpret the time-resolved absorption spectra of benzophenone.[66] The failure to observe emission from the *cis*-enol in hydrocarbon solvents suggests that the proton-transfer rate in the excited state is extremely rapid with an estimated lower limit of $3\times10^{12}\ s^{-1}$ at room temperature and $4\times10^{10}\ s^{-1}$ at 77 K. Thus, while proton transfer appears to be a necessary step in the mechanism for photochromism, it does not play a role in the rate-determining step.

Molecules such as 7-azaindole and methyl salicylate are excellent for the study of proton translocation over short distances. As can be seen from the study of 7-azaindole, a great deal of information can be learned about proton tunneling through relatively low-lying barriers. In contrast, the study of methyl salicylate results in a unique way of observing the dynamics of molecular geometry changes. In the next few years the application of both picosecond spectroscopy and high-resolution laser spectroscopy should lead to a much better understanding of proton translocation over short distances. This type of information is very important not only in understanding this basic and important chemical phenomena, but will have an impact on the study of the photochemistry of vision[67] and the chemical mechanism for genetic mutation.[68] While intramolecular proton-transfer reactions are excellent models for the study of the encounter complex, they do not furnish any information about proton motion within the solvent. Since a vast number of chemical and biological reactions take place in water, where proton interaction with the solvent is very important, it is necessary to also study intermolecular proton-transfer reactions.

IV. INTERMOLECULAR PROTON TRANSFER

The earliest studies of excited-state proton reactions were concerned with the measurement of excited-state *pK* values both in the singlet and the triplet manifold. These studies are thoroughly reviewed in Refs. 13 and 18. Since the excited singlet state of most aromatic molecules lasts only a few nanoseconds, it was difficult to measure directly the dynamics of intermolecular proton transfer. Instead, as outlined in the introduction, steady-state measurements were used. Picosecond spectroscopy has made it possible to probe in real time the intermolecular transfer process.

The most thoroughly studied compounds are derivatives of 2-naphthol. The emission spectra of 2-naphthol and its derivatives are shown in Fig. 12. The parent compound was found to have a rate of proton ejection ($\vec{k}$) of about 5×10^{7} s^{-1} using phase fluorimetry,[26] nanosecond emission

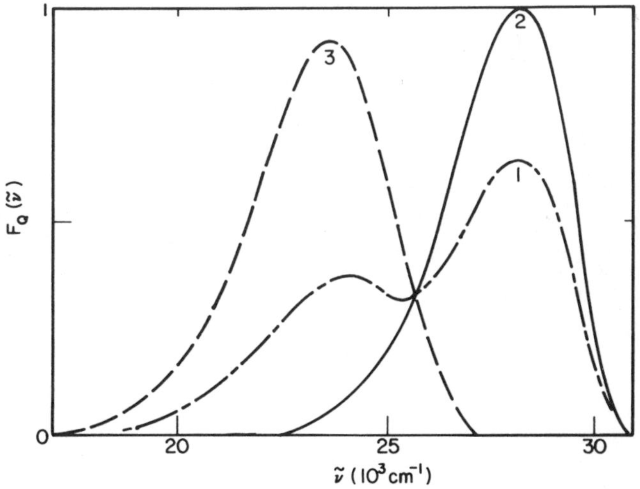

Fig. 12. Fluorescence spectra of *β*-naphthol in water at 25°. (1): pH 5–6; (2): 0.15 M $HClO_4$; (3): 0.02 M NaOH. (From Ref. 13.)

spectroscopy[24, 25, 69] and

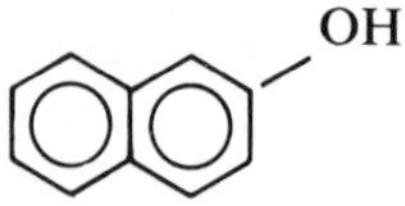

picosecond emission spectroscopy.[22, 70] The rate constants obtained by each of these techniques are summarized in Table III. They are in good agreement with estimates made from fluorescence titration studies. Excited state *pK* values for 2-naphthol-6-sulfonate are lower than those of the parent compound 2-naphthol. As expected from the p*K**'s the rate of proton discharge for 2-naphthol-6-sulfonate is about 20 times faster. The solutions to equations (2) and (3) is given by equations (12) and (13).[28]

$$[\mathrm{AH}(S_1)] = \frac{[\mathrm{AH}(S_1)]_0}{\gamma_2-\gamma_1}[(\gamma_2-x)e^{-\gamma_1}+(x-\gamma_1)e^{-\gamma_2}] \qquad (12)$$

$$[\mathrm{A}^-(S_1)] = \frac{\vec{\mathrm{k}}[\mathrm{AH}_1(S_1)]_0}{\gamma_2-\gamma_1}[e^{-\gamma_1}-e^{-\gamma_2}] \qquad (13)$$

TABLE III
Excited State Behavior of Several Aromatic Alcohols

	p*K**	Fluorescence lifetimes at pH=0	pH=7	pH=12	Deprotonation rate (s^{-1})
2-Naphthol-3,6-disulfonate	0.5±1.1	14	21	21	$3.1\pm1.1\ 10^{10}$
8-Hydroxypyrene-1,3,6-trisulfonate	0.4±0.1	5.6	6.3	6.2	$2.0\pm1.0\ 10^{10}$
2-Naphthol-6-sulfonate	1.66[a]				$1.02\pm0.2\ 10^{10}$
2-Naphthol	2.5,2.8[a]				4.6±0.5[c, d] $7\pm2\ 10^{7}$[b]

[a]Taken from Ref. 28.
[b]Taken from Ref. 22.
[c]Taken from Ref. 25.
[d]Taken from Ref. 24.

where

$$\gamma_{1,2}=\tfrac{1}{2}\left\{(x+y)\pm\left[(y-x)^2+4\vec{k}\overleftarrow{k}[H_3O^+]\right]^{1/2}\right\}$$

and

$$x=k_r+k_{nr}+\vec{k} \qquad y=k_r'+k_{nr}'+\overleftarrow{k}[H_3O^+]$$

They predict that as the $[H_3O^+]$ is increased, so that $\overleftarrow{k}[H_3O^+]$ becomes greater than $\vec{k}$, the rate of appearance of emission from the deprotonated anion will no longer be controlled by the rate of proton ejection $\vec{k}$ but instead by the rate of reprotonation $\overleftarrow{k}[H_3O^+]$. The intensity of the anion emission is of course diminished due to the reprotonation reaction. By monitoring the apparent rise of the anion fluorescence as a function pH, it is possible to determine the rate constant $\overleftarrow{k}$. For 2-naphthol $\overleftarrow{k}$ was found to be $5\pm1\times10^{10}$ liter-mole^{-1} s^{-1} and for 2-naphthol-6-sulfonate it was found to be $9\pm3\times10^{10}$ liter-mole^{-1} s^{-1}.[22, 70] The primary contribution to both these rates is from the proton diffusion in water. They are in reasonably good agreement with other determinations of the proton diffusion rate.[38] However, by comparison with 8-hydroxy-1,3,6-pyrene trisulfonate in which the $\overleftarrow{k}$ is slowed by a factor of 4 in the excited state due to the equivalent reactivity of the $-SO_3^-$ and $-O^-$ groups in the excited state, one would have expected the 2-naphthol-6-sulfonate to have a slower rather than a faster reprotonation rate than 2-naphthol.

Addition of a second sulfonate group to form the compound 2-naphthol-3-6-disulfonate results in an even smaller excited state pK^* than is found for 2-naphthol or 2-naphthol-6-sulfonate. The deprotonation rate was found to be 3.1×10^{10} s^{-1}.[71] The pyrene derivative HPS had a deprotonation of about 3.2×10^{10} s^{-1}. The excited state pK^*'s are related directly to the standard free energy, ΔG_R° for the protonation–deprotonation reaction. When the rate of deprotonation for the naphthol derivatives and the 8-hydroxy-1,3,6-pyrene trisulfonate are plotted as a function of the excited state pK^*, a linear relationship is found to exist over four orders of magnitude (Fig. 13). The existence of this linear free energy relationship would suggest that the mechanism for proton ejection is similar for the four compounds. More careful measurements will be needed to determine exactly how valid this relationship is, especially in light of the difference in the rate for reprotonation for 2-naphthol and 2-naphthol-6-sulfonate.

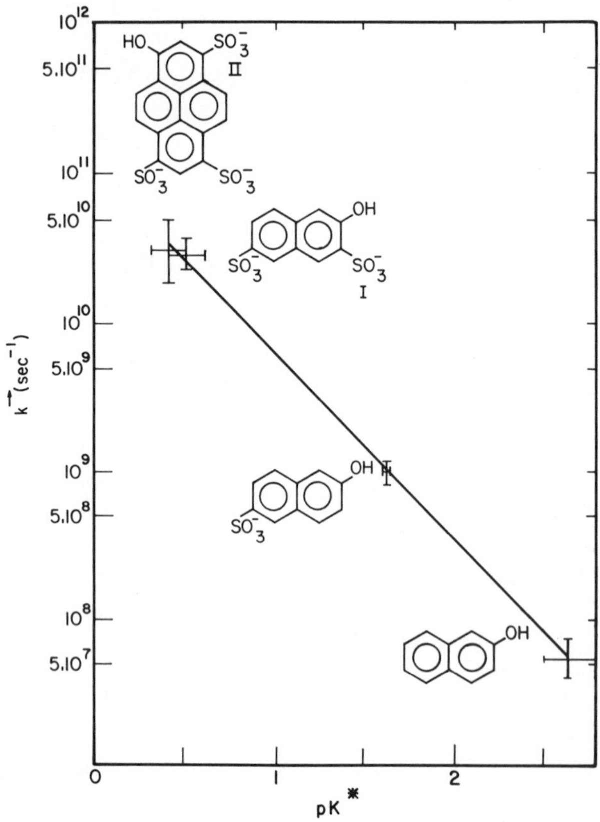

Fig. 13. The colinearity of the measured rate of acid dissociation of excited naphthol and pyrene derivatives with the acid dissociation constant of the excited compound. Data taken from Table II.

At low pH (<2) aminopyrene (AP), α-naphthylamine (αNA), and β-naphthylamine (βNA)

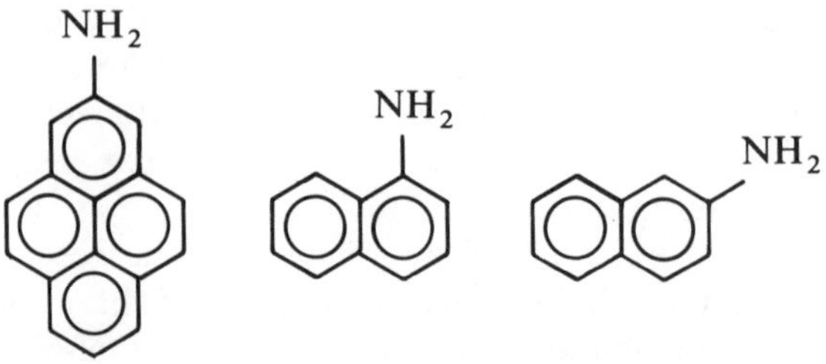

AP αNA βNA

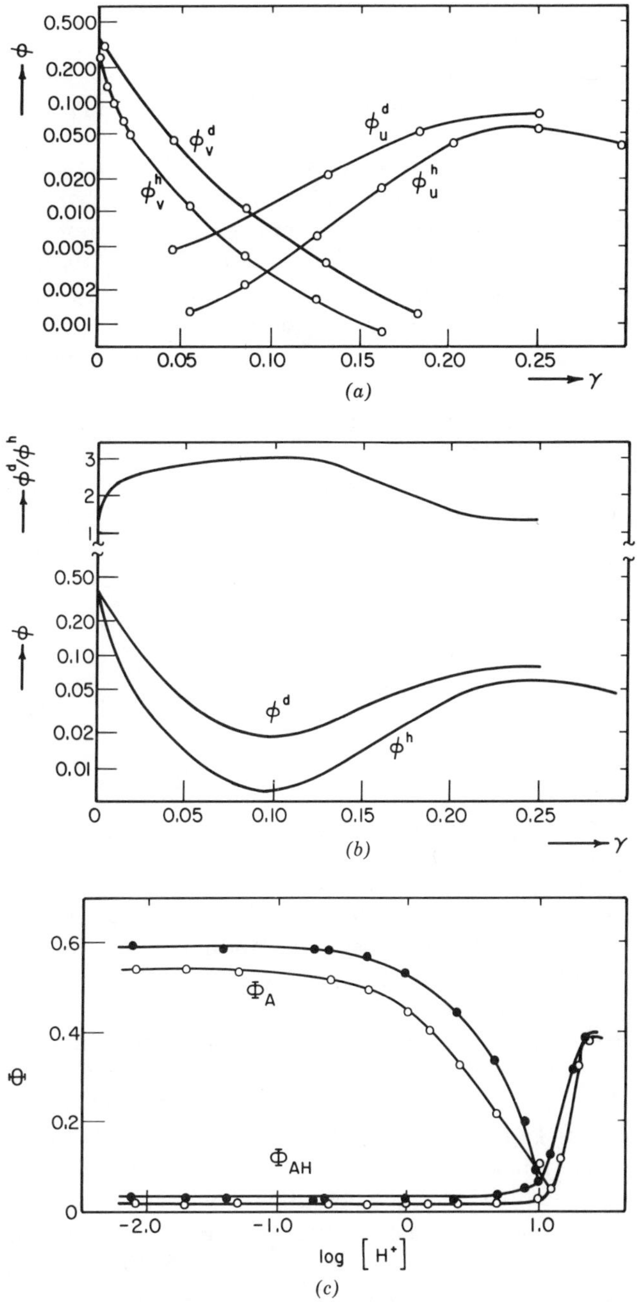

Fig. 14. (a) Fractional quantum yields of separate fluorescence components of 2-naphthylamine. γ: molar fraction of acid in sulphuric acid–water mixture; Φ_v^h: molecular (violet) component in H_2SO_4/H_2O; Φ_v^d: same in D_2SO_4/D_2O, Φ_u^h: cationic (UV) component in H_2SO_4/H_2O; Φ_u^d: same in D_2SO_4/D_2O. (b) Total fluorescence quantum yields of 2-naphthylamine. Φ^h: in H_2SO_4/H_2O; Φ^d: in D_2SO_4/D_2O. Upper curve: isotopic ratio Φ^d/Φ^h. (From Ref. 72). (c) Semilogarithmic plots of the fluorescence quantum yields of neutral 1-aminopyrene Φ_A and cation Φ_{AH} as a function of $[H^+]$ in 20% acetonitrile systems (○: H_2SO_4/H_2O; ●: D_2SO_4/D_2O). (Data taken from Refs. 72 and 74.)

become protonated in the ground state.[72–74] Absorption of a photon results in a lowering of the pK accompanied by the release of a proton. Titration curves for the fluorescence from AP and β-NA are shown in Fig. 14.[72, 74] The midpoints in the titration curves of the cation emission and the neutral emission do not cross at the same $[H_3^+O]$. This indicates that the molecules do not behave as simple two-component systems. Time-resolved emission studies showed that the neutral lifetimes were about 27 ns for the naphthylamines and 5.9 ns for aminopyrene at low $[H_3^+O]$ and decreased as the $[H_3^+O]$ was increased. At very high proton concentrations equations (12) and (13) would predict that in the absence of a quenching process the intensity of neutral emission would be reduced, but its lifetime would approach that of the cation. At moderate proton concentration the cation did have a short lifetime; however, as the proton concentration was increased the lifetime was lengthened to 167 nsec for AP and 47 ns for α-NA. This is in contrast to what (12) and (13) would predict. The change in lifetimes with the addition of protons excluded the possibility of a static quenching mechanism being responsible for the typical titration curves. Instead the following kinetic model was proposed:

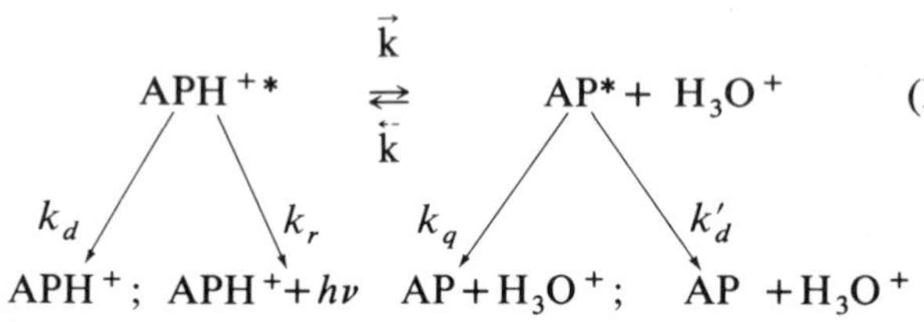

(IV)

where k_q is the second-order rate constant for quenching of the excited neutral by the protons. For aminopyrene, the deprotonation rate ($\vec{k}$) could be directly measured by observing the formation of the neutral AP* absorption spectrum. $\vec{k}$ was found to be 4.2×10^9 s^{-1} for protons and 3.4×10^9 s^{-1} for deuterons[74] in H_2O : acetonitrile (4 : 1) and 3.7 and 3.4×10^9 s^{-1} in a H_2O : acetonitrile solution (1 : 1). k_q was obtained from a Stern–Volmer plot. Analysis of the fluorescence lifetimes resulted in a determination of $\overleftarrow{k}$. The values of these rate constants for AP, α-NA and β-NA are summarized in Table IV. From $\vec{k}$ and $\overleftarrow{k}$ the excited state could be kinetically determined. As can be seen from Table IV there is poor agreement between the determination of pK* by kinetic measurements, the Förster cycle and fluorescence titration. This disagreement is almost certainly due to the effect of the dynamic quenching by the protons. Since the reprotonation rates are considerably smaller than a diffusion-controlled rate, changes of the molecular geometry most likely occur during proton uptake. Such a geometry change could account for the difference in the

TABLE IV
Excited State Behavior of Several Aromatic Amines

	$\vec{k}$ (s^{-1})	$\overleftarrow{k}$ (liter mole^{-1} s^{-1})	k'_q (liter mole^{-1} s^{-1})	pK^* (Kinetic)	pK^* (Förster)	pK^* (Titration)
Aminopyrene	4.2×10^9	2.7×10^8	3.6×10^7	−1.2	−5.5	−0.55
α-Napthylamine	1.3×10^9	1.2×10^8	8.9×10^9	−1.0		1.9
β-Napthylamine	1×10^9	1.5×10^8	3.3×10^8	−0.8	−8.1	0.8

pK^* measured by kinetic means and through the Förster cycle. For the dynamic measurements, the aminopyrene solubility was enhanced by the addition of 20–50% acetonitrile. This could account in part for the difference in pK^* determined by the two methods, since the Förster cycle measurement was made in water. The pK^* measured in 20% acetonitrile was not very different than that measured in 50% acetonitrile, thus the solvent did not affect the data enough to account for the very large difference between the two methods.

V. SOLVENT CAGE EFFECT

Upon dissociation, the molecular fragments are formed within the solvent cage that surrounded the parent molecule. While still within the solvent cage, the radicals or ions that were formed can undergo geminate recombination.[75] Several techniques have been used to study the rate of geminate recombination and the dynamics of escape from the cage. Picosecond absorption has been used to probe the geminate recombination of iodine atoms formed by photolysis of I_2.[76] Magnetic field effects on the triplet quantum yield have been used to follow geminate recombination of radical ions formed during electron transfer reactions.[77] For radical ions, there is also an electrostatic effect in addition to the solvent cage. For the proton and anion formed by the excitation of an aromatic alcohol like 8-hydroxy-1,3,6-pyrene trisulfonate the solvent cage may have no effect on the dynamics, since the proton is directly transferred several solution layers away from the donor. The electrostatic effect can be very large. In order to probe this electrostatic cage effect, the amount of ground-state HPS anion present 20 ns after excitation was measured.[78, 79] It was assumed that all the protons within the electrostatic cage formed by the interaction of the excited anion and the proton would react instantly when the anion returned to the ground state. Those protons that escaped the electrostatic cage would react much more slowly. The size of the anion absorption detected immediately after the flash would be due to the slowly reacting protons. By quenching the singlet excited HPS anion, it was hoped to decrease the probability of escape. This would result in a decrease in the ground-state anion absorbance. From an analysis of the data, a cage lifetime of about 2 nsec was obtained. The local $[H_3O^+]$ predicted by such a long lifetime for the cage is much greater than that needed to quench anion fluorescence. A cage lifetime of 2 ns is also not compatible with the large diffusion rate of protons in solution. As was suggested by the authors time-resolved experiments will be necessary to determine the dynamics of any cage effect.

Preliminary time-resolved fluorescence measurements give a somewhat better indication of what the cage lifetime might be.[80] Both the rise in the fluorescence of the anion and the decay of neutral emission of 8-hydroxy-1,3,6-pyrene were measured. Careful examination of both the rise of the anion and the decay of the neutral indicate the presence of biphasic behavior. The biphasic kinetics can more readily be seen in D_2O. It is quite possible that the faster rate constant is due to protons that immediately escape the cage, while the slower decay is due to the destruction of the cage.

VI. GENERATION OF A pH JUMP

The picosecond studies above (Fig. 14) have indicated that the rate of proton ejection can be estimated from the pK^*. For compounds with a pK^* less than about 2 it is possible to generate a transient proton population within a few nanoseconds. Pulsed lasers such as the solid-state ruby or the gas-phase exciplex laser are capable of producing more than a 100 mJ in the ultraviolet. This corresponds more than to 10^{17} photons/pulse. The laser pulse can be focused to a very small cross-section. If the optical density of the ground-state acid is large enough to absorb 10% of the incident radiation, a final pH of 4 can be in principle be easily attained. Since the ground-state pK of most of the acids studied is greater than 7, a pH jump of at least 3 can be obtained. Such a pH jump would be ideal for the study of biological reactions such as ATP synthesis.[3] Chemical reactions such as acid catalyzed tautomerization and cis–trans isomerization can in principle also be studied with this technique. Such a catalysis with conventional light sources to generate a pH change has been reported. However, it has been shown to be due to artifacts.[81, 82] When intense laser pulses are used additional artifacts can appear. One of these is the generation of solvated electrons and the products obtained through their reaction with the solution.[83]

Two studies of the nanosecond pH jump has been carried out with pulsed lasers. The rate of reprotonation of 8-hydroxypyrene-1,3,6 trisulfonate has been studied as a function of ionic strength.[38] The rate

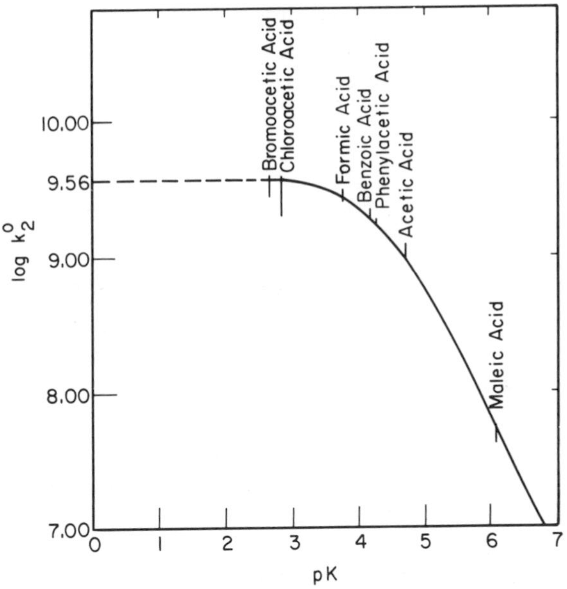

Fig. 15. Logarithm of the rate constant for reprotonation of the ground state anion of HPS (k_{21}^0) for the weak acids as a function of their pK values. (Data from ref. 39).

constant depended on the square root of the ionic strength as is predicted by the semiempirical Deby–Hückel–Brönsted–Davies equation. The rate of reprotonation in the presence of buffers depended strongly on the pK of the acid (Fig. 15).

The change in pH obtained by laser excitation was monitored directly by exciting 2-naphthol 6-sulfonate or 2-naphthol-3,6-disulfonate in the presence of pH indicator dyes.[84] The appartus used in this study is shown in Fig. 16. The acid–base reactions in aqueous solution of 2-naphthol-3,6-disulfonate following a laser pulse at 347.2 nm are given in Scheme V, where In is a pH indicator.

$$
\begin{gathered}
\text{naphthol}\ (S_o) \xrightarrow{h\nu} \text{naphthol*}\ (S_1) \\
\text{naphthol*}\ (S_1) + H_2O \rightleftarrows \text{naphtholate*}\ (S_1) + H_3O^+ \\
H_3O^+ + In^- \rightleftarrows HIn + H_2O
\end{gathered}
\tag{V}
$$

The rapid bleaching of the basic form of 71 μM Bromocresol Green (pK=4.7) is completed within approximately 200 ns. Observation of a decrease in absorbance due to protonation of the indicator dye occurred only after a short lag. This delay can be accounted for by the combined

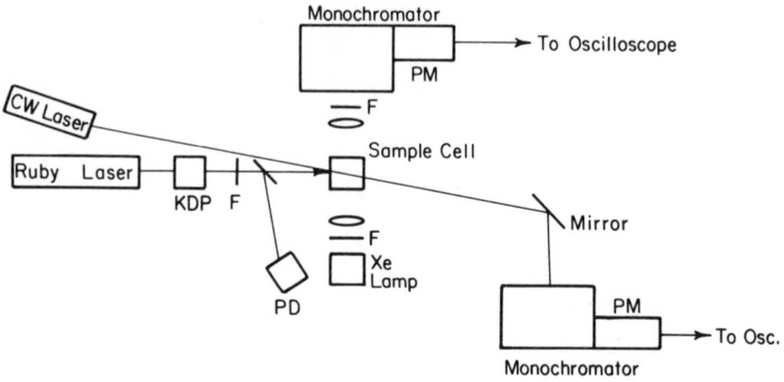

Fig. 16. Nanosecond laser apparatus. F=filters; PD=photodiode, PM=photomultiplier.

effect of an absorbance decrease (corresponding to protonation of indicator) superimposed on an absorbance increase reflecting the formation of hydrated electrons. To verify this idea, the experiment was repeated in the absence of the indicator: Hydrated electrons were observed with a decay time of 150 ns. Considering the fact that the extinction coefficient of hydrated electrons at this wavelength is about half that of Bromocresol Green ($1.5 \cdot 10^4$ and $3.9 \cdot 10^4$ M^{-1} cm^{-1}, respectively), accurate estimation of the rate of protonation of the indicator is not possible. However, it is complete within 200 ns. Decay of the protonated indicator occurs with a first-order rate constant of 4×10^5 s^{-1}.

The protonation of the indicator was also investigated at 442 nm, close to the absorbance maximum of the indicator's acid form. In the absence of an indicator, laser excitation of 2-naphthol-3,6-disulfonate generates some long-living intermediates absorbing at 442 nm.[84] If 100 μM Bromocresol Green is also present, the absorbance at 442 nm is much greater, indicating the formation of the protonated form of the indicator. The same experiments were repeated using Phenol Red (pK=8.0), Bromothymol Blue (pK=7.4), Neutral Red (pK=7), Bromocresol Purple (pK=6.3), and Bromocresol Green (pK=4.7). In all these cases, the rapid decay (50–200 ns) of the basic form of the indicator and the rapid (100–200 ns) formation of the acidic form of the indicator were accelerated as the pK value of the indicators increased.

A pH jump could also be detected when 2-naphthol-6-sulfonate was excited. This compound is somewhat better than 2-naphthol-3,6-disulfonate, since hydrated electrons formed by the laser excitation were hardly detectable. On the other hand, due to the relatively small extinction coefficient at 347.2 nm, much higher concentrations (10–20 nM) had to be employed in order to detect the discharge of protons.

VII. SUMMARY

As we have seen, the application of lasers has resulted in a revival of the study of proton-transfer reactions. Important features of these reactions are now accessible. Within the next few years, studies of light-generated proton transfer will provide new insight into chemical reaction mechanisms. Thus the simple proton will continue to provide us with valuable information on how and why reactions take place as well as the mechanism for mass transport in important liquids such as water.

Acknowledgments

We thank Professor Joshua Jortner for helpful discussions. This work was supported by the Research Corporation, the National Science Foundation, and the donors of the Petroleum Research Fund administered by the American Chemical Society at the University of Illinois, by the Binational Science Foundation—Grant Number 1404 and the Israel Commission for Basic Research at the University of Tel-Aviv. K. J. Kaufmann thanks the Max Planck Institute for Biophysical Chemistry and D. Huppert thanks Bell Laboratories for their hospitality during the preparation of this manuscript.

References

1. J. N. Brönsted and K. J. Pedersen, *Z. Phys. Chem.*, **108**, 185 (1924).
2. M. Eigen and L. DeMayer, *Z. Electrochem.*, **59**, 986 (1955).
3. P. Mitchell, *Biol. Rev. Cambridge Philos. Soc.*, **41**, 445 (1966).
4. D. Oesterhelt and W. Stockenius, *Proc. Natl. Acad. Sci.* (*USA*), **70**, 2853 (1973).
5. A. J. Kresge, *Acct. Chem. Res.*, **8**, 354 (1975).
6. E. Buncel, and C. C. Lee, eds., *Isotopes in Organic Chemistry*, Elsevier, Amsterdam, 1976.
7. R. A. Marcus, *Faraday Symp.*, **10**, 60 (1975).
8. M. Eigen, *Angew. Chem. Int. Ed. Eng.*, **3**, 1 (1964), and references therein.
9. G. Briere and F. Gaspard, *J. Chim. Phys.*, **64**, 1071 (1967).
10. S. Meiboom, *J. Chem. Phys.*, **34**, 375 (1961).
11. M. D. Newton, *J. Chem. Phys.*, **67**, 5535 (1977).
12. E. Grunwald and E. K. Ralph, *Acct. Chem. Res.*, **4**, 107 (1971).
13. A. Weller, *Prog. Reaction Kinetics*, **1**, 189 (1961).
14. K. Weber, *Z. Physik. Chem.*, **77**, 1207 (1931).
15. T. Förster, *Naturwissenschaften*, **36**, 186 (1949).
16. R. P. Bell, *The Proton in Chemistry*, 2nd ed., Cornell University Press, Ithaca, N.Y., 1973.
17. T. A. Godfrey, G. Porter, and P. Suppan, *Disc. Faraday Soc.*, **39**, 194 (1965).
18. J. F. Ireland and P. A. H. Wyatt, *J. Chem. Soc. Faraday 1*, **68**, 1053 (1972); **69**, 161 (1973).
19. T. Förster, *Z. Electrochem.*, **54**, 42 (1950).
20. T. C. Werner and D. M. Hercules, *J. Phys. Chem.*, **74**, 1030 (1970).
21. S. G. Schulman and I. Pace, *J. Phys. Chem.*, **76**, 1996 (1972).
22. A. J. Campillo, J. H. Clark, S. L. Shapiro, and K. R. Winn, *Proceedings of the First International Conference on Picosecond Phenomena*, C. V. Shank, E. P. Ippen, and S. L. Shapiro, eds., Springer-Verlag, Berlin, 1978.
23. A. Weller, *Z. Physik. Chem. N.F.*, **15**, 438 (1958).
24. M. R. Loken, J. W. Hayes, J. R. Gohlke, and L. Brand, *Biochemistry*, **11**, 4779 (1972).

25. M. Ofran and J. Feitelson, *Chem. Phys. Lett.*, **19**, 427 (1973).
26. A. B. Demjaschkewitch, N. K. Zaitsev, and M. G. Kuzmin, *Chem. Phys. Lett.*, **55**, 80 (1978).
27. A. Weller, *Z. Electrochem.*, **56**, 662 (1952).
28. J. F. Ireland and P. A. H. Wyatt, *Adv. Phys. Org. Chem.*, **12**, 131 (1976).
29. G. Jackson and G. Porter, *Proc. Roy. Soc. (London)*, **A260**, 13 (1961).
30. R. A. Marcus, *J. Am. Chem. Soc.*, **91**, 7224 (1969).
31. M. M. Kreevoy and D. E. Konasewich, *Adv. Chem. Phys.*, **21**, 243 (1971).
32. P. R. Wells, *Linear Free Energy Relationships*, Academic Press, London, 1968.
33. M. G. Evans and M. Polanyi, *Trans. Faraday Soc.*, **34**, 11 (1938).
34. T. Förster and S. Völker, *Chem. Phys. Lett.*, **34**, 1 (1975).
35. A. Weller, *Disc. Faraday Soc.*, **27**, 28 (1959).
36. P. Debye, *Trans. Electrochem. Soc.*, **82**, 265 (1942).
37. A. Weller, *Z. Phys. Chem. N.F.*, **13**, 335 (1957).
38. J. N. Brönsted, *Chem. Rev.*, **5**, 231 (1928).
39. K. Sölc and W. H. Stockmayer, *Int. J. Chem. Kinetics*, **5**, 733 (1973).
40. K. C. Ingham and M. A. El-Bayoumi, *J. Am. Chem. Soc.*, **96**, 1674 (1974).
41. M. A. El-Bayoumi, P. Avouris, and W. R. Ware, *J. Chem. Phys.*, **62**, 2499 (1975).
42. W. M. Hetherington III, R. H. Micheels, and K. B. Eisenthal, *Chem. Phys. Lett.*, **66**, 230 (1979).
43. A. Weller, *Z. Elektrochem.*, **60**, 1144 (1956).
44. H. Beens, K. H. Grellmann, M. Gurr, and A. H. Weller, *Disc. Faraday Soc.*, **39**, 183 (1965).
45. K. Sandros, *Acta Chem. Scandia*, **A30**, 761 (1976).
46. W. Klöpffer and G. Naundorf, *J. Limin.*, **8**, 457 (1974).
47. U. V. Naiboken, E. N. Pavlova, and B. A. Zadorozhnyi, *Opt. Spectr.*, **6**, 312 (1156).
48. J. E. Del Bene, *J. Am. Chem. Soc.*, **95**, 6517 (1973).
49. B. A. Zadorozhnyi and I. K. Ischenko, *Opt. Spectr.*, **19**, 306 (1965).
50. J. H. Caudet, G. C. Newland, H. W. Peters, and J. W. Tamblyn, *Soc. Plastics Eng. Trans.*, **1**, 26 (1961).
51. K. K. Smith and K. J. Kaufmann, *J. Phys. Chem.*, **82**, 2286 (1978).
52. D. J. Bradley and G. H. C. New, *Proc. IEEE*, **62**, 313 (1974).
53. P. J. Thistlethwaite and G. J. Woolfe, *Chem. Phys. Lett.*, **63**, 401 (1979).
54. E. M. Kosower and H. Dodiuk, *J. Lumin.*, **11**, 249 (1975).
55. J. Goodman and L. E. Brus, *J. Am. Chem. Soc.*, **100**, 7472 (1978).
56. A. A. Lamola and L. J. Sharp, *J. Phys. Chem.*, **70**, 2534 (1966).
57. W. Klopffer, *J. Polymer Sci.*, **57**, 205 (1976).
58. S.-Y. Houe, W. M. Hetherington III, G. M. Korenowski, and K. B. Eisenthal, *Chem. Phys. Lett.*, **68**, 282 (1979).
59. H. Shizuka, K. Matsui, T. Okamura, and I. Tanaka, *J. Phys. Chem.*, **79**, 2731 (1975).
60. H. Shizuka, K. Matsui, Y. Hirata, and I. Tanaka, *J. Phys. Chem.*, **80**, 2079 (1976).
61. H. Shizuka, K. Matsui, Y. Hirata, and I. Tanaka, *J. Phys. Chem.*, **81**, 2243 (1977).
62. P. M. Rentzepis, *Science*, **202**, 174 (1978).
63. R. Nakagaki, T. Kobayashi, and S. Nagakura, *Bull. Chem. Soc. Japan*, **51**, 1671 (1978).
64. R. Nakagaki, T. Kobayashi, J. Nakamura, and S. Nagakura, *Bull. Chem. Soc. Japan*, **50**, 1909 (1977).
L. A. Brey, G. B. Schuster, and H. G. Drickamer, *J. Chem. Phys.*, **67**, 2648 (1977).
65. T. Rosenfeld, M. Ottolenghi, and A. Y. Meyer, *Mol. Photochem.*, **5**, 39 (1973).
66. B. I. Greene, R. M. Hochstrasser, and R. B. Weisman, *J. Chem. Phys.*, **70**, 1247 (1979).

67. K. Peters, M. L. Appleburry, and P. M. Rentzepis, *Proc. Natl. Acad. Sci. USA*, **74**, 3119 (1977).
68. P. O. Löwdin, *Adv. Quantum Chem.*, **2**, 213 (1965).
69. A. Gafni, R. L. Modlin, and L. Brand, *J. Phys. Chem.*, **80**, 898 (1976).
70. J. H. Clark, S. L. Shapiro, A. J. Campillo, and K. R. Winn, *J. Am. Chem. Soc.*, **101**, 746 (1979).
71. K. K. Smith, D. Huppert, M. Gutman, and K. J. Kaufmann, *Chem. Phys. Lett.*, **64**, 522 (1979).
72. Th. Förster, *Chem. Phys. Lett.*, **17**, 309 (1972).
73. K. Tsutsumi and H. Shizuka, *Chem. Phys. Lett.*, **52**, 485 (1977).
74. H. Shizuka, K. Tsutsumi, H. Takeuchi, and I. Tanaka, *Chem. Phys. Lett.*, **62**, 408 (1979).
75. R. M. Noyes, *J. Am. Chem. Soc.*, **77**, 2042 (1955).
76. T. J. Chuang, G. W. Hoffmann, and K. B. Eisenthal, *Chem. Phys. Lett.*, **25**, 201 (1974).
77. J. Werner, Z. Schulten, and K. Schulten, *J. Chem. Phys.*, **67**, 646 (1977).
78. M. Hauser, H. P. Haar, and U. K. A. Klein, *Ber. Bunsenges. Physik. Chem.*, **81**, 27 (1977).
79. H. P. Haar, U. K. A. Klein, and M. Hauser, *Chem. Phys. Lett.*, **58**, 525 (1978).
80. K. K. Smith, D. Huppert, and K. J. Kaufmann, unpublished data.
81. F. D. Saeva and G. R. Olin, *J. Am. Chem. Soc.*, **97**, 5631 (1975).
82. E. J. Chandross, *J. Am. Chem. Soc.*, **98**, 1053 (1976).
83. U. Lachish, M. Ottolenghi, and G. Stein, *Chem. Phys. Lett.*, **48**, 402 (1977).
84. M. Gutman and D. Huppert, *J. Biochem. Biophys. Methods*, **1**, 19 (1979).

AUTHOR INDEX

Numbers in parentheses are reference numbers and indicate that the author's work is referred to although his name is not mentioned in the text. Numbers in *italics* show the pages on which the complete references are listed.

SUBJECT INDEX